IELD PRECISION RATIO : $\dfrac{\text{CLOSURE}}{\text{AVG. DIST.}}$

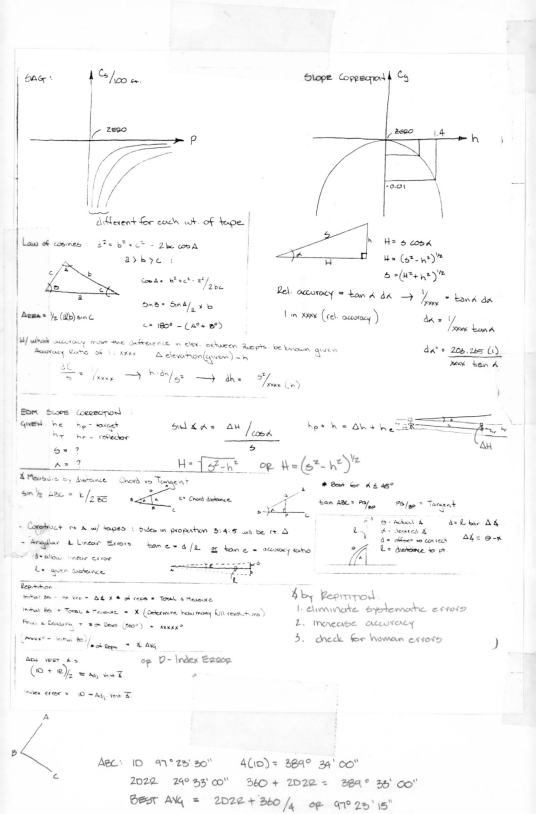

SAG:

$C_s / 100$ ft.

ZERO

P

different for each wt. of tape

SLOPE CORRECTION C_g

ZERO   1.4

h

-0.01

Law of cosines: $a^2 = b^2 + c^2 - 2bc \cos A$

$a > b > c$:

$\cos A = \dfrac{b^2 + c^2 - a^2}{2bc}$

$\sin B = \dfrac{\sin A}{a} \times b$

$c = 180° - (A° + B°)$

AREA $= \frac{1}{2}(a)(b) \sin C$

$H = S \cos \alpha$

$H = (S^2 - h^2)^{1/2}$

$S = (H^2 + h^2)^{1/2}$

Rel. accuracy $= \tan \alpha \, d\alpha \rightarrow \dfrac{1}{xxxx} = \tan \alpha \, d\alpha$

1 in xxxx (rel. accuracy)

$d\alpha = \dfrac{1}{xxxx \tan \alpha}$

W/ what accuracy must the difference in elev. between 2 pts. be known given
Accuracy Ratio of 1: xxxx Δ elevation (given) - h

$\dfrac{dC}{S} = \dfrac{1}{xxxx} \rightarrow h \cdot dh/S^2 \rightarrow dh = \dfrac{S^2}{xxxx}(h)$

$d\alpha'' = \dfrac{206.265 \,(1)}{xxxx \tan \alpha}$

EDM SLOPE CORRECTION :

GIVEN: he h_p - target
        h_T h_r - reflector

$S = ?$
$\alpha = ?$

$\sin \angle \alpha = \dfrac{\Delta H / \cos \alpha}{S}$

$H = \sqrt{S^2 - h^2}$ or $H = (S^2 - h^2)^{1/2}$

$h_p + h = \Delta h + h_e$

ΔH

$\angle$ Measure by distance   Chord vs Tangent

$\sin \frac{1}{2} \angle BC = k/2 \overline{BO}$

$k =$ Chord distance

* Best for $\alpha \le 45°$

$\tan \angle BC = PG/BP$ $PG/BP =$ Tangent

- Construct rt $\angle$ w/ tapes : sides in proportion 3:4:5 will be rt. Δ

$\theta =$ Actual $\angle$ $d = R \tan \Delta \angle$
$\alpha =$ desired $\angle$
$d =$ offset to correct $\Delta \angle = \theta - \alpha$
$R =$ distance to pt

- Angular & Linear Errors $\tan e = d/R$ or $\tan e =$ accuracy ratio

$d =$ allow. linear error
$R =$ given distance

Repetition
Initial BS - BS Rel - $\Delta \& \alpha \times \#$ of reps = TOTAL $\angle$ MEASURE
Initial BS + TOTAL $\angle$ MEASURE = X (Determine how many full revolutions)
Final $\angle$ Reading + $\angle$ of Revol (360°) - xxxxx°

$(xxxx° - $ Initial BS$) / \#$ of Reps = $\angle$ AVG.

ADJ VERT $\angle$ S
$(ID + IR)/2 = $ Adj Vert $\overline{I}$

or D - INDEX ERROR

Index error = $ID - $ Adj Vert $\overline{I}$.

$\angle$ by REPITITION:
1. eliminate systematic errors
2. increase accuracy
3. check for human errors

A
B
C

ABC: ID 97° 23' 30"   4(ID) = 389° 34' 00"
      2D2R 29° 33' 00"   360 + 2D2R = 389° 33' 00"
      BEST AVG = $\dfrac{2D2R + 360}{4}$ or 97° 23' 15"

CBA : ID 262° 36' 30"   BEST AVG = $\dfrac{1D1R + 360}{2}$ or 262° 36' 45"
       1D1R 165° 13' 30"

Surveying

SEVENTH EDITION

Francis H. Moffitt
University of California—Berkeley

Difference in Stationing between the PC & first station on curve (c_1) the angle to be subtending
a full station is equal to D $d_1 = c_1 (D)/100$

$c_1 = 2100 - 2054$ or $46'$

$c_1 =$ actual chord distance $\to$ $C_1 = 2R (\sin \frac{1}{2} d_1)$

- Difference in Stationing between last Full Station & PT (c'_2)

$c'_2 = 2234 - 2200$ or $34'$ $d_2° = c'_2 (D)/100$

$C_{2} =$ actual chord distance $\to$ $C_2 = 2R (\sin \frac{1}{2} d_2)$

- Difference between full stations (c) $d = c'D/100$ $C = 2R \sin \frac{1}{2} D$
  "   "   "   ½ stations $C_{50} = 2R \sin \frac{°}{4}$
  "   "   "   ¼ stations $C_{25} = 2R \sin \frac{°}{8}$

Ex. $PI = P_C + T$ GIVEN: Δ, D

$R = 5729.5780/D$ $T = R \tan (\Delta/2)$ $P_C = PI - T$ $L = \Delta/D (100)$ in feet $PT = PC + L$

- determine c_1, c'_2 and C_{100} (or C_{50}, C_{25})

- determine $d_1/2 = c_1 (D)/2(100)$ $d_2/2 = c'_2 (D)/2(100)$ $D/2 = C_{100}(D)/2(100)$ $D°/2 = C_{50}(D)/2(100)$
  (subdeflection $\times$'s)

- determine chords $C_1 = 2R \sin d_1/2$ $C_2 = 2R \sin d_2/2$ $C_{100} = 2R \sin D/2$ $C_{50} = 2R \sin D°/$

STA	SUBDEFLECTION $\times$	CHORDS	CURVE DATA			
PI STA			$\Delta \times =$	$D =$	$T =$	$L =$
			PI STA $=$	$R =$	$P_C @$	$PT @$
PC STA	$0° \ 00' \ 00'$					

* FOR STAKING A CURVE, BEGIN @ BOTTOM
OF FIELD NOTES

VERTICAL CURVES - constant change in elevation
g (gradient) %

1. $BVC = V_{STA} - L/2$ $EVC = V_{STA} + L/2$

  BVC elev. $= V_{STA}$ elev. $- g_1 (L/2)$ EVC elev. $= V_{STA}$ elev. $+ g_2 (L/2)$

2. ELEV. OF A PT. ON GRADE LINES $Y = BVC$ elev. $+ Y'$ $Y' = g_1 (x')$ | x' - in stations |

3. ELEV. OF ANY POINT ON CURVE $Y = BVC$ elev. $+ \underbrace{g_1 (x') + r/2 [(x')^2]}_{Y'}$ $r = g_2 - g_1 / L$ (in stations)

4. ELEV. OF TURNING POINT" $Y = BVC$ elev. $+ g_1 (x) + r/2 (x^2)$ where: $x = -g_1/r$

Sponsoring Editor: Cliff Robichaud
Project Editor: Pamela Landau
Designer: Robert Sugar
Production Manager: Marion Palen
Compositor: Composition House Ltd.
Printer & Binder: Halliday Lithograph Corp.
Art Studio: J & R Technical Services, Inc.

Surveying, Seventh Edition

Copyright © 1982 by Harper & Row, Publishers, Inc.

Library of Congress Cataloging in Publication Data

Moffitt, Francis H.
  Surveying.

  (The Harper & Row series in civil engineering)
  Includes bibliographies and index.
  1. Surveying. I. Bouchard, Harry, 1889–1954.
II. Title. III. Series.
TA545.B7 1982   526.9   81-6826
ISBN 0-06-044559-9   AACR2

Contents

Chapter 4 The Measurement of Angles 131

Chapter 5 Random Errors 181

Chapter 6 Field Operations with the Transit or Theodolite 218

Chapter 7 The Direction of a Line 234

Chapter 8 Traverse Surveys and Computations 261

Chapter 9   Traversing with Inertial Surveying Systems   341

Chapter 10   Horizontal Control Networks   360

Chapter 11   State Plane Coordinate and Universal Transverse Mercator Systems   434

Chapter 12   Practical Astronomy 478

Chapter 13   Horizontal and Vertical Curves 519

Chapter 14   Tacheometry 553

Appendix A   Adjustment of Elementary Surveying Measurements by the Method of Least Squares   745

Appendix B   The Adjustment of Instruments   808

Appendix C   Tables   818

Preface

Since the publication of the sixth edition of this textbook in 1975, the practice of surveying has been greatly influenced by spaceage technology. This technology has been applied to the further development, refinement, and miniaturization of electronic distance meters (EDM's), the incorporation of electronic angular measurement capabilities in these instruments, and the automatic storage and processing of survey data by built-in microprocessors. Application of refined inertial guidance and navigation systems to surveying problems has resulted in revolutionary surveying systems which permit large survey projects to be executed with speeds unheard of just a decade ago.

The Navy Navigation Satellite System referred to as Transit now allows the surveyor to determine the position and elevation of any point on earth to an accuracy of one-half metre with the appropriate ground receivers which are constantly being improved. This system of orbiting satellites will become increasingly important to the surveyor in the future as more and more satellites are placed in orbit to give even better positioning accuracy. Positioning by satellite observations has been introduced in Chapter 10, which discusses horizontal control networks.

This seventh edition introduces these new concepts of surveying to students in order they may keep abreast of new technology and be prepared for future developments. The latest EDM's with their capabilities of measuring horizontal and vertical angles have been introduced in Chapters 2 and 4. The principle of the gyroscope as it relates to the determination of directions using the gyrotheodolite is explained in Chapter 7. The gyroscope is also discussed in a new chapter (Chapter 9) on inertial surveying along with the concept of accelerometers for measuring horizontal and vertical displacements.

Discussion of many of the tacheometric instruments described in previous editions of this textbook has been eliminated because the author feels that modern lightweight EDM's have practically displaced these instruments for the rapid measurement of distances.

All logarithmic calculations have been eliminated, in addition to the stadia tables and the tables of trigonometric functions. Modern handheld calculators are capable of generating logarithms where needed, and the trigonometric functions required in surveying computations. They are much more efficient than table lookup. All of the calculations required in the main part of the text can be performed on handheld calculators likely used by the student. Some of the more advanced handheld calculators are required for the reduction of matrix equations found in the appendix on least-squares adjustments (Appendix A).

Minor additions found in this edition include expansion on the use of the metric system, coordinate transformations for use in running random traverses, locating slope stakes by inversing construction lines, the universal transverse Mercator system of rectangular coordinates, modern photogrammetric instruments, descriptions by metes and bounds, subdivision surveys, and the adjustment of intersection and resection surveys by least squares.

The bibliographies at the ends of the chapters have been brought up to date (1980) to reflect the rapidly changing techniques employed in surveying. It is through the selected reading of articles listed in the bibliography that students can expand their knowledge of the principles and practice of surveying. The listings rely quite heavily on publications of the American Congress on Surveying and Mapping[1] and the Canadian Institute of Surveying.[2] These publications are most likely to be found in the college library, or can be obtained through the two societies.

The author wishes to express thanks to all of the teachers and students who have offered comment, criticism, and suggestions relating to the sixth edition. Special acknowledgment is given to Dr. James M. Anderson, Mr. Leslie F. Gregerson, Mr. Sean Curry, Mr. Chris Crawford, Mr. Moir D. Hoag, and Mr. James R. Barr for their help during the preparation of the manuscript of this edition. Finally, the author wishes to thank the instrument manufacturers and the federal agencies who generously furnished many of the photographs and illustrations used in this edition.

Francis H. Moffitt

[1] American Congress on Surveying and Mapping, 210 Little Falls Street, Falls Church, Virginia 22046. (703) 241-2446.
[2] Canadian Institute of Surveying, Box 5378, Station F, Ottawa, Ontario, Canada, K2C, 3J1.

Surveying

Chapter 1
Introduction

1-1. SURVEYING   The purpose of surveying is to locate the positions of points on or near the surface of the earth. Some surveys involve the measurement of distances and angles for the following reasons: (1) to determine horizontal positions of arbitrary points on the earth's surface; (2) to determine elevations of arbitrary points above or below a reference surface, such as mean sea level; (3) to determine the configuration of the ground; (4) to determine the directions of lines; (5) to determine the lengths of lines; (6) to determine the positions of boundary lines; and (7) to determine the areas of tracts bounded by given lines. Such measurements are data-gathering measurements.

In other surveys it is required to lay off distances and angles to locate construction lines for buildings, bridges, highways, and other engineering works, and to establish the positions of boundary lines on the ground. These distances and angles constitute layout measurements.

A survey made to establish the horizontal or vertical positions of arbitrary points is known as a *control survey.* A survey made to determine the lengths and directions of boundary lines and the area of the tract bounded by these lines, or a survey made to establish the positions of boundary lines on the ground is termed a *cadastral, land, boundary,* or *property survey.* A survey conducted to determine the configuration of the ground is termed a *topographic survey*. The determination of the configuration of the bottom of a body of water is a *hydrographic survey*. Surveys executed to locate or lay out engineering works are known as *construction surveys*. A survey performed by means of aerial photography is called an *aerial survey* or a *photogrammetric survey*.

The successful execution of a survey depends on surveying instruments of a rather high degree of precision and refinement, and also on the proper use and handling of these instruments in the field. All surveys involve some computations, which may be made directly in the field, or performed in the office, or in both places.

Some types of surveys require very few computations, whereas others involve lengthy and tedious computations. In the study of surveying, therefore, the student not only must become familiar with the field operational techniques, but must also learn the mathematics applied in surveying computations.

1-2. BASIC DEFINITIONS   In order to gain a clear understanding of the procedures for making surveying measurements on the earth's surface, it is necessary to be familiar with the meanings of certain basic terms. The terms discussed here have reference to the actual figure of the earth.

An *oblate spheroid*, also called an *ellipsoid of revolution*, is a solid obtained by rotating an ellipse on its shorter axis. The idealized figure of the earth is an oblate spheroid in which the earth's rotational axis serves as the shorter or minor axis. Because of its relief, the earth's surface is not a true spheroid. However, an imaginary surface representing a mean sea level extending over its entire surface very nearly approximates a spheroid. This imaginary surface is used as the figure on which surveys of large extent are computed.

A *vertical line* at any point on the earth's surface is the line that follows the direction of gravity at that point. It is the direction which a string will assume if a weight is attached to the string and the string is suspended freely at the point. At a given point there is only one vertical line. The earth's center of gravity cannot be considered to be located at its geometric center, because vertical lines passing through several different points on the surface of the earth do not intersect in that point. In fact, all vertical lines do not intersect in any common point. A vertical line is not necessarily normal to the surface of the earth, nor even to the idealized spheroid. The angle between the vertical line and the normal to the spheroid at a point is called the *deflection of the vertical*.

A *horizontal line* at a point is any line that is perpendicular to the vertical line at the point. At any point there are an unlimited number of horizontal lines.

A *horizontal plane* at a point is the plane that is perpendicular to the vertical line at the point. There is only one horizontal plane through a given point.

A *vertical plane* at a point is any plane that contains the vertical line at the point. There are an unlimited number of vertical planes at a given point.

A *level surface* is a continuous surface that is at all points perpendicular to the direction of gravity. It is exemplified by the surface of a large body of water at complete rest (unaffected by tidal action).

A *horizontal distance* between two given points is the distance between the points projected onto a horizontal plane. The horizontal plane, however, can be defined at only one point. For a survey the reference point may be taken as any one of the several points of the survey.

A *horizontal angle* is an angle measured in a horizontal plane between two vertical planes. In surveying this definition is effective only at the point at which the measurement is made or at any point vertically above or below it.

A *vertical angle* is an angle measured in a vertical plane. By convention, if the angle is measured upward from a horizontal line or plane, it is referred to as a *plus* or *positive* vertical angle, and also as an *elevation* angle. If the angle is measured downward, it is referred to as a *minus* or *negative* vertical angle, and also as a *depression* angle.

A *zenith angle* is also an angle measured in a vertical plane, except that, unlike a vertical angle, the zenith angle is measured down from the upward direction of the plumb line.

The *elevation* of a point is its vertical distance above or below a given reference level surface (see Section 3-1).

The *difference in elevation* between two points is the vertical distance between the two level surfaces containing the two points.

Plane surveying is that branch of surveying wherein all distances and horizontal angles are assumed to be projected onto one horizontal plane. A single reference plane may be selected for a survey where the survey is of limited extent. For the most part this book deals with plane surveying.

Geodetic surveying is that branch of surveying wherein all distances and horizontal angles are projected onto the surface of the reference spheroid that represents mean sea level on the earth.

The surveying operation of leveling takes into account the curvature of the spheroidal surface in both plane and geodetic surveying. The leveling operation determines vertical distances and hence elevations and differences of elevation.

1-3. UNITS OF MEASUREMENT In the United States, the linear unit most commonly used at the present time is the foot and the unit of area is the acre, which is 43,560 ft^2. In most other countries throughout the world distances are expressed in metres. The metre is also used by the National Geodetic Survey of the United States Department of Commerce as well as other federal and state agencies in the United States engaged in establishing control. However, the published results of some of these control survey operations are given in both units, or are available to the user in both units.

On all United States government land surveys, the unit of length is the Gunter's chain, which is 66 ft long and is divided into 100 links, each of which is 0.66 ft or 7.92 in. long. A chain, therefore, equals $\frac{1}{80}$ mile. This is a convenient unit where areas are to be expressed in acres, since 1 acre = 10 square chains. A distance of 2 chains 18 links can also be written as 2.18 chains. Any distance in chains can be readily converted into feet, if desired, by multiplying by 66.

In those portions of the United States that came under Spanish influences another unit, known as the vara, has been used. A vara is about 33 in. long. The exact length varies slightly in different sections of the Southwest, where the lengths of property boundaries are frequently expressed in this unit.

For purposes of computation and plotting, decimal subdivisions of linear units are the most convenient. Most linear distances are therefore expressed in feet and tenths, hundredths, and thousandths of a foot. The principal exception to this practice is in the layout work on a construction job, where the plans of the structures are dimensioned in feet and inches. Tapes are obtainable graduated either decimally or in feet and inches.

Volumes are expressed in either cubic feet or cubic yards.

Angles are measured in degrees (°), minutes ('), and seconds ("). One circumference $= 360°; 1° = 60'; 1' = 60''$. In astronomical work some angles are expressed in hours (^{h}), minutes (^{m}), and seconds (^{s}). Since one circumference $= 24^h = 360°$, it follows that $1^h = 15°$ and $1° = \frac{1}{15}^h = 4^m$ (see Section 12-4).

Although surveying instruments which measure angles in the sexagesimal system are graduated in degrees, minutes, and seconds, it is necessary, in computations with hand calculators, to convert to degrees and decimal degrees (unless the calculator has provision for this conversion) in order to compute the trigonometric functions of the angles, and then if necessary to convert back to degrees, minutes, and seconds. The number of decimal places to be retained in the decimal degree is a function of the least reading of the instrument or of the given angle. Equivalents of the decimal part of a degree are:

$$0.00001° = 0.036''$$

$$0.0001° = 0.36''$$

$$0.001° = 3.6''$$

$$0.01° = 36''$$

$$0.1° = 6'$$

Thus, if the angle is given to the nearest minute, two decimal places must be retained; if the angle is given to the nearest second, then four decimal places must be retained.

Example 1-1

Convert 153° 43′ 17.2″ to decimal form.

Solution: $153° \, 43′ \, 17.2″ = 153° \left(43 \dfrac{17.2}{60} \right)′ = 153° \, 43.287′$

$$153° \, 43.287′ = \left(153 \dfrac{43.287}{60} \right)° = 153.72144°$$

which is also written 153°.72144.

Example 1-2

Convert 24.4652° to degrees, minutes, and seconds.

Solution: $0.4652° \times 60 = 27.912′$

$0.912′ \quad \times 60 = 55″$

Answer: 24° 27′ 55″

In order to help visualize the physical size of small dimensions, consider that 0.01 ft is very nearly equal to $\frac{1}{8}$ in., and that 0.10 ft is therefore about $1\frac{1}{4}$ in. Also, the sine or tangent of 1′ of arc is approximately 0.0003 (three zeroes 3). Thus 1′ of arc subtends about 0.03 ft or $\frac{3}{8}$ in. in 100 ft. At 1000 ft, 1′ of arc subtends 0.30 ft or $3\frac{1}{2}$ in.

The sine or tangent of 1″ of arc is approximately 0.000005 (five zeroes 5). Thus at 1000 ft, 1″ subtends 0.005 ft or about $\frac{1}{16}$ in. Another relationship is embodied in the expression "a second is a foot in 40 miles." Taking the radius of the earth to be 4000 miles, a second of arc at the center of the earth subtends approximately 100 ft on the earth's surface.

Quite frequently it becomes necessary to convert a small angle from its arc or radian value to its value expressed in seconds, and vice versa. We find from trigonometry that the sine, the tangent, and the arc or radian value of 1″ are all equal within the limits of computational practicality. As a consequence of this, and designating θ as a small angle,

$$\text{arc } \theta \simeq \sin \theta \simeq \tan \theta$$

If the small angle is expressed in seconds, its arc, or sine, or tangent can be determined by the following relationships

$$\text{arc } \theta = \theta″ \times \text{arc } 1″ \qquad \text{(exact)}$$

$$\sin \theta = \theta″ \times \sin 1″ \qquad \text{(approximate)}$$

$$\tan \theta = \theta″ \times \tan 1″ \qquad \text{(approximate)}$$

The value of the arc, the sine, and the tangent of 1″ is 0.0000048481 to 10 decimal places. If the arc or radian value of a small angle is known, or if the sine or tangent is known, then its value in seconds can be determined by rearranging the three preceding expressions to give

$$\theta'' = \frac{\text{arc } \theta}{\text{arc } 1''} = \frac{\text{arc } \theta}{48481 \times 10^{-10}} = 206265 \text{ arc } \theta$$

$$\theta'' = \frac{\sin \theta}{\sin 1''} = \frac{\tan \theta}{\tan 1''}$$

Example 1-3

A target 4 in. wide is placed on the side of a building 1600 ft away from a survey point. What angle is subtended at the point between the left and right edge of the target?

Solution: The target width is 0.333 ft, and thus the arc of the small angle is 0.333/1600 = 0.000208. The angle in seconds is then 0.000208/48481 × 10^{-10} = 42.9″ or 0.000208 × 206265 = 42.9″

1-4. METRIC EQUIVALENTS The metric system, used extensively throughout the world, is finding a more widespread adoption in the United States. This adoption has come about primarily to facilitate exchange of scientific and technical data with countries using the metric system.

The basis of distance measurement in the metric system is the metre, which is defined as 1,650,763.73 wavelengths of orange-red krypton gas with an atomic weight of 86 and at a specified energy level in the spectrum. This natural standard was adopted in lieu of a man-made physical standard, such as the International Prototype Metre Bar, because of the ever-present danger that the latter might be destroyed by accident or by a hostile act.

The metre is more commonly subdivided into the following units:

$$1 \text{ centimetre (cm)} = 0.01 \text{ metre}$$

$$1 \text{ millimetre (mm)} = 0.001 \text{ metre}$$

$$1 \text{ micrometre } (\mu m) = 0.001 \text{ mm} = 10^{-6} \text{ metre}$$

$$1 \text{ nanometre (nm)} = 0.001 \ \mu m = 10^{-9} \text{ metre}$$

$$1 \text{ angström (Å)} = 0.1 \text{ nm} = 10^{-10} \text{ metre}$$

It was not until 1959 that the technical and scientific agencies of the United States and the United Kingdom adopted the following equivalence:

$$1 \text{ U.S. inch} \quad = \quad 2.54 \text{ centimetres}$$

$$1 \text{ British inch} = \quad 2.54 \text{ centimetres}$$

$$1 \text{ U.S. foot} \quad = 30.48 \text{ centimetres}$$

The previous United States equivalent was 1 metre = 39.37 U.S. inches, or 1 U.S. inch = 2.540005+ centimetres. Since plane and geodetic surveys may involve very long distances, sometimes on the order of 200 miles, and since results of important nationwide surveys going back more than 50 years are still used, inconsistencies would be introduced if the 1959 equivalence was adopted for surveying operations. Therefore the previous equivalence of 1 metre = 39.37 U.S. inches is used to define the *U.S. survey foot.* Thus 1 survey foot = 12/39.37 metre. This differs from the new definition of the foot (1 ft = 30.48 cm) by about two parts per million (2 ppm). To get an appreciation of this difference consider that in a distance of 100,000 metres, which is slightly more than 60 miles, the difference between the two values is 0.66 ft or 8 in.

The kilometre (km) is 1000 m = 3280.833 ft which is approximately $\frac{5}{8}$ of a mile. One mile is approximately 1.6 km.

The unit of area in many countries which use the metric system is the *hectare* (ha) which is 10,000 m^2. Thus one hectare is approximately $2\frac{1}{2}$ acres.

Volumes are expressed in cubic metres (m^3) which is approximately 1.3 cubic yard (yd^3).

1-5. CENTESIMAL SYSTEM   Many surveying and mapping instruments with angular scales are graduated in the centesimal system. In this system one circumference is divided into 400 grads$^{(g)}$. The grad is divided into 100 centesimal minutes$^{(c)}$, and the minute is subdivided into 100 centesimal seconds$^{(cc)}$. The degree value of the grad is determined by the following equivalences:

$$400^g = 360°$$

$$1^g = 0.9°$$

$$0.01^g = 1^c = 0.009° = 0° \, 00' \, 32.4''$$

$$0.0001^g = 1^{cc} = 0.00009° = 0° \, 00' \, 0.32''$$

An angle expressed as 324.4625^g could also be expressed as 324^g 46^c 25^{cc}, but this latter notation would be much more awkward and entirely unnecessary.

If trigonometric tables based on centesimal units are available, many computations are greatly simplified.

In order to convert an angle given in grads to its equivalent value in degrees, multiply the grad value by 0.9.

Example 1-4

Convert 324.4625^g to degrees, minutes, and seconds.

Solution: Multiplying by 0.9 gives $324.4625 \times 0.9 = 292.01625°$. Conversion from decimal degrees to degrees, minutes, and seconds then follows.

$$0.01625° \times 60 = 0.975' \quad \text{(minutes)}$$

$$0.975 \times 60 = 58.5'' \quad \text{(seconds)}$$

The value of the conversion is thus $292° 00' 58.5''$.

When converting from degrees, minutes, and seconds to grads, the angle must be converted to decimal degrees and the results then divided by 0.9.

Example 1-5

Convert $142° 22' 14.5''$ to grads.

Solution: $\quad 142° 22' 14.5'' = 142°\left(22 \dfrac{14.5}{60}\right)' = 142° 22.242'$

$$142° 22.242' = 142\left(\dfrac{22.242}{60}\right)° = 142.37070°$$

$$142.37070° = \left(\dfrac{142.37070}{0.9}\right)^g = 158.18967^g$$

1-6. TABLES Appendix C contains three tables that are useful in surveying computations. Table A is useful for converting an angle given in degrees, minutes, and seconds into its radian value.

Example 1-6

Determine the radian value of $37° 14' 53.6''$.

Solution: From Table A,

37°	= 0.645 77 182
14′	= 0.004 07 243
53″	= 0.000 25 695
0.6″	= 0.000 00 291

$$37° \ 14′ \ 53.6″ = 0.650 \ 10 \ 411 \ \text{radian}$$

Example 1-7

Compute the length of a circular arc of radius 2500 m that is subtended by an angle of 37° 14′ 53.6″.

Solution: The length of the arc is equal to the radius multiplied by the subtended angle in radians. Thus

$$\text{arc length} = 2500 \times 0.650 \ 10 \ 411 = 1625.260 \ \text{m}$$

Tables B and C give trigonometric formulas for the solution of right triangles and oblique triangles, respectively.

1-7. FIELD NOTES When a survey is performed for the purpose of gathering data, the field notes become the record of the survey. If the notes have been carelessly recorded and documented, falsified, lost, or made grossly incorrect in any way, the survey or a portion of the survey is rendered useless. Defective notes result in tremendous waste of both time and money. Furthermore, it will become obvious that, no matter how carefully the field measurements are made, the survey as a whole may be useless if some of those measurements are not recorded or if the meaning of any record is ambiguous.

Some surveying instruments (see Section 4-23) are capable of automatic storage of field measurements. However, they do not preclude the necessity of recording these data in the field notes because they are usually accompanied by field sketches and diagrams.

The keeping of neat, accurate, complete field notes is one of the most exacting tasks. Although several systems of note keeping are in general use, certain principles apply to all. The aim is to make the clearest possible notes with the least expenditure of time and effort. Detailed examples of forms of notes suited to the principal surveying operations will be given later.

All field notes should show when, where, for what purpose, and by whom the survey was made. The signature of the recorder should appear. Habits are formed by constant repetition. One of the best habits is to make each page of notes complete. The field notes of a surveyor or engineer are often presented as evidence in court cases, and such things as time and place and members of the field party may be of extreme importance.

A hard pencil—4H or harder—should be used to prevent smearing. The notebook should be of good quality, since it is subjected to hard usage. No erasures should be made, because such notes will be under suspicion of having been altered. If an error is made, a line should be drawn through the incorrect value and the new value should be inserted above. In some organizations the notes are kept in ink, but this is rather inadvisable unless a waterproof ink is used or unless there is no possible chance of the notes becoming wet.

Clear, plain figures should be used, and the notes should be lettered rather than written. The record should be made in the field book at the time the work is done, and not on scraps of paper to be copied into the book later. Copied notes are not original notes, and there are too many chances for making mistakes in copying or of losing some of the scraps. All field computations should appear in the book so that possible mistakes can be detected later.

It should always be remembered that the notes are frequently used by other than the person who makes them. For this reason they should be so clear that there can be no possible chance of misinterpreting them.

If the numbers stamped on the field equipment are recorded, it may be possible to explain at some later date apparent discrepancies which were caused by some imperfection in the equipment. Thus errors in linear measurements can sometimes be explained by a broken tape being improperly repaired.

1-8. METHODS OF KEEPING NOTES There are four general methods of keeping notes: (1) by a written description of what has been done; (2) by means of a sketch on which all numerical values are shown; (3) by a tabulation of the numerical values; and (4) by a combination of these methods.

1. A detailed written description of what has been done is given in the notes for surveys made in connection with the subdivision of the public lands. However, for the usual surveys such a description would likely be long and involved, and it would be difficult to pick out the numerical values that are to be used in the office computations. By means of sketches or by a proper tabulation of the field measurements, the notes are greatly simplified and yet the field operations will be perfectly apparent to one who has a knowledge of surveying.
2. In the case of a relatively simple survey, such as that of a piece of property with few sides, a sketch that is roughly to scale can be made, and all linear and angular values can be shown directly on the sketch.

3. Where many angles and distances are measured from the same point, as in the case of a topographic survey, a sketch showing all observed values would be hopelessly complicated. For this reason the angles and distances are recorded in tabular form, care being taken to show clearly between what points the measurements are made. The notes for most leveling operations are recorded in tabular form.

4. On extensive surveys a combination of tabulated numerical values and sketches is used. Whenever there may be any doubt concerning field conditions, a sketch accompanies the numerical values. This sketch need not be drawn to scale; in fact, the doubt can usually be cleared up more easily by a distorted scale. The notes of most route surveys, as surveys for railroads, highways, canals, and transmission lines, are usually kept in this fashion. The numerical values are recorded on the left-hand page of the notebook, and the right-hand page is used for the sketch. If the notes start from the bottom of the page, and the various points and lines on the sketch are placed opposite the numerical values relating to them, the right-hand page will be a normally oriented map of what has been found on the ground when looking in the forward direction.

1-9. INDEXING OF NOTES   Since many isolated surveys may appear in the same notebook, an index in the front pages of the book will assist in locating any desired survey. Each notebook should bear a number. The owner's name and address should also appear to aid in its recovery in case of loss. Though desirable, it is not necessary that the notes for a given survey appear on consecutive pages. If separated, they should be cross-referenced. Some type of card index or loose-leaf index to all the books is extremely valuable. Property surveys should be indexed according to the name of the owner of the property and also according to the location of the property.

1-10. ERRORS AND MISTAKES   The value of a distance or an angle obtained by field measurements is never exactly the true value, except by chance. The measured value approaches the true value as the number and size of errors in the measurements become increasingly small. An error is the difference between the true value of a quantity and the measured value of the same quantity. Errors result from instrumental imperfections, personal limitations, and natural conditions affecting the measurements. Examples of instrumental errors are (1) a tape that is actually longer or shorter than its indicated length; (2) errors in the graduations of the circles of an engineer's transit; and (3) a defect in adjustment of a transit or level. Examples of personal limitations are the observer's inability to bisect a target or read a vernier exactly, inability to maintain a steady tension on the end of a tape, and failure to keep a level bubble centered at the instant at which a leveling observation is taken. Examples of natural conditions affecting a measurement are temperature changes, wind, refraction of a line of sight because of atmospheric conditions, and magnetic attraction.

An error is either a *systematic error* or a *random error.* A systematic error is one the magnitude and algebraic sign of which can theoretically be determined. If a tape is found to measure 99.94 ft between the 0-ft mark and the 100-ft mark when compared with a standard, then the full tape length introduces a systematic error of +0.06 ft each time it is used to measure the distance between two given points. If a tape is used at a temperature other than that at which it was compared with a standard, then the amount by which the non-standard temperature increases or decreases the length of the tape can be computed from known characteristics of the material of which the tape is made.

A random error is one the magnitude and sign of which cannot be predicted. It can be plus or minus. Random errors tend to be small and tend to distribute themselves equally on both sides of zero. If an observer reads and records a value of, say, 6.242 ft when the better value is 6.243 ft, a random error of −0.001 ft has been introduced. When an individual is holding a signal on which a transitman is sighting, failure to hold the signal directly over the proper point will cause a random error of unknown size and algebraic sign in the measured angle. If, however, he *fixes* the signal eccentrically, the resultant error will be systematic.

A *mistake* is not an error, but is a blunder on the part of the observer. Examples of mistakes are failure to record each full tape length in taping, misreading a tape, interchanging figures, and forgetting to level an instrument before taking an observation. Mistakes are avoided by exercising care in making measurements, by checking readings, by making check measurements, and to a great extent by common sense and judgment. If, for example, a leveling rod is read and the reading is recorded as 7.13 ft, whereas the levelman knows that this is absurd since he is very nearly at the top of a 14-ft rod, then he is exercising common sense in suspecting a mistake.

The subject of random errors is considered in more detail in Chapter 5. Systematic errors and methods for their elimination are discussed in the appropriate sections throughout the book.

1-11. ACCURACY AND PRECISION Since surveying is after all a measurement science, it is necessary to distinguish between the two terms *accuracy* and *precision* which, if not understood, cause needless confusion. The accuracy of a measurement is an indication of how close it is to the true value of the quantity that has been measured. In order to obtain an accurate measurement, one must have calibrated the measuring instrument by comparison with a standard. This allows for the elimination of systematic errors.

The precision of a measurement has to do with the refinement used in taking the measurement, the quality (but not necessarily the accuracy) of an instrument, the repeatability of the measurement, and the finest or least count of the measuring device.

As an illustration of the difference between these two terms, suppose that two different taping parties (Chapter 2) measure the same line five times, each using a different 100-ft tape. The first party reports the following measurements: 736.80, 736.70, 736.75, 736.85, and 736.65 ft. The second party reports the following measurements: 736.42, 736.40, 736.40, 736.42, and 736.41 ft. Further suppose that the correct or true length of the line is 736.72 ft. Obviously, from an examination of the spread of the results, the measurements reported by the second party are more precise. However, those reported by the first party are more accurate because they tend to group around the true value. Thus the tape used by the second party has some kind of systematic error that has not been accounted for in the reported measurements.

PROBLEMS

1-1. How many acres are contained in a rectangular area which measures 1000 by 2000 ft?

1-2. How many hectares are contained in the area given in Problem 1-1?

1-3. A distance of 79 chains 45 links shows on a map between two boundary markers. What is the corresponding length in feet?

1-4. What is the length of the line in Problem 1-3 in metres?

1-5. Convert the following decimal degrees to their corresponding values in the sexagesimal system: (a) 144.4°; (b) 217.13°; (c) 146.244°; (d) 10.5022°.

1-6. Convert the following angles to decimal degree form: (a) 26° 04′; (b) 182° 14′ 11″; (c) 44° 17′ 52.4″.

1-7. Convert the angles given in Problem 1-5 to their equivalent values in the centesimal system of grads.

1-8. Convert the angles given in Problem 1-6 to grads.

1-9. Convert the following angles to decimal degrees: (a) 14.42^g; (b) 262.422^g; (c) 171.3663^g; (d) 57.42266^g.

1-10. Convert the angles given in Problem 1-9 to their values in the sexagesimal system.

1-11. Convert the following lengths to metres: (a) 1285 ft; (b) 1406.2 ft; (c) 19805.02 ft; (d) 52.264 ft.

1-12. Convert the following lengths to feet: (a) 125 m; (b) 1466.8 m; (c) 5245.63 m; (d) 1048.224 m.

1-13. Convert the following volumes to cubic metres: (a) $14,626 \text{ yd}^3$; (b) $29,800 \text{ yd}^3$; (c) 2126.4 yd^3; (d) 1286.25 yd^3.

1-14. Convert 55,163.68 m to the equivalent length: (a) in survey feet; (b) in feet defined by the relationship 1 in. = 25.4 mm.

1-15. The sides of a triangle measure 725.55, 708.62, and 87.90 ft. Using the formulas given in Table C of Appendix C, compute the three angles in the triangle.

1-16. Compute the area of the triangle in square feet given in Problem 1-15.

1-17. The sides of a triangle are 1065.15, 1810.38, and 1320.95 m. Compute the three angles in the triangle.

1-18. Compute the area of the triangle in square metres given in Problem 1-17.

1-19. Two sides and the included angle of a triangle are 1818.20 m, 1474.85 m, and 37°14′20″, respectively. Compute the length of the remaining side and the two remaining angles.

1-20. Compute the area of the triangle given in Problem 1-19.

1-21. In triangle ABC, $A = 35° 14' 22''$; $B = 49° 18' 16''$; side $AB = c = 276.143$ m. Compute sides a and b.

1-22. Compute the area of the triangle given in Problem 1-21.

BIBLIOGRAPHY

Lyddan, R. H. 1977. U.S.G.S. Plans for conversion to the metric system. *Proc. Ann. Meet. Amer. Cong. Surv. & Mapp.* p. 1.

Strasser, Georg. 1975. The toise, the yard and the metre—the struggle for a universal unit of length. *Surveying and Mapping* 35:25.

Chapter 2
Measurement of
Horizontal Distances

2-1. HORIZONTAL DISTANCES   One of the basic operations of surveying is the determination of the distance between two points on the surface of the earth. In surveys of limited extent the distance between two points at different elevations is reduced to its equivalent horizontal distance either by the procedure used to make the measurement or by computing the horizontal distance from a measured slope distance. Distances are measured by scaling from a map, by pacing, by using an odometer, tacheometry, electronic distance meters (EDM's), or by taping. This chapter will emphasize the use of the tape and the EDM's.

2-2. PACING   Where approximate results are satisfactory, distances can be obtained by pacing. A person can best determine the length of his pace by walking over a line of known length several times, maintaining a natural walking stride. No particular advantage is obtained by developing a pace of, say, 3 ft. The natural stride is reproducible from day to day, whereas an artificial pace is not. The number of paces can be counted with a tally register or by the use of a pedometer, which is carried like a watch in a vertical position in the pocket.

2-3. ODOMETER The odometer of a motor vehicle will give fairly reliable distances along highways, provided that the odometer is periodically checked against a known distance. This method can often be used to advantage on preliminary surveys where precise distances are not necessary. The odometer is used to obtain distances for writing descriptions of the locations of survey control points and markers.

2-4. TACHEOMETRY Distances can be measured indirectly by the use of optical surveying instruments in conjunction with measuring bars or rods. The measurements are performed quite rapidly and are sufficiently accurate for many types of survey operations. Their use, however, is limited to providing check measurements in high-accuracy surveys, in which the measurements are performed either by direct taping or by EDM. Methods and operations used in this kind of measurement are referred to generally as *tacheometry.*

The telescope of an engineer's transit or theodolite usually contains three horizontal cross hairs that are used for tacheometry. The top and bottom hairs are called *stadia* hairs. These are usually so spaced that each unit intercepted between them on a graduated rod held vertically at a point some distance from the theodolite represents a distance of 100 units from the theodolite. Most distances obtained in topographic surveys conducted in the field are determined by stadia measurements. Both horizontal and vertical distances can be obtained by means of stadia tacheometry. This method is described in Chapter 14.

The self-reducing telemeter, which is discussed in Chapter 14, is another form of tacheometric instrument and performs generally the same function as the stadia hairs of the theodolite.

A somewhat more accurate method of tacheometry than the use of the transit and its variations is that of *subtense.* This consists of setting up a short base at one end of the line to be measured and then measuring the small horizontal angle subtended by the base with a precise theodolite at the other end of the line.

The short base, consisting of either a graduated rod or a rod containing a target on each end, is oriented so that it is in a horizontal position and is also perpendicular to the line being measured. The rod is called a *subtense bar.* The distance is computed from the isosceles triangle formed by the known base on the rod and the sides of the opposite measured angle. The subtense bar is discussed in greater detail in Chapter 14.

2-5. ELECTRONIC SURVEYING In recent years several ingenious electronic systems have been developed for the express purpose of measuring distances in surveying with a high degree of accuracy. They are based on the invariant velocity of light or electromagnetic waves in a vacuum. It is of interest to note that the first of these instruments, which is called the *geodi-meter*,† was the outgrowth of instrumentation developed to determine an accurate value of the velocity of light. The value adopted by the International Union of Geodesy and Geophysics in 1957 is 299,792.5 km/s, and its differs by only 0.4 km from that determined by geodimeter instrumentation in 1951.

A discussion of the use of the various EDM instruments for determining distances is presented following the sections that describe the use of the tape. This newer system has practically displaced the tape in modern surveys except for relatively short measurements and for layout work in construction surveys.

2-6. CHAINS Two kinds of chains were formerly used in surveying, namely, the 100-ft engineer's chain and the 66-ft Gunter's chain. Each kind is divided into 100 links. A link of an engineer's chain is therefore 1 ft long, and a link of a Gunter's chain is only 0.66 ft or 7.92 in. long. Such chains are seldom, if ever, used at present. Although the 66-ft chain is still employed as a *unit* of length in the survey of the public lands and of some farm lands, the actual field measurements are made with steel tapes graduated in chains and links. If the distances recorded are in chains, a chain of 66 ft is implied.

2-7. TAPES Steel tapes for most surveying operations are graduated in feet or metres together with decimal parts of these units. Their lengths vary from 50 to 300 ft and from 15 to 100 m although the 100-ft and the 30-m tape are the most common length. Lightweight tapes may be graduated to hundredths of a foot or centimetre for the entire length. However, the usual foot tape is graduated to feet, with an end foot divided to hundredths of a foot. Metric tapes are usually graduated to decimetres throughout with an end decimetre divided to millimetres.

As most engineering and architectural plans show dimensions in feet and inches, a tape graduated in feet and inches is an advantage on construction work for layout purposes.

Tapes graduated in metres are used in most countries outside of the United States. These tapes are also used on most geodetic work in all countries.

Tapes of cloth, or of cloth containing threads of bronze or brass, are sometimes used where low precision is permissible and where a steel tape might be broken, as in cross sectioning for a railroad or a highway.

For extreme precision an invar tape, made of an alloy of steel and nickel, is used. The advantage of a tape of this material is that its coefficient of thermal expansion is about one-thirtieth that of steel, and hence its length is not so seriously affected by temperature changes. However, since such a tape is

† Geodimeter is a registered trade mark of AGA Aktiebolag.

expensive and must be handled very carefully to prevent kinking, invar tapes are not used for ordinary work.

Tapes are calibrated by comparison with a standard which is maintained by the National Bureau of Standards and by certain state, county, and city agencies. A few universities and state sections of the American Congress on Surveying and Mapping also maintain standard lengths by which tapes can be calibrated. The owner of the tape specifies under what conditions the tape should be calibrated, that is, what tension should be applied to the tape, and whether it is to be supported throughout its entire length or whether it is to be supported only at the ends. The calibration report then gives the length of the tape under the specified conditions and at some standard temperature, usually 68°F or 20°C.

2-8. EQUIPMENT USED FOR TAPING For the direct measurement of a line several hundred feet or metres long, the equipment used consists of a 100-ft or 30-m steel tape, two plumb bobs, one or more line rods, a set of taping pins, and, if the ground is hilly, a hand level. These items are shown in Fig. 2-1. A spring balance, which can be seen to the extreme right in Fig. 2-3, is used to apply the desired pull or tension to the tape during the measurement. The tension is expressed in pounds or kilograms.

Line rods, also called "range poles," are from $\frac{3}{8}$ in. to more than 1 in. in diameter and from 6 to 8 ft (2 m) long. They are pointed at one end and are painted with alternate bands of white and red. The rods are used to sight on and thus keep the forward and rear ends of the tape on the line that is being measured.

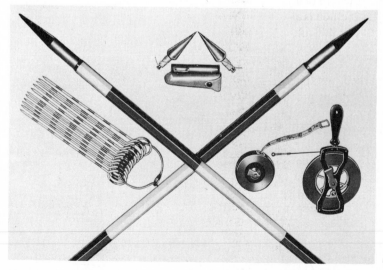

Figure 2-1. Equipment used for taping. Courtesy of W. & L. E. Gurley Co.

Taping pins are used to mark the positions of the ends of the tape on the ground while a measurement is in progress. A taping pin may be a heavy spike, but is usually a piece of No. 10 wire that is 10 to 18 in. long, is sharpened at one end, and is provided with an eye at the other end. Pieces of colored cloth can be tied to the eyes to make the pins more visible in tall grass or weeds.

The hand level is used to keep the two ends of the tape in the same horizontal plane when a measurement is made over rough or sloping ground.

The plumb bobs are used to project a point on the ground up to the tape, or to project a point on the tape down to the ground.

Some tapes are kept on reels when not in use. But a metal tape must be entirely removed from the reel when any length other than a few feet or metres is to be measured. If such a tape is not supplied with thongs on both ends with which to hold the tape, a taping pin can be slipped through the eye at the end of the tape and used as a handle. A tape that is thrown together in the form of a series of loops when not in use must be carefully unwrapped and checked for short kinks before it can be used for measurement. As long as a tape is stretched straight, it will stand any amount of tension that two people can apply. If kinked or looped, however, a very slight pull is sufficient to break it.

2-9. MEASUREMENTS WITH TAPE HORIZONTAL   The horizontal distance between two points can be obtained with a tape either by keeping the tape horizontal or by measuring along the sloping ground and computing the horizontal distance. For extreme precision, such as is required in the length of a baseline in a triangulation system, the latter method is used. This method is also advantageous where steep slopes are encountered and it would be difficult to obtain the horizontal distance directly.

For moderate precision where the ground is level and fairly smooth, the tape can be stretched directly on the ground, and the ends of the tape lengths can be marked by taping pins or by scratches on a paved area. Where the ground is level but ground cover prevents laying the tape directly on the ground, both ends of the tape are held at the same distance above the ground by the forward tapeman and the rear tapeman. The tape is preferably held somewhere between knee height and waist height. The graduations on the tape are projected to the ground by means of the plumb bobs. The plumb-bob string is best held on the tape graduation by clamping it with the thumb, so that the length of the string can be altered easily if necessary (this can be seen in Fig. 2-9). When a tape is supported throughout its length on the ground and subjected to a given tension, a different value for the length of a line will be obtained than when the tape is supported only at the two ends and subjected to the same tension (see Section 2-18). Where fairly high accuracy is to be obtained, the method of support must be recorded in the field notes, provided both methods of support are used on one survey. Experienced tapemen should obtain as good results by plumbing the ends of the tape over the marks as they will obtain by having the tape supported on the ground.

When the ground is not level, either of two methods may be used. The first is to hold one end of the tape on the ground at the higher point, to raise the other end of the tape until it is level, either by estimation or with the aid of the hand level, and then to project the tape graduation over the lower point to the ground by means of a plumb bob. The other method is to measure directly on the slope as described in Sections 2-11 and 2-12. These methods are shown in Fig. 2-2.

For high precision, a taping tripod or taping buck must be used instead of a plumb bob. Such a tripod is shown in Fig. 2-3. Taping tripods are usually used in groups of three, the rear tripod then being carried to the forward position. A pencil mark is scribed at the forward tape graduation, and on the subsequent measurement the rear tape graduation is lined up with this mark in order to carry the measurement forward. Since taping is usually done on the slope when tripods are used, the elevations of the tops of the tripods must be determined at the same time the taping proceeds. The elevations, which are determined by leveling (Chapter 3), give the data necessary to reduce the slope distances to horizontal distances as discussed in Section 2-12.

The head tapeman carries the *zero* end of the tape and proceeds toward the far end of the line, stopping at a point approximately a tape length from the point of beginning. The rear tapeman lines in the forward end of the tape by sighting on the line rod at the far end of the line. Hand signals are used to bring the head tapeman on line. The rear tapeman takes a firm stance and holds the tape close to his body with one hand, either wrapping the thong around his hand or holding a taping pin which has been slipped through the

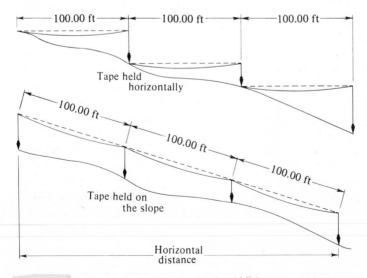

Figure 2-2. Taping over sloping ground using 100-ft tape.

Figure 2-3. Taping tripod.

eye of the tape as shown in Fig. 2-4. Standing to the side of the tape, he plumbs the end graduation over the point on the ground marking the start of the line. The tip of the plumb bob should be less than $\frac{1}{8}$ in. or about 3 mm above the ground point.

The head tapeman applies the tension to be used, either by estimation or by means of a spring balance fastened to the zero or forward end of the tape. At approximately the correct position on the ground, he clears a small area where the taping pin will be set. After again applying the tension, the head tapeman waits for a vocal signal from the rear tapeman indicating that the latter is on the rear point. When the plumb bob has steadied and its tip is less than $\frac{1}{4}$ in. or about 5 mm from the ground, the head tapeman dips the end of the tape slightly so that the plumb bob touches the ground. Then he, or a third member of the taping party, sets a taping pin at the tip of the plumb bob to mark the end of the first full tape length as shown in Fig. 2-5. The pin is set at right angles to the line and inclined at an angle of about 45° with the ground away from the side on which the rear tapeman will stand for the next measurement. The tape is then stretched out again to check the position of the pin. The notekeeper records the distance, 100.00 ft, or 30.000 m, in the field notes. The tape is advanced another tape length, and the entire process is repeated.

Figure 2-4. Plumbing over point.

If the taping advances generally downhill, the head tapeman checks to see that the tape is horizontal by means of the hand level. If the taping advances generally uphill, the rear tapeman checks for level.

When the end of the line is reached, the distance between the last pin and the point at the end of the line will usually be a fractional part of a tape length. The rear tapeman holds the particular full-foot or decimetre graduation that will bring the subgraduations at the zero end of the tape over the point marking the end of the line. The head tapeman rolls the plumb-bob string

Figure 2-5. Setting taping pin to mark forward position of tape.

along the subgraduations with his thumb until the tip of the plumb bob is directly over the ground point marking the end of the line.

Two types of end graduations of a tape that reads in feet are shown in Fig. 2-6. In view (a) the subgraduations are outside the zero mark, and the fractional part of a foot is added to the full number of feet. Hence the distance is 54 + 0.44 = 54.44 ft. This type is referred to as an *add* tape. In view (b) the subgraduations are between the zero and the 1-ft graduation, and the fractional part of a foot must be subtracted from the full number of feet. So the distance is 54 − 0.28 = 53.72 ft. This type is called a *cut* tape. Because of the variation in the type of end graduations, the rear tapeman must call out the

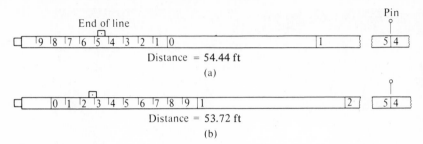

Figure 2-6. Graduations at end of foot-graduated tape.

actual foot mark he holds, and both the head tapeman and the notekeeper must agree that the value recorded in the field notes is the correct value.

Add tapes are more convenient to use than cut tapes simply because it is easier to add than to subtract the decimal part of the whole unit. The design of an add metric tape which is divided in decimetres throughout its length is shown in Fig. 2-7. The outside decimetre is further subdivided to centimetres and then to millimetres. In the illustration, the rear tapeman holds the 14.7-m mark at the pin, and the head tapeman reads 0.072 m at the end of the line. The distance is thus $14.7 + 0.072 = 14.772$ m.

Where the slope is too steep to permit bringing the full length of the tape horizontal, the distance must be measured in partial tape lengths, as shown in Fig. 2-8. It is then necessary to enter a series of distances in the field notes. Some or all of them will be less than a full tape length. For a partial tape length, the head tapeman holds the zero end and the rear tapeman holds a convenient whole foot or decimetre mark which will allow the selected length of tape to be horizontal. When the forward pin is set, this partial tape length is recorded in the field notes. The head tapeman than advances with the zero end of the tape, and the rear tapeman again picks up a convenient whole foot or decimetre mark and plumbs it over the pin. Each partial tape length is recorded as it is measured or as the forward pin is set. In Fig. 2-9 is illustrated the use of a device called a *tape clamp* for holding a tape at any place other than at an end.

If a tape clamp is not available, the rear tapeman must then hold the tape in one hand in such a manner that it neither injures his hand nor damages the tape. At the same time he must be able to sustain a tension of between 10 and

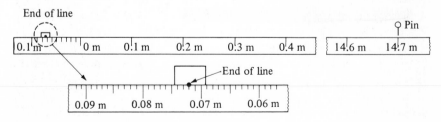

Figure 2-7. Graduations at end of metric tape.

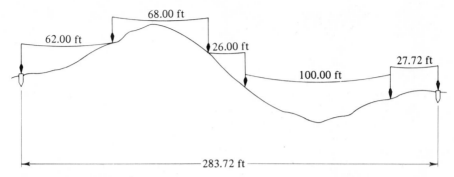

68.00 ft

62.00 ft

26.00 ft

27.72 ft

100.00 ft

283.72 ft

Figure 2-8. Breaking tape.

20 lb, or between 5 and 10 kg. The technique shown in Fig. 2-10 is a satisfactory solution to this problem. The tape is held between the fleshy portion
of the fingers and that of the palm. Enough friction is developed to sustain a
tension upward of 25 to 30 lb (10 to 15 kg) without injury or discomfort to
the tapeman. He must not turn his hand too sharply, however, otherwise the
tape may become kinked.

All distances should be taped both forward and backward, to obtain a
better value of the length of the line and to detect or avoid mistakes. When the
backward measurement is made, the new positions of the pins should be

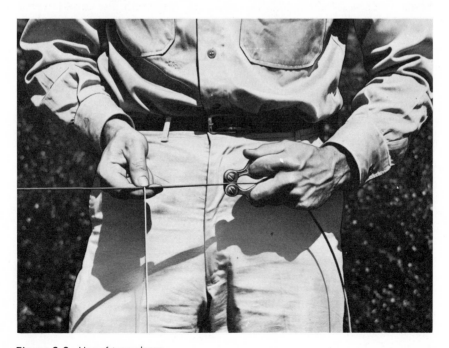

Figure 2-9. Use of tape clamp.

Figure 2-10. Method of holding tape when clamp is not available.

completely independent of their previous positions. This practice eliminates the chance of repeating a mistake.

2-10. TENSION Most steel tapes are correct in length at a temperature of 68°F (20°C) when a tension of 10 to 12 lb or 5 kg is used and the tape is supported throughout the entire length. If this same tension is used when the tape is suspended from the two ends, the horizontal distance between the ends of the tape will be shorter than the nominal length. The amount of the shortening depends on the length and the weight of the tape. A light 100-ft tape weighs about 1 lb. Such a tape, when suspended from the two ends, would be shortened about 0.042 ft. A heavy 100-ft tape weighs about 3 lb and would be shortened about 0.375 ft. Some engineers attempt to eliminate this error by increasing the tension used. The tension for the light tape is then increased to about 18 lb. It is practically impossible to eliminate the error in the heavy tape by this method, as the tension would have to be increased to about 50 lb. Generally, less tension is used and a correction is applied to each measured length.

2-11. REDUCTION OF SLOPE MEASUREMENTS BY VERTICAL ANGLES Where slopes are considerable or where high accuracy is required, or where both these conditions occur, taped measurements are made directly along the slopes as stated in Section 2-9. The slope distances are then reduced to the corresponding horizontal distances by computations. In order that these computations can be performed, either the vertical angle of the slope measurement or the difference in elevation between the two ends of the tape or the line must be known.

In Fig. 2-11 a transit or theodolite (Chapter 4) is set up over one end of the line to be measured. The inclination of the telescope is set so that the line of sight of the transit will be generally parallel with the slope of the ground. The instrumentman guides the two tapemen in order to keep the tape on line and on a constant slope. He then reads the vertical angle α in order to compute the horizontal distance. If the distance measured along the slope is s, and the slope makes a vertical angle α with the horizontal, the corrected horizontal distance is then

$$H = s \cos \alpha \qquad (2\text{-}1)$$

Example 2-1

A measurement is made along a line that is inclined by a vertical angle of $3° \, 22'$ as measured using an engineer's transit. The slope measurement is 3236.86 ft. What is the corresponding horizontal distance?

Solution: Using Eq. (2-1),

$$H = 3236.86 \times 0.998274 = 3231.27 \text{ ft}$$

Example 2-2

A horizontal distance of 745.00 ft is to be established along a line that slopes at a vertical angle of $5° \, 10'$. What slope distance should be measured off?

Solution: Using Eq. (2-1),

$$s = \frac{745.00}{0.99594} = 748.04 \text{ ft}$$

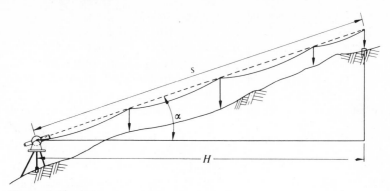

Figure 2-11. Slope taping using transit or theodolite.

The accuracy with which a vertical angle must be measured in order to reduce a slope distance to the corresponding horizontal distance will now be investigated. Accuracy in distances can be considered in one of two ways: an absolute accuracy and a relative accuracy. If a 10,000-ft distance is measured with an accuracy of a foot, then the implied absolute accuracy is 1 ft. The relative accuracy in this instance is 1 ft divided by 10,000 ft or 1/10,000 or 1 : 10,000. A statement of relative accuracy is preferred because it is independent of the lengths of lines in the survey.

Equation (2-1) expresses the horizontal distance H in terms of the measured slope distance s and the angle of inclination α. To determine the accuracy with which the vertical angle must be measured in order to meet a given relative accuracy in the resulting horizontal distance, differentiate Eq. (2-1) with respect to α, thus

$$dH = -s \sin \alpha \, d\alpha \qquad (2\text{-}2)$$

Disregarding the minus sign, the relative accuracy is then

$$\frac{dH}{H} = \frac{s \sin \alpha \, d\alpha}{s \cos \alpha} = \tan \alpha \, d\alpha \qquad (2\text{-}3)$$

The small angle $d\alpha$ can be computed for any desired relative accuracy and a particular slope angle.

Example 2-3

To what accuracy must the slope angle of Example 2-1 be measured if the relative accuracy of the horizontal distance is to be 1/25,000?

Solution: By Eq. (2-3),

$$\tan \alpha \, d\alpha = \frac{1}{25,000} \quad \text{and} \quad d\alpha = \frac{1}{25,000 \tan 3° \, 22'}$$

Then (see Section 1-3)

$$d\alpha'' = \frac{206,265}{25,000 \times 0.05883} = 140'' = 2'\ 20''$$

Example 2-4

To what accuracy must the slope angle of Example 2-1 be measured if the horizontal distance is to be accurate to 0.02 ft?

Solution: By Eq. (2-2),

$$s \sin \alpha\ d\alpha = 0.02 \text{ ft} \quad \text{and} \quad d\alpha'' = \frac{0.02 \times 206,265}{3236.86 \times 0.05873} = 21.7''$$

The same results would be obtained by defining the relative accuracy as 0.02/3236.86 and then using Eq. (2-3) as shown in Example 2-3.

Equation (2-3) can be used to compute a curve showing the necessary vertical angle accuracy required for different slope angles if the relative accuracy is known. Figure 2-12 shows three such curves for three different relative accuracies. An examination of these curves shows that 1′ accuracy in measuring vertical angles is adequate for 1 : 10,000 relative accuracy except for angles of more than 20°, which is fairly infrequent. On the other hand, in order to meet the 1 : 100,000 criterion the accuracy of vertical angles must be better than 10″ for slopes of more than 10°. An investigation of the accuracy

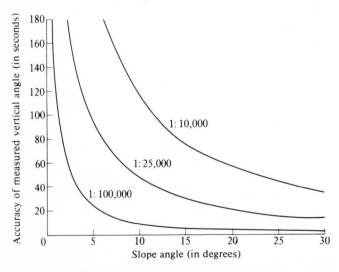

Figure 2-12. Accuracy required in measuring vertical angles when slope taping.

required in measuring slope angles becomes important when one considers the very high accuracies of distance measurement using EDM's (Section 2-24). Poorly measured vertical angles degrade the overall accuracy of the distance measurements.

2-12. REDUCTION OF SLOPE MEASUREMENTS BY DIFFER-ENCE IN ELEVATION Measurements made on the slope can be reduced to their corresponding horizontal distances if the differences in elevation between the two ends of the tape have been measured by leveling (see Chapter 3). One method is to set a hub (a 1 × 1 in. or 2 × 2 in. stake driven flush with the ground) or a stake at each point marking the end of the tape after the taping pin has been pulled, and then to run a line of levels over these points in order to compute successive differences in elevation between the points. This technique assumes that the two tapemen hold their end of the tape the same height above the ground, so that the measured difference in elevation is in fact the difference in elevation between the two ends of the tape.

In Fig. 2-13, s is the tape length along the sloping line, h is the difference in elevation between the two ends of the tape, and H is the corresponding horizontal distance to be determined. In this right-angled triangle,

$$H = (s^2 - h^2)^{1/2} \qquad (2\text{-}4)$$

In precise measurement of baselines (see Section 10-11) or in high-order control surveys, in which the taping tripod described in Section 2-9 is employed, a level party either immediately precedes or follows the taping party in order to measure the elevations of the successive positions of the tripod tops before the tripods are moved forward along the line. Section 10-11 discusses the technique of setting taping posts into the ground along the baseline in lieu of taping tripods. The elevations of the tops of the posts are determined just as in the case of tripods. Because of the permanency of the posts, however, there is no need for both the taping party and the level party to operate at the same time.

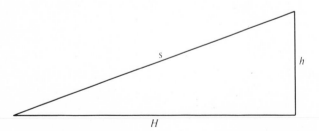

Figure 2-13. Reduction of slope distance using difference in elevation.

If a long sloping line is to be measured using the plumb bobs and taping pins, and if the ground slope is fairly uniform, the two tapemen can then parallel the ground with the tape by holding the two ends an equal distance above the ground for each tape length. The difference in elevation between the two ends of the line is then measured using the level. However, if the slope of the line is irregular, hubs must be set at each tape length as discussed above. Each tape length is then reduced separately.

Example 2.5

A line measures 1446.25 ft along a constant slope. The difference in elevation between the two ends of the line is 57.24 ft. What is the horizontal length of the line?

Solution: By Eq. (2-4),

$$H = (1446.25^2 - 57.24^2)^{1/2} = 1445.12 \text{ ft}$$

Example 2-6

What slope distance must be laid out along a line that rises 6 ft/100 ft in order to establish a horizontal distance of 750 ft?

Solution: The difference in elevation between the two ends of the line is $6 \times 7.50 = 45.00$ ft. Then by Eq. (2-4), $s = (H^2 + h^2)^{1/2}$ or

$$s = (750^2 + 45^2)^{1/2} = 751.35 \text{ ft}$$

The difference C between the slope distance and the horizontal distance can be determined by expanding the right side of Eq. (2-4) by the binomial expansion to give

$$H = s - \frac{h^2}{2s} - \frac{h^4}{8s^3} - \cdots \qquad (2\text{-}5)$$

Thus, neglecting the last term in the above expansion, gives

$$s - H = C = \frac{h^2}{2s} \qquad (2\text{-}6)$$

An analysis of the accuracy required in determining the difference in elevation between the two ends of a tape in slope measurements can be made by differentiation of Eq. (2-6) with respect to h, thus,

$$dC = \frac{h \, dh}{s}$$

If both sides are divided by s, the relative accuracy expression is obtained, thus

$$\frac{dC}{s} = \frac{h\,dh}{s^2} \tag{2-7}$$

Example 2-7

A baseline is measured with a 50-m tape. The error in leveling over the tops of the taping posts is not to contribute more than one part in 100,000 error to the relative accuracy of the measurement. The largest difference in elevation between any two successive posts is 3.60 m. With what accuracy must this difference in elevation be determined?

Solution: By Eq. (2-7), the value of dC/s must be 1/100,000 which in turn is $(h\,dh)/s^2$. Thus

$$dh = \frac{s^2}{100{,}000h} = \frac{50^2}{100{,}000 \times 3.60} = 0.007 \text{ m}$$

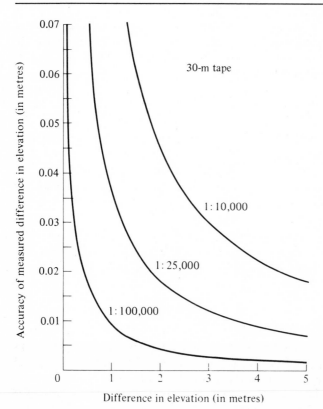

Figure 2-14. Accuracy required in measuring difference in elevation between two ends of a 30-m tape when slope taping.

Equation (2-7) can be used to compute curves for a given tape length that show the necessary accuracy in determining elevations for various differences of elevation and for different accuracy ratios. The curves shown in Fig. 2-14 are for a 30-m tape at three different accuracy ratios.

2-13. SYSTEMATIC ERRORS IN TAPING

The principal systematic errors in linear measurements made with a tape are (1) incorrect length of tape, (2) tape not horizontal, (3) fluctuations in the temperature of the tape, (4) incorrect tension or pull, (5) sag in the tape, (6) incorrect alignment, and (7) tape not straight.

2-14. INCORRECT LENGTH OF TAPE

At the time of its manufacture, a tape is graduated when under a tension of about 10 lb or 4.5 kg. A steel tape will maintain a constant length under a considerable amount of handling and abuse. This is not true, however, of an invar tape which must be handled with great care. In either case, if a tape is compared with a standard length under specific conditions of temperature, tension, and method of support, the distance between the two end graduations will seldom equal the nominal length indicated by the graduation numbers. The correction to be applied to any measurement made with the tape in order to account for this discrepancy is called the *absolute correction* C_a, and is given by

$$C_a = \text{true length} - \text{nominal length} \tag{2-8}$$

The true length is the value determined by calibration under specific conditions. The calibration or standard tension will range anywhere from 10 to 30 lb or 5 to 15 kg, and is specified by the user of the tape. The calibration comparison can be made with the tape supported throughout its length, or supported only at the two ends, or supported at the two ends and at one or more intermediate points.

The absolute error $-C_a$ is usually assumed to be distributed uniformly throughout the length of the tape. Thus the absolute correction in a measured distance is directly proportional to the number and fractional parts of the tape used in making the measurement.

2-15. TAPE NOT HORIZONTAL

If the tape is assumed to be horizontal but actually is inclined, an error is introduced. The amount of this error C can be computed from Eq. (2-6) or may be taken as $h^2/2s$. If one end of a 100-ft tape is 1.41 ft higher or lower than the other, the error will amount to 0.01 ft. It should be noted that the error is proportional to the square of the vertical distance. When one end is 2.82 ft above or below the other, the error increases to 0.04 ft.

Errors from this source are cumulative and may be considerable when measuring over hilly ground. The error can be kept at a minimum by using a hand level to determine when the tape is horizontal.

2-16. CHANGES IN TEMPERATURE The coefficient of thermal expansion of steel is about 0.0000065/°F or 1.15×10^{-5}/°C. As a difference of 15°F in temperature will cause a change in length of nearly 0.01 ft in a 100-ft tape, a measurement made with a tape on an extremely hot day in the summer cannot be expected to agree with one made during zero weather. A difference of 3°C changes the length of a 30-m tape by 1 mm.

If the temperature remains constant and is different from that at which the tape was standardized, the resulting error will be cumulative and in direct proportion to the number of tape lengths measured. When the field temperature differs considerably from the standard, the temperature should be determined and a correction should be applied to the measured length.

The amount of the correction for temperature is determined by the formula

$$C_t = L\alpha(T - T_s) \qquad (2\text{-}9)$$

in which C_t is the correction to be applied, in feet or metres; L is the length of the tape actually used, in feet or metres; α is the coefficient of thermal expansion; T is the temperature at which the measurement is made; T_s is the temperature at which the tape was standardized.

Since the temperature of the tape may be considerably different from that of the surrounding air on a bright day, important measurements should be made on a cloudy day, or early in the morning, or late in the afternoon.

The advantage of an invar tape over the steel tape is its very low coefficient of thermal expansion, which is 0.0000002 or less per degree Fahrenheit or 3.60×10^{-7}/°C. Because the effective temperature of a tape is very difficult to determine, even with the best thermometers, the invar tape will give better results than steel and should be used in surveys of high order. Even considering this advantage, however, the best results with an invar tape will be obtained on an overcast day or even at night.

The tape thermometer attached to the tape is fairly responsive to temperature changes. If taping is performed with the tape supported on the ground, the tape either takes on or gives off heat from or to the supporting material. Since different portions of the tape can be supported by different types of material—for example, blacktop pavement, grassy sod, concrete—the only place where the observed temperature is reasonably correct is that portion of the tape to which the thermometer is fastened. If the tape is suspended in air during the measurements, it reaches a state of thermal equilibrium in which the tape temperature is uniform throughout its length (except for spotty shaded areas). If the tape has been lying on the ground and then picked up for stretching, whoever reads the thermometer must wait for a minute or so for the tape to come to thermal equilibrium. Otherwise the effect of the tape being in contact with the ground will cause a considerable error in the observed temperature, sometimes as much as 30°F or 16°C.

2-17. INCORRECT TENSION

If a tape is used in the field under a tension different from the standard tension used in calibration, the tape will change length a slight amount according to the relationship between stress and strain. The amount to be added to or subtracted from the measured length in order to account for the difference in tension is a function of the measured length, the field tension, the standard tension, the cross-sectional area of the tape, and the modulus of elasticity of the material of which the tape is made. This is referred to as the *correction for pull* or tension and is given by

$$C_p = \frac{(P - P_s)L}{AE} \tag{2-10}$$

in which C_p is the correction per tape length, in feet or metres; P is the tension applied, in pounds or kilograms; P_s is the standard tension, in pounds or kilograms; L is the length in feet or metres; A is the cross-sectional area, in square inches or square centimetres; E is the modulus of elasticity of the steel. The modulus of elasticity of steel is about 28,000,000 to 30,000,000 psi (pounds per square inch) or 2,100,000 kg/cm².

A light tape, weighing about 1 lb, will have a cross-sectional area of about 0.0030 in.² An increase of 10 lb in tension will stretch a 100-ft tape about 0.011 ft. The same increase with a heavy tape, weighing about 3 lb, will produce a change in length of but 0.004 ft.

On careful work, a spring balance is used for maintaining a constant tension. If the spring balance is not used, errors from incorrect tension will tend to be compensating, since the tensions applied to the tape may be either greater or less than the standard tension. The tendency is usually to underpull, rather than to exceed the standard tension.

2-18. SAG

If a tape has been standardized when supported throughout its entire length, a correction must be applied to every measurement that is made with the tape suspended from the two ends, or the tension must be increased. A tape suspended in this way will take the form of a catenary, and the horizontal distance between the two ends will be less than when the tape is supported throughout its entire length. If the amount of the sag is such that the center of a 100-ft tape is about $7\frac{3}{8}$ in. below the two ends, the measured distance has been shortened by 0.01 ft. The correction for sag is given by the formula

$$C_s = n\frac{w^2 l^3}{24P^2} \tag{2-11}$$

in which C_s is the correction for sag, in feet or metres; n is the number of unsupported lengths; w is the weight per foot of tape, in pounds or kilograms per metre; l is the unsupported length, in feet or metres; P is the tension applied, in pounds or kilograms.

When a tape is suspended from the two ends and subjected to the standardized tension for the tape supported throughout its length, the observed distance is always greater than the true distance, and the error due to sag is constant.

If the total weight of the tape is W and the total length is L, then Eq. (2-11) can be expressed as

$$C_s = \frac{W^2 L}{24 P^2} \tag{2-12}$$

in which C_s is then the difference in distance between the ends of the tape when supported throughout and when supported only at the two ends.

Error from sag can be eliminated by applying the necessary corrections to the observed distances or, in the case of a light tape, by increasing the tension sufficiently to compensate for the effect of sag. When a tape having a total weight of W lb or kg is supported at the two ends, this required tension P_n, which is called the *normal tension* for the tape, may be found by solving the following formula by trial:

$$P_n = \frac{0.204 W \sqrt{AE}}{\sqrt{P_n - P_s}} \tag{2-13}$$

For a light tape, the tension would be increased from 10 to about 18 lb. For a heavy tape, the normal tension would be about 50 lb, which would be difficult to apply in practice.

Example 2-8

A 30-m tape weighing 0.900 kg has a cross-sectional area of 0.0485 cm². The tape measures 30.000 m when supported throughout under a tension of 5 kg. The modulus of elasticity is 2.1×10^6 kg/cm². What tension is required to make the tape measure 30 m when supported only at the two ends?

Solution: Using Eq. (2-13), try 10 kg as a first approximation.

$$10 = \frac{0.204 \times 0.900 \sqrt{0.0485 \times 2,100,000}}{\sqrt{10 - 5}} = \frac{58.594}{\sqrt{5}} = 26.204 \text{ kg}$$

Try 18 kg.

$$18 = \frac{58.594}{\sqrt{13}} = 16.25 \text{ kg}$$

Try 17 kg.

$$17 = \frac{58.594}{\sqrt{12}} = 16.91 \text{ kg}$$

Try 16.95 kg.

$$16.95 = \frac{58.594}{\sqrt{11.95}} = 16.95 \text{ kg} \quad (\text{checks})$$

For extremely accurate work the tape is supported at enough points to render the error due to sag negligible; or the tape is standardized when it is supported in the manner in which it is to be used in the field.

2-19. INCORRECT ALIGNMENT   If a field measurement comprises more than one tape length and the points marking the ends of the various lengths are not along a straight line, the measured length will be too great. The amount of error for any given tape length may be computed from Eq. (2-6), where h is the amount one end of the tape is off line. To produce an error of 0.01 ft in a 100-ft measurement, one end of the tape would have to be 1.41 ft off line. Under normal circumstances the rear tapeman should be able to keep the forward tapeman much closer to the true line than this, and the error from this source should be practically negligible. For measurements of high precision the tapemen can be kept on line with a theodolite.

2-20. TAPE NOT STRAIGHT   If the tape is not stretched straight, as when it is being bent either horizontally or vertically around trees or bushes or is blown by a strong wind, the measured length will be too great. The error is the least when the obstruction is near the middle of the tape. If the middle point of a 100-ft tape is 0.71 ft or about $8\frac{1}{2}$ in. off line, the resulting error in length is 0.01 ft. Errors from this source are cumulative.

2-21. APPLYING CORRECTIONS TO TAPE MEASUREMENTS
In order to illustrate the application of corrections to tape measurements, the following examples are given.

Example 2-9

A steel tape is standardized at 68°F under a tension of 15 lb when supported throughout its entire length, and the distance between the zero mark and the 100-ft mark is 99.98 ft. The tape weighs 0.013 lb/ft and has a cross-sectional area of 0.0040 in.^{2} In the field this tape is used under a 15-lb tension and is supported at the ends only, and the temperature of the tape is recorded as 88°F throughout the measurement. A series of distances are measured with the tape held horizontal, and the observed distances are recorded as 100.00, 100.00, 100.00, and 57.22 ft. What is the total actual or true distance after corrections for systematic errors have been applied?

Solution: The total observed distance is 357.22 ft. The errors for incorrect length, for temperature, and for incorrect tension are all in direct proportion to the length of the measurement. Therefore the total correction for incorrect length is

$$-0.02 \times 3.57 = -0.071 \text{ ft}$$

and the total correction for temperature is

$$C_t = 357.22 \times 0.0000065 \times (88 - 68) = +0.046 \text{ ft}$$

The correction for incorrect tension is zero, since the standard pull was used when the measurements were made.

For computing the correction for sag, the three 100-ft measurements are treated together and the 57-ft measurement is treated separately. For the three 100-ft measurements, by Eq. (2-11),

$$C_s = 3 \times \frac{0.013^2 \times 100^3}{24 \times 15^2} = -0.094 \text{ ft}$$

For the 57-ft measurement,

$$C_s = \frac{0.013^2 \times 57.2^3}{24 \times 15^2} = -0.006 \text{ ft}$$

The total sag correction is −0.100 ft.

The total correction to be applied to the measured distance is −0.071 + 0.046 − 0.100 = −0.125 ft. To the nearest hundredth, it is −0.13 ft. So the corrected distance is 357.09 ft.

Example 2-10

A steel tape is standardized at 68°F under a tension of 15 lb when supported at the two ends only, and the distance between the zero mark and the 100-ft mark is 99.972 ft. The total weight of the tape is 2.80 lb, and its cross-sectional area is 0.0088 in.² For steel E is assumed to be 30,000,000 psi. What is the distance between the zero and 100-ft marks when the tape is supported at the two ends only, the tension is 25 lb, and the standard temperature prevails?

Solution: First determine the distance when the tape is supported throughout and the tension is 15 lb. Then determine the distance for support throughout and a 25-lb pull. Finally, determine the distance when the tape is supported only at the two ends under a 25-lb pull. When the tape is supported throughout under a 15-lb pull, the distance between the end graduations will be greater than that with the tape supported only at the ends by the following amount:

$$C_s = \frac{0.028^2 \times 100^3}{24 \times 15^2} = 0.145 \text{ ft}$$

The distance for support throughout and a 15-lb pull is

$$99.972 + 0.145 = 100.117 \text{ ft}$$

When the tension is increased to 25 lb with the tape supported throughout, the additional increase in length is

$$C_p = \frac{(25 - 15) \times 100}{0.0088 \times 30,000,000} = 0.004 \text{ ft}$$

The distance for support throughout and a 25-lb pull is 100.121 ft.

Finally, when the support is taken away and the tape is supported only at the ends under a 25-lb pull, the distance will be less than that for support throughout. The correction is

$$C_s = \frac{0.028^2 \times 100^3}{24 \times 25^2} = 0.052 \text{ ft}$$

The distance between end graduations when the tape is supported at the ends only under a tension of 25 lb is 100.069 ft.

Example 2-11

A steel tape with a cross-sectional area of 0.039 cm^2 and weighing 0.909 kg has a length of 30.009 m between the zero and the 30-m marks when supported throughout at 20°C and subject to a tension of 5.5 kg. This tape is used to measure a distance along a uniform 4% grade and is supported throughout during the measurement. The tension applied is 12 kg. The temperature of the tape is 2°C. The measured slope distance is recorded as 381.615 m. What is the corrected horizontal distance?

Solution: All errors affecting this measurement are in direct proportion to the distance considered since there is no sag correction. For each tape length,

$$C_a = 30.009 - 30.000 = +0.0090 \text{ m}$$

$$C_t = 30 \times 1.15 \times 10^{-5}(2 - 20) = -0.0062 \text{ m}$$

$$C_p = \frac{(12 - 5.5) \times 30}{0.039 \times 2.1 \times 10^6} = +0.0024 \text{ m}$$

The length of the tape as used in the field is thus $30.0000 + 0.0090 - 0.0062 + 0.0024 = 30.0052$ m. The correct slope distance is thus

$$s = 381.615 + \frac{381.615}{30} \times 0.0052 = 381.681 \text{ m}$$

The slope angle is $\tan^{-1} 0.04 = 2° 17'$, and the correct horizontal distance is $381.681 \cos 2° 17' = 381.376$ m.

2-22. RANDOM TAPING ERRORS Random errors are introduced into taping measurements from several causes. Some of these are the following: (1) error in determining temperature of tape; (2) failure to apply the proper tension; (3) wind deflecting the plumb bob; (4) taping pin not set exactly where the plumb bob touched the ground; (5) inability of the tapeman to steady the plumb bob; (6) inability of the observer to estimate the last place in reading between graduations. Random errors and their effects on measurements will be discussed more fully in Chapter 5.

2-23. SPECIFICATIONS FOR LINEAR MEASUREMENTS In the preceding articles the various sources of error in linear measurements have been discussed. The principal sources of error in any tape length are those due to the incorrect length of the tape, variations in temperature and tension, incorrect determination of slope, and inaccurate marking of the tape end. These errors may either increase or decrease the length of the measurement. The total random error in any tape length may be assumed equal to the square root of the sum of the squares of the maximum individual errors.

Specifications for any desired accuracy can be prepared from a knowledge of these errors. Thus if the actual error is not to exceed 1/5000, the following limits for the various errors in 100 ft would suffice:

1. Length of tape to be known within 0.01 ft: maximum error, ± 0.01 ft.
2. Temperature to be known within 10°F: maximum error, ± 0.006 ft.
3. Tension to be known within 2 lb: maximum error, ± 0.001 ft.
4. For slopes averaging 5%, vertical distance to be known within 0.2 ft: maximum error, ± 0.01 ft.
5. Tape lengths to be marked within 0.01 ft: maximum error, ± 0.01 ft.
6. Alignment errors to be eliminated by keeping tape on line by transit.
7. Effect of sag to be eliminated by having tape standardized when suspended from the two ends: if the tension is known within 2 lb, the maximum sag error is about ± 0.006 ft.

2-24. ELECTRONIC DISTANCE MEASURING INSTRUMENTS (EDM'S)

The electronic distance measuring instruments referred to as EDM's in this text are a relatively new development in the field of surveying The instrument sends out a beam of light or high-frequency microwaves from one end of a line to be measured, and directs it toward the far end of the line. A reflector or transmitter-receiver at the far end reflects the light or microwaves back to the instrument where they are analyzed electronically to give the distance between the two points. The first such instrument was the geodimeter mentioned in Section 2-5. It became available to the general surveying and engineering profession in the early 1950s. The early geodimeter as well as all subsequent models employs a modulated light beam for determining distance. It was followed in the late 1950s by the tellurometer, an instrument that employs modulated microwaves for determining distances. Instruments similar to the tellurometer were introduced in subsequent years, and these along with the geodimeter became the standard instruments for measuring over long distances. The advantage of the microwave instruments is their operability in fog or moderate rain, day or night, as well as their generally longer range.

The development and perfection of small light-emitting diodes in the mid-1960s as well as a general miniaturization of electronics using solid-state components caused a revolution in the design of EDM's. This second generation of instruments is much more portable, takes less power to operate, and is much simpler to operate and read. However, these EDM's do not have the range that the earlier ones have.

A third generation of EDM's employing highly coherent laser light has been brought to perfection in recent years. This type of instrument has the distinct advantage of long-range, low-power requirements, and fairly good portability as well as ease of operation and readout.

Modern EDM's are fully automatic to such an extent that, after the instrument and its transponder (a light reflector or a microwave receiver-transmitter) have been set over the two ends of the line to be measured, the operator need only depress a button, and the slope distance is automatically displayed. More complete EDM's also have the capability of measuring horizontal and vertical angles as well as the slope distance (see Section 4-23). These instruments, referred to as *total station* instruments, display horizontal and vertical distances between the ends of the line, and horizontal and vertical angles to a survey point. These displayed outputs can be recorded on either paper or magnetic tape for further calculations in a computer.

Because of the proliferation of EDM's available to the surveyor or engineer, no attempt is made in this textbook to catalog or describe all of them. Only the principles on which these instruments operate will be presented. The student is urged to consult the readings at the end of this chapter and to obtain descriptive brochures from instrument manufacturers in order to appreciate the wide variety of EDM's.

Because of their suitability, durability, and reliability, and also because they represent a fairly substantial investment, the older geodimeters and microwave instruments are still in use throughout the world. The trend in instrument development, however, is toward the second generation short-range and intermediate-range more portable instruments incorporating full

Figure 2-15. Wild DI4 Distomat mounted atop a theodolite. Courtesy of Wild Heerbrugg Instruments, Inc.

Figure 2-16. Geodimeter 120 mounted atop a theodolite. Data collection unit is attached to tripod. Courtesy of AGA Geodimeter, Inc.

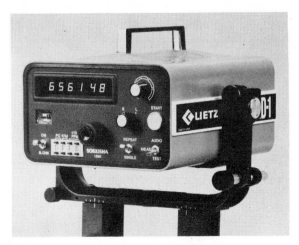

Figure 2-17. Lietz RED-1 distance meter. Courtesy of Lietz Co.

Figure 2-18. H-P 3800 distance meter in operation showing retroreflector set up at other end of line. Courtesy of Hewlett Packard Co.

automation. The laser light instruments are gradually displacing the older instruments for intermediate- to long-range operations.

The classification of instruments can be made by their range capabilities, although this is somewhat arbitrary. A short-range EDM is one that is capable of measuring distances up to about 2 miles. This is not to say that all short-range instruments have a 2-mile range, because the upper range of some EDM's is only 1000 to 1500 m. The very short-range instruments, however, incorporate features such as automatic reading, light weight, very low power requirements, and adaptability to angle-measuring surveying instruments.

The instruments shown in Figs. 2-15 through 2-18 are all classed as short to intermediate range. The range can be extended by adding reflectors to the far end of the line in order to get a strong enough return signal at the longer distances. The light carrier for all of these instruments is infrared light in the 900- to 930-nm wavelength region. Note that this is out of the visible portion of the spectrum.

The instrument shown in Fig. 2-19 uses high-frequency microwaves as the carrier. It has an upper range of about 20 miles using special antennae for transmission and reception.

The instrument shown in Fig. 2-20 uses a laser light beam as the carrier. This is a typical long-range instrument, and under favorable atmospheric conditions can measure lines 40 miles long.

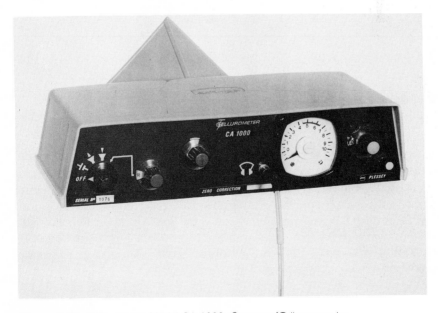

Figure 2-19. Tellurometer Model CA 1000. Courtesy of Tellurometer, Inc.

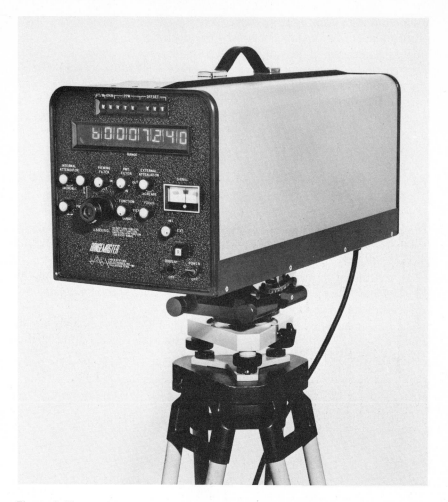

Figure 2-20. K & E Rangemaster. Courtesy of Keuffel & Esser Co.

2-25. MEASUREMENT PRINCIPLE OF EDM USING LIGHT
WAVES   In all EDM's using tungsten, mercury, laser, or infrared light as the carrier, a continuous beam of light is generated in the instrument. Before it enters the aiming optics and is directed to a reflector such as shown in Fig. 2-21 which is placed at the other end of the line to be measured, this continuous beam is intensity modulated at a very high frequency. The modulation, in effect, chops the beam up into wavelengths that are a direct function of the modulating frequency. This wavelength is given by

$$\lambda = \frac{V_a}{f} \tag{2-14}$$

Figure 2-21. Single and triple reflectors. Courtesy of Keuffel & Esser Co.

in which λ is the wavelength of modulation, in metres; V_a is the velocity of the light through the atmosphere, in metres per second; f is the modulating frequency, in hertz. The value of V_a is a function of air temperature, pressure, and partial pressure of water vapor, as discussed in Section 2-27.

The intensity of the modulated light varies from zero at the beginning of each wavelength to a maximum at 90°, back to zero at 180°, to a second maximum at 270°, and back to zero at 360°. The distance between every other zero point is thus one full wavelength. The EDM therefore provides a measuring tape of light whose length is equal to the wavelength of the modulated light. For example, if the modulating frequency is 10 MHz and the velocity of the light is approximately 300,000 km/s, the modulated wavelength is about 30 m, or about 100 ft.

In Fig. 2-22 the EDM is located to the left, at one end of a line to be measured, and the reflector R to the right, occupying the other end of the line. The reflector is a corner of a cube of glass in which the sides of the cube

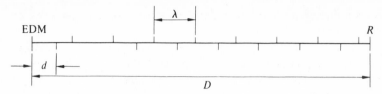

Figure 2-22. Measurement principle of EDM.

are perpendicular to one another within very close tolerances. This perpendicularity causes incoming light to be reflected internally and to emerge parallel with itself. The corner cube thus constitutes a retroreflector. An integral number of wavelengths plus a partial distance designated *d* make up the total distance from the EDM to the reflector back to the EDM. Note that if the reflector (or the EDM) were to be moved either forward or backward along the line by a distance of one-half wavelength, or any number of half wavelengths, the value of the partial distance *d* would be the same in each case. This partial distance is measured in the instrument by some type of a phase meter. The desired distance *D* between the two ends of the line is then given by

$$D = \tfrac{1}{2}(n\lambda + d) \tag{2-15}$$

in which *n* is the integral number of wavelengths in the double distance. One way in which this equation could be solved would be to have prior knowledge of the length of the double path to the nearest half wavelength, which requires that the length of the line be known to the nearest quarter wavelength. Since this is not practical, the ambiguity of *n* can be resolved by employing the multiple-frequency technique. If the measurement is made at one known frequency and it is then repeated using a slightly different frequency, two different values of *d* will be read on the phase meter. Knowing the two values of the wavelengths, two equations given by Eq. (2-15) can now be solved simultaneously to give the value of the unknown *n* and thus the desired distance *D*.

The multiple-frequency technique for resolving ambiguity is incorporated directly into the modern EDM's. One such system in common use is the decade-modulation technique. Assume that a modulation frequency of 15 MHz is set up in the instrument, resulting in a half wavelength of 10 m. Let a full sweep of the phase meter represent this 10-m distance. The phase-meter reading then gives the unit metre and decimal part of the metre in the measured distance from 0 to 9.999 m. In a distance of, say, 3485.276 m this frequency would give the 5.276 part. Switching down to 1.5 MHz, the half wavelength is now 100 m, which is resolved by the phase meter to give the tens of metres—in this instance 80 (8 tens). The next frequency is then 0.15 MHz, which in conjunction with the phase meter, gives the hundreds of metres, which in this instance is 400 (4 hundreds). Finally, a 15-kHz frequency will give the number of thousand metres in the distance, which in this instance is 3000 (3 thousands).

The instrument shown in Fig. 2-23, which reads in feet, employs the decade-modulation technique. The operator first aims the EDM optics at the reflector at the other end of the line using the sighting scope, and manipulates a pair of slow-motion screws to perfect the horizontal and vertical alignment. The best alignment is determined by observing a return-signal strength meter. This is a characteristic of most of the EDM's. The operator then balances the outgoing and incoming signal strengths in order to ensure proper functioning of the electronic components. Using a slide switch, he moves to the first frequency and dials in the appropriate number of feet and decimal parts.

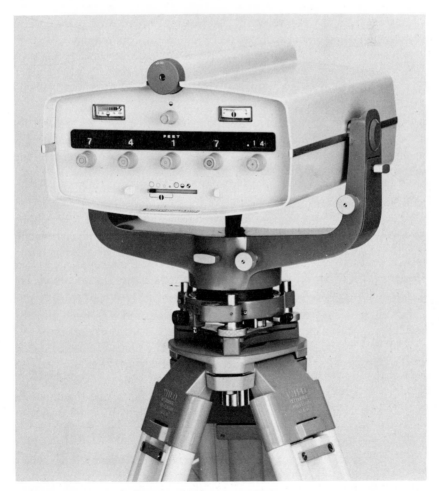

Figure 2-23. H-P 3800 distance meter showing controls and distance reading. The slide switch at lower center is correlated with the control knobs and electrical meters by means of symbols. Courtesy of Hewlett Packard Co.

This is seen to be 7.14 ft. He then slides the switch to the next lower frequency and dials in 10 ft; the next lower frequency gives 400 ft; and the lowest frequency gives 7000 ft. The measured distance is thus 7417.14 ft.

In the later models of this instrument, and in most modern EDM's, the different frequencies are sequentially changed in the instrument itself without intervention by the operator. The slope distance is then automatically displayed.

2-26. MEASUREMENT PRINCIPLES OF EDM USING MICRO-WAVES
The microwave instruments generate superhigh-frequency (SHF) or extremely high-frequency (EHF) electromagnetic waves in the range of 3 to 35 GHz as the carrier waves. These, in turn, are modulated to frequencies varying from 10 to 75 MHz, depending on the type of instrument. The length of the modulated wave is given by

$$\lambda = \frac{V_r}{f} \tag{2-16}$$

in which λ is the wavelength of modulation, in metres or feet; V_r is the velocity of the microwave through the atmosphere, in metres or feet per second; f is the modulating frequency, in hertz. The value of V_r depends on the prevailing temperature, atmospheric pressure, and the partial pressure of water vapor in the atmosphere, as discussed in Section 2-27. Two similar instruments are involved in measurement with the microwave instruments, one at each end of the line to be measured. These are referred to as the master and the remote. Observations are made at the master station, whereas the remote instrument, which must also have an operator, serves like a reflector in the light-generating EDM's.

The operator at the master station selects a modulating frequency at which the microwaves are transmitted toward the remote instrument. He indicates to the remote operator by voice communication, which is built into the instrument, the transmitting frequency being used. The remote operator sets his instrument to correspond to that frequency. The signal is received by the remote instrument and is retransmitted back to the master with no delay. A phase meter at the master gives the total amount by which the outgoing and incoming signals are out of phase. This, in effect, gives the fractional or decimal part of the wavelength by which the double path from master to remote to master departs from an integral number of wavelengths. It is equivalent to distance d in Fig. 2-22.

If either the master or the remote instrument were to be moved toward or away from one another by one-half of the modulated wavelength, the phase meter would go through one complete cycle and display the same value as before. Thus the ambiguity that exists in the light-wave instruments is also present in the microwave instruments.

The technique for resolving the ambiguity in the number of whole wavelengths contained in the double distance is generally the same as that used in the light-generating EDM that is, by the multiple frequency technique.

2-27. EFFECT OF ATMOSPHERIC CONDITIONS ON WAVE VELOCITY
The conditions of the atmosphere which have an effect on the velocity of propagation of light and microwaves are the air temperature, the atmospheric pressure, and the relative humidity. The temperature and relative humidity, in turn, define the vapor pressure in the atmosphere. A knowledge of these conditions allows a determination of the *refractive index* of the air, which must be known to compute the velocity of light or microwaves under given meteorological conditions.

For light waves the refractive index n_g of standard air is given by

$$n_g = 1 + \left(287.604 + \frac{4.8864}{\lambda_c^2} + \frac{0.068}{\lambda_c^4}\right)10^{-6} \qquad (2\text{-}17)$$

in which λ_c is the wavelength of the light carrier in micrometres. For the kinds of light used in EDM's, the values of λ_c are as follows:

CARRIER	$\lambda_c (\mu m)$
Mercury vapor	0.5500
Incandescent	0.5650
Red laser	0.6328
Infrared	0.900–0.930

The refractive index n_a for light waves at conditions departing from the standard air can then be computed by

$$n_a = 1 + \frac{0.359474(n_g - 1)p}{273.2 + t} - \frac{1.5026e \times 10^{-5}}{273.2 + t} \qquad (2\text{-}18)$$

in which p is the atmospheric pressure, in millimetres of mercury (torr); t is the air temperature, in degrees Celsius; e is the vapor pressure, in torr.

For practically all distance measurements with light, the last term of Eq. (2-18) involving vapor pressure can be neglected since relative humidity has very little effect on light waves.

The velocity of the light wave in air, V_a, is related to the velocity of light in a vacuum by

$$V_a = \frac{c}{n_a} \qquad (2\text{-}19)$$

The value for c is 299,792.5 km/s.

Example 2-12

The modulating frequency of a red laser beam is 24 MHz. This beam travels through the atmosphere, the temperature of which is 26°C and the atmospheric pressure 759 torr. What is the modulated wavelength of the light?

Solution: By Eq. (2-17), the refractive index of standard air for the laser carrier is

$$n_g = 1 + \left(287.604 + \frac{4.8864}{0.6328^2} + \frac{0.068}{0.6328^4}\right)10^{-6} = 1.0003002$$

By Eq. (2-18), the refractive index of the air under the given atmospheric conditions, neglecting the last term, is

$$n_a = 1 + \frac{0.359474(1.0003002 - 1) \times 759}{273.2 + 26} = 1.0002738$$

The velocity of laser light through this atmosphere is, by (Eq. 2-19),

$$V_a = \frac{299,792.5}{1.0002738} = 299,710.4 \text{ km/s}$$

Finally, the modulated wavelength is given by Eq. (2-14),

$$\lambda = \frac{299,710.4}{24 \times 10^6} = 0.01248793 \text{ km} = 12.48793 \text{ m}$$

The effect of water vapor pressure, which can be neglected when working with light, is quite large when using microwave EDM's. Consequently, the relative humidity must be very carefully determined in the field at the time of measurement. A high-quality psychrometer which gives wet and dry bulb thermometer readings must be employed for the determination of vapor pressure.

The refractive index of microwaves n_r is given by

$$(n_r - 1)10^6 = \frac{103.49}{273.2 + t}(p - e) + \frac{86.26}{273.2 + t}\left(1 + \frac{5748}{273.2 + t}\right)e \quad (2\text{-}20)$$

where p is the atmospheric pressure, in millimetres of mercury; e is the vapor pressure, in millimetres of mercury; t is the air temperature (dry bulb), in degrees Celsius.

The velocity of propagation of microwaves V_r through the atmosphere is then given by

$$V_r = \frac{c}{n_r}$$ (2-21)

and the modulated wavelength is given by Eq. (2-16).

Example 2-13

What is the wavelength in metres of microwaves modulated at a frequency of 10 MHz if the atmospheric pressure is 643 torr, temperature is 23.9°C, and vapor pressure is 3.5 torr?

Solution: By Eq. (2-20),

$$(n_r - 1)10^6 = \frac{103.49}{297.1}(643.0 - 3.5) + \frac{86.26}{297.1}\left(1 + \frac{5748}{297.1}\right)3.5 = 243.4$$

and $n_r = 1.0002434$. By Eq. (2-21),

$$V_r = \frac{299,792.5}{1.0002434} = 299,719.5 \text{ km/s}$$

Finally, by Eq. (2-16),

$$\lambda = \frac{299,719.5}{10 \times 10^6} = 0.02997195 \text{ km} = 29.97195 \text{ m}$$

The equations given above for determining refractive index are presented in different forms in various articles and publications; and there are slight but insignificant variations in results obtained by the different expressions. These particular forms are presented here because they allow the reader to appreciate the relative significance of the temperature, pressure, and vapor pressure on both light and microwaves.

The effects of atmospheric conditions are handled in various ways by different EDM systems. The corrections are of little consequence for short distances using the light-generating EDM's. For longer lines an error of 10°C in the effective temperature of the beam path will introduce a relative error of 10 ppm; and an error of 25 millimetres of mercury in measuring atmospheric pressure will also introduce a relative error of 10 ppm. Corrections are either computed based on meteorological data determined at the time of measurement, or else the instrument circuitry is modified to take the atmospheric conditions into account. In the instrument shown in Fig. 2-18, for example, environmental corrections are dialed into the power unit. This

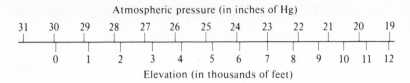

Figure 2-24. Relationship between elevation in feet and atmospheric pressure in inches.

changes the modulating frequency in order to maintain a constant wavelength at any temperature and pressure.

In the case of the microwave EDM's, the partial pressure of water vapor obtained by the wet and dry bulb thermometer readings must be determined quite accurately. A 2-mm error in vapor pressure or a 1.5°C error in the difference between the dry and wet bulb temperatures will produce a relative error of about 10 ppm at normal temperatures. This relative error increases with an increase in air temperature. Assuming that the meteorological conditions have been measured satisfactorily, corrections to measured distances are easily made with the aid of the various charts, tables, or nomographs that are furnished with the instruments.

If an aneroid barometer (Section 3-6) is used to determine atmospheric pressure, the elevation reading in feet or metres must be converted to the appropriate value of pressure in inches or millimetres of mercury. This conversion is made using charts furnished by the instrument maker. Figure 2-24 shows a simplified chart that can be used to convert feet of elevation to inches of mercury. Altimeter dials used in surveying are usually graduated so that a reading of, say, 1000 ft will be obtained at sea level under standard atmospheric conditions. This is done to eliminate negative altimeter readings at or near sea level when the atmospheric pressure increases above standard. It is necessary to know this zero reading because it must be subtracted from the altimeter reading before using a chart such as that shown in Fig. 2-24.

2-28. INSTRUMENTAL ERRORS IN EDM If a modern EDM is properly tuned, there will be very few sources of instrumental errors that need correcting. One error, referred to as the "reflector constant," is caused by not having the effective center of the reflector plumbed over the far end of the line. This type of error is shown in Fig. 2-25 for a corner cube reflector. The distance through which the light travels in the glass cube during retroreflection is $a + b + c$, which in turn is equal to $2t$. The distance t is measured from the face of the reflector to the corner of the glass cube. The equivalent air distance through which the light travels is $1.517 \times 2t$ on account of the refractive index of the glass. The effective corner of the cube is at R and represents the end of the line. If the plumb line passes vertically in front of point R, then an error c_R would be introduced into the measured length of the line. Then distance c_R would have to be subtracted in this instance. The reflector constant is a relatively small value, most commonly 30 to 40 mm. The reflector constant

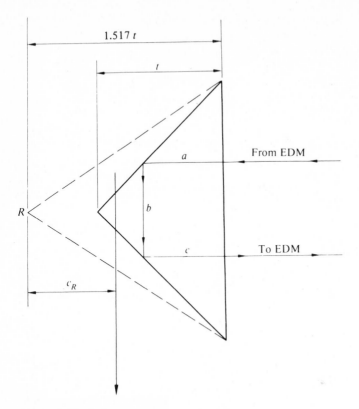

Figure 2-25. Reflector constant.

is effectively eliminated by advancing the electrical center of the EDM by a corresponding amount during manufacture.

Since reflectors are made and mounted by different manufacturers, the reflector constants of all of the reflectors used with any one given instrument may not all be the same. Thus, it is necessary to determine a combined instrument-reflector constant C_I for each combination to be used by the surveyor. Once this constant has been determined, it is added to any measurement made with that particular combination. The constant C_I can be entered into most of the modern EDM's by the operator via a keyboard and is then automatically added to the measured slope distance.

Two methods can be employed to determine the combined constant. In the first, a line as long as practicable is established and measured using a steel or invar tape. Because of the inherent accuracy of the EDM's, this line must be measured to a high degree of accuracy. If the known distance is then measured with the EDM, the distance corrected for meterological conditions and the slope of the line should agree with the known (taped) distance. Otherwise

$$C_I = \text{taped distance} - \text{EDM measured distance} \qquad (2\text{-}22)$$

Figure 2-26. Layout for determining combination instrument-reflector constant.

If a reliable base line of known length is not available and if it is not feasible to measure a line, a second method can be employed to determine the instrument constant. In Fig. 2-26 three points A, B, and C are located on a straight line. The EDM occupies point A, and the distances AB and AC are measured. The EDM is moved to B, and the distance BC is measured. These three measurements are corrected for slope and meterological conditions. The constant is then computed as (measured $AB + C_I$) + (measured $BC + C_I$) = measured $AC + C_I$ giving

$$C_I = \text{measured } AC - (\text{measured } AB + \text{measured } BC) \qquad (2\text{-}23)$$

These two methods for determining the combination instrument-reflector constant are also applied to microwave instruments. In this case, the remote instrument is considered as the reflector.

A determination of the value of C_I is made under the assumption that the instrument is properly tuned to give the correct modulation frequencies. An error in the frequency produces a scale error such as the incorrect tape length discussed in Section 2-14. For example, if the correct modulation frequency is 10 MHz and if the actual frequency departs from this by 100 Hz, a relative error of 10 ppm affects each measurement. The frequencies can be checked by means of a frequency counter, which can be obtained from an electronic laboratory or through the instrument manufacturer or service outlet. A frequency check should be performed at regular intervals, particularly if high-order surveys or surveys with very long lines are being preformed. Alternatively, if the EDM is checked regularly against a known distance, applying corrections for instrument and reflector constants, meteorological conditions, and slope, a frequency shift can be detected.

2-29. GROUND REFLECTIONS OF MICROWAVES The microwave EDM's have a relatively wide beam. Consequently, the waves traveling from one end of a line to the other may have strong reflections off the intervening terrain particularly if it is smooth and free of coarse vegetation. These reflected waves can also be quite serious when measurements are made over water. Such reflected waves will cause a faulty distance to be displayed because they travel over longer paths than the desired direct rays. If a series of fine readings are taken, each with a different frequency, and if there are strong reflections, these readings will vary in a cyclic manner. If the readings are plotted as a function of carrier frequency, they ideally take the shape of a sine wave. The cyclic variation in the fine readings is termed *swing*. The interpretation of the swing curve that will extract the best value is a matter of considerable experience and judgment and is outside the scope of the book. However, generally a straight average of the fine readings will be of sufficient accuracy for most measurements.

2-30. REDUCTION OF SLOPE MEASUREMENTS IN EDM

If the lines measured by EDM's are not too long, the slope distances obtained after the readings are corrected for meteorological and instrumental effects can be reduced to the corresponding horizontal distances by the methods outlined in Sections 2-11 and 2-12. In order to obtain the correct vertical angle used in the slope-reduction formulas, the height of the EDM above the ground station, the height of the reflector or remote EDM, the height of the instrument used to measure the vertical angle, and the height of the target on which the instrument was sighted when measuring the vertical angle must all be measured and recorded at the time of measurement.

In Fig. 2-27 the slope distance ER is measured with the EDM at E and the reflector at R. The vertical angle α is measured with the transit or theodolite at T to the target at P. In this instance the slope angle along the EDM line is larger than that along the line from T to P by the small angle $\Delta\alpha$.

Denoting AE as the height of the EDM, AT as the height of the theodolite, BR as the height of the reflector (or the remote instrument), and BP as the height of the target, the difference between the heights of the two instruments at A is $AE - AT$; and the difference between the heights of the reflector and the target is $BR - BP$. Note that $BR - BP$ is greater than $AE - AT$ in this instance. The difference, denoted ΔH, is then $(BR - BP) - (AE - AT)$ and is shown in Fig. 2-28. This value of $\Delta\alpha$ in radians is then $\Delta H \cos \alpha/ER$, and $\Delta\alpha'' = \Delta H \cos \alpha/ER$ arc $1''$. The vertical angle of the measured line is then $\alpha' = \alpha + \Delta\alpha$.

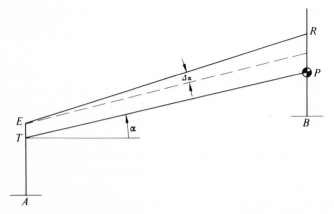

Figure 2-27. Difference between vertical angle of measured slope distance and vertical angle of line of sight.

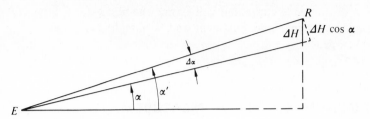

Figure 2-28. Reduction to correct vertical angle of slope measurement.

Example 2-14

A distance is measured along a slope with an EDM which, when corrected for meteorological conditions and instrumental constants, is 714.652 m. The EDM is 1.750 m above the ground, and the reflector is located 1.922 m above ground. A theodolite is used to measure a vertical angle of $+4° 25' 15''$ to a target sitting 1.646 m above ground. The theodolite itself is 1.650 m above ground. What is the horizontal length of the line?

Solution: From the analysis given above, the situation is as shown in Fig. 2-27. The value of ΔH in Fig. 2-28 is

$$(1.922 - 1.646) - (1.750 - 1.650) = 0.276 - 0.100 = 0.176 \text{ m}$$

The small angle $\Delta\alpha$ in seconds is then

$$\frac{0.176 \cos 4° 25' 15''}{714.652 \times 0.000004848} = 51''$$

The slope of the measured line is then

$$+4° 25' 15'' + 51'' = +4° 26' 06''$$

The horizontal distance is computed using Eq. (2-1). Thus

$$H = 714.652 \cos 4° 26' 06'' = 714.652 \times 0.997006 = 712.512 \text{ m}$$

If the slope distance obtained by EDM is to be reduced by difference in elevation between the two ends of the line, then the height of the EDM and the height of the reflector or remote EDM must be measured and recorded. In Fig. 2-29 the EDM is set up at A and the reflector at B. The height of the EDM is AE and that of the reflector is BR. The difference in elevation between A and B is h. But the difference in elevation between E and R along which the

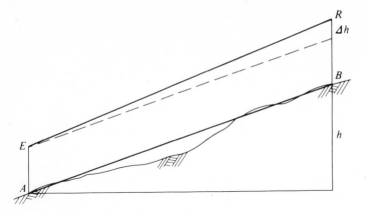

Figure 2-29. Reduction to correct difference in elevation.

measurement is made is $h + \Delta h$. In this instance $\Delta h = BR - AE$. The corrected value of h is then used in Eq. (2-4) in order to obtain the horizontal distance from A to B.

If the measured lines are very long, the slope distance cannot be considered as the hypotenuse of a right triangle as assumed in short slope distances. When using a steel tape to make measurements, if the tape is held horizontal each time it is stretched out, the measurements are automatically made along a curved line that generally conforms with the curvature of the earth. This is shown ideally in Fig. 2-30 in which the curvature of the earth is exaggerated out of proportion to individual tape lengths. This curved distance, if reduced to sea level as explained in Chapter 10, is then called a *geodetic distance.*

When using an EDM, a single measurement is made of the entire line rather than in tape-length sections. Thus in Fig. 2-31 the EDM-measured slope distance is AB', and the desired distance is AB. The reduction required to compute AB is somewhat more complex than the simple expressions given by Eqs. (2-1) and (2-4). This is thoroughly discussed in Sections 10-29 and 10-30.

If a line measures 2 miles using the EDM and the vertical angle is $5°$, the difference between the horizontal distance computed by Eq. (2-1) and that determined as outlined in Section 10-29 will be in error by no more than one part in 25,000. This relative accuracy is sufficient for all but high-order control surveys. Thus if the maximum range of the EDM is 2 miles, then the more elaborate slope-reduction calculations are needed only on lines that slope more than $5°$. Figure 2-32 is a curve prepared on the basis of the slope

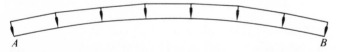

Figure 2-30. Taping along curved surface.

Figure 2-31. Measuring long line using EDM.

reductions given in Section 10-29 for lines 10,000 ft or less in length in which the desired relative accuracy is not to drop below one part in 25,000. The curve indicates the maximum size of vertical slope angle as a function of distance. If the vertical angle exceeds this limit, then the result of Eq. (2-1) is not sufficiently accurate to give 1/25,000, and the more elaborate reduction of Section 10-29 must be employed.

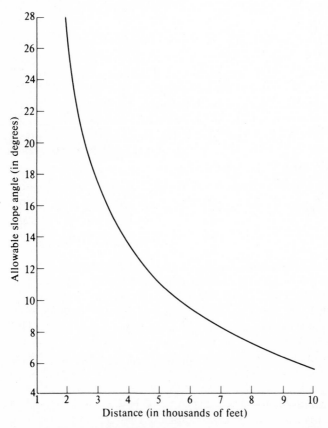

Figure 2-32. Allowable slope angle to permit using simple reduction equation in order to maintain relative accuracy of 1:25,000.

2-31. LAYOUT MEASUREMENTS USING EDM   In construction surveying, precalculated dimensions taken from plans and drawings must be laid out on the ground along well-defined construction lines. The steel tape is well suited to this type of surveying because the construction lines are usually rather short and the tape is convenient to handle. If the construction project is fairly extensive and involves the laying out of several long lines, however, the EDM can be employed to advantage. The technique to be used depends on the features of the instrument at hand.

If the EDM is designed to make a measurement along a direct line between it and a stationary reflector, which can be referred to as a static measurement, the technique to be used can take the following form. The approximate distance can be established within one or two feet using stadia hairs in the sighting telescope (see Sections 2-4 and 14-2) and two points are then set on line a few feet in front of and behind the tentative point. The reflector is then set over the tentative point, and the distance is measured with the EDM. If there is significant slope, this distance is reduced to a horizontal (plan) distance using either Eq. (2-1) or (2-4). The presumably small difference between the measured distance and the distance indicated on the construction plans is then quickly laid off with a tape along the line defined by the two points previously set.

If the EDM is designed to follow a moving reflector in order to get the continuous slope distance, which can be referred to as a dynamic measurement, the positioning of the point is somewhat simplified. If the slope is significant, the correct slope distance to be laid off can be computed as shown in Example 2-2 or 2-6, in which the construction-plan distance is used for H in the examples. Alignment of the point can be made at the same time using the cross hairs of the sighting telescope.

The fully automatic EDM computes and displays horizontal distance from the slope distance measured with the instrument. After a tentative horizontal distance has been established, the reflector is moved forward or backward. Then continuous or near-continuous readout of horizontal distance is obtained. The point is set on line where the displayed distance is equal to the construction plan distance.

2-32. ACCURACY OF EDM MEASUREMENTS   At close range the accuracy of the EDM is limited by a constant uncertainty such as 1 cm, 5 mm, or 0.01 ft. As the measured distance is increased, this constant value becomes relatively inconsequential. Beyond, say, 500 to 1000 m, all of the currently available EDM's will give relative accuracies of better than one part in 25,000. This is very difficult to obtain by taping and requires taping tripods and very careful attention to systematic errors. On the other hand this accuracy is practically assured using the EDM's. The factors that limit both the relative and the absolute accuracies of EDM measurements are the meteorological conditions at the time of measurement. If these are known with sufficient accuracy, then all but the very short-range instruments are capable of relative accuracies of one part in 100,000 or better.

If very long lines must be measured with the maximum accuracy, then the uncertainty of the meteorological conditions along the entire beam path becomes important. For the majority of measurements in the short to inter-mediate range, meteorological measurements taken only at the instrument end of the line are sufficient to obtain the desired accuracy. This can be enhanced by taking a mean of the readings at both ends of the line. Improve-ments can further be made by sampling the meteorological conditions at intermediate points along the line, which of course is complicated usually by the necessity of elevating the meteorological instruments to considerable heights above the intervening terrain. The ultimate solution at present is to fly along the line and record the temperature and pressure profile all along the line. This technique has been employed in California along lines used to measure very small earthquake fault displacements over great distances.

PROBLEMS

2-1. A surveyor paces a 100-ft length six times with the following results: $35\frac{1}{3}$, 34, $34\frac{2}{3}$, 35, $34\frac{2}{3}$, and 35 paces. How many paces must he step off in order to lay off a distance of 10 chains?

2-2. A surveyor paces a 50-m length six times with the following results: $56\frac{1}{2}$, 57, $56\frac{1}{2}$, 58, 57, and $56\frac{2}{3}$ paces. How many paces must he step off in order to lay off a dis-tance of 300 m?

2-3. A slope distance of 1421.20 ft is measured between two points with a slope angle of 3° 32′. Compute the horizontal distance between the points.

2-4. In Problem 2-3 if the vertical angle is in error by 2′, what error is produced in the calculated horizontal distance?

2-5. The difference in elevation between two points is 16.264 m. The measured slope distance is 383.804 m. Compute the horizontal distance.

2-6. A measurement of 148.866 m is made between two points on a 4° 02′ slope. Compute the horizontal distance between the two points.

2-7. In Problem 2-5 if the difference in elevation is in error by 0.020 m, what error is produced in the corresponding horizontal distance?

2-8. A line was measured along sloping ground with a 100-ft tape, and the following results were recorded:

SLOPE DISTANCE (ft)	DIFFERENCE IN ELEVATION (ft)
100.00	5.86
100.00	3.04
63.00	3.16
100.00	10.08
27.44	2.15

What is the horizontal length of the line?

2-9. A tape that measures 99.92 ft between the zero and 100-ft mark is used to lay out foundation walls for a building 350.00 × 640.00 ft. What observed distances should be laid out?

2-10. A 100-ft steel tape weighs 1.62 lb and measures 100.003 ft when supported throughout under a 10-lb tension. It is used in the field under the same tension, but supported only at the ends. The recorded length is 585.28 ft. What is the correct length of the line?

2-11. What distance should be laid out with a tape that measures 99.988 ft under the prevailing field conditions if a $\frac{1}{2}$-mile distance is to be established?

2-12. What distance on a 5% grade should be laid out with a tape that measures 29.992 m under field conditions if the horizontal distance is to be 550.00 m?

2-13. A 100-ft steel tape measures correctly when supported throughout its length under a tension of 10 lb and at a temperature of 72°F. It is used in the field at a tension of 18 lb and supported only at the ends. The temperature throughout the measurement averages 84°F. The measured length is 748.25 ft (the tape is suspended between the 48-ft mark and the zero end for the last measurement). The tape weighs 2.00 lb and has a cross-sectional area of 0.0060 in.². Assuming that E is 28,000,000 psi for steel, what is the actual length of the line?

2-14. A 30-m tape weighs 12 g/m and has a cross section of 0.020 cm². It measures correctly when supported throughout under a tension of 8.5 kg and at a temperature of 20°C. When used in the field, the tape is only supported at its ends, under a tension of 8.5 kg. The temperature is 13.5°C. What is the distance between the zero and 30-m mark under these conditions?

2-15. A 100-ft steel tape weighs 1.80 lb and has a cross-sectional area of 0.0056 in.². The tape measures 100.00 ft when supported throughout under a tension of 10 lb. Assume that $E = 28 \times 10^6$ psi. What tension, to the nearest $\frac{1}{4}$ lb, must be applied to overcome the effect of sag when the tape is supported only at the two ends?

2-16. A 50-m tape weighs 24 g/m and has a cross section of 0.038 cm². It measures 49.9862 m under a tension of 2.20 kg when supported only at the two ends. What does the tape measure if it is supported throughout and under a tension of 6 kg? $E = 2.1 \times 10^6$ kg/cm².

2-17. With what accuracy must the difference in elevation between the two ends of a 100-ft tape be known if the difference in elevation is 8.60 ft and the accuracy ratio is to be at least 1:10,000?

2-18. With what accuracy must a difference in elevation between the two ends of a 30-m tape be known if the difference in elevation is 2.280 m and the accuracy ratio is to be at least 1:25,000?

2-19. Compute the refractive index of mercury vapor light at a temperature of 68°F and barometric pressure of 27.50 in. Hg. Neglect the effect of vapor pressure.

2-20. What is the refractive index of red laser light at a temperature of 22°C and barometric pressure of 710 torr? Neglect the effect of vapor pressure.

2-21. What is the velocity of mercury vapor light at a temperature of 10°C and barometric pressure of 720 torr.

2-22. Microwaves are propagated through an atmosphere of 76°F, atmospheric pressure of 714 torr, and a vapor pressure of 12.6 torr. If the modulating frequency is 30 MHz, what is the modulated wavelength?

2-23. What is the wavelength of the modulated light of Problem 2-19 if the frequency of modulation is 30 MHz?

2-24. Microwaves are modulated at a frequency of 75 MHz. They are propagated through an atmosphere at a temperature of 12°C, atmospheric pressure of 712 torr, and a vapor pressure of 7.6 torr. What is the modulated wavelength of these waves?

2-25. Determine the velocity of red laser light through an atmosphere at 80°F and elevation 5280 ft.

2-26. Referring to Fig. 2-26, AB measures 1162.28 ft; BC measures 925.50 ft; AC measures 2087.66 ft using a particular EDM-reflector combination. A line measures 1752.84 ft with this instrument-reflector combination. What is the correct length of the line?

2-27. The height of an EDM set up at A is 5.06 ft. The height of the reflector set up at B is 4.20 ft. The height of the theodolite set up at A and used to measure a vertical angle is 5.25′. The height of the target at B on which the vertical angle sight is taken is 5.00 ft. The vertical angle is $-4° 22′ 35″$. The slope distance, after meteorological corrections, is 2645.63 ft. What is the horizontal distance between A and B?

2-28. The height of an EDM set up at M is 1.500 m. The height of the reflector set up at P is 1.300 m. The height of the theodolite at M used to measure the vertical angle is 1.615 m. The height of the target at P on which the vertical angle sight is taken is 1.385 m. The slope distance, after meterological corrections, is 1676.442 m. The measured vertical angle is $+3° 02′ 32″$. What is the horizontal distance between M and P?

2-29. With what accuracy must an angle of 5° be measured in a slope distance of 10,000 ft in order to obtain an accuracy ratio of 1 : 50,000 when computing the horizontal distance?

BIBLIOGRAPHY

American Congress on Surveying and Mapping-Control Surveys Division. 1971. *Electronic Distance Measuring Instruments*, Technical Monograph No. CS-2. Washington, D.C.

Bell, T. P. 1978. A practical approach to electronic distance measurement. *Surveying and Mapping* 38:335.

Brinker, D. M. 1971. Modern taping practice versus electronic distance measuring. *Proc. Fall Conv. Amer. Cong. Surv. & Mapp.* p. 354.

Buckner, R. B. 1970. Survey distance measurements and when to use what. *Proc. 30th Ann. Meet. Amer. Cong. Surv. & Mapp.* p. 1.

Bullock, M. L. 1975. HP 3810A—The Layout Machine. *Proc. Fall Conv. Amer. Cong. Surv. & Mapp.* p. 14.

Carrol, C. L., Jr. 1974. Field comparison of steel surveyor's tapes. National Bureau of Standards Report, NBSIR 73-408. Washington, D.C.: U.S. Government Printing Office.

Gort, A. F. 1977. The Hewlett-Packard 3820A Electronic Total Station. *Proc. Fall Conv. Amer. Cong. Surv. & Mapp.* p. 308.

Greene, J. R. 1977. Accuracy Evaluation in Electro-Optic Distance-Measuring Instruments. *Surveying and Mapping* 37:247.

Greulich, G., and Schellens, D. F. 1973. Surveying with Zeiss Reg Elta 14 in densely populated areas. *Proc. 33d Ann. Meet. Amer. Congr. Surv. & Mapp.* p. 451.

Hodges, D. J. 1970. Electro optical distance measuring instruments. *Canadian Surveyor* 24:47.

International Association of Geodesy. 1967. *Electromagnetic Distance Measurement.* Toronto: Univ. of Toronto Press.

Jones, H. E. 1971. Systematic errors in tellurometer and geodimeter measurements. *Canadian Surveyor* 25:406.

Kelly, M. L. 1979. Field calibration of electronic distance measuring devices. *Proc. Ann. Meet. Amer. Cong. Surv. & Mapp.* p. 425.

Kesler, J. M. 1973. EDM slope reduction and trigonometric leveling. *Surveying and Mapping* 33:61.

Lesley, G. B. 1967. Laser geodimeter. *Surveying and Mapping* 27:625.

Meade, B. K. 1972. Precision in electronic distance measuring. *Surveying and Mapping* 32:69.

Moffitt, F. H. 1958. The tape comparator at the University of California. *Surveying and Mapping* 18:441.

Moffitt, F. H. 1971. Field evaluation of the Hewlett Packard Model 3800 distance meter. *Surveying and Mapping* 31:79.

Moffitt, F. H. 1975. Calibration of EDM's for precision measurement. *Surveying and Mapping* 35:147.

Mogg, M. I. 1970. The MA100—a high-precision infrared tellurometer. *Canadian Surveyor* 24:66.

Rick, J. O. 1975. ACSM Student Chapter offers tape comparison service to surveyors. *Surveying and Mapping* 35:239.

Robertson, K. D. 1975. A method for reducing the index of refraction errors in length measurement. *Surveying and Mapping* 35:115.

Romaniello, C. G. 1977. EDM 1976. *Surveying and Mapping* 37:25.

Rosenberg, P. 1975. Corrections for EDM equipment mounted on surveying instruments. *Proc. Fall Conv. Amer. Cong. Surv. & Mapp.* p. 25.

Strasser, G. 1973. A reducing electronic tacheometer—the Wild DI 3 distomat. *Proc. 33rd Ann. Meet. Amer. Cong. Surv. & Mapp.* p. 441.

Tomlinson, R. W. 1971. Short-range electronic distance measuring instruments. *Proc. Fall Conv. Amer. Cong. Surv. & Mapp.* p. 235.

Chapter 3
Leveling

3-1. GENERAL Leveling is the operation in surveying performed to determine and establish elevations of points, to determine differences in elevation between points, and to control grades in construction surveys. The elevation of a point has been defined as its vertical distance above or below a given reference level surface. The reference level surface most commonly used in the United States is the National Geodetic Vertical Datum of 1929, formerly called the Sea Level Datum of 1929. This surface was established by connecting all of the major level lines (see Section 3-24) in the country to 26 tidal benchmarks along the Atlantic, Gulf of Mexico, and Pacific Coasts. Other vertical datums are used locally by engineers and surveyors, but this practice causes confusion when tying together level lines which originate on different datums.

A benchmark is a permanent or semipermanent physical mark of known elevation. It is set as a survey marker in order to provide a point of beginning for determining elevations of other points in a survey. A good benchmark is a bronze disk set either in the top of a concrete post or in the foundation of a structure. Other locations for benchmarks are the top of a culvert headwall, the top of an anchor bolt, or the top of a spike driven into the base of a tree. The elevations of benchmarks are determined to varying degrees of accuracy

66

by the field operations to be described in this chapter. Benchmarks established throughout the country by the National Geodetic Survey (NGS) to a high order of accuracy define the National Vertical Geodetic Datum of 1929.

The basic instrument used in leveling is a spirit level which establishes a horizontal line of sight by means of a telescope fitted with a set of cross hairs and a level bubble. The level is described in later sections of this chapter. Other instruments used for determining vertical distances are the engineer's transit, the theodolite, the EDM, the aneroid barometer, the hand level and the telescopic alidade. The use of the transit, the theodolite, and the telescopic alidade are explained in Chapters 4 and 14.

Modern inertial survey systems described in Chapter 9 and satellite Doppler receivers described in Chapter 10 can be employed to determine elevations quite rapidly. However the accuracy of the measured elevations is not as high as that obtained by most of the methods to be discussed in this chapter.

3-2. CURVATURE AND REFRACTION The measurements involved in leveling take place in vertical planes. Consequently, the effects of earth curvature and atmospheric refraction must be taken into account because these effects occur in the vertical direction. They are allowed for either by the appropriate calculations or else by the measuring techniques designed to eliminate these effects.

In order to examine the effect of earth curvature, consider the amount by which the level surface passing through point A of Fig. 3-1 departs from a horizontal plane at the distance AB from the point of tangency. This departure is shown to be the vertical distance CB. If the effect of earth's curvature is designated c, and the distance from the point of tangency to the point in question is designated K, then the curvature effect can be shown to be

$$c = 0.667K^2 \qquad\qquad (3\text{-}1)$$

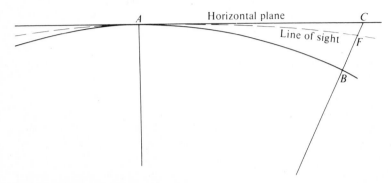

Figure 3-1. Curvature and refraction.

where c is the curvature effect in feet and K is the distance in miles. This expression for earth curvature is based on a mean radius of the earth, or 3959 miles. It should be noted that the error due to curvature is proportional to the square of the distance from the point of tangency to the point in question. Thus for a distance of 10 miles the value of c is about 67 ft, whereas for a distance 100 ft it diminishes to 0.00024 ft. In the metric system

$$c_M = 0.0785K_M^2 \qquad (3\text{-}2)$$

in which c_M is the curvature effect, in metres; K_M is the distance, in kilometres.

Rays of light passing through the earth's atmosphere in any direction other than vertical are refracted or bent from a straight path. This bending usually takes place in a direction toward the earth's surface under normal conditions of temperature and pressure gradients. Such bent rays of light tend to diminish the effect of curvature a slight amount, normally about 14% of the curvature effect. This effect is shown by the dashed line of sight in Fig. 3-1. The effect of both curvature and refraction is reduced to the distance FB. Thus under normal conditions the combined effect of curvature and refraction can be computed from one of the following:

$$(c + r) \quad = 0.574K^2 \qquad (3\text{-}3)$$

$$(c + r) \quad = 0.0206M^2 \qquad (3\text{-}4)$$

$$(c + r)_M = 0.0675K_M^2 \qquad (3\text{-}5)$$

in which $(c + r)$ is the combined effect of curvature and refraction, in feet; $(c + r)_M$ is the combined effect of curvature and refraction, in metres; K is the distance, in miles; M is the distance, in thousands of feet; K_M is the distance, in kilometres.

When leveling with the spirit level, the line of sight is ordinarily no more than about 2 m or 6 ft above the ground throughout its length. Depending on the weather, the temperature gradient near the ground can be the reverse of what it is, say, at 20 ft above ground on account of the heat radiating from the ground. In this zone, therefore, the effect of atmospheric refraction is uncertain. Fortunately, however, because of the relatively short lines of sight that are encountered in spirit leveling and also because of balanced sight distance (see Section 3-28), this uncertainty is of minor consequence except in very precise leveling.

3-3. TRIGONOMETRIC LEVELING   The difference in elevation between two points can be determined by measuring the vertical angle of the line from one point to the other and then computing the difference in elevation from a knowledge of either the slope distance or the horizontal distance between the two points. In Fig. 3-2, DE is a horizontal line, and the vertical

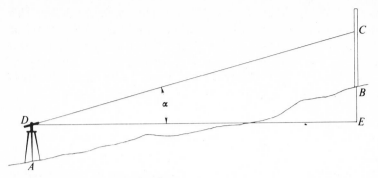

Figure 3-2. Trigonometric leveling.

distance CE is $DC \sin \alpha$ or $DE \tan \alpha$. So the difference in elevation between A and B is $AD + CE - CB$, where AD is the height of the instrument above point A, and CB is the height of the line of sight above point B measured with a leveling rod or with some other type of graduated rod. This method of determining difference in elevation is limited to horizontal distances less than 1000 ft when moderate precision is sufficient, and to proportionately shorter distances as higher precision is desired.

Many of the EDM's discussed in Chapter 2 contain built-in micro-processors which compute the distance CE from the measured slope distance and the vertical angle. The vertical angle is either measured directly in some instruments, or else must be entered into the keyboard of the EDM.

For distances beyond 1000 ft the effects of curvature and refraction must be considered and applied. The situation shown in Fig. 3-3 represents a line of intermediate length, up to perhaps 2 miles. The angle at C is shown to be 90° and is so assumed in this range. At 2 miles it is actually about 90° 01′ 44″. The vertical distance CE is equal to $DE \sin \alpha$ or $DC \tan \alpha$ depending on whether the slope distance obtained by EDM or horizontal distance from triangulation (Chapter 10) is known. The vertical distance from a level line through D to point E is FE, which is $CE + CF$. But CF is the total effect of curvature and refraction which, by Eq. (3-4), is $0.0206(DC/1000)^2$ or $0.0206(DE \cos \alpha/1000)^2$. The difference in elevation between A and B is then $AD + CF + CE - EB$. The values of AD and EB are recorded at the time of measurement.

In order to eliminate the uncertainty in the curvature and refraction correction, vertical-angle observations are made at both ends of the line as close in point of time as possible. This pair of observations is termed *reciprocal vertical-angle observation*. The correct difference in elevation between the two ends of the line is then the mean of the two values computed both ways without taking into account curvature and refraction.

The accuracy of the determination of difference in elevation over a long distance is basically a function of the uncertainty of the atmospheric refraction and of the accuracy of the vertical angles. The slope distance obtained with EDM will be so accurate that no appreciable error will be introduced from

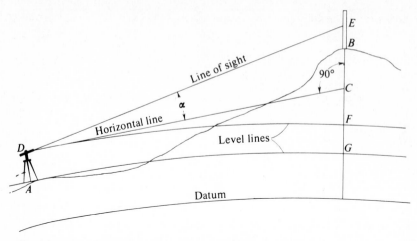

Figure 3-3. Curvature and refraction in trigonometric leveling.

this source. The effect of uncertainty in atmospheric refraction is held to a minimum by reciprocal observations. Bearing in mind that a second is a foot in 40 miles, then in a distance of say 2 miles the error of difference in elevation can be held to 0.20 ft if the vertical-angle accuracy is about 4″, which is obtainable with the precise theodolites discussed in Chapter 4. This error is reduced in proportion to a reduction in distance. If reciprocal observations are taken say every 1000 m and the accuracy of vertical-angle measurements is of the order of 3″ to 4″, the error can be held to $1\frac{1}{2}$ to 2 cm or about 0.05 to 0.07 ft.

When applying trigonometric leveling to very long lines, the slope distance is measured using the intermediate or long-range EDM's, and reciprocal vertical angles are measured using precise theodolites. If possible, the reciprocal angles should be made simultaneously in order to eliminate the refraction uncertainty. If this proves unfeasible, then more than one set should be observed at different times in order to average out the errors due to this uncertainty.

Example 3-1

The slope distance between two mountain peaks determined by EDM measurement is 76,963.54 ft. The vertical angle at the lower of the two peaks to the upper peak is +3°02′05″. The reciprocal vertical angle at the upper peak is −3°12′55″. Compute the difference in elevation between the two peaks first by using only the single vertical angle and applying correction for curvature and refraction, and second by using the average obtained by both vertical angles.

Solution: The difference in elevation using only the single vertical angle is

$$\Delta H = 76{,}963.54 \sin 3°\,02'\,05'' + 0.0206\left(\frac{76{,}963.54}{1000}\cos 3°\,02'\,05''\right)^2$$

$$76{,}963.54 \sin 3°\,02'\,05'' = 76{,}963.54 \times 0.05294113 = \quad 4074.54$$

$$0.0206\left(\frac{76{,}963.54}{1000}\cos 3°\,02'\,05''\right)^2 = 0.0206 \times 5906 \qquad = \; +121.66$$

$$\Delta H = +4196.20 \text{ ft}$$

The difference in elevation using the average is

$$76{,}963.54 \sin 3°\,02'\,05'' = \quad 4074.54$$

$$76{,}963.54 \sin 3°\,12'\,55'' = \quad 4316.71$$

$$\overline{}$$

$$\Delta H = \text{average} = +4195.63 \text{ ft}$$

3-4. DIRECT DIFFERENTIAL LEVELING The purpose of differential leveling is to determine the difference of elevation between two points on the earth's surface. The most accurate method of determining differences of elevation is with the spirit level and a rod, in the manner illustrated in Fig. 3-4. It is assumed that the elevation of point *A* is 976 ft and that it is desired to determine the elevation of point *B*. The level is set up, as described in Section 3-22, at some convenient point so that the instrument is higher than both *A* and *B*. A leveling rod is held vertically at the point *A*, which may be on the top of a stake or on some solid object, and the telescope is directed toward the rod. The vertical distance from *A* to a horizontal plane can be read on the rod where the horizontal cross hair of the telescope appears to coincide. If the rod reading is 7.0 ft, the plane of the telescope is 7.0 ft above the point *A*. The elevation of this horizontal plane is 976 + 7 = 983 ft. The leveling rod is next held vertically at *B* and the telescope is directed toward the rod. The vertical distance from *B* to the same horizontal plane is given by the rod reading with which the horizontal cross hair appears to coincide. If

Figure 3-4. Direct differential leveling.

the rod reading at B is 3.0 ft, the point B is 3.0 ft below this plane and the elevation of B is $983 - 3 = 980$ ft. The elevation of the ground at the point at which the level is set up need not be considered.

The same result may be obtained by noting that the difference in elevation between A and B is $7 - 3 = 4$ ft, and that B is higher than A. The elevation of B equals the elevation of A plus the difference of elevation between A and B, or $976 + 4 = 980$ ft.

3-5. STADIA LEVELING   Stadia leveling combines features of trigonometric leveling with those of direct differential leveling. In stadia leveling, vertical angles are read by using the transit or theodolite, and horizontal distances are determined at the same time by means of the stadia hairs mentioned in Section 2-4. As in direct differential leveling, the elevation of the ground at the point at which the instrument is located is of no concern in the process. Stadia leveling is a rapid means of leveling when moderate precision is sufficient. It is described in detail in Section 14-10.

3-6. LEVELING WITH ANEROID BAROMETER   The fact that atmospheric pressure, and hence the reading of a barometer, decreases as the altitude increases is utilized in determining differences of elevation. Because of transportation difficulties, the mercurial barometer is not used for survey purposes. Instead, the aneroid barometer or altimeter is used in surveys in which errors of 5 to 10 ft are of no consequence. Altimeters vary in size from that of an ordinary watch to one that is 10 or 12 in. in diameter. The dials sometimes have two sets of graduations, namely, feet or metres of elevation and inches or millimetres of mercury. The smaller altimeters can be read by estimation to about 10 ft. A larger type, one of which is shown in Fig. 3-5, is much more sensitive, and differences of elevation of 2 or 3 ft can be detected.

An altimeter is, of course, subject to natural changes in atmospheric pressure due to weather, and it is also subject to effects of temperature and humidity. For this reason altimeters should be used in groups of three or more.

In Fig. 3-6, an altimeter is maintained at L, which is a point of known elevation and is designated the low base. A second altimeter, whose reading has been compared with the one kept at the low base, is taken to H, which is a second point of known elevation and is designated the high base. At regular intervals, say every 5 min, the altimeters at the low and high bases are read. A third altimeter, which is referred to as the field altimeter or roving altimeter, is initially compared to the low-base altimeter and then taken to various points whose elevations are to be established. At each point the reading of the roving altimeter and the time are recorded. The elevation of any point can be determined from the reading of the roving altimeter at the point and the readings

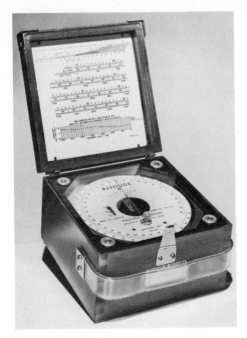

Figure 3-5. Precision altimeter.

of the altimeters at the high and low bases at that time. In order to maintain accuracy, limitations must be imposed on the elevation difference between the two bases and on the area covered by the roving altimeter. These limitations tend to eliminate the effects of changing atmospheric pressure due to the weather and also the effects of differences in temperature and humidity.

In the accompanying tabulation the readings of the three altimeters indexed at the low base are shown. The high-base altimeter is taken to the

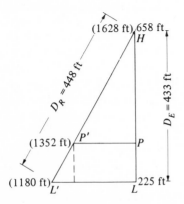

Figure 3-6. Two-base altimetry.

high base and the roving altimeter is taken to various field points. Three readings of the roving altimeter are shown in Table 3-1.

TABLE 3-1. Two-Base Altimeter Readings. Indexing on Low Base; Low-Base Altimeter 1180; High-Base Altimeter 1189; Roving Altimeter 1188

TIME	LOW-BASE READING (ft)	LOW-BASE ELEVATION (ft)	HIGH-BASE READING (ft)	HIGH-BASE ELEVATION (ft)	ROVING-ALTIMETER READING (ft)
1:55 P.M.	1180	225	1637	658	1360
2:15 P.M.	1186	225	1650	658	1425
2:20 P.M.	1184	225	1646	658	1452

Before either the high-base readings or the roving-altimeter readings are used for computing elevations, they must be corrected for index error. Thus the high-base readings must be reduced by 9 ft and the roving-altimeter readings by 8 ft. It is possible to make all three altimeters read the same at the time of indexing by physically adjusting the pointers, but this practice is not recommended. The corrected readings are shown here for convenience.

TIME	LOW-BASE READING (ft)	CORRECTED HIGH-BASE READING (ft)	CORRECTED ROVING-ALTIMETER READING (ft)	DIFFERENCE HIGH-LOW (ft)	DIFFERENCE ROVING-LOW (ft)	Δh (ft)
1:55 P.M.	1180	1628	1352	448	172	166
2:15 P.M.	1186	1641	1417	455	231	220
2:20 P.M.	1184	1637	1444	453	260	249

The known difference in elevation between the high base and the low base is $658 - 225 = 433$ ft. This value is represented by the length of the vertical line LH in Fig. 3-6. At the time of the first field reading, the difference between the corrected readings at the high and low bases is 448 ft. This value is represented by the length of the sloped line $L'H$. Also, at the same time, the difference between the corrected readings at the field point and the low base is 172 ft, which is represented by the distance $L'P'$. Then, by proportion, the difference in elevation Δh between the low base and the field point, or the distance LP, is $(172)(433)/448 = 166$ ft. Similarly, the differences in elevation between the low base and the other two points are, respectively, $(231)(433)/455 = 220$ ft and $(260)(433)/453 = 249$ ft.

3-7. TYPES OF SPIRIT LEVELS
The instrument most extensively used in leveling is the engineer's level. It consists essentially of a telescope to

which a very accurate spirit level is attached longitudinally. The telescope is supported at the ends of a straight bar that is firmly secured at the center to the perpendicular axis on which it revolves. The whole is supported on a tripod.

There are three general types of engineer's levels. These are the dumpy level; the tilting level; and the automatic or self-leveling level. The design features will vary in any given type of level, but the operating principle is the same.

3-8. DUMPY LEVEL   The engineer's dumpy level is shown in Fig. 3-7. It consists of five basic parts: the telescope *a* which is rigidly attached to the cross bar *b*; the level vial *c* encased in a housing which is also attached to the cross bar; the leveling head *d*; and the tripod *e*. The vertical axis of rotation is a tapered spindle rigidly attached to the lower side of the cross bar. This tapered spindle fits into a tapered bore through the center of the leveling head. A clamp *f* either permits (when loose) or prevents (when clamped) rotation of the spindle inside the leveling head. It is used to control the rotation of the telescope in a horizontal plane when sighting. A *slow motion* or tangent screw *g* is used after the clamp has been clamped in order to make a fine adjustment to the direction of the telescope when taking a rod reading.

The leveling head contains either three or four leveling screws (the latter shown in Fig. 3-7) which tilt the leveling head against the top of the plate *h*. This also tilts the vertical spindle to the vertical position, using the level bubble for reference as discussed in Section 3-22. The tripod consists of three legs shod with steel and connected by hinge joints to the metal tripod head.

Figure 3-7. Dumpy level.

When the level has been unscrewed from the top of the tripod for storage, it should be replaced by a threaded tripod cap in order to prevent the threads from being damaged.

3-9. THE TELESCOPE A longitudinal section of a typical telescope used in a surveying instrument is diagrammed in Fig. 3-8. The essential parts are the barrel, the positive objective lens, the negative focusing lens, the reticle which holds a set of cross hairs, and the eyepiece system. The objective lens is a compound lens composed of both crown and flint glass to eliminate chromatic aberration. Its function is to form an image of the object sighted. The image would be formed ahead of the cross hairs. However, the focusing lens diverges the light rays and brings them into focus in the plane of the cross hairs as shown in Fig. 3-8. The eyepiece lenses magnify the image together with the cross hairs in order to give the instrument man the ability to sight accurately and to be able to accurately read the leveling rod graduations. The image formed by the objective and the focusing lens is inverted. Some eyepiece systems erect the image to give a normal view. Other eyepiece systems do not, so the image is seen upside down.

The cross hairs used in some surveying instruments are very fine threads, taken from the cocoon of a brown spider. Many instrument makers use finely drawn platinum wires, some use fine glass threads, and others use a glass diaphragm on which lines have been etched. Levels intended for ordinary work have two cross hairs, one horizontal and one vertical. Those instruments that are intended for precise leveling have two additional horizontal hairs, one above and one below the usual horizontal cross hair. These additional hairs are called *stadia hairs* and are found in most transits and theodolites.

The cross hairs are attached to a reticle, or cross-hair ring. As shown in Fig. 3-9, this reticle is practically a heavy brass washer that is somewhat smaller in diameter than the inner surface of the telescope barrel. It is thick enough to be tapped and threaded for the capstan-headed screws by which it is held in place in the telescope tube. The holes through the telescope tube are large enough to permit the cross-hair ring to be rotated through a small angle and are therefore covered by beveled washers. The ring can be moved vertically or horizontally by turning the proper capstan screws.

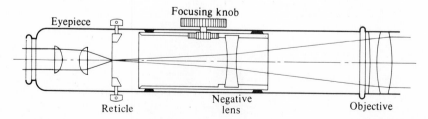

Figure 3-8. Longitudinal section of telescope.

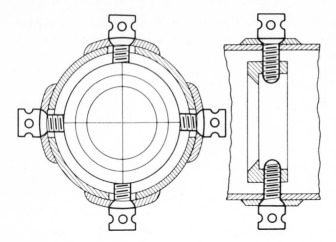

Figure 3-9. Cross-hair reticle.

Focusing of the eyepiece is accomplished by changing the distance between it and the cross hairs. On most telescopes this is done by twisting the end of the eyepiece.

3-10. LEVEL TUBE The spirit level c in Fig. 3-7 is the part of the instrument on which the accuracy chiefly depends. As indicated in Fig. 3-10, in plan in view (a) and in side elevation in view (b), it consists of a sealed glass tube nearly filled with alcohol or with a mixture of alcohol and ether. The upper surface of the tube, and sometimes also the lower surface, is ground to form a longitudinal circular curve. The sensitiveness of the bubble tube is dependent on the radius of this circular curve, and is usually expressed in terms of the angle through which the tube must be tilted to cause the bubble to move over one division of the scale etched on the glass. This angle may vary from 1″ or 2″, in the case of a precise level, up to 10″ to 30″ on an engineer's level. The radius to which the tube for an engineer's level is ground is usually between 75 and 150 ft.

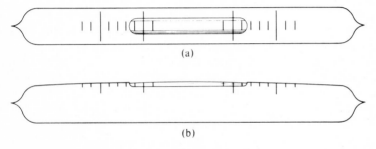

Figure 3-10. Level tube.

In a properly designed level there should be a definite relation between the sensitiveness of the bubble and the magnifying power of the telescope. If, with the telescope focused on a distant object, the leveling screws are turned just enough to cause a barely perceptible movement of the horizontal cross hair, a corresponding displacement of the bubble should be apparent. Conversely, when the bubble is changed very slightly, a movement of the horizontal cross hair should be detected.

The angular value of one division of the level tube can be determined by turning the leveling screws so as to cause the bubble to move over several divisions and by observing the corresponding vertical distance the horizontal cross hair is raised or lowered on a leveling rod held at a known horizontal distance from the level. Thus if the bubble is moved over five divisions and the vertical movement of the cross hair is 0.102 ft on a rod 300 ft from the level, the angular value S of one division is given by

$$\tan S = \frac{0.102}{5 \times 300}$$

or

$$S'' = \frac{\tan S}{\tan 1''} = \frac{0.102}{5 \times 300 \times 0.00000485} = 14''$$

The metal case containing the level tube is adjustable vertically at one end.

3-11. SIGHTING THROUGH THE TELESCOPE When sighting through the telescope of a surveying instrument, whether its a level, a transit, a theodolite, or an alidade, the observer must first focus the eyepiece system for *his individual eye*. This is most easily done by holding an opened field book about six inches in front of the objective lens and on a slant in order to obscure the view ahead of the telescope and to allow light to enter the objective lens. This is shown in Fig. 3-11. The *eyepiece* is now twisted in or out until the cross hairs are sharp and distinct. Now, with the eyepiece system focused, the telescope is pointed at the object to be sighted, with the observer looking along the top of the telescope barrel (some telescopes are provided with peep sights with which to make this initial alignment). The rotational motion is then clamped. The object to be sighted should now be in the field of view. The tangent screw is then used to bring the line of sight directly on the point.

The *focusing lens* is now focused in order to bring the image of the point into the plane of the cross hairs. This can be checked by moving the eye up and down or sideways a small amount to detect whether or not the cross hairs appear to move with respect to the object sighted. If they do not appear to move, the telescope is properly focused. If they *do* appear to move, the

Figure 3-11. Focusing the eyepiece.

condition called *parallax* exists between the image and the cross hairs. This condition is shown in Fig. 3-12. If the image formed by the lens lies either in front of or behind the plane of the cross hairs as shown in Fig. 3-12(a) and (b) and the observer's eye is raised or lowered slightly, the cross hairs seem

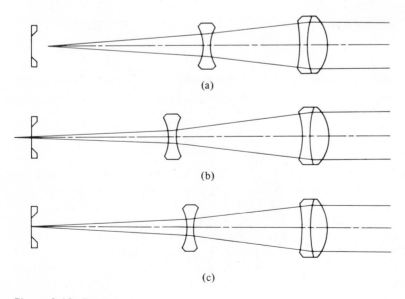

(a)

(b)

(c)

Figure 3-12. Parallax.

to move up and down on the face of a leveling rod. Refocusing while observing the cross hairs will bring the focusing lens in the proper position, and the parallax is eliminated as shown in Fig. 3-12(c).

It is advisable to occasionally check the eyepiece focus by the method shown in Fig. 3-11, particularly when the observer must make a lengthy series of observations.

3-12. TILTING LEVEL When a level is used, the level bubble must be centered at the instant of observation. Otherwise serious random errors are introduced. If he is using a conventional dumpy level, the observer must constantly check the level bubble and readjust the leveling screws. The tilting level, shown in Fig. 3-13, is brought to approximate level by means of a circular type, or bull's-eye, level. When a reading is to be taken, the observer rotates a tilting knob, which moves the telescope through a small vertical angle. With his head in the normal position for viewing the rod through the telescope, he can at the same time look through a window to the left of the eyepiece of the telescope and observe the more-sensitive, main-level bubble as two half images of opposite ends of the bubble. These half images are brought into superposition and made visible to the observer by a prismatic arrangement directly over the bubble. The observer then tilts the telescope until the two half images are made to coincide, in which position the bubble

Figure 3-13. Tilting level. Courtesy of Keuffel & Esser Co.

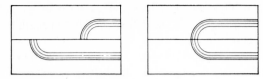

Figure 3-14. Coincidence bubble.

is centered. The split bubble before and after coincidence is shown in Fig. 3-14. Note that the apparent discrepancy between the two half images is actually twice the centering error. Because of the fine tilting adjustment and the ability to view both ends of the bubble simultaneously, the observer can center the bubble to within 0.3 mm.

In Fig. 3-15 is shown a tilting level of European design. Its operation is identical with that of the level shown in Fig. 3-13. The eyepiece, however,

Figure 3-15. European tilting level. Courtesy of Kern Instruments, Inc.

does not erect the image of the leveling rod, and this difference accounts for the relatively shorter telescope. In this instrument the image of the coincidence bubble appears directly in the field of view of the telescope rather than through an adjacent eyepiece, and the observer may view the bubble, the cross hairs, and the image of the leveling rod with one eye and from one position.

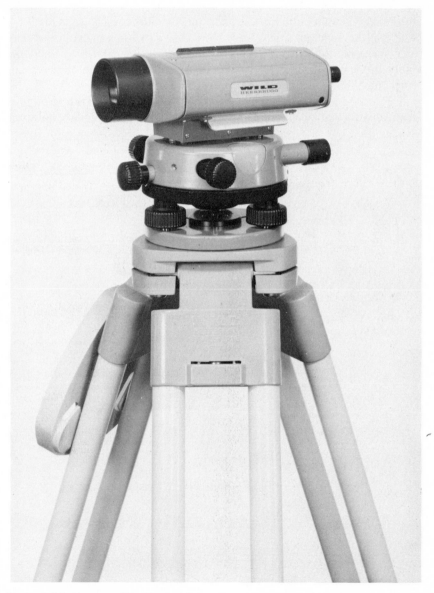

Figure 3-16. European tilting level with provision for rotating telescope about longitudinal axis. Courtesy of Wild Heerbrugg Instruments, Inc.

The tilting level shown in Fig. 3-16 contains a coincidence bubble which is viewed through a separate window like that of the level shown in Fig. 3-13. Also, the level contains a horizontal circle which makes it convenient for layout and construction surveys. The unique feature of this level is the capability of rotating the telescope and the level tube 180° about the telescope's longitudinal axis. This feature permits a direct and reverse reading on a level rod, the mean value of which is free from the error of an inclined line of sight discussed in Section 3-28.

3-13. SELF-LEVELING OR AUTOMATIC LEVEL The level shown in Fig. 3-17 is said to be self-leveling. When the bull's-eye bubble has been centered, a prism carried on a pendulum supported by two pairs of wires reflects the light rays entering the objective lens on back to the eyepiece end of the telescope. The lengths of the supporting wires and the positions of the points of suspension are so designed that the only rays of light reflected back to the intersection of the cross hairs by the swinging prism are the horizontal rays passing through the optical center of the objective lens. Hence, as long as the prism is free to swing, a horizontal line of sight is maintained, even

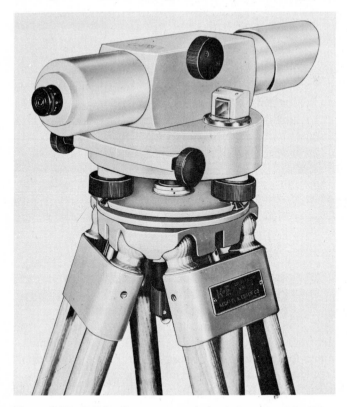

Figure 3-17. Self-leveling or automatic level. Courtesy of Keuffel & Esser Co.

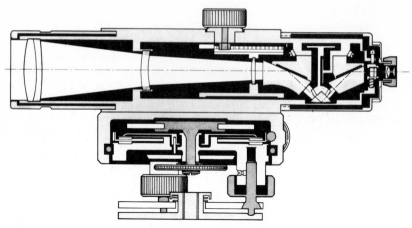

Figure 3-18. Section through self-leveling level.

though the telescope barrel itself is not horizontal. A damping device brings the pendulum to rest quite rapidly, so that the observer does not have to wait until it settles of its own accord. This type of level has the advantage of offering very rapid instrument setups and of eliminating random errors in centering the level bubble.

The self-leveling level shown in Fig. 3-17 is shown in cross section in Fig. 3-18. A small triangular-shaped prism can be seen slightly forward of the eyepiece near the bottom of the telescope. Light on entering the objective lens passes through the focusing lens and then through a fixed prism. It is then reflected off the triangular swinging prism and enters a second fixed prism, from which it is deflected into the eyepiece system.

The automatic level shown in Fig. 3-19 operates by means of a gravity-actuated compensator similar in general principle to that described above. This level contains an optical micrometer, the function of which is to allow very precise readings of graduated level rods as described in Section 3-21.

All automatic levels operate generally on the same principle as that described in this section.

3-14. GEODETIC LEVEL The geodetic level, used for extreme precision required in first-order leveling conducted by the National Geodetic Survey, is a tilting level made of invar to lessen the effect of temperature. The magnification of the objective-eyepiece combination is about 42 diameters. The geodetic level is equipped with stadia hairs for determining three rod readings at one observation and for determining the lengths of the sights. The tripod on which the geodetic level rests is of such height that the observer may stand erect while observing. The unusually high tripod brings the line of sight somewhat farther from the intervening ground than does the ordinary tripod, helping to lessen differential refraction of the line of sight.

Figure 3-19. Self-leveling level with optical micrometer. Courtesy of Wild Heerbrugg Instruments, Inc.

The precise tilting level shown in Fig. 3-20 has a 42-power telescope and contains a very sensitive coincidence-type bubble in addition to an optical micrometer. This level, together with a precise leveling rod, is quite suitable for precise geodetic leveling.

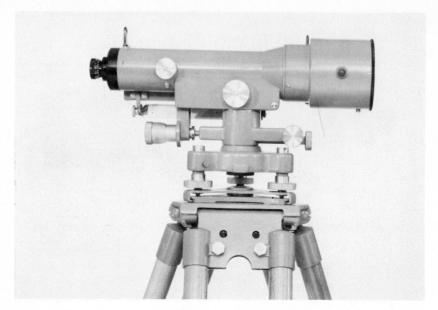

Figure 3-20. Precise tilting level with optical micrometer used for geodetic leveling. Courtesy of Wild Heerbrugg Instruments, Inc.

3-15. HAND LEVEL AND CLINOMETER The hand level, shown in Fig. 3-21, is a brass tube with a small level tube mounted on the top. A 45° mirror on the inside of the main tube enables the user to tell when it is being held horizontally. As the rod viewed through the level is not magnified, the length of sight is limited by the visibility of rod readings with the naked eye.

The hand level is used on reconnaissance surveys where extreme accuracy is unnecessary and in taping to determine when the tape is being held horizontally. It is also used to advantage for estimating how high or how low the engineer's level must be set in order to be able to read the leveling rod.

Figure 3-21. Hand level. Courtesy of Keuffel & Esser Co.

Figure 3-22. Clinometer. Courtesy of Keuffel & Esser Co.

The clinometer, shown in Fig. 3-22, can be used in the same manner as the hand level. In addition it can be employed for measuring vertical angles where approximate results are sufficient.

3-16. LEVELING RODS There are two general classes of leveling rods, namely, self-reading and target rods. A self-reading rod has painted graduations that can be read directly from the level. When a target rod is used, the target is set by the rodman as directed by the levelman, and the reading is then made by the rodman. Some types of rods can be used either as self-reading rods or as target rods.

The graduations on self-reading rods should appear sharp and distinct for the average length of sight. In the United States the rods are ordinarily graduated so as to indicate feet and decimals, the smallest division usually being 0.01 ft. Metric rods are graduated to centimetres, and rod readings are estimated to millimetres.

3-17. PHILADELPHIA ROD The Philadelphia rod, front and rear views of which are shown in Fig. 3-23, is made in two sections that are held together by the brass sleeves *a* and *b*. The rear section slides with respect to the front section, and it can be held in any desired position by means of the clamp screw *c* on the upper sleeve *b*. The rod is said to be a *short rod* when the rear portion is not extended, and a *long* or *high rod* when it is extended. The short rod is used for readings up to 7 ft. For readings between 7 and 13 ft, the long rod must be used. When the rod is fully extended, the graduations on the face are continuous.

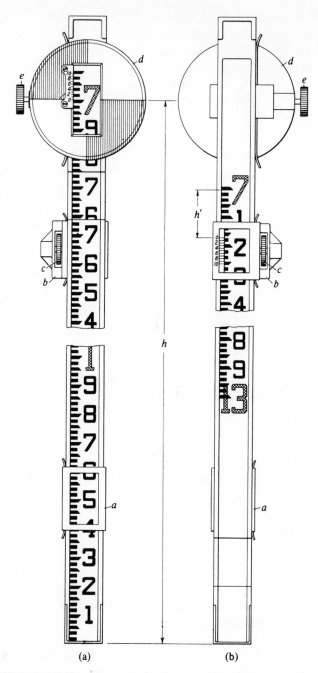

Figure 3-23. Philadelphia rod.

The rod shown in Fig. 3-23 is graduated to hundredths of a foot by alternate black and white spaces painted on the rod. Each fifth hundredth is indicated by a longer graduation mark, so that an acute angle is formed at one corner of the black space of which the graduation is a part. The tenths are marked by large black figures, half above and half below the graduation mark, and the feet are shown in a similar manner by red figures (shaded in the illustrations). The graduations can be seen distinctly through the telescope of a level at distances up to about 300 ft, and the rod can thus be used as a self-reading rod.

In case the graduations cannot be read directly from the telescope with sufficient accuracy, the target d is used. The target is a circular or elliptical metal plate divided into quadrants alternating red and white. There is an opening in the face of the target in order that the graduations on the face of the rod can be seen through it. One side of this opening is beveled to a thin edge, and a scale is marked along this edge so that it is close to the face of the rod. This scale is used for determining readings between graduation marks. One of its ends is exactly on the horizontal line dividing the colors on the target.

When the clamp e is loose, the target can be moved over the face of the rod until the line dividing the colors coincides with the horizontal cross hair of the level. The target can then be fixed in that position by tightening the clamp e.

When the target is used on a high rod, it is first set exactly at 7 ft on the extension part of the rod, as shown in Fig. 3-23(a). The extension with the target is then raised until the line dividing the colors coincides with the horizontal cross hair of the telescope, and the rod is held in that position by tightening the clamp c. To determine the reading for a high rod, the graduations on the back of the sliding part and a scale on the back of the upper sleeve are used. As shown in view (b) these graduations begin at 7 ft at the top and increase downward to 13 ft.

The reading of a high rod is the distance from the base of the rod to the target. Thus for the position shown in Fig. 3-23(b), the rod reading represents the distance h. When the target is set at 7 ft on the extension part of the rod while the rod is closed, the reading of the high rod is 7 ft, and the 7-ft graduation on the back of the rod is opposite the zero of the scale on the sleeve. As the rod is raised the 7-ft mark moves upward, whereas the zero mark of the scale remains stationary since it is attached to the lower portion of the rod. The distance h' between these two marks therefore increases as the rod is extended further. Consequently, the distance h' is equal to the amount by which the target is raised above 7 ft, and for any high-rod reading the distance h is equal to 7 ft plus the distance h'. To obviate actual addition, the foot graduations on the back of the rod are numbered downward from 7 to 13. Thus when the rod is extended 1 ft, the reading on the back is $7 + 1 = 8$ ft, and so on. It is seen therefore that the numbers must increase downward in order that the rod readings may become greater as the target is raised.

If the rod has been damaged by allowing the upper portion to slide down with a bang, it is possible that the reading on the back of the rod will be less than 7 ft when the rear section is in the lowered position. In this case the target should be set at the corresponding reading on the face of the rod before extending the rod.

3-18. PRECISE LEVELING RODS A precise rod is graduated on an invar strip which is independent of the main body of the rod except at the shoe at the bottom of the rod. The graduations are in yards, in feet, or in metres, and the smallest graduations are 0.01 yd, 0.01 ft, and 1 cm, respectively. A yard rod or a metre rod is also graduated in feet on the back of the rod. These foot graduations are painted directly on the main body of the rod. They serve as a check on the more precise readings taken on the invar strip, and help to prevent mistakes in rod readings. The precise rod is equipped with either a bull's-eye level or a pair of level vials at right angles to each other to show when the rod is vertical, and also with a thermometer. Front and back views of a precise rod are shown in Fig. 3-24.

3-19. READING THE ROD DIRECTLY If the target is not used, the reading of the rod is made directly from the telescope. The number of feet is given by the red figure just below the horizontal cross hair when the level has an erecting telescope, or just above the horizontal cross hair when it has an inverting telescope. The number of tenths is shown by the black figure just below or above the hair, the position depending on whether the telescope is erecting or inverting. If the reading is required to the nearest hundredth, the number of hundredths is found by counting the divisions between the last tenth and the graduation mark nearest to the hair. If thousandths of a foot are required, the number of hundredths is equal to the number of divisions between the last tenth and the graduation mark on the same side of the hair as that tenth, and the number of thousandths is obtained by estimation.

The readings on the rod for the positions x, y, and z in Fig. 3-25(a) are determined as follows: For x, the number of feet below the cross hair is 4, the number of tenths below is 1, and the cross hair coincides with the first graduation above the tenth mark; consequently, the reading is 4.11 ft to the nearest hundredth, or 4.110 ft to the nearest thousandth. For y, the feet and tenths are again 4 and 1, respectively; also, the hair is just midway between the graduations indicating 4 and 5 hundredths, and therefore the reading to the nearest hundredth can be taken as either 4.14 or 4.15 ft. In determining the hundredths it is convenient to observe that the hair is just below the acute-angle graduation denoting the fifth hundredth, and it is therefore unnecessary to count up from the tenth graduation. If thousandths are required, the number of hundredths is the lower one, or 4; and since the hair is midway between two graduation marks on the rod and the distance between the

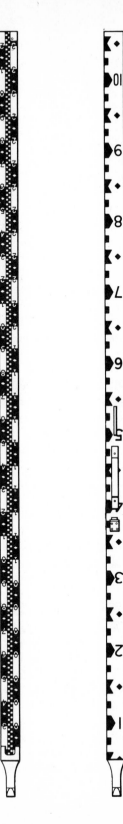

Figure 3-24. Precise leveling rod.

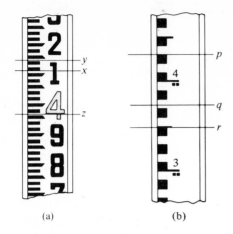

(a)   (b)

Figure 3-25. Direct reading of rod. (a) Rod graduated to 0.01 ft. (b) Rod graduated to 1 cm.

graduations is 1 hundredth or 10 thousandths of a foot, the number of thousandths in the required reading is $\frac{1}{2} \times 10$, or 5. Hence the reading to the nearest thousandth is 4.145 ft. For z, the reading to the nearest hundredth is 3.96 ft and that to the nearest thousandth is 3.963 ft.

Direct high-rod readings are made with the rod fully extended, as the graduations on the face of the rod then appear continuous.

The metric rod shown in Fig. 3-25(b) is numbered every decimetre and graduated in centimetres. The double dot shown below the decimetre numbers indicates the readings are in the 2-m range. The readings for positions p, q, and r are, respectively, 2.430 m, 2.374 m, and 2.349 m.

3-20. VERNIERS   The vernier is a device by means of which readings closer than the smallest division of a scale can be made with more certainty than they can be obtained by estimation. The verniers found on a leveling rod are shown in Fig. 3-26. The front view in (a) shows the target and vernier for a short rod, and the rear views in (b) and (c) show the vernier on the back of the rod, which is employed in making high-rod readings when the target is used.

The principle of the vernier is dependent on the fact that the lengths of the divisions on the vernier are slightly less than the lengths of the divisions on the scale. A vernier of the type found on leveling rods is shown in Fig. 3-27. On this vernier a length corresponding to nine of the smallest divisions on the scale, or 0.09 ft, is divided into 10 parts on the vernier. The length of one of these vernier divisions is $\frac{1}{10} \times 0.09$ or 0.009 ft. The difference between the length of one division on the scale and one division on the vernier is equal to $0.01 - 0.009 = 0.001$ ft. Hence, if the vernier is shifted from the position shown in Fig. 3-27(a), where the zero of the vernier is coinciding with the 0.3-ft mark, to the position shown in view (b), where the first division on the vernier coincides with the 0.31-ft mark, the distance the vernier has been

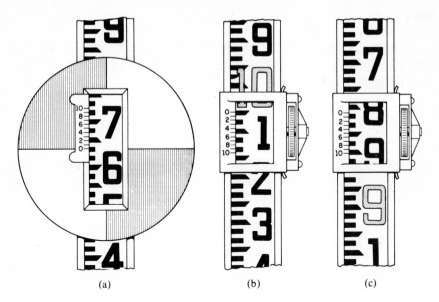

Figure 3-26. Verniers on leveling rod.

moved must be the difference between the length of a main-scale division and the length of a vernier division, or 0.001 ft. In Fig. 3-27(c) the eighth division on the vernier is coinciding with a mark on the main scale, and hence the vernier must have been moved 0.008 ft from the original position in view (a). The number of thousandths of a foot must always be the same numerically as the number of the vernier graduation that coincides with a mark on the main

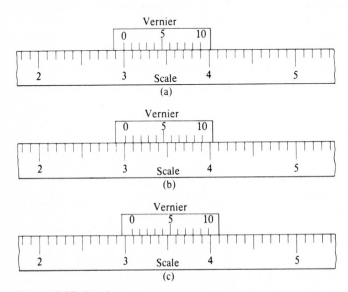

Figure 3-27. Vernier.

scale. The readings shown in the three positions in Fig. 3-27 are 0.300, 0.301, and 0.308 ft, respectively.

If s represents the spaces on the main scale, and n represents the number of divisions on the vernier, the vernier formula used to determine the least reading l is given as

$$l = \frac{s}{n} \qquad (3\text{-}6)$$

The least reading is thus determined by dividing the length of the smallest division on the scale by the number of divisions on the vernier between two marks that coincide with marks on the main scale. The least count of the vernier shown in Fig. 3-27 is 0.01/10 = 0.001 ft.

The readings on the rods shown in Fig. 3-26 are as follows: (a) 0.635 ft, (b) 10.053 ft, (c) 8.807 ft. Mistakes can often be averted by first estimating the number of thousandths and then looking at that part of the vernier where the coincidence must occur.

3-21. OPTICAL MICROMETER The optical micrometer, the principle of which is shown in Fig. 3-28, measures the vertical distance on the rod from the point at which a horizontal line of sight strikes the rod to the next lower graduation mark. It consists of a thick piece of optical glass, the front and rear surfaces of which are ground flat and parallel to one another, together with a micrometer drum graduated so as to indicate the amount by which the plano-parallel element raises or lowers the line of sight. The unit either fits over the barrel of the telescope directly in front of the objective lens or else is an integral part of the instrument. A complete revolution of the drum, or that part of a revolution between the zero graduation and the last graduation, represents a fixed value, such as 0.01 ft, 1 cm, or 5 mm. The rod graduations must therefore correspond to the particular optical micrometer being used.

The micrometer eliminates the necessity for estimating between graduations, and it furthermore subdivides the distance between graduations very

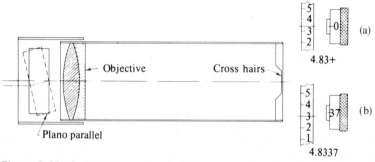

Figure 3-28. Optical micrometer principle.

precisely into 50 or 100 parts. The position of the horizontal line of sight on the rod shown in Fig. 3-28(a) is between 4.83 and 4.84 ft, and the micrometer drum reads zero. Also the drum is divided into 100 parts. When the micrometer drum is turned and the plano-parallel element is rotated so that the line of sight is lowered to the 4.83-ft rod graduation, [Fig. 3-28(b)] the micrometer drum reads 37. This reading means that the line of sight has been lowered 37/100 of 0.01 ft, or 0.0037 ft. The rod reading is therefore 4.8337 ft.

The optical micrometer seen on the level shown in Fig. 3-29 both raises and lowers the line of sight 0.01 ft from an undeviated position by means of the plano-parallel element. When at the zero or undeviated position, the micrometer drum reads 10; when lowered the maximum amount, it reads 20; and when raised the maximum amount, it reads 0. Three readings using this type of optical micrometer are shown in Fig. 3-30. If the micrometer is set to read 10.0, the cross hair is located about one-fourth of the distance from the 5.24- and the 5.25-ft graduation. This is the undeviated position of the line of sight. If the micrometer is now rotated in order to lower the cross hair to the whole graduation at 5.24, the micrometer drum reading will be, for example 12.4. This represents 0.0124 ft. The complete reading with this setting is then 5.2400 + 0.0124 = 5.2524 ft. If the micrometer is now rotated in order to

Figure 3-29. Optical micrometer drum.

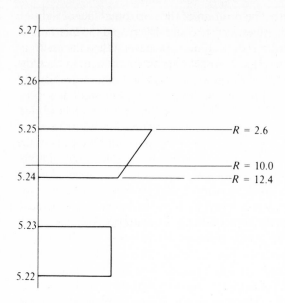

Figure 3-30. Micrometer drum readings for different positions of line of sight.

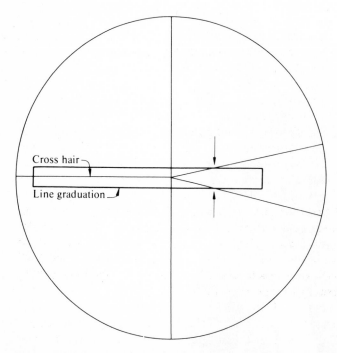

Figure 3-31. Wedge-type reticle. The thickness of the line graduation is greatly exaggerated. Symmetry is observed at the arrows.

bring the cross hair up to the 5.25-ft graduation, the drum reading will be, say, 2.6, which represents 0.0026 ft. The complete reading with this setting is then 5.2500 + 0.0026 = 5.2526 ft. The mean of the two readings is thus 5.2525 ft.

When using an optical micrometer, the instrument man must constantly check the coincidence of the level bubble, because manipulation of the micrometer knob may cause the telescope to go slightly out of level.

The graduations shown in Fig. 3-30 are characteristic of the Philadelphia rod graduations as well as those of many of the precise rods in use for high-order surveys. Their design requires that a single horizontal cross hair be set exactly on the upper or lower edge of the black bar before the micrometer drum is read. A somewhat more satisfactory design for use with the optical micrometer is the line graduation shown in Fig. 3-31. The level used with this type of design contains a horizontal cross hair that splits into a wedge half-way across the field of view. This wedge can be centered on the line graduation with a high degree of accuracy by the principle of symmetry.

3-22. SETTING UP THE LEVEL The purpose of direct leveling, as explained in Section 3-4, is to determine the difference of elevation between two points by reading a rod held on the points. These rod readings can be made by the levelman without setting the target, or the target can be set as directed by the levelman and the actual reading made by the rodman. At the instant the readings are made, it is necessary that the line of sight determined by the intersection of the cross hairs and the optical center of the objective be horizontal. In a properly adjusted instrument this line will be horizontal only when the bubble is at the center of the bubble tube.

The first step in setting up the level is to spread the tripod legs so that the tripod head will be approximately horizontal. The legs should be far enough apart to prevent the instrument from being blown over by a gust of wind, and they should be pushed into the ground far enough to make the level stable. Repairs to a damaged instrument are always expensive. For this reason, neither a level nor a transit should be set up on a pavement or a sidewalk if such a setup can possibly be avoided. When the instrument must be so set up, additional care should be exercised to protect it from possible mishaps.

If the level contains four leveling screws, the telescope is turned over either pair of opposite leveling screws as shown in Fig. 3-32(a). The bubble is then brought approximately to the center of the tube by turning the screws in *opposite* directions. The level bubble moves in the direction of the left thumb, a point well worth remembering. No great care should be taken to bring the bubble exactly to the center the first time.

The next step is to turn the telescope over the other pair of screws and to bring the bubble exactly to the center of the tube by means of these screws. This is shown in Fig. 3-32(b). The telescope is now turned over the first pair of screws once more, and this time the bubble is centered accurately. The telescope is then turned over the second pair of screws and if the bubble has

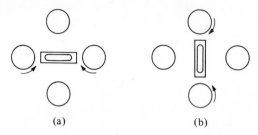

Figure 3-32. Manipulation of four leveling screws.

moved away from the center of the tube, it is brought back to the center. When the instrument is finally leveled up, the bubble should be in the center of the tube when the telescope is turned over either pair of screws. If the instrument is in adjustment, the bubble should remain in the center as the telescope is turned in any direction.

The beginner will need considerable practice in leveling up the instrument. It is by practice alone that he is able to tell how much to turn the screws to bring the bubble to the center. The more sensitive the bubble, the more skill is required to center it exactly. For the final centering when the bubble is to be moved only a part of a division, only one screw need be turned. The screw that has to be tightened should be turned if both are a little loose, and the one that has to be loosened should be turned when they are tight. When the telescope is finally leveled up, all four screws should be bearing firmly but should not be so tight as to put a strain in the leveling head. If the head of the tripod is badly out of horizontal, it may be found that the leveling screws turn very hard. The cause is the binding of the ball-and-socket joint at the bottom of the spindle. The tension may be relieved by loosening both screws of the other pair.

When a three-screw instrument is to be leveled, the level bubble is brought parallel with a line joining any two screws, such as a and b in Fig. 3-33(a). By rotating these two screws in opposite directions, the instrument is tilted about the axis l–l, and the bubble can be brought to the center. The level bubble is now brought perpendicular to the line joining these first two screws, as shown in Fig. 3-33(b). Then only the third screw c is rotated to bring the bubble to the center. This operation tilts the instrument about the axis m–m. The procedure is then repeated to bring the bubble exactly to the center in both directions.

When leveling a three-screw tilting level or an automatic level equipped with a bull's-eye bubble, as shown in Fig. 3-33(c), opposite rotations of screws a and b cause the bubble to move in the direction of the axis m–m. Rotation of screw c only causes the bubble to move in the direction of the axis l–l.

In walking about the instrument the levelman must be careful not to step near the tripod legs, particularly when the ground is soft. Neither should any part of the level be touched as the readings are being made, because the bubble can be pulled off several divisions by resting the hand on the telescope or on

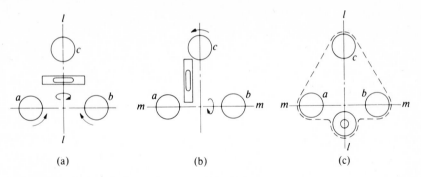

Figure 3-33. Manipulation of three leveling screws.

a tripod leg. The bubble will not remain in the center of the tube for any appreciable length of time. The levelman should form the habit of always checking the centering of the bubble just before and just after making a reading. Only in this way can he be sure that the telescope was actually horizontal when the reading was made.

3-23. SIGNALS In running a line of levels it is necessary for the levelman and the rodman to be in almost constant communication with each other. As a means of communication, certain convenient signals are employed. It is important that the levelman and the rodman understand these in order to avoid mistakes. When the target is used, it is set by the rodman according to signals given by the levelman. Raising the hand above the shoulder, so that the palm is visible, is the signal for raising the target; lowering the hand below the waist is the signal for lowering the target. The levelman, viewing the rod and the rodman through the telescope, should remember that he can see them much more distinctly than he can be seen by the rodman. Hence his signals should be such that there is no possible chance of misunderstanding. A circle described by the hand is the signal for clamping the target, and a wave of both hands indicates that the target is properly set, or all right. The signal for plumbing the rod is to raise one arm above the head and then to lean the body in the direction in which the rod should be moved.

3-24. RUNNING A LINE OF LEVELS In the preliminary example of direct leveling, given in Section 3-4, it was assumed that the difference of elevation between the two points considered could be obtained by a single setting of the level. This will be the case only when the difference in elevation is small and when the points are relatively close together. In Fig. 3-34, rods at the points A and K cannot be seen from the same position of the level. If it is required to find the elevation of point K from that of A, it will be necessary to set up the level several times and to establish intermediate points such as

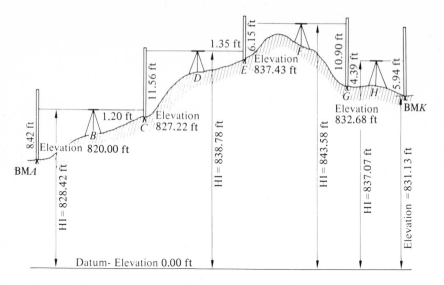

Figure 3-34. Direct leveling.

C, E, and G. These are the conditions commonly encountered in the field and may serve as an illustration of the general methods of direct leveling.

Let the elevation of the benchmark (abbreviated BM) at A be assumed as 820.00 ft. This is recorded as shown in the leveling notes of Fig. 3-35. The level is set up at B, near the line between A and K, so that a rod held on the BM will be visible through the telescope; the reading on the rod is found to be 8.42 ft. This reading is called a backsight reading, or simply a backsight (abbreviated BS), and is recorded as such in the notes. A backsight is the rod reading taken on a point of known elevation to determine the *height of instrument* (abbreviated HI). If the BS of 8.42 ft is added to the elevation of A, the HI is obtained. Thus HI = 820.00 + 8.42 = 828.42 ft. This is shown in the notes.

After the HI has been established, a point C called a *turning point* is selected that is slightly below the line of sight. This point should be some stable unambiguous object, so that the rod can be removed and put back in the same place as many times as may be necessary. For this purpose, a sharp-pointed solid rock or a well-defined projection on some permanent object is preferable. If no such object is available, a stake or a railroad spike can be driven firmly in the ground and the rod held on top of it. After the turning point at C, designated TP-1, is set or selected, a reading is taken on the rod held on TP-1. If this reading is 1.20 ft, TP-1 is 1.20 ft below the line of sight, and the elevation of TP-1 is 828.42 − 1.20 = 827.22 ft as shown in the notes. This rod reading is called a foresight (abbreviated FS). An FS is taken on a point of unknown elevation in order to determine its elevation from the height of instrument.

LEVELING, BMA to BMK SHATTUCK AVE. SEWER PROJECT AKRON, OHIO						MAY 18, 1981	
					LEVEL # 4096	LEVEL J. BROWN	
STA	BS	HI	FS	ELEV	ROD # 18	ROD F. SMITH	
BMA	8.42	828.42		820.00			
TP₁	11.56	838.78	1.20	827.22	BMA is top of iron pipe, S.E.cor.		
TP₂	6.15	843.58	1.35	837.43	Shattuck + Maple Sts.		
TP₃	4.39	837.07	10.90	832.68			
BMK			5.94	831.13	BMK is Bronze disk in sidewalk		
					N.W. Cor. Shattuck + Vincent Sts.		
ΣBS	+30.52		ΣFS	-19.39			
	-19.39						
	+11.13				831.13		
					-820.00		
					+ 11.13 checks		

Figure 3-35. Level notes.

Occasionally successive foresights and backsights are taken on an overhead point such as on a point in the roof of a tunnel. The foresight taken on such a point is added to the HI to obtain the elevation of the point. The backsight taken on the point is subtracted from the elevation of the point to determine the HI. Such readings must be carefully noted in the field notes.

While the rodman remains at C, the level is moved to D and set up as high as possible but not so high that the line of sight will be above the top of the rod when it is again held at C. This can be quickly checked by means of a hand level. The reading 11.56 ft is taken as a backsight. Hence, the HI at D is 827.22 + 11.56 = 838.78 ft. When this reading is taken, it is important that the rod be held on exactly the same point that was used for a foresight when the level was at B.

After the backsight on C has been taken, another turning point E is chosen and a foresight of 1.35 ft is obtained. The elevation of E is 838.78 − 1.35 = 837.43 ft. The level is then moved to F and the backsight of 6.15 ft taken on E. The new HI is 837.43 + 6.15 = 843.58 ft. From this position of the level a foresight of 10.90 ft is taken on G, the elevation of which is 843.58 − 10.90 = 832.68 ft. The level is then set up at H, from which position a backsight reading of 4.39 ft is taken on G and a foresight reading of 5.94 ft is taken on the new BM at K. The final HI is 832.68 + 4.39 = 837.07 ft, and the elevation of K

is 837.07 − 5.94 = 831.13 ft. As the starting elevation was 820.00 ft, the point K is 11.13 ft higher than A.

3-25. CHECKING LEVEL NOTES To eliminate arithmetical mistakes in the calculation of HI's and elevations, the arithmetic should be checked on each page of notes. Adding backsights gives $\sum$ BS; adding foresights gives $\sum$ FS. Then $\sum$ BS − $\sum$ FS should equal the difference in elevation (DE) between the starting point to the last point on the page. This is shown in the notes of Fig. 3-35. $\sum$ BS − $\sum$ FS = +11.13 ft, and the calculated DE is also +11.13 which checks the arithmetic. The last point on the page should then be carried to the following page before the BS on that point is recorded in the notes.

3-26. CHECK LEVELS Although the arithmetic in the reduction of the field notes may have been verified, there is no guarantee that the difference of elevation is correct. The difference of elevation is dependent on the accuracy of each rod reading and on the manner in which the field work has been done. If there has been any mistake in reading the rod or in recording a reading, the difference of elevation is incorrect.

The only way in which the difference of elevation can be checked is by carrying the line of levels from the last point back to the original benchmark or to another benchmark whose elevation is known. This is called "closing a level circuit." If the circuit closes on the original benchmark, the last point in the circuit, BMK in Fig. 3-34, must be used as a turning point; that is, after the foresight has been read on the rod at K from the instrument set up at H, the level must be moved and reset before the backsight is taken on K in order to continue the circuit to closure. Otherwise, if a mistake was made in reading the rod on the foresight to K from the setup at H, this mistake will not be discovered when checking the notes. A plan view of the level circuit between A and K in which the circuit is closed back on A is shown in Fig. 3-36(a). Note that the level has been reset between the FS taken on K and the BS taken on K. In Fig. 3-36(b), the level circuit has been continued to a known benchmark P to close the circuit. BMK is used as a turning point in this instance.

If the line of levels is carried back to BMA in the above example, on return the measured elevation of A should be 820.00 ft. The difference represents the error of closure of the circuit and should be very small. If a large discrepancy exists, the mistake may have been made in adding and subtracting backsight and foresight readings. This will be discovered on checking the notes. Otherwise there was a wrong reading of the rod or a wrong value was entered in the field notes. The adjustment of a level circuit based on the error of closure is discussed in Section 5-11.

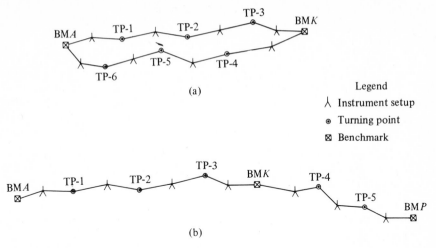

(a)

Legend

λ Instrument setup

$\circledcirc$ Turning point

$\boxtimes$ Benchmark

(b)

Figure 3-36. Closing on known benchmark.

3-27. SOURCES OF ERROR IN LEVELING The principal sources of error in leveling are instrumental defects, faulty manipulation of the level or rod, settling of the level or the rod, errors in sighting, mistakes in reading the rod or in recording or computing, errors due to natural sources, and personal errors.

3-28. INSTRUMENTAL ERRORS The most common instrumental error is caused by the level being out of adjustment. As has been previously stated, the line of sight of the telescope is horizontal when the bubble is in the center of the tube, provided the instrument is in perfect adjustment. When it is not in adjustment, the line of sight will either slope upward or downward when the bubble is brought to the center of the tube. The various tests and adjustments of the level are given in Appendix B.

Instrumental errors can be eliminated or kept at a minimum by testing the level frequently and adjusting it when necessary. Such errors can also be eliminated by keeping the lengths of the sights for the backsight and foresight readings nearly equal at each setting of the level. Since it is never known just when an instrument goes out of adjustment, this latter method is the more certain and should always be used for careful leveling.

In Fig. 3-37 the line of sight with the level at B should be in the horizontal line $EBGK$. If the line of sight slopes upward as shown and a sight is taken on a rod at A, the reading is AF, instead of AE. This reading is in error by the amount of $EF = e_1$. When the telescope is directed toward a rod held at C or D, the line of sight will still slope upward through the same vertical angle if it is assumed that the bubble remains in, or is brought to, the center of the tube.

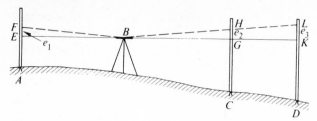

Figure 3-37. Errors caused by imperfect adjustment of level.

The rod reading taken on C will be CH, which is in error by an amount $GH = e_2$. If the horizontal distances BE and BG are equal, the errors e_1 and e_2 will be alike, and the difference between the two rod readings AF and CH will be the true difference of elevation between A and C. A rod reading taken at D will be DL, with an error $KL = e_3$. Since BK is longer than BE, the difference in elevation between A and D will be in error by an amount equal to the difference between e_3 and e_1. Similar reasoning applies if the line of sight slopes downward instead of upward. This error also exists in an automatic level.

It is not always feasible or even possible to balance a foresight distance with a backsight distance. This situation might occur on account of the terrain or when a series of foresights must be taken from a single instrument setup as in checking grade over a large area. It is advisable in such cases to check the level and make the necessary adjustments (see Section 3-38 and also Appendix B). An alternative is to employ a level like the one shown in Fig. 3-16 in which the rod can be read with the telescope right-side up and upside down, giving a mean reading free of error from a sloping line of sight. Another alternative is the technique of reciprocal observations discussed in Section 3-36.

Extremely long sights should also be avoided. The further the rod is from the level, the greater will be the space covered on the rod by the cross hair and the more difficult it will be to determine the reading accurately. For accurate results, sights with the engineer's level should be limited to about 300 ft.

3-29. ERRORS DUE TO MANIPULATION As has been previously stated, the careful levelman will form the habit of checking the position of the bubble just before and just after making each rod reading. This is the only way in which he can be certain that he is getting the proper reading.

The amount of the error due to the bubble being off center will depend on the sensitiveness of the bubble. A very convenient way to determine this error for any given bubble and for any given distance is to remember that an error of 1' in angle causes an error of about 1 in. or 0.08 ft at a distance of 300 ft (3 cm in 100 m). Thus if a 30" bubble is off one division at the instant the reading is made, the resulting error will be about 0.04 ft when the rod is 300 ft away (15 mm in 100 m). This type of error does not exist in an automatic level.

A common mistake in handling the rod is in not being careful to see that the target is properly set before a high-rod reading is made with the target. Many rods have been damaged by allowing the upper portion to slide down rapidly enough to affect the blocks at the bottom of the lower section and the top of the upper section. If this has been done, it is probable that the reading on the back of a Philadelphia rod will not be exactly 7 ft when set as a low rod. In this case the target should be set at that reading, rather than at exactly 7 ft, for a high-rod reading with the target. When a high rod is being read directly from the level, the rodman should make sure that the rod is properly extended and has not slipped down.

When the target is being used, the levelman should check its position after it has been clamped in order to make sure that it has not slipped. The beginner will be astonished at the care that must be exercised in making the target coincide exactly with the horizontal cross hair. Readings taken on the same point about 300 ft from the level may vary by several hundredths of a foot if the bubble is not exactly centered and if the target is carelessly set.

3-30. ERRORS DUE TO SETTLEMENT If any settlement of the level takes place in the interval between the reading of the backsight and the reading of the foresight, the resulting foresight will be too small, and all elevations beyond that point will be too high by the amount of the settlement. Also, if the turning point should settle while the level is being moved forward after the foresight has been taken, the backsight on the turning point from the new position of the level will be too great and all elevations will be too high by the amount of the settlement.

Errors due to the settlement of the level can be avoided by keeping the level on firm ground. If this is impossible, stakes can be driven in the ground and the tripod legs can be set on these instead of directly on the ground. In precise leveling, two rods and two rodmen are used in order that the backsight and foresight readings from a setup can be made more quickly. The error is still further diminished by taking the backsight first at one setup of the level and the foresight first at the next setup.

A proper choice of turning points should eliminate error from their settlement. If soft ground must be crossed, long stakes should be used as turning points.

3-31. ERRORS IN SIGHTING If parallax exists between the image formed by the focusing lens and the plane of the cross hairs, an error will be introduced in the rod reading. Parallax is eliminated by the procedure discussed in Section 3-11.

The rod should be plumb when the reading is made. The levelman can tell whether or not the rod is plumb in one direction by noting if it is parallel to

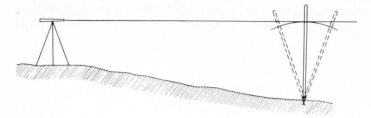

Figure 3-38. Waving the rod.

the vertical cross hair. He cannot tell, however, if it is leaning toward or away from him. The leveling rods used on precise work are equipped with circular levels, so that the rodman can tell when he is holding the rod vertically. For less accurate work, the rodman can balance the rod between his fingers if the wind is not blowing, or he can wave it slowly toward and then away from the level as indicated in Fig. 3-38. The least reading obtainable is the proper one. If the target is being used, the line dividing the colors should just coincide with the cross hair and then drop away from it. Errors from failure to hold the rod plumb will be much greater on readings near the top of the rod than for those near the bottom. For this reason more care should be exercised when making high-rod readings.

For careful work the lengths of the backsight and the foresight from the same setup should be kept nearly equal (see Section 3-28). If in ascending a steep hill the level is always kept on the straight line between the turning points, the distance to the backsight will be about twice as great as the distance to the foresight, and considerable error may result if the instrument is not in good adjustment. If there are not obstructions, these two distances can be kept nearly equal by setting the level some distance away from the straight line between the turning points. By thus zigzagging with the level, this source of error can be eliminated.

3-32. MISTAKES IN READING ROD, RECORDING, AND COMPUTING
A common mistake in reading the rod is to misread the number of feet or tenths or metres and decimetres. The careful levelman observes the foot and tenth marks both above and below the cross hair. On close sights, no foot mark may appear within the field of the telescope. In this case the reading can be checked by directing the rodman to place his finger on the rod at the cross hair or if the reading is a high one, by having him slowly raise the rod until a foot mark appears in the telescope. In case of doubt the target can always be used.

Some instruments for precise leveling are equipped with three horizontal cross hairs. All three hairs are read at each sighting. If the hairs are evenly spaced, the difference between the readings of the upper and the middle hairs should equal the difference between the readings of the middle and lower hairs. This comparison is always made before the rodman leaves a turning point.

Where readings to thousandths of a foot are being made with the target, a common mistake in recording is to omit one or more ciphers from such readings as 5.004, and to record instead 5.04 or 5.4. Such mistakes can be avoiding by making sure that there are three decimal places for each reading. Thus the second reading, if correct, should be recorded as 5.040, and the third as 5.400. If the values are not so recorded, the inference would be that the levelman was reading only to hundredths of a foot on the second reading and only to tenths on the third.

Other common mistakes of recording are the transposition of figures and the interchanging of backsight and foresight readings. If the levelman will keep the rodman at the point long enough to view the rod again after recording the reading, mistakes of the first type can often be detected. To prevent the interchange of readings, the beginner should remember that ordinarily the first reading taken from each position of the level is the backsight reading and that only one backsight is taken from any position of the level. Any other sights taken are foresights.

Mistakes in computations, as far as they affect the elevations of turning points and benchmarks, can be detected by checking the notes, as described in Section 3-25. This should be done as soon as the bottom of a page is reached, so that incorrect elevations will not be carried forward to a new page.

3-33. ERRORS DUE TO NATURAL SOURCES

One error due to natural sources is that caused by curvature and refraction as described in Section 3-2. The error from this source amounts to but 0.0002 ft in a 100-ft sight (0.01 mm/30 m) and to about 0.002 ft in a 300-ft sight (0.7 mm/100 m). So for ordinary leveling it is a negligible quantity. It can be practically eliminated by keeping the backsight and foresight distances from the same setup equal. In precise leveling if the backsight and foresight distances are not substantially equal, a correction is applied to the computed difference of elevation.

The familiar heat waves seen on a hot day are evidence of refraction and when they are seen, refraction may be a fruitful source of error in leveling. When they are particularly intense, it may be impossible to read the rod unless the sights are much shorter than those usually taken. Refraction of this type is much worse close to the ground. For careful work it may be necessary to discontinue the leveling for 2 or 3 hr during the middle of the day. It may be possible to keep the error from this source at a low figure by taking shorter sights and by so choosing the turning points that the line of sight will be at least 3 or 4 ft or 1 m above the ground.

Better results will usually be obtained when it is possible to keep the level shaded. If the sun is shining on the instrument, it may cause an unequal expansion of the various parts of the instrument; or if it heats one end of the bubble tube more than the other, the bubble will be drawn to the warmer end of the tube. For precise work the level must be protected from the direct rays of the sun.

TABLE 3-2. Classification, Standards of Accuracy, and General Specifications for Vertical Control

CLASSIFICATION	FIRST-ORDER CLASS I, CLASS II	SECOND-ORDER CLASS I	SECOND-ORDER CLASS II	THIRD-ORDER
PRINCIPAL USES Minimum standards; higher accuracies may be used for special purposes	Basic framework of the National Network and of metropolitan area control. Extensive engineering projects. Regional crustal movement investigations. Determining geopotential values	Secondary control of the National Network and of metropolitan area control. Large engineering projects. Local crustal movement and subsidence investigations. Support for lower-order control	Control densification, usually adjusted to the National Net. Local engineering projects. Topographic mapping. Studies of rapid subsidence. Support for local surveys	Miscellaneous local control; may not be adjusted to the National Network. Small engineering projects. Small engineering projects. Drainage studies and gradient establishment in mountainous areas
RECOMMENDED SPACING OF LINES National Network	Net A; 100 to 300 km *Class I* Net B; 50 to 100 km *Class II*	Secondary Net; 20 to 50 km	Area Control; 10 to 25 km	As needed
Metropolitan control; other purposes	2 to 8 km As needed	0.5 to 1 km As needed	As needed As needed	As needed As needed
SPACING OF MARKS ALONG LINES	1 to 3 km	1 to 3 km	Not more than 3 km	Not more than 3 km
INSTRUMENT STANDARDS	Automatic or tilting levels with parallel plate micrometers; invar scale rods	Automatic or tilting levels optical micrometers or three-wire levels; invar scale rods	Geodetic levels and invar scale rods	Geodetic levels and rods

FIELD PROCEDURES	Double-run; forward and backward, each section 1 to 2 km	Double-run; forward and backward, each section 1 to 2 km	Double- or single-run 1 to 3 km for double-run	Double- or single-run 1 to 3 km for double-run
Section length				
Maximum length of sight	50 m *Class I*; 60 m *Class II*	60 m	70 m	90 m
Max. difference in lengths Forward & backward sights per setup	2 m *Class I*; 5 m *Class II*	5 m	10 m	10 m
per section (cumulative)	4 m *Class I*; 10 m *Class II*	10 m	10 m	10 m
Max. length of line between connections	Net A; 300 km Net B; 100 km	50 km	50 km double-run 25 km single-run	25 km double-run 10 km single-run
MAXIMUM CLOSURES[a]				
Section: fwd. and bkwd.	3 mm $\sqrt{K}$ *Class I*: 4 mm $\sqrt{K}$ *Class II*	6 mm $\sqrt{K}$	8 mm $\sqrt{K}$	12 mm $\sqrt{K}$
Loop or line	4 mm $\sqrt{K}$ *Class I*: 5 mm $\sqrt{K}$ *Class II*	6 mm $\sqrt{K}$	8 mm $\sqrt{K}$	12 mm $\sqrt{K}$

[a] Check between forward and backward runnings where K is the distance in kilometres.

To guard against changes in the length of the leveling rod from variations in temperature, the graduations on rods used for precise leveling are placed on strips of invar, which has an extremely small coefficient of expansion. For ordinary leveling, errors from this source are negligible.

3-34. PERSONAL ERRORS   Some levelmen consistently tend to read the rod too high or too low. This may be caused by defective vision, or by an inability to decide when the bubble is properly centered. In general, personal errors tend to be compensating, although some individuals may manipulate the level in such a way as to cause them to be cumulative. Such a tendency can be discovered by leveling over a line that has been previously checked by several different parties, or by running a number of circuits of levels each of which begins and ends on the same point.

3-35. LIMITS OF ERROR   If care is used in leveling, most of the errors will tend to be random. For this reason the error in any line can be expected to be proportional to the number of setups. Since the number of setups per mile of levels will be nearly constant, the error will also be proportional to the distance in miles. The precision that is being attained can be determined by comparing the two differences of elevation obtained by running levels in both directions over a line.

Level lines are classified as first order, second order, third order, or fourth order in accordance with the field procedures used and also with the agreement between the results of leveling in both directions over a line. Table 3-2 outlines the standards for all but fourth-order accuracy in vertical control surveys and gives the principal uses to which these levels of accuracy are applied. Trigonometric and barometric leveling are considered as being of fourth-order accuracy or less.

3-36. RECIPROCAL LEVELING   In leveling across a river or a deep valley, it is usually impossible to keep the lengths of the foresight and the backsight nearly equal. In such cases reciprocal leveling is used, except where approximate results are sufficient. The difference of elevation between points on the opposite sides of the river or valley is then obtained from two sets of observations.

The method is illustrated by Fig. 3-39. The level is first set up at L_1 and rod readings are taken on the two points A and B. From these readings a difference of elevation is obtained. The level is then taken across the stream and set at such a position L_2 that $L_2 B = L_1 A$ and $L_2 A = L_1 B$. From this second position, readings are again taken on A and B, and a second difference of elevation is obtained. It is probable that these two differences will not agree. Both may be incorrect because of instrumental errors and curvature and

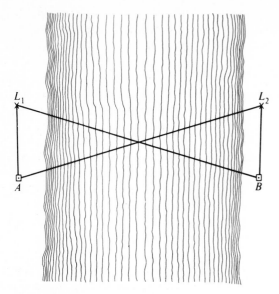

Figure 3-39. Reciprocal leveling.

refraction. However, the true difference of elevation should be very close to the mean of the two differences thus obtained.

The accuracy of this method will be increased if two leveling rods can be used, so that no appreciable time will elapse between the backsight and the foresight readings. It will be further increased by taking a number of rod readings on the more distant point, and using the average of these readings, rather than to depend on a single observation. If the distance AB is very great, it is important that the atmospheric conditions be the same for both positions of the level. Otherwise a serious error may be introduced by a changed coefficient of refraction.

3-37. THREE-WIRE LEVELING Three-wire leveling, also referred to as precise leveling, is a process of direct leveling wherein three cross hairs, referred to as threads, are read and recorded rather than the single horizontal cross hair. The cross hairs are spaced so as to represent a horizontal distance of about 100 ft for every full foot intercepted on the rod between the upper and lower cross hairs. Any tilting level equipped with three horizontal cross hairs can be used for three-wire leveling.

In the geodetic level discussed in Section 3-14, the cross hairs are spaced so as to represent a horizontal distance of about 350 ft for every foot of rod interval. This spacing has the advantage in that with greater magnification long sights can be made, while at the same time all three hairs can be observed on the rod. Suppose, for example, that the sight length is 400 ft, the reading of the center cross hair is 1.500 ft, and the ratio distance to rod intercept is 100 : 1.

The upper cross hair would fall on the 3.500-ft mark, but the lower cross hair would fall 0.500 ft below the bottom of the rod. If the ratio is 350 : 1, the upper cross hair would fall on the 2.072-ft mark whereas the lower cross hair would fall on the 0.928-ft mark.

The rods used are graduated in feet, yards, or metres. Since three-wire leveling is used for great precision, the precise leveling rods described in Section 3-18 are preferred for use in leveling of this type. Furthermore, the rods are used in pairs.

The direction diagram and the form of notes for three-wire leveling are given in Figs. 3-40 and 3-41. The left-hand page is for the backsight readings, and the right-hand page is for the foresight readings. It is to be noted that a station refers to an instrument setup and not to a point on the ground. The procedure is explained with reference to the notes of Fig. 3-41. The levels are carried from BM23 to BM24 in the forward direction. The sun is shining from direction 5, and the wind is blowing from direction 2.

The forward rodman, carrying rod No. 4, selects a turning point and the rear rodman is at the previous turning point. The level is now set up, and the observer backsights on the back turning point, reading 2487, 2416, and 2345 on the three cross hairs. He must take care that the level bubble is centered at each reading. These readings are made on a metre rod and are actually 2.487, 2.416, and 2.345 m, respectively. The back of the rod is read by using only the middle cross hair and the reading is 7.93 ft. The upper interval and the lower interval of 71 and 71 are computed and found to agree within 2 mm. So the mean of the three cross-hair readings is computed. This is compared with the equivalent reading in feet as a check against a mistake.

The observer then takes readings on rod No. 4 which is held on the forward turning point. The readings are 0519, 0444, and 0369 from the front of the rod and 1.46 ft from the back of the rod. The half intervals are then computed for agreement, and the mean reading is computed and recorded. The total backsight interval of 142 mm, representing the backsight distance, differs from the total foresight interval of 150 mm, and the sum of the backsight

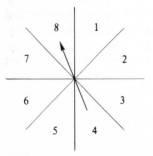

Figure 3-40. Direction diagram.

intervals from the point of beginning (BM23) is out of balance with the sum of the foresight intervals.

The back rodman advances to locate the next forward turning point. The next instrument setup is at station 43. At this setup the foresight readings are taken before the backsight readings. Alternating the sequence of readings at each setup tends to eliminate the effect of instrument settlement. Note that rod No. 7 will be sighted on first at each setup. The accumulated sum of the backsight intervals is still out of balance with the sum of the foresight intervals, so the foresight distances must be made shorter. After station 46, the sums of intervals are out of balance by only 1 mm.

As a check on computing the mean readings, all the readings are added, the sum is divided by three, and the result is compared with the sum of the means. This check is shown on the backsight page. A further check is made by adding the middle-thread readings in feet and converting the sum to metres as shown.

3-38. COLLIMATION CORRECTION In the example of Section 3-37 the sum of the backsight intervals is 725 mm and the sum of foresight intervals is 724 mm. If the distance-to-intercept ratio is assumed to be 350 : 1, the total of the backsight lengths is $725 \times 350 = 253{,}750$ mm or about 254 m. The total foresight length is 253 m. This is a very close balance, requiring no measurable correction for the error discussed in Section 3-28 and shown in Fig. 3-37. Usually, however, such a close balance is not possible, and it becomes necessary to apply a collimation correction before the correct difference in elevation can be determined. This correction is commonly referred to as the C-factor correction, in which C represents the inclination of the line of sight when the level bubble is centered.

In Fig. 3-42 it is assumed that the level bubble is centered but the line of sight is inclined by the small angle α. The reading of a rod would be R, instead of R'. The error introduced is $R - R' = D \tan \alpha = D\alpha$, where α is a small angle expressed in radians. The correction to R is then $-D\alpha = DC$, in which C is the value of the C factor. Thus $C = -\alpha$ and it follows that if C is positive, the line of sight is inclined downward.

In order to correct for an inclined line of sight when precise levels are run, the C factor must be determined from time to time, perhaps as frequently as twice a day. It takes only a few minutes to determine C, and this factor should be evaluated not only for the tilting level but also for the self-leveling level, since the same type of error always exists in this type of instrument.

To compute C, two points at least 200 ft apart are chosen in fairly level terrain, and two independent differences of elevation between the points are determined from two setups. In Fig. 3-43(a) the instrument is set up about 20 ft from A. The distance to A is d_1, and the distance to B is D_1. These distances are measured. The rod reading at A is n_1, and that at B is N_1. The line of sight is shown sloping downward, or C is assumed to be positive. The

THREE-WIRE LEVELING

DATE : 8-30-59

Sun : 5

FORWARD – ~~BACKWARD~~
(Strike out one word)

No. of Station	Thread Reading Back Sight	Mean	Middle Thread Reading Feet	Thread Interval	Sum of Intervals
	2487			71	
42	2416	2416.0	7.93	71	
	2345			142	142
	2800			83	
43	2717	2716.7	8.92	84	
	2633			167	309
	3242			61	
44	3181	3181.0	10.43	61	
	3120			122	431
	2744			72	
45	2672	2672.3	8.77	71	
	2601			143	574
	2602			75	
46	2527	2526.7	8.29	76	
	2451			151	725
	40538	13512.7	44.34		
	−10417	3472.3	11.40		
	3 ⟌30121	+10040.4	+32.94 ft.	= +10.04	meters
	+10040.3				

Figure 3-41. Notes for three-wire levels.

THREE-WIRE LEVELING

From B.M.: *23*   To B.M.: *24*

Wind : *2*   Time : *345*

Rod and Temp.	Thread Reading Fore Sight	Mean	Middle Thread Reading Feet	Thread Interval	Sum of Intervals
4	0519			75	
22°C	0444	0444.0	1.46	75	
	0369			150	150
7	0748			82	
	0666	0665.3	2.19	84	
	0582			166	316
4	0663			57	
	0606	0606.3	1.99	56	
	0550			113	429
7	1066			74	
	0992	0991.7	3.25	75	
	0917			149	578
4	0838			73	
24°C	0765	0765.0	2.51	73	
	0692			146	724
	10417	3472.3	11.40		

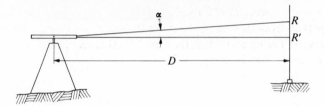

Figure 3-42. Inclination of line of sight of level.

difference in elevation (abbreviated DE) found from this setup is the corrected backsight at A minus the corrected foresight at B. Thus

$$DE_{AB} = (n_1 + Cd_1) - (N_1 + CD_1) \qquad (3\text{-}7)$$

The instrument is then set up about 20 ft from B, as indicated in Fig. 3-43(b). Rod readings N_2 and n_2 are taken at A and B, and the distances D_2 and d_2 from the instrument to the two rods are measured. The difference in elevation found from the second setup is

$$DE_{AB} = (N_2 + CD_2) - (n_2 + Cd_2) \qquad (3\text{-}8)$$

Equating the right-hand sides of Eqs. (3-7) and (3-8) gives

$$(n_1 + Cd_1) - (N_1 + CD_1) = (N_2 + CD_2) - (n_2 + Cd_2)$$

The result obtained by solving for C is

$$C = \frac{(n_1 + n_2) - (N_1 + N_2)}{(D_1 + D_2) - (d_1 + d_2)} \qquad (3\text{-}9)$$

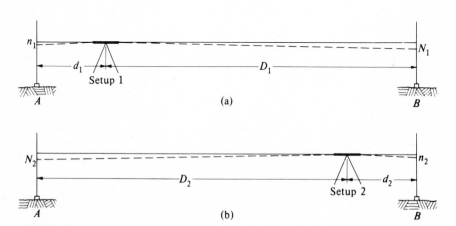

Figure 3-43. Setups to determine C factor.

Since, as will be shown, corrections for collimation errors are computed from intervals rather than from distances, the two terms in the denominator are expressed in intervals. The quantity C is then expressed in feet per foot of interval or in millimetres per millimetre of interval.

Notes for the determination of C are shown in Fig. 3-44. The leveling rod was graduated in feet and thread readings were recorded in feet. The conditions are represented in Fig. 3-43. When the instrument was set up near A, the readings of the three threads with the rod held at A were 4.600, 4.448, and 4.298, as shown in column 2 of Fig. 3-44. Their mean, which is shown in column 3, is 4.449 ft. This is n_1 in Fig. 3-43. The value of d_1 expressed as an interval is 0.302 ft, as shown in column 4 of Fig. 3-44 and again in column 5. The rod is now taken to B, and the three thread readings are 4.237, 3.515, and 2.790, as shown in column 6. The mean, which is 3.514, is shown in column 7. This is N_1 in Fig. 3-43. The value of D_1 expressed as an interval is 1.447, as shown in columns 8 and 9 of Fig. 3-44.

With the instrument set up near B in Fig. 3-43, the rod readings taken on B are recorded on the left-hand page of the notes in Fig. 3-44 in column 2. These rod readings give n_2 and d_2, as shown. In column 5 the quantity $(d_1 + d_2)$, or the sum of the intervals, is shown as 0.692 ft. From this setup, the rod readings taken at A are recorded on the right-hand page in column 6. These readings give N_2 and D_2, as shown. In column 9 the quantity $(D_1 + D_2)$

					DETERMINATION OF C-FACTOR					
		BACKSIGHT					FORESIGHT			
No. of Station	Thread Reading Feet	Mean Feet		Thread Interval Feet	Sum of Intervals Feet		Thread Reading Feet	Mean Feet		Thread Interval Feet	Sum of Intervals Feet
(1)	(2)	(3)		(4)	(5)		(6)	(7)		(8)	(9)
	4.600			0.152			4.237			0.722	
1	4.448	4.449=n_1		0.150			3.515	3.514=N_1		0.725	
	4.298		d_1=	0.302	0.302·d_1		2.790		D_1=	1.447	1.447·D_1
	4.400			0.195			5.863			0.698	
2	4.205	4.205=n_2		0.195			5.165	5.164·N_2		0.700	
	4.010		d_2=	0.390	0.692=d_1+d_2		4.465		D_2=	1.398	2.845=D_1+D_2
		8.654·n_1+n_2						8.678·N_1+N_2			−0.692=d_1+d_2
		−8.678=N_1+N_2									$(D_1$+$D_2)$−$(d_1$+$d_2)$=2.153
		−0.024=$(n_1$+$n_2)$−$(N_1$+$N_2)$									
		$C=\frac{.0024}{2.153}$=−0.0112 ft /ft of Interval									

Figure 3-44. Notes for determining C factor.

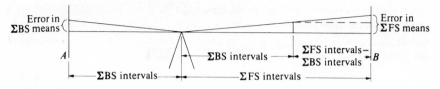

Figure 3-45. Imbalance of sum of backsight and sum of foresight intervals.

is shown as 2.845. The sum of the two means in column 3 gives $(n_1 + n_2)$ in Eq. (3-9); the sum of the two means in column 7 gives $(N_1 + N_2)$. The remaining computations needed to determine C are shown directly in the notes.

Assume that the results obtained when a line of levels was run between two points A and B in Fig. 3-45 are as follows: sum of backsight means, 31.422 ft; sum of foresight means, 12.556 ft; sum of backsight intervals, 22.464 ft; and sum of foresight intervals, 27.845 ft. The difference between $\sum$ FS intervals and $\sum$ BS intervals is $27.845 - 22.464 = 5.381$ ft. Therefore the observed value of $\sum$ FS means must be corrected by an amount equal to the C factor times the difference between the intervals, that is, by $-0.0112 \times 5.381 = -0.060$ ft. The corrected value of $\sum$ FS means is thus $12.556 - 0.060 = 12.496$ ft. The corrected difference in elevation is then $\sum$ BS means $-\sum$ FS means $= 31.422 - 12.496 = +18.926$ ft.

3-39. PROFILE LEVELS The purpose of profile leveling is to determine the elevations of the ground surface along some definite line. Before a railroad, highway, transmission line, sidewalk, canal, or sewer can be designed, a profile of the existing ground surface is necessary. The route along which the profile is run may be a single straight line, as in the case of a short sidewalk; a broken line, as in the case of a transmission line or sewer; or a series of straight lines connected by curves, as in the case of a railroad, highway, or canal. The data obtained in the field are usually employed in plotting the profile. This plotted profile is a graphical representation of the intersection of a vertical surface or a series of vertical surfaces with the surface of the earth, but it is generally drawn so that the vertical scale is much larger than the horizontal scale in order to accentuate the differences of elevation. This is called vertical exaggeration.

3-40. STATIONS The line along which the profile is desired must be marked on the ground in some manner before the levels can be taken. The common practice is to set stakes at some regular interval—which may be 100, 50, or 25 ft, or 30, 20, or 10 m depending on the regularity of the ground surface and the accuracy required—and to determine the elevation of the ground surface at each of these points. The beginning point of the survey is designated as station 0. Points at multiples of 100 ft or 100 m from this point are termed *full stations*. Horizontal distances along the line are most conveniently reckoned by the station method. Thus points at distances of 100, 200, 300, and 1000 ft from the starting point of the survey are stations 1, 2, 3, and 10, respectively. Intermediate points are designated as *pluses*. A point that is 842.65 ft from the beginning point of the survey is station 8 + 42.65. If the plus sign is omitted, the resulting figure is the distance, in feet, from station 0.

In the remainder of this chapter, reference will be made to stations of 100 ft. However, 100-m stations are handled in exactly the same way if the leveling is performed in metric units.

When the stationing is carried continuously along a survey, the station of any point on the survey, at a known distance from any station or plus, can be calculated. Thus, a point that is 227.94 ft beyond station 8 + 42.65 is 842.65 + 227.94 = 1070.59 ft from station 0 or at station 10 + 70.59. The distance between station 38 + 66.77 and station 54 + 43.89 is 5443.89 − 3866.77 = 1577.12 ft.

In the case of a route survey, the stationing is carried continuously along the line to be constructed. Thus if the survey is for a highway or a railroad, the stationing will be carried around the curves and will not be continuous along the straight lines which are eventually connected by curves. For the method of stationing that is used in surveys of this sort, see Chapter 13.

3-41. FIELD ROUTINE OF PROFILE LEVELING

The principal difference between differential and profile leveling is in the number of foresights, or −S readings, taken from each setting of the level. In differential leveling only one such reading is taken, whereas in profile leveling any number can be taken. The theory is exactly the same for both types of leveling. A backsight, or +S reading, is taken on a benchmark or point of known elevation to determine the height of instrument. The rod is then held successively on as many points, whose elevations are desired, as can be seen from that position of the level, and rod readings, called *intermediate foresights* (IFS), are taken. The elevations of these points are calculated by subtracting the corresponding rod readings from the height of the instrument (HI). When no more stations can be seen, a foresight is taken on a turning point, the level is moved forward, and the process is repeated.

The method of profile leveling is illustrated in Fig. 3-46. The level having been set up, a sight is taken on a benchmark, not shown in the sketch. Intermediate foresights are then taken on stations 0, 1, 2, 2 + 65, 3, and 4. The sight is taken at station 2 + 65 because there is a decided change in the ground slope at that point. The distance to this point from station 2 is obtained either by pacing or by taping, the better method depending on the precision required. To determine the elevation of the bottom of the brook between stations 4 and 5, the level is moved forward after a foresight reading has been taken on the

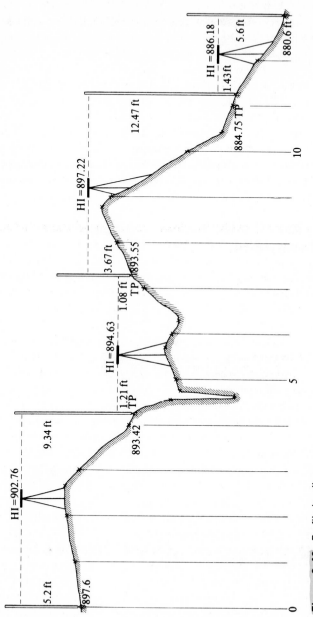

Figure 3-46. Profile leveling.

turning point just beyond station 4. With the level in the new position, a backsight is taken on the turning point and intermediate foresights are taken on stations 4 + 55, 4 + 63, 4 + 75, 5, 5 + 70, 6, 6 + 25, and 7, and lastly a foresight is taken on a turning point near station 7. From the third setup, a backsight is taken on the turning point and intermediate foresights are taken on stations 8, 8 + 75, 9, 10, 10 + 40, and 11, and a foresight is taken on a turning point near station 11. From the final setup shown in the figure, a backsight is taken on this turning point and intermediate foresights are taken on stations 12 and 13. Finally, a foresight is taken on a benchmark not shown in the sketch.

Readings have thus been taken at the regular 100-ft stations and at intermediate points wherever there is a decided change in the slope. The level has not necessarily been set on the line between the stations. In fact, it is usually an advantage to have the level from 30 to 50 ft away from the line, particularly when readings must be taken on intermediate points. More of the rod will then be visible through the telescope and the reading can be made more easily and quickly.

Benchmarks are usually established in the project area by differential leveling prior to running the profile leveling. When running the profile leveling, backsights and foresights on benchmarks and turning points must be taken with the same accuracy as that used to establish the elevations of the project benchmarks, usually to the hundredth of a foot (0.001 m). This is necessary in order to maintain the overall accuracy of the profile leveling. If intermediate foresights along profiles are taken along bare ground, they need only be read to the nearest tenth of a foot (1 cm). However, if the entire profile is a paved surface, it may be required to read the intermediate foresights to the hundredth of a foot, depending on the purpose of the profile. The profile leveling is then adjusted between previously established project benchmarks.

If benchmarks have not been established in advance, they should be established as the work progresses. Benchmarks may be from 10 to 20 stations apart when the differences of elevation are moderate, but the vertical intervals between benchmarks should be about 20 ft where the differences of elevation are considerable. These benchmarks should be so located that they will not be disturbed during any construction that may follow. Their elevations should be verified by running check levels.

The notes for recording the rod readings in profile leveling are the same as those for differential leveling except for the addition of a column for intermediate foresights. The notes and calculations for a portion of the profile leveling shown in Fig. 3-46 are given in Fig. 3-47.

3-42. PLOTTING THE PROFILE To facilitate the construction of profiles, paper prepared especially for the purpose is commonly used. This has horizontal and vertical lines in pale green, blue, or orange, so spaced as to represent certain distances to the horizontal and vertical scales. Such paper is

STA	BS	HI	FS	IFS	ELEV	LEVEL # 4096	LEVEL J. BROWN
						PROFILE LEVELING	MAY 21, 1981
BM	4.18	902.76			898.58	ROD # 18	ROD F. SMITH
0				5.2	897.6		
1				4.6	898.2		
2				3.9	898.9		
2+65				3.8	899.0	⟨ BS = 9.06	884.75
3				4.9	897.9	⟨ FS = -22.89	- 898 · 58
4				9.2	893.6	- 13.83	- 13·83
TP-1	1.21	894.63	9.34		893.42	checks	
4+55				4.4	890.2		
+63				9.9	884.7		
+75				5.3	889.3		
5				5.0	889.6		
5+70				3.9	890.7		
6				4.5	890.1		
6+25				5.4	889.2		
7				2.2	892.4		
TP-2	3.67	897.22	1.08				
8				2.4	894.8		
8+75				1.1	896.1		
9				1.9	895.3		
10				8.4	888.8		
10+40				11.3	885.9		
11				12.2	885.0		
TP-3			12.47		884.75		
⟨BS =	+9.06		⟨FS= 22.89				

Figure 3-47. Profile-level notes.

called *profile paper.* If a single copy of the profile is sufficient, a heavy grade of paper is used. When reproductions are necessary, either a thin paper or tracing cloth is available. The common form of profile paper is divided into $\frac{1}{4}$-in. squares by fairly heavy lines. The space between each two such horizontal lines is divided into five equal parts by lighter horizontal lines, the distance between these light lines being $\frac{1}{20}$ in. In order to accentuate the differences of elevation, the space between two horizontal lines can be considered as equivalent to 0.1, 0.2, or 1.0 ft, and the space between two vertical lines as 25, 50, or 100 ft, according to the total difference of elevation, the amount of vertical exaggeration desired, the length of the line, and the requirements of the work.

To aid in estimating distances and elevations, each tenth vertical line and each fiftieth horizontal line are made extra heavy. A piece of profile paper showing the profile for the level notes given in Fig. 3-47 is illustrated in Fig. 3-48. The elevation of some convenient extra-heavy horizontal line is assumed to be 900 ft and a heavy vertical line is taken as station 0. Each division between horizontal lines represents 1 ft and each division between vertical lines represents 100 ft, or 1 station. As the elevation or station of each printed line is known, the points on the ground surface can be plotted easily. When these points are connected with a smooth line, an accurate representation of that ground surface should result.

It is often a convenience to have several related profiles plotted on the same sheet. Thus in designing a pavement for a city street three profiles may appear, namely, those of the center line and of the two curb lines. This is practically a necessity when there is any considerable difference in elevation between the two sides of the street. When only a single copy is being made, different colored inks can be used to distinguish one profile from another. When reproductions are to be made, different kinds of lines, such as different combinations of dots and dashes, are used.

As indicated in Section 17-12, it is possible to prepare a profile from a topographic map.

When it is desirable to have both the plan and the profile appear on the same sheet, paper which is half plain and half profile-ruled is used. Plans for highways and sewers often are prepared in this manner, the location plan appearing at the top of the sheet and the profile below it.

Any information that may make the profile more valuable should be added. Thus the names of the streets or streams and the stations at which they are crossed should appear. The locations and elevations of benchmarks may appear as notes on the map. There should be a title giving the following information: what the profile represents, its location, its scales, the date of the survey, and the names of the surveyor and the draftsman.

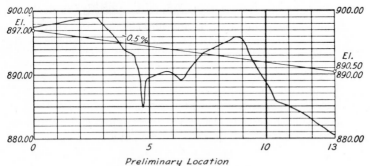

Preliminary Location
Saline-Manchester Highway
Washtenaw County, Mich.
Scales: Horiz. 1in.=400 ft.
Vert. 1in.=15 "
Surveyed Sept. 14,1959 by J.Brown
Platted Sept. 28,1959 by A.Wishnefski

Figure 3-48. Profile.

3-43. GRADE LINES The irregular line in Fig. 3-48 represents the original ground surface along which the profile has been taken. The line to which this surface is brought by grading operations is called the *grade line*. When planning a railroad, the grade line represents the proposed position of the base of the rail. For a street or highway the grade line is the finished surface at the center line. For construction purposes it is usually more convenient to let the grade line represent the subgrade, that is, the base of the ballast on a railroad or the bottom of the pavement slab on a paved highway.

The principal purpose for which a profile is constructed is to enable the engineer to establish the grade line. The aim of the engineer is to keep the volume of earthwork at a minimum, to have the grade line straight for considerable distances and its inclination within the allowable limit, and to have the excavation and the embankment balance over reasonably short stretches. To determine the best grade line may require considerable study, but the saving of even a few hundred cubic yards of grading will pay for many hours of such study.

Where changes in the inclination of the grade line occur, the straight grades are connected by vertical curves (see Section 13-14).

3-44. RATE OF GRADE The inclination of the grade line to the horizontal can be expressed by the ratio of the rise or fall of the line to the corresponding horizontal distance. The amount by which the grade line rises or falls in a unit of horizontal distance is called the *rate of grade* or the *gradient*. The rate of grade is usually expressed as a percentage—that is, as the rise or fall in a horizontal distance of 100 ft. If the grade line rises 2 ft in 100 ft, it has an ascending grade of 2%, which is written +2%. If the grade line falls 1.83 ft in 100 ft, it has a descending grade of 1.83%, which is written −1.83%. The sign + indicates a rising grade line and the sign − indicates a falling grade line.

The rate of grade is written along the grade line on the profile. The elevation of grade is written at the extremities of the line, and also at each point where the rate of grade changes. It is common practice to enclose in small circles the points on the profile where the rate of grade changes.

The rate of grade, in percent, is equal to the total rise or fall in any horizontal distance divided by the horizontal distance expressed in stations of 100 ft. The total rise or fall of a grade line in any given horizontal distance is equal to the rate of grade, in percent, multiplied by the horizontal distance in stations. The horizontal distance, in stations of 100 ft, in which a given grade line will rise or fall a certain number of feet is equal to the amount of the required rise or fall divided by the rate of grade in percent.

Example 3-2

The grade elevation in a construction project at station 40 + 50 is to be 452.50 ft. The elevation along this same grade line at station 52 + 00 is to be 478.80 ft. What is the percent grade? What is the grade elevation at station 48 + 25? Refer to Fig. 3-49.

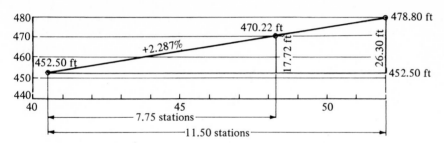

Figure 3-49. Calculation of grade and elevation.

Solution: The difference in elevation between the two points is 478.80 − 452.50 = +26.30 ft. The difference in stations is 52 − 40.50 = 11.50 stations. The percent grade is thus +26.30/11.50 = +2.287%. Station 48 + 25 lies 7.75 stations beyond station 40 + 50. The rise of the grade line between these two stations is 7.75 × 2.287 = +17.72 ft. The grade elevation at station 48 + 25 is thus 452.50 + 17.72 = 470.22 ft.

PROBLEMS

3-1. A backsight of 3.0455 m is taken on a point 60 m from the level. A foresight of 1.1508 m is then taken on a point 220 m from the level. Compute the correct difference in elevation between the two points, taking into account the effect of curvature. Neglect the effect of refraction on the line of sight.

3-2. Sighting across a lake 15 miles wide through a pair of binoculars, what is the height of the shortest tree on the opposite shore whose tip the observer can see if his eyes are 5 ft 6 in above the shore line on which he stands.

3-3. A backsight of 3.865 ft is taken on a point 20 ft from the level. A foresight of 2.680 ft is then taken on a point 220 ft from the level. Compute the correct difference in elevation between the two points, taking into account the effect of curvature. Neglect the effect of refraction.

3-4. A vertical angle of −2° 42′ 30″ is read to a target which is 12.0 ft above ground station *B*. The telescope of the instrument is 5.2 ft above ground station *A*. The horizontal distance *AB* is 23,200 ft. The elevation of station *B* is 175.52 m. Compute the elevation of station *A* in metres.

3-5. A slope distance between two points *P* and *Q* is measured using EDM, giving a corrected slope distance of 47,572.60 ft. The elevation of the lower point *P* is 1986.58 ft. Reciprocal vertical angles are measured at *P* = +2° 12′ 18″, and at *Q* = −2° 19′ 20″. Assume the instrument and the target are at the same height above ground at the two stations. What is the elevation of point *Q*?

3-6. In Problem 3-5, using only the measured vertical angle at *Q*, together with the normal effect of curvature and refraction, compute the elevation of *Q*.

3-7. A vertical angle of +18° 20′ 30″ is measured to the tip of a church spire from an instrument set up 1920 ft away from the spire, measured horizontally. The telescope of the instrument is 5.25 ft above the ground station, whose elevation is 415.65 ft. What is the elevation of the tip of the spire? Assume normal refraction conditions.

3-8. The slope distance between two points R and S is measured as 14,752.25 m. A vertical angle of $+3°\ 53'\ 40''$ is measured from R to S. The instrument is 1.65 m above ground at R; the target is 1.65 m above ground at S. Compute the difference in elevation between R and S.

3-9. Three altimeters A, B, and C read as follows when set on a benchmark whose elevation is 334 ft: $R_A = 1282$; $R_B = 1300$; and $R_C = 1290$. Altimeter A is kept at the benchmark; altimeter B is taken to a benchmark at an elevation of 1725 ft; altimeter C is used as a field altimeter. The following readings were taken:

TIME	ALTIMETER A	ALTIMETER B	ALTIMETER C	POINT
10:30	1290	2674	1600	1
10:55	1288	2680	1824	2
11:00	1296	2688	1820	3
11:45	1290	2684	1445	4
12:00	1288	2678	1566	5

Determine the elevation of the five field points.

3-10. If the sensitiveness of a bubble of an engineer's level is $24''$ and the graduations on the tube are 2 mm apart, what is the radius of curvature of the bubble tube in metres?

3-11. A level is set up, and the end of the level bubble is carefully brought to a graduation on the tube by manipulating the leveling screws. A reading of 1.4655 m is taken on a rod held 100 m from the level. The level is then tilted so as to cause the end of the bubble to move over six divisions, and a second rod reading is taken. This second reading is 1.5436 m. What is the sensitiveness of the bubble?

3-12. A level is set up 180 ft from a leveling rod. One end of the level bubble is carefully brought to a graduation on the tube by means of the leveling screws. A reading of 4.735 ft is taken on the rod. The bubble is moved through five divisions and a second rod reading of 4.586 ft is taken. What is the sensitiveness of the bubble?

3-13. Complete the accompanying set of notes recorded in running differential levels between two benchmarks. Check the notes to detect arithmetical mistakes.

BM 31 ELEV : 197.28 ft.

STATION	BS	HI	FS	ELEVATION (ft)
BM28	5.18			176.42
TP1	11.62		0.86	
TP2	10.96		1.44	
TP3	12.02		2.18	
BM29	8.75		4.45	
TP4	4.46		6.84	
TP5	4.31		6.38	
TP6	3.20		11.76	
TP7	3.91		4.11	
BM30	6.60		7.86	
TP8	6.73		6.78	
BM31			4.25	

3-14. The line of sight of a level rises at the rate of 0.145 ft/100 ft when the level bubble is centered. A backsight of 5.620 ft is taken on point A, which is 80 ft from the level. The elevation of A is 363.723 ft. A foresight of 4.156 ft is taken on point B, which is 200 ft from the level. If the bubble was centered for both the backsight and the foresight, what is the elevation of B? Neglect curvature and refraction.

3-15. Assume that an area is to be graded to a level surface, and that the grading is to be controlled by grade stakes. The line of sight of the level to be used falls at the rate of 0.102 ft/100 ft when the bubble is centered. A backsight is taken on a point that defines the grade elevation, and the rod reads 4.46 ft. The backsight point is 44 ft from the level. A series of grade stakes are set at the following distances from the level:

STAKE	DISTANCE (ft)	STAKE	DISTANCE (ft)	STAKE	DISTANCE (ft)
1	20	6	114	11	15
2	68	7	137	12	70
3	36	8	190	13	190
4	102	9	166	14	230
5	128	10	202	15	144

Compute the correct rod readings which will put the foot of the rod at grade at each of the 15 points.

3-16. Assume the graduations of a 13-ft Philadephia rod to be in the plane of the front of the rod and that the shoe or foot of the rod is $1\frac{5}{8}$ in. wide front to back. What is the effect on a rod reading of 11.00 ft if the rod rests on a flat surface and is waved forward toward the instrument such that the top of the rod moves 8 in.? What is the effect if the rod is waved backward away from the level by the same amount?

3-17. Complete the accompanying set of notes in running differential levels between two benchmarks. Check the notes for arithmetical mistakes.

STATION	BS	HI	FS	ELEVATION (m)
BM101A	2.087			47.466
TP1	2.110		0.884	
TP2	1.846		1.462	
TP3	0.484		1.598	
TP4	0.256		1.778	
BM102A	1.411		2.324	
TP5	1.798		1.780	
TP6	2.024		1.752	
TP7	3.198		1.046	
TP8	3.148		2.353	
TP9	2.862		3.046	
BM103A	2.140		3.021	
TP10	0.898		2.106	
BM104A			0.989	

3-18. The instrument man observes a low reading of 11.260 ft and a high reading of 11.280 ft when the rodman waves the rod. How far is the rod out of plumb at the 13-ft mark when the high reading is taken?

3-19. Complete and check the accompanying set of profile level notes.

STATION	BS	HI	FS	IFS	ELEVATION (ft)
BM3	3.09				976.28
TP1	4.18		2.90		
0 + 00				3.2	
1 + 00				1.9	
1 + 20				1.4	
2 + 00				3.3	
2 + 50				3.2	
3 + 00				4.9	
3 + 80				7.2	
TP2	2.02		7.08		
4 + 00				2.9	
5 + 00				3.0	
6 + 00				5.0	
TP3	6.84		4.91		
6 + 42				7.7	
6 + 46				9.9	
6 + 60				9.7	
7 + 00				6.8	
8 + 00				1.3	
BM4			1.44		

976.06 BM4 ELEV.

3-20. Using a horizontal scale of $1'' = 1$ station and a vertical scale of $1'' = 5$ ft, plot the profile of the line to which the notes of Problem 3-19 apply.

3-21. The grade elevation at station $25 + 00.00$ is to be 2456.70 ft. What will be the grade elevation at station $39 + 14.25$, if the grade line falls at the rate of 2.25% between these two points.

3-22. The grade elevation of station $153 + 50$ is 763.60 ft, and the grade elevation of station $159 + 75$ is 785.44 ft. What is the rate of grade, in percent, of a slope joining these two stations?

3-23. Complete and check the accompanying set of profile level notes.

STATION	BS	HI	FS	IFS	ELEVATION (ft)
BM38	6.94				491.29
22 + 00				6.4	
23 + 00				6.0	
23 + 71				3.2	
24 + 00				2.4	
24 + 18				2.0	
TP1	0.92		8.72		

Continued

STATION	BS	HI	FS	IFS	ELEVATION (ft)
25 + 00				4.6	
26 + 00				12.9	
26 + 80				9.8	
27 + 00				9.5	
27 + 78				12.4	
28 + 00				12.4	
TP2	5.63		10.10		
28 + 27				3.3	
28 + 47				0.9	
29 + 00				2.8	
29 + 48				6.1	
30 + 00				7.3	
30 + 60				8.6	
31 + 00				10.2	
32 + 00				9.7	
BM39			9.83		

3-24. Using a horizontal scale of $1'' = 1$ station and a vertical scale of $1'' = 10$ ft, plot the profile of the line from the notes of Problem 3-23.

3-25. Compute the slope of the line of sight of a level from the following observations taken in each instance with the level bubble centered. From the first instrument setup, the reading on a rod at A is 3.729 ft and that on a rod at B is 4.286 ft. The distance from the level to A is 30 ft, and the distance to B is 220 ft. From the second instrument setup, the reading on the rod at A is 4.368 ft and that at B is 4.908 ft. The distance to A is 190 ft, and that to B is 15 ft.

BIBLIOGRAPHY

Berry, R. M. 1969. Experimental techniques for levels of high-precision using the Zeiss Ni 2 automatic level. *Proc. 29th Ann. Meet. Amer. Cong. Surv. & Mapp.* p. 309.

Berry, R. M. 1976. History of geodetic leveling in the United States. *Surveying and Mapping* 36:137.

Berry, R. M. 1977. Observational techniques for use with compensator leveling instruments for first-order levels. *Surveying and Mapping* 37:17.

Bray, J. A. 1978. Sea level variation on the Southeast Pacific Coast. *Proc. Ann. Meet. Amer. Cong. Surv. & Mapp.* p. 224.

Cheng-Yeh Hou, Veress, S. A., and Colcord, J. E. 1972. Refraction in precise leveling. *Surveying and Mapping* 32:231.

de Jong, S. H., and Siebenhuener, H. F. W. 1972. Seasonal and secular variations of sea level on the Pacific Coast of Canada. *Canadian Surveyor* 26:4.

Ellingwood, C. F., and Holdahl, S. R. 1972. The precise leveling test network of the National Geodetic Survey. *Proc. Ann. Meet. Amer. Cong. Surv. & Mapp.* p. 275.

Franceschini, G. J., and Mitchell, R. J. 1969. Southern California cooperative leveling program. *Proc. 29th Ann. Meet. Amer. Cong. Surv. & Mapp.* p. 296.

Guth, J. E. 1976. Survey procedures for determining mean high water. *Proc. Fall Conv. Amer. Cong. Surv. & Mapp.* p. 405.

Hradileh, L. 1972. Refraction in trigonometric and three-dimensional terrestrial networks. *Canadian Surveyor* 26:59.

Kesler, J. M. 1973. EDM slope reduction and trigonometric leveling. *Surveying and Mapping* 33:61.

Kivioja, L. A. 1979. New mercury leveling instruments. *Surveying and Mapping* 39:61.

Lechapelle, G. 1979. Redefinition of National Geodetic Networks. *Canadian Surveyor* 33:273.

Lippold, H. R., Jr. 1980. Readjustment of the National Geodetic Vertical Datum. *Surveying and Mapping* 40:155.

Maltby, C. S., and Doskins, E. 1968. Bench mark monumentation. *Surveying and Mapping* 28:45.

O'Connor, D. C. 1968. Some meteorological factors affecting the accuracy of barometric altimetry. *Surveying and Mapping* 28:477.

Poetzschke, H. 1979. Motorized leveling. *Proc. Ann. Meet. Amer. Cong. Surv. & Mapp.* p. 321.

Rowland, J. B. 1969. Atmospheric refraction in long-line vertical angles. *Surveying and Mapping* 29:59.

Selley, A. D. 1977. A trigonometric level crossing of the Strait of Belle Isle. *Canadian Surveyor* 31:249.

Swanson, R. L., Thurlow, C. I., and Hicks, S. D. 1976. A proposed uniform tidal datum system for the United States. *Proc. Fall Conv. Amer. Cong. Surv. & Mapp.* p. 411.

Whalen, C. T., and Balazs, E. I. 1977. Test results of first-order class III leveling. *Surveying and Mapping* 37:45.

Chapter 4
The Measurement
of Angles

4-1. HORIZONTAL ANGLE A horizontal angle is the angle formed by two intersecting vertical planes. The vertical planes intersect along a vertical line which contains the vertex of the angle. In surveying, an instrument for measuring angles occupies this vertex, and the vertical line formed by the two vertical planes coincides with the vertical axis of the instrument. In Fig. 4-1 two vertical planes, OAZ and OBZ, intersect along the vertical line OZ and form the horizontal angle AOB. Point C lies in the vertical plane containing line OA. Therefore the direction of OC is the same as that of OA. Point D lies in the vertical plane containing line OB, and so the direction of OD is the same as that of OB. Point O, being the vertex of the angle, is the *at* station. The angle at O from A to B is the angle AOB, point A being the *from* or backsight station and point B being the *to* or foresight station. In Fig. 4-1 the horizontal angle at O from C to D is exactly the same as angle AOB. The horizontal projection of OC is OA, and the horizontal projection of OD is OB. The horizontal angle between two lines therefore is the angle between the projections of the lines onto a horizontal plane.

A horizontal angle in surveying has a direction or sense; that is, it is measured or designated to the right or to the left, or it is considered clockwise or counterclockwise. In Fig. 4-1 the angle at O from A to B is counterclockwise and the angle at O from D to A is clockwise.

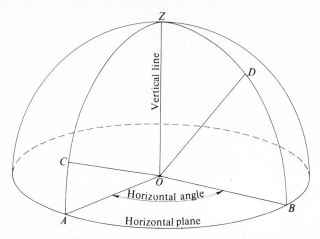

Figure 4-1. Horizontal angle.

The common methods of measuring horizontal angles are by measuring distances, by transit, by theodolite, or by total station instruments.

4-2. ANGLES BY MEASURED DISTANCES
It is possible to make simple surveys by using just a tape, the angles being computed from the linear measurements. One method of doing this is to divide the survey area into a series of connected triangles, the sides of which are measured. From the lengths of the resulting triangle sides, the angles at all the vertices are computed by using the formulas of plane trigonometry, and the various separate angles are added to obtain the whole angle at each vertex. If all the sides in Fig. 4-2 are measured, all the angles in the tract can be computed by solving triangles *ABD*, *CDB*, and *EAD*. To obtain the angle at *D* from *E* to *C*, for example, the three angles at *D* from the individual triangles are added.

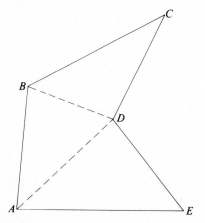

Figure 4-2. Survey with a tape.

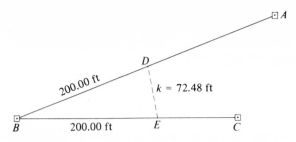

Figure 4-3. Measuring angle with tape using chord method.

If the figure is large and if the lines are clear, the distances can be measured using the EDM. The field procedure known as *trilateration* (discussed in Chapter 10) employs the EDM measurements for computing angles in large triangles.

To avoid measuring the interior lines in a tract, as in Fig. 4-2, the angle at each vertex may be determined directly by the chord method or by the tangent method. In the chord method (Fig. 4-3) equal distances are laid off from B on the two lines BA and BC, and points D and E are set. These points should be lined in carefully. The chord distance DE, or k, is then measured. The angle is computed from the relation $\sin \frac{1}{2} ABC = k/2BD$. Thus if BD and BE are each 200 ft and k is 72.48 ft, then $\sin \frac{1}{2} ABC = 72.48/(2 \times 200) = 0.1812$, $\frac{1}{2} ABC = 10° 26\frac{1}{2}'$, and $ABC = 20° 53'$.

In the tangent method (Fig. 4-4) a perpendicular to BC is established at P. (See Section 4-3 for a method of laying out a right angle.) Point Q is set on the line BA. The distances BP and PQ are measured with the tape, and the relation $\tan ABC = PQ/BP$ gives the angle at B. This method becomes weaker as the angle increases and it is not satisfactory for angles much over 45° since the line representing the perpendicular may not be a true perpendicular and its intersection with the line, as at Q in Fig. 4-4, will be doubtful. An extremely small angle can be measured quite accurately with the tape by using the tangent method since the perpendicular will be relatively short.

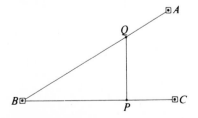

Figure 4-4. Measuring angle with tape using tangent method.

The accuracy of the values obtained by these methods is dependent on the size of the angle, on the care with which the points are set on line, on the accuracy of the measured lengths, and when the tangent method is used, on the care taken in erecting the perpendicular. With reasonable care the value of an angle determined by one of these methods would agree with the value obtained with the transit within 1′ or 2′.

4-3. LAYING OFF ANGLES WITH THE TAPE It is sometimes necessary to stake out an angle more accurately than it can be done with the ordinary transit. Since it is impossible to lay off directly with the transit the angle of 67° 42′ 14″ shown in Fig. 4-5, the angle is laid off as exactly as possible with the instrument and a point G, 1000 ft from E, is established. The angle DEG is measured by the repetition method described in Section 4-17. A point H is next located from G at a distance $GH = 1000 \tan (67° 42′ 14″ − DEG)$. The line through E and H will make the required angle with the line ED. This method is used in land surveying, where a line is frequently extended several miles and where a small error in an angle may cause a considerable discrepancy in the location of a point at the end of the line.

A perpendicular to a line can be constructed with a tape by the 3-4-5 method. Since a triangle is right angled if its sides are proportional to the numbers 3, 4, and 5, a right angle can be laid out by constructing a triangle whose sides are in these proportions. Thus in Fig. 4-6 let it be required to erect a perpendicular to the line AD at A. First, a temporary point B is established on the line AD and 30 ft or 9 m from A. Then, by using A and B as centers and swinging arcs with radii of 40 and 50 ft, or 12 and 15 m, respectively, the point C is located at the intersection of these two arcs. This can

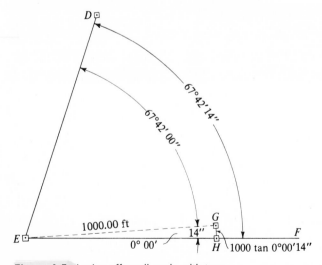

Figure 4-5. Laying off small angle with tape.

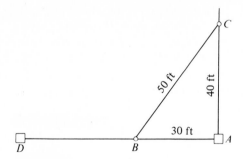

Figure 4-6. Constructing right angle with tape.

be most easily done if two tapes are available. If only a single 100-ft tape is at hand, the zero mark can be held at B and the 100-ft mark held at A, while the point C is located by holding the 50- and 60-ft marks together and stretching the tape tight. Alternatively, if a 30-m tape is used, the 30-m mark is held at A, holding the 15- and 18-m marks at C. After a stake has been driven at C, the exact point on the stake can be found by striking two arcs on the top of the stake. One arc is struck at a time, and the intersection of the two arcs is the required point. Instead of the 30-, 40-, and 50-ft lengths, any other convenient lengths that are in the same proportions can be used.

A perpendicular to a given line through a point off the line can be constructed by first estimating where the perpendicular will intersect the line. To drop a perpendicular from point P to line MN in Fig. 4-7, first select point F on MN. The line FL is established by the 3-4-5 method, points being set at A and at B. The offset distance k from FL to P is then measured, and this distance is laid off on line MN from F to establish point G. Thus the line GP is perpendicular to the line MN.

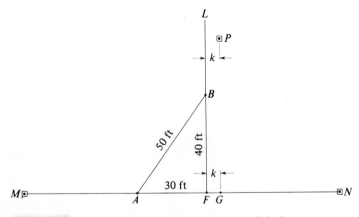

Figure 4-7. Perpendicular to a line through a point off the line.

4-4. ANGLES WITH TRANSIT-THEODOLITE The engineer's or surveyor's transit, more commonly referred to as a theodolite (the two terms are used interchangeably), is an instrument of great versatility. By means of a telescope, the line of sight is well defined. As a result long sights and accurate sighting are possible.

The theodolite contains a horizontal circle of either glass or silver. The circle can be read to the nearest 1′ down to an arc as small as 0.2″ using appropriate circular verniers or auxiliary glass scales. The horizontal circle

Figure 4-8. Engineers transit. Courtesy of Keuffel & Esser Co.

is used when measuring or laying off horizontal angles. The theodolite also contains a vertical circle which is used to measure or lay off vertical angles or zenith angles (see Section 4-18). The standard engineer's transit shown in Fig. 4-8 contains silver circles and circular verniers. It usually contains a compass circle by which magnetic bearings may be observed (see Section 7-11). The optical reading glass-circle theodolites usually do not have an integral compass, but some theodolites have provision for attaching a separate compass needle.

The reticle containing the cross hairs is equipped with auxiliary stadia hairs which may be used to determine distances by using the optical constants of the telescope. The telescope is provided with a level bubble that is sensitive enough to allow the transit to be used for spirit leveling and to eliminate uncertainties in measuring vertical angles. The telescope can be rotated about its horizontal axis through a complete circle. This last feature gives the transit its name. The word "transit" means to pass over or cross over, and the line of sight of the transit can be made to cross over from one side to the other by rotating the telescope about its horizontal axis.

4-5. REPEATING VERSUS DIRECTION INSTRUMENTS   Two

classes of instruments are used to measure horizontal angles. These are the repeating theodolite and the direction theodolite. The engineers transit shown in Fig. 4-8 is a repeating instrument in that it is capable of accumulating angles on the horizontal circle by means of two separate rotations to be described in Section 4-8. The measurement of an angle by repetition is described in Section 4-17. A direction theodolite essentially contains only one rotation, and it therefore does not permit angles to accumulate on the horizontal circle. The direction theodolite is discussed in Section 4-21. There is no difference in the vertical circles of the two classes of instruments.

4-6. PARTS OF THE ENGINEER'S TRANSIT   A modern engi-

neer's transit is made up of three subassemblies. These are shown in Fig. 4-9. The upper part is the *alidade*. This contains a circular cover plate that is equipped with two level vials at right angles to one another and is rigidly connected to a solid conical shaft called the inner spindle. The vernier for the horizontal circle is contained in the cover plate. The alidade also contains the frames supporting the telescope which are called the *standards*, the vertical circle and its vernier, the compass box, and finally the telescope and its level vial.

The middle part of a transit contains the *horizontal limb*, which is rigidly connected to a hollow conical shaft called the "outer spindle or intermediate center." The inside of the outer spindle is bored with a taper to receive the inner spindle of the alidade. The limb assembly also contains a clamp, called

Figure 4-9. Transit showing three subassemblies. Courtesy of W. & L. E. Gurley Co.

the *upper clamp*, which allows or prevents rotation of the inner spindle within
the outer spindle. The graduated horizontal circle is attached to the upper
face of the limb.

The lower part of the transit is the *leveling-head assembly*. It contains the
leveling screws, a bottom plate that screws onto the tripod, a shifting device
that permits the transit to be moved about $\frac{1}{4}$ to $\frac{3}{8}$ in. in any direction while
the tripod remains stationary, a half ball that allows the transit to tilt while
being leveled; and the four-arm piece, or the spider, into the center of which
fits the outer spindle. The assembly also contains a clamp, called the *lower*

clamp, which allows or prevents rotation of the outer spindle, and a hook for attaching a plumb-bob string to the center of the transit. Some of the parts can be seen in Fig. 4-8.

4-7. TRANSIT TELESCOPE The telescope on the transit shown in Fig. 4-8 is similar to that on an engineer's level, but it is shorter in length. Its main parts are the objective, the internal focusing lens, the focusing wheel, the cross hairs, and the eyepiece. The telescope is attached to the transverse axis, or horizontal axis, which rests on the standards and revolves in bearings at the tops of the standards. The telescope is held at any desired inclination by means of a clamp, called the "vertical clamp," and can be rotated slowly in a vertical plane by means of a tangent screw, or slow-motion screw, called the "vertical tangent screw," attached to one standard. The telescope in Fig. 4-8 is shown with a sunshade which shades the objective when the instrument is in use, but this sunshade is replaced by a cap when the transit is not being used.

4-8. ROTATION OF SPINDLES The inner and outer spindles, together with the tapered bore in the four-arm piece, provide for rotation of the alidade and the limb about the vertical axis of the instrument. With the instrument set up on a tripod and with both the lower clamp and the upper clamp in a clamped position, neither the alidade which contains the vernier nor the limb which contains the graduated horizontal circle can move. If the upper clamp remains clamped and the lower clamp is loosened, then the outer spindle is free to rotate in the four-arm piece, but both the alidade and the limb must rotate together about the vertical axis. As a result the vernier will appear opposite the same horizontal circle graduation during rotation since they move together.

If the lower clamp is clamped and the upper clamp is loosened, the horizontal circle is fixed in position, but the alidade is allowed to rotate about the vertical axis and the vernier moves with it. As the vernier moves around the edge of the fixed graduated circle, the reading on the circle changes correspondingly.

To permit a small rotation of the alidade or the inner spindle in the outer spindle, a tangent screw, or slow-motion screw, is provided. Every tangent screw operates against a spring, which reduces play or backlash in the motion. A small rotation of the outer spindle in the four-arm piece is permitted by a tangent screw similar to that controlling the slow motion of the alidade. Neither of these tangent screws will work unless the corresponding clamp is clamped.

Rotation of the inner spindle in the outer spindle is known as the *upper motion* of the transit. Such rotation is controlled by the upper clamp and the upper tangent screw. Rotation of the outer spindle in the four-arm piece is known as the *lower motion* of the transit. This rotation is controlled by the lower clamp and the lower tangent screw. An upper motion of the transit changes the reading on the horizontal circle. A lower motion of the transit does not change the reading on the horizontal circle because when the lower motion is used, the vernier on the alidade moves with the horizontal circle on the limb. In other words either of the motions will rotate the alidade and thus the line of sight about the vertical axis of the instrument, but only the upper motion will change the circle reading.

In practice the terms *upper motion* and *lower motion* are also used to refer to the clamps and tangent screws controlling the motions. For instance, when the lower clamp is tightened, it is said, "the lower motion is clamped"; and when the upper clamp is loosened, it is said, "the upper motion is unclamped." Or when a sight is taken with the upper clamp tightened and the lower clamp loosened, it is said, "a sight is taken by means of the lower motion."

4-9. HORIZONTAL CIRCLE AND VERNIERS The silver horizontal circle of a transit is located on the periphery of the upper surface of the horizontal limb. It is graduated in various ways, three of which are shown in Fig. 4-10. Each small division on the circle shown in views (a) and (c) is $\frac{1}{2}°$ or 30'. A division in view (b) is $\frac{1}{3}°$ or 20'. One in view (d) is $\frac{1}{6}°$ or 10'. On the circles in views (a), (b), and (c) the graduations for every 10° are numbered both clockwise and counterclockwise from 0° through 360°. This method of numbering is convenient when measuring angles clockwise or counterclockwise. The circle in view (d) is numbered clockwise only. However, it can also be used for measuring counterclockwise angles.

There are two verniers for the horizontal circle of a transit. They are intended to be exactly 180° apart. These are called the A vernier and the B vernier. Each vernier is located in a window in the cover plate and is covered with glass. When the inner spindle is in its proper position in the outer spindle, the outer edge of each vernier comes almost in contact with the inner edge of the graduated circle. Although the scale on the horizontal limb is circular and is graduated in angular units, the principle of a vernier used with this scale is the same as that of a vernier used on a leveling rod and described in Section 3-20.

The vernier shown in Fig. 4-10 (a) is one of the two verniers of a 1' transit. It is actually a double direct vernier, the part to the left of zero being graduated clockwise and the part to the right of zero being graduated counterclockwise. The main scale, or the circle, is graduated to 30' and there are 30 spaces in each part of the vernier. The least count of the combination is therefore 30'/30 or 1'. The clockwise part of the vernier is used with the numbers on the circle increasing clockwise. The counterclockwise part of the vernier is used with the numbers on the circle increasing counterclockwise. The letter A at the zero graduation of the vernier signifies which vernier is being used. The letter B would appear on the vernier on the opposite side of the circle.

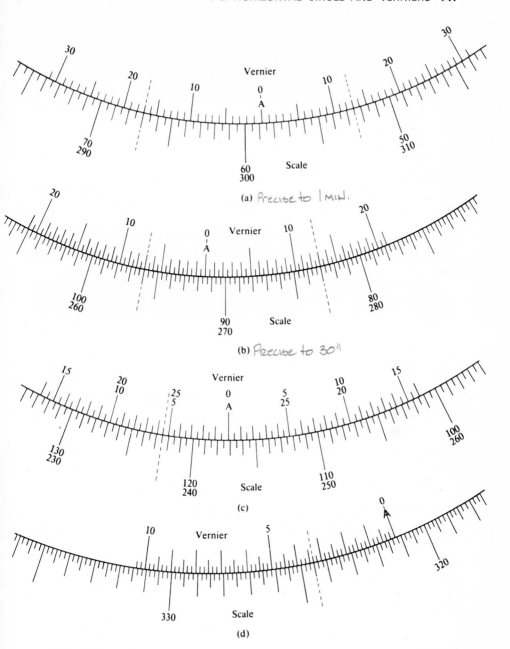

(a) Precise to 1 min.

(b) Precise to 30"

(c)

(d)

Figure 4-10. Transit verniers.

In Fig. 4-10 (b) the circle is divided to 20' and there are 40 spaces in either the clockwise part of the vernier or the counterclockwise part. Hence the least count of the combination is 20'/40 or $\frac{1}{2}'$ or 30". A transit containing such a circle and vernier is known as a 30" transit. Except for the least count, the verniers in views (a) and (b) are of the same type.

A folded vernier is represented in Fig. 4-10 (c). This type of vernier is used where space for a double vernier is not available. It is to be noted that there are two sets of numbers on the vernier, those of one set increasing clockwise and those of the other increasing counterclockwise. The numbers closest to the circle increase toward the left from 0 at the center up to 15; and from the extreme right, where 15 is repeated, they continue to increase going toward the left to the center mark, which corresponds to 30. The least count of this combination is 30", since a division on the circle is 30' and the complete vernier contains 60 spaces. In Fig. 4-10 (d) is shown a single vernier with 60 spaces. Since a division on the circle is 10', the least count is $10'/60 = \frac{1}{6}'$ or 10".

Any vernier is read by first noting the graduation on the circle beyond which the vernier index lies, and then adding to the value of that graduation the value of the vernier graduation that coincides with a circle graduation. When the circle in Fig. 4-10 (a) is read clockwise, the vernier index lies beyond the 58° 30' graduation, and the vernier shows coincidence at the 17' mark. Hence the reading is 58° 30' + 17', or 58° 47'. The corresponding counterclockwise reading is 301° 13'. For the clockwise reading in Fig. 4-10 (b) the vernier index lies beyond the 91° 20' graduation, and the vernier shows coincidence at the 7' mark. Hence the reading is 91° 20' + 7', or 91° 27' 00". The corresponding counterclockwise reading is 268° 33' 00". The readings of the other verniers are as follows: in view (c), 117° 05' 30" or 242° 54' 30", and in view (d), 321° 13' 20".

The length of a division on the vernier is slightly smaller than that of a division on the circle. So when an exact coincidence occurs, the two vernier marks on either side of the graduation that coincides with a circle graduation should fall just inside the corresponding circle graduations. If the transit does not have provision for magnification of the graduations by means of an attached microscope, a hand magnifying glass or reading glass must be used. When reading the circle, the observer should look radially along the graduations to eliminate the effect of parallax between the circle and vernier graduations.

4-10. VERTICAL CIRCLE AND VERNIER The vertical circle is rigidly connected to the transverse axis of the telescope and moves as the telescope is raised or depressed. It is clamped in a fixed position by the clamp that controls the vertical motion of the telescope, and after being clamped, it can be moved through a small vertical angle by means of the tangent screw. The vertical circle can be either a full circle or a half circle. The former is graduated from 0° to 90° in both directions and then back to 0; the latter is graduated from 0° to 90° in both directions. Each division on either is $\frac{1}{2}$°, or 30'. On most transits the vernier for reading the vertical circle is mounted on the standard directly beneath the circle and almost comes in contact with the circle. It is a double direct vernier with the graduations numbered from 0 to 30 in both directions. The least count of the vertical circle and its vernier on almost every modern transit is 1'.

When the line of sight, and thus the telescope, of a well-adjusted transit is horizontal, the vernier will read 0° on the vertical circle. If some part of the transit is out of adjustment or if the transit is not carefully leveled, an initial reading other than 0° may be obtained. This is called the *index error*. On a transit with the vernier rigidly attached to the standard, this initial reading must be determined before a vertical angle can be measured correctly (see Section 4-18). On some transits the vernier can be moved slightly on the standard by means of a tangent screw, and its correct position is determined by a level bubble called a *control bubble* or index bubble. With this arrangement the circle will read 0° when the index bubble is centered and the line of sight is horizontal, so that the correct vertical angle can be measured directly.

4-11. OPTICAL READING THEODOLITE The mechanical motions of the optical reading repeating theodolite are essentially the same as those of the engineer's transit. That is, they have an upper motion, a lower motion, and a vertical motion together with the appropriate clamps and tangent screws. The telescope contains either an erecting or an inverting eyepiece, but many of the optical reading theodolites do not contain a telescope bubble. Instead, a horizontal line of sight is established by first centering the vertical circle control bubble and then setting off a 90° zenith angle using the vertical clamp and tangent screw. Some theodolites contain a pendulum compensator which eliminates the need for a control bubble. However the instrument must be leveled up quite closely in order to allow the compensator to operate properly.

The inner spindle of most theodolites is hollow in order to provide a line of sight for the *optical plummet* which takes the place of the plumb bob used to center the theodolite over the point to be occupied. A schematic diagram of the optical plummet is shown in Fig. 4-11. After the instrument has been leveled up over the ground station, the observer loosens the clamp which holds the tribrach to the top of the tripod, and shifts the instrument until the line of sight of the optical plummet, defined by a set of cross hairs as seen through the optical plummet eyepiece, intersects the point marking the ground station. Since the top of the tripod may not be perfectly level, the instrument must be shifted parallel with itself. If it is rotated while being shifted, then the line of sight of the optical plummet will no longer be vertical. Both the leveling and the centering must be checked after this shifting operation.

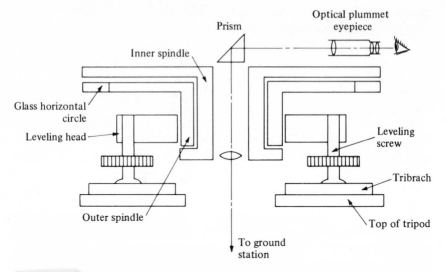

Figure 4-11. Schematic diagram of optical plummet.

The glass circles of the optical reading theodolite are much simpler to read than the silver circles of the engineer's transit described in Section 4-9. Figure 4-12 shows a typical repeating theodolite with glass circles. A mirror located on the side of one of the standards directs light through a window into the optical system, thereby illuminating the circles. A light bulb can be attached to the window in lieu of the reflected light in case of poor natural

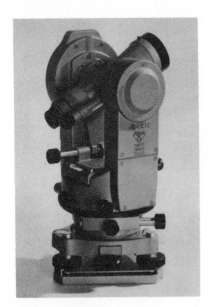

Figure 4-12. Lietz T60D repeating theodolite. Courtesy of Lietz Co.

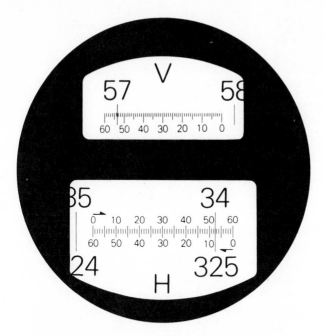

Figure 4-13. Circles and scales of Lietz T60D theodolite.

Figure 4-14. Wild T1 repeating theodolite. Courtesy of Wild Heerbrugg Instruments, Inc.

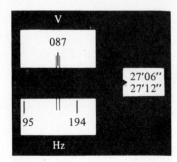

Figure 4-15. Circle reading of Wild T1 theodolite.

lighting or nighttime operation. The microscope eyepiece located to the side of the telescope eyepiece is used to view both the horizontal and the vertical circle under great magnification through a series of prisms and lenses located inside the housing of the instrument. Figure 4-13 shows the circles of the theodolite shown in Fig. 4-12 as viewed through the microscope eyepiece. The upper part of the field of view is the vertical circle and glass scale showing a zenith angle reading of 57° 53.4′. The lower part is the horizontal circle graduated both clockwise (the upper figures) and counterclockwise (the lower figures) together with the appropriate glass scales containing the minute graduations. The arrows indicate the direction, clockwise or counterclockwise. The clockwise circle reading is 34° 52.6′, and the counterclockwise circle reading is 325° 07.4′.

Different methods are employed in various instruments to enable the observer to make the circle reading. The field of view through the reading micrometer of the instrument shown in Fig. 4-14 is shown in Fig. 4-15. The observer sees the horizontal circle marked Hz, the vertical circle marked V, and the optical micrometer which replaces the glass scale. The micrometer allows an estimation to 3″ of arc. After sighting on the station being observed, the observer turns the micrometer operating screw located on the side of the standard (see Fig. 4-22) and sets the double index lines symmetrical with the degree graduation on either the horizontal or the vertical circle. He then reads the circle, which in this example shows a vertical circle reading of 87° 27′ 09″.

4-12. INSTRUMENT STATIONS　The point over which a transit or theodolite is set up is called an *instrument station.* Such a point should be marked as accurately as possible on some firm object. On many surveys each station is marked by a wooden stake or hub driven flush with the ground, and a tack or small nail marks the exact point on the stake. When the survey is intended to be relatively permanent, each point can be marked by a monument of stone or concrete or of iron pipe set in concrete.

Triangulation stations and *traverse stations* of many of the federal, state, county, and city agencies engaged in survey work are marked by bronze disks set in the tops of firmly planted concrete monuments. A point on rock can be marked by a chiseled cross. For identifying such points, the necessary facts can be written on the stone with keel or lumber crayon. To identify a hub and to indicate its location, a projecting stake, which is called a *guard stake* or *witness stake*, is placed near the hub. On this stake are marked in keel the station number of the hub and any other necessary information, such as the name or purpose of the survey.

4-13. SETTING UP INSTRUMENT In setting up a theodolite it is necessary to have the center of the instrument, which is the point of intersection of the transverse axis and the vertical axis of the instrument, directly over a given point on the ground. Also, the circle and the transverse axis must be horizontal. The position of a transit over a point on the ground may be indicated by a plumb bob suspended from a hook or ring attached to the lower end of the centers. The plumb-bob string should be held by a sliding knot in order that the height of the bob can be adjusted. The point of the bob should almost touch the mark over which it is desired to set the transit. The optical plummet is used in place of the plumb bob if the instrument contains this feature.

In setting up a theodolite the tripod legs are spread and their points are so placed that the top of the tripod is approximately horizontal and the telescope is at a convenient height for sighting. The instrument should be within a foot of the desired point, but no extra care is taken to set it closely at once. When setting up on rough ground, two legs of the tripod should be set at about the same elevation and the top of the tripod should be made approximately horizontal by shifting the third leg. If the instrument is more than a few inches from the given point, the tripod is lifted bodily without changing the inclinations of the legs and the instrument is set as near as possible to the point. By pressing the legs firmly into the ground, the plumb bob or the cross hairs of the optical plummet can usually be brought to within $\frac{1}{4}$ in. of the point. This brings the point within the range of the shifting head of the instrument.

When the tripod head is approximately horizontal and the plumb bob (or cross hairs) is very near the point, the bubbles of the plate levels are brought almost to the centers of the tubes by means of the leveling screws. If there are four screws, the plates are rotated until one level is parallel to a line through each pair of opposite screws. Then each bubble is brought to the center separately by turning the screws of the corresponding pair. When the plates are finally leveled up, all four screws should be bearing firmly but should not be so tight as to put a strain in the leveling head. If the leveling screws turn very hard, the cause may be the binding of the ball-and-socket joint at the bottom of the spindle. The tension may be relieved by loosening both screws of the other pair. If there are three leveling screws, the plates are rotated so that one level is parallel to a line through any pair of screws. Both bubbles are then brought to the center in succession by first turning the screws of that pair in opposite directions and then turning the third screw.

After the bubbles are approximately centered, the instrument is loosened on the tripod plate by loosening two *adjacent* leveling screws, or the retaining clamp, and the plumb bob or optical plummet cross hairs is brought exactly over the point on the ground by means of the shifting head. This arrangement allows the head of the theodolite to be moved laterally without disturbing the tripod. Then the instrument is held securely in position on the tripod plate by tightening the leveling screws that were previously loosened or by clamping the retaining ring. The bubbles of the plate levels are brought exactly to the centers of the tubes, and the position of the instrument over the point is observed. If the instrument has moved off the point in leveling up, it is brought back to the proper position by means of the shifting head, and the instrument is leveled again.

4-14. SIGHTING WITH THE THEODOLITE When an observer sights with the telescope of a theodolite, he must first focus the cross hairs so that they appear sharp to his eye (see Section 3-11). Then, provided the cross hairs are in proper focus, he brings the image of the object sighted on into focus by manipulating the focusing knob which moves the objective lens or the internal focusing lens. The focus should be tested for movement between cross hairs and image by moving the eye back and forth slightly. If parallax exists, it is removed by refocusing the objective or focusing lens. If the cross hairs go out of focus, then both the eyepiece and the objective or focusing lens must be refocused.

When making a pointing with the telescope, it is best to sight directly over the top of the telescope and then clamp the upper or lower motion, whichever is being used. The telescope is then raised or lowered and held in position by means of the vertical clamp. When the object to be sighted appears in the field of view, the vertical cross hair is centered on the object with the upper or lower tangent screw, the final motion of the tangent screw being made *against* the spring of the tangent screw. The vertical tangent screw is then used to bring the object sighted in the center of the field of view, that is, near the center horizontal cross hair. This final movement of the telescope may cause the vertical cross hair to go off the point, in which case it is brought on by the upper or lower tangent screw, whichever is being used.

The sight should be made on as small an object as is feasible or is consistent with the length of the sight. If the station is marked on the ground, the ideal sight is, of course, directly on the marked point. If the station is obscured, a pencil or a taping pin held vertically on the station makes a good target. If these are still obscured, a plumb bob centered over the station, either hand

Figure 4-16. Use of a commercial string target.

held or suspended from a tripod, is satisfactory. It is often difficult to see a plumb-bob string on a sight more than about 500 ft. To make the string visible, a string target can be used. It may be either a commercial type as shown in Fig. 4-16 or one fashioned on the spot. A sheet of white paper folded twice makes a good string target. The plumb-bob string is held inside and along the second fold, and the paper is held diagonally so that one corner points downward along the freely suspended plumb-bob string. This is shown in Fig. 4-17. A line rod, which is from $\frac{3}{8}$ to $\frac{1}{2}$ in. in diameter, is a good target for sights between 500 and 1000 ft in length. It must be straight and it must be carefully plumbed. The observer should sight as far down the rod as possible. A range pole from $\frac{3}{4}$ to $1\frac{1}{4}$ in. in diameter is satisfactory for sights beyond 500 ft. It, too, must be straight and plumbed over the station. Various sighting targets are used which sit on top of a tripod centered over the point to be observed.

Figure 4-17. Use of string target fashioned on the spot.

4-15. MEASURING A HORIZONTAL ANGLE With the theodolite set up, centered, and leveled over the *at* station, the A vernier or the glass circle is usually set to read 0° for convenience. This is done by loosening both the upper and lower clamps and rotating the horizontal limb by slight pressure applied with the finger of one hand to its underside while holding the standards in place with the other hand, until the 0° graduation of the circle is brought almost opposite the index of the vernier or glass scale. The upper motion is then clamped, and coincidence is obtained between the 0° graduation and the vernier or glass scale index by using the upper tangent screw. If the instrument contains a silver circle, the setting must be made with the aid of the magnifying glass, as this operation constitutes a reading of the circle.

The lower motion being free, the line of sight is directed toward the *from* or backsight station. The lower motion is now clamped, and the line of sight is brought exactly on the backsight station by turning the lower tangent screw. For this sight the circle reading is observed as 0° 00′ and it is recorded as the initial circle reading.

The final reading is obtained by loosening the upper clamp, rotating the alidade about the vertical axis, and directing the line of sight toward the *to* or foresight station. If the alidade has been turned in a clockwise direction, the reading on the clockwise graduations will increase by the angle of rotation. The upper clamp is tightened, and the line of sight is brought exactly on the foresight station by turning the upper tangent screw. This motion of the upper tangent screw will change the circle reading correspondingly. The reading of the circle for this sight is recorded as the final circle reading.

The difference between the initial reading and the final reading is the angle through which the line of sight was turned in going from the backsight station to the foresight station. As stated before, the A vernier or glass scale is usually set to read 0° before the backsight is made. However, it need not be set at any particular reading before the backsight is taken, since it is immaterial what the initial reading is. As an example, suppose it is desired to measure the angle at *L* from *M* to *P* in Fig. 4-18, and when a backsight is taken to station *M* the circle reading is 122° 43′. The upper clamp is loosened,

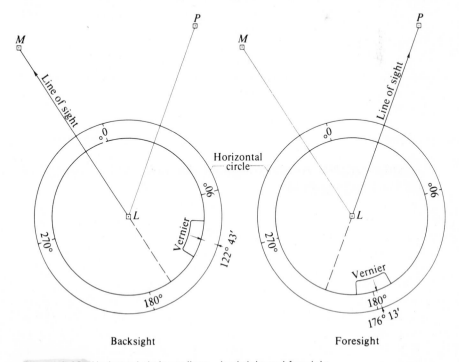

Figure 4-18. Horizontal circle readings—backsight and foresight.

the line of sight is rotated so that it is directed to P, and the upper clamp is then tightened. The line of sight is brought exactly on station P by using the upper tangent screw. The circle reading is now $176° 13'$. The angle at L from M to P is therefore the difference between the initial reading and the final reading, which is $176° 13' - 122° 43' = 53° 30'$.

The foregoing is the basic procedure in measuring a horizontal angle with the theodolite. The principles involved are as follows: (1) The backsight is taken by a lower motion, and the backsight reading is the initial circle reading; (2) the foresight is taken by an upper motion, and the foresight reading is the final circle reading; (3) the horizontal angle is obtained by taking the difference between the initial and final circle readings. The procedure as presented is not sufficient in itself to obtain a reliable measurement of a horizontal angle for the reasons discussed in the following section.

4-16. DOUBLE CENTERING If the various components of a theodolite are considered as lines and planes, one finds that the graduated horizontal and vertical circles together with their verniers constitute planes. The line of sight, the transverse or horizontal axis of the telescope, the vertical axis, and the axis of any bubble tube are all lines in space. These lines and planes bear definite relations to one another in a well-adjusted instrument. The line of sight is normal to the horizontal axis; the horizontal axis is normal to the vertical axis; the vertical axis is normal to the plane containing the horizontal circle; the line of sight is parallel to the axis of the telescope bubble tube; the axes of the plate levels lie in a plane parallel to the horizontal circle; and the movement of the focussing lens in and out when it is focused is parallel to the line of sight. The inner spindle and the outer spindle must be concentric, and the horizontal circle and the verniers must be concentric. The line joining the indices of the A and B verniers must pass through the center of the horizontal circle. A maladjustment of the theodolite in one or more of these respects is to be expected. In fact, a theodolite is never in perfect adjustment.

Two procedures for eliminating almost all of the errors due to maladjustment of the transit are (1) double centering and (2) reading both the A vernier and the B vernier or diametrically opposite sides of the circle (see Section 4-21). Of the two, double centering is the more important. Double centering or double sighting consists of making a measurement of a horizontal or a vertical angle once with the telescope in the direct or erect position and once with the telescope in the *reversed*, *inverted*, or *plunged* position. The act of turning the telescope upside down, that is, rotating it about the transverse axis, is called "plunging" or "transiting" the telescope. When the telescope is in its plunged position, the telescope bubble is on top of the telescope.

In Section 4-15, the instrument was set up at station L. A backsight was taken on station M with the telescope direct, and the circle reading was

122° 43′. A foresight was then taken on station *P* with the telescope still direct, and the circle reading was 176° 13′. The value of the angle by the direct measurement is 53° 30′.

To measure the angle with the telescope reversed or plunged, the vertical clamp is loosened and the telescope is plunged about its transverse axis. The circle reading is still 176° 13′. The lower clamp is loosened to allow the circle to move with the vernier or glass scale and thus preserve the reading on the circle. The line of sight is now directed toward station *M*, the lower clamp is tightened, and the line of sight is brought exactly on the station by using the lower tangent screw. The initial reading for this second measurement is thus 176° 13′, which was the final reading for the first measurement. The upper clamp is loosened, the vernier or glass scale being allowed to move along the circle so as to increase the circle reading, and the line of sight is directed to station *P*. The upper clamp is tightened and the line of sight is brought exactly on the foresight station with the upper tangent screw. The final circle reading is now observed and recorded. Say this reading is 229° 44′. The second value of the angle at *L* from *M* to *P* is thus 229° 44′ − 176° 13′, or 53° 31′. The mean of the values with the telescope direct and reversed is 53° 30′ 30″, and this mean value is free from errors due to practically all the maladjustments of the instrument.

When an angle is measured with the telescope direct and reversed and the resulting two values disagree by more than the least count of the circle, the whole measurement should be repeated. If double centering is repeated and the discrepancy is always of the same amount and in the same direction, the discrepancy indicates bad adjustment of the theodolite, but the values can be used. A mean of the two readings should be taken as the best value of the angle.

4-17. ANGLES BY REPETITION Study of Section 4-16 indicates that an angle can be measured to the nearest 30″ with a 1′ repeating theodolite by double centering, that an angle can be measured to the nearest 15″ with a 30″ instrument by double centering, and so on. This is true. Moreover, an experienced instrumentman with a good 1′ theodolite is capable of measuring an angle to the nearest 10″ by repeating the angle several times, accumulating the successive readings on the circle, and then dividing by the number of repetitions.

To illustrate the principle of repetition, Table 4-1 gives in the left-hand column the successive values of the angle 25° 10′ 23″ multiplied one, two, three, four, five, and six times. In the middle column the tabulation gives the corresponding values of the angle that would be read on a 1′ transit after one, two, three, four, five, and six repetitions if the initial backsight reading were 0° and if there were no errors in pointing or mistakes in using the wrong motion. The right-hand column gives the values of the angles in the middle column divided by one, two, three, four, five, and six, respectively.

TABLE 4-1. Angle by Repetition

NUMBER	MULTIPLE VALUES OF TRUE ANGLE	SUCCESSIVE TRANSIT READINGS	VALUES OBTAINED FROM TRANSIT READINGS
1	25° 10′ 23″	25° 10′	25° 10′ 00″
2	50° 20′ 46″	50° 21′	25° 10′ 30″
3	75° 31′ 09″	75° 31′	25° 10′ 20″
4	100° 41′ 32″	100° 42′	25° 10′ 30″
5	125° 51′ 55″	125° 52′	25° 10′ 24″
6	151° 02′ 18″	151° 02′	25° 10′ 20″

If the repetitions could go on indefinitely, the transit reading on the twelfth repetition divided by 12 would be 25° 10′ 25″, division after the twenty-fourth repetition would give 25° 10′ 22.5″, and so on. The practical limit is reached, however, somewhere between the sixth and twelfth repetition because of errors of graduations, eccentricities of centers, play in the instrument, and inability of the observer to point with sufficient speed and accuracy to preserve precision.

The procedure for repeating an angle six times with a theodolite is to measure it three times with the telescope direct and three times with the telescope reversed, in that order. It is understood that the observer is experienced, that his personal errors will be small, and that he will make no mistakes. Otherwise there is no justification for repeating angles.

The instrument is set up, centered, and leveled. The initial backsight reading is taken on both the A and B verniers with a transit so equipped because it is assumed that a great degree of refinement is necessary. The backsight is taken with a lower motion and with the telescope direct. Then with the telescope still in the direct position, the upper clamp is loosened and the pointing on the foresight station is made by using the upper clamp and tangent screw. The first foresight reading is made on the circle to determine the approximate value of the angle. Next, with the telescope direct, the lower clamp is loosened and the line of sight is brought back on the backsight station by using the lower clamp and tangent screw. The upper clamp is loosened and the line of sight is swung around to the foresight station. The circle reading is thus increased. A third backsight is taken on the *from* station and the third foresight is taken on the *to* station. Each repetition increases the circle reading. After the third repetition, the telescope is transited to the reversed position, and the angle is turned off three more times with the telescope reversed. The final readings on both the A vernier and the B vernier are taken. The accompanying three sets of notes will illustrate how the final angle is obtained.

In the first set the approximate angle is the difference between the initial backsight and foresight readings and is 28° 10′. When this angle is multiplied by six and the product is added to the initial reading of 0°, the sum is less than 360°. So the vernier or glass scale has not gone completely around the circle.

Readings for Set No. 1

			AT STATION B (30″ TRANSIT)			
FROM	TO	REPETITION	TELESCOPE	CIRCLE	VERNIER A	VERNIER B
A	C	0	D	0° 00′	00″	00″
		1	D	28° 10′	00″	
		6	R	169° 00′	00″	30″

The mean of the initial readings on the A and B verniers is 0° 00′ 00″. The mean of the final readings on the A and B verniers is 169° 00′ 15″. The difference is 169° 00′ 15″. This divided by six is 28° 10′ 02.5″, which is the accepted value.

Readings for Set No. 2

			AT STATION B (30″ TRANSIT)			
FROM	TO	REPETITION	TELESCOPE	CIRCLE	VERNIER A	VERNIER B
A	C	0	D	232° 41′	00″	00″
		1	D	260° 51′		
		6	R	41° 41′	00″	30″

In the second set of notes the difference between the initial backsight and foresight readings is 28° 10′. If the angle is multiplied by six and the product is added to the initial reading of 232° 41′, the sum is greater than 360°. Consequently, the final reading must be increased by 360°. (The number of complete revolutions is determined by inspection.) The mean of the initial readings on the A and B verniers is 232° 41′ 00″, and that of the final readings on the A and B verniers *increased by* 360° is 401° 41′ 15″. The difference is 169° 00′ 15″. This divided by six is 28° 10′ 02.5″, which is the accepted value.

Readings for Set No. 3

			AT STATION 40 + 12.50 (1′ TRANSIT)			
FROM	TO	REPETITION	TELESCOPE	CIRCLE	VERNIER A	VERNIER B
16 + 22.60	51 + 02.81	0	D	114°	20′	19′
		1	D	274°	52′	
		6	R	357°	33′	33′

In the third set of notes the approximate angle being measured is 274° 52′ − 114° 20′ = 160° 32′. When this angle is multiplied by six and the product is added to the initial reading, the sum is 1077° 32′. Obviously, the vernier has passed the 360° mark twice. Consequently, the final reading must be increased by 720°. The mean of the initial readings of the A and B verniers is 114° 19′ 30″; that of the final readings of the A and B verniers *increased by* 720° is 1077° 33′. The difference is 963° 13′ 30″, which is divided by six. The result, 160° 32′ 15″, is the adopted value.

Many repeating theodolites with glass circles and scales do not have provision for reading diametrically opposite sides of the circle (equivalent to an A- and B-vernier reading), but the principle of repetition applies just as well to these instruments.

4-18. VERTICAL AND ZENITH ANGLES   A vertical angle is an angle measured in a vertical plane from a horizontal line upward or downward to give a positive or a negative value, respectively. Positive and negative vertical angles are sometimes referred to as elevation or depression angles, respectively. A vertical angle thus lies between 0° and ±90°.

A zenith angle is an angle measured in a vertical plane downward from an upward directed vertical line through the instrument. It is thus between 0° and 180°.

Some instruments, such as the engineer's transit, measure vertical angles while most optical theodolites measure zenith angles. Referring to the engineer's transit, centering the plate levels brings the vertical axis of the transit truly vertical, provided that the plate levels are in adjustment. Centering the telescope level brings the telescope, and hence the line of sight, into a truly horizontal position, provided that the telescope level and the line of sight are both in adjustment. When the plate-level bubbles are centered and the telescope is brought to a horizontal position by centering the telescope-level bubble, the vernier should read 0° on the vertical circle. If the plate levels are out of adjustment or are not centered at the time a vertical angle is to be measured, then the vertical circle will not read 0° when the telescope is brought horizontal. It will have an initial reading, which may be positive or negative. Furthermore, as the alidade is rotated in azimuth, that is, rotated about its vertical axis, this initial reading will vary. The initial reading is termed the *index error* of the vertical circle. It is determined for a given vertical angle by pointing the telescope in the direction of the desired line, centering the telescope-level bubble, and reading the vertical circle. This method assumes that the line of sight is parallel to the axis of the telescope bubble, and that the vernier is in proper position on the standard.

To measure the vertical angle of the line of sight from one point to another, the transit is set up over the first point, centered, and leveled. The line of sight is brought in the direction of the other point, and the telescope bubble is centered. The reading of the vertical circle is recorded as the initial reading or as the index error. The line of sight is then raised or lowered and is directed accurately to the second point by means of the vertical clamp and tangent screw. The reading of the vertical circle is recorded as the final reading. The correct vertical angle is the difference between the initial and final readings. In recording both the initial and final readings, it is necessary to give each value its correct algebraic sign—the + for an elevation angle and the − for a depression angle. As an example, suppose that when the telescope bubble is centered, the vertical circle reads +0° 02′. When the line of sight is now raised and directed to the desired point, the vertical circle reads +14° 37′. The correct vertical angle is the difference between the two readings, or +14° 37′ − (+0° 02′) = +14° 35′. As a further example, suppose the initial reading is +0° 02′ and the final reading is −21° 14′. The difference is −21° 14′ − (+0° 02′) = −21° 16′, which is the correct vertical angle.

When the initial reading of the vertical circle is recorded as an index error and the final reading is recorded directly as the vertical angle, this angle is corrected by applying the index correction, which is equal to the index error but of opposite sign. Thus if the index error is +0° 03′ and the recorded vertical angle is −16° 48′, the index correction −0° 03′ is applied to −16° 48′ to obtain −16° 51′, which is the correct vertical angle.

If the plate bubbles are out of adjustment, the transit can be leveled very accurately for measuring vertical angles by using the more sensitive telescope bubble. The procedure is as follows. Level the transit initially using the plate bubbles. Now bring the telescope over a pair of leveling screws and center the telescope bubble by using the vertical clamp and tangent screw. Rotate the telescope 180° in azimuth. If the telescope bubble does not come to the center, bring it halfway to the center by turning the vertical tangent screw, and then center the bubble by using the two leveling screws. Rotate the telescope again 180° in azimuth. The bubble should remain centered. If it does not, repeat the process of bringing it halfway to center with the tangent screw and the rest of the way with the leveling screws. When the bubble remains centered after the telescope is rotated through 180° in azimuth, bring the telescope over the other pair of leveling screws and bring the bubble to center by using the leveling screws only. This operation makes the vertical axis of the transit truly vertical.

If the axis of the transit is truly vertical but the line of sight is not parallel to the axis of the telescope level, or the vertical-circle vernier is displaced, or there is a combination of these maladjustments, an index error will result. This error can be detected in a transit having a full vertical circle by reading the vertical angle with the telescope in the direct position and in the reversed position and comparing the two values. If the two disagree, the index error is one-half the difference between the direct and reversed readings. The mean of the direct and reversed readings is free from index error. Measuring a vertical angle with the telescope direct and reversed will not, however, eliminate the error introduced by not having the vertical axis truly vertical. It is therefore necessary to have the transit leveled carefully when vertical angles are to be measured.

The procedure for measuring a vertical angle with a transit equipped with a movable vertical-circle vernier is to sight directly at the point, center the control-level bubble on the vernier by using the control-level tangent screw, and then read the vertical angle. This is free of index error.

One important difference between the horizontal circle and the vertical circle of the transit is in the placement of the vernier with respect to the graduated circle. On the horizontal circle the vernier is located inside the circle. On the vertical circle the vernier is located outside the circle. When horizontal and vertical angles are measured concurrently, the observer is alternately reading the horizontal-circle and vertical-circle verniers, and he must be on guard against mistakes in reading the circle for the vernier because of the difference in placement of the verniers.

The vertical circle of an instrument that measures zenith angles is graduated from 0° clockwise to 360° (back to 0°). With the telescope in the direct position, the zenith angle equals the vertical circle reading. With the telescope in the reversed position, the zenith angle is 360° minus the circle reading. After a zenith angle Z is determined with the instrument, the vertical angle V is obtained as $V = 90° - Z$. This gives the correct algebraic sign to V.

4-19. SOURCES OF ERROR IN THEODOLITE ANGLES   The sources of error in theodolite angles are instrumental, personal, and natural. Most of the instrumental errors have been discussed in preceding articles.

1. If the line of sight of a telescope is not parallel with the telescope level, an error is introduced in the measurement of vertical angles. This error is eliminated by reading the vertical angle with the telescope both direct and reversed.

2. If the vertical-circle vernier is displaced on the standard from its correct position, a corresponding error will result in the measured value of a vertical angle. This error is eliminated by measuring the vertical angle with the telescope both direct and reversed.

3. If the optical axis of the telescope is not parallel with the movement of the focussing lens when it is being focused, an error will be introduced into the value of a horizontal angle when the backsight and foresight stations are at considerably different distances from the transit. This error is eliminated by measuring with the telescope direct and reversed. There is no error from this source if the two other stations are at about the same distance from the transit.

4. If the line of sight is not perpendicular to the horizontal axis of the transit and the telescope is rotated on that axis, the line of sight generates a cone whose axis is horizontal. In Fig. 4-19, the instrument center is at O which is the apex of the cone whose base is the dashed circle HAH'. The line of sight to a point S is an element of that cone. The line OZ is a vertical line through the instrument. The amount by which the line of sight is out of perpendicular with the horizontal axis is designated by the small angle e.

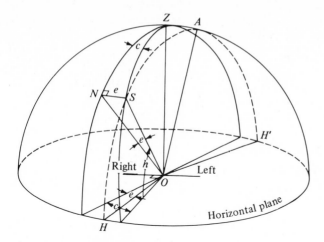

Figure 4-19. Line of sight not perpendicular to horizontal axis.

The error in the horizontal direction to point S is the angle c. In the spherical triangle ZNS, side NS is perpendicular to side ZN. Angle h is the vertical angle to S. Then by the law of sines for spherical triangles,

$$\frac{\sin c}{\sin N} = \frac{\sin e}{\sin ZS} \quad (N = 90°)$$

or

$$\sin c = \frac{\sin e}{\sin ZS} = \frac{\sin e}{\cos h} \quad (h = 90° - ZS) \tag{4-1}$$

Since both c and e are small angles, Eq. (4-1) can be expressed as

$$c'' = e'' \sec h \tag{4-2}$$

If the line of sight is to the left of the normal to the horizontal axis (as depicted in Fig. 4-19), then the clockwise circle reading is too large by the angle c. If a horizontal angle is measured between two stations to which the vertical angles are about the same, there will be no error from this source. If however, the vertical angles to the backsight and the foresight stations are different from one another, the total error c_t in the horizontal angle is

$$c_t = e(\sec h_F - \sec h_B) \tag{4-3}$$

This error is eliminated by double centering.

Example 4-1

A backsight is taken on the FROM station with the telescope inclined downward at a vertical angle of $-12°\,22'$. A foresight is taken on the TO station with the telescope inclined upward $+42°\,50'$. The line of sight departs to the left of the normal to the horizontal axis by $25''$. The horizontal angle measured clockwise is $76°\,15'\,10''$. Compute the correct horizontal angle.

Solution: This situation is depicted in Fig. 4-19. Thus, both the backsight and the foresight readings are too large, and the total error in the measured angle is, by Eq. (4-3),

$$c_t = 25''(\sec 42°\,50' - \sec 12°\,22') = 8.50''$$

The correct angle is thus $76°\,15'\,10'' - 8.5'' = 76°\,15'\,01.5''$.

---

5. If the horizontal axis is not perpendicular to the vertical axis and the telescope is rotated on the horizontal axis, the line of sight will generate a plane that is not vertical, as it would be if the instrument were in perfect adjustment. In Fig. 4-20, the right side of the horizontal axis is seen to be inclined upward by small angle e. The line of sight sweeps through the tilted plane $Z'SH'$ instead of the vertical plane ZSH. This produces an error c in the horizontal direction of line OS.

In the spherical triangle $Z'SH$, $c = Z'$. Then by the law of sines for spherical triangles,

$$\frac{\sin Z'}{\sin e} = \frac{\sin HS}{\sin Z'S} = \frac{\sin h}{\sin Z'S} = \frac{\sin h}{\cos(90° - Z'S)} \tag{4-4}$$

but since $90° - Z'S = h$ (approximately) and $c = Z'$, Eq. (4-4) becomes

$$\sin c = \sin e \tan h$$

or since c and e are both small angles

$$c'' = e'' \tan h \tag{4-5}$$

If the right side of the horizontal axis is inclined upward (as depicted in Fig. 4-20), and if h is positive, then the clockwise circle reading is too large by the angle c. The total error in a measured angle is thus

$$c_t'' = e''(\tan h_F - \tan h_B) \tag{4-6}$$

Errors from this source are eliminated by double centering.

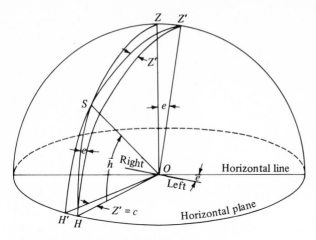

Figure 4-20. Inclined horizontal axis.

Example 4-2

Given the same data as in Example 4-1, except that the left side of the horizontal axis is inclined upward by 25″. Compute the correct horizontal angle.

Solution: Since the left side is high (opposite to that depicted in Fig. 4-20), the backsight reading will be too large since h is negative. The foresight reading will be too small. The total error in the measured angle is thus, by Eq. (4-6),

$$c_t = -25''[\tan 42° 50' - \tan(-12° 22')] = -28.7''$$

The correct angle is thus $76° 15' 10'' + 28.7'' = 76° 15' 38.7''$.

---

6. If the plate bubbles are out of adjustment, the vertical axis is not truly vertical when the bubbles are centered. This defect causes an error in the horizontal direction of each pointing that varies both with the direction of pointing and with the vertical angle. When the direction of pointing is in the same direction as the inclination of the vertical axis, no error will result, no matter what the vertical angle may be. If the direction of pointing is at right angles to the direction of inclination of the vertical axis, the error will be greatest. Furthermore the greater the vertical angle becomes, the greater will be the error of pointing. This error is not of great consequence in ordinary transit work. It *cannot* be eliminated by double centering. It *can* however be eliminated by proper use of the plate bubble (or bubbles). If the instrument contains two plate bubbles at right angles to one another, when one bubble is brought over a pair of leveling screws, the other bubble is over the opposite pair (or over the third leveling screw if the instrument contains only three

leveling screws). Both bubbles are brought to the center by turning the leveling screws. The instrument is then rotated 180° about its vertical axis. If both bubbles return to the center of the tube, the vertical axis is truly vertical within the sensitivity of the plate levels. If they do not return to the center, each bubble is brought *halfway* back by turning the corresponding leveling screws. This makes the vertical axis truly vertical. The leveling screws should not be moved after this step while angles are being measured. If the instrument contains only one plate level, the above reversal procedure is repeated for directions at right angles to one another. If the theodolite contains a telescope level, the vertical axis can be made truly vertical by the technique described in Section 4-18.

7. If the A and B verniers are not set 180° apart, the two readings will be in disagreement. If the inner and outer spindles do not cause the verniers and the horizontal circle to move concentrically with one another, an error will be introduced that varies from zero at one pointing to a maximum at right angle to the original pointing. These two errors are eliminated by reading both verniers and adopting a mean of the two readings. Since errors due to this source are very small, the A vernier alone is read in most ordinary transit work.

8. Except in work of high precision, the use of an instrument with very small inequalities in the spacing of the graduations around the horizontal circle or around the verniers will not affect the accuracy of ordinary transit work. If the precision of the work warrants it, the error in an angle at a station can be reduced by measuring the angle several times and distributing the initial backsight reading uniformly around the circle.

It is evident from the foregoing discussion that the error caused by defects in adjustment of the theodolite will be greatest when the survey is being made on hilly terrain with varying vertical angles, and when some points are very close to the instrument and others are far away. Practically all instrumental errors can be eliminated by taking the mean of two angles, one of which is observed with the telescope direct and the other with the telescope inverted.

Personal errors are introduced into theodolite work by the instrumentman and his signalmen. Some of these errors are serious if short sights are encountered, but they are usually negligible on long sights.

Failure to have the instrument centered over the occupied station, bad pointing of the telescope on the target, or failure to have the target directly over the station sighted are all personal errors that affect accuracy in a similar manner.

If the theodolite is not set up exactly over the station, all angles measured at that point will be in error. The error decreases as the length of the sight increases. The diagram in Fig. 4-21 shows the effect of measuring an angle from B' instead of B, the actual station. The correct angle ABC equals the measured angle $AB'C + e_1 + e_2$. The maximum value of e_1 will be obtained when BB' is perpendicular to BC, in which case $\tan e_1 = BB'/BC$. Likewise the value of e_2 will be greatest when BB' is at right angles to BA, or when

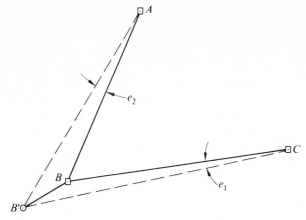

Figure 4-21. Instrument not properly centered.

$\tan e_2 = BB'/BA$. A convenient relation to remember is that the error in a sight will be about 1′ when the length of the sight is 300 ft and the actual point of setup is about 1 in. off the theoretical line of sight. Thus if the distances BA and BC are each about 300 ft and the distance BB' is 1 in., both e_1 and e_2 will be less than 1′ and the measured angle will be in error by less than 2′. Consequently, when precision is required and when the lengths of the sights are short, the transit must be set up more carefully than when the sights are long or when small errors are of no importance.

The error in an angle due to errors of pointing can also be determined from the relationship just given. Thus when the pointing is made to a point 1 in. to the right or left of a station 300 ft from the transit, the resulting error in the measured angle will be 1′. For a point 100 ft away, this error would be increased to about 3′.

Parallax caused by not having the image of the object sighted focused in the plane of the cross hairs introduces a personal error of varying amounts in the measurement of both horizontal and vertical angles. Its elimination has been discussed in Section 3-11.

Faulty centering of level bubbles is a personal error different from maladjustment of the bubbles. Proper manipulation of the level bubbles has been discussed in previous sections.

If the instrument contains verniers, errors are introduced when the observer reads the verniers. If the observer does not use a reading glass, or if he does not look radially along the graduations when reading the verniers, he may be one or two graduations in either direction from the correct coincidence. Errors in reading the verniers are not to be confused with mistakes in reading the verniers to be discussed in Section 4-20.

The instrument man should be very careful in walking about the theodolite. The tripod is easily disturbed, particularly when it is set up on soft ground. If a tripod leg is accidentally brushed against, the backsight should be checked at once. Of course the plates should be releveled first, if necessary.

Some causes of natural errors are gusts of wind, heat from the sun shining on the instrument, settling of the tripod legs, and horizontal and vertical refraction of the line of sight due to atmospheric conditions. The errors are of little consequence in all but extremely precise work. The theodolite can be sheltered from the wind and from the rays of the sun if work under unfavorable weather conditions is necessary and accuracy demands such protection. Stakes may be driven to receive the tripod legs in unstable ground. Horizontal refraction is avoided by not allowing transit lines to pass close to such structures as buildings, smokestacks, and stand pipes, which will radiate a great deal of heat. The effect of vertical refraction can be reduced to a minimum when vertical angles are being measured if the vertical angle can be measured at both ends of the line. Otherwise its effect can be determined by methods discussed in Section 3-2 or by reference to prepared tables discussed in Chapter 12.

4-20. MISTAKES IN THEODOLITE ANGLES

All mistakes are of a personal nature, but they are not to be confused with personal errors. They are blunders on the part of the surveyor, and are known in the field as "busts." A mistake can have any magnitude and will usually render any observation useless. Some of the more frequent mistakes are forgetting to level the instrument; turning the wrong tangent screw; transposing digits in recording a reading, as writing 291° instead of 219°; dropping a full 20′, 30′, or 40′ from a vernier reading; reading the wrong vernier; reading the wrong circle (clockwise or counterclockwise); reading the wrong side of the vernier; recording a small vertical angle with the wrong algebraic sign; failing to center the control bubble on a vertical circle containing one before reading the circle; sighting on the wrong target; recording the wrong reading on a leveling rod when measuring a vertical angle.

4-21. DIRECTION THEODOLITE

The direction theodolite, one example of which is shown in Fig. 4-22, does not provide for a lower motion as is contained in a repeating instrument. When it is set up over a point, the horizontal circle is essentially fixed in position. Both the backsight reading and the foresight reading are taken by means of one motion, which is the movement of the reading device with respect to the graduated horizontal circle. The difference between the backsight and the foresight readings is the value of the angle.

The graduations on both the horizontal and the vertical glass circles are further subdivided by means of a circular optical micrometer. This permits directions to be read to the nearest second or smaller.

Different methods are employed to enable the observer to make a circle reading.

The direction theodolite shown in Fig. 4-23 has two different kinds of graduations. In the earlier and very widespread model, the view seen through

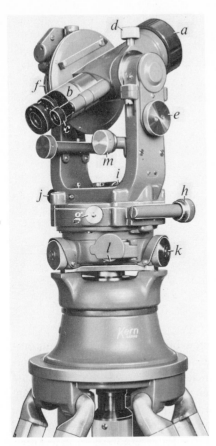

Figure 4-22. Optical reading theodolite. (a) Main telescope; (b) focusing ring; (c) reading microscope; (d) clamp of vertical circle; (e) micrometer operating screw; (f) lighting mirror; (g) clamp of horizontal circle; (h) azimuth tangent screw; (i) horizontal level; (j) optical plummet; (k) leveling screw; (1) circle orienting gear; and (m) altitude tangent screw. Courtesy of Kern Instruments, Inc.

the reading microscope is shown in Fig. 4-24. In the upper part of the field of view can be seen the diametrically opposite graduations of either the horizontal circle or the vertical circle. These are displaced 180° apart and are brought together in the field of view by internal optics. The ultimate reading is thus a mean of two positions on the circle just like the combined A and B vernier readings of a transit. The lower part of the field of view shows the optical micrometer that can be read directly to 1″ and estimated to $\frac{1}{10}$″. The left part of Fig. 4-24 shows the view the observer sees after he has sighted on

Figure 4-23. Wild T2 direction theodolite. Courtesy of Wild Heerbrugg Instruments, Inc.

his signal using the clamp and tangent screw but before he sets the micrometer operating screw. He now turns the micrometer screw until the diametrically opposite graduations are coincident, as shown in the right part
of the figure. After coincidence, it is seen that 265° has its corresponding
reading of 85° *lying to the right*. Thus the reading is 265°. The index line also
indicates the reading to lie somewhere between 265° and 266°. The circle is
graduated every 20′. However, if the bottom set of graduations are caused
to move 10′ to the right, the upper graduations will then move 10′ to the left,
and coincidence would occur once more. Thus each 20′ graduation represents 10′ of absolute movement. In order to determine the number of tens of

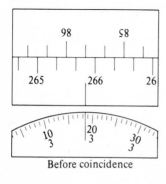

Before coincidence

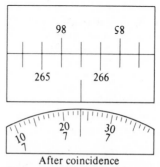

After coincidence

Circle reading.. 265° 40′

Drum reading.. ___7′ 23.6″__

265° 47′ 23.6″

Figure 4-24. Circle reading on earlier Wild T2.

minutes, the observer counts the number of spaces between 265° and 85° which in the figure is four. The reading thus far is 265° 40′. On the micrometer drum below, the number that repeats itself is the number of minutes to be added to the multiple of 10′, and this is seen to be 7′. The graduations give the number of seconds as 23.6″. The reading is therefore 265° 40′ + 7′ + 23.6″ = 265° 47′ 23.6″.

The most common mistake in reading the micrometer shown in Fig. 4-24 is miscounting the number of spaces between the direct degree number and its corresponding inverted number to get the tens of minutes. This must be carefully guarded against. In order to eliminate this source of mistake, the newer model of this same instrument employs the technique shown in Fig. 4-25. After the telescope sighting has been made, the instrumentman makes the two sets of three graduations at the top of the field of view coincide using the micrometer screw. The small triangle below the degree number points to the appropriate number of tens of minutes. The optical micrometer drum is then read to obtain the remainder of the reading.

A switching knob located on the standard opposite the vertical circle of the theodolite shown in Fig. 4-23 is used to present to the observer either the horizontal circle or the vertical circle. Note that only one or the other appears in the field of view as indicated in Figs. 4-24 and 4-25.

The view of the circle and optical micrometer of the direction theodolite shown in Fig. 4-26 is presented in Fig. 4-27. The technique for reading is similar to that of Fig. 4-24. Note that the circle is graduated each 4′. Therefore the absolute value of each space between the direct and inverted numbers is 2′. In this example, the reading is 73° 26′ plus 1′ 56.7″, or 73° 27′ 56.7″.

The circle graduations shown in Fig. 4-28 appear in the direction theodolite of Fig. 4-29. This circle is also graduated in grads. The two pairs of graduation opposite either the letter V or the letter H are bisected by the two index lines to give a mean of the positions of diametrically opposite parts of the circle. The tens of minutes are indicated by the small squares shown in the figure. The horizontal circle reading here is 56^g 53^c 34^{cc} or 56.5334^g.

The vertical circles of the direction theodolites, which measure zenith angles, are equipped with either an automatic pendulum compensator or a

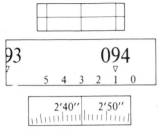

Horizontal or vertical circle
94° 12′ 44″

Figure 4-25. Circle reading on later Wild T2.

Figure 4-26. Wild T3 direction theodolite. Courtesy of Wild Heerbrugg Instruments, Inc.

control bubble which eliminates any index error from the vertical circle reading. The control bubble is usually the coincidence type discussed in Section 3-12 and shown in Fig. 3-14.

The direction theodolite contains an optical plummet which allows the instrument to be precisely centered over the instrument station.

An altogether different method for centering the instrument over the ground station is provided by the telescoping plumbing rod, the upper part of which can be seen under the instrument in Fig. 4-29. The bottom of the rod is pointed and is set into the station mark. The top of the rod is moved laterally to bring the rod into a plumb position by means of the tribrach which moves over the top face of the tripod. The plumb position is indicated by a bull's-eye bubble attached to the rod. The upper end of the rod and,

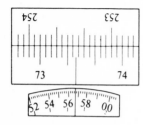

Horizontal circle 73° 26'

$$\frac{1' 56.7''}{73° 27' 56.7''}$$

Figure 4-27. Circle reading on Wild T3.

```
        124
9 8 7 6 5 4|3|2 1 0
     [⫿⫿ ⫿⫿] V
     [⫿⫿ ⫿⫿] H
         56
9 8 7 6|5|4 3 2 1 0
      |⫿⫿⫿⫿|⫿⫿⫿⫿|
      20  30  40
```

Horizontal circle 400^g
56^g 53^c 34^{cc}

Figure 4-28. Circle reading on Kern DKM2-A.

consequently, the tribrach is then locked into position by means of a knurled clamping nut at the upper end of the rod. This method of centering the instrument has the distinct advantage of allowing the theodolite to be removed and replaced by an EDM, a reflector, or a target, without disturbing the tribrach. This technique is referred to as *forced centering,* since each subsequent instrument or target placed on the tripod is automatically centered over the ground station.

Forced centering can be accomplished without the use of the plumbing rod if the tribrach which is secured to the top of the tripod is provided with

Figure 4-29. Kern DKM2-A direction theodolite. Courtesy of Kern Instruments, Inc.

a convenient release of the instrument. Nearly all of the modern theodolites incorporate the forced-centering feature, which allows rapid interchangeability of instruments and targets over the same point.

A cautionary note is needed regarding the optical reading theodolites. Although they are extremely rugged, they should be treated with care in handling. When one of these instruments is being removed from its case to be set up, or is removed from the tripod to be placed back into its case, it should be lifted with one hand by the standard on the side opposite that of the vertical circle and supported underneath by the other hand. This helps to preserve the critical alignment of the internal optical components. Also, as soon as the instrument is set on its tripod or into the tribrach, it should be secured. Otherwise the tripod might inadvertently be moved or even picked up, allowing the instrument to fall to the ground.

4-22. ANGLES FROM MEASURED DIRECTIONS As an example of recording and reducing the field notes when using the direction instrument, refer to Fig. 4-30. A direction theodolite, reading to 1″, is set up at *A* for the purpose of measuring the angles between *B* and *C*, *C* and *D*, and *D* and *E*. The telescope in its direct position is pointed at *B*, the motion is clamped, and the pointing is perfected by means of the tangent screw. The circle is read as 132° 20′ 36″ and recorded as shown in Fig. 4-31 opposite D for "direct." The telescope is then turned to *C*, and the reading of 175° 46′ 21″ is recorded as shown. The telescope is next turned to *D*, and a reading of 192° 06′ 04″ is recorded. Finally, a reading of 232° 15′ 32″ is obtained by a sight to *E*.

In order to eliminate instrumental systematic errors, the telescope is now reversed, and the line of sight is directed to *B*. With the telescope in this reverse position, the circle reading is displaced 180° from the direct reading. It is 312° 20′ 42″, which is recorded opposite R for "reverse." The discrepancy between the 36″ in D and the 42″ in R reflects sighting errors and instrumental

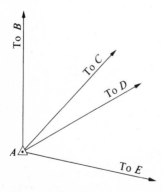

Figure 4-30. Angles measured by direction instrument.

STATION	D or R	CIRCLE READING	MEAN D & R
	⊼ ℗ STATION A		
B	D	132° 20 ′36°	132° 20′ 39″
	R	312° 20′ 42″	
C	D	175° 46′ 21″	175° 46′ 23″
	R	355° 46′ 25″	
D	D	192° 06′ 04″	192° 06′ 05″
	R	12° 06′ 06″	
E	D	232° 15′ 32″	232° 15′ 35″
	R	52° 15′ 38″	

Figure 4-31. Direction instrument notes.

errors. The mean of the two values, or 39″, is suffixed to the direct reading of 132° 20′ to give the mean of the D and R values as shown. This mean is free from instrumental errors, but not from random errors. Sights are then taken to points *C*, *D*, and *E* to obtain the reversed readings, which are recorded as shown in the notes.

The directions to the four points are now used to obtain the angles. The angle at *A* from *B* to *C* is equal to the direction to *C* minus the direction to *B*, and so on. Thus

$$\angle BAC = 175° \, 46' \, 23'' - 132° \, 20' \, 39'' = 43° \, 25' \, 44''$$

$$\angle CAD = 192° \, 06' \, 05'' - 175° \, 46' \, 23'' = 16° \, 19' \, 42''$$

$$\angle DAE = 232° \, 15' \, 35'' - 192° \, 06' \, 05'' = 40° \, 09' \, 30''$$

$$\angle BAD = 192° \, 06' \, 05'' - 132° \, 20' \, 39'' = 59° \, 45' \, 26''$$

The direction instrument has provision for advancing the circle for a new set of readings as just discussed. For the next set the initial reading would then be 132° 20′ 36″ plus a value determined from the number of sets, or *positions*, to be read. This added value is $180°/n$, in which n is the number of positions to be observed. Thus if three positions are to be observed, the initial reading for the second set would be 132° 20′ 36″ + 60° = 192° 20′ 36″, and the initial reading for the third set would be 252° 20′ 36″. For each position a set of angles can be computed as just shown for the first position. The final angles are then the means obtained from all the sets.

4-23. TOTAL STATION INSTRUMENTS A total station instrument is one which contains an EDM unit with which to measure distances, and horizontal and vertical circles for measuring horizontal and vertical

Figure 4-32. Circular binary code disk.

Figure 4-33. Zeiss Reg Elta 14 Electronic Recording Tacheometer-Rangefinder. Courtesy of Keuffel & Esser Co.

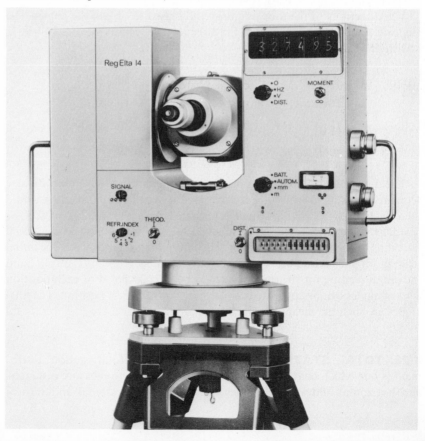

angles. The horizontal and vertical rotations of the telescope are converted from mechanical angular displacements to direct digital readouts. The conversion is accomplished by various binary codes like the one shown in Fig. 4-32 which are printed on glass circles. The codes are read by either photoelectric, magnetic, or direct contact pickup. The signals are then sorted electronically, and the angular displacements can be displayed on small Nixie tubes in the form of horizontal and vertical angles. The angular as well as distance outputs can also be stored on magnetic tape and printed in hard copy for further processing by a computer. The instrument shown in Fig. 4-33 displays either the horizontal angle or the vertical angle or the slope distance on six-digit Nixie tubes. These three values together with the identifying information can be punched on tape.

The instrument shown in Fig. 4-34 contains a microprocessor which computes the horizontal distance and the vertical distance to a point based on slope distance input from the EDM unit and the vertical circle reading. Horizontal angles in a survey are measured in a manner identical to that using the theodolites previously described. A data collector attached to the instrument automatically stores the measurements which are in turn fed directly to a calculator in the field or taken to the office for survey calculations.

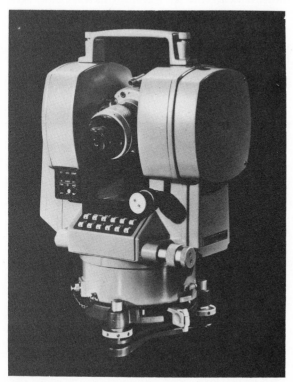

Figure 4-34. H-P 3820A Total Station Instrument.
Courtesy of Hewlett Packard Co.

A unique feature of the HP 3820A is an automatic level detector which senses the amount and direction that the instrument is out of level. The output from the level detector is automatically entered into the micro-processor to compute corrections to both the measured horizontal and vertical angles before they are displayed and recorded on magnetic tape.

The digital-readout total station theodolites are a relatively recent development. Advances made in the development of microprocessors and coded disks are certain to have an accelerating effect on the acceptance of these novel instruments. To attempt to describe all of the different total station instruments in detail would be outside the scope of this text. The student should consult the selected readings at the end of this chapter and should obtain descriptive brochures from the instrument manufacturers in order to gain more insight into the operation of these new instruments.

4-24. ACCURACY REQUIRED IN MEASURING ANGLES

The accuracy required in the measurement of horizontal angles is largely determined by the purpose of the survey and by the accuracy with which the linear measurements are made. A balance should be maintained between the accuracy of the angular measurements and the accuracy of the linear mea-surements.

The location of a point is frequently determined by a horizontal angle from some line of reference and a distance from the vertex of the angle. Thus in Fig. 4-35 a point P can be located by the angle NAP and the distance AP, neither of which can be measured exactly. If the angular error is $\pm e$ and it is

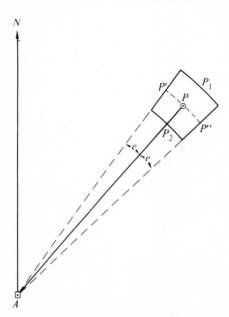

Figure 4-35. Relation between angular and linear errors.

assumed that there is no linear error, the position of P would be between P' and P''. On the other hand, if it is assumed that there is no angular error, the effect of linear error would be to locate P somewhere between P_1 and P_2. If the accuracy of the angular measurements is consistent with the accuracy of the linear measurements, the values of $P'P$, $P_2 P$, $P''P$, and $P_1 P$ should all be equal, and the located position of P will fall somewhere within an approximate square.

When the allowable linear error in a distance l is assumed to be d, the corresponding allowable angular error e can be computed from the relation $\tan e = d/l$. If the linear error is expressed as a ratio, this ratio will also be $\tan e$. Where the angular error e is assumed, the allowable error in distance will be the length of the line times the tangent of the angle e.

In Table 4-2 the allowable errors in angles and the allowable errors in linear measurements are given for various assumed accuracies of linear and angular measurements. From this tabulation it is evident that if the distances are being obtained by the stadia method with a relative accuracy of 1/500, the angles need be measured only to the nearest 5′, whereas if a relative accuracy of 1/10,000 is required, the angles should be known to 20″. If EDM's are used to obtain distances, then the comparable accuracy in angular measurements can only be obtained using an instrument that is accurate to 1″ or 2″. This angular accuracy, however, is not required in all surveys even when EDM's are used. The high accuracy of the EDM's is inherent in the instruments and is obtained with no additional effort.

The accuracy required in the measurement of vertical angles for reducing slope distances and for trigonometric leveling are discussed in Sections 2-11 and 3-3.

TABLE 4-2. Relations between Linear and Angular Errors

ALLOWABLE ANGULAR ERROR FOR GIVEN LINEAR ACCURACY		ALLOWABLE LINEAR ERROR FOR GIVEN ANGULAR ACCURACY					
ACCURACY OF LINEAR MEASURE-MENTS	ALLOW-ABLE ANGULAR ERROR (″)	LEAST READING IN ANGULAR MEASURE-MENTS	ALLOWABLE LINEAR ERROR IN				RATIO
			100′	500	1000′	5000′	
1/500	6′ 53	5′	0.145	0.727	1.454	7.272	1/688
1/1000	3′ 26	1′	0.029	0.145	0.291	1.454	1/3440
1/5000	0′ 41	30″	0.015	0.073	0.145	0.727	1/6880
1/10,000	0′ 21	20″	0.010	0.049	0.097	0.485	1/10,300
1/50,000	0′ 04	10″	0.005	0.024	0.049	0.242	1/20,600
1/100,000	0′ 02	5″	0.002	0.012	0.024	0.121	1/41,200
1/1,000,000	0′ 00.2	2″	0.001	0.005	0.010	0.048	1/103,100
		1″		0.002	0.005	0.024	1/206,300

4-25. ACCURACY OF ANGLES USED IN TRIGONOMETRIC
COMPUTATIONS   If the measured angles are to be used in trigonometric computations, the values in Table 4-2 may be somewhat misleading. It appears from the table that with an angular error of 1′ the corresponding linear accuracy should be 1/3440. If the measured angle is used in computing the length of an unknown side, the accuracy of the computed length may be considerably less than this, the exact error depending on the size of the angle and the particular function used.

As an example, let it be assumed that in Fig. 4-36 the distance AC has been measured with no error and the angle at A is accurate within 1′. The distance BC is $AC \sin A$. With $A = 30° \pm 01'$ and $AC = 500.00$ ft, BC will be 250.00 ± 0.126 ft. The accuracy of this computed value therefore is 0.126/250 or 1/1980. If the base AB of the triangle had been measured and BC had been computed from the relation $BC = AB \tan A$, the accuracy of the computed result would be still less than when the sine is used. With $AB = 500.00$ and $A = 30° \pm 01'$, $BC = 288.68 \pm 0.194$ ft and the accuracy of the result is 0.194/288.68 or 1/1490.

TABLE 4-3. Accuracy of Values Calculated by Trigonometry

SIZE OF ANGLE AND FUNCTION		ANGULAR ERROR				
		1′	30″	20″	10″	5″
SINE	COSINE	ACCURACY OF COMPUTED VALUE USING SINE OR COSINE				
5° or	85°	1/300	1/600	1/900	1/1800	1/3600
10°	80°	1/610	1/1210	1/1820	1/3640	1/7280
20°	70°	1/1250	1/2500	1/3750	1/7500	1/15,000
30°	60°	1/1990	1/3970	1/5960	1/11,970	1/23,940
40°	50°	1/2890	1/5770	1/8660	1/17,310	1/34,620
50°	40°	1/4100	1/8190	1/12,290	1/24,580	1/49,160
60°	30°	1/5950	1/11,900	1/17,860	1/35,720	1/71,440
70°	20°	1/9450	1/18,900	1/28,330	1/56,670	1/113,340
80°	10°	1/19,500	1/39,000	1/58,500	1/117,000	1/234,000
TAN OR COT		ACCURACY OF COMPUTED VALUE USING TAN OR COT				
5°		1/300	1/600	1/900	1/1790	1/3580
10°		1/590	1/1180	1/1760	1/3530	1/7050
20°		1/1100	1/2210	1/3310	1/6620	1/13,250
30°		1/1490	1/2980	1/4470	1/8930	1/17,870
40°		1/1690	1/3390	1/5080	1/10,160	1/20,320
45°		1/1720	1/3440	1/5160	1/10,310	1/20,630
50°		1/1690	1/3390	1/5080	1/10,160	1/20,320
60°		1/1490	1/2980	1/4470	1/8930	1/17,870
70°		1/1100	1/2210	1/3310	1/6620	1/13,250
80°		1/590	1/1180	1/1760	1/3530	1/7050
85°		1/300	1/600	1/900	1/1790	1/3580

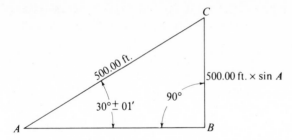

Figure 4-36. Accuracy of angles in trigonometric computations.

In Table 4-3 is shown the accuracy of computed results involving the sines and cosines and the tangents and cotangents of various angles for angular errors of 1′, 30″, 20″, 10″, and 5″. From this table it is evident that when the sine of an angle of approximately 45° is to be used in a computation and the accuracy of the computed result is to be 1/10,000, the angle must be measured to the nearest 20″. When the tangent of 45° is used, the angle must be measured to 10″ to maintain the same accuracy. The accuracy of computations involving the sines, tangents, or cotangents of small angles or the cosines, tangents, or cotangents of angles near 90° will be very low, unless these angles are measured much more exactly than is usually the case.

PROBLEMS

4-1. The sides of the triangle shown in Fig. 4-2 were measured as follows: $AB = 213.48$ m; $AD = 282.04$ m; $AE = 316.84$ m; $CB = 322.18$ m; $CD = 232.56$ m; $BD = 192.23$ m; $DE = 238.80$ m. Compute the angles in the triangles to the nearest 10″.

4-2. An angle ABC in Fig. 4-3 is determined by linear measurements in the manner described in Section 4-2. If $BD = BE = 100$ ft and $k = 63.80$ ft, what is the value of the angle?

4-3. A signal has been displaced 0.25 ft perpendicular to the line of sight of a transit. If the signal is 650 ft away from the transit, what angular error is produced? Express the answer to the nearest second.

4-4. A line rod is $\frac{1}{2}$ in. in diameter. What angle would be subtended between the two sides at a distance of 250 ft? Answer to the nearest second.

4-5. In Problem 4-4 if the instrument is 300 m away, what angle would the line rod subtend?

4-6. An 8-ft range pole is plumbed over a station on which a sight is made with a transit. If the sight is made 6 ft up from the ground and if the top of the pole is out of plumb 2 in. in a direction normal to the line of sight, what angular error is produced in a 400-ft sight?

4-7. A 1′ transit has a 6-in. diameter horizontal circle. What is the difference in the width of spacings on the circle and those on the vernier? Take the answer to the nearest 0.00001 in.

4-8. A horizontal angle is measured at Q from P to R by the method of double centering. A backsight on P gives a circle reading of 91° 52′ 20″, and a foresight on R gives a circle reading of 146° 18′ 10″. After plunging the telescope, the backsight is made to *P* and a foresight is made to *R*. The final circle reading is 200° 43′ 40″. What is the value of the angle?

4-9. A horizontal angle is measured at *F* from *E* to *G* by the method of double centering. A backsight on *E* gives a circle reading of 302° 16′ 30″, and a foresight on *G* gives a circle reading of 20° 20′ 00″. After plunging the telescope, the backsight is made to *E* and a foresight is made to *G*. The final circle reading is 98° 23′ 00″. What is the value of the angle?

4-10. An angle is measured by repetition with a 1′ transit. The mean of the A and B verniers on the initial backsight is 102° 50′ 00″. After the first repetition, the A vernier reads 262° 18′. After the sixth repetition, the mean of the A and B verniers is 339° 36′ 30″. Compute the value of the angle.

4-11. An angle is measured by repetition using a 30″ transit. The initial mean backsight reading is 359° 59′ 45″ on the clockwise circle. After the first repetition, the A vernier of the clockwise circle reads 72° 29′ 00″. After the twelfth repetition, the mean of A and B verniers of the clockwise circle is 149° 47′ 00″. Compute the value of the angle.

4-12. An angle is measured six times direct and six times reverse using a 5″ theodolite. The initial backsight reading is 0° 00′ 15″. After the first repetition, the circle reads 77° 20′ 00″. After the twelfth repetition, the circle reads 208° 01′ 45″. Compute the value of the angle.

4-13. A vertical angle is measured to a signal on top of a hill. With the telescope in the direct position, the circle reads +18° 32′. With the telescope in the reversed position, the circle reads +18° 37′. If the vernier reads zero when the transit is truly leveled and the telescope bubble is centered, what causes the apparent discrepancy? What is the size of the actual discrepancy?

4-14. In laying off a vertical angle, the surveyor sights the transit in the direction of the point to be established. He brings the telescope horizontal, centers the telescope bubble and reads −0° 03′ on the vertical circle. What reading should be set on the vertical circle in order to establish a +20° 20′ vertical angle?

4-15. The line of sight of a theodolite falls at the rate of 0.035 m per 100 m when the telescope bubble is centered. Assuming that the instrument is in proper adjustment in all other respects, what is the value of the vertical angle to a point 1000 m away if the vertical circle reads −3° 52′?

4-16. The distance between the upper ends of the standards which provide the points of support for the horizontal axis of the transit is 5.150 inches. The left standard is 0.0020 in high. What error will this produce in the measurement of a clockwise horizontal angle if the backsight vertical angle is −10° 00′ and the foresight vertical angle is +45°? Is the measured angle too large or too small?

4-17. The right side of the telescope axis of a transit is 0.0015 in. high. The distance between points of support is 6.600 in. On measuring a clockwise horizontal angle, the backsight vertical angle is 0° 00′ and the foresight vertical angle is −32° 30′. What error will this produce in the horizontal angle? Is the measured angle too large or too small?

4-18. The line of sight of a transit makes an angle of 89° 55′ with the transverse axis. What error will this produce in the measurement of a clockwise horizontal angle if the backsight is level and the foresight is inclined 30° 00′?

4-19. In laying off construction lines an engineer lays off a line approximately at right angles with a construction base line, and sets a point 185.00 m away from the transit. The angle is measured by repetition and found to be 90° 00′ 20″. What offset should be measured perpendicular to the construction line at the set point (see Fig. 4-5)?

4-20. If the diameter of the glass circle shown in Fig. 4-27 is 150.00 mm, what is the spacing of the circle graduations (nearest tenth of micrometre)?

4-21. A direction theodolite is used to measure angles about a point. Four positions of the circle are used. The resulting directions are listed below.

STATION SIGHTED	MEAN D AND R POSITION 1	MEAN D AND R POSITION 2	MEAN D AND R POSITION 3	MEAN D AND R POSITION 4
E	135° 21′ 25.6″	180° 21′ 21.0″	225° 21′ 28.9″	270° 21′ 26.5″
F	172° 59′ 06.0″	217° 59′ 06.6″	262° 59′ 12.9″	307° 59′ 09.6″
G	203° 10′ 59.2″	248° 10′ 58.4″	293° 11′ 05.1″	338° 11′ 06.1″
H	245° 34′ 47.4″	290° 34′ 44.4″	335° 34′ 55.3″	20° 34′ 57.7″

Compute the three angles for each position. Compute the mean of each of the three angles from the four positions.

4-22. The following directions were observed using a direction theodolite that reads in grads. Compute the three angles for each position. Compute the mean of each of the angles.

STATION SIGHTED	MEAN D AND R POSITION 1	MEAN D AND R POSITION 2	MEAN D AND R POSITION 3	MEAN D AND R POSITION 4
P	20.1150	70.1130	120.1120	170.1135
Q	34.2676	84.2640	134.2634	184.2655
R	132.9838	182.9800	232.9780	282.9825
S	255.4482	305.4440	355.4412	5.4455

4-23. Convert the mean angles computed in Problem 4-21 to their grad values.

4-24. Convert the mean angles computed in Problem 4-22 to their degree-minute-second values.

4-25. A theodolite is used to measure directions. If it can be assumed that the error in the directions is within a spread of ± 3″, how closely must the length of a line 3 km long be measured in order to be consistent with the angular accuracy?

4-26. A 30″ transit is used to measure angles by the method of double centering. If the measured angle is in error by no more than ± 10″, with what accuracy must a line 4000 ft long be measured in order that the accuracies be consistent?

4-27. Assume that side BC of Fig. 4-36 is measured with no error and found to be 375.55 ft. If angle A is measured as 13° 52′ 40″ ± 10″, what will be the accuracy of the computed distance AC?

4-28. Assuming no error in side AC or Fig. 4-36 which is 514.264 m long, angle A is measured as 26.2624^g ± 30^{cc}. What will be the accuracy of the computed side BC?

4-29. In Problem 4-28, what will be the accuracy of the computed side AB?

BIBLIOGRAPHY

Burnett, W. H. 1967. The encoder for automatic angle readout. *Surveying and Mapping* 27:649.

Colcord, J. E. 1971. Error analysis in angulation design. *Proc. Fall Conv. Amer. Cong. Surv. & Mapp.* p. 200.

Erickson, K. E. 1977. Electronic Surveying System. *Proc. Ann. Meeting Amer. Cong. Surv. & Mapp.* p. 209.

Gort, A. F. 1977. The Hewlett–Packard 3820A Electronic Total Station. *Proc. Fall Conv. Amer. Cong. Surv. & Mapp.* p. 308.

Greulich, G., and Schellens, D. F. 1973. Surveying with Zeiss Reg Elta 14 in densely populated areas. *Proc. Ann. Meet., Amer. Cong. Surv. & Mapp.,* p. 451.

Leitz, H. 1979. Ten years of electronic tacheometry—Zeiss: 1968, Reg Elta-14—1978, Elta-2 and Elta-4. *Proc. Ann. Meet. Amer. Cong. Surv. & Mapp.* p. 547.

Schellens, D. F. 1968. New design elements in surveying instruments. *Proc. 28th Ann. Meet. Amer. Cong. Surv. & Mapp.* p. 473.

Chapter 5
Random Errors

5-1. NATURE OF RANDOM ERRORS In the processes of measuring distances, elevations, and angles described in Chapters 2, 3, and 4, several sources of errors and mistakes were discussed. Mistakes are eliminated when the observer takes such precautions as measuring a distance in both directions, checking level notes, repeating angles direct and reversed, and using other procedures that will help to avoid or detect the mistakes. Systematic errors can be practically eliminated by using calibrated tapes, by correcting taped distances for temperature, tension, sag, and slope, by determining instrumental constants and making frequency checks in EDM, by balancing backsight and foresight distances in leveling, by alternating the sequence of taking backsight and foresight readings, by having the leveling rod plumb when it is read, and by double centering to measure angles.

Random errors, often called *accidental errors*, are unpredictable in regard to both size and algebraic sign. They are truly accidental and cannot be avoided. The statements in the articles to follow are based on three important principles of random errors: (1) A plus error will occur as frequently as will a minus error; (2) small errors will occur more frequently than large errors; and (3) very large errors do not occur at all, or the chance for a large error to occur is remote.

5-2. PROBABILITY OF AN ERROR OCCURRING If serious thought is given to some of the examples of sources of random errors discussed in Chapters 2, 3, and 4, it will be seen that any one single error can be considered to be in itself the result of an indefinite number of very small elementary errors introduced together at any one time. The size and algebraic sign of the resultant random error are determined by the manner in which these elementary errors build up when combined in a measurement. Each element of error is thought of as having an equal chance of being plus or minus, and each element is further considered as having the same size as each other element. For a given number of these elements, or units, of error acting together, we can determine the possible number of combinations of the units, the size of the resultant error from each combination, and the probability of each such size occurring.

Assume that the size of each unit of error is unity. If one unit acts alone, the size of the resultant error is one and the possible number of combinations is two, since the error may be either $+1$ or -1. The probability of either combination occurring is one-half or one out of two. Probability is defined as the number of times a given value should occur divided by the total number of times all values can occur. If two units act together, there are four possible combinations. These are $+1$ and $+1$, $+1$ and -1, -1 and -1, and -1 and $+1$. The sizes of the resultant errors are, respectively, $+2, 0, -2$, and 0. The probability of $+2$ occurring is one-quarter, that of 0 occurring is two-quarters, and that of -2 occurring is one-quarter.

Each time the number of units acting at any one time is increased by one, the number of possible combinations will double. Three units of error acting together give 8 possible combinations; four units give 16 combinations, and so on. Table 5-1 shows the relationship between the number of combinations and the probability that an error of a given size will occur.

Let m designate the number of units acting at any one time. If C is the number of combinations with q specified positive $(+)$ units, then C is given by

$$C = \frac{m(m-1)(m-2)\cdots(m-q+1)}{q!} \tag{5-1}$$

The total number of combinations N which could possibly occur is 2^m. Thus, the probability P of q positive units occurring at any one time is

$$P = \frac{C}{N} = \frac{C}{2^m} \tag{5-2}$$

The quantities m, C, N, and P are shown in the headings of Table 5-1.

The results of Table 5-1 are plotted in Fig. 5-1, views (a) through (f). The shaded rectangle represents the following probability. In (a), $\frac{1}{2}$; in (b), $\frac{1}{4}$; in (c), $\frac{1}{8}$; in (d), $\frac{1}{16}$; in (e), $\frac{1}{32}$; and in (f), $\frac{1}{64}$. It is to be noted that the total

TABLE 5-1. Characteristics of Random Errors

NUMBER OF UNITS COMBIN- ING AT ONE TIME (m)	UNITS IN COMBINATIONS (q SPECIFIED)	NUMBER OF COMBI- NATIONS (C)	TOTAL NUMBER OF COMBI- NATIONS (N)	SIZE OF RESULTANT ERROR	PROBABILITY OF OCCURRENCE OF RESULTANT ERROR (P)
1	$+1$	1		$+1$	$\frac{1}{2}$
	-1	1	2	-1	$\frac{1}{2}$
2	$+1, +1$	1		$+2$	$\frac{1}{4}$
	$+1, -1$	2		0	$\frac{2}{4}$
	$-1, -1$	1	4	-2	$\frac{1}{4}$
3	$+1, +1, +1$	1		$+3$	$\frac{1}{8}$
	$+1, +1, -1$	3		$+1$	$\frac{3}{8}$
	$+1, -1, -1$	3		-1	$\frac{3}{8}$
	$-1, -1, -1$	1	8	-3	$\frac{1}{8}$
4	$+1, +1, +1, +1$	1		$+4$	$\frac{1}{16}$
	$+1, +1, +1, -1$	4		$+2$	$\frac{4}{16}$
	$+1, +1, -1, -1$	6		0	$\frac{6}{16}$
	$+1, -1, -1, -1$	4		-2	$\frac{4}{16}$
	$-1, -1, -1, -1$	1	16	-4	$\frac{1}{16}$
5	$+1, +1, +1, +1, +1$	1		$+5$	$\frac{1}{32}$
	$+1, +1, +1, +1, -1$	5		$+3$	$\frac{5}{32}$
	$+1, +1, +1, -1, -1$	10		$+1$	$\frac{10}{32}$
	$+1, +1, -1, -1, -1$	10		-1	$\frac{10}{32}$
	$+1, -1, -1, -1, -1$	5		-3	$\frac{5}{32}$
	$-1, -1, -1, -1, -1$	1	32	-5	$\frac{1}{32}$
6	$+1, +1, +1, +1, +1, +1$	1		$+6$	$\frac{1}{64}$
	$+1, +1, +1, +1, +1, -1$	6		$+4$	$\frac{6}{64}$
	$+1, +1, +1, +1, -1, -1$	15		$+2$	$\frac{15}{64}$
	$+1, +1, +1, -1, -1, -1$	20		0	$\frac{20}{64}$
	$+1, +1, -1, -1, -1, -1$	15		-2	$\frac{15}{64}$
	$+1, -1, -1, -1, -1, -1$	6		-4	$\frac{6}{64}$
	$-1, -1, -1, -1, -1, -1$	1	64	-6	$\frac{1}{64}$

area in each case is one. In Fig. 5-1(f), the vertical scale has been reduced to one-half of that in the other five views.

If the number of units of error were increased indefinitely, the graphical representation of the size of the resultant errors and the frequency of their occurrence would approach a bell-shaped curve called the *probability curve*, or the *normal error distribution curve*. Such a curve is shown in Fig. 5-2. This curve portrays the three principles of random errors stated in Section 5-1. The equation of this curve is

$$y = ke^{-h^2x^2} \tag{5-3}$$

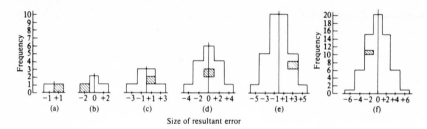

Figure 5-1. Plot of size of resultant error versus frequency based on plus and minus unit errors.

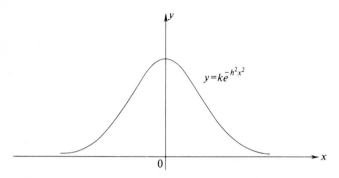

Figure 5-2. Probability curve.

in which y is the relative frequency of occurrence of an error of a given size, x is the size of the error, k and h are constants that determine the shape of the curve, and e is the base of natural logarithms. The values of k and h under a given set of circumstances will be discussed in Section 5-4.

5-3. CONDITION OF LEAST SQUARES In Section 1-10 an error is defined as the difference between the true value of a quantity and the measured value of the same quantity. Since the true value of a quantity can never be determined, the errors in the measurements can be determined only to the extent that systematic errors can theoretically be computed and eliminated. Let it be presumed that a quantity has been measured several times and that a slightly different value has been obtained from each measurement. Then by an extended investigation of Eq. (5-3) the value of the quantity that has the most frequent chance of occurrence or that has the *maximum probability* of occurrence, is the one that will render the sum of the squares of the errors a minimum.

Suppose, for example, that in a given set of measurements, $M_1, M_2, \ldots,$ M_n of a quantity, some value is adopted to represent the *best possible value* of the measured quantity obtainable from the set. Let this be denoted as M. The following subtractions are then made:

$$M_1 - M = v_1$$

$$M_2 - M = v_2$$

$$\vdots$$

$$M_n - M = v_n$$

The v's on the right-hand side are termed *residuals*. They are similar to errors (x's), but errors are obtained by subtracting the true value, rather than the best possible value, from the measurements. The equation of the probability curve, or Eq. (5-3), can then be written in terms of the residuals as follows:

$$y = ke^{-h^2v^2} \tag{5-4}$$

in which v is the size of the residual and the other terms are defined as for Eq. (5-3).

The probability of the occurrence of v_1, for example, is then the area of the probability curve at this value. This area is obtained by multiplying the ordinate by an arbitrary increment Δv, as shown in Fig. 5-3. If a probability is denoted by P, the probability that $v_1, v_2, \ldots, v_n$ will occur in the set of measurements is as follows:

$$P_{v_1} = y_1 \, \Delta v = ke^{-h^2v_1^2} \, \Delta v$$

$$P_{v_2} = y_2 \, \Delta v = ke^{-h^2v_2^2} \, \Delta v$$

$$\vdots$$

$$P_{v_n} = y_n \, \Delta v = ke^{-h^2v_n^2} \, \Delta v$$

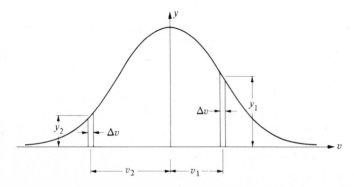

Figure 5-3. Probability of occurrence of a residual v_i.

Referring back to Section 5-2, if for example, four units of error are imagined to be acting at any one time, what is the probability that all four are positive (+) errors at the same time? The probability that any one single error will be positive is obviously $\frac{1}{2}$. The probability that all four will be positive is $\frac{1}{2} \times \frac{1}{2} \times \frac{1}{2} \times \frac{1}{2} = \frac{1}{16}$ as shown in Table 5-1. Thus, the probability that a set of events will occur simultaneously is the product of their separate probabilities. Applying this concept to the set of residuals above, the probability that all of the v's will occur simultaneously in the set of measurements (the M_i's) is equal to the product of their separate probabilities. Thus,

$$P_{(v_1, v_2, \ldots, v_n)} = (ke^{-h^2v_1^2}\,\Delta v)(ke^{-h^2v_2^2}\,\Delta v)\cdots(ke^{-h^2v_n^2}\,\Delta v)$$

or

$$P_{(v_1, v_2, \ldots, v_n)} = k^n(\Delta v)^n e^{-h^2(v_1^2 + v_2^2 + \cdots + v_n^2)} \tag{5-5}$$

The quantity M must be selected so as to give the maximum probability of the simultaneous occurrence of the v's. According to Eq. (5-5), that unique set of v's which has the highest probability of occurring is the one that will make $P_{(v_1, v_2, \ldots, v_n)}$ a maximum. This condition exists when the negative exponent of e in Eq. (5-5) has a minimum value; that is, for maximum probability,

$$v_1^2 + v_2^2 + \cdots + v_n^2 = \text{minimum}$$

or

$$\sum v^2 = \text{minimum} \tag{5-6}$$

The expression $\sum v^2$ meaning "sum of the squares of the v's" is also frequently written $[vv]$. The square brackets signify "sum of" and vv is v^2.

Equation (5-6) expresses the *condition of least squares* that is imposed on a set of measurements when each is made with the same reliability (see Section 5-10) and the resulting set of residuals conforms to the principles relating to random errors stated in Section 5-1.

When a quantity is measured directly several times, the best value or most probable value, of the quantity is the arithmetic mean of the measured values. This can be shown as follows:

$$v_1 = M_1 - M;\, v_1^2 = (M_1 - M)^2$$
$$v_2 = M_2 - M;\, v_2^2 = (M_2 - M)^2$$
$$\vdots$$
$$v_n = M_n - M;\, v_n^2 = (M_n - M)^2$$
$$\sum v^2 = (M_1 - M)^2 + (M_2 - M)^2 + \cdots + (M_n - M)^2$$

In order to make $\sum v^2$ a minimum, in accordance with Eq. (5-6), the procedure is to differentiate with respect to M and to set the derivative equal to zero. Thus

$$\frac{d}{dM}\left(\sum v^2\right) = -2(M_1 - M) - 2(M_2 - M) - \cdots - 2(M_n - M) = 0 \quad (5\text{-}7)$$

Then

$$\frac{d^2}{dM^2}\left(\sum v^2\right) = 2 + 2 + 2 + \cdots + 2 = \text{a positive number}$$

indicating a minimum. Also, from Eq. (5-7),

$$M_1 - M + M_2 - M + \cdots + M_n - M = 0$$

or

$$nM = M_1 + M_2 + \cdots + M_n$$

Hence,

$$M = \frac{M_1 + M_2 + \cdots + M_n}{n} = \text{arithmetic mean}$$

5-4. STANDARD ERROR   A residual v is treated as a random error in every respect. Suppose that 100 measurements of the length of a line were made and that all systematic errors have been eliminated. The mean of all the measurements is the most probable value. If the mean is now subtracted from each measured value, 100 residuals of varying size would result. If each size of residual is plotted against the frequency of occurrence of that size, the resulting curve will be similar to that of Fig. 5-3. This curve is called the *normal distribution curve*. Its equation is

$$y = \frac{1}{\sigma\sqrt{2\pi}} e^{(-1/2\sigma^2)v^2} \quad (5\text{-}8)$$

in which y is the relative frequency of the occurrence of a residual of a given size, σ is a constant to be determined from the measurements, and v is the size of the residual. The value of σ is obtained from the measurements by the formula†

$$\sigma = \sqrt{\frac{\sum v^2}{n-1}} \quad (5\text{-}9)$$

† Some handheld and desk calculators have provision for computing standard errors based on Eq. (5-9) without first having to compute all of the residuals. Only the measurements themselves are entered into the calculator. The student is cautioned, however, that if the numbers involved are large, internal truncation may take place, giving an erroneous value of σ. In order to avoid this difficulty, if a series of measurements are, for example, 546.55, 546.58, . . . , 546.50, they should be entered as 0.55, 0.58, . . . , 0.50, and the value 546 added after the mean has been determined in the calculator.

in which $\sum v^2$ is the sum of the squares of the residuals and n is the number of measurements. The quantity σ is referred to as the *standard error* of the set of measurements. It indicates the precision of the measurements relative to any other set of measurements. The smaller σ becomes, the greater is the precision. The term standard error used in this book is also called *standard deviation*.

The quantity $(n - 1)$ in the denominator under the radical in Eq. (5-9) represents the number of extra measurements taken to determine a value. For example, if a line was measured five times, or $n = 5$, there are four extra measurements, since one would have been sufficient although it would not be very reliable. As the number of extra measurements increases, the standard error becomes smaller. In statistics the quantity $(n - 1)$ is sometimes referred to as the number of degrees of freedom, and σ is the *estimated* standard deviation. Since a residual is obtained by subtracting the mean value of n measurements from any one of these measurements, the following statement must be true. If the mean of n measurements and any $n - 1$ of the measurements are given, the one measurement that was left out can be determined. In other words, it is a dependent measurement.

Example 5-1

The length of a line is measured five times. Four of the measured lengths are 154.26, 154.29, 154.26, and 154.21 ft. Also, the mean of all the measured lengths is 154.25 ft. What is the value of the dependent measurement?

Solution: The sum of the five measurements is obviously five times the mean, or $5 \times 154.25 = 771.25$ ft. The sum of the four given measurements is 617.02 ft. The dependent measurement is $771.25 - 617.02 = 154.23$ ft.

A comparison of the probability curve given by Eq. (5-3) and the curve representing the distribution of errors derived from an actual set of measurements given by Eq. (5-4) and Eq. (5-8) shows that a residual v is treated as an error x. Also, the two constants k and h are obtained in terms of values derived from actual measurements by the relations

$$k = \frac{1}{\sigma\sqrt{2\pi}}$$

and

$$h^2 = \frac{1}{2\sigma^2}$$

It is seen that as k and h increase, so does the precision of the measurements increase. The quantity h is known as the *precision modulus* of the measurements.

5-5. HISTOGRAMS   A *histogram* is a visual representation of the distribution of either a set of measurements or a set of residuals. The representations showing the distribution of unit errors in Fig. 5-1 are histograms. A histogram shows by examination whether the measurements or the residuals are symmetrical about some central value, such as the mean value or the zero residual; they show the total spread of values of the measurements or residuals; they show frequencies of the different values; and they show how peaked or how flat the distribution of the values is.

As an example of the development of a histogram, consider a set of 100 measurements of the number of revolutions of a planimeter wheel when measuring the area of a figure. The results of these measurements, arranged in ascending order were as shown in Table 5-2. The arithmetic mean of these values, rounded off to the nearest 0.001 revolution (rev), is 1.667. Subtracting the mean from each measured value gives the set of 100 residuals associated with these measurements. The residuals are also arranged in ascending order in Table 5-3 and the unit is the revolution of the wheel.

A histogram can be drawn for each set of tabulated values. In Fig. 5-4, which is the histogram for the first set, the abscissas are the measurements in the range from 1.649 to 1.685 rev plotted at intervals of 0.001 rev. Thus

TABLE 5-2. Planimeter Measurements in Revolutions of the Wheel

1.651	1.662	1.665	1.669	1.674
51	63	65	69	74
53	63	66	69	74
53	63	66	69	74
56	63	66	69	74
56	63	66	1.670	75
57	63	66	70	75
57	63	66	70	76
58	63	67	70	76
58	64	67	71	77
58	64	67	71	77
59	64	67	71	78
59	64	67	71	1.680
1.660	64	67	71	80
61	64	67	71	80
61	65	68	71	81
61	65	68	72	83
61	65	68	72	84
61	65	68	73	84
61	65	68	73	85

TABLE 5-3. Residuals in Revolutions of the Wheel

−0.016	−0.005	−0.002	+0.002	+0.007
16	04	02	02	07
14	04	01	02	07
14	04	01	02	07
11	04	01	02	07
11	04	01	03	08
10	04	01	03	08
10	04	01	03	09
09	04	0	03	09
09	03	0	04	+0.010
09	03	0	04	10
08	03	0	04	11
08	03	0	04	13
07	03	0	04	13
06	03	0	04	13
06	02	+0.001	04	14
06	02	01	05	16
06	02	01	05	17
06	02	01	06	17
06	02	01	06	18

each interval is given a discrete value. The ordinates show the frequency of occurrence of each given abscissa. In Fig. 5-5, which is the histogram for the set of 100 residuals, the abscissas are the sizes of the residuals in the range from −0.018 to +0.018 rev.

An examination of the two histograms—that of the measurements and that of the residuals—shows that they are identical except for the numbering of the abscissa scales. This means that if the values in a set of measurements of a

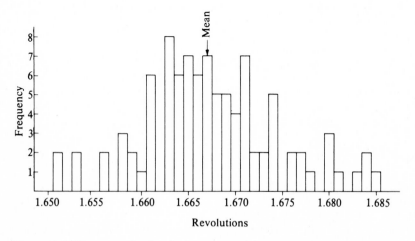

Figure 5-4. Histogram of polar planimeter measurements.

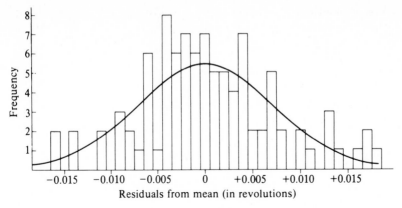

Figure 5-5. Histogram and normal distribution curve of residuals.

quantity are plotted in the form of a histogram, the residuals can be analyzed visually without also having been plotted as a histogram.

The value of σ for the residuals given in Table 5-3 can be computed by Eq. (5-9). The normal distribution curve can then be plotted to some arbitrary scale by applying Eq. (5-8) to compute the ordinates. However, it is usually desirable to plot the normal distribution curve to the same scale as the residual histogram in Fig. 5-5. In order that this may be done Eq. (5-8) must be modified as follows: Let the abscissa interval, which is designated the *class interval*, be denoted by I, and let n denote the number of measurements or residuals. Then Eq. (5-8) takes the form

$$y = \frac{nI}{\sigma\sqrt{2\pi}} e^{(-1/2\sigma^2)v^2} \tag{5-10}$$

After the value of σ has been determined from the residuals, it is possible to compute the following constant quantities for the set:

$$\frac{nI}{\sigma\sqrt{2\pi}} = K_1 \quad \text{and} \quad \frac{1}{2\sigma^2} = K_2$$

When these constants are substituted in Eq. (5-10), that equation becomes

$$y = \frac{K_1}{e^{K_2 v^2}} \tag{5-11}$$

Example 5-2

For the normal distribution curve for the set of residuals shown in histogram form in Fig. 5-5, compute the ordinates corresponding to abscissas that are multiples of 0.002 rev. Plot the curve onto the histogram.

Solution: The sum of the squares of the residuals, or $\sum v^2$, is found by computation to be 0.005235. These computations are not shown, but the procedure is similar to that indicated in Example 5-3. The standard error is then

$$\sigma = \sqrt{\frac{0.005235}{100 - 1}} = \pm 0.0073 \text{ rev}$$

Since $n = 100$ and $I = 0.001$ rev, the value of K_1 in Eq. (5-11) is

$$K_1 = \frac{100 \times 0.001}{0.0073\sqrt{2\pi}} = 5.46$$

Also,

$$K_2 = 9383$$

v	v^2	$K_2 v^2$	$e^{K_2 v^2}$	$y = \dfrac{K_1}{e^{K_2 v^2}}$
0	0	0	1.000	5.46
± 0.002	0.000004	0.0375	1.038	5.26
± 0.004	0.000016	0.1501	1.162	4.70
± 0.006	0.000036	0.338	1.403	3.88
± 0.008	0.000064	0.601	1.82	3.00
± 0.010	0.000100	0.938	2.56	2.13
± 0.012	0.000144	1.351	3.86	1.41
± 0.014	0.000196	1.839	6.30	0.87
± 0.016	0.000256	2.40	11.0	0.50
± 0.018	0.000324	3.04	20.9	0.26

The remainder of the computations are arranged in tabular form, as shown. The computed values of y has been plotted on the histogram in Fig. 5-5, and the normal distribution curve has been drawn as a continuous line through the plotted points. Note that the area under the curve is essentially the same as the area bounded by the histogram, as it should be. Note also that if the histogram of residuals had not been prepared, the normal distribution curve could have been plotted on the histogram of the measurements shown in Fig. 5-4 by renumbering the abscissa scale so that $v = 0$ occurs at the mean measured value of 1.667 rev.

The class interval I of the data given in the above tables and shown in Figs. 5-4 and 5-5 is 0.001 rev. This class interval is somewhat arbitrary and could be taken as some other value, say, 0.005 rev. A limit to the size of the

TABLE 5-4. Planimeter Measurements Grouped Using a Class Interval of 0.005 Rev

(1)	(2)	(3)	(4)	(5)	(6)	(7)
			FREQUENCY	RESIDUAL AT		
		CENTRAL	× CENTRAL	CENTRAL		
INTERVAL	FREQUENCY	VALUE	VALUE	VALUE	v^2	$\sum v^2$
1.651–1.655	4	1.653	6.612	−0.014	0.000196	0.000784
1.656–1.660	10	1.658	16.580	−0.009	0.000081	0.000810
1.661–1.665	28	1.663	46.564	−0.004	0.000016	0.000448
1.666–1.670	27	1.668	45.036	+0.001	0.000001	0.000027
1.671–1.675	18	1.673	30.114	+0.006	0.000036	0.000648
1.676–1.680	8	1.678	13.424	+0.011	0.000121	0.000968
1.681–1.685	5	1.683	8.415	+0.016	0.000256	0.001280
	100		166.745			0.004965

Mean = 1.667 rev
$\sigma = \pm 0.0071$ rev

class interval is reached when the plotted histograms no longer reflect the actual distribution of either the measurements or the residuals. In order to examine the effect of a change of class interval, the planimeter measurements given in Table 5-2 are grouped using a class interval of 0.005 rev in Table 5-4.

The first two columns are plotted in the form of a histogram to give the same area as the histogram of Fig. 5-4. Note that the abscissa scale remains the same but the frequency scale must be reduced by a factor of five. The resulting histogram shown in Fig. 5-6 is considerably more coarse than that of

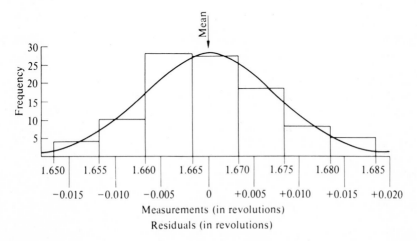

Figure 5-6. Histogram of measurements and residuals together with normal distribution curve for *l* = 0.005 rev.

Fig. 5-4, but the picture of the distribution of the measurements is generally the same.

A continuation of the tabulation gives a simplified method for computing the standard error when a great number of measurements are involved. The central value for each class interval is listed in the third column. This value, which represents each measurement in the interval, is multiplied by the corresponding frequency to give the products appearing in column 4. The sum of the products in column 4 is virtually the same as the sum of the 100 individual measurements. The mean, 1.667 rev, is the same in this instance as before. If the mean is now subtracted from the central value of each class interval, the residual for each central value is obtained and is entered in column 5. This is a representative residual for all those measurements falling in the class interval.

Squaring each residual to give the values in column 6 and then multiplying each squared value by the frequency give the $\sum v^2$ for each class interval. The value is shown in column 7. The values in column 7 are added to give the total $\sum v^2 = 0.004965$. This differs slightly from the value 0.005235 given in Example 5-2. This slight difference is accounted for by the fact that the measurements are not distributed uniformly in each interval.

In order to plot the normal distribution curve for this distribution, the values of K_1 and K_2 are first computed. Thus

$$K_1 = \frac{100 \times 0.005}{0.0071\sqrt{2\pi}} = 28.1 \qquad K_2 = 9919$$

The ordinates to the curve are then computed the same way as shown in Example 5-2.

v	v^2	$K_2 v^2$	$e^{K_2 v^2}$	$y = \dfrac{K_1}{e^{K_2 v^2}}$
0	0	0	1.000	28.1
±0.005	0.000025	0.2480	1.282	21.9
±0.010	0.000100	0.9919	2.680	10.5
±0.015	0.000225	2.232	9.30	3.02
±0.020	0.000400	3.968	53.2	0.53
±0.025	0.000625	6.199	490	0.06

The curve is now superimposed on the measurement histogram of Fig. 5-6. Note that the zero residual falls at the 1.667-rev point on the abscissa scale, since the mean value gives zero residual. A close comparison between the normal distribution curves of Figs. 5-5 and 5-6 would show that they are practically identical in shape and amplitude. Thus a reasonable change in the class interval chosen to make an analysis of a set of measurements will have little effect on the results.

5-6. MEASURES OF PRECISION

Both the standard error σ and the precision modulus h are measures of precision of a set of measurements. If the curve for Eq. (5-8) were to be integrated between the limits $-\sigma$ and $+\sigma$, the area between these limits would be 0.6826 times the total area under the curve, the total area being unity. In other words approximately 68% of all the residuals for the set of measurements should be equal to or less than the standard error. In Example 5-2, 72% of the residuals fall between -0.0073 and $+0.0073$ rev. For further analysis the curve plotted in Fig. 5-5 is redrawn in Fig. 5-7.

The area under the curve between the limits $-\sigma$ and $+\sigma$ is indicated. Also, as can be seen on this curve, the standard error occurs at the point of inflection on each side of the center.

Let it be desired to find the size of a residual, plus or minus, within which one-half, or 50%, of the residuals should fall. This size would be the fractional part of σ for which the area under the curve will be 0.50. The limits of this area are the dashed lines in Fig. 5-7. The error of a size embracing 50% of all residuals is called the *probable error*. It is obtained by the formula

$$E = 0.6745\sigma = 0.6745\sqrt{\frac{\sum v^2}{n-1}} \qquad (5\text{-}12)$$

in which E is the probable error of any one of the measurements, and the other symbols have the meanings previously given. Another way of indicating the meaning of probable error is to say that the residual of any one measurement has an equal chance of being either less than E or greater than E. In Example 5-2 the probable error is ±0.0048 rev. It is seen that 55% of the residuals fall in the range between -0.0048 and $+0.0048$ rev.

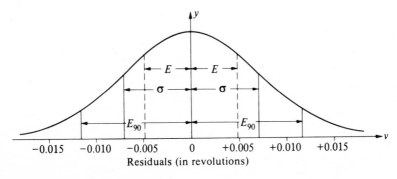

Figure 5-7. Measures of precision.

Example 5-3

The measurements of the difference in elevation between two points by three-wire leveling are listed in the accompanying tabulation. What are the standard error and the probable error of any one of these measurements?

MEASUREMENT NUMBER	MEASURED DIFFERENCE IN ELEVATION (ft)	RESIDUAL v (ft)	v^2
1	21.232	−0.001	0.000001
2	21.235	+0.002	0.000004
3	21.233	0	0
4	21.225	−0.008	0.000064
5	21.233	0	0
6	21.221	−0.012	0.000144
7	21.239	+0.006	0.000036
8	21.244	+0.011	0.000121
9	21.232	−0.001	0.000001
10	21.234	+0.001	0.000001
11	21.229	−0.004	0.000016
12	21.235	+0.002	0.000004
13	21.237	+0.004	0.000016
14	21.230	−0.003	0.000009
15	21.235	+0.002	0.000004
	Sum = 318.494	$\sum v^2 = 0.000421$	
	Mean = 21.233		

Solution: First find the mean of the measured differences in elevation. Then compute each residual and its square, and get $\sum v^2$. The standard error of any one of these measurements is, by Eq. (5-9),

$$\sigma = \sqrt{\frac{0.000421}{15 - 1}} = \pm 0.0055 \text{ ft}$$

The probable error of any one of these measurements is, by Eq. (5-12),

$$E = 0.6745(\pm 0.0055) = \pm 0.0037 \text{ ft}$$

Theoretically, one-half of the residuals should lie between +0.0037 and −0.0037 ft.

The probable error E, as well as the quantities h and σ, is a measure of the precision of a given set of measurements and can be computed after the measurements have been made. The quantity E is sometimes referred to as the

50% *error*, because 50% of all the residuals are theoretically expected to fall between $+E$ and $-E$. This measure of precision was used extensively in the past for defining the reliability of a set of measurements. However, its use has been almost completely replaced by the concept of standard error.

Another very useful measure of precision is the 90% *error*, or the value below which 90% of all residuals of a set of measurements is expected to fall. The 90% error is determined from the standard error by the relation

$$E_{90} = 1.6449\sigma \qquad (5\text{-}13)$$

This measure of precision is valuable in deciding the worth of a given set of measurements, since it indicates the size of the residual that is exceeded by only 10% of the residuals. In Example 5-2, E_{90} is ±0.0118 rev. Actually, a total of 88% of the residuals fall in the range between -0.0118 and $+0.0118$ rev.

5-7. PROPAGATION OF RANDOM ERRORS

Thus far in this chapter consideration has been given to the measures of precision of directly measured quantities. In Example 5-1 a line has been repeatedly measured; in Example 5-2 the area of a figure was measured several times; in Example 5-3 a difference in elevation was measured 15 times. In these three instances the measurement was presumed to have been obtained by direct application of the measuring instrument. It could be argued correctly, of course, that a difference in elevation as illustrated by Example 5-3 is the result of the difference between two direct observations of the level rod. The question then arises: If the standard error of each rod reading is known, what then *is* the standard error of the difference in elevation obtained from the rod readings?

The difference in elevation discussed above is what is called a "derived quantity," or a computed quantity based on directly observed quantities. In order to investigate how random errors propagate and what the standard error of a derived quantity is, consider a very simple case of the measurement of the line AC of Fig. 5-8 in two sections AB and BC. Letting $X = AB$ and $Y = BC$, then the total length is $U = AC$. But $U = X + Y$, and U is considered to be a derived quantity. Suppose that X and Y are both measured several times and that a mean value for each, designated $\bar{X}$ and $\bar{Y}$, is obtained. Now denoting the difference between the mean and the measured values of X_i

Figure 5-8. Sum of two directly measured quantities.

as x_i, and the difference between the mean and the measured values of Y_i as y_i, the x's and y's are seen to be residuals. Then

$$X_1 = \bar{X} + x_1 \qquad\qquad Y_1 = \bar{Y} + y_1$$
$$X_2 = \bar{X} + x_2 \quad \text{and} \quad Y_2 = \bar{Y} + y_2$$
$$\vdots \qquad\qquad\qquad\qquad \vdots$$
$$X_n = \bar{X} + x_n \qquad\qquad Y_n = \bar{Y} + y_n$$

The subscript i above indicates any one of a series of quantities from 1 to n. If $\bar{U}$ denotes the most probable value of the sum of the two distances obtained by adding $\bar{X} + \bar{Y}$, and if u_i denotes the difference between $\bar{U}$ and the value U_i obtained by adding the measurements $X_i + Y_i$, then

$$U_1 = \bar{U} + u_1 = X_1 + Y_1 = \bar{X} + x_1 + \bar{Y} + y_1$$
$$U_2 = \bar{U} + u_2 = X_2 + Y_2 = \bar{X} + x_2 + \bar{Y} + y_2$$
$$\vdots$$
$$U_n = \bar{U} + u_n = X_n + Y_n = \bar{X} + x_n + \bar{Y} + y_n$$

or

$$\bar{U} + u_1 = \bar{X} + x_1 + \bar{Y} + y_1$$
$$\bar{U} + u_2 = \bar{X} + x_2 + \bar{Y} + y_2$$
$$\vdots$$
$$\bar{U} + u_n = \bar{X} + x_n + \bar{Y} + y_n$$

But since

$$\bar{U} = \bar{X} + \bar{Y}$$

then

$$\left.\begin{aligned} u_1 &= x_1 + y_1 \\ u_2 &= x_2 + y_2 \\ &\vdots \\ u_n &= x_n + y_n \end{aligned}\right\} \tag{5-14}$$

Equation (5-14) states that if AB of Fig. 5-8 is measured with a residual of, say, $+0.06$ ft and BC is measured with a residual of $+0.03$ ft, the sum of AB and BC will have a residual of $+0.09$ ft; that is, $x_i = +0.06$ ft, $y_i = +0.03$ ft, and $u_i = +0.09$ ft.

Squaring both sides of Eq. (5-14) and adding gives

$$u_1^2 = x_1^2 + 2x_1 y_1 + y_1^2$$
$$u_2^2 = x_2^2 + 2x_2 y_2 + y_2^2$$
$$\vdots$$
$$u_n^2 = x_n^2 + 2x_n y_n + y_n^2$$
$$\overline{\sum u^2 = \sum x^2 + 2\sum xy + \sum y^2} \tag{5-15}$$

Equation (5-15) contains the sum of the products term $2\sum xy$, which must be examined. If x_i is a residual, $\sum x = 0$ and likewise $\sum y = 0$. If $\sum xy = 0$, then this term of Eq. (5-15) can be dropped. In making two field measurements, one of $X_i = AB$ and the other of $Y_i = BC$, the first measurement should have no effect on the second. In other words the two measurements are independent and uncorrelated. If the measurements are independent and uncorrelated, so are their residuals. Thus x_i and y_i are assumed to be uncorrelated, and $\sum xy$ tends toward zero. Equation (5-15) can then be written

$$\sum u^2 = \sum x^2 + \sum y^2 \tag{5-16}$$

Dividing both sides by $n - 1$ gives

$$\frac{\sum u^2}{n - 1} = \frac{\sum x^2}{n - 1} + \frac{\sum y^2}{n - 1}$$

or by Eq. (5-9),

$$\sigma_U^2 = \sigma_X^2 + \sigma_Y^2$$

and

$$\sigma_U = \sqrt{\sigma_X^2 + \sigma_Y^2} \tag{5-17}$$

Thus the standard error of the sum of two quantities is equal to the square root of the sum of the squares of the standard errors of the individual quantities. This concept can be extended to the sum of any number of quantities that are not correlated.

Example 5-4

A line is measured in six sections, and the lengths of the sections together with the standard error in the length of each section are given in the accompanying tabulation. What is the standard error of the length of the line?

SECTION	LENGTH (ft)	σ (ft)
1	961.22	± 0.044
2	433.12	± 0.031
3	1545.90	± 0.060
4	355.40	± 0.021
5	1252.54	± 0.100
6	320.40	± 0.075

Solution: The sum of the lengths is 4868.58 ft, and the standard error of the total length of the line is, by Eq. (5-17),

$$\sigma_s = \sqrt{0.044^2 + 0.031^2 + 0.060^2 + 0.021^2 + 0.100^2 + 0.075^2}$$
$$= \pm 0.150 \text{ ft}$$

Example 5-5

A line is laid out using a 200-ft tape. The standard error of each tape length laid out is assumed to be ± 0.03 ft. If 12 tape lengths are laid out, giving a total length of 2400 ft, what is the standard error of the length of the line?

Solution: Since each tape length has the same standard error, then Eq. (5-17) can be stated as

$$\sigma_s = \sigma\sqrt{n} \tag{5-18}$$

Hence the standard error of the 2400-ft distance is $0.03\sqrt{12} = \pm 0.104$ ft.

One measure of precision that is of importance in surveying is the standard error of the mean of a set of like measurements. The standard error of each measurement of the set is given by Eq. (5-9). The standard error of the sum of all the measurements of the set is given by Eq. (5-18) of Example 5-5. Then, since the mean is equal to the sum divided by the number of measurements, the standard error of the mean is given by

$$\sigma_M = \frac{\sigma\sqrt{n}}{n} = \frac{\sigma}{\sqrt{n}} \tag{5-19}$$

Equation (5-19) can be combined with Eq. (5-9) to give a different form of σ_M. Thus

$$\sigma_M = \sqrt{\frac{\sum v^2}{n(n-1)}} \tag{5-20}$$

Example 5-6

Compute the standard error of the mean of Example 5-3.

Solution: The standard error of the measurements is ± 0.0055 ft. Then by Eq. (5-19), the standard error of the mean is $0.0055/\sqrt{15} = \pm 0.0014$ ft.

The standard error of the mean indicates what spread would be expected if the set of measurements were to be repeated several times and the means were analyzed. If the set of measurements of Example 5-3 were to be repeated several times under the same field conditions, then, as shown in Example 5-6, about 68% of the means should be expected to fall within ± 0.0014 ft of the mean of the means.

The analysis employed to investigate the propagation of standard error through a very simple function $U = X + Y$ can be applied to any function of a set of independent uncorrelated variables whose standard errors are known.

Let U be some function of measured quantities $X, Y, Z, \ldots, Q$; that is,

$$U = f(X, Y, Z, \ldots, Q)$$

If each independent variable is allowed to change by a small amount dX, $dY, dZ, \ldots, dQ$, then the quantity U will change by an amount dU given by the expression from the calculus

$$dU = \frac{\partial U}{\partial X} dX + \frac{\partial U}{\partial Y} dY + \frac{\partial U}{\partial Z} dZ + \cdots + \frac{\partial U}{\partial Q} dQ$$

in which $\partial U/\partial X$ denotes the partial derivative of U with respect to X, and likewise for the other variables. Applying this to a set of measurements, let $x_i = dX_i$, $y_i = dY_i$, $z_i = dZ_i, \ldots, q_i = dQ_i$ be a set of residuals of the measured quantities. Then, letting $u_i = dU_i$,

$$\left. \begin{aligned} u_1 &= \frac{\partial U}{\partial X} x_1 + \frac{\partial U}{\partial Y} y_1 + \frac{\partial U}{\partial Z} z_1 + \cdots + \frac{\partial U}{\partial Q} q_1 \\ u_2 &= \frac{\partial U}{\partial X} x_2 + \frac{\partial U}{\partial Y} y_2 + \frac{\partial U}{\partial Z} z_2 + \cdots + \frac{\partial U}{\partial Q} q_2 \\ &\;\;\vdots \\ u_n &= \frac{\partial U}{\partial X} x_n + \frac{\partial U}{\partial Y} y_n + \frac{\partial U}{\partial Z} z_n + \cdots + \frac{\partial U}{\partial Q} q_n \end{aligned} \right\} \tag{5-21}$$

Squaring both sides of Eq. (5-21) and adding gives

$$u_1^2 = \left(\frac{\partial U}{\partial X}\right)^2 x_1^2 + 2\left(\frac{\partial U}{\partial X}\right)\left(\frac{\partial U}{\partial Y}\right)x_1 y_1 + \cdots + \left(\frac{\partial U}{\partial Y}\right)^2 y_1^2 + \cdots + \left(\frac{\partial U}{\partial Q}\right)^2 q_1^2$$

$$u_2^2 = \left(\frac{\partial U}{\partial X}\right)^2 x_2^2 + 2\left(\frac{\partial U}{\partial X}\right)\left(\frac{\partial U}{\partial Y}\right)x_2 y_2 + \cdots + \left(\frac{\partial U}{\partial Y}\right)^2 y_2^2 + \cdots + \left(\frac{\partial U}{\partial Q}\right)^2 q_2^2$$

$$\vdots$$

$$u_n^2 = \left(\frac{\partial U}{\partial X}\right)^2 x_n^2 + 2\left(\frac{\partial U}{\partial X}\right)\left(\frac{\partial U}{\partial Y}\right)x_n y_n + \cdots + \left(\frac{\partial U}{\partial Y}\right)^2 y_n^2 + \cdots + \left(\frac{\partial U}{\partial Q}\right)^2 q_n^2$$

$$\sum u^2 = \left(\frac{\partial U}{\partial X}\right)^2 \sum x^2 + 2\left(\frac{\partial U}{\partial X}\right)\left(\frac{\partial U}{\partial Y}\right)\sum xy + \cdots + \left(\frac{\partial U}{\partial Y}\right)^2 \sum y^2 + \cdots + \left(\frac{\partial U}{\partial Q}\right)^2 \sum q^2$$

$$(5\text{-}22)$$

in which some of the squared and cross-product terms have been omitted for the sake of simplicity. If the measured quantities are independent and are not correlated with one another, the cross products tend toward zero. Dropping these cross products and then dividing both sides of Eq. (5-22) by $n - 1$ gives

$$\frac{\sum u^2}{n-1} = \left(\frac{\partial U}{\partial X}\right)^2 \frac{\sum x^2}{n-1} + \left(\frac{\partial U}{\partial Y}\right)^2 \frac{\sum y^2}{n-1}$$

$$+ \left(\frac{\partial U}{\partial Z}\right)^2 \frac{\sum z^2}{n-1} + \cdots + \left(\frac{\partial U}{\partial Q}\right)^2 \frac{\sum q^2}{n-1}$$

or

$$\sigma_U^2 = \left(\frac{\partial U}{\partial X}\right)^2 \sigma_X^2 + \left(\frac{\partial U}{\partial Y}\right)^2 \sigma_Y^2 + \left(\frac{\partial U}{\partial Z}\right)^2 \sigma_Z^2 + \cdots + \left(\frac{\partial U}{\partial Q}\right)^2 \sigma_Q^2 \quad (5\text{-}23)$$

The general expression for propagation of standard error through any function expressed by Eq. (5-23) can then be applied to any special function.

Let $U = X + Y + Z$, in which X, Y, and Z are three measured quantities. Then, from Eq. (5-23),

$$\sigma_U = \sqrt{\sigma_X^2 + \sigma_Y^2 + \sigma_Z^2}$$

which is identical with Eq. (5-17) for three variables.

Let $U = X - Y$, in which X and Y are two measured quantities. Then, from Eq. (5-23),

$$\sigma_U = \sqrt{\sigma_X^2 + \sigma_Y^2} \quad (5\text{-}24)$$

Let $U = XY$, in which X and Y are two measured quantities. Then, from Eq. (5-23),

$$\sigma_U = \sqrt{Y^2\sigma_X^2 + X^2\sigma_Y^2} \qquad (5\text{-}25)$$

Let $U = AX$, in which A is a constant and X is a measured quantity. Then, from Eq. (5-23),

$$\sigma_U = A\sigma_X \qquad (5\text{-}26)$$

Example 5-7

If the standard error of a backsight reading from a setup of a level is ±0.02 ft, and the standard error of the foresight reading from the same setup is also ±0.02 ft, what is the standard error of the difference in elevation between the two turning points on which the readings are taken?

Solution: The difference in elevation between the two turning points is equal to the backsight minus the foresight. Then by Eq. (5-24) the standard error of the difference in elevation is

$$\sigma = \sqrt{0.02^2 + 0.02^2} = \pm0.028 \text{ ft}$$

Example 5-8

The dimensions of a rectangular field are measured with a steel tape and found to be 550.00 ft with a standard error of ±0.07 ft and 800.00 ft with a standard error of ±0.12 ft. What are the area of the field and the standard error of the area?

Solution: The area of the field is $550.00 \times 800.00 = 440,000$ ft^2. The standard error of the area is, by Eq. (5-25),

$$\sigma = \sqrt{550^2 \times 0.12^2 + 800^2 \times 0.07^2} = \pm86.6 \text{ ft}^2$$

Example 5-9

What is the standard error of the determination of the area in Example 5-8, in acres?

Solution: Since one acre equals 43,560 ft^2, the standard error of the area in acres is, by Eq. (5-26),

$$\sigma = \left(\frac{1}{43,560}\right)(\pm86.6) = \pm0.00199 \text{ acre}$$

Example 5-10

The diameter D of the base of a cone is measured as 3.002 in., with a standard error of ± 0.0005 in. The height h of the cone is measured as 5.45 in. with a standard error of ± 0.01 in. What are the volume of the cone and the standard error of the volume?

Solution: The volume of a cone is

$$V = \frac{\pi D^2 h}{12} = \frac{\pi \times 3.002^2 \times 5.45}{12} = 12.86 \text{ in.}^3$$

The standard error of the volume can be expressed, according to Eq. (5-23), as follows:

$$\sigma_V = \sqrt{\left(\frac{\partial V}{\partial D} \sigma_D\right)^2 + \left(\frac{\partial V}{\partial h} \sigma_h\right)^2}$$

Then

$$\frac{\partial V}{\partial D} = \frac{\pi D h}{6} = \frac{\pi \times 3.002 \times 5.45}{6} = 8.567 \text{ in.}^3/\text{in.}$$

and

$$\frac{\partial V}{\partial h} = \frac{\pi D^2}{12} = \frac{\pi \times 3.002^2}{12} = 2.359 \text{ in.}^3/\text{in.}$$

Hence the standard error of the volume is

$$\sigma_V = \sqrt{(8.567 \times 0.0005)^2 + (2.359 \times 0.01)^2} = \pm 0.024 \text{ in.}^3$$

Example 5-11

Two sides and the included angle of a triangle were measured with the following results: $a = 472.58$ ft and $\sigma_a = \pm 0.09$ ft; $b = 214.55$ ft and $\sigma_b = \pm 0.06$ ft; $C = 37° 15'$ and $\sigma_C = \pm 30''$. Compute the area of the triangle, in square feet, and compute the standard error of the area.

Solution: The area A of the triangle is given by the relationship

$$A = \tfrac{1}{2}ab \sin C = \tfrac{1}{2} \times 472.58 \times 214.55 \sin 37° 15' = 30{,}686 \text{ ft}^2$$

The standard error of the area is, by Eq. (5-23),

$$\sigma_A = \sqrt{\left(\frac{\partial A}{\partial a} \sigma_a\right)^2 + \left(\frac{\partial A}{\partial b} \sigma_b\right)^2 + \left(\frac{\partial A}{\partial C} \sigma_C\right)^2}$$

Then

$$\frac{\partial A}{\partial a} = \frac{1}{2} b \sin C = \frac{1}{2} \times 214.55 \sin 37° 15' = 64.93 \text{ ft}^2/\text{ft}$$

$$\frac{\partial A}{\partial b} = \frac{1}{2} a \sin C = \frac{1}{2} \times 472.58 \sin 37° 15' = 143.03 \text{ ft}^2/\text{ft}$$

and

$$\frac{\partial A}{\partial C} = \frac{1}{2} ab \cos C = \frac{1}{2} \times 472.58 \times 214.55 \cos 37° 15' = 40,354 \text{ ft}^2/\text{rad}$$

Since σ_C expressed in radians (rad) is $30 \times 0.00000485 = 0.0001455$ rad, the standard error of the computed area is

$$\sigma_V = \sqrt{(64.93 \times 0.09)^2 + (143.03 \times 0.06)^2 + (40,354 \times 0.0001455)^2}$$
$$= \pm 12 \text{ ft}^2$$

5-8. WEIGHTED MEASUREMENTS The *weight* of a measurement can be thought of as the value or the worth of that measurement relative to any other measurement. In order to investigate the relationship between the weight of a given observation and the measures of precision σ and h, consider the length of a line to have been measured 15 times by using the same measuring technique, the same degree of refinement, the same experienced personnel, and so on. Suppose that the 15 values are as shown in the accompanying tabulation.

MEASUREMENT NUMBER	VALUE (ft)	MEASUREMENT NUMBER	VALUE (ft)	MEASUREMENT NUMBER	VALUE (ft)
1	432.33	6	432.33	11	432.36
2	432.33	7	432.36	12	432.33
3	432.36	8	432.31	13	432.33
4	432.33	9	432.33	14	432.36
5	432.31	10	432.36	15	432.33

Each of the listed values has equal weight, because all conditions of measurement were identical. The most probable. value of the length of the line may therefore be obtained by adding all the values and dividing by 15, that is, by finding the mean value. Since the value 432.33 appears eight times, the value 432.36 appears five times, and the value 432.31 appears twice, the mean can be computed as follows:

$$8 \times 432.33 = 3458.64$$

$$5 \times 432.36 = 2161.80$$

$$\frac{2 \times 432.31 = \quad 864.62}{15} \qquad 6485.06/15 = 432.337 \text{ ft}$$

The 15 measurements can be considered equivalent to 3 measurements with the relative worths, or weights, of 8, 5, and 2. By analysis of the arithmetic in the assumed example, an expression for a *weighted mean* can be formed as follows:

$$\overline{M} = \frac{p_1 M_1 + p_2 M_2 + \cdots + p_n M_n}{p_1 + p_2 + \cdots + p_n} = \frac{\sum (pM)}{\sum p} \qquad (5\text{-}27)$$

in which $\overline{M}$ is the weighted mean of several measurements of a quantity; M_1, M_2, and so on are the measurements; and p_1, p_2, and so on are the relative weights of the measurements.

Now suppose that the standard error of each of the 15 measurements of the line in the assumed example is ± 0.02 ft. The mean of the eight values of 432.33 ft has a standard error of $\pm 0.02/\sqrt{8}$ ft by Eq. (4-12); the mean of the five values of 432.36 ft has a standard error of $\pm 0.02/\sqrt{5}$ ft, and the mean of the two values of 432.31 ft has a standard error of $\pm 0.02/\sqrt{2}$ ft. For convenient reference the values of the three means, their weights, their standard errors, and the squares of the standard errors are given in the following tabulation:

VALUE (ft)	WEIGHT p	STANDARD ERROR (ft)	σ^2 (ft^2)
432.33	8	$\pm 0.02/\sqrt{8}$	0.0004/8
432.36	5	$\pm 0.02/\sqrt{5}$	0.0004/5
432.31	2	$\pm 0.02/\sqrt{2}$	0.0004/2

The assumed example demonstrates the following very important relationship between relative weights and measures of precision: The weight of a measured value is inversely proportional to the square of the standard error of the measurement, and directly proportional to the square of the precision modulus, that is,

$$p \propto \frac{1}{\sigma^2} \propto h^2 \qquad (5\text{-}28)$$

In the preceding discussion the value ± 0.02 is assumed to be the standard error of each individual measurement of the set, and each measurement has the same weight. This quantity is sometimes referred to as the standard error of unit weight, and is designated σ_0.

The standard error of unit weight of a set of n measurements of the same quantity, each having different weights, is given by the expression

$$\sigma_0 = \sqrt{\frac{\sum (pv^2)}{n - 1}} \qquad (5\text{-}29)$$

in which the v's are the differences between the weighted measurement and the weighted mean; that is, $v_i = M_i - \overline{M}$. The standard error of each weighted value can then be expressed in terms of its weight and the standard error of unit weight, as follows:

$$\sigma_p = \sigma_0 \sqrt{\frac{1}{p}} \qquad (5\text{-}30)$$

in which σ_p is the standard error of a weighted value, and p is the weight. Of course, if the weight of a measurement is unity, then $\sigma = \sigma_0$. The quantity $1/p$, which is the reciprocal of the weight, is referred to as a *weight number* or *cofactor*.

Example 5-12

Three lines of levels were run between BM20 and BM21 by different routes, but each setup and each rod reading were made with the same degree of care. On the first line, 17 setups were required and the measured difference in elevation was 29.492 ft; on the second line, 9 setups were required and the measured difference in elevation was 29.440 ft; on the third line, 10 setups were required and the measured difference in elevation was 29.480 ft. What is the weighted mean of the measured differences in elevation between BM20 and BM21?

Solution: According to Eq. (5-18) in Example 5-5, the standard error of the determination of the difference in elevation between two points is proportional to the square root of the number of setups required. Therefore the standard errors of lines 1, 2, and 3 are in proportion to the quantities $\sqrt{17}$, $\sqrt{9}$, and $\sqrt{10}$. The relative weights of the values of the difference in elevation obtained from lines 1, 2, and 3, are, by Eq. (5-28), $\frac{1}{17}$, $\frac{1}{9}$, and $\frac{1}{10}$, respectively. Then by Eq. (5-27) the weighted mean of the measured differences in elevation is

$$\overline{M} = \frac{(\frac{1}{17})(29.492) + (\frac{1}{9})(29.440) + (\frac{1}{10})(29.480)}{\frac{1}{17} + \frac{1}{9} + \frac{1}{10}} = 29.466 \text{ ft}$$

If the weighted mean is computed by Eq. (5-27), the standard error of the weighted mean is given by the expression

$$\overline{\sigma} = \sqrt{\frac{\sum (pv^2)}{\sum p(n - 1)}} \qquad (5\text{-}31)$$

Example 5-13

Determine the standard error of the weighted mean difference in elevation in Example 5-12.

Solution: The calculations may be arranged as follows:

$$v_1 = 29.492 - 29.466 = +0.026 \qquad v_1^2 = 0.000676$$
$$v_2 = 29.440 - 29.466 = -0.026 \qquad v_2^2 = 0.000676$$
$$v_3 = 29.480 - 29.466 = +0.014 \qquad v_3^2 = 0.000196$$

$$p_1 v_1^2 = \frac{1}{17} \times 0.000676 = 0.0000398 \qquad \sum p = 0.2699$$

$$p_2 v_2^2 = \frac{1}{9} \times 0.000676 = 0.0000751 \qquad \bar{\sigma} = \sqrt{\frac{0.0001345}{0.2699(3-1)}} = \pm 0.0158 \text{ ft}$$

$$p_3 v_3^2 = \frac{1}{10} \times 0.000196 = 0.0000196$$

$$\sum (pv^2) = \overline{0.0001345}$$

Example 5-14

What is the standard error of each of the measured differences in elevation in Example 5-12?

Solution: By Eq. (5-29),

$$\sigma_0 = \sqrt{\frac{0.0001345}{2}} = \pm 0.0082 \text{ ft}$$

Then by Eq. (5-30),

$$\sigma_1 = 0.0082\sqrt{17} = \pm 0.0338 \text{ ft}; \qquad \sigma_2 = 0.0082\sqrt{9} = \pm 0.0246 \text{ ft};$$
$$\sigma_3 = 0.0082\sqrt{10} = \pm 0.0259 \text{ ft}$$

5-9. SIMPLE ADJUSTMENTS OF MEASUREMENTS If the three angles in a triangle were to be measured by methods described in Chapter 4 and each angle was measured with the same degree of precision, then each of the three measured angles would carry the same weight. If the sum of the three measured angles is not 180°, then it is expected that each measured angle would be corrected by the same amount. The reasonableness of this

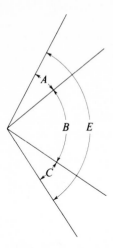

Figure 5-9. Angles measured about a point.

adjustment is proved by an extended investigation of the principle of least squares developed in Section 5-3.

If three angles are measured about a point, as the angles A, B, and C in Fig. 5-9, and the total angle, as E, is also measured and if all measurements are made with the same precision, $A + B + C$ should equal E. If the sum of the measured values of A, B, and C does not equal the measured value of E, then each of the measured angles must receive the same amount of correction, because their weights are identical.

Example 5-15

In Fig. 5-9, the measured angles are $A = 22° 10' 10''$, $B = 71° 05' 50''$, $C = 24° 16' 00''$, and $E = 117° 31' 00''$. What are the corrected angles?

Solution: The computations may be arranged as in the accompanying tabulation.

ANGLE	MEASURED VALUE	CORRECTION	ADJUSTED VALUE
A	$22° 10' 10''$	$-15''$	$22° 09' 55''$
B	$71° 05' 50''$	$-15''$	$71° 05' 35''$
C	$24° 16' 00''$	$-15''$	$24° 15' 45''$
Sum $=$	$117° 32' 00''$		$117° 31' 15''$
E	$117° 31' 00''$	$+15''$	$117° 31' 15''$
Error $=$	$1' 00''$		

If a series of angles closing the horizon about a point are measured, their total should be 360°. If the total is not 360° and all angles were measured with the same degree of refinement, each should be corrected by an equal amount. Obviously this amount is found by taking the difference between 360° and the measured total and dividing that difference by the number of angles.

5-10. ADJUSTMENT OF WEIGHTED MEASUREMENTS The corrections to be applied to uncorrelated weighted measurements are inversely proportional to their weights. Equation (5-6) is the statement of the condition of least squares which applies to measurements of equal or unit weight. Assume now a set of measurements $M_1, M_2, \ldots, M_n$ with varying weights $p_1, p_2, \ldots, p_n$ and the corresponding residuals $v_1, v_2, \ldots, v_n$. By Eq. (5-28), p is directly proportional to h^2. The probability P that $v_1, v_2, \ldots, v_n$ will occur in the set is as follows:

$$P_{v_1} = y_1 \, \Delta v = k_1 e^{-h_1^2 v_1^2} \, \Delta v$$

$$P_{v_2} = y_2 \, \Delta v = k_2 e^{-h_2^2 v_2^2} \, \Delta v$$

$$\vdots$$

$$P_{v_n} = y_n \, \Delta v = k_n e^{-h_n^2 v_n^2} \, \Delta v$$

Also, since the probability that residuals will occur simultaneously in the set is equal to the product of their separate probabilities,

$$P_{(v_1, v_2, \ldots, v_n)} = (k_1 e^{-h_1^2 v_1^2} \, \Delta v)(k_2 e^{-h_2^2 v_2^2} \, \Delta v) \cdots (k_n e^{-h_n^2 v_n^2} \, \Delta v)$$

or

$$P_{(v_1, v_2, \ldots, v_n)} = (k_1 k_2 \cdots k_n) \Delta v^n e^{-(h_1^2 v_1^2 + h_2^2 v_2^2 + \cdots + h_n^2 v_n^2)}$$

Obviously, if P is to be a maximum in order to give the most probable value from the measurements, the negative exponent of e must be made a minimum; that is,

$$h_1^2 v_1^2 + h_2^2 v_2^2 + \cdots + h_n^2 v_n^2 = \text{a minimum}$$

Since p is proportional to h^2, then

$$p_1 v_1^2 + p_2 v_2^2 + \cdots + p_n v_n^2 = \text{a minimum}$$

or

$$\sum (pv^2) = \text{minimum} \tag{5-32}$$

or

$$[pvv] = \text{minimum}$$

The residual v for any value is defined as the measured value minus the most probable value, or $v_i = M_i - M$, in which M_i is the ith measured value and M is the most probable value. Thus the most probable value, or the adjusted value, is $M = M_i - v_i$. It can be seen that if the quantity $-v$ is added to the measured value, the adjusted value is obtained. But, $-v$ must be the correction to the measured value. Thus

$$c = -v \qquad\qquad (5\text{-}33)$$

in which c is the correction.

Note that c^2 has both the same magnitude and the same algebraic sign as v^2. Then, if Eq. (5-32) is to be satisfied, the corrections must be inversely proportional to the weights. For example, if one measurement is, say, five times as good as another like measurement, then the correction to the first measurement should be only one-fifth of the correction to the second measurement. As an illustration of the adjustment of weighted measurements, consider the three angles of the triangle in Fig. 5-10. Assume that the mean value of six measurements of angle A is $32° 40' 15''$, the mean value of three measurements of angle B is $84° 33' 15''$, and a single measurement of angle C is $62° 46' 00''$. Also assume that each measurement of an angle is made with the same degree of refinement, and that the standard error of each single measurement is $\pm 1'$. Then, by Eq. (5-19), the standard error of the mean of the six measurements of A is $\pm 1'/\sqrt{6}$; the standard error of the mean of the three measurements of B is $\pm 1'/\sqrt{3}$; and the standard error of the measurement of C is $\pm 1'/\sqrt{1}$. So by Eq. (5-28) the relative weights are $p_A = 6$, $p_B = 3$, and $p_C = 1$. In general the number of repetitions of an angle indicates its relative weight.

The computations for determining the adjusted angles may be arranged as shown in the accompanying tabulation. Since the sum of the observed angles is $179° 59' 30''$, the total of the three corrections must be $30''$. Also, the

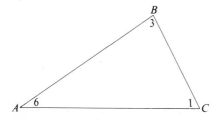

Figure 5-10. Angles in a triangle.

ANGLE	NUMBER OF MEASURE-MENTS	MEAN VALUE OF ANGLE	RELATIVE CORRECTIONS	CORRECTIONS	ADJUSTED VALUE
A	6	32° 40′ 15″	$\frac{1}{6}$ or $\frac{1}{9}$	$\frac{1}{9} \times 30'' = 3.3''$	32° 40′ 18.3″
B	3	84° 33′ 15″	$\frac{2}{6}$ or $\frac{2}{9}$	$\frac{2}{9} \times 30'' = 6.7''$	84° 33′ 21.7″
C	1	62° 46′ 00″	$\frac{6}{6}$ or $\frac{6}{9}$	$\frac{6}{9} \times 30'' = 20.0''$	62° 46′ 20.0″
		179° 59′ 30″	$\frac{9}{6}$ $\frac{9}{9}$		180° 00′ 00.0″

correction to the mean value of angle A is only one-half that of the correction to the mean value of angle B and is only one-sixth that of the correction to the mean value of angle C. The relative corrections to the three angles A, B, and C are therefore $\frac{1}{6}$, $\frac{1}{3}$, and $\frac{1}{1}$ or $\frac{1}{6}$, $\frac{2}{6}$, and $\frac{6}{6}$. Now the sum of these relative corrections should be found, and the numerator of this sum, or 9, should be used as the denominator of each of a second set of relative corrections. The sum of this second set should be unity. The corrections to the measured angles are then obtained by multiplying the second set of relative corrections by the total error, and the adjusted angles are determined.

5-11. ADJUSTMENT OF A LEVEL CIRCUIT

The adjustment of a level circuit is a special case of adjusting weighted observations. When several lines of levels between two points are compared, the weight of each line is inversely proportional to the number of setups, as shown in Example 5-12. It was also pointed out that the standard error of a line is proportional to the square root of the length of the line. Then, by Eq. (5-28) the weight of a line of levels is inversely proportional to the distance along the line. This fact is the basis of the following principle for adjusting a line or circuit of levels: Corrections to measured elevations of points are proportional to the distances from the beginning of the line or circuit to the points, or are proportional to the numbers of setups between the beginning of the line or circuit and the points.

Example 5-16

A line of levels is run from BM30 to BM33, the elevations of which have been previously established. As shown in Fig. 5-11, two intermediate bench marks, BM31 and BM32 are established. The fixed elevation of BM30 is 453.52 ft, and the fixed elevation of BM33 is 482.10 ft. As a result of the field work the eleva-

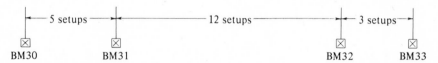

Figure 5-11. Line of levels between two points.

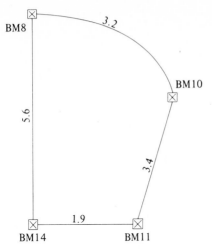

BM8

3.2

5.6

BM10

3.4

1.9

BM14   BM11

Figure 5-12. Circuit of level lines.

tion of BM31 is 440.98 ft; that of BM32 is 464.25 ft; and that of BM33 is 482.23 ft. The error of closure is therefore +0.13 ft. There are a total of 20 setups distributed as shown in Fig. 5-11. The relative corrections for BM's 31, 32, and 33 are therefore $\frac{5}{20}$, $\frac{17}{20}$, and $\frac{20}{20}$, respectively. The correction to the elevation of BM31 is $\frac{5}{20} \times 0.13 = 0.03$ ft; that to the elevation of BM32 is $\frac{17}{20} \times 0.13 = 0.11$ ft; and that to the elevation of BM33 is $\frac{20}{20} \times 0.13 = 0.13$ ft. The adjusted elevations of BM's 31, 32, and 33 are, respectively, 440.95, 464.14, and 482.10 ft.

Example 5-17

A circuit of levels is shown in Fig. 5-12. It starts and ends at BM8 and passes through BM10, BM11, and BM14. The distances between the several bench-marks, in thousands of feet, are given on the diagram. Also the known eleva-tion of BM8 is 445.108 ft, and the measured elevations of the benchmarks are as shown in the accompanying tabulation. What are the adjusted elevations of BM's 10, 11, and 14?

BENCHMARK	MEASURED ELEVATION (ft)	DISTANCE FROM BM8 (1000) (ft)	CORRECTION (ft)	ADJUSTED ELEVATION (ft)
8	445.108	0	0	445.108 (fixed)
10	430.166	3.2	$3.2/14.1 \times 0.074 = 0.017$	430.183
11	433.421	6.6	$6.6/14.1 \times 0.074 = 0.035$	433.456
14	440.380	8.5	$8.5/14.1 \times 0.074 = 0.045$	440.425
8	445.034	14.1	$14.1/14.1 \times 0.074 = 0.074$	445.108

Solution: The closing error on BM8 is 445.108 − 445.034 = 0.074 ft. The corrections to the measured elevations of the benchmarks are computed as shown in the table. Since the measured elevation of BM8 is too small, the corrections are added to the measured elevations of the other benchmarks.

5-12. OTHER ADJUSTMENTS The necessity for applying corrections to observations that are subject to random errors enters into all types of surveying measurements. The adjustment of the various types of measurements will be discussed in the separate chapters where the need arises. Appendix A shows the adjustment of elementary surveying measurements by the principle of least squares.

PROBLEMS

5-1. If eight coins are tossed in the air at random, what is the probability that half of the coins will come up heads?

5-2. In Problem 5-1, what is the probability that three coins will come up tails?

5-3. Sixteen coins are tossed in the air at random. What is the probability that half of the coins will come up heads?

5-4. In Problem 5-3, what is the probability that five coins will come up tails?

5-5. Plot a histogram similar to those shown in Fig. 5-1 for the coins in Problem 5-1.

5-6. Four dice are rolled at random. Determine the probability that the sum of the marks on the upper face of the dice will be: (a) four; (b) five; and (c) six.

5-7. Plot a histogram to show the theoretical frequency distribution of the sum of the marks on three dice if they are rolled a total of 800 times.

5-8. The difference in elevation between two fixed points was measured repeatedly for a total of 12 times with the following results:

MEASUREMENT NUMBER	DIFFERENCE IN ELEVATION (m)	MEASUREMENT NUMBER	DIFFERENCE IN ELEVATION (m)
1	7.3522	7	7.3530
2	7.3532	8	7.3520
3	7.3526	9	7.3520
4	7.3514	10	7.3521
5	7.3522	11	7.3531
6	7.3510	12	7.3528

Compute the mean and standard error of the mean of these observations.

5-9. Compute the probable error and the 90% error of the set of measurements of Problem 5-8.

5-10. As an experiment in reaction time, five periods of a pendulum were timed by a stop watch. This measurement was repeated 100 times with the following results, given in seconds of time.

8.08	8.08	8.02	7.90	7.92
8.00	8.05	8.10	8.01	7.94
7.86	8.18	8.19	7.98	8.07
7.79	7.94	7.98	8.23	7.81
8.04	8.00	8.02	8.15	8.12
7.87	7.95	7.75	7.93	7.89
7.93	8.00	8.05	7.99	8.00
8.12	8.00	8.00	7.98	8.01
8.08	8.15	8.02	7.98	8.05
7.92	7.98	8.02	7.86	8.00
7.91	7.90	8.00	8.15	8.10
7.95	8.00	7.91	7.82	7.80
8.00	8.08	7.98	8.10	7.98
7.93	7.93	8.11	8.09	8.00
8.01	7.98	8.00	8.04	7.98
8.15	8.14	8.15	8.05	7.90
8.05	8.15	7.97	8.25	8.08
8.01	8.01	7.99	8.02	7.95
8.02	7.89	7.88	8.02	8.05
7.85	8.07	8.10	7.95	8.05

(a) Plot these observations in the form of a histogram with a class interval of 0.05 s, beginning with 7.75 s. (b) Compute the mean of the observed values, the standard error of the set of observations, the standard error of the mean, and the probable error of the mean. (c) Using the standard error of the set of observations, compute ordinates to the normal distribution curve representing the set. Plot the curve on the histogram in (a). (d) Locate the positions of $+\sigma$ and $-\sigma$ on the curve in (c).

5-11. Assume that the standard error of a rod reading in leveling due to multiple sources of random errors is ± 0.005 ft. The distance between a fixed benchmark and one whose elevation is to be determined is 2800 ft. If the average length of sight is 120 ft, what is the standard error of the measured value of the elevation of the new benchmark?

5-12. A line was measured to be 1170 m using a 30-m tape. If the standard error of each tape length is ± 0.003 m, what is the standard error of the measurement?

5-13. A line AE is broken into sections for measurement with a tape. The results are $AB = 440.00$ ft ± 0.040 ft, $BC = 1052.00$ ft ± 0.055 ft, $CD = 570.00$ ft ± 0.056 ft, and $DE = 527.75$ ft ± 0.050 ft. Compute the length of the line and the standard error of the length.

5-14. If the line AE in Problem 5-13 is measured with an EDM and the length is 2589.50 ft ± 0.060 ft, what is the most probable length of the line?

5-15. The difference in elevation between two benchmarks was measured by each of three field parties using different kinds of leveling instruments. The results are as follows: Party 1, $DE = 17.42$ ft ± 0.07 ft; Party 2, $DE = 17.40$ ft ± 0.04 ft; Party 3, $DE = 17.46$ ft ± 0.05 ft. What is the most probable difference in elevation between the two benchmarks.

5-16. What is the standard error of the most probable difference in elevation computed in Problem 5-15.

5-17. The volume of a pyramid is $\frac{1}{3}bh$. The base of a pyramid is a rectangle with measured sides 10.2 ft ± 0.2 ft and 24.5 ft ± 0.3 ft. The height of the pyramid is measured as 6.32 ft ± 0.15 ft. Compute the volume of the pyramid and the standard error of the volume.

5-18. Three sides of a triangle are measured with the following results: $a = 1552.65$ m ± 0.025 m, $b = 682.22$ m ± 0.030 m, $c = 1888.60$ m ± 0.040 m. Compute the angles in the triangle together with the standard errors of the angles.

5-19. What is the standard error of the area of the triangle of Problem 5-18?

5-20. Two sides and the included angle of a triangle are measured as follows: $a = 376.85$ ft ± 0.045 ft, $b = 646.80$ ft ± 0.060 ft, and angle $C = 47° 14' ± 30''$. Compute the area of the triangle and the standard error of the area.

5-21. One side and two adjacent angles of a triangle are measured in order to determine the lengths of the other two sides, because the vertex opposite the measured side is inaccessible. The side c measures 372.624 m ± 0.022 m, angle A measures $72° 26' 10'' ± 10''$, angle B measures $60° 16' 30'' ± 15''$. Compute angle C, side a, and side b. Compute the standard error of each quantity.

5-22. A line of levels is carried from BM A whose elevation is 146.522 m, to a new BM P, requiring 10 setups. The measured difference in elevation is -3.436 m. A line is carried from BM B, whose elevation is 146.851 m, to BM P, requiring six setups. The measured difference in elevation is -3.755 m. A line is carried from BM C whose elevation is 132.768 m, to BM P, requiring four setups. The measured difference in elevation is $+10.312$ m. Compute the weighted elevation of BM P and the standard error of this elevation.

5-23. What is the standard error of each of the three measured differences in elevation in Problem 5-22?

5-24. The weight of an angle is assumed to be proportional to the number of times it has been repeated (see Section 4-17). Five angles in a five-sided figure are measured with the following results:

ANGLE	OBSERVED VALUE	NUMBER OF REPETITIONS
1	86° 15′ 20″	6
2	134° 44′ 35″	2
3	75° 48′ 50″	8
4	167° 02′ 05″	6
5	76° 08′ 50″	4

Compute the adjusted values of the angles (nearest tenth of a second).

5-25. Three contiguous angles were measured separately at O, and then the total angle was measured. The results are as follows:

FROM STATION	TO STATION	OBSERVED VALUE	NUMBER OF REPETITIONS
P	Q	14° 15′ 52″	6
Q	R	26° 25′ 40″	4
R	S	60° 54′ 26″	4
P	S	101° 36′ 10″	12

Compute the adjusted angles (nearest tenth of a second).

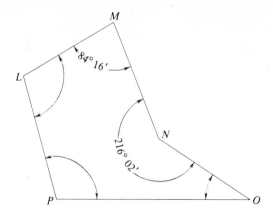

Figure 6-1. Interior angles.

first instance the graduation numbers of the horizontal circle and vernier or scale that increase in a counterclockwise direction would be used; in the second instance those that increase in a clockwise direction would be used.

The form of notes used to record the interior angles is a matter of importance, since the notes are the record of the measurements. The form of notes shown in Fig. 6-2 is a record of interior angles measured at stations M and N of Fig. 6-1, each angle being measured once with the telescope direct and once with it reversed.

The occupied station appears under the column headed *Station*. The backsight station and the foresight station appear respectively above and below under the column headed $\frac{From}{To}$ *Station*. The three readings under the column headed *Circle* are the initial backsight reading, the initial foresight reading, and the second foresight reading. The first value of the interior angle at station M from L to N, measured with the telescope direct, is obtained by subtracting $0°\,00'$ from $84°\,16'$ and is recorded under the column headed *Interior Angle* as

STATION	FROM STATION TO	CIRCLE	INTERIOR ANGLE
	L	$0°\,00'$	—
M		$84°\,16'$	$84°\,16'$
	N	$168°\,32'$	$84°\,16'$
		mean	$84°\,16'\,00''$
	M	$0°\,04'$	—
N		$216°\,06'$	$216°\,02'$
	O	$72°\,09'$	$216°\,03'$
		mean	$216°\,02'\,30''$

Figure 6-2. Notes for recording interior angles.

STATION	FROM TO	STATION	CIRCLE	INTERIOR ANGLE
		L	0° 00′	—
M			275° 44′	84° 16′
		N	191° 28′	84° 16′
			mean	84° 16′ 00″

Figure 6-3. Notes for recording interior angles with transit having only clockwise numbering.

84° 16′. The second value, measured with the telescope reversed, is obtained by subtracting 84° 16′ from 168° 32′, giving 84° 16′. This is entered under the *Interior Angle* column. The first value of the interior angle at N from M to O, measured with the telescope direct, is obtained by subtracting 0° 04′ from 216° 06′ and is recorded under the column headed *Interior Angle* as 216° 02′. The second value, measured with the telescope reversed, is obtained by subtracting 216° 06′ from (72° 09′ + 360°) and is recorded as 216° 03′. The mean of the two angles is recorded as 216° 02′ 30″. At both M and N the counterclockwise circle was read because the angles are counterclockwise interior angles.

If interior angles are measured in a counterclockwise direction with a transit on which the circle is graduated in a clockwise direction only, then the circle readings would decrease when the angles are turned. The angle would be obtained by subtracting the lower reading from the upper reading. As an example, if the angle at M in Fig. 6-1 was measured from L to N with such a transit, the notes would be as shown in Fig. 6-3.

The first value of the interior angle is obtained by subtracting 275° 44′ from (0° 00′ + 360° 00′). The second value is obtained by subtracting 191° 28′ from 275° 44′. In order to avoid confusion, interior angles should be measured clockwise if the circle does not contain a counterclockwise system of graduation numbers.

The sum of the interior angles in a closed plane figure must equal $(n - 2) \times 180°$, in which n is the number of sides in the figure. This relation furnishes a check on the accuracy of the field measurements of the angles and a basis for distributing the errors in measurement as discussed in Sections 5-9 and 5-10.

6-3. MEASURING ANGLES TO THE RIGHT Most surveys performed to establish horizontal control by traversing between fixed control points are made by measuring all the angles from the backsight to the foresight station in a clockwise direction. Angles to the right are shown in Fig. 6-4, in which a traverse is run from station Park to station Church through stations 1, 2, 3, and 4. The form for recording the notes is the same as those for recording readings for interior angles.

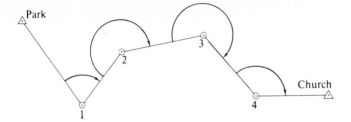

Figure 6-4. Angles to the right.

The check on the angles measured to the right is made by comparing the final azimuth, as computed from the angles, with a fixed azimuth. This check is fully discussed in Section 8-6 in Chapter 8.

6-4. MEASURING DEFLECTION ANGLES

A deflection angle, as shown in Fig. 6-5, is the angle made by the foresight with the prolongation of the backsight. Thus at station 17 + 42.86, a deflection to the *left* of 18° 26′ is shown; at station 38 + 14.53, the deflection is 40° 12′ 30″ to the *right*.

To measure the deflection angle with the transit occupying station 17 + 42.86, a backsight is taken on station 0 with the telescope direct. The circle is read and the value is recorded as the initial circle reading. Say the reading is 0° 00′. The telescope is plunged to the reversed position so that the line of sight is now directed along the prolongation of the line from station 0 to station 17 + 42.86. The circle still reads 0° 00′ since no angle has been turned.

The upper clamp is loosened and the line of sight is turned to the left and directed to station 38 + 14.53. When the line of sight is brought exactly on the point with the upper tangent screw, the circle is read. If the clockwise circle is read, the reading will have decreased from 0° (or 360°) to 341° 34′, the difference being 18° 26′. If the counterclockwise circle is read, the reading will have increased from 0° to 18° 26′, the difference being 18° 26′.

The angle is now measured a second time. With the telescope *still in the reversed position*, the lower clamp is loosened and the line of sight is directed toward the backsight station. The line of sight is brought exactly on station 0 by means of the lower tangent screw. The clockwise circle still reads 341° 34′, and the counterclockwise circle still reads 18° 26′. The telescope is plunged

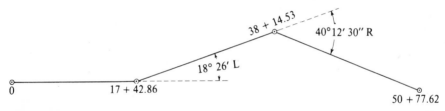

Figure 6-5. Deflection angles.

STATION	FROM STATION TO	CIRCLE	DEFLECTION ANGLE
	0 + 00.00	0° 00′	—
17 + 42.86		341° 34′	18° 26′
	38 + 14.53	323° 08′	18° 26′
		mean	18° 26′ 00″ left
	17 + 42.86	0° 00′	—
38 + 14.53		40° 13′	40° 13′
	50 + 77.62	80° 25′	40° 12′
		mean	40° 12′ 30″ right

Figure 6-6. Deflection-angle notes with clockwise circle.

again (it is thus brought back to its direct position) so that the line of sight is now directed along the prolongation of the backsight. The clockwise circle still reads 341° 34′, and the counterclockwise circle still reads 18° 26′.

The upper clamp is loosened and the line of sight is turned to the left and brought on station 38 + 14.53. The circle is read to obtain the final reading. If the clockwise circle is read, the reading will have decreased from 341° 34′ to 323° 08′, the difference being 18° 26′. If the counterclockwise circle is read, the reading will have increased from 18° 26′ to 36° 52′, the difference being 18° 26′. The direction of the deflection angle, in this case *left*, is as important as the magnitude of the angle.

The two angles shown in Fig. 6-5 would be recorded from the clockwise circle readings as shown in the form in Fig. 6-6.

If the counterclockwise circle is read, the readings would be recorded as shown in the form in Fig. 6-7.

When a deflection angle is measured only once, the more serious instrumental errors are introduced into the measurement. If a deflection angle is measured by double centering, most of the instrumental errors are eliminated, and the second value of the angle affords a check on the first value.

STATION	FROM STATION TO	CIRCLE	DEFLECTION ANGLE
	0 + 00	0° 00′	—
17 + 42.86		18° 26′	18° 26′
	38 + 14.53	36° 52′	18° 26′
		mean	18° 26′ 00″ left
	17 + 42.86	0° 00′	—
38 + 14.53		319° 47′	40° 13′
	50 + 77.62	279° 35′	40° 12′
		mean	40° 12′ 30″ right

Figure 6-7. Deflection-angle notes with counterclockwise circle.

Deflection angles are used frequently in surveys for highways, railroads, transmission lines, and similar routes, which extend in the same general direction, although they are also employed by many surveyors and engineers in running traverse lines. The principal advantages of using deflection angles are (1) azimuths are easily computed from deflection angles (see Section 8-5); (2) the deflection angles are used in the computation of circular curves for highway and railroad work; and (3) deflection angles are easily plotted. One distinct disadvantage is the likelihood of the notekeeper making a mistake in recording the direction of the deflection angle.

In a closed figure in which the sides do not cross one another, the difference between the sum of the right deflection angles and the sum of the left deflection angles should equal 360°. In a closed survey in which the lines cross once, as in a figure eight, the sum of the right deflection angles should equal the sum of the left deflection angles. If the lines do not cross or cross any *even* number of times, the difference between the sums should equal 360°. If the lines cross any *odd* number of times, the sums should equal one another.

6-5. LAYING OFF ANGLES In all types of construction work when laying out highways, bridges, buildings, and other structures, it becomes necessary to establish lines in given directions and thus to lay off both horizontal and vertical angles. A line called a *base line* is established in the vicinity of the work. The base line may or may not be a portion of a control traverse, its position depending on circumstances. Nevertheless, it acts as the control for the construction. Stations along the base line are occupied by the theodolite, and backsights are made to other base line stations. The angles called for on construction plans and drawings are laid off with respect to the base line.

The principle of double centering is employed in turning angles from the base line. If construction lines are quite long, the principle of repetition may be employed. For example, to lay off an angle of 55° 12′ 30″ by double centering with a 1′ transit, a backsight is taken on a baseline station with the circle reading 0° 00′. The upper clamp is loosened and the telescope is turned until the circle reading is approximately 55° 12′. The upper clamp is tightened and the vernier is made to read 55° 12′ as closely as possible by turning the upper tangent screw. A point is set on the line of sight, preferably being marked by a tack in a stake or a hub. The telescope is inverted and a second backsight is taken on the baseline station by a lower motion. When the backsight has been made, the upper clamp is loosened and the telescope is turned until approximately 2 × 55° 12′ 30″, or 110° 25′, is read. The upper clamp is tightened and the exact reading of 110° 25′ is made by turning the upper tangent screw. A second point is set on the line of sight alongside the first point. The second angle is 55° 13′. A third point halfway between the other two points is the correct point, and the line joining this point and the transit station makes the desired angle with the base line.

To lay off an angle by repetition, see Sections 4-3 and 4-17.

If a direction theodolite is used to lay off a horizontal angle, the backsight is made along the base line and the circle is read. The angle to be laid out is added to or subtracted from the backsight circle reading (depending on whether the angle is to be laid off clockwise or counterclockwise) to obtain the circle reading for directing the line of sight to the point to be set. The optical micrometer is then set to the correct number of minutes and seconds, and the clamp and tangent screw are used to make the remainder of the circle reading. The set point should be checked by measuring the angle with the telescope in the reversed position.

To lay off a vertical angle with a theodolite having a full vertical circle, carefully level the instrument by using either the plate levels as described in Section 4-19 or the telescope level as described in Section 4-18. Rotate the telescope in the direction of the point to be set and incline the telescope until the vertical circle reads approximately the angle to be laid off. Clamp the vertical clamp and set off the exact reading by using the vertical tangent screw. The point is marked by reference to the center horizontal cross hair. Invert the telescope and rotate the transit on its vertical axis until the line of sight is in the desired direction. Sight the first point and clamp the vertical clamp. If the vertical circle reads the desired angle when the horizontal cross hair is brought on the point previously set, it may be concluded that the index error was zero and the point has been correctly set. If the vertical circle does not read the desired angle, the vertical tangent screw is turned until it does. A second point is set above or below the first point. A third point halfway between the two will be in very nearly the correct position.

To lay off a vertical angle with a transit having a half vertical circle, level the instrument as before and rotate the telescope until the line of sight is in the direction of the desired point. Bring the telescope horizontal and center the telescope bubble by using the vertical tangent screw. Read the vertical circle to obtain the index error. Add the index error algebraically to the desired vertical angle to obtain the correct circle reading. Set off the correct circle reading by using the vertical clamp and tangent screw. The line of sight is now directed along the desired vertical angle and the point is set on line.

If the theodolite reads zenith angles, then any vertical angle to be laid off must be converted to the equivalent zenith angle beforehand.

6-6. STRAIGHT LINE BY DOUBLE CENTERING Whenever a straight line is to be accurately extended for any considerable distance, the method of double centering should be used. If in Fig. 6-8 the line AB is to be extended to the right, the instrument is set up at B, and a backsight is taken on A with the telescope in the normal or direct position. The telescope is then plunged. If the theodolite is in adjustment, the line of sight will then be the line BC. If the theodolite is not in adjustment, it may be either the line BC' or the

Figure 6-8. Extending straight line by double centering.

line BC''. A stake is driven on the line of sight and the exact line is marked by a point on the stake. The backsight having first been checked to see that the telescope has not moved off line, either the upper clamp or the lower clamp is loosened and a second backsight is taken on A, this time with the telescope in the inverted position. The telescope is again plunged and the line of sight is again directed toward C. If the instrument is in adjustment, the line of sight will strike the point previously set. If it is out of adjustment, the line of sight will now miss that point and a second point should be set on this line. A third point, midway between the two points already set, will be on the line AB extended. Thus if C' is the first point set, C'' will be the second one and the correct line will be marked by C, which is midway between C' and C''.

The line can be extended further by moving the theodolite to C and repeating the procedure, a fourth point D being located midway between D' and D''. The first backsight might, of course, be made with the telescope in the inverted position. If this is done, the second backsight is taken with it normal. In Fig. 6-8 the scale is very much distorted. With an instrument in reasonably good adjustment, the three points C', C'', and C should all fall on the same 2-in. stake at a distance of 1200 to 1500 ft from the instrument. If they do not, the instrument should be adjusted. Otherwise when this kind of work is being done, unnecessary time will be spent in driving stakes.

Establishing a straight line by the method of double centering is nothing more than laying off a zero deflection angle twice, the first backsight being taken with the telescope direct and the second one with it reversed.

6-7. ESTABLISHING POINTS ON A STRAIGHT LINE The simplest case of establishing points on a straight line is when the two ends of the line are marked and the entire line is visible from one end or the other. Thus if it is necessary to establish additional points between A or B in Fig. 6-8, the transit can be set up at either A or B and a sight taken to the other end of the line. The rodman is then directed by motions to the right or the left to the point that is on the line. The rodman may use a line rod if the lengths are considerable. When they are short and the ground is clear, he may have the instrumentman line in the stake or other object on which the point is to be marked. A point on the top of the stake may be obtained by lining in the line rod again or a pencil or a plumb-bob string.

If the ground surface is so shaped that the telescope does not have to be raised or lowered appreciably after taking the sight on the opposite end of the line and if the focus of the telescope is likewise unchanged, this work can be done with a theodolite that is not in perfect adjustment. If the inclination of the telescope must be changed considerably and the focus altered, important points should be checked by taking two sights, one with the telescope normal and another with it inverted.

If points are to be established between B and C, the distant point C should first be established by the method of double centering and a foresight should be taken on C before line is given for the intermediate points. Where points are being set at regular intervals, as in a route survey, the instrument is moved only when the foresight distance becomes so great that it is impossible to give line accurately or when the ground conditions prevent further accurate sighting. The forward station should be set by the method of double centering. The careful instrumentman will use this procedure as a check, even when he is reasonably certain that his instrument is in good adjustment.

6-8. BALANCING IN   It often happens that the two ends of a line are not intervisible but some intermediate point can be found from which the ends can be seen. Thus in Fig. 6-9 both A and B can be seen from the top of an embankment. The method of finding, by trial, the location of a point C that will be on the line AB is called *balancing in*. The transit is first set up at C', which is as nearly on the line as can be estimated. A backsight is taken on A and the telescope is plunged. From the sketch it is evident that the line of sight strikes considerably to the left of B. The amount by which it fails to strike B is measured or estimated, and the transit is moved a proportionate amount (in the sketch about one-third) of this distance and the procedure is repeated. After a few trials, the line of sight should strike B. The final movement of the instrument can often be made with the shifting head of the transit. Until the transit is near the correct point, it is unnecessary to level the instrument carefully.

To check the position of the transit, backsight on A with the telescope in the inverted position, plunge the telescope, and check to see whether or not

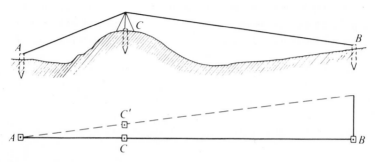

Figure 6-9. Balancing in.

the line of sight is directed on B. If it is, then the transit is on line. If it is to one side of B, measure the distance by which it misses B and move the transit in the opposite direction a proportionate amount of this distance. A final check is made by double centering as described in Section 6-6. Point B should fall at the midpoint of the two points B' and B'' (analogous to C' and C'' of Fig. 6-8).

After the correct position of the transit at C has been established, any additional points between C and A or between C and B can be set by the method described in the preceding section.

6-9. RANDOM LINES   In Fig. 6-10 the distance from A to B may be very great and the ground may be so thickly wooded or so hilly that it is impossible to see directly from A to B or to find an intermediate point from which both A and B can be seen. Under any of these conditions, a trial straight line, or *random line*, is run as near as can be estimated to the true line. This line, as $ACDEF$, is run by the method of double centering to a point from which B can be seen. The length AF of the random line, the distance FB, and the angle at F from A to B are then measured. In the triangle ABF, two sides and the included angle are known, and the angle at A can be computed. The instrument is set up again at A, the computed angle is turned off from the line AF by using the method described in Sections 4-3 or 6-5, and the line AB is run. If this new line fails to strike B exactly, points along the line can be shifted slightly to bring them on line, the amounts they are to be moved being obtained by proportion.

A second method of locating points on the required line is to set up the instrument at points on the random line, turn off angles equal to the measured angle AFB, and measure distances obtained by proportion. Thus in Fig. 6-10 if the angle at E is made equal to the angle at F, then $EG = FB \times (AE/AF)$. Points adjacent to C and D may be set in a similar manner.

Another method of locating points along the line AB is discussed in Chapter 8. In this method a closed random traverse is run, instead of a straight line, and points on the traverse are located with respect to the line AB by computation. From these traverse points computed angles and distances are laid off to determine additional points on AB. Furthermore, these additional points may be set on AB at any desired distances from A or B.

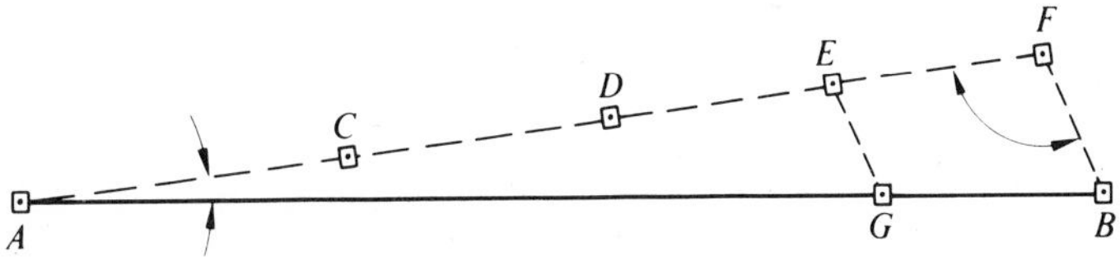

Figure 6-10. Random line.

6-10. INTERSECTION OF TWO STRAIGHT LINES   Two examples of the intersection of straight lines are shown in Fig. 6-11. To find the intersection of lines *AB* and *CD* in view (a), the transit is set up at either *A* or *B* and a sight is taken along the line *AB*. Stakes, known as *straddle stakes*, are then set at *E'* and *E"*, being so located that the point of intersection will fall somewhere between them. These stakes should be set so close that a string can be stretched between them. If it is difficult to estimate where the intersection is to be, more than two such stakes can be set.

The transit is next moved to *C* or *D* and a sight is taken along the line *CD*. The point *E* is then found at the intersection of the line of sight and the string stretched between *E'* and *E"*. If two transits are available, one can be set up on each line and *E* can be found at the intersection of the two lines of sight.

In Fig. 6-11 (b) the intersection of the lines *AB* and *CD* is on the prolongations of both lines. If the lines are short, the stakes at *E'* and *E"* can be set by placing the transit at *A*, sighting on *B*, and extending this line to *E'* and *E"*. When the lines are long, the transit is set up at *B* and points on the straddle stakes are set by double centering. The intersection is determined by moving the transit to either *C* or *D* and finding where the new line of sight strikes

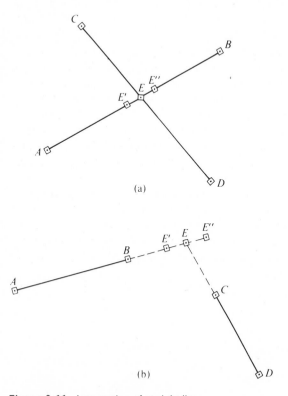

(a)

(b)

Figure 6-11. Intersection of straight lines.

the string stretched between E' and E''. If the transit is at C, the method of double centering should be used.

6-11. OBSTACLES ON A LINE

When a boundary line is being extended by the method developed in Section 6-6 or when the center line for a highway or railroad is being staked out in preparation for construction, the problem of extending the line through heavily wooded or congested areas is invariably faced. Where timber cannot be cut, as in a boundary survey, or where buildings are not removed until immediately before construction begins and an obstacle is on the line, it becomes necessary to leave the line in order to pass the obstacle. Two methods for passing an obstacle are given here.

The first method, illustrated in Fig. 6-12, consists of establishing two equal perpendicular offsets from the desired line AF, as BB' and CC'. To provide a check, an equal offset from A to A' is also laid off. The line $A'B'C'$ is parallel to the line ABC and is extended beyond the obstacle to points D', E', and F' by the methods described in Sections 6-6 and 6-7, the point F' being set as a check. Perpendicular offsets equal in length to BB' and CC' are established at these points to locate points D, E, and F. If measurements are to be carried along the line being run, then the offsets at C and D' should be made parallel by using the theodolite, so that the unmeasurable distance CD will be equal to $C'D'$, which can be measured.

The second method, illustrated in Fig. 6-13, is used where there are many obstacles on the desired line and where running an offset parallel line as just described is not feasible. This method consists of laying off a small deflection angle α from the desired line at a convenient point as A, setting point E at a measured distance from A, then laying off a deflection angle 2α from line AE in the opposite direction, and setting point B on the new line of sight so that $EB = AE$. If the deflection angle α is now turned from line EB at B, the line of sight will be directed along the desired line. If there is an obstacle between B and C, the line EB may be extended to a point at F. The angle 2α is then laid off at F and the distance FC is laid off equal to BF. The procedure is repeated until the obstructed area has been traversed. Finally, the direction of the desired line beyond C can be established by laying off the angle α at C.

This method is convenient in that isosceles triangles are laid off, and the angles α and 2α can have the same values throughout. A random traverse discussed in Chapter 8 will accomplish the same results with perhaps more speed in the field, but more computations will be needed. Using a random

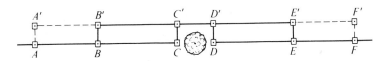

Figure 6-12. Prolonging line past obstacle.

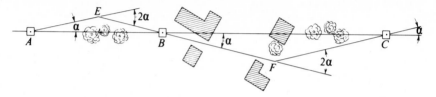

Figure 6-13. Passing obstacles on line.

traverse also requires computed distances and angles to be laid out, necessitating additional field work.

6-12. PARALLEL LINES The principles of plane geometry are used in the establishment of parallel lines. In Fig. 6-12 the lines AF and $A'F'$ are made parallel by making the distance between them the same at B, C, D, and E. Equal offsets at two points are sufficient to establish the position of the parallel line, but the field work can be checked if offsets are made at three or more points.

In Fig. 6-14 the line CD is made parallel to the line AB by making the opposite interior angles at B and C equal. In the field the angle ABC would be measured and the direction of CD determined by laying off the angle BCD equal to the measured angle ABC. If the length of CD is more than a very few feet, these angles should be measured by repetition.

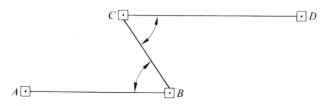

Figure 6-14. Parallel lines.

6-13. LOCATION OF A POINT The field location of a point involves the measurement of enough angles and distances to enable a person in the office to plot the point in its correct position, either with respect to two fixed points or with respect to a fixed point and a fixed line through that point. As the two fixed points and the required point are the vertexes of some triangle, the position of the required point will also be fixed when the triangle is determined.

A triangle is determined by (1) two sides and the included angle; (2) a side and the two adjacent angles; (3) three sides; (4) two sides and the angle opposite one of these sides. The field applications of these methods of locating a point are shown in Fig. 6-15. In view (a) the point P is located from A by

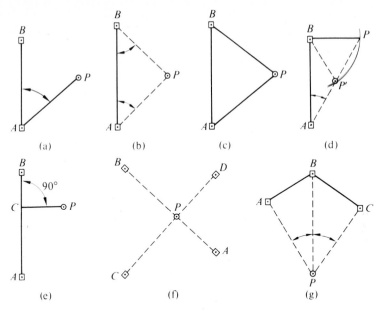

Figure 6-15. Locating a point.

the measurement of an angle from the known line AB and a distance. In view (b) two angles are measured at the known points A and B, and in view (c) two distances from A and B are required. In view (d) an angle at the known point A and the side opposite from the known point B are measured. In view (e) the measurements are the perpendicular distance CP from P to the line AB and the distance BC or AC. In view (f) the point P is at the intersection of the known lines AB and CD. The location of P in view (g) is determined by measuring the angles at P between the lines to three fixed points A, B, and C.

Although each of these methods has its particular uses, the first one shown—the measurement of an angle and a distance—is used much more frequently than the others. The shape of the figure may sometimes preclude the use of some particular method. Thus in view (b) where two angles are measured, the location will be very uncertain if the angles at A and B are so small as to give an indefinite intersection when the lines AP and BP are, for example, drawn on a map. This method is used when the measurement of distances is difficult or impossible. It is also the basis of triangulation where, with one side and all the angles of a triangle known, the unknown sides are computed.

The method in view (c) can be used when the survey is being made with only a tape. The location of P will be best when the angle at P is near $90°$ and will be uncertain when this angle approaches $0°$ or $180°$. In such cases the method shown in view (d) will give a better location. This method can also be used when the measurement of the side AP is impossible. When this method is adopted, the approximate shape of the triangle must be known as there generally will be two solutions of a triangle when two sides and the angle opposite one of them are known. In every case it must be known whether the located point is to the right or left of the line of reference.

The method represented in view (e) is frequently used in route surveying to locate objects with respect to the center line. It is most advantageously used when extreme precision is unnecessary, as considerable field work is involved in locating the point C exactly. On construction work, important points that are liable to be disturbed during the process of the work can be replaced if their location is determined by the intersection of lines, as shown in view (f). The location will be more certain if the distances from P to A, B, C, and D are also measured.

In view (g) is shown a *three-point resection* to determine the position of the point P occupied by an instrument when this position is not known before the instrument is set up. The positions of points A, B, and C are known, and the angles at P from A to B and from B to C are measured. The computations necessary to determine the position of P are described in Section 10-35.

PROBLEMS

6-1. The interior angles in a seven-sided figure are measured as follows: a, 83° 22′ 00″; b, 193° 04′ 30″; c, 88° 58′ 00″; d, 149° 16′ 00″; e, 116° 41′ 30″; f, 133° 40′ 00″; g, 134° 54′ 30″. All angles have equal weights. Compute the adjusted interior angles.

6-2. The interior angles in a six-sided figure are measured as follows: p, 126° 10′; q, 43° 32′; r, 222° 58′; s, 91° 04′; t, 116° 17′; u, 119° 56′. All angles have equal weight. Compute the adjusted interior angles.

6-3. The interior angles in a six-sided figure are measured as follows: h, 90° 00′ 15″; i, 89° 58′ 00″; j, 216° 30′ 30″; k, 55° 14′ 45″; l, 147° 00′ 30″; m, 121° 16′ 30″. All angles have equal weight. Compute the adjusted interior angles.

6-4. The following equal-weight deflection angles were measured in a five-sided traverse that closes on itself. A, 88° 14′ L; B, 70° 52′ L; C, 121° 30′ L; D, 36° 03′ R; E, 115° 37′ L. Compute the adjusted deflection angles.

6-5. The following equal-weight deflection angles were measured in a seven-sided traverse that closes on itself: P, 55° 14′ 15″ R; Q, 52° 10′ 00″ R; R, 93° 38′ 30″ R; S, 72° 53′ 30″ L; T, 107° 40′ 00″ L; U, 111° 14′ 30″ L; V, 90° 47′ 00″ R. Compute the adjusted deflection angles.

6-6. The line of sight of a transit makes an angle of 89° 57′ 10″ with the transverse axis. What would be the angular error in the line CD in Fig. 6-8 if each extension of the line AB was made with only one plunging and each backsight was made with the telescope in the direct position?

6-7. If each of the distances AB, BC, and CD in Problem 6-6 is 350 ft, how far off the true prolongation of AB will point D be set?

6-8. In Problems 6-6 and 6-7 if the backsight at B were made with the telescope direct and the backsight at C were made with the telescope inverted, and if only one plunging was performed at each point, what will be the resulting angular error in the direction of CD? What will be the offset error of point D?

6-9. In Fig. 6-10 AC = 580.00 ft; CD = 445.00 ft; DE = 430.00 ft; EF = 305.00 ft; and FB = 6.97 ft. The angle at F from B to E is measured as 93° 53′. What is the angle at A from F to B? What is the distance EG, if EG is parallel to FB?

6-10. Compute the length of line AB of Problem 6-9.

6-11. In Fig. 6-10 AC = 137.00 m; CD = 130.00 m; DE = 125.00 m, EF = 225.00 m; and FB = 3.352 m. The angle at F from B to E is measured as 115° 32′. Compute the angle at A from F to B. What is the distance EG, if EG is parallel to FB?

6-12. Compute the length of the line AB of Problem 6-11.

6-13. In Fig. 6-10 AC = 295.50 ft; CD = 280.00 ft; DE = 280.00 ft, EF = 250.00 ft; and FB = 8.02 ft. The angle at F from B to E is 100° 22′. Compute the angle at A from F to B.

6-14. What is the distance EG in Problem 6-13?

6-15. What is the length of the line AB to Problem 6-13?

6-16. In Fig. 6-13 AE = EB = 180.00 ft, BF = FC = 425.00 ft, and angle α = 1° 10′. What is the length of the line AC?

6-17. In Fig. 6-13 AE = EB = 150.00 m; BF = FC = 315.00 m; and angle α is 2° 25′. What is the length of AC?

Chapter 7
The Direction of a Line

7-1. ASTRONOMICAL MERIDIAN A plane passing through a point on the surface of the earth and containing the earth's axis of rotation defines the *astronomical meridian* at the point. The direction of this plane may be established by observing the position of the sun or a star, as described in Chapter 12, or by gyrocompassing as discussed in Section 7-16. By popular usage the intersection of this meridian plane with the surface of the earth is known as the *true meridian*.

7-2. MAGNETIC MERIDIAN The earth acts very much like a bar magnet with a north magnetic pole located considerably south of the north pole defined by the earth's rotational axis. The magnetic pole is not fixed in position, but rather changes its position continually. A magnetized needle freely suspended on a pivot will come to rest in a position parallel to the magnetic lines of force acting in the vicinity of the needle. Generally the greatest component of the magnetic force at a point is that created by the earth's magnetic field, but other components may be created by other magnetic fields such as those around electric-power lines, reinforcing bars in roads and structures, and iron deposits. The direction of the magnetized needle defines the *magnetic meridian* at the point at a specific time. Unlike the true meridian, whose direction is fixed, the magnetic meridian varies in direction.

A gradual shift in the earth's magnetic poles back and forth over a great many years causes a secular change, or variation, which amounts to several degrees in a cycle. An annual variation of negligible magnitude is experienced by the earth's magnetic field. A daily variation causes the needle to swing back and forth through an angle of not much more than one-tenth of a degree each day.

Local attraction, the term applied to the magnetic attractions other than that of the earth's magnetic field, may change at a given location. This change may be caused by a variation in the voltage carried by a power line or by a gradual increase or decrease of a magnetic field in reinforcing bars, wire fences, underground utility pipes, or other metal parts.

7-3. ASSUMED MERIDIAN For convenience in a survey of limited extent, any line of the survey may be assumed to be a meridian or a line of reference. An assumed meridian is usually taken to be in the general direction of the true meridian.

7-4. CONVERGENCE OF MERIDIANS True meridians on the surface of the earth are lines of geographic longitude, and they converge toward each other as the distance from the equator toward either of the poles increases. The amount of convergence between two meridians in a given vicinity depends on (1) its distance north or south of the equator and (2) the difference between the longitudes of the two meridians. Magnetic meridians tend to converge at the magnetic poles but the convergence is not regular and it is not readily obtainable.

7-5. GRID MERIDIAN Another assumption is convenient in a survey of limited extent. When a line through one point of the survey has been adopted as a reference meridian, whether true or assumed north, all the other meridians in the area are considered to be parallel to the reference meridian. This assumption eliminates the necessity for determining convergence. The methods of plane surveying assume that all measurements are projected to a horizontal plane and that all meridians are parallel straight lines. These are known as *grid meridians*.

Two basic systems of grids are used in the United States to allow plane surveying to be carried statewide without any appreciable loss of accuracy. In each separate grid one true meridian is selected. This is called the *central meridian*. All other north-south lines in that grid are parallel to this line. The two systems, which are known as the *Lambert conformal projection* and the *transverse Mercator projection*, are the subject of Chapter 11.

7-6. AZIMUTH OF A LINE The azimuth of a line on the ground is the horizontal angle measured from the plane of the meridian to the vertical plane containing the line. Azimuth gives the direction of the line with respect to the meridian. It is usually measured in a clockwise direction with respect to either the north meridian or the south meridian. In astronomical and geodetic work azimuths are measured from the south meridian. In plane surveying azimuths are generally measured from north.

Since the advent of the state grid systems mentioned in Section 7-5, engineers and surveyors are using more and more the control monuments established by the large federal agencies engaged in control surveys. These agencies publish the azimuths of lines in their control networks as measured from the south. Because of this practice, an increasing number of engineers engaged in state-wide, county-wide, and city-wide control surveying are using azimuths measured from the south.

A line may have an azimuth between 0° and 360°. In Fig. 7-1 the line *NS* represents the meridian passing through point *O*, with *N* toward the north. The azimuth of the line *OA* measured from the north is 70°; that of *OB* is 145°; that of *OC* is 235°; and that of *OD* is 330°. Azimuths are called *true* azimuths when measured from the true meridian, *magnetic* azimuths when measured from the magnetic meridian, *assumed* azimuths when referred to an arbitrary north-south line, and *grid* azimuths when referred to the central meridian in a grid system.

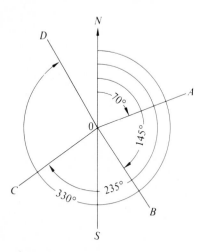

Figure 7-1. Azimuths.

7-7. BACK AZIMUTH When the azimuth of a line is stated, it is understood to be that of the line directed from an original point to a terminal point. Thus a line *LP* has its origin at *L* and its terminus at *P*. If the azimuth of *LP* is stated as 85°, then the azimuth of the line *PL* must have some other value

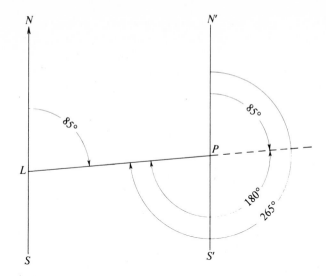

Figure 7-2. Relationship between azimuth and back azimuth.

since *LP* and *PL* do not have the same direction. One direction is the reverse of the other. For the purpose of discussing back azimuth, consider the line *LP* in Fig. 7-2. In the diagram, *NS* is the meridian through *L* and *N'S'* is the meridian through *P*. According to the assumption in plane surveying, the two

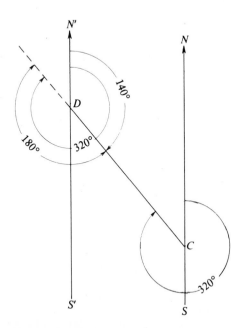

Figure 7-3. Azimuth and back azimuth.

meridians are parallel to each other. If the azimuth of *LP* is 85°, then the azimuth of *PL* is 85° + 180°, or 265°. Thus, the back azimuth of *LP* is the same as the azimuth of *PL*. In Fig. 7-3 the azimuth of *CD* is 320° and the back azimuth of *CD*, which is the azimuth of *DC*, is 320° − 180°, or 140°.

From the preceding explanation, it is seen that the back azimuth of a line can be found from its forward azimuth as follows: If the azimuth of the line is less than 180°, add 180° to find the back azimuth. When the azimuth of the line is greater than 180°, subtract 180° to obtain the back azimuth.

7-8. BEARING OF A LINE   The bearing of a line also gives the direction of the line with respect to the reference meridian. Unlike an azimuth, which is always an angle measured in a definite direction from a definite half of the meridian, a bearing angle is never greater than 90°. The bearing states whether the angle is measured from the north or the south and also whether the angle is measured toward the east or toward the west. For example, if a line has a bearing of S 35° E (called south 35° east), the bearing angle 35° is measured from the south meridian eastward. In Fig. 7-4 are shown four lines, the origin in each case being point *O*. Each line lies in a different quadrant. The bearing of *OA* is N 70° E; that of *OB* is S 35° E; that of *OC* is S 55° W; and that of *OD* is N 30° W.

It is apparent from Fig. 7-4 that the bearing angle of a line must be between 0° and 90°. A stated bearing is a *true* bearing, a *magnetic* bearing, an *assumed* bearing, or a *grid* bearing, according to whether the reference meridian is true, magnetic, assumed, or grid. In land surveying and in the conveyance of title to property, references are made to maps, notes, and plats of previous surveys which are recorded with the various counties throughout the United States. When reference is made to a bearing in such a previous survey, the

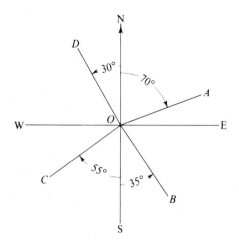

Figure 7-4. Bearings.

term *record bearing* is used. If reference is made to a property deed, the term *deed bearing* is used. The terms "deed bearing" and "record bearing" are commonly used interchangeably.

7-9. BACK BEARING The back bearing of a line is the bearing of a line running in the reverse direction. In Fig. 7-5 the bearing of the line AB is N 68° E, and the bearing of BA, which is in the reverse direction, is S 68° W.

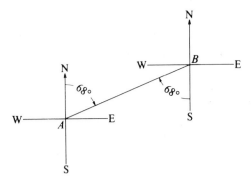

Figure 7-5. Bearing and back bearing.

The bearing of a line in the direction in which a survey containing several lines is progressing is called the *forward* bearing, whereas the bearing of the line in the direction opposite to that of progress is the *back* bearing. The back bearing can be obtained from the forward bearing by simply changing the letter N to S or S to N and also changing E to W or W to E.

7-10. RELATION BETWEEN AZIMUTHS AND BEARINGS To simplify computations based on survey data, bearings may be converted to azimuths and azimuths may be converted to bearings. The instances when such conversion is convenient will become apparent in later chapters. The conversion itself is quite simple.

An inspection of Fig. 7-1 will show that the line OA, whose azimuth from north is 70°, lies in the northeast quadrant since the angle eastward from the meridian is less than 90°. Furthermore it is apparent that the bearing angle and the azimuth are identical. Therefore the bearing of OA is N 70° E. The line OB is 145° from the north meridian. It lies south of a due-east line and is therefore in the southeast quadrant. The problem in this case is to determine the angle from the south meridian. Since the north meridian and the south meridian are 180° apart, the problem is solved by subtracting the azimuth, or 145°, from 180° to arrive at the bearing angle, which is 35°. Therefore the bearing of the line OB is S 35° E. The line OC is 235° from the north meridian in a clockwise direction and is beyond the south meridian in a westerly direction by an angle of 235° − 180°, or 55°. Therefore if the azimuth is 235°, the bearing is S 55° W. For the line *OD* the angle from the north meridian is 330° in a clockwise direction. The angle from the north meridian in a counterclockwise or westerly direction is 360° − 330°, or 30°. The line *OD* lies in the northwest quadrant, and its bearing is N 30° W.

The rules to observe in converting from azimuths to bearings are very quickly established in a person's mind after the rules have been put to practice a few times. They are as follows: (1) If an azimuth from north is between 0° and 90°, the line is in the northeast quadrant and the bearing angle is equal to the azimuth, (2) if an azimuth from north is between 90° and 180°, the line is in the southeast quadrant, and the bearing angle is 180° minus the azimuth; (3) if the azimuth from north is between 180° and 270°, the line is in the southwest quadrant, and the bearing angle is the azimuth minus 180°; (4) if the azimuth from north is between 270° and 360°, the line is in the northwest quadrant, and the bearing angle is 360° minus the azimuth. These conversions should be performed mentally. For example, to determine the bearing of a line whose azimuth is 142° 29′ 54″, mentally subtract 142° from 179° and write 37°; mentally subtract 29′ from 59′ and write 30′; subtract 54″ from 60″ and write 06″. The bearing is S 37° 30′ 06″ E. Such mental subtraction is highly efficient in all phases of surveying computations, and should be practiced at the outset.

To convert from bearings to azimuths it is only necessary to reverse the foregoing rules, and the computations should be made with the same mental ease. The azimuth of a line in the northeast quadrant is equal to the bearing angle; that of a line in the southeast quadrant is 180° minus the bearing angle; that of a line in the southwest quadrant is 180° plus the bearing angle; that of a line in the northwest quadrant is 360° minus the bearing angle. As an example, the azimuth of a line whose bearing is N 77° 43′ 16″ W is 282° 16′ 44″.

7-11. THE MAGNETIC COMPASS

Since a freely suspended magnetized needle will lie in the magnetic meridian, the direction of a line can be determined with respect to the needle, and thus to the magnetic meridian, by measuring the angle between the line and the needle. The magnetic compass is constructed so as to allow a needle to swing freely on a pivot when in use, and to allow a line of sight to be directed from the occupied point to a terminal point. As shown in Fig. 7-6 a graduated circle is rotated as the line of sight is rotated. The north-seeking end of the compass needle is read against the circle to obtain the angle between the magnetic meridian and the line of sight.

The circle and the needle are encased in a metal compass box and are covered with a glass plate. The line of sight normally is fixed in line with the zero mark or the north graduation on the circle. Thus if a line of sight is directed along the north magnetic meridian, the needle will point to the zero mark or to the north graduation. As the line of sight is turned clockwise from magnetic north, the needle remains in the magnetic meridian, but the graduated circle is turned clockwise through the corresponding angle. When the

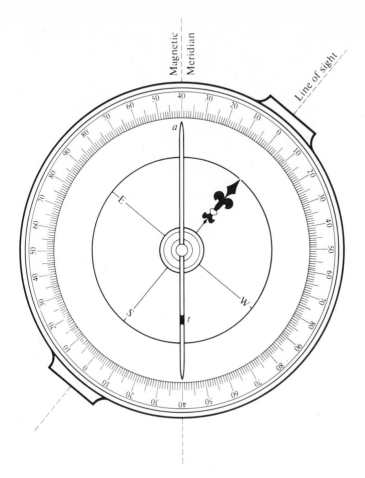

Figure 7-6. Compass circle.

line of sight is turned exactly 90° east of north, then the letter E is brought opposite the north end of the compass needle. The circle thus indicates that the magnetic bearing is due east, as it should. If the line of sight is turned exactly 180° from north, the letter S is brought opposite the north end of the needle and the magnetic bearing of the line of sight is shown to be due south.

Three general types of compasses are used in surveying. The pocket compass of various designs is used where only rough estimates of directions are needed. It is easily carried and is held in the hand when used. The line of sight is established by a combination of a peepsight and a slotted vane, whereby the observer sights from the origin toward the terminus of the line whose bearing he wishes to determine. In order that the compass needle may swing freely on its pivot, the pivot must be kept very sharp. If the compass needle is allowed to jostle around inside the compass box, the constant jarring against the pivot will in time dull the pivot point and render the compass

Figure 7-7. Brunton compass. Courtesy of Keuffel & Esser Co.

useless. The pocket compass is constructed so that when the sights are folded down, the needle is lifted off its pivot and held against the glass cover. Some types of pocket compasses, such as the Brunton compass shown in Fig. 7-7, contain a pin that when depressed will clamp the compass needle. Thus when the observer has made a sight, he depresses the pin and can then conveniently read the bearing.

The surveyor's compass (Fig. 7-8) is the instrument that was used in the past to run surveys of reasonable accuracy. Its main part is the compass box containing the graduated circle between 4 and 6 in. in diameter, the magnetic needle, one or two level bubbles to allow the circle to be brought horizontal, long sight vanes to allow fairly accurate pointings along rather steeply inclined lines, and a screw for lifting the needle off the pivot and against the glass cover plate. The compass box is fitted with a spindle that rotates in a socket. The socket fits in a leveling head consisting of a ball-and-socket joint, and the whole compass assembly is leveled about the ball-and-socket against friction applied by a clamping nut. The leveling head is fastened to a tripod or to a staff and is used in the field in this fashion. The circle is usually graduated in half degrees. Just as on the pocket compass, the sight vanes are aligned so that the line of sight passes through the north graduation on the circle.

The transit compass is similar to the surveyor's compass in all respects save for the method of leveling and for the line of sight. The compass box sets in the center of the horizontal limb of the transit, astride of which are the standards that hold the telescope (see Fig. 4-8). The method of leveling the transit, and thus the compass box, was described in Section 4-13. The line of sight for the transit compass is the collimation line of the telescope defined by the center of the objective lens and the intersection of the cross hairs. The telescope of the transit is capable of being raised or depressed for sighting along inclined lines. The letter N on the compass circle is normally under the objective end of the telescope when the telescope is in the direct

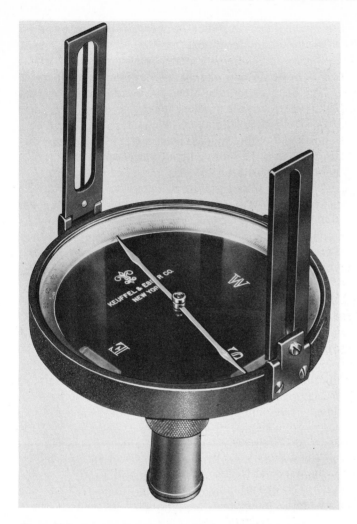

Figure 7-8. Surveyor's compass. Courtesy of Keuffel & Esser Co.

position (see Section 4-16). The letter S is under the eyepiece. Therefore this arrangement keeps the north graduation directed along the line of sight just as in the case of the pocket compass and the surveyor's compass.

7-12. DIP OF THE COMPASS NEEDLE   The lines of force created by the earth's magnetic field are directed toward the north and south magnetic poles and are horizontal only at points about halfway between the poles. From the halfway point toward the poles these lines of force become increasingly steep, and at the poles they are practically vertical.

A magnetic needle pivoted at its center of gravity would dip down on one end in order to remain oriented parallel to the lines of force in the vicinity.

In the northern hemisphere the north end of the needle would dip down. To overcome the dip of the needle and to cause it to come to rest about its pivot in a horizontal position, a coil of wire or a small clip of nonmagnetic metal is fastened to the end of the needle opposite that of the direction of dip to act as a counterweight. The metal is adjusted along the needle until, when the compass is level, the top surface of the needle coincides, or nearly coincides, with the top surface of the graduated circle. The counterbalance tells at a glance which is the north end of the compass needle. In the northern hemisphere the weight is usually on the south end, although with some needles this is not true. The observer must determine the north end for himself.

If a given compass is used in the same general vicinity, the position of the weight rarely needs to be changed. If, however, the compass is used in areas widely separated in a north-south direction, then the weight must be shifted accordingly. Although shifting the weight is a simple operation in itself, it must be done with great delicacy and with a fine touch so as not to dull the pivot in taking up or replacing the needle and so as to keep from bending the needle.

7-13. DETERMINING DIRECTIONS WITH THE MAGNETIC COMPASS   When using a pocket compass, the observer occupies one end of the line whose magnetic bearing he wishes to obtain. He holds the compass level, releases the needle by lifting the sights to their sighting position, sights the other end of the line, and allows the needle to come to rest. He then depresses the needle clamp and reads the circle at the north end of the compass needle. This reading gives him the magnetic bearing of the line directed from his position to the point sighted.

When using a surveyor's compass or a transit, the observer occupies one end of the line, centers the instrument over the point, and levels the instrument. He releases the clamp holding the needle, allowing it to rest on the pivot. He then directs the line of sight toward a point at the other end of the line and brings it on that point. When the needle comes to rest, he reads the circle at the north end of the needle to obtain the bearing of the line from the occupied point to the point sighted on. He then clamps the needle before disturbing the instrument.

To obtain the back magnetic bearing as a check, he occupies the second point and sights back to the first point, reading the north end of the needle as before. The bearing angles should show reasonable agreement, and the letters N and S as well as E and W should be reversed on the back bearing. This check should be made before leaving the point.

7-14. MAGNETIC DECLINATION   The magnetic poles do not coincide with the poles defined by the earth's rotational axis, and certain irregularities in the earth's magnetic field cause local and regional variations in the position of the needle. Therefore, except in a very narrow band around the earth, the magnetic needle does not point in the direction of true north. In some areas the needle points east of true north; in other areas the needle points west of true north. The amount and direction by which the magnetic needle is off the true meridian is called the *magnetic declination* (formerly called *variation*). Declination is positive or plus when the needle points east of true north, and it is negative or minus when the needle points west of true north. The declination varies from $+23°$ in the state of Washington to $-22°$ in the state of Maine, the total range over the United States being $45°$.

The declination can be determined for a given locality at a certain date by referring to the Isogonic Chart for Magnetic Declination prepared by the United States Geological Survey about every five years. The Isogonic Chart for 1980 is shown in Fig. 7-9. As seen on the chart, a narrow band, or line, running through portions of the states of Michigan, Wisconsin, Illinois, Indiana, Kentucky, Tennessee, Alabama, and Florida shows zero declination. This is called the *agonic line*. All portions to the east of this line have a west or minus declination. Those portions to the west of this line have an east or plus declination.

As noted in Section 7-2 the direction of the magnetic meridian at a given point is changing continually. The change to be considered when the compass is used for surveying is the secular variation, which causes the declination to change slowly in one direction in a given locality over a great many years and then change in the other direction. The cycles through which the change in declination occurs vary in length. Furthermore, this variation cannot be predicted because not enough is known about secular variation.

If the declination in a given vicinity for a previous year is desired, the declination for a year for which an isogonic chart has been published should be determined. Then by referring to the dashed lines on the chart, the annual change is determined. This annual change is multiplied by the difference between the desired year and the charted year to give the total change in declination from the charted year to the desired year. The proper correction is applied to the declination at the time of charting. As an example, suppose that the declination at Denver in 1948 is desired. Consult the isogonic chart for either 1945 or 1950. The procedure with the 1945 chart follows:

$$
\begin{aligned}
\text{annual change at Denver in 1945} &= -2' \\
\text{lapse of time covered} &= +3 \text{ years} \\
\text{total change in declination} &= (+3) \times (-2') = -6' \\
\text{declination at Denver in 1945} &= +14°\,16' \\
&\quad\;\; -06' \\
\hline
\text{declination at Denver in 1948} &= +14°\,10'
\end{aligned}
$$

If the 1950 chart is used, this is the work:

$$
\begin{aligned}
\text{annual change at Denver in 1950} &= -1.8' \\
\text{lapse of time covered} &= -2 \text{ years} \\
\text{total change in declination} &= (-2) \times (-1.8') = +3.6'' \\
\text{declination at Denver in 1950} &= +14°\,00.0' \\
&\quad\;\; +3.6' \\
\hline
\text{declination at Denver in 1948} &= +14°\,03.6'
\end{aligned}
$$

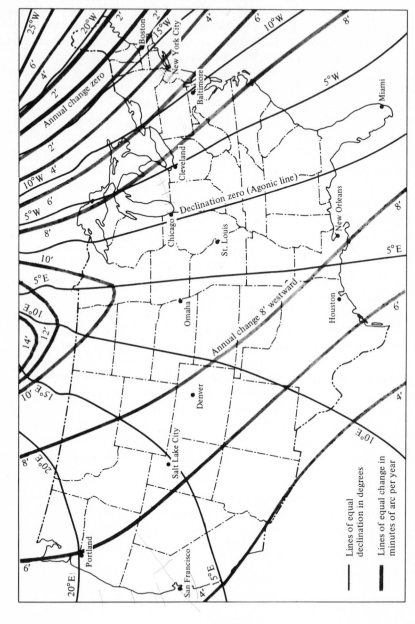

Figure 7-9. Lines of equal magnetic declination and of equal annual change in the United States for 1980. Courtesy of U.S. Geological Survey.

The difference between the two results is due in part to interpolation and in part to inaccuracy in the charts. If the declination at Denver for 1860 is required, the approximate declination could be obtained by working backward from a recent chart, but the better way would be to obtain the values from a chart prepared as near to 1860 as possible. This refinement is not always justified, since magnetic bearings shown on charts in many instances are only approximate at best and local irregularities can be several times as great as any error caused by using a chart for a different year.

7-15. RELATION BETWEEN TRUE AND MAGNETIC BEARINGS AND AZIMUTHS
True azimuths differ from magnetic azimuths by the magnitude of the magnetic declination at the time. If the declination is east of north, then all magnetic azimuths will be less than corresponding true azimuths by the amount of declination; if the declination is west of north then magnetic azimuths will be larger than corresponding true azimuths by the amount of declination. Conversion between true and magnetic directions is best made by means of azimuths.

Example 7-1

The magnetic declination on a certain date at a given place is 15° 15′ W. The magnetic azimuth of a line is 124° 20′. What is the true azimuth of the line?

Solution: The solution to this problem is shown in Fig. 7-10. The magnetic azimuth is seen to be larger than the true azimuth by 15° 15′, and the true azimuth is therefore 124° 20′ − 15° 15′, or 109° 05′.

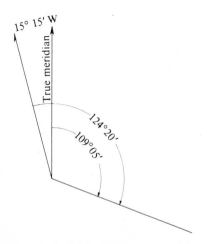

Figure 7-10. Relationship between true and magnetic azimuth.

Example 7-2

The magnetic bearing of a line was observed as S 12° 30′ E when the magnetic declination was 22° E. What is the true bearing of the line?

Solution: This problem is solved best by making the conversion by azimuths. The work follows:

$$
\begin{aligned}
\text{magnetic bearing} &= \text{S } 12°\,30′\,\text{E} \\
\text{magnetic azimuth} &= 167°\,30′ \\
\text{declination} &= +\ 22° \\
\hline
\text{true azimuth} &= 189°\,30′ \\
\text{true bearing} &= \text{S } 9°\,30′\,\text{W}
\end{aligned}
$$

Example 7-3

The magnetic bearing of a line was recorded as S 80° 15′ W in 1880 at a place which had a declination of 16° E in that year. What is the present magnetic bearing if the declination is now 4° 30′ E?

Solution: The reduction to true azimuth in this problem is made in the following manner:

$$
\begin{aligned}
\text{magnetic bearing in 1880} &= \text{S } 80°\,15′\,\text{W} \\
\text{magnetic azimuth in 1880} &= 260°\,15′ \\
\text{declination in 1880} &= +\ 16° \\
\hline
\text{true azimuth} &= 276°\,15′ \\
\text{declination at present} &= -4°\,30′ \\
\hline
\text{magnetic azimuth at present} &= 271°\,45′ \\
\text{magnetic bearing at present} &= \text{N } 88°\,15′\,\text{W}
\end{aligned}
$$

Example 7-4

The true azimuth, from north, of a line is given as 144° 15′. This line is to be traced by using a compass when the declination is 3° 30′ W. What should be the reading on the compass to observe this line?

Solution: The computations follow:

$$
\begin{aligned}
\text{true azimuth} &= 144°\,15′ \\
\text{declination} &= +\ 3°\,30′ \\
\hline
\text{magnetic azimuth} &= 147°\,45′ \\
\text{magnetic bearing} &= \text{S } 32°\,15′\,\text{E}
\end{aligned}
$$

From the foregoing examples, the process might be considered unnecessarily long. This may be true if the conversion of only one or two lines is involved. When the survey involves several lines, however, the method of converting directions from one meridian to another through azimuths is consistent throughout, and its use tends to reduce mistakes involving algebraic signs. Also, in making conversions, a diagram like that in Fig. 7-10 should be drawn showing the angles to relatively correct size before deciding in each case whether a declination is to be added to or subtracted from an azimuth or a bearing angle.

Computations involving azimuths and bearings will be treated in more comprehensive examples in Chapter 8.

7-16. GYROCOMPASSING   The difficulties in using the magnetic compass to determine true directions are the uncertainty in the magnetic declination at the point of observation, the amount and direction of local attraction, and the inability to read the compass circle much better than 10' of arc. These difficulties are alleviated if the earth's rotation is used to define the true meridian. This method employs the *gyrocompass* which contains a high-precision pendulum gyroscope that orients its spin axis to true north by the angular rotational force of the earth.

The principle of the gyroscope is described with reference to Fig. 7-11. The *spin* axis is designated *s* in Fig. 7-11(a). The angular velocity vector of the spinning wheel is represented by the vector *S* in Fig. 7-11(b). The line perpendicular to the spin axis and designated *f* is called the *input* axis of the gyroscope. If an input force couple or torque *c–c* is applied to the spin axis, it tends to rotate the wheel about the input axis and out of its plane. The angular velocity vector *F* set up by the force couple *c–c* is shown in Fig. 7-11(b). The axis designated *p* is the *precession* axis about which the plane of

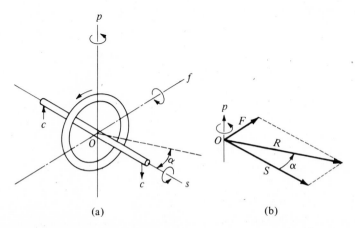

(a)   (b)

Figure 7-11. Principle of gyroscope.

the gyroscope wheel rotates. The three axes are all perpendicular to one another.

Referring to Fig. 7-11(b), the resultant of the two vectors S and F is R. This forms the angle α with vector S. Angle α is the angle of precession of the plane of the wheel about axis p. Consequently, the gyroscope will precess until its spin axis coincides with the direction of vector R. However, if the force couple continues to be applied, this rotation or precession will continue indefinitely. If the force couple is reduced to zero, then F goes to zero and the precession will stop.

In Fig. 7-12, the earth's angular velocity vector is represented by the vector E. Three different points on the meridian are shown. Point A lies on the equator where the horizontal component H of E is the full vector E. Vector H lies in the meridian. The vertical component of E is zero. At point B, whose latitude is shown as 40° N, the horizontal component of E is $H = E \cos 40°$, and the vertical component $V = E \sin 40°$. C lies at the pole where the horizontal component of E is zero and the vertical component is

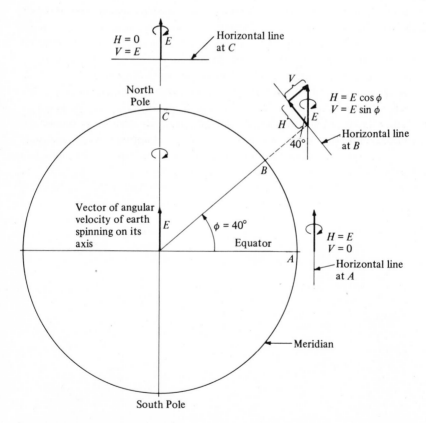

Figure 7-12. Horizontal and vertical components of earth's angular velocity vector at different latitudes.

E. Thus the horizontal component *H* of the earth's angular velocity vector decreases from a maximum at the equator to zero at the pole.

Imagine a gyroscope to be suspended on a wire in such a manner that its spin axis is constrained to lie in a horizontal plane, and making an angle *A* with the meridian. Figure 7-13(a) is an overhead view of such an arrangement, which is incorporated in an instrument called a *gyrocompass*. Point *p* is the precessional axis represented by the wire. The precessional axis is thus vertical. The angular velocity vector of the spinning wheel is designated *S* as in Fig. 7-11. The earth's rotation causes a force couple to be applied to the gyroscope's input axis causing the gyroscope spin axis to precess. The input angular velocity vector is the horizontal component *H*. The angle *A*, between the original direction of the spin axis and the vector *H* is shown in Fig. 7-13(b). The spin axis precesses to vector R_1. When it arrives at this new position as shown in Fig. 7-13(c), a new resultant R_2 is formed. The spin axis then precesses to this new position R_2. The new resultant R_3 is shown in Fig. 7-13(d). When the spin axis finally aligns itself with the meridian as shown in Fig. 7-13(e), the torque set up by the earth's rotation no longer acts on the spin axis, and the precession of the gyroscope comes to a stop. The spin axis thus points to true north. Actually, the precession causes the spin axis to overshoot the meridian because of mass momentum until the strength of the earth's rotational torque causes it to slow down and stop (at some angle to the west of true north in Fig. 7-13). This point is called the *west reversal* or *turning point* of the spin axis. Precession then takes place to the east until the east turning point is reached. This oscillation, whose period increases with latitude, would continue with a relatively large amplitude unless the oscillations are physically dampened by some braking arrangement imposed on the gyroscope precession axis. The dampening action limits the amplitude, but the period of oscillation remains constant for any given latitude.

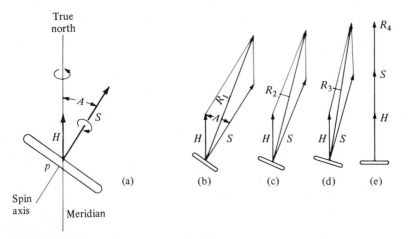

Figure 7-13. Gyroscope spin axis precessing to north.

The gyrocompass operates most efficiently at the equator where the horizontal component of E is a maximum. The horizontal component decreases rapidly toward the pole, and the gyrocompass rotates freely at the pole since no horizontal component exists.

If the gyroscope spin axis is constrained to remain horizontal, the force couple set up by the earth's rotation represented by the vertical vector V cannot act on the gyroscope spin axis. This is shown in Fig. 7-14. The vertical vector V acts upward to produce the force couple C_v-C_v, which would cause the spin axis to precess about axis p, tilting it out of the horizontal position through vertical angle β. But, since the spin axis is constrained in a horizontal position, no precession can take place. Thus, the vertical component V of the earth's angular velocity vector has no effect on the gyroscope.

The instrument attached to the top of the theodolite shown in Fig. 7-15 is a gyrocompass in which the gyroscope motor is suspended on a thin tape similar to a plumb bob. The upper end of the tape is connected to the upper end of the tubular housing at the top. A cross-sectional view of the gyrocompass is shown in Fig. 7-16. The moving mark oscillates with the gyroscope. This mark is projected down to a V-shaped index which is viewed through the eyepiece. Since the eyepiece index is attached to the instrument, it is fixed with respect to the alidade of the theodolite.

After the instrument is set up over the survey station, the theodolite telescope is aligned approximately to true north, by means of a magnetic compass (taking into account the declination of the compass needle). The gyrocompass is then attached to the theodolite. The gyro motor is now turned on and allowed to run up to full operating speed, after which the gyroscope is released. The projected moving mark can then be observed through the eyepiece as moving across the graduations contained in the eyepiece scale. This scale, together with the V-shaped index, is shown in Fig. 7-17.

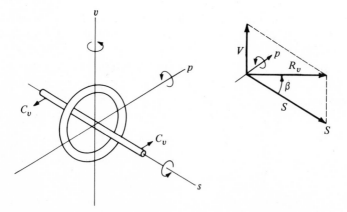

Figure 7-14. Vertical component of earth's angular velocity vector acting on gyroscope.

Figure 7-15. Wild GAK1 Gyrocompass. Courtesy of Wild Heerbrugg Instruments, Inc.

Either of two methods can be used to determine the precise direction of true north with an accuracy of 30″ or better. The first is called the *Turning Point Method.* The observer follows the oscillations of the moving mark, keeping it centered in the V-shaped index by turning the horizontal motion tangent screw of the theodolite (the upper tangent screw when using a repeating theodolite). When a turning point is reached, the moving mark will appear to remain stationary in the V-shaped index. A horizontal circle reading of the theodolite is now made and recorded. The observer then follows the movement of the moving mark, keeping it centered in the V-index until the opposite turning point is reached. The horizontal circle is again read and recorded. Three or more turning points are observed, from which the horizontal circle reading representing true north is computed by

$$N_1 = \frac{1}{2}\left(\frac{u_1 + u_3}{2} + u_2\right); \quad N_2 = \frac{1}{2}\left(\frac{u_2 + u_4}{2} + u_3\right), \quad \text{etc.} \quad (7\text{-}1)$$

and

$$N = \frac{\sum N_i}{n} \quad (7\text{-}2)$$

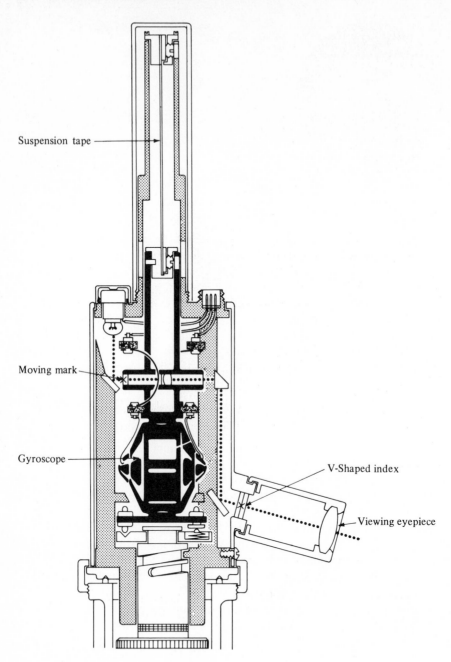

Figure 7-16. Cross section of gyrocompass.

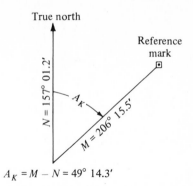

$A_K = M - N = 49°\ 14.3'$

Figure 7-17. Gyrocompass scale and V-shaped index.

Figure 7-18. Azimuth from gyrocompass and circle readings.

in which N is the horizontal circle reading indicating true north, $u_1, u_3, \ldots,$ are, say, the west turning point circle readings, and $u_2, u_4, \ldots$ are the east turning point readings.

The azimuth of a reference line is established by sighting on a reference mark direct and reversed, obtaining the mean of the D and R readings, and comparing this circle reading with the value of N computed by Eq. (7-2). Let M designate the mean circle reading to the reference mark and N represent the circle reading indicating true north as diagrammed in Fig. 7-18. The azimuth of the line determined by the gyrocompass is then

$$A_K = M - N \qquad (7\text{-}3)$$

Example 7-5

A series of clockwise circle readings were taken using the turning point method as follows. The direct reading to the reference mark was made before the north determination, and the reversed reading was taken afterwards.

WEST TURNING POINT	EAST TURNING POINT	REFERENCE MARK
u_1 154° 56.0′		D 206° 15.2′
	u_2 159° 04.8′	R 26° 15.8′
u_3 154° 58.6′		Mean 206° 15.5′
	u_4 159° 03.6′	
u_5 154° 59.0′		
	u_6 159° 02.8′	

Compute the azimuth of the line from the occupied point to the reference mark.

Solution: By Eqs. (7-1) and (7-2),

$$N_1 = \frac{1}{2}\left(\frac{154° 56.0' + 154° 58.6'}{2} + 159° 04.8'\right) = 157° 01.0'$$

$$N_2 = \frac{1}{2}\left(\frac{159° 04.8' + 159° 03.6'}{2} + 154° 58.6'\right) = 157° 01.4'$$

$$N_3 = \frac{1}{2}\left(\frac{154° 58.6' + 154° 59.0'}{2} + 159° 03.6'\right) = 157° 01.2'$$

$$N_4 = \frac{1}{2}\left(\frac{159° 03.6' + 159° 02.8'}{2} + 154° 59.0'\right) = \underline{157° 01.1'}$$

$$N = 157° 01.2'$$

The azimuth of the reference mark is by Eq. (7-3), $206° 15.5' - 157° 01.2' = 49° 14.3'$.

If the direction of the line of sight of the telescope of the theodolite does not agree with the point of exact symmetry of the oscillating mark by an angle E, then the azimuth determined as in Example 7-5 is not correct. This angle, which is usually small, can be determined and checked from time to time by observing the azimuth A_K of a line of known azimuth using the gyrocompass. Then, if the known azimuth is A,

$$E = A - A_K \qquad (7\text{-}4)$$

This small angle is shown in Fig. 7-19. The correct azimuth of a line computed as in Example 7-5 is obtained by adding the calibration value E. Thus

$$A = A_K + E = M - N + E \qquad (7\text{-}5)$$

The second method for determining azimuths using the gyrocompass is called the *Transit Method*. After the theodolite has been set up and oriented to approximate north, the clockwise horizontal circle is read. A stop watch with a trailing hand is used to measure the time it takes for the moving mark to travel from the V-index to the west and back to the V-index (considered positive) and the time it takes for the mark to travel from the V-index to the east and back again to the V-index (considered negative). Three or more such transit times of the moving mark across the V-index must be observed. It is also necessary to measure the amplitude of the moving mark in scale units

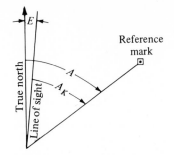

Figure 7-19. Calibration angle E.

of the index scale. This is done by noting the scale readings of the moving mark at the west and the east turning point and taking the mean of these two readings.

The time difference Δt is the difference between the time t_L it takes for the moving mark to move from the V-index to the left and back to the V-index (positive swing time) and the time t_R it takes for the moving mark to travel from the V-index to the right and back to the V-index (negative swing time). Thus

$$\Delta t = t_L - t_R \qquad (7\text{-}6)$$

If the average t_L and the average t_R are the same, then the line of sight of the theodolite points to true north (except for E). If they are not, then a correction ΔN must be applied to the initial theodolite circle reading N' in order to obtain the circle reading N which represents true north. The value of ΔN is computed by

$$N = ca\,\Delta t \qquad (7\text{-}7)$$

in which c is a proportionality factor relating scale readings to time, a is the mean of the amplitudes left and right, and Δt is given by Eq. (7-6). Then

$$N = N' + \Delta N \qquad (7\text{-}8)$$

The proportionality factor c is determined by making two separate measurements with different initial orientation directions N' of the telescope. The directions N'_1 and N'_2 should each be about 15' to the west and east, respectively, of the middle of the oscillations of the moving mark. For the true north readings of the theodolite circle, the following equations are solved

$$N = N'_1 + c\,\Delta t_1 a_1$$

$$N = N'_2 + c\,\Delta t_2 a_2$$

from which

$$c = \frac{N_2' - N_1'}{\Delta t_1 a_1 - \Delta t_2 a_2} \qquad (7\text{-}9)$$

Example 7-6

A theodolite is set up at a point, and a sight is taken to a reference mark. The gyrocompass is then attached, the telescope is pointed to approximate north, and the horizontal circle is read as 188° 22.6′ for this orientation [N' of Eq. (7-8)]. The following readings are taken, using the transit method. The swing time to the left is measured first.

TIME OF TRANSIT	AMPLITUDE	REFERENCE MARK
L $4^m\ 45.7^s$ R $8^m\ 24.4^s$ L $11^m\ 57.7^s$ R $15^m\ 36.8^s$ $19^m\ 10.2^s$	L 8.0 R 7.7 Av 7.85	D 46° 16.6′ R 226° 16.0′ Mean 46° 16.3′

The value of E for the gyrocompass is given as $+2.7'$, and the factor c is $0.046'/s$. Compute the azimuth to the reference mark.

Solution: The first swing time from the V-index to the left and back to the V-index is $8^m\ 24.4^s - 4^m\ 45.7^s = (+) 3^m\ 38.7^s$. The first swing time to the right is $11^m\ 57.7^s - 8^m\ 24.4^s = (-) 3^m\ 33.3^s$. The second swing time to the left is $15^m\ 36.8^s - 11^m\ 57.7^s = (+) 3^m\ 39.1^s$, and to the right is $19^m\ 10.2^s - 15^m\ 36.8^s = (-)3^m\ 33.4^s$. Then

$$\Delta t_1 = 3^m\ 38.7^s - 3^m\ 33.3^s = +5.4^s; \qquad \Delta N_1 = 0.046 \times 7.85 \times 5.4 = +2.0'$$

$$\Delta t_2 = -3^m\ 33.3^s + 3^m\ 39.1^s = +5.8^s; \qquad \Delta N_2 = 0.046 \times 7.85 \times 5.8 = +2.1'$$

$$\Delta t_3 = 3^m\ 39.1^s - 3^m\ 33.4^s = +5.7^s; \qquad \Delta N_3 = 0.046 \times 7.85 \times 5.7 = +2.1'$$

$$\text{Mean } \Delta N = +2.1'$$

By Eq. (7-8), $N = 188° 22.6' + 2.1' = 188° 24.7'$. By Eq. (7-5), $A = 46° 16.3' - 188° 24.7' + 2.7' = -142° 05.7'$ or $217° 54.3'$ clockwise from north.

Reliable azimuths are ordinarily determined in survey work by measuring angles from existing reference azimuths as explained in Chapter 8, or by making observations on the sun or the stars as explained in Chapter 12. However, in many situations such as in underground mine surveys or where the area is perpetually overcast with clouds, these methods would not be possible. The gyrocompass on the other hand permits the determination of reliable azimuths under these adverse conditions.

PROBLEMS

7-1. Convert the following bearings to azimuths: (a) S 15° 25′ 20″ E; (b) N 18° 16′ 30″ E; (c) N 18° 16′ 30″ W; (d) S 27° 20′ 15″ W; (e) S 88° 16′ 40″ E.

7-2. Convert the following bearings to azimuths: (a) N 45° 27′ 45″ W; (b) N 27° 20′ 35″ E; (c) S 27° 20′ 35″ E; (d) S 77° 20′ 15″ W; (e) N 89° 10′ 15″ W.

7-3. Convert the following azimuths to bearings: (a) 106° 22′ 10″; (b) 57° 18′ 50″; (c) 327° 14′ 20″; (d) 267° 53′ 15″; (e) 178° 14′ 20″.

7-4. Convert the following azimuths to bearings: (a) 20° 12′ 30″; (b) 90° 04′ 50″; (c) 166° 14′ 40″; (d) 270° 14′ 15″; (e) 354° 26′ 15″.

7-5. The true azimuth of a line is 187° 10′. The observed magnetic bearing of the line is S 2° 15′ E. What is the magnetic declination at the point of observation?

7-6. In 1890 the magnetic bearing of a line was observed to be S 2° 30′ E, at which time the magnetic declination was 4° 10′ W. This year the observed magnetic bearing of the same line is S 8° 30′ W. What is the present magnetic declination?

7-7. In 1888 the magnetic declination was 7° 15′ W. The magnetic bearing of a line in that year was N 79° 45′ W. The present magnetic bearing of the line is N 87° 30′ W. What is the present magnetic declination?

7-8. Given the following magnetic bearings: (a) S 1° 40′ W; (b) S 1° 10′ E; (c) N 20° 30′ W; (d) N 3° 10′ W; (e) N 45° 00′ W; (f) S 38° 10′ E. The magnetic declination is 16° 10′ W. Compute the true bearings.

7-9. Given the following magnetic bearings: (a) N 18° 30′ W; (b) S 2° 20′ E; (c) S 6° 14′ W; (d) N 25° 50′ E; (e) N 88° 20′ E; (f) N 86° 14′ W. The magnetic declination is 17° 20′ W. Compute the true bearings.

7-10. Using the magnetic declination chart for 1980, determine the magnetic declination: (a) at New York City in 1945; (b) at San Francisco in 1955; (c) at Chicago in 1990; (d) at Miami in 1880; (e) at St. Louis in 1890; (f) at Denver in 1910; (g) at Boston in 2010.

7-11. Using the magnetic declination chart for 1980, determine the magnetic declination: (a) at Houston in 1880; (b) at Portland in 1930; (c) at Baltimore in 1900; (d) at New Orleans in 1990; (e) at Cleveland in 1940; (f) at Omaha in 1950; (g) at Salt Lake City in 1925.

7-12. At the time a survey was run, the magnetic declination was 6° 50′ E. The magnetic bearings of several lines observed at the time were as follows: AB = N 26° 20′ W; BC = S 4° 40′ E; CD = N 2° 15′ E; DE = S 58° 00′ E; EF = N 88° 30′ W. These lines are to be retraced using a compass when the declination is 0° 30′ W. What bearings should be set off on the compass?

7-13. A retracement survey is to be run by the compass at the present time when the magnetic declination is 3° 20′ W. When the original survey was run, the magnetic declination was 1° 50′ E. The original survey magnetic bearings were as follows: AB = S 1° 14′ E; BC = S 48° 12′ E; CD = N 88° 30′ E; DE = N 10° 40′ W; EF = S 88° 10′ W. What bearings should be set off on the compass?

BIBLIOGRAPHY

Bowker, O. W. 1972. Lightweight gyro azimuth surveying instrument. *Proc. Ann. Meet. Amer. Cong. Surv. & Mapp.* p. 379.

Dawson, E., and Dalgetty, L. C. 1967. The march of the compass in Canada. *Canadian Surveyor* 21:380.

Gregerson, L. F. 1973. Report on gyroscopic experiments in Canada, 1969–1972. *Proc. Ann. Meet. Amer. Cong. Surv. & Mapp.* p. 144.

Gregerson, L. F. 1974. The gyroscope as a tool for azimuth determination. *Canadian Surveyor* 28:37.

Mining surveying and rock deformation measurements—Symposium. 1970. *Canadian Surveyor* 24:23–46, 86–135.

Chapter 8
Traverse Surveys
and Computations

8-1. TRAVERSE A traverse is a series of connected lines of known length related to one another by known angles. The lengths of the lines are determined by direct measurement of horizontal distances, by slope measurement, or by indirect measurement based on the methods of tacheometry. These methods are discussed in Chapters 2 and 14. The angles at the traverse stations between the lines of the traverse are measured with the instruments discussed in Chapter 4. They can be interior angles, deflection angles, or angles to the right.

The results of field measurements related to a traverse will be a series of connected lines whose lengths and azimuths, or whose lengths and bearings, are known. The lengths are horizontal distances; the azimuths or bearings are either true, magnetic, assumed, or grid.

A completely different method of traversing can be performed using an inertial system which is described in Chapter 9. This system incorporates a gyro-stabilized platform containing three accelerometers oriented at right angles to one another, a precise clock, and an on-board computer. The accelerometers sense accelerations in the north–south, east–west, and vertical directions from which displacements in these three directions are computed.

In general, traverses are of two classes. One of the first class is an open traverse. It originates either at a point of known horizontal position with respect to a horizontal datum or at an assumed horizontal position, and terminates at an unknown horizontal position. A traverse of the second class is a closed traverse, which can be described in any one of the following three ways: (1) It originates at an assumed horizontal position and terminates at that same point; (2) it originates at a known horizontal position with respect to a horizontal datum and terminates at that same point; (3) it originates at a known horizontal position and terminates at another known horizontal position. A known horizontal position is defined by its geographic latitude and longitude, by its Y and X coordinates on a grid system, or by its location on or in relation to a fixed boundary.

Traverse surveys are made for many purposes and types of projects, some of which follow:

1. To determine the positions of existing boundary markers.
2. To establish the positions of boundary lines.
3. To determine the area encompassed within the confines of a boundary.
4. To determine the positions of arbitrary points from which data may be obtained for preparing various types of maps, that is, to establish *control* for mapping.
5. To establish ground control for photogrammetric mapping.
6. To establish control for gathering data regarding earthwork quantities in railroad, highway, utility, and other construction work.
7. To establish control for locating railroads, highways, and other construction work.

8-2. OPEN TRAVERSE   An open traverse is usually run for exploratory purposes. There are no arithmetical checks on the field measurements. Since the figure formed by the surveyed lines does not close, the angles cannot be summed up to a known mathematical condition. None of the positions of the traverse stations can be verified, since no known or assumed position is included except that of the starting station. To strengthen an open traverse, that is, to render it more reliable, several techniques may be employed. Each distance can be measured in both directions and can be roughly checked by using the stadia hairs of the theodolite (see Chapter 14). The measurements of the angles at the stations can be repeated by using the methods of Chapters 4 and 6 and checked approximately by observing magnetic bearings. The directions of the lines can be checked by observing the sun or the stars to determine true azimuths or bearings of selected lines in the traverse. An open traverse should not be run for any permanent project or for any of the projects indicated in Section 8-1 because it does not reveal mistakes or errors and the results are always open to doubt.

8-3. CLOSED TRAVERSE   A traverse that closes on itself immediately affords a check on the internal accuracy of the measured angles, provided that the angle at each station has been measured. As will be discussed later, a traverse that closes on itself gives an indication of the consistency of measuring distances as well as angles by affording a check on the position closure of the traverse. Unless astronomical observations for azimuths have been made at selected station of a traverse that closes on itself, the only provision for verifying the directions of the lines is that afforded by the angular closure. In this type of traverse there is no check on the systematic errors introduced into the measurement of lengths. Therefore when this type of traverse is executed for a major project, the measuring apparatus must be carefully calibrated to determine the systematic errors and to eliminate them.

A traverse that originates at a known position and closes on another known position is by far the most reliable, because a check on the position of the final point checks both the linear and angular measurements of the traverse. When a point of known position is referred to, it is understood that such a point has been located by procedures as precise as, or more precise than, those used in the traverse being executed. These procedures are the methods of either traversing, triangulation, or trilateration. Triangulation and trilateration are the subjects of Chapter 10.

8-4. INTERIOR-ANGLE TRAVERSE An interior-angle traverse is shown in Fig. 8-1. The azimuth or bearing of the line AP is known. The lengths of the traverse lines are measured to determine the horizontal distances. With the transit or theodolite at A, the angle at A from P to B is measured to

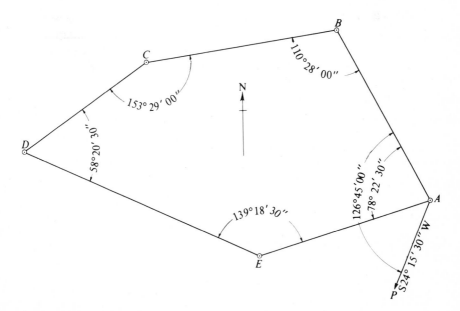

Figure 8-1. Interior-angle traverse.

determine the azimuth of line *AB*. The angle at *A* from *E* to *B* is also measured, as this is one of the interior angles in the figure. The instrument is then set up at *B*, *C*, *D*, and *E* in succession, and the indicated angles are measured. The notes for the angles are recorded as indicated in Chapter 4. Had the azimuth of one of the traverse sides been known, there would have been no necessity for measuring the angle at *A* from *P* to *B*.

To test the internal angular closure, the interior angles are added and their sum is compared to $(n - 2)180°$, which in this example is $(5 - 2)180° = 540°$. The total angular error, or the closure, is $1'30''$. In the tabulation in Table 8-1 the measured angles are adjusted by assuming that the angular error is of the same amount at each station. This assumption may not be valid, because an error in angular measurement, all other things being equal, will increase as the lengths of the adjacent sides decrease. In the traverse of Fig. 8-1 the interior angles are presumed to have been measured to the nearest 30″. The manner of adjusting the $1'30''$ closure can be somewhat arbitrary in this instance. For example, instead of applying an 18″ correction to each angle, the discrepancy could be distributed by correcting the angles at *A*, *B*, and *C* by 30″ each, or by correcting the angles at *A*, *C*, and *D* by 30″ each.

After the adjusted angles are computed, they should always be added to see whether their sum is, in fact, the proper amount. A mistake in arithmetic either in adding the measured angles or in applying the corrections will become apparent.

The azimuth of the line *AP* in Fig. 8-1 is known, and the azimuths of all the traverse sides can be determined by using the measured angle at *A* from *P* to *B* and the adjusted interior angles in the closed figure. Since the bearing of *AP* is S 24° 15′ 30″ W, the azimuth of *AP* reckoned from north is 204° 15′ 30″. As indicated in Fig. 8-2 the azimuth of *AB* is equal to the azimuth of *AP* plus the angle at *A* from *P* to *B*, or 204° 15′ 30″ + 126° 45′ 00″ = 331° 00′ 30″. The azimuths of the other lines are computed systematically by applying the adjusted angle at each station to the azimuth of each backsight line in turn. To determine the azimuth of *BC*, compute the azimuth of *BA* by subtracting 180° from the azimuth of *AB*, and then add the adjusted angle at *B* from *A* to *C*. A similar procedure is adopted at each station (see Table 8-2).

TABLE 8-1. Adjustment of Angles in an Interior Angle Traverse

STATION	MEASURED ANGLE	CORRECTION	ADJUSTED ANGLE
A	78° 22′ 30″	+18″	78° 22′ 48″
B	110° 28′ 00″	+18″	110° 28′ 18″
C	153° 29′ 00″	+18″	153° 29′ 18″
D	58° 20′ 30″	+18″	58° 20′ 48″
E	139° 18′ 30″	+18″	139° 18′ 48″
	539° 58′ 30″		540° 00′ 00″
	−540° 00′ 00″		
	closure = −1′ 30″		

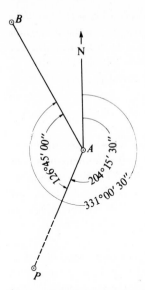

Figure 8-2. Determining azimuth of line *AB*.

TABLE 8-2. Computation of Azimuths and Bearings Using Adjusted Interior Angles

LINE	AZIMUTH	BEARING
AP	204° 15′ 30″	S 24° 15′ 30″ W
	(+ ∠*A*) + 126° 45′ 00″	
AB	331° 00′ 30″	N 28° 59′ 30″ W
BA	151° 00′ 30″	
	(+ ∠*B*) + 110° 28′ 18″	
BC	261° 28′ 48″	S 81° 28′ 48″ W
CB	81° 28′ 48″	
	(+ ∠*C*) + 153° 29′ 18″	
CD	234° 58′ 06″	S 54° 58′ 06″ W
DC	54° 58′ 06″	
	(+ ∠*D*) + 58° 20′ 48″	
DE	113° 18′ 54″	S 66° 41′ 06″ E
ED	293° 18′ 54″	
	(+ ∠*E*) + 139° 18′ 48″	
EA	432° 37′ 42″	
EA	72° 37′ 42″	N 72° 37′ 42″ E
AE	252° 37′ 42″	
	(+ ∠*A*) + 78° 22′ 48″	
AB	331° 00′ 30″ check	N 28° 59′ 30″ W

The azimuths of the lines can also be computed by applying the angles at
A, E, D, C, and B in that order. Then each interior angle would be *subtracted*
from the azimuth of the proper back line. This is apparent from a study of
Fig. 8-1.

The bearing of each line, if desired, is determined from its azimuth. Since
the conversion from azimuths to bearings is easily made, it is unwise to
attempt to compute the bearing of each line directly by applying the proper
angle in the traverse to the bearing of the adjacent line.

In a traverse of n sides or stations the closure of the measured angles should
not exceed the least count of the vernier or scale of the instrument times the
square root of n. With care in sighting, in centering the instrument, and in
reading the verniers or scales, the angular closure can be expected to be half
this allowable amount. Of course, with few traverse stations, this precision
may not be obtainable since the opportunities for random errors to compensate are few.

One very important point to observe in the preceding example of an
interior-angle traverse is that there is no check on the angle at A from P to
B. If this angle is in error, then the error affects the azimuth of each line in
the traverse. To avoid the possibility of making a mistake in measuring this
angle, the clockwise angle at A from B to P should also be measured. This
measurement immediately affords a check at station A since the angle at A
from P to B and the angle at A from B to P should total 360°. If the azimuth
of one of the sides of the traverse were known, then this uncertainty would
have been avoided, because the test of the accuracy in this case is simply
that the sum of the interior angles should equal $(n - 2)$ 180°.

8-5. DEFLECTION-ANGLE TRAVERSE   A deflection-angle traverse
that originates at station 0 + 00 and closes on station 22 + 20 is shown in
Fig. 8-3. The azimuth of the line FG is fixed as 196° 35′ and the azimuth of
EM is fixed as 306° 43′, these directions having been established by previous
surveys. The traverse is made to fit between these two fixed azimuths. After the
deflection angles and the lengths of the several courses have been measured,
the check on angular closure is made by carrying azimuths through the traverse
by applying the deflection angles. In Fig. 8-4 the azimuth of GF is 16° 35′,
obtained by subtracting 180° from the azimuth of FG. So the azimuth of the
line FA is 16° 35′ plus 44° 28′, which is the right deflection angle at F from G
to A. The addition gives 61° 03′ as the azimuth of FA. Now to determine
the azimuth of the next line AB, the right deflection angle at A from F to B
is added to the azimuth of FA prolonged. This sum is 61° 03′ + 57° 42′ =
118° 45′. Observe that no back azimuths need be computed as was the case
for the interior-angle traverse. To compute an azimuth by using a deflection
angle, simply add a right deflection angle to the *forward* azimuth of the backsight to obtain the *forward* azimuth of the foresight; or subtract a left deflection angle from the *forward* azimuth of the backsight to obtain the *forward*

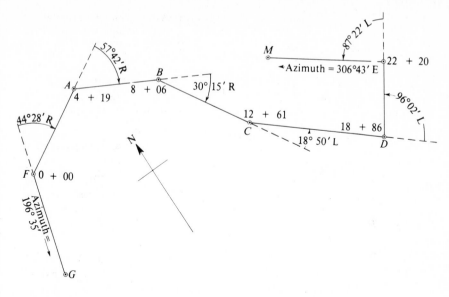

Figure 8-3. Deflection-angle traverse between fixed azimuths.

azimuth of the foresight. A computation of azimuths in deflection-angle traverse is given in Table 8-3.

In the example in Fig. 8-3, the computed azimuth of *EM* failed to check by 3′. Since six deflection angles were measured, the correction to each angle is 30″. Instead of applying this correction to each angle and recomputing the azimuths, the azimuths themselves are adjusted. The azimuth of *FA* receives

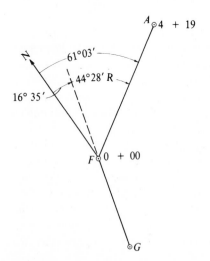

Figure 8-4. Determining azimuth of line *FA*.

TABLE 8-3. Computation of Azimuths and Bearings Using Deflection Angles

LINE	AZIMUTH	CORRECTION	ADJUSTED AZIMUTH	ADJUSTED BEARING
GF	16° 35′	fixed		N 16° 35′ 00″ E
	(+ ∠*F*) + 44° 28′			
FA	61° 03′	−0′ 30″	61° 02′ 30″	N 61° 02′ 30″ E
	(+ ∠*A*) + 57° 42′			
AB	118° 45′	−1′ 00″	118° 44′ 00″	S 61° 16′ 00″ E
	(+ ∠*B*) + 30° 15′			
BC	149° 00′	−1′ 30″	148° 58′ 30″	S 31° 01′ 30″ E
	(− ∠*C*) − 18° 50′			
CD	130° 10′	−2′ 00″	130° 08′ 00″	S 49° 52′ 00″ E
	(− ∠*D*) − 96° 02′			
DE	34° 08′	−2′ 30″	34° 05′ 30″	N 34° 05′ 30″ E
DE	394° 08′			
	(− ∠*E*) − 87° 22′			
EM	306° 46′	−3′ 00″	306° 43′ 00″	N 53° 17′ 00″ W
	306° 43′	fixed		
	closure = +03′			

a 30″ correction, since this azimuth was obtained by considering only one measured angle; the azimuth of *AB* receives a 1′ 00″ correction, since this azimuth was obtained by using two angles, and so on. The correction to the last azimuth is 6 × 30″ = 3′ 00″, since this azimuth was obtained by using all six deflection angles.

In Fig. 8-5 is illustrated a deflection-angle traverse that closes on the point of origin at *L*. The bearing of the line *OP* is known to be N 66° 02′ W. Before the bearings of the remaining sides are computed, the angles must be adjusted so that the difference between the sum of the right deflection angles and the sum of the left deflection angles is 360°. It is found that the sum of the right deflection angles must be reduced and the sum of the left deflection angles must be increased. The correction in this case is distributed equally among all of the angles. Had the angular closure been, for example, only 01′, then 30″ could have logically been added to the angle at *N* and subtracted from the angle at *Q*, since these two angles are formed by the shortest lines of sight. The tabulation of Table 8-4 shows this adjustment of deflection angles.

Bearings of the sides of the traverse are computed by first converting the bearing of *OP* to an azimuth, then applying the adjusted deflection angles to obtain azimuths, and finally converting the azimuths to bearings as shown in Table 8-5.

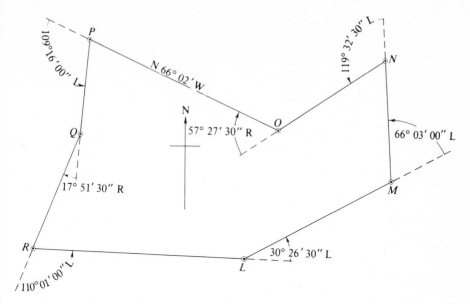

Figure 8-5. Deflection-angle traverse that closes on point of origin. (Adjusted angles are shown.)

The angular closure of a deflection-angle traverse should be no more than the least count of the vernier of the instrument times the square root of the number of angles in the traverse. A deflection angle should never be measured without double centering the instrument, because the error caused by the line of sight not being normal to the horizontal axis of the instrument may be too large to be tolerated.

TABLE 8-4. Adjustment of Deflection Angles

STATION	DEFLECTION ANGLE	CORRECTION	ADJUSTED DEFLECTION ANGLE
L	30° 26′ 00″ L	+30″	30° 26′ 30″ L
M	66° 02′ 30″ L	+30″	66° 03′ 00″ L
N	119° 32′ 00″ L	+30″	119° 32′ 30″ L
O	57° 28′ 00″ R	−30″	57° 27′ 30″ R
P	109° 15′ 30″ L	+30″	109° 16′ 00″ L
Q	17° 52′ 00″ R	−30″	17° 51′ 30″ R
R	110° 00′ 30″ L	+30″	110° 01′ 00″ L

$\sum$ right 75° 20′ 00″
$\sum$ left 435° 16′ 30″
difference 359° 56′ 30″
closure 3′ 30″

$\sum$ right 75° 19′ 00″
$\sum$ left 435° 19′ 00″
360° 00′ 00″ check

TABLE 8-5. Computation of Azimuths and Bearings Using Adjusted Deflection Angles

LINE	AZIMUTH	BEARING
OP	293° 58′ 00″	N 66° 02′ 00″ W
	(− ∠P) − 109° 16′ 00″	
PQ	184° 42′ 00″	S 4° 42′ 00″ W
	(+ ∠Q) + 17° 51′ 30″	
QR	202° 33′ 30″	S 22° 33′ 30″ W
	(− ∠R) − 110° 01′ 00″	
RL	92° 32′ 30″	S 87° 27′ 30″ E
	(− ∠L) − 30° 26′ 30″	
LM	62° 06′ 00″	N 62° 06′ 00″ E
LM	422° 06′ 00″	
	(− ∠M) − 66° 03′ 00″	
MN	356° 03′ 00″	N 3° 57′ 00″ W
	(− ∠N) − 119° 32′ 30″	
NO	236° 30′ 30″	S 56° 30′ 30″ W
	(+ ∠O) + 57° 27′ 30″	
OP	293° 58′ 00″ check	N 66° 02′ 00″ W

8-6. ANGLE-TO-THE-RIGHT TRAVERSE   Either an open traverse or a closed traverse can be executed by measuring angles to the right. The method of measuring the angles is described in Section 6-3. The method of computing azimuths from a given fixed azimuth is similar to that employed in an interior-angle traverse. A forward azimuth is always obtained by *adding* the angle to the right to the azimuth of the backsight. Angles to the right are always employed when a traverse is executed with a direction instrument.

As an illustration of an angle-to-the-right traverse executed by use of a direction instrument refer to Fig. 8-6. Station P is occupied, a backsight is taken on station T, and the circle is read. A foresight is taken on station Q, and the circle is read. This procedure is repeated at each station along the traverse, each backsight and foresight being observed with the telescope both direct and reversed. Notes for a 10″ direction instrument are given in Table 8-6.

Assume that the azimuth of PT is known to be 88° 20′ 10″. The azimuth of the line PQ is the azimuth of PT plus the angle to the right at P. The angle to the right at P from T to Q can be obtained by subtracting the mean circle reading to T from the mean circle reading to Q. This angle, as computed from notes, is the 92° 16′ 00″ − 51° 54′ 50″ = 40° 21′ 10″. Then the azimuth of PQ is 88° 20′ 10″ + 40° 21′ 10″ = 128° 41′ 20″. This computation can be simplified if the mean circle reading to Q is added to the azimuth of PT, and then

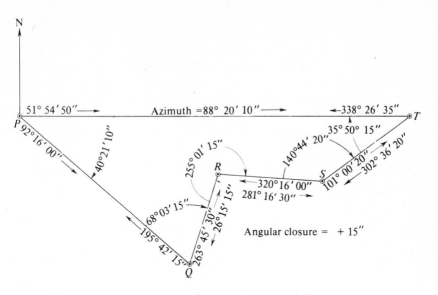

Figure 8-6. Angles-to-the-right from direction instrument readings.

TABLE 8-6. Notes for Measuring Angles to the Right When Using a Direction Instrument

STATION		D OR R	READING	MEAN
At P	T	D	51° 54′ 50″	51° 54′ 50″
		R	231° 54′ 50″	
	Q	D	92° 15′ 50″	92° 16′ 00″
		R	272° 16′ 10″	
At Q	P	D	195° 42′ 10″	195° 42′ 15″
		R	15° 42′ 20″	
	R	D	263° 45′ 20″	263° 45′ 30″
		R	83° 45′ 40″	
At R	Q	D	26° 15′ 10″	26° 15′ 15″
		R	206° 15′ 20″	
	S	D	281° 16′ 30″	281° 16′ 30″
		R	101° 16′ 30″	
At S	R	D	320° 16′ 00″	320° 16′ 00″
		R	140° 16′ 00″	
	T	D	101° 00′ 10″	101° 00′ 20″
		R	281° 00′ 30″	
At T	S	D	302° 36′ 10″	302° 36′ 20″
		R	122° 36′ 30″	
	P	D	338° 26′ 30″	338° 26′ 35″
		R	158° 26′ 40″	

the mean circle reading to T is subtracted from the sum. In this manner the circle reading for the foresight is always added and the circle reading for the backsight is always subtracted.

The computation is shown in Table 8-7 as it would be made without the aid of a calculating machine designed for adding and subtracting angles on the sexagesimal system. If such a calculating machine is used, only the forward azimuth of each line of the traverse need be recorded on the computation sheet. The resulting azimuths are then adjusted to the fixed azimuth, and bearings may be obtained from the adjusted azimuths.

TABLE 8-7. Computations of Azimuths Using Angle-to-the-Right Notes

LINE	AZIMUTH	CORRECTION	ADJUSTED AZIMUTH
PT	88° 20′ 10″ fixed		
	+ 92° 16′ 00″		
	180° 36′ 10″		
	− 51° 54′ 50″		
PQ	128° 41′ 20″	−3″	128° 41′ 17″
QP	308° 41′ 20″		
	+263° 45′ 30″		
	572° 26′ 50″		
	−195° 42′ 15″		
QR	376° 44′ 35″		
QR	16° 44′ 35″	−6″	16° 44′ 29″
RQ	196° 44′ 35″		
	+281° 16′ 30″		
	478° 01′ 05″		
	− 26° 15′ 15″		
RS	451° 45′ 50″		
RS	91° 45′ 50″	−9″	91° 45′ 41″
SR	271° 45′ 50″		
	+101° 00′ 20″		
	372° 46′ 10″		
	−320° 16′ 00″		
ST	52° 30′ 10″	−12″	52° 29′ 58″
TS	232° 30′ 10″		
	+338° 26′ 35″		
	570° 56′ 45″		
	−302° 36′ 20″		
TP	268° 20′ 25″	−15″	268° 20′ 10″
PT	88° 20′ 25″		
	88° 20′ 10″ fixed		
	closure + 15″		

8-7. TRAVERSE-BY-AZIMUTH METHOD

In running a traverse for the purpose of establishing a lower order of control for mapping and for locating the positions of ground objects with respect to such a supplementary traverse, the transit or theodolite can be handled so that the clockwise circle reading at all times indicates the azimuth of the line of sight. This procedure eliminates the need for computing azimuths from interior angles, deflection angles, or angles to the right.

In Fig. 8-7 is shown a portion of a supplementary traverse in which the line *DE* has an azimuth of 96° 22′. This may be a true, magnetic, assumed, or grid azimuth. The instrument is set up at station *E* and the clockwise circle is set to read the azimuth of the line *ED*, which is 276° 22′. A lower motion is used to take a backsight along the line *ED*. The theodolite and the horizontal circle are now oriented. In other words, when any sight is taken by an upper motion, the clockwise circle reading will always indicate the azimuth of the line of sight. If a pointing is made on station *F* and the clockwise circle reads 68° 02′, this is the azimuth of the line *EF* and is recorded as such.

When the transit is set up at *F*, the circle is oriented by setting the azimuth of *FE*, which is 248° 02′, on the clockwise circle and backsighting along the line *FE* by a lower motion. A sight taken on point *P* by an upper motion will give the azimuth of the line *FP* directly on the clockwise circle. The readings to *P*, *Q*, and *G* are shown to be, respectively, 57° 25′, 192° 52′, and 113° 10′.

The advantages of executing a traverse by observing azimuths are that the transit or theodolite is allowed to do the work of adding and subtracting angles, and that the field notes are easily reduced to map form by having all lines related to the same meridian directly in the notes.

The disadvantage of such a procedure is in not realizing the benefit of double centering, which eliminates instrumental errors and which makes mistakes in reading the circle quite obvious. The purpose of the traverse, however, is usually such that small errors are of little consequence (see Section 14-8).

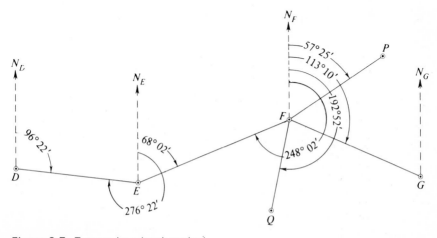

Figure 8-7. Traverse by azimuth method.

If a traverse is to be run on magnetic azimuths and the directions of the lines are to be consistent with one another, the initial setup determines the specific magnetic meridian to which all other lines are referred. To orient on the magnetic meridian, set the clockwise circle to read zero, unclamp the compass needle, loosen the lower clamp, and rotate the transit until the compass needle points to the north point of the compass circle. Tighten the lower clamp, and make an exact setting by using the tangent screw. The transit is then oriented for measuring magnetic azimuths. The procedure for carrying azimuths through the remaining lines in the traverse is the same as that previously described.

8-8. AZIMUTH TRAVERSE An azimuth traverse is a continuous series of lines of sight related to one another by measured angles only. The distances between the instrument stations are not measured. An azimuth traverse serves one of two purposes. The first purpose is to permit the determination of directions far removed from a beginning azimuth without the necessity of measuring distances. As an example, consider a pair of intervisible stations, such as Ridge 1 and Ridge 2 in Fig. 8-8, situated high on a ridge, and assume that the azimuth of the line joining these stations is known. A traverse is to be run in an adjacent valley, and the basis of azimuths of this traverse is to be the same as that for the line along the ridge.

To carry the azimuth down into the valley, one end of the known line, as Ridge 1, is occupied and the other end of the line, Ridge 2, is used as a back-sight for measuring an angle to the right, a deflection angle, or, if the traverse

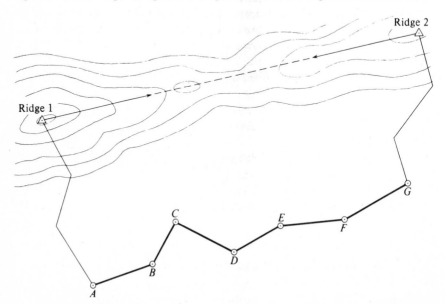

Figure 8-8. Azimuth traverse.

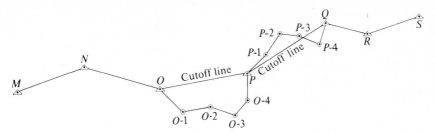

Figure 8-9. Azimuth traverse by cutoff lines.

is to close by occupying station Ridge 2, an interior angle. All the other points are occupied by the instrument, and angles are measured. That portion of the survey from Ridge 1 to A and the portion from G to Ridge 2 are azimuth traverses. The lengths of the lines from AB through FG are measured along with the angles. The angular closure can be computed, and the traverse from A to G will be on the same basis of azimuths as is the straight line from Ridge 1 to Ridge 2.

The second purpose of an azimuth traverse is to avoid carrying azimuths through extremely short traverse sides. Since the angular error will increase as the lines of sight become shorter, the desirable traverse is one with long sights. Such sights may not be practical, however, for several reasons. The ground over which the traverse must follow may be rough and the sides may have to be short to alleviate difficult taping; the traverse may run through a city where every street corner or every point at a break in the street center line is a transit station; the traverse may be run along a curving right of way where intervening brush would interfere with the measurement of the lengths of long traverse sides. In these situations the angular errors accumulating from short lines of sight can be isolated by employing an azimuth traverse of one or more stations through which the azimuth is computed. This use of an azimuth traverse is illustrated in Fig. 8-9. The azimuth is carried through lines MN, NO, OP, PQ, QR, and RS, although the azimuths of the lines forming the loops that have been cut off must also be determined. The lengths of the cutoff lines OP and PQ would not be measured. All the remaining distances would be measured.

The angular closure along the route $MNOPQRS$ is adjusted by methods discussed in the previous sections. This adjustment fixes the directions of the lines OP and PQ. The loop O, O-1, O-2, O-3, O-4, P is then adjusted to the azimuth of OP. The loop P, P-1, P-2, P-3, P-4, Q is then adjusted to the azimuth of PQ.

8-9. REFERENCING A TRAVERSE STATION
As an aid in relocating a point that may become hidden by vegetation or buried beneath the surface of the ground, or as a means of replacing a point that may have been destroyed, measurements are made to nearby permanent or semipermanent

objects. This process is known as *referencing* or *witnessing* the point. Property corners and, on important surveys, all instrument stations are usually referenced.

In Fig. 8-10 are shown the locations of the witness points with respect to station *A*. If the stake at *A* cannot be found at a later time, its position can be determined by locating the intersection of arcs struck with the trees as centers. The point would, of course, be determined by two arcs, but a third measurement is taken to serve as a check. If it is likely that any of the witness points will be destroyed, additional witnesses are located.

The method of recording the witnesses is the same as that used in notes for all United States land surveys (see Chapter 18). First, the object is described. If it is a tree, its diameter is given. Next, the bearing from the station to the witness point is given. Last, the distance is recorded. If the measurement is made to any definite point, such as a nail driven in the root or side of a tree, that fact should be stated. To be of the most value in replacing a missing station, the witnesses should be less than 100 ft from the station and, if possible, the arcs should intersect approximately at right angles.

On many surveys no permanent objects may be available as witnesses. In such cases additional stakes can be driven. The method illustrated in Fig. 6-15(f) is the most satisfactory way of referencing a point so that it can be replaced in its original position. In this figure the point *P* is at the intersection of the lines *AB* and *CD*. If the distances from *P* to the four stakes are measured carefully, the point can be replaced if any two of the stakes remain, and the relocation can be checked if three of them can be found. This method is commonly used in referencing transit stations on route surveys, where it is known that all centerline stakes will be destroyed as soon as grading operations are begun.

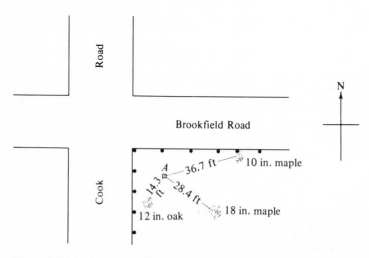

Figure 8-10. Referencing a point.

In addition to the specific witnesses, a general description of the location of the traverse station should be given, so that a person searching for it will have a fairly good idea of where to begin his search.

8-10. TRAVERSE COMPUTATION The result of the field work in executing a traverse of any kind is a series of connected lines whose directions and lengths are known. The angular closure is distributed to give a series of preliminary adjusted azimuths or bearings. Errors in the measured lengths of the traverse sides, however, will tend to alter the shape of the traverse. The steps involved in adjusting a traverse whose preliminary adjusted azimuths have been determined are as follows:

1. Determine the distance that each line of the traverse extends in a north or south direction, and the distance that each line extends in an east or west direction. These distances are called, respectively, *latitudes* and *departures*.
2. Determine the algebraic sum of the latitudes and the algebraic sum of the departures, and compare them with the fixed latitude and departure of a straight line from the origin to the closing point. This presumes a closed traverse.
3. Adjust the discrepancy found in step (2) by apportioning the closure in latitudes and the closure in departures on a reasonable and logical basis.
4. Determine the adjusted position of each traverse station with respect to some origin. This position is defined by its Y coordinate and its X coordinate with respect to a plane rectangular coordinate system, the origin being the intersection of the Y axis and the X axis with the Y axis being in the direction of the meridian.

The above sequence of computations can be altered to one in which the plane rectangular coordinates are computed after step (1) above, and then the coordinates of the last point of the traverse are compared with the fixed coordinates of this point. This comparison establishes the closure error that is used to adjust the computed coordinates. Both of these sequences are presented in the sections to follow.

When the computations have been performed, the position of each traverse station is known with respect to any other traverse station. Furthermore, each station is related in position to any other point that is defined on the same coordinate system, even though the point is not included in the traverse.

If the purpose of the survey is merely to control a map of a limited area and the positions of the traverse stations are to be located on the map sheet by distances and deflection angles, or by distances and bearings, then there is no need for computing the latitudes and departures or the coordinates, since the ultimate position will be determined graphically.

Surveying computations are performed by electrically operated hand-held or desk calculators, graphical methods, or electronic computers. The selection of the method depends on several factors, which include the accuracy of the field work, the number of significant figures necessary, the availability of electric power, the size of the project and the extent of the computations, the complexity of the computations, the cost, the time restriction, the availability of trained personnel, and the availability of computing equipment.

Very few traverses are computed by longhand means. Electrically operated handheld or desk calculators are used most for performing traverse computations. Where electric power is not available, battery-operated computers are used. Many large mapping agencies that do extensive traverse work employ calculating machines designed specifically for traverse computations. They also employ electronic computers of small or medium capacity.

8-11. LATITUDES AND DEPARTURES The latitude of a line is the distance the line extends in a north or south direction. A line running in a northerly direction has a plus latitude; one running in a southerly direction has a minus latitude.

The departure of a line is the distance the line extends in an east or west direction. A departure to the east is considered plus; a departure to the west is minus.

In Fig. 8-11 the bearing of the line MP is N 23° 34′ 20″ E and its length is 791.28 ft. The bearing of PR is S 51° 05′ 10″ E and its length is 604.30 ft. The line MP has a latitude of $+725.25$ ft and a departure of $+316.44$ ft. The line PR has a latitude of -379.60 ft and a departure of $+470.20$ ft. From Fig. 8-11 it is seen that the latitude of each line is the length of the line times the cosine of the bearing angle, and that the departure of each line is the length of the line times the sine of the bearing angle. If D represents the length of the line and B is the bearing angle, then

$$\text{latitude} = D \cos B \qquad (8\text{-}1)$$

$$\text{departure} = D \sin B \qquad (8\text{-}2)$$

If the direction of the line is given in terms of its azimuth from north and A represents that azimuth, then

$$\text{latitude} = D \cos A \qquad (8\text{-}3)$$

$$\text{departure} = D \sin A \qquad (8\text{-}4)$$

When bearings are used, the sine and cosine of the bearing angle are always considered positive, and the algebraic signs of the latitude and departure are

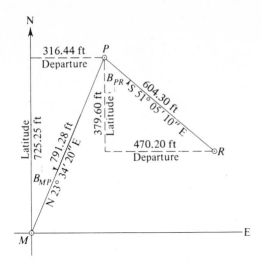

Figure 8-11. Latitudes and departures.

obtained from the quadrant. The signs of the sine and cosine of an azimuth in each of the four quadrants (northeast, southeast, southwest, and northwest) are shown in Fig. 8-12, in which Roman numerals are used for the quadrants. These signs agree, respectively, with the algebraic signs of the functions of angles lying in the four quadrants defined in trigonometry. Thus, the correct algebraic signs are automatically generated in the calculator or computer.

The calculation of latitudes and departures for the traverse of Fig. 8-13 is shown in Table 8-8 as it would be set up for using a desk or handheld calculating machine. The bearings given in the tabulation are those computed after an adjustment was made for the angular closure of the traverse. Several variations of the form of computation will be found in practice. The form can be extended to include the calculation of coordinates. If the computer generates the sine and cosine function internally, the trigonometric functions shown in Table 8-8 are not necessary.

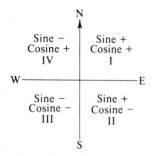

Figure 8-12. Algebraic signs of azimuth-angle functions.

TABLE 8-8. Calculation of Latitudes and Departures by Desk or Handheld Computer

STATION	BEARING	LENGTH	COSINE	SINE	LATITUDE +	LATITUDE −	DEPARTURE +	DEPARTURE −
A								
	N 47° 28′ 00″ E	483.52	0.676019	0.736884	326.87		356.30	
B								
	S 8° 27′ 30″ W	392.28	0.989123	0.147090		388.01		57.70
C								
	S 56° 27′ 00″ W	886.04	0.552665	0.833404		489.68		738.43
D								
	N 26° 16′ 30″ E	452.66	0.896680	0.442680	405.89		200.39	
E								
	N 39° 18′ 00″ W	279.33	0.773840	0.633381	216.16			176.92
F								
	S 80° 20′ 30″ E	421.97	0.167773	0.985826		70.79	415.99	
A								
		2915.80			948.92	948.48	972.68	973.05
					−948.48		−973.05	
					+0.44		−0.37	

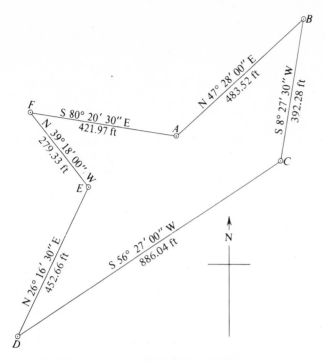

Figure 8-13. Traverse showing lengths and bearings.

8-12. CLOSURE IN LATITUDES AND DEPARTURES
In a traverse that closes on the point of origin, the algebraic sum of the latitudes should be zero and the algebraic sum of the departures should be zero. In the traverse of Fig. 8-13 the sum of the plus latitudes is 948.92 ft, and the sum of the minus latitudes is 948.48 ft. The closure in latitude is therefore $+0.44$ ft. The sum of the plus departures is 972.68 ft and the sum of the minus departures is 973.05 ft. Thus the closure in departure is -0.37 ft. These closures result from errors in measuring angles and distances when executing the traverse. Even though an adjustment has been made to eliminate the angular closure, each angle has not necessarily received the correct amount of adjustment.

In a traverse that originates at one known position and closes on another known position, the algebraic sum of the latitudes and the algebraic sum of the departures must equal, respectively, the latitude and departure of the line joining the origin and the closing point. The latitude and departure of this line are usually given as a difference in Y coordinates and a difference in X coordinates, respectively, between the point of beginning and the closing point. An example of a traverse computation between two known positions will be given following a discussion of coordinates.

If the error in latitudes or the error in departures is large, a mistake is indicated. The first place to search for the mistake is in the computations.

Sources of mistakes are, of course, very numerous. If no mistakes can be found in the computations, then a large discrepancy in latitudes or departures, or both, indicates a mistake in the field work of measuring either an angle or a distance. If the length of one line contains a mistake, the line itself may be selected from the remainder of the traverse by analyzing the traverse closure.

8-13. TRAVERSE CLOSURE The closure of a traverse is the line that will exactly close the traverse. In Fig. 8-14 the line $A'A$ is the closure of a traverse that theoretically closes on itself. The angle B that this line makes with the meridian is determined by the relation

$$\tan B = \frac{\text{closure in departure}}{\text{closure in latitude}} \qquad (8\text{-}5)$$

The length D of the line is given by the equation

$$D = \sqrt{(\text{closure in latitude})^2 + (\text{closure in departure})^2} \qquad (8\text{-}6)$$

In the traverse computed in Section 8-11 and shown in Fig. 8-13, the bearing angle of the traverse closure is $\tan^{-1}(0.37/0.44) = 40°$.

Since the algebraic sum of the latitudes is positive, the latitude of the closure must be negative; and since the algebraic sum of the departures is negative, the departure of the closure must be positive. So the bearing of the closure must be S 40° E. The length of the closure is $\sqrt{0.37^2 + 0.44^2} = 0.57$ ft.

A *closure precision* for a given traverse may be obtained by dividing the length of the traverse closure by the total length of the traverse. It is to be expected that the length of the traverse closure will be greater in long surveys

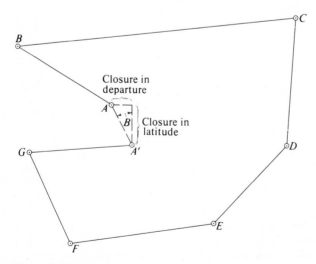

Figure 8-14. Traverse closure.

than in short ones if the same precautions are taken in both cases. Therefore the lengths of the traverse closures of two surveys of different lengths do not indicate the relative degrees of accuracy of the surveys. The closure precisions, however, do indicate the relative degrees of precision of the two surveys. The length of the closure of the traverse computed in Section 8-11 is 0.57 ft and the total length of the traverse is 2915.80 ft. The closure precision is therefore 0.57/2915.80 or about 1/5120. If a traverse that was twice as long had the same length of closure, its closure precision would be about 1/10,240. Traverse work is often rated as to order of accuracy on the basis of the closure precision.

It should be noted that if the traverse begins and closes on the same point, the traverse closure is the result of random errors rather than of systematic errors. Thus if a traverse of this kind is run with a tape of incorrect length or with an EDM containing a frequency error, the actual error in the length of each line in the traverse can be considerable, even through the traverse closure and the closure precision are very small. Also, if the beginning azimuth or bearing is wrong and if the azimuths or bearings of the remaining sides have been determined by using the adjusted angles in the figure, then the azimuths or bearings of all the sides are wrong by approximately the same amount. If a traverse begins on one known position and closes on another known position, then the amounts by which it fails to close on the basis of the computations reflects both systematic and random errors in the field work. The amount by which it reflects the systematic errors depends on the shape of the traverse. If the traverse is run in a direct line between the two known points, then all the systematic errors are represented. If the two points lie on an east-west line, for example, and if the traverse is run north for a considerable distance, then east, and then south for a considerable distance to the point of closure, the systematic errors accumulated in traveling north will be canceled by those accumulated in traveling south, insofar as they affect the traverse closure.

8-14. BALANCING A TRAVERSE   Before the results of a traverse are usable for determining areas or coordinates, for publishing the data, or for computing lines to be located from the traverse stations, the traverse must be mathematically consistent; that is, the closures in latitudes and departures must be adjusted out. Applying corrections to the individual latitudes and departures so that they will sum up to a given condition is called *balancing a traverse*.

At this point it will be well to examine Table 4-2 to appreciate the relationship between the accuracy in angular measurements and the accuracy in linear measurements. For example, according to Table 4-2, an accuracy of one part in 5000 in linear measurements requires that the error in angular measurements must not be greater than about 0′ 41″. This accuracy is obtained by using a 1′ transit if the angles are turned once direct and once reversed. On the other hand if angles are measured within $1'\,00''$, the distances should be measured to an accuracy of one part in 3440 which is about $1\frac{1}{2}$ ft/mile. In each of these examples the two types of measurements will be consistent with one another.

Only three basic conditions can exist. (1) The angular accuracy is higher than the linear accuracy; (2) the angular accuracy is the same as the linear accuracy; (3) the angular accuracy is lower than the linear accuracy. An extreme example of the first condition is a traverse, executed in rugged terrain, in which angles are measured directly to $1''$ and distances are determined by holding the tape horizontally throughout the measurements. The angular accuracy is expected to be high, whereas the linear accuracy is expected to be low because it is necessary to break tape. The second condition is exemplified by a traverse, executed in fairly level terrain, in which angles are measured with a $30''$ or $1'$ transit and the tape is handled by experienced personnel; or a similar traverse in which angles are measured with a $1''$ instrument and the tape is supported on posts or taping bucks and is handled by experienced personnel; or else the distances are measured with an EDM. The third condition is represented by a traverse in which directions are obtained by compass bearings and distances are measured with an EDM. There are, of course, various degrees of these three basic conditions.

In order to balance a traverse properly, the conditions should be known because the conditions themselves govern the selection of the method employed. In general, traverses are executed so as to satisfy the second condition as a matter of policy, economy, and common sense. The method of balancing based on this condition is known as the *compass rule*. In using the compass rule to balance or adjust a traverse, the errors, and, consequently, the corrections, are presumed to be in direct proportion to distance. Note that this is the basis for the adjustment of a level circuit.

8-15. BALANCING BY THE COMPASS RULE According to the compass rule, the correction to the latitude of a side is to the length of that side as the closure in latitude of the traverse is to the total length of the traverse; and the correction to the departure of a side is to the length of that side as the closure in departure of the traverse is to the total length of the traverse. For example, in the traverse computed in Section 8-11 the compass rule states that the correction c_L to the latitude of the line CD may be found from the proportion $c_L/886.04 = 0.44/2915.80$; and the proportion for finding the correction c_D to the departure of the line CD is $c_D/886.04 = 0.37/2915.80$. Both c_L and c_D have algebraic signs opposite to those of the closure in latitude of the traverse and the closure in departure of the traverse. Thus for the line CD, $c_L = -0.13$ ft and $c_D = +0.11$ ft. The corrections to the latitudes and departures for each line in the traverse are shown in Table 8-9.

After the corrections have been computed, they should be added to check whether, in fact, their sums equal the closures in latitude and departure. In

TABLE 8-9. Corrections to Latitudes and Departures

STATION	LENGTH	LATITUDES +	LATITUDES −	DEPARTURES +	DEPARTURES −	c_L	c_D	BALANCED LATITUDES +	BALANCED LATITUDES −	BALANCED DEPARTURES +	BALANCED DEPARTURES −
A	483.52	326.87		356.30		−0.07	+0.06	326.80		356.36	
B	392.28		388.01		57.70	−0.06	+0.05		388.07		57.65
C	886.04		489.68		738.43	−0.13	+0.11		489.81		738.32
D	452.66	405.89		200.39		−0.07	+0.06	405.82		200.45	
E	279.33	216.16			176.92	−0.04	+0.04	216.12			176.88
F	421.97		70.79	415.99		7 −0.04	+0.05		70.86	416.04	
A	2915.80	948.92	948.48	972.68	973.05	−0.43 −0.44	+0.37	948.74	948.74	972.85	972.85
		−948.48		−973.05							
		+0.44		−0.37							

the example, the sum of the latitude corrections as first computed was less than the closure in latitude by 0.01 ft. This is a rounding-off error and is corrected by adding the 0.01 ft to that correction which lies nearest to the next higher 0.01 ft. Thus the correction to the latitude of the line FA has been changed from -0.06 to -0.07.

After the balanced latitudes and departures have been computed, they should be added in order to provide a check on the computations.

8-16. LEAST-SQUARES ADJUSTMENT   The least-squares method of adjustment of a traverse applies corrections to the latitudes and departures in such a manner that the weighted sum of the squares of the adjustments to the measured angles and the measured lengths is a minimum according to the principles developed in Chapter 5. If the assumption can be made that the measured lengths and the measured angles are equally reliable, that is, their weights are the same, then an adjustment of the traverse by the method of least squares will give essentially the same results as the compass rule. This is according to the second condition discussed in Section 8-14. If the weights are not equal, however, then the method of least squares can accommodate to this fact and give theoretically better results.

The vast majority of traverses executed in the practice of surveying can be adjusted by the compass rule. When traverses of great extent and very high accuracy are run and when the lines in the traverse join one another at junction points, the overall adjustment should be made by the least-squares method. A method of adjusting a traverse net of crossed lines and junction points by least squares is given in Appendix A. Adjustment between junction points is then performed by the compass rule.

The development of the rigorous methods of least squares for the adjustment of traverses and traverse nets is outside the scope of this book. The interested reader is referred to the readings at the end of this chapter.

8-17. REMARKS ON ADJUSTMENTS   A traverse is balanced to make the figure formed by the traverse geometrically consistent. If the traverse is run to establish control points for other traverses of equal or lower order, the adjusted positions of the points furnish checks on the work to follow. If the traverse is run to determine areas, the consistency in the figure affords a check on the calculations. In many surveys, however, no balancing will be required, particularly when the latitudes and departures are to be used only in plotting the positions of the stations on a map and when the closure is too small to be scaled on the map.

8-18. RECTANGULAR COORDINATES   The rectangular coordinates of a point are the distances measured to the point from a pair of mutually perpendicular axes. As in analytic geometry the distance from the X axis is the Y coordinate of the point and the distance from the Y axis is the X coordinate

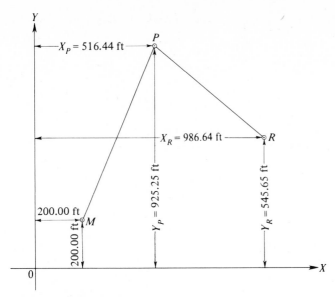

Figure 8-15. Rectangular coordinates.

of the point. In plane surveying conducted in the United States, the north part of the meridian on which the survey is based is the positive Y axis, and a line perpendicular to this meridian through an assumed origin is the X axis. The X axis is positive in an easterly direction. Also, the Y and X coordinates are expressed in feet or metres. In Fig. 8-15 the Y and X coordinates of M are, respectively, $+200.00$ ft and $+200.00$ ft; those of P are $+925.25$ ft and $+516.44$ ft; and those of R are $+545.65$ ft and $+986.64$ ft.

In general the origin of coordinates is situated far enough south and west of the area to make the coordinates of all points in the survey positive quantities. This practice reduces the chances of making mistakes in computations involving the coordinates. Furthermore there is no need to assign an algebraic sign to the coordinates of a point, since it is understood that all coordinates are positive. In subsequent examples in this chapter the origin is selected so as to render all the coordinates positive.

Before the coordinates of the stations of a traverse can be computed, the coordinates of at least one station must be known or else they must be assumed. In Fig. 8-15 the coordinates of station M were assumed to be 200.00 and 200.00. The Y coordinate of station P is determined by adding the latitude of the line MP to the Y coordinate of station M; the X coordinate of P is determined by adding the departure of MP to the X coordinate of M. If the subscript 1 is used for the first station and the subscript 2 is used for the second station of the line 1-2, then

$$Y_2 = Y_1 + \text{latitude of 1-2} \tag{8-7}$$

$$X_2 = X_1 + \text{departure of 1-2} \tag{8-8}$$

TABLE 8-10. Computation of Coordinates with Desk Calculating Machine

STATION	BALANCED LATITUDES		BALANCED DEPARTURES		Y COORDINATES	X COORDINATES
	+	−	+	−		
A					4166.20	6154.22
	326.80		356.36			
B					4493.00	6510.58
		388.07		57.65		
C					4104.93	6452.93
		489.81		738.32		
D					3615.12	5714.61
	405.82		200.45			
E					4020.94	5915.06
	216.12			176.88		
F					4237.06	5738.18
		70.86	416.04			
A					4166.20	6154.22

These equations can be applied to obtain Y and X coordinates all the way around a traverse. If a traverse originates and closes on the same point, then the final Y and X coordinates of the point of beginning computed by using the latitudes and departures of the traverse sides should be the same as the beginning Y and X coordinates. This will be true provided the traverse has been balanced. An addition to the tabulation of the computations for the traverse of Section 8-11 is given in Table 8-10 to show the calculation of co-ordinates of the points in the traverse. The columns headed Balanced Latitudes and Balanced Departures are repeated for continuity. The coordinates of station A are known to be $Y = 4166.20$ and $X = 6154.22$.

A check on the arithmetic is provided by computing the coordinates of station A from those of station F to see whether or not they are the same as the given coordinates of A.

8-19. ADJUSTMENT OF TRAVERSE BY COORDINATE AD-JUSTMENT The coordinates of traverse stations may be computed directly from the measured lengths of the sides of the traverse, the preliminary azimuths or bearings of the sides, and the coordinates of the point of beginning. This work is done conveniently with a desk or hand calculator by first putting the Y coordinate of the point of beginning into the calculator, adding the product of the distance and the cosine of the bearing angle of the first side, and using the result as the Y coordinate of the second point. This result is simply the sum of the latitude of the first line and the Y coordinate of the point of be-ginning, as called for by Eq. (8-7). The sum is combined with the product D cos B of the second line, and the procedure is repeated for successive lines. Due regard must be paid to algebraic signs. Each intermediate result is the Y coordinate of a point in the traverse. In a closed traverse the computed Y

coordinate of the point on which the traverse closes will differ from its known Y coordinate by the closure in latitude.

Next, the X coordinate of the point of beginning is put into the calculator, and to this is added the product $D \sin B$ for the first line. The result is the X coordinate of the second point, as the operation is simply the application of Eq. (8-8), or the addition of the departure of the first line to the X coordinate of the point of beginning. Similarly, the product $D \sin B$ for each successive line is used, due regard being paid to algebraic signs, and each intermediate result is taken as the X coordinate of a point in the traverse. The computed X coordinate of the point on which the traverse closes will differ from the known X coordinate by the closure in departure.

After the preliminary coordinates have been computed, the correction to be applied to each Y coordinate is found by multiplying the closure in latitude by the ratio of the distance of the point from the point of beginning to the total length of the traverse. An adjustment is then made to each X coordinate in the same manner; that is, the correction to a Y coordinate is to the discrepancy in Y coordinates (closure in latitude) as the distance from the point of beginning is to the total length of the traverse. Similarly, the correction to an X coordinate is to the discrepancy in X coordinates (closure in departure) as the distance from the point of beginning is to the total length of the traverse. Since corrections are proportionate to length, this is to be recognized as a compass-rule adjustment. If a discrepancy is plus, the corrections to all coordinates are minus; if a discrepancy is minus, the corrections to all coordinates are plus.

Tables 8-11 and 8-12 show the computation and the adjustment of coordinates of traverse stations. The traverse originates on station Richmond, whose coordinates are $Y = 28,221.34$ and $X = 20,370.66$, and closes on station Kenney, whose coordinates are $Y = 18,972.77$ and $X = 17,146.82$. The azimuths are preliminary azimuths obtained after an adjustment of the measured angles in the traverse. The column headed Total Distance is the distance from the point of beginning to the station in question. It is needed only to the nearest foot, because it is used to compute the corrections that are, or should be, relatively small.

The computed Y coordinate of station Kenney is too large by 1.18 ft. This is comparable to a closure in latitude of $+1.18$ ft. The computed X coordinate of station Kenney is too large by 0.29 ft. Thus it is apparent that the Y coordinate of each point must be reduced by an amount proportionate to the distance from the beginning point to the point under consideration, and that the X coordinate of each point must be reduced in the same proportion.

The Y coordinate of station A is reduced by an amount equal to $(1.18)(2872)/16,101 = 0.21$ ft; the Y coordinate of B is reduced by $(1.18)(4850)/16,101 = 0.35$ ft; and so on. The X coordinate of A is reduced by $(0.29)(2872)/16,101 = 0.05$ ft; the X coordinate of B is reduced by $(0.29)(4850)/16,101 = 0.09$ ft; and so on.

TABLE 8-11. Computation of Coordinates with Desk or Handheld Computer

STATION	DISTANCE	TOTAL DISTANCE	AZIMUTH	COSINE	SINE	Y	X
Richmond						28,221.34	20,370.66
A	2872.45	2,872	110° 10′ 20″	−0.3448432	+0.9386603	27,230.80	23,066.91
B	1977.14	4,850	184° 22′ 15″	−0.9970916	−0.0762115	25,259.41	22,916.23
C	1440.40	6,290	177° 49′ 54″	−0.9992841	+0.0378355	23,820.04	22,970.73
D	1635.97	7,926	167° 02′ 11″	−0.9745127	+0.2243322	22,225.77	23,337.73
E	3230.84	11,157	212° 18′ 28″	−0.8451893	−0.5344671	19,495.10	21,610.95
F	1491.48	12,648	284° 50′ 23″	+0.2561160	−0.9666461	19,877.09	20,169.22
G	2231.98	14,880	270° 42′ 25″	+0.0123382	−0.9999239	19,904.63	17,937.41
Kenney	1220.96	16,101	220° 20′ 12″	−0.7622542	−0.6472777	18,973.95	17,147.11
				fixed		18,972.77	17,146.82
					closure	+1.18	+0.29

$$\text{closure precision} = \frac{\sqrt{1.18^2 + 0.29^2}}{16,101} = \frac{1}{13,200}$$

TABLE 8-12. Adjustment of Coordinates

STATION	TOTAL DISTANCE	Y CORRECTION	X CORRECTION	ADJUSTED Y	ADJUSTED X
Richmond				28,221.34	20,370.66
A	2,872	−0.21	−0.05	27,230.59	23,066.86
B	4,850	−0.35	−0.09	25,259.06	22,916.14
C	6,290	−0.46	−0.11	23,819.58	22,970.62
D	7,926	−0.58	−0.14	22,225.19	23,337.59
E	11,157	−0.82	−0.20	19,494.28	21,610.75
F	12,648	−0.93	−0.23	19,876.16	20,168.99
G	14,880	−1.09	−0.27	19,903.54	17,937.14
Kenney	16,101	−1.18	−0.29	18,972.77	17,146.82

The corrections to all the coordinates, together with the adjusted coordinates, are shown in Table 8-12. Actually these results can be combined with the previous computations of Table 8-11 in one tabulation, which would show the unadjusted and adjusted coordinates. Each adjusted coordinate may be entered directly below the corresponding unadjusted coordinate or alongside it in a separate column; or the last few digits of the unadjusted coordinate may be struck out, and the adjusted figures entered above them. The two separate tabulations are shown here for the sake of clarity.

8-20. THE USE OF RECTANGULAR COORDINATES

The rectangular coordinates of a point uniquely define its position with respect to a known horizontal datum. Therefore the use of coordinates is the most convenient method of publishing the horizontal positions of points that have been located in the field. The coordinates of a point give its position with respect to any other point located on the same coordinate system or horizontal datum. Coordinates are used to plot the positions of survey points on a map sheet as explained in Section 8-34.

If the coordinates of two stations 1 and 2 are known, the latitude and departure of the line from station 1 to station 2 are obtained by transposing terms of Eqs. (8-7) and (8-8) to give the following relations:

$$\text{latitude of } 1\text{-}2 = Y_2 - Y_1 \tag{8-9}$$

$$\text{departure of } 1\text{-}2 = X_2 - X_1 \tag{8-10}$$

Referring to Fig. 8-16, the tangent of the bearing angle B of the line joining the two stations is given by the equation

$$\tan B = \frac{X_2 - X_1}{Y_2 - Y_1} \tag{8-11}$$

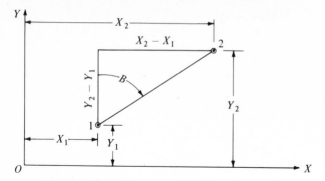

Figure 8-16. Calculation of bearing angle.

The distance D between the two points is given by the following equation:

$$D = \sqrt{(Y_2 - Y_1)^2 + (X_2 - X_1)^2} \qquad (8\text{-}12)$$

If the numerator of Eq. (8-11) is positive, the line from 1 to 2 bears east, and if the denominator is positive, the line bears north.

The computation made to determine the bearing and length of a line from the coordinates of the two points defining the line is called *inversing* the line; this is one of the most frequent computations made in surveying.

The azimuth A_N of a line 1-2 measured from the north meridian is given by the equation

$$\tan A_N = \frac{X_2 - X_1}{Y_2 - Y_1} \qquad (8\text{-}13)$$

The azimuth A_S of the same line measured from the south meridian is given by the equation

$$\tan A_S = \frac{X_1 - X_2}{Y_1 - Y_2} \qquad (8\text{-}14)$$

In Eqs. (8-9), (8-10), (8-11), (8-13), and (8-14), due consideration must be given to the algebraic signs of the coordinates and the differences in coordinates.

The equations of straight lines and second-degree curves given by the methods of analytic geometry are equally applicable to problems in surveying when coordinates are used. The slope of a straight line in analytic geometry is given as $m = \tan \alpha$, in which α is the angle measured from the positive X axis. As seen in Fig. 8-17, the slope of the line OP is $m = \tan \alpha$. The angle α is the complement of the bearing angle B of the line or of its azimuth from north. Therefore, in surveying, the slope of the line is given as $\cot B$ or $\cot A$, the choice depending on whether bearings or azimuths are used. The

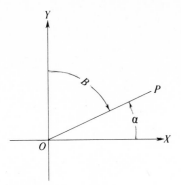

Figure 8-17. Slope of a line.

following equations from analytic geometry, applicable to surveying, are given here for convenience.

1. Equation of a straight line: When the coordinates (X_1, Y_1) and (X_2, Y_2) of two points are given,

$$\frac{Y - Y_1}{Y_2 - Y_1} = \frac{X - X_1}{X_2 - X_1} \qquad (8\text{-}15)$$

When the coordinates (X_1, Y_1) of one point and the bearing angle B of the line are given,

$$Y - Y_1 = \cot B(X - X_1) \qquad (8\text{-}16)$$

When the line passes through the origin and its bearing angle is given,

$$Y = X \cot B \qquad (8\text{-}17)$$

Equations (8-15) and (8-16) can be reduced to the general form

$$aX + bY + c = 0 \qquad (8\text{-}18)$$

Then, referring to Fig. 8-18 the azimuth of the line is given by

$$\tan A = -\frac{b}{a} \qquad (8\text{-}19)$$

The intercepts of the line on the X and Y axes are

$$X \text{ intercept} = -\frac{c}{a} \qquad (8\text{-}20)$$

$$Y \text{ intercept} = -\frac{c}{b} \qquad (8\text{-}21)$$

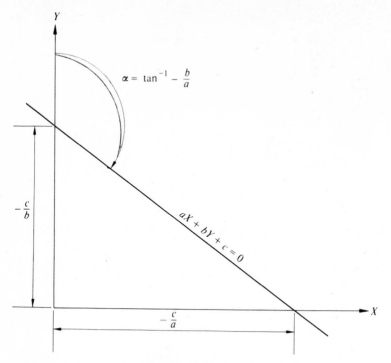

Figure 8-18. General form of equation of a line.

One of the most convenient forms of the equation of a line in solving for the distance from the line to a point off the line is the normal form given in terms of the coefficients a, b, and c of Eq. (8-18). If these are known, the normal form is then given by

$$\frac{a}{\pm\sqrt{a^2 + b^2}} X + \frac{b}{\pm\sqrt{a^2 + b^2}} Y + \frac{c}{\pm\sqrt{a^2 + b^2}} = 0 \qquad (8\text{-}22)$$

The algebraic sign of the radical in the denominator is chosen to be the same as that of b. In order to determine the distance d shown in Fig. 8-19 between the line and a point P off the line, the coordinates of P are substituted into Eq. (8-22) the left-hand side of which becomes the desired distance.

2. Equation of a circle of radius R: When its center is at the origin,

$$X^2 + Y^2 = R^2 \qquad (8\text{-}23)$$

When coordinates of its center are $X_c = H$ and $Y_c = K$,

$$(X - H)^2 + (Y - K)^2 = R^2 \qquad (8\text{-}24)$$

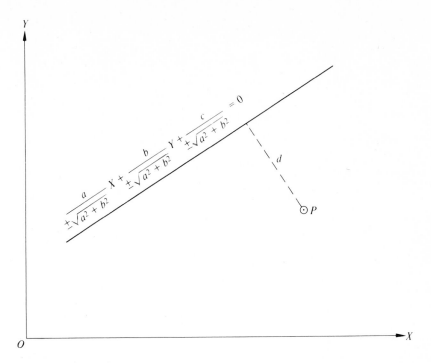

Figure 8-19. Normal form of equation of a line.

3. Area of polygon:

$$2A = X_1(Y_2 - Y_n) + X_2(Y_3 - Y_1) + \cdots$$
$$+ X_{n-1}(Y_n - Y_{n-2}) + X_n(Y_1 - Y_{n-1}) \qquad (8\text{-}25)$$

or

$$2A = Y_1(X_2 - X_n) + Y_2(X_3 - X_1) + \cdots$$
$$+ Y_{n-1}(X_n - X_{n-2}) + Y_n(X_1 - X_{n-1}) \qquad (8\text{-}26)$$

4. Dividing a line into two parts in the proportion R_1/R_2:

$$X_0 = \frac{X_1 R_2 + X_2 R_1}{R_1 + R_2} \qquad Y_0 = \frac{Y_1 R_2 + Y_2 R_1}{R_1 + R_2} \qquad (8\text{-}27)$$

in which the coordinates of the points at the ends of the line are (X_1, Y_1) and (X_2, Y_2) and the coordinates of the division point are (X_0, Y_0).

The foregoing equations are but a few of many that can be taken from analytic geometry and applied to surveying computations. Engineers and surveyors in the past have been prone to apply the more cumbersome methods of trigonometry to the solution of surveying problems, but when a programmable desk or hand computer is available, the methods of analytic geometry are much more efficient than are trigonometric methods. The following examples are given to show how easily some of the methods of analytic geometry are applied.

Example 8-1

For the traverse computed in Section 8-11, determine the Y and X coordinates of the intersection of the line joining points E and C with the line joining points D and A.

Solution: The equation of each of these lines may be written from the coordinates computed in Section 8-18 by using Eq. (8-15). The equation of the line EC is

$$\frac{Y - 4020.94}{4104.93 - 4020.94} = \frac{X - 5915.06}{6452.93 - 5915.06} \tag{1a}$$

and the equation of the line DA is

$$\frac{Y - 3615.12}{4166.20 - 3615.12} = \frac{X - 5714.61}{6154.22 - 5714.61} \tag{2a}$$

Because the actual coordinates are relatively large numbers which give terms with many significant figures in the calculations, the values of the coordinates should be reduced by subtracting a constant from each Y and X coordinate. In this problem each Y coordinate is reduced by 3000 ft, and each X coordinate is reduced by 5000 ft. The two line equations then become

$$\frac{Y' - 1020.94}{1104.93 - 1020.94} = \frac{X' - 915.06}{1452.93 - 915.06} \tag{1b}$$

$$\frac{Y' - 615.12}{1166.20 - 615.12} = \frac{X' - 714.61}{1154.22 - 714.61} \tag{2b}$$

Rearranging and collecting terms, we get

$$83.99X' - 537.87Y' = -472{,}277.11 \tag{1c}$$

$$551.08X' - 439.61Y' = +123{,}394.38 \tag{2c}$$

These two equations are solved simultaneously to find the coordinates of the intersection. If Eq. (2c) is multiplied by 537.87/439.61, and Eq. (1c) is subtracted from the new equation, the work is as follows:

$$647.26X' - 537.87Y' = +150,975.04$$
$$\underline{83.99X' - 537.87Y' = -472,277.11}$$
$$590.27X' \qquad\quad = +623,252.15$$
$$X' \qquad\qquad = +1055.876$$

Substituting the value of X' in Eqs. (1c) and (2c) to provide a check gives Y' as 1042.93. The coordinates of the point of intersection therefore, are,

$$Y = 1042.93 + 3000 = 4042.93$$
$$X = 1055.88 + 5000 = 6055.88$$

Example 8-2

For the traverse computed in Section 8-11, determine the coordinates of the intersection of a line passing through point B and parallel to line CD with the line AF.

Solution: The equations of these two lines are written and then solved simultaneously to obtain the coordinates of the point of intersection. The bearing of the line passing through B is the same as that of CD. Therefore $\tan B_{CD} = (5714.61 - 6452.95)/(3615.12 - 4104.93)$ or $\cot B = (3615.12 - 4104.93)/(5714.61 - 6452.95) = +0.663393$, and the equation of the line through B is, by Eq. (8.16),

$$Y - 4493.00 = +0.663393(X - 6510.58) \tag{1a}$$

The equation of the line AF is, by Eq. (8-15),

$$\frac{Y - 4166.20}{4237.06 - 4166.20} = \frac{X - 6154.22}{5738.18 - 6154.22} \cdot \tag{2a}$$

In order to keep the number of significant figures in the calculations as small as possible, each Y coordinate is reduced by 4000 ft and each X coordinate by 5000 ft. The two line equations then become

$$Y' - 493.00 = +0.663393(X' - 1510.58) \tag{1b}$$

$$\frac{Y' - 166.20}{237.06 - 166.20} = \frac{X' - 1154.22}{738.18 - 1154.22} \tag{2b}$$

Rearranging and collecting terms gives

$$0.663393X' - Y' = +509.11 \tag{1c}$$

$$70.86X' + 416.04Y' = +150,933.88 \tag{2c}$$

These equations are solved simultaneously to give the coordinates of the intersection. When Eq. (1c) is multiplied by 70.86/0.663393 and the new equation is subtracted from Eq. (2c), the work is as follows:

$$
\begin{array}{r}
70.86X' - 106.81Y' = + 54,380.34 \\
70.86X' + 416.04Y' = +150,933.88 \\
\hline
+ 522.85Y' = + 96,553.54 \\
Y' = +184.668
\end{array}
$$

From Eqs. (1c) and (2c) it is found that $X' = 1045.81$ and $X' = 1045.79$. The adopted value is 1045.80. The final coordinates therefore, are,

$$Y = 184.67 + 4000 = 4184.67$$

$$X = 1045.80 + 5000 = 6045.80$$

Example 8-3

Where does the line CD of Example 8-2 intersect the X and Y axes?

Solution: See Fig. 8-20. The equation of CD is written in the form of Eq. (8-15),

$$\frac{Y - 4104.93}{3615.12 - 4104.93} = \frac{X - 6452.93}{5714.61 - 6452.93}$$

which reduces to the general form given by Eq. (8-18); thus

$$489.81X - 738.32Y - 129,957.73 = 0$$

Then by Eqs. (8-20) and (8-21),

$$X \text{ intercept} = -\frac{-129,957.73}{489.81} = +265.32 \text{ ft}$$

$$Y \text{ intercept} = -\frac{-129,957.73}{-738.32} = -176.02 \text{ ft}$$

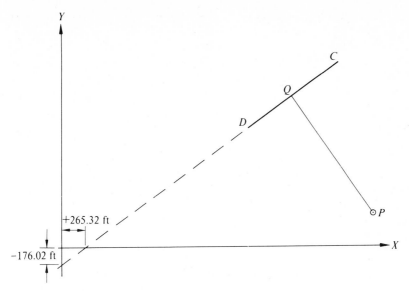

Figure 8-20. Distance from a point to a line.

Example 8-4

Compute the distance PQ from the line CD of Example 8-3 to a point P whose coordinates are $X_P = 6{,}802.52$ and $Y_P = 2654.88$.

Solution: Point P is shown in relation to the line CD in Fig. 8-20. Either of two methods can be used to solve this problem. The first is to substitute the coordinates of P into the normal form of the equation of CD given by Eq. (8-22) and solve for the distance. The general form of the equation of CD is already given in Example 8-3. Then, since b is negative,

$$-\sqrt{a^2 + b^2} = -\sqrt{489.81^2 + 738.32^2} = -886.02$$

giving

$$\frac{489.81}{-866.02} X + \frac{-738.32}{-886.02} Y + \frac{-129{,}957.73}{-886.02} = 0$$

or

$$-0.552820X + 0.833299Y + 146.68 = 0$$

which is the normal form of the equation of CD. The distance from CD to P is then obtained by substituting the coordinates of P into the normal form. Thus

$$PQ = (-0.552820)(6802.52) + (0.833299)(2654.88) + 146.68 = -1401.58$$

The negative sign is not significant because the desired quantity is the absolute distance between the line and the point.

The second method for solving this problem is to keep the line CD in the two-point form given by Eq. (8-15), then write the equation of a perpendicular line through P in the point-slope form. The slope of this line is 90° different from that of CD. These two equations are solved simultaneously to give a point of intersection Q on line CD. PQ is then solved by Eq. (8-12).

Example 8-5

The coordinates of the center of a circular curve defining the center line of a highway are $Y = 2655.10$ and $X = 17,255.35$. The radius of the curve is 750.00 ft. Determine the coordinates of the intersection of this center line with a line passing through a point whose coordinates are $Y = 3844.20$ and $X = 17,000.00$ and having a bearing S 30° E (see Fig. 8-21).

Solution: The equation of the circular curve is, by Eq. (8-24),

$$(X - 17,255.35)^2 + (Y - 2655.10)^2 = 750.00^2 \tag{1a}$$

The equation of the straight line is, by Eq. (8-16),

$$Y - 3844.20 = -\cot 30° (X - 17,000.00) \tag{2a}$$

The cotangent is minus because the bearing is in the southeast quadrant. The calculations may be simplified by subtracting 2000 ft from each Y coordinate and 17,000 ft from each X coordinate before the equations are solved, and then adding these amounts to the computed coordinates. Therefore

$$(X' - 255.35)^2 + (Y' - 655.10)^2 = 750.00^2 \tag{1b}$$

$$Y' - 1844.20 = -1.732051 (X' - 0) \tag{2b}$$

To solve these simultaneous equations, first the value of Y' obtained from Eq. (2b) is substituted in Eq. (1b). Thus

$$Y' = -1.732051X' + 1844.20 \tag{2c}$$

$$(X' - 255.35)^2 + (-1.732051X' + 1844.20 - 655.10)^2 = 750.00^2 \tag{1c}$$

Expanding, rearranging, and collecting terms gives

$$4X'^2 - 4629.86X' + 916,662.43 = 0$$

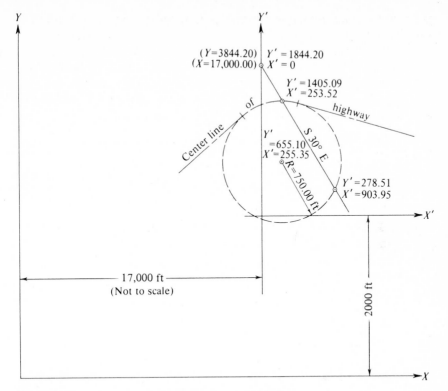

Figure 8-21. Intersection of circular curve and straight line.

This is a quadratic equation of the form $ax^2 + bx + c = 0$ whose solution is $x = (-b \pm \sqrt{b^2 - 4ac})/2a$, and its solution will give two values of X', as follows:

$$X' = \frac{4629.86 \pm \sqrt{21{,}435{,}603.6196 - 14{,}666{,}598.8800}}{8}$$

$$X' = 903.95 \quad \text{or} \quad X' = 253.52$$

Substituting the values of X' in Eq. (2c) gives

$$Y' = 278.51 \quad \text{or} \quad Y' = 1405.09$$

As a check on the computation, both values of X' and Y' may be substituted in Eq. (1b) to see whether both values satisfy the equation. An inspection of Fig. 8-21 shows that the values obtained by using the smaller value of X' are the coordinates of the desired point. The final coordinates of the point are obtained by adding 2000 ft to Y' and 17,000 ft to X'. The results are $Y = 3405.09$ ft and $X = 17,253.52$ ft.

8-21. COORDINATES OF UNOCCUPIED POINTS In order to determine the coordinates of points that cannot be included in the traverse because it is not possible to set the transit up over such points, spur lines or tie lines must be run from the traverse to these points. For example, in Fig. 8-22 the boundaries of a parcel of land are defined by an iron pipe, a nail in a tree, and two fence posts. The only corner that can be occupied is the one marked by the iron pipe. The problem is to determine the lengths and bearings of the property lines.

It will be assumed that the coordinates of the iron pipe at point A and the bearing of the line AB are known. The procedure is then as follows:

1. Measure the lengths of the traverse sides AB, BC, CD, and DA, and also the lengths of the three tie lines BP, CR, and DS.
2. Measure the angles in the traverse $ABCD$, and also the angles between traverse lines and the tie lines to P, R, and S, as shown in Fig. 8-22.

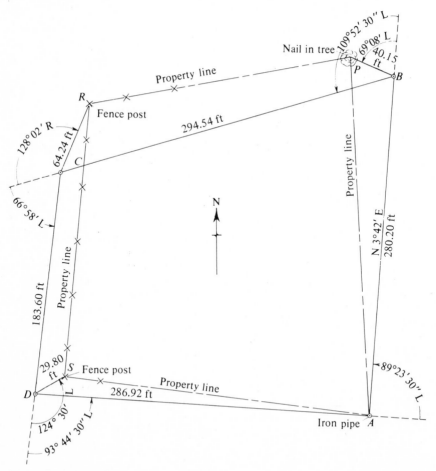

Figure 8-22. Traverse to determine boundary.

3. Adjust the angles in the traverse $ABCD$, and compute the bearings of lines BC, CD, and DA based on the known bearing of the line AB.
4. Balance the traverse $ABCD$.
5. Compute the coordinates of points B, C, and D.
6. Determine the bearings of the lines BC, CD, and DA based on the coordinates computed in (5).
7. Compute the bearings of the tie lines from the bearings computed in (5) and the measured angles.
8. Compute the latitudes and departures of the tie lines.
9. Compute the coordinates of the unoccupied stations from the coordinates of traverse stations B, C, and D and the latitudes and departures of the lines BP, CR, and DS, respectively.

When the coordinates of P, R, and S have been computed, the bearings and lengths of the boundary lines may be computed by Eq. (8-11) and (8-12).

Example 8-6

The following deflection angles were measured at stations A, B, C, and D of the traverse in Fig. 8-22.

STATION	FROM	TO	DEFLECTION ANGLE	ADJUSTED DEFLECTION ANGLE
A	D	B	89° 23′ 30″ L*	89° 24′ 00″ L
B	A	C	109° 52′ 00″ L*	109° 52′ 30″ L
B	A	P	69° 08′ 00″ L	—
C	B	D	66° 58′ 00″ L*	66° 58′ 30″ L
C	B	R	128° 02′ 00″ R	—
D	C	A	93° 44′ 30″ L*	93° 45′ 00″ L
D	C	S	124° 30′ 00″ L	—
			check	360° 00′ 00″

The lengths as determined in the field are as follows:

LINE	LENGTH (ft)
AB	280.20
BC	294.54
BP	40.15
CD	183.60
CR	64.24
DA	286.92
DS	29.80

Determine the lengths and bearings of the boundary lines AP, PR, RS, and SA.

Solution: The sum of the four angles in the closed traverse shown by an asterisk (*) is 359° 58′. These angles are adjusted to total 360°, as shown in the last column. The bearing of the line *AB* is known to be N 3° 42′ E. The bearings of the other traverse sides (not including the tie lines) are computed by subtracting each left deflection angle from the forward azimuth of each backsight as outlined in Section 8-5. When these azimuths are converted to bearings, the results are as follows:

LINE	ADJUSTED BEARING
AB	N 3° 42′ 00″ E
BC	S 73° 49′ 30″ W
CD	S 6° 51′ 00″ W
DA	S 86° 54′ 00″ E

From the adjusted bearings and the measured lengths of the traverse sides, the latitudes and departures are computed and balanced. From the balanced latitudes and departures, the coordinates of *B*, *C*, and *D* are determined, a check being made by computing the coordinates of *A* from point *D*. For the sake of brevity Table 8-13 shows only unbalanced and balanced latitudes and departures of the traverse sides, together with the computed coordinates of the traverse stations.

From the coordinates of the traverse stations, the azimuths of the lines *AB*, *BC*, and *CD* are computed, since these azimuths are needed to determine the azimuths of the tie lines. Then for line *AB* by Eq. (8-13),

$$\tan A_N = \frac{+18.13}{+279.68} \qquad A = 3° 42′ 30″$$

For line *BC*,

$$\tan A_N = \frac{-282.82}{-81.98} \qquad A = 253° 50′ 00″$$

For line *CD*,

$$\tan A_N = \frac{-21.86}{-182.25} \qquad A = 186° 50′ 30″$$

The azimuths and bearings of the tie lines are determined as shown in Table 8-14.

From the bearings and lengths of the tie lines, the latitudes and departures and then the coordinates are computed, as shown in Table 8-15.

TABLE 8-13. Computation of Coordinates of Traverse Stations

STATION	LATITUDE +	LATITUDE −	DEPARTURE +	DEPARTURE −	BALANCED LATITUDE +	BALANCED LATITUDE −	BALANCED DEPARTURE +	BALANCED DEPARTURE −	Y COORDINATE	X COORDINATE
A	279.62		18.08		279.68		18.13		1000.00	1000.00
B		82.05		282.88		81.98		282.82	1279.68	1018.13
C		182.29		21.90		182.25		21.86	1197.70	735.31
D		15.52	286.50			15.45	286.55		1015.45	713.45
A									1000.00	1000.00
	279.62	279.86	304.58	304.78	279.68	279.68	304.68	304.68		

TABLE 8-14. Computation of Azimuths of Tie Lines

LINE	AZIMUTH	LINE	AZIMUTH	LINE	AZIMUTH
AB	3° 42′ 30″	BC	253° 50′ 00″	CD	186° 50′ 30″
AB	363° 42′ 30″	$(+ \angle C)$	$+128° 02′ 00″$	$(- \angle D)$	$-124° 30′ 00″$
$(- \angle B)$	$-69° 08′ 00″$	CR	381° 52′ 00″	DS	62° 20′ 30″
BP	294° 34′ 30″	CR	21° 52′ 00″		
BP	N 65° 25′ 30″ W	CR	N 21° 52′ 00″ E	DS	N 62° 20′ 30″ E

Since the coordinates of each boundary corner are known, the bearings and lengths of the boundary lines may now be computed by Eq. (8-11) and Eq. (8-12).

For line AP,

$$\tan B = \frac{981.62 - 1000.00}{1296.38 - 1000.00}; \quad B = \text{N } 3° 33′ 00″ \text{ W}$$

$$AP = \sqrt{(981.62 - 1000.00)^2 + (1296.38 - 1000.00)^2} = 296.95 \text{ ft}$$

For line PR,

$$\tan B = \frac{759.24 - 981.62}{1257.32 - 1296.38}; \quad B = \text{S } 80° 02′ 20″ \text{ W}$$

$$PR = \sqrt{(759.24 - 981.62)^2 + (1257.32 - 1296.38)^2} = 225.78 \text{ ft}$$

For line RS,

$$\tan B = \frac{739.84 - 759.24}{1029.28 - 1257.32}; \quad B = \text{S } 4° 51′ 50″ \text{ W}$$

$$RS = \sqrt{(739.84 - 759.24)^2 + (1029.28 - 1257.32)^2} = 228.86 \text{ ft}$$

For line SA,

$$\tan B = \frac{1000.00 - 739.84}{1000.00 - 1029.28}; \quad B = \text{S } 83° 34′ 40″ \text{ E}$$

$$SA = \sqrt{(1000.00 - 739.84)^2 + (1000.00 - 1029.28)^2} = 261.80 \text{ ft}$$

The above calculation of lengths and bearings from coordinates can be done quite conveniently on most hand calculators by means of the rectangular to polar coordinate conversion function.

TABLE 8-15. Computation of Coordinates of Boundary Corners

STATION	LENGTH	BEARING	LATITUDE		DEPARTURE		Y COORDINATE	X COORDINATE
			+	−	+	−		
B	40.15	N 65° 25′ 30″ W	16.70			36.51	1279.68	1018.13
P							1296.38	981.62
C	64.24	N 21° 52′ 00″ E	59.62		23.93		1197.70	735.31
R							1257.32	759.24
D	29.80	N 62° 20′ 30″ E	13.83		26.39		1015.45	713.45
S							1029.28	739.84

8-22. COORDINATE TRANSFORMATIONS It frequently becomes necessary in surveying to transform rectangular coordinates in one coordinate system to their positions in another coordinate system. This can be done if the coordinates of at least two points are known in both systems. In Fig. 8-23(a), point A has coordinates X_A and Y_A in the XY coordinate system. The azimuth AB is designated α, obtained by inversing AB. In Fig. 8-23(b) point A has coordinates X'_A and Y'_A in the $X'Y'$ coordinate system. The azimuth of AB in this second system is α'. If the $X'Y'$ axes are rotated through angle $\theta = \alpha' - \alpha$, a new set of coordinates (X'_A) and (Y'_A) relative to the dashed (X'), (Y') coordinate axes are obtained. These coordinates are computed by the rotation equations from analytical geometry. Thus

$$(X'_A) = X'_A \cos \theta - Y'_A \sin \theta$$
$$(Y'_A) = X'_A \sin \theta + Y'_A \cos \theta \tag{8-28}$$

If the dashed axes are now translated in the X-direction by $X_A - (X'_A) = c$ and in the Y-direction by $Y_A - (Y'_A) = d$, X_A and Y_A are obtained. Thus

$$X_A = X'_A \cos \theta - Y'_A \sin \theta + c$$
$$Y_A = X'_A \sin \theta + Y'_A \cos \theta + d \tag{8-29}$$

The value of θ is obtained by inversing the line in each system. Then Eq. (8-28) is solved to determine (X'_A) and (Y'_A). The c and d terms are then obtained by

$$c = X_A - (X'_A)$$
$$d = Y_A - (Y'_A) \tag{8-30}$$

The coordinates of any other point such as point C in the $X'Y'$ system can be determined by Eq. (8-29).

If the scale of the two coordinate systems are different, then the correct coordinates would be obtained by multiplying the terms on the right side of Eq. (8-28) by some scale s giving

$$(X'_A) = (s \cos \theta)X'_A - (s \sin \theta)Y'_A$$
$$(Y'_A) = (s \sin \theta)X'_A + (s \cos \theta)Y'_A \tag{8-31}$$

And Eq. (8-29) becomes

$$X_A = (s \cos \theta)X'_A - (s \sin \theta)Y'_A + c$$
$$Y_A = (s \sin \theta)X'_A + (s \cos \theta)Y'_A + d \tag{8-32}$$

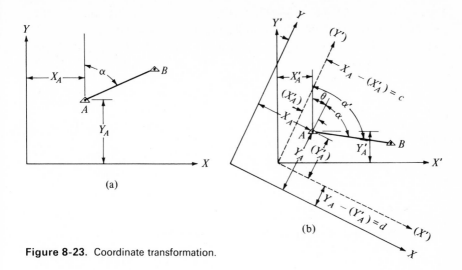

Figure 8-23. Coordinate transformation.

Let $a = s \cos \theta$ and $b = s \sin \theta$. Then for any point, Eq. (8-32) becomes

$$
X = aX' - bY' + c
$$
$$
Y = bX' + aY' + d
$$

(8-33)

Equation (8-33) is called a similarity transformation or a four-parameter transformation, the four parameters being a, b, c, and d. The parameters are solved by substituting the known coordinates of two points in both systems into Eq. (8-33) as shown by the following example.

Example 8-7

The coordinates of A and B are given for two systems. The coordinates of two additional points M and N in the second system are also given.

POINT	FIRST SYSTEM		SECOND SYSTEM	
	X	Y	X'	Y'
A	2120.00	1180.50	1145.25	1345.20
B	3570.80	1925.70	2595.35	599.82
M			2095.18	1278.48
N			2502.55	1296.80

Compute the coordinates X_M, Y_M, X_N, Y_N in the first system.

Solution: Substituting the known coordinates of A and B in the first system in the left side of Eq. (8-33) and their coordinates in the second system in the right side gives four equations in the four unknowns $a, b, c,$ and d. These are solved simultaneously.

$$2120.00 = 1145.25a - 1345.20b + c$$
$$1180.50 = 1345.20a + 1145.25b \qquad + d$$
$$3570.80 = 2595.35a - 599.82b + c$$
$$1925.70 = 599.82a + 2595.35b \qquad + d$$

The solution of these equations gives $a = +0.582440$; $b = +0.813281$; $c = +2546.99$; $d = -534.41$. Substituting the coordinates of M and N in the second system into the right side of Eq. (8-33) gives

$$X_M = 0.582440 \times 2095.18 - 0.813281 \times 1278.48 + 2546.99 = 2727.54$$

$$Y_M = 0.813281 \times 2095.18 + 0.582440 \times 1278.48 - 534.41 = 1914.20$$

$$X_N = 0.582440 \times 2502.55 - 0.813281 \times 1296.80 + 2546.99 = 2949.91$$

$$Y_N = 0.813281 \times 2502.55 + 0.582440 \times 1296.80 - 534.41 = 2256.17$$

If the scale is identical between the two systems, then $a = \cos\theta$ and $b = \sin\theta$, and $\sqrt{a^2 + b^2} = 1$. In this example, $\sqrt{a^2 + b^2} = 1.000331$, indicating that the scale of the first system is larger than that of the second system.

If there are more than two points known in both systems, the parameters $a, b, c,$ and d can be solved by the method of least squares, using the method of observation equations given in Sections A-3(a) and (b) of Appendix A.

8-23. LOCATION OF A LINE BASED ON COMPUTATIONS

The problem of locating a desired line in the field by measurement from stations in a control traverse is constantly faced in location surveys, construction surveys, and boundary surveys. In Sections 6-9 and 6-11 two methods were discussed for laying out a straight line in the field by means of a straight random traverse and by a traverse such as that illustrated in Fig. 6-13. Either of these problems can be solved by running a control random traverse between the two ends of the line. Latitudes and departures can be computed for the lines in the random traverse, and coordinates can be computed for the traverse stations. From the coordinates of the two ends of the desired line, its length and bearing can be computed. Furthermore, the coordinates of any point on the desired line can be computed.

If the control random traverse is on the same system of coordinates as that of the two ends of the line along which points are to be set, the calculated closure error at the far end of the line is used as the basis of a compass rule adjustment of the intermediate points on the random traverse.

If the control random traverse begins at one end of the line and some assumed bearing of the starting line is used to compute the azimuths of the lines in the control traverse, then the two coordinate systems will be different, and a coordinate transformation is made. This transformation determines the coordinates of the random traverse stations in the system of the given line on which points are to be set.

By use of the computed coordinates at selected points on the desired line, the lengths and bearings of tie lines from the control-traverse stations can be computed. The angle to be laid off from a control-traverse line to run a tie line is computed from the known bearing of the tie line and that of the traverse line. In this manner all the necessary angles are computed, and the points are set on the desired line by reoccupying the traverse stations and running the tie lines on the computed bearings and for the computed distances.

Example 8-8

The coordinates of the two ends of line AP of Fig. 8-24 are $X_A = 1452.60$ m, $Y_A = 2086.12$ m; $X_P = 2533.14$ m, $Y_P = 2461.85$ m. Points are to be set on this line at 1, 2, . . . every 200 m beginning at A. A random traverse $ABCDEP$ is run by deflection angles and distances on an assumed meridian with the following results:

STATION	DEFLECTION ANGLE	LINE	MEASURED DISTANCE (m)
B	21° 06′ 10″ R	AB	271.40
C	26° 42′ 30″ L	BC	295.35
D	36° 02′ 10″ R	CD	239.50
E	49° 36′ 10″ R	DE	307.65
		EP	150.80

No backsight was available at A, and thus no deflection angle was measured. The approximate direction of AB was determined by observing a compass bearing at B and applying the magnetic declination. The approximate bearing of AB is N 50° 15′ E. Compute the necessary distances and azimuths to be run

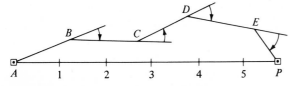

Figure 8-24. Random traverse run to establish points on line AP.

from B to points 1 and 2; from C to point 3; from D to point 4; and from E to point 5.

Solution: The approximate directions of BC, CD, DE, and EP (which are consistent with one another to within the accuracy of the measured angles) are computed as discussed in Section 8-5. The bearings are shown in Table 8-16. The calculations to determine the tentative coordinates are also shown.

The coordinates of the points given in Table 8-16 must now be transformed into the system in which line AP lies by means of Eq. (8-33). The values of a, b, c, and d are found using points A and P. The fixed coordinates of A and P are placed on the left side of Eq. (8-33), and the tentative coordinates of A and P from Table 8-16 are placed on the right side giving

$$1452.60 = 1452.60a - 2086.12b + c$$

$$2086.12 = 2086.12a + 1452.60b \qquad + d$$

$$2533.14 = 2528.04a - 2476.82b + c$$

$$2461.85 = 2476.82a + 2528.04b \qquad + d$$

Solving gives $a = +0.999\,7225$; $b = -0.013\,8195$; $c = -28.43$; and $d = +20.66$. The value of $\sqrt{a^2 + b^2}$ is $0.999\,8180$ which gives an indication of the closure precision of the random traverse, assuming that the fixed coordinates of A and P are without error. Using the values of the tentative coordinates of the random traverse points B, C, D, and E from Table 8-16 in the right side of Eq. (8-33) gives their coordinates in the system of the fixed line AB.

$$X_B = 0.999\,7225 \times 1661.26 + 0.013\,8195 \times 2259.66 - 28.43 = 1663.60$$

$$Y_B = 0.999\,7225 \times 2259.66 - 0.013\,8195 \times 1661.26 + 20.66 = 2256.74$$

$$X_C = 0.999\,7225 \times 1941.11 + 0.013\,8195 \times 2354.10 - 28.43 = 1944.67$$

$$Y_C = 0.999\,7225 \times 2354.10 - 0.013\,8195 \times 1941.11 + 20.66 = 2347.28$$

$$X_D = 0.999\,7225 \times 2109.41 + 0.013\,8195 \times 2524.50 - 28.43 = 2115.31$$

$$Y_D = 0.999\,7225 \times 2524.50 - 0.013\,8195 \times 2109.41 + 20.66 = 2515.31$$

$$X_E = 0.999\,7225 \times 2413.00 + 0.013\,8195 \times 2574.32 - 28.43 = 2419.48$$

$$Y_E = 0.999\,7225 \times 2574.32 - 0.013\,8195 \times 2413.00 + 20.66 = 2560.92$$

The transformed coordinates of the points in the random traverse are now inversed in order to obtain the correct azimuths of the lines. This is shown in Table 8-17. These azimuths are needed to compute the necessary angles to turn when running the tie lines.

TABLE 8-16. Computation of Tentative Coordinates

STATION	BEARING	LATITUDE	DEPARTURE	Y'	X'
A				2086.12	1452.60
	N 50° 15′ 00″ E	+173.54	+208.66		
B				2259.66	1661.26
	N 71° 21′ 10″ E	+94.44	+279.85		
C				2354.10	1941.11
	N 44° 38′ 40″ E	+170.40	+168.30		
D				2524.50	2109.41
	N 80° 40′ 50″ E	+49.82	+303.59		
E				2574.32	2413.00
	S 49° 43′ 00″ E	−97.50	+115.04		
P				2476.82	2528.04

The coordinates of points 1 through 5 of Fig. 8-24 are now computed as shown in Table 8-18. The bearing of AP is found by inversing to give N 70° 49′ 30″ E.

The tie lines can now be inversed to determine their azimuths and lengths. These inverse computations are shown in Table 8-19.

TABLE 8-17. Inversing to Obtain Azimuths of Lines in Random Traverse

LINE	ΔX	ΔY	AZIMUTH
AB	+211.00	+170.62	51° 02′ 20″
BC	+281.07	+90.54	72° 08′ 40″
CD	+170.64	+168.03	45° 26′ 30″
DE	+304.17	+45.61	81° 28′ 20″
EP	+113.66	−99.07	131° 04′ 40″

TABLE 8-18. Calculation of Coordinates of Points on Line *AP*

STATION	LENGTH	LATITUDE[a]	DEPARTURE[a]	Y	X
A				2086.12	1452.60
	200.00	65.691	188.904		
1				2151.81	1641.50
	200.00	65.691	188.904		
2				2217.50	1830.41
	200.00	65.691	188.904		
3				2283.19	2019.31
	200.00	65.691	188.904		
4				2348.88	2208.22
	200.00	65.691	188.904		
5				2414.58	2397.12

[a] Carried an extra place to avoid accumulated rounding-off errors.

TABLE 8-19. Inversing Tie Lines

LINE	ΔX	ΔY	AZIMUTH	DISTANCE
B-1	−22.10	−104.93	191° 53′ 40″	107.23
B-2	+166.81	−39.24	103° 14′ 20″	171.36
C-3	+74.64	−64.09	130° 39′ 00″	98.38
D-4	+92.91	−166.43	150° 49′ 40″	190.61
E-5	−22.36	−146.34	188° 41′ 10″	148.04

The angles necessary to lay off in the field from the random traverse to the points on the desired line are obtained from the azimuths involved. For example in Fig. 8-25, with the theodolite at B, the right deflection angle from A to point 2 is 103° 14′ 20″ − 51° 02′ 20″ = 52° 12′ 00″ R. This angle should be laid off by double centering. The calculated distance of 171.36 m is measured off to locate the desired point.

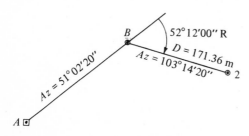

Figure 8-25. Deflection angle to tie line.

8-24. RADIATION TRAVERSE   A radiation traverse is one that is executed at a single station. It is applicable in an open area with no obstructed line of sight to the other points in the survey. It can be used either to determine the positions of existing points or to lay out points as in a construction project. Assume that the position of some point in the area, defined by its plane coordinates, is known and also that the direction to some other point is known. For example, the position of point A in Fig. 8-26 and the azimuth of the line AP are both known. Point A is occupied with the transit or theodolite and a backsight is made to point P. The angles between all the lines AB through AH are measured with respect to line AP. The theodolite is then replaced by the EDM (unless both features are incorporated into the same instrument). The distances to points B through H are then measured and reduced to horizontal distances. The angles are used to determine the azimuths of all the lines with reference to the known azimuth of AP. The azimuths (or bearings) together with the horizontal distances are used to compute latitudes and departures of the lines AB through AH. These values are then

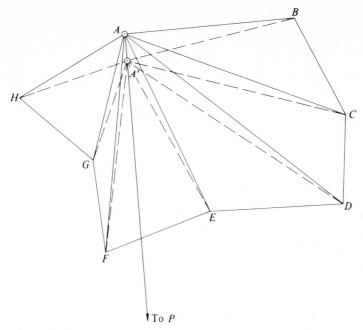

Figure 8-26. Radiation traverse.

added to the coordinates of A to get the coordinates of B through H by Eqs. (8-7) and (8-8).

In order to provide a check on the measurements and discover any possible blunders, a second point A' is carefully set on the line AP. This can be done at the time the initial backsight is made on P. The distance AA', which can be fairly short, say 10 to 12 ft, is then measured. The coordinates of A' can be computed later by applying the latitude and departure of AA' to the coordinates of A using Eqs. (8-7) and (8-8). After the measurements are made at A, the instruments are moved to A', a backsight is made along the line $A'P$, and the angles between this line and the lines $A'B$ through $A'H$, shown by dashed lines in Fig. 8-26, are measured. The EDM is then used to measure the lengths of the lines. These lengths, reduced to horizontal distances, are then used together with the azimuths of $A'B$ through $A'H$ to obtain the latitudes and departures of all the lines. The coordinates of the points computed from A' are then compared with those computed from A. If they are in reasonable agreement, the mean of each pair is adopted. Otherwise the mistake must be found in the field.

The main drawback to the application of the radiation traverse is the weak angular relationship between the lines joining the perimeter points such as B through H of Fig. 8-26. For example, if the line EF is relatively short, then a small uncertainty in the length of AE or AF, or both, will have a relatively large effect on the direction of EF. Consequently, the lines on the perimeter cannot be used as starting lines in subsequent surveys in the area.

When using the radiation traverse to set out points, it is presumed that the coordinates of the points are known, having been either derived from the geometry of the construction lines or scaled from the construction plans. Each radiating line in the layout scheme can then be inversed to determine its length and bearing, using Eqs. (8-11) and (8-12). The inversed values of the bearings are used to compute the angles or circle readings to be set off in the field with respect to the reference line, such as AP in the discussion above. The distances along the radiating lines are then laid out according to methods discussed in Section 2-31.

After all the points have been set, their coordinates can be checked from an auxiliary station such as A' as outlined above. This check precludes the possibility of setting out a point in the wrong place. It also verifies the inverse calculations made to determine distances, angles, and bearings.

8-25. AREA FROM RECTANGULAR COORDINATES The area of a figure for which coordinates have been computed is determined by either Eq. (8-25) or (8-26) of Section 8-20. In evaluating each term of this equation, some of the terms will be positive and some negative giving the possibility of a net positive area or a net negative area. This is of no consequence unless the result is subject to further analytical treatment (see Section 8-29). The application of Eq. (8-25) to the area enclosed within the *boundary* of Fig. 8-22, the coordinates of which were computed in Section 8-21, is shown in Table 8-20.

TABLE 8-20. Computation of Area Within a Boundary by Coordinates

	+	−
$X_1(Y_2 - Y_4) = 1000.00(1296.38 - 1029.28) =$	267,100	
$X_2(Y_3 - Y_1) = 981.62(1257.32 - 1000.00) =$	252,590	
$X_3(Y_4 - Y_2) = 759.24(1029.28 - 1296.38) =$		202,793
$X_4(Y_1 - Y_3) = 739.84(1000.00 - 1257.32) =$		190,376
	519,690	393,169
	−393,169	
	2)126,521	
area $=$	63,260 ft^2	

8-26. AREAS FROM MAPS When a map of a required area is available, the area can be obtained by dividing the figure into geometrical shapes (triangles, trapezoids, and rectangles), and computing the area of these figures from the dimensions scaled from the map. This method is limited to figures that are bounded by approximately straight lines. When the three sides a, b, and c of a triangle have been scaled from the map, the area may be found by the formula

$$\text{area} = \sqrt{s(s-a)(s-b)(s-c)} \qquad (8\text{-}34)$$

where $s = \frac{1}{2}(a + b + c)$. These lengths are sometimes the result of direct field measurements.

When two sides a and b and the included angle C of a triangle are known, the formula for the area is

$$\text{area} = \frac{1}{2}ab \sin C \qquad (8\text{-}35)$$

If the figure is very irregular in shape, as is generally the case when the boundaries are bodies of water, its area can be obtained most accurately by the use of the planimeter. The planimeter can be used also to check an area that has been computed from the field measurements.

The larger planimeters having adjustable arms are generally more carefully constructed and hence are more accurate than the smaller ones. The precision of the resulting area is dependent on the scale to which the map is drawn, and on the skill of the operator of the planimeter. With care, areas that are accurate within 1 % can be obtained by means of the planimeter.

8-27. AREA WITH ONE CURVED BOUNDARY Two examples of areas with one curved boundary are shown in Fig. 8-27. In view (a) one side of the figure is a circular arc with radius R and central angle FGE. This area can be calculated in two parts, namely, the sector EFG and the part $ABCDEGFA$ bounded by straight lines. The area of the sector is $\pi R^2 \times$ angle $FGE/360°$.

In Fig. 8-27 (b) the boundary AF is an irregular curve. The area of this figure can be divided into two parts by the straight line GH, and computations for the portion bounded by straight lines can be made by any of the ordinary methods. If perpendicular offsets from the straight line GH to the curved boundary are measured, either on the ground or on a map, the irregular portion can be considered as a series of triangles and trapezoids. These offsets should be measured at all breaks in the curved boundary. The area of each triangle and each trapezoid can be computed separately. When the boundary is sufficiently regular, the offsets can be taken at equal intervals along the line GH, and the computations can thus be simplified considerably. The area of the regular portion $AJLK$ in Fig. 8-27 (b) can be obtained by a single multiplication by applying the trapezoidal rule, which is

$$\text{area} = W\left(\frac{h_1 + h_n}{2} + h_2 + h_3 + \cdots + h_{n-1}\right) \qquad (8\text{-}36)$$

in which W is the common spacing of the offsets and n is the number of offsets.

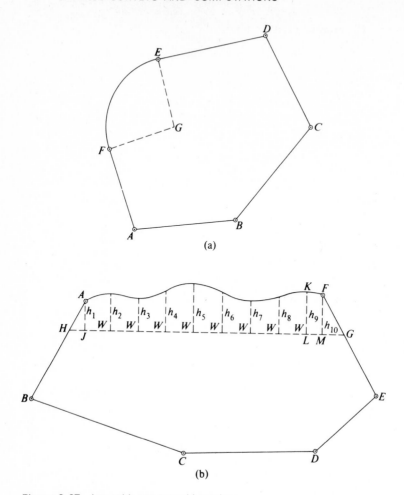

Figure 8-27. Area with one curved boundary.

Example 8-9

Determine the area $AFGH$ in Fig. 8-27 (b) by using the trapezoidal rule for the regular portion. The distances are as follows:

$h_1 = 22.6$ ft $h_6 = 36.9$ ft $HJ = 14$ ft

$h_2 = 28.0$ ft $h_7 = 30.0$ ft $LM = 13.5$ ft

$h_3 = 27.1$ ft $h_8 = 31.5$ ft $MG = 21$ ft

$h_4 = 30.6$ ft $h_9 = 34.8$ ft $W = 25$ ft = common spacing

$h_5 = 38.5$ ft $h_{10} = 35.0$ ft

Solution: The computations follow:

$$\text{area } AJH \quad = \frac{(14.0)(22.6)}{2} = 158 \text{ ft}^2$$

$$\text{area } KFML = \frac{(34.8 + 35.0)(13.5)}{2} = 471 \text{ ft}^2$$

$$\text{area } FGM \quad = \frac{(21.0)(35.0)}{2} = 368 \text{ ft}^2$$

$$\text{area } AJLK \ = 25\left(\frac{22.6 + 34.8}{2} + 28.0 + 27.1 + 30.6 + 38.5 \right.$$

$$\left. + 36.9 + 30.0 + 31.5\right)$$

$$= 6283 \text{ ft}^2$$

The total area $AFGH$ is therefore 7280 ft^2.

The assumption made in using the trapezoidal rule is that the curved boundary is composed of chords connecting the ends of the offsets. When the offsets are taken closely enough together and when the curves are flat, no considerable error is introduced by this assumption.

A more accurate value for the area between a straight line and an irregular boundary may be obtained by taking offsets at regular intervals and applying Simpson's one-third rule to that portion lying between an *odd* number of offsets. Simpson's one-third rule assumes that the curve through each successive three points is a portion of a parabola. In Fig. 8-28 the curve through points A, B, and C is assumed to be a segment of a parabola cut off by the

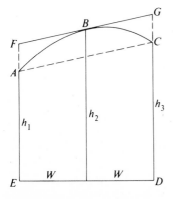

Figure 8-28. Area by Simpson's one-third rule.

chord AC. The area between the chord and the parabola, found by the calculus, is equal to two-thirds the area of the parallelogram $AFGC$. The area of $ABCDE$ is therefore

$$\text{area} = 2W\left(\frac{h_1 + h_3}{2}\right) + \frac{2}{3} \times 2W\left(h_2 - \frac{h_1 + h_3}{2}\right)$$

Expanding and collecting terms gives

$$\text{area} = \frac{W}{3}(h_1 + 4h_2 + h_3)$$

For the next three offsets, the area would be, by the same reasoning,

$$\text{area} = \frac{W}{3}(h_3 + 4h_4 + h_5)$$

By extending this reasoning to any *odd* number of offsets and summing up the results, we obtain the following relation:

$$\text{area} = \frac{W}{3}[h_1 + 2(h_3 + h_5 + \cdots + h_{n-2}) + 4(h_2 + h_4 + \cdots + h_{n-1}) + h_n]$$

in which h_1 is the first offset; h_n is the last odd-numbered offset; W is the common interval at which the offsets are taken.

In its condensed form Simpson's one-third rule may be expressed as follows:

$$\text{area} = \frac{W}{3}(h_1 + h_n + 2\sum h_{\text{odd}} + 4\sum h_{\text{even}}) \qquad (8\text{-}37)$$

Example 8-10

Determine the area $AFGH$ in Fig. 8-27 (b) by using Simpson's one-third rule where it applies.

Solution: Obviously Simpson's one-third rule applies only between the first and the ninth offsets. The area of the remaining portion will be computed by the trapezoidal rule. Thus,

$$\text{area } AJLK = \tfrac{25}{3}[22.6 + 34.8 + 2(27.1 + 38.5 + 30.0)$$

$$+ 4(28.0 + 30.6 + 36.9 + 31.5)] = 6305 \text{ ft}^2$$

The remaining area, as computed in the preceding example, is 997 ft². Therefore the total area $AFGH$ is 7302 ft², which is 22 ft² more than that obtained by the trapezoidal rule. This difference is due to the fact that the irregular boundary is generally concave toward the line *HG*. If the irregular boundary is concave away from the straight line from which offsets have been measured, Simpson's one-third rule will give a smaller area than will the trapezoidal rule.

8-28. PROBLEMS IN OMITTED MEASUREMENTS The common examples of omitted measurements are shown in Fig. 8-29. In view (a) the length or azimuth, or both the length and the azimuth, of the side *EA* may be unknown. The latitudes and departures of the known sides are computed. Coordinates are then computed, and *EA* is inversed to obtain its bearing and length.

In Fig. 8-29 (b) two adjacent sides are involved in the missing measurements. Both lengths, both azimuths, or the azimuth of one line and the length of the other may be unknown. From the calculated coordinates of *A* and *D*, the azimuth and length of the closing line *DA* are computed. The triangle *ADE* can now be solved. When the lengths of *DE* and *EA* are missing, the triangle is solved from the computed length *DA* and the three angles, which

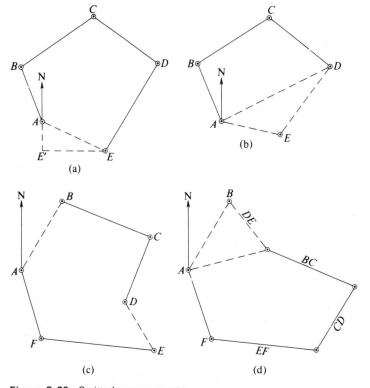

Figure 8-29. Omitted measurements.

can be calculated from the known azimuths. When the azimuths of *DE* and *EA* are unknown, the three sides of the triangle are known and the angles are computed. When the azimuth of one of the sides and the length of the other are unknown, the triangle is solved from the two known sides and the angle opposite one of the known sides. Frequently there will be two solutions in this last case, and the approximate shape of the figure must then be known. The trigonometric formulas for the solution of oblique triangles are given in Table C in Appendix C.

In Fig. 8-29 (c) the omitted measurements occur on sides that are not adjacent. Since the latitudes and departures of equal parallel lines are equal, this problem can be solved by shifting the line *DE* until it is adjacent to *AB*, so as to form the closed figure shown in view (d). The solution of this problem is the same as that outlined for view (b).

8-29. PARTING OFF LAND

One of the common problems of land surveying is the division of an irregular polygon into two or more parts with known areas. The division may be made by a line of known direction, or it may be made by a line from a given starting point on the perimeter of the polygon.

In Fig. 8-30 let the problem be the division of the polygon *ABCDEF* into two equal parts by a line parallel to *AF*. The azimuths and lengths of the

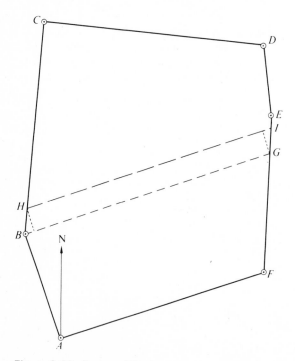

Figure 8-30. Parting off land.

TABLE 8-21. Computation of Area by Coordinates

STATION	Y	X			+	−
A(=1)	0	108.5	108.5	(301.5 − 197.0) =	11,338	
B(=2)	301.5	0	0	(917.1 − 0) =		
C(=3)	917.1	60.7	60.7	(843.8 − 301.5) =	32,918	
D(=4)	843.8	710.9	710.9	(644.1 − 917.1) =		194,076
E(=5)	644.1	729.9	729.9	(197.0 − 843.8) =		472,099
F(=6)	197.0	710.8	710.8	(0 − 644.1) =		457,826

$$+44,256 \qquad -1,124,001$$
$$+44,256$$
$$2) - 1,079,745$$
$$\text{area} = -539,873$$

sides being known, the first step is the calculation of the entire area. If a rough sketch is made from the field measurements, the approximate position of the line HI, which is to divide the area into two equal parts, can be determined. From the figure it is evident that H will be somewhere on the line BC. The exact location of HI is determined most easily by first including an imaginary line BG parallel to AF and calculating the area of the part $ABGE$. This area will probably be less than the required area and the line must be shifted up.

The solution to this problem is obtained by the methods of analytic geometry. The rectangular coordinates of the corners are determined from the latitudes and departures of the sides. For convenience the Y axis should be chosen through B and the X axis through A. The coordinates are as shown in Table 8-21 along with the determination of the area of the entire field found by Eq. (8-25).

This area is $-539,870$ ft². The algebraic sign is significant in this solution, since one-half this area, or $-269,935$ ft², is a term in one of the equations to be used in solving the problem analytically.

From Fig. 8-30 and from the conditions in the problem, the following four statements are apparent: (1) Point H lies on the line BC; (2) point I lies on the line EF; (3) the area of $ABHIF$ is one-half the total area, or $-269,935$ ft²; and (4) the line HI has the same bearing as does the line AF. If the coordinates of the unknown point H are denoted by Y_H and X_H and the coordinates of the unknown point I by Y_I and X_I, the following four equations may be written: By Eq. (8-15),

$$\frac{Y_H - Y_B}{Y_C - Y_B} = \frac{X_H - X_B}{X_C - X_B}$$

By Eq. (8-15),

$$\frac{Y_I - Y_E}{Y_F - Y_E} = \frac{X_I - X_E}{X_F - X_E}$$

By Eq. (8-25),

$$X_A(Y_B - Y_F) + X_B(Y_H - Y_A) + X_H(Y_I - Y_B)$$
$$+ X_I(Y_F - Y_H) + X_F(Y_A - Y_I) = 2(-269,935)$$

By Eq. (8-11),

$$\frac{X_F - X_A}{Y_F - Y_A} = \tan B_{AF} = \tan B_{HI} = \frac{X_I - X_H}{Y_I - Y_H}$$

Substituting the known values of the coordinates in these equations gives four equations which may be solved simultaneously to obtain the values of the four unknowns X_H, Y_H, X_I, and Y_I. These equations are

$$\frac{Y_H - 301.5}{615.6} = \frac{X_H - 0}{60.7} \tag{1a}$$

$$\frac{Y_I - 644.1}{-447.1} = \frac{X_I - 729.9}{-19.1} \tag{2a}$$

$$(108.5)(104.5) + 0(Y_H - 0) + X_H(Y_I - 301.5)$$
$$+ X_I(197.0 - Y_H) + 710.8(0 - Y_I) = -539,870 \tag{3a}$$

$$\frac{X_I - X_H}{Y_I - Y_H} = \frac{602.3}{197.0} \tag{4a}$$

The results obtained by cross multiplying and rearranging and collecting terms follow:

$$615.6X_H - 60.7Y_H + 18,301.05 = 0 \tag{1b}$$

$$447.1X_I - 19.1Y_I - 314,035.98 = 0 \tag{2b}$$

$$301.5X_H - 197.0X_I - X_H Y_I + X_I Y_H + 710.8Y_I - 551,208 = 0 \tag{3b}$$

$$X_H - X_I - 3.05736Y_H + 3.05736Y_I = 0 \tag{4b}$$

Expressing Y_H in terms of X_H from Eq. (1b) gives

$$Y_H = 10.14168X_H + 301.5 \tag{1c}$$

Expressing Y_I in terms of X_I from Eq. (2b) gives

$$Y_I = 23.40838X_I - 16,441.7 \tag{2c}$$

When these values of Y_H and Y_I are substituted in Eqs. (3b) and (4b), the results are

$$16,743.2X_H + 16,743.2X_I - 13.26670X_H X_I - 12,237,968 = 0 \quad (3c)$$

$$30.00677X_H - 70.56784X_I + 51,190 = 0 \quad (4c)$$

or

$$X_I = 0.42522X_H + 725.40$$

Substituting the value of X_I from Eq. (4c) in Eq. (3c) gives

$$5.64127X_H^2 - 14,239X_H + 92,451 = 0 \quad (3d)$$

From this quadratic equation, $X_H = 6.5588$. By Eq. (4c), $X_I = 728.19$. By Eq. (1c), $Y_H = 368.0$. By Eq. (2c), $Y_I = 604.0$. By Eq. (8-12), $BH = 66.8$ ft; $EI = 40.1$ ft; $HI = 759.2$ ft.

In Fig. 8-31 let the problem be the division of the polygon $ABCDEF$ of Fig. 8-30 into two equal parts by a line GH from the middle point of the side AF. This problem is solved by analytic geometry by first computing co-ordinates of the traverse stations and then computing the area from the

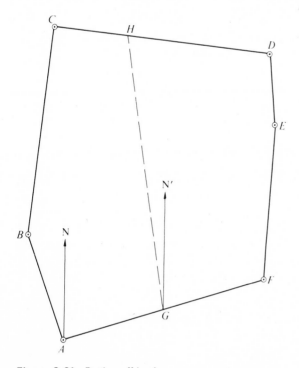

Figure 8-31. Parting off land.

coordinates. The coordinates of point G are $Y = 98.5$ and $X = 409.6$, which are equal to the means of the coordinates of A and F since G is the middle point of the line AF. From Fig. 8-31 and from the conditions in the problem, two statements are apparent: (1) Point H lies on line CD and (2) the area of $ABCHG$ is one-half the total area. Two equations suffice to compute the coordinates X_H and Y_H of the unknown point H. The equations are as follows: By Eq. (8-15),

$$\frac{Y_H - Y_C}{Y_D - Y_C} = \frac{X_H - X_C}{X_D - X_C}$$

By Eq. (8-25),

$$X_A(Y_B - Y_G) + X_B(Y_C - Y_A) + X_C(Y_H - Y_B)$$
$$+ X_H(Y_G - Y_C) + X_G(Y_A - Y_H) = -2(269,935)$$

It is only necessary to substitute the values of the known coordinates in these equations and solve the simultaneous equations thus obtained. The work follows:

$$73.3X_H + 650.2Y_H - 600,747.7 = 0 \tag{1a}$$
$$818.6X_H + 348.9Y_H - 543,594.4 = 0 \tag{2a}$$

$$73.3X_H + 650.2Y_H - 600,747.7 = 0 \tag{1b}$$
$$1525.5X_H + 650.2Y_H - 1,013,026.9 = 0 \tag{2b}$$

$$\overline{\quad 1452.2X_H \qquad\qquad -412,279.2 = 0 \quad}$$

$$X_H = 283.9 \qquad\qquad Y_H = 891.9$$

By Eq. (8-12), $CH = 224.5$ ft and $GH = 803.3$ ft. By Eq. (8-11), the bearing of GH is N 9° 00′ W.

8-30. METHODS OF PLOTTING A SURVEY Traverses as well as calculated figures such as parcels of land can be plotted to suitable scale by using the computed coordinates of the traverse stations or the points in the figures, or by laying off distances and bearings. The latter method is used to advantage when the traverse has not been computed and balanced. The method of coordinates is the more reliable of the two methods, however, because most of the work in plotting has been arrived at numerically.

8-31. PLOTTING BY DISTANCE AND BEARINGS The most convenient approach to plotting traverse lines by their lengths and bearings is to reference all the work on the map sheet to a very carefully constructed square which measures for convenience 10 in. on a side. In Fig. 8-32 a 10-in.

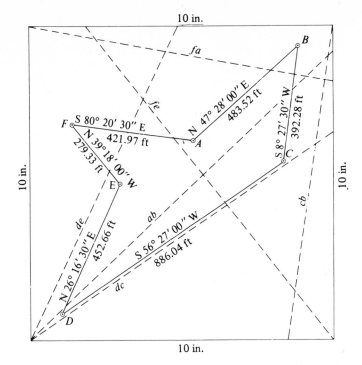

Figure 8-32. Plotting traverse by tangents or cotangents of bearing angles.

square has been constructed on the map sheet. This will be used to plot the traverse from Section 8-11 and shown in Fig. 8-13. The lines, their lengths and bearings, and either the tangent or the cotangent of the bearing angle are listed (see Table 8-22). If the bearing angle is less than 45°, the tangent is listed; if greater than 45°, the cotangent is listed.

The first traverse station is plotted at some tentative point on the map sheet, the position of which depends on the size of the sheet, the scale of the plotting, and the shape of the traverse. This tentative position may need repositioning if the traverse begins to run off the limits of the map sheet. In the present instance station A is plotted in the position shown in Fig. 8-32.

TABLE 8-22. Data for Plotting a Traverse by Lengths and Bearings

LINE	LENGTH	BEARING	TANGENT B	COTANGENT B
AB	483.52	N 47° 28′ 00″ E		0.91740
BC	392.28	S 8° 27′ 30″ W	0.14871	
CD	886.04	S 56° 27′ 00″ W		0.66314
DE	452.66	N 26° 16′ 30″ E	0.49369	
EF	279.33	N 39° 18′ 00″ W	0.81849	
FA	421.97	S 80° 20′ 30″ E		0.17018

Next, the direction of the line AB is established. Since its bearing is N 47° 28′ 00″ E, if we imagine commencing at the southwest corner of the square, going eastward 10 in. and then northward 10 cot 47° 28′ 00″ = 9.174″ along the right edge of the square, a point is plotted 9.174″ up from the lower right southeast corner of the square. A line joining this point with the southwest corner of the square defines the dashed line ab which bears N 47° 28′ 00″ E. This direction is transferred to station A and a line through A of indefinite length is drawn. The length AB is now laid off along this line according to the plotting scale. This defines the position of B.

Next the direction of BC is established. The bearing of BC is S 8° 27′ 30″ W. Thus commencing at the northeast corner of the square, we go south 10″ along the right side of the square and then west 10 tan 8° 27′ 30″ = 1.487″ along the lower edge of the square. When joined with the northeast corner, a point plotted at this position gives the dashed line cb which is in the proper direction. A line is drawn through B parallel with cb and the length BC is laid off to scale in order to locate point C.

The above procedure is used with each succeeding line to lay off the correct direction and distance. The end of the last line FA should fall on the initial point A. If the survey is of reasonable accuracy, even though it had not been balanced, it should close on paper with no detectable error. Any discrepancy between the beginning and ending position of the plotted, closed traverse therefore is due to either errors or mistakes in plotting or else a combination of plotting errors and mistakes.

A graphical application of the compass rule is used to adjust a misclosure. The amount of closure should first be scaled to examine whether a unit such as 10 or 100 ft has been misplotted. The direction of closure can be an indication of which line might be plotted with the wrong length, because the closure line and the faulty line should be parallel with one another. If the closure is relatively small, then each plotted point can be shifted by an amount proportionate to its distance from the point of beginning and in the direction required to move the last plotted point into correct position (see line $a'a$ of Fig. 14-24).

8-32. PLOTTING BY RECTANGULAR COORDINATES The method of plotting a traverse or a figure by rectangular coordinates is more accurate than plotting by lengths and bearings and has advantages not possessed by the latter method. It is known in advance that the traverse closes. The erroneous plotting of any point does not affect the location of any of the succeeding points, as the plotting of each point is independent of the rest. Each point can be checked as soon as plotted by comparing the scaled distance to the preceding station with the length measured in the field. The proper size of sheet for the map may be determined by an inspection of the coordinates.

A series of *grid* lines is carefully constructed. The lines of the two sets are perpendicular to one another, and those of each set are spaced at intervals of 50, 100, 500, or 1000 ft, or at some other regular interval, the distance depending on the scale to which the traverse or map is to be plotted. All these grid lines should be plotted in as brief a period of time as possible, so that the measurements are consistent with one another under the given ambient conditions of temperature and humidity. Each grid line is labeled with its proper Y coordinate or X coordinate and, finally, the several points are plotted, according to their coordinates, from the nearest grid lines. Plotting by coordinates is shown in Fig. 8-33, where the stations of the traverse of Section 8-11 are plotted by using a 200-ft grid spacing. When all the points have been plotted, their positions are checked by scaling the distances between successive points and comparing them with the known distances.

Grid lines and coordinates can be plotted mechanically by means of an instrument called a *coordinatograph*. One is shown in Fig. 8-34. A girder travels in the X direction, and its motion is capable of being read or set by means of a scale and rotating dial to an accuracy of approximately ± 0.003 in. or less than 0.1 mm. A pencil, pen, or needle holder travels along the girder in the Y direction and its motion also can be read or set to the same accuracy. Grid lines are ruled in both directions by appropriate settings of the scales and dials, which are determined by the choice of the plotting scale. Each point is then plotted, according to its coordinates, by making appropriate settings of the scales and dials.

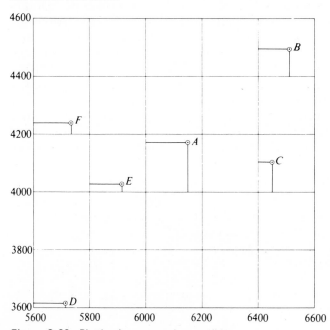

Figure 8-33. Plotting by rectangular coordinates.

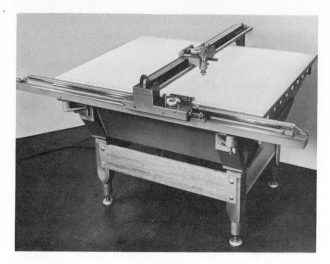

Figure 8-34. Coordinatograph. Courtesy of Unitech Corp.

The plotting unit of the coordinatograph shown in Fig. 8-35 is driven by command from a preprogrammed control unit through the medium of punched cards. In modern computer-driven graphics the plotter unit containing a pencil or pen can be instructed to draw grid lines, plot points, draw straight lines and curves, letter or number points, or both; that is, it does anything that can be performed by a draftsman. This kind of a system is most advantageous when the plotting of grids and control points is a continuous, everyday operation. It is not feasible when drafting of this nature is limited.

The value of a map may be increased considerably by retaining the grid lines on the finished map, either as complete lines or as tick marks at the edges of the sheet and cross marks at the grid intersections in the interior portion. The map user can then detect any change in scale in any or all portions of the map by measuring between grid lines or between grid ticks. Where accurate lengths are to be determined between well-defined points, the coordinates of points defining the ends of the lines are scaled from the nearest grid lines and

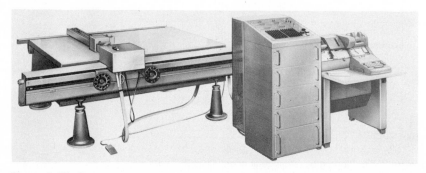

Figure 8-35. Computer-controlled coordinatograph. Courtesy of Coradi A. G.

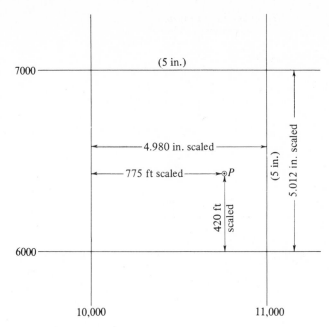

Figure 8-36. Scaling coordinates.

the scaled distances are either increased or decreased to allow for the scale change. The lengths may then be computed by Eq. (8-12). In Fig. 8-36 the grid was originally plotted with 5-in. spacings representing 1000 ft. The scaled distances to point P from the grid lines representing $Y = 6000$ ft and $X = 10,000$ ft are 420 ft and 775 ft, respectively. The actual distance between the grid lines in the Y direction is 5.012 in. and the distance in the X direction is 4.980 in. The Y coordinate of the point P is therefore $6000 + (420)(5)/5.012 = 6419$ ft and the X coordinate is $10,000 + (775)(5)/4.980 = 10,777$ ft.

8-33. SCALES The scales commonly used on maps intended for engineering purposes are those found on the edges of an engineer's scale. Thus 1 in. on the paper represents some multiple of 10, 20, 30, 40, 50, or 60 ft on the ground. Maps are generally classed as large scale when the scale is greater than 1 in. = 100 ft, as intermediate scale when the scale is between 1 in. = 100 ft and 1 in. = 1000 ft, and as small scale when the scale is less than 1 in. = 1000 ft. The large scales are used when a large amount of detail is to be shown, or where the area covered by the map is small. In general, the scale should be as small as possible and still represent the detail with sufficient precision. Maps intended for the design of engineering projects are commonly plotted to scales between 1 in. = 20 ft and 1 in. = 800 ft, the exact scale depending on the detail to be shown and the area covered.

Surveys made for architects are often plotted to the scales used by architects, namely, 1 in. = 4 ft, 1 in. = 8 ft, and 1 in. = 16 ft.

TABLE 8-23. Approximate Equivalents Between Engineer's Scale and Metric Representative Fractions

ENGINEER'S SCALE	APPROXIMATE RF FOR METRIC SYSTEM	
1 in. = 20 ft	$\frac{1}{200}$ or $\frac{1}{250}$	1:200 or 1:250
1 in. = 40 ft	$\frac{1}{500}$	1:500
1 in. = 50 ft	$\frac{1}{500}$	1:500
1 in. = 100 ft	$\frac{1}{1000}$	1:1000
1 in. = 200 ft	$\frac{1}{2000}$ or $\frac{1}{2500}$	1:2000 or 1:2500
1 in. = 250 ft	$\frac{1}{2000}$ or $\frac{1}{2500}$	1:2000 or 1:2500
1 in. = 1000 ft	$\frac{1}{10000}$	1:10000
1 in. = 2000 ft	$\frac{1}{20000}$ or $\frac{1}{25000}$	1:20000 or 1:25000

The topographic maps prepared by the U. S. Geological Survey are plotted to so-called natural scales. These scales are expressed as ratios. On a map drawn to a scale of 1/20,000, one unit on the map represents 20,000 units on the ground. Thus 1 in. on the paper represents 20,000 in. on the ground, 1 ft represents 20,000 ft, and 1 m represents 20,000 m on the ground.

By the decimal nature of the metric system, ratios or *representative fractions* (RF) are used exclusively to express scale. Some approximate equivalents between engineer's scales and representative fractions used in the metric system are given in Table 8-23. It is to be noted that these are not exact equivalents by any means.

As previously stated, the scale of a map may change slightly because of the shrinkage or stretching of the drawing paper. The scale is changed also when the map is reproduced. The change may be slight in blueprinted reproductions, or it may be considerable when photographic reductions or enlargements are made. Since the original scale no longer applies to the duplicates, a graphical scale, similar to one of those shown in Fig. 8-37, should appear on the map.

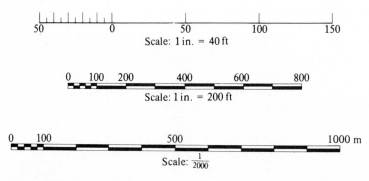

Figure 8-37. Graphical scales.

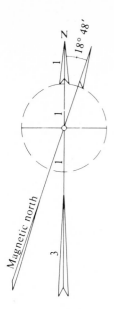

Figure 8-38. Meridian arrows.

8-34. MERIDIAN ARROW   A meridian arrow should appear on the map. It should be simple in design and although it should be of sufficient length to permit the reasonably accurate scaling of directions on any part of the map, it should not be so conspicuous as to convey the impression that the map is a drawing of an arrow. The proportions of an arrow that can be constructed easily with drawing instruments are shown in Fig. 8-38. The dotted circle does not appear on the finished drawing. When both true and magnetic directions are known, two arrows are combined as in Fig. 8-38—a full-headed arrow usually indicating true north and a half-headed arrow indicating magnetic north. The magnetic declination is shown in figures. If the map contains grid lines, or grid ticks, then grid north is indicated by the grid lines running in the *Y* direction.

8-35. TITLES   On an engineering drawing the title appears in the lower right-hand corner of the sheet. This same convention is followed as far as possible on maps, although on many maps the shape of the area covered may be such that the title is placed elsewhere on the sheet to give the drawing a more balanced appearance.

As maps may pass through many hands and may be intended as permanent records, more time is spent on the lettering than would be justified on the shop drawings for an engineering project. The very ornate lettering found on older maps is no longer used, carefully constructed Roman or Gothic capitals being used instead. Legends and notes, separated from the main title, may be shown in single-stroke Reinhardt letters.

The title should be sufficiently complete to permit the identification of the map. It should include the purpose of the survey, the name of the owner or the organization for whom the survey is made, the location (city or township, county, and state), the date of the survey, the name of the engineer or surveyor responsible for the work, and the scale. The name of the draftsman may also appear.

PROBLEMS

8-1. The adjusted *clockwise* interior angles in a traverse which originates and closes at point A are as follows:

STATION	FROM	TO	ADJUSTED CLOCKWISE INTERIOR ANGLE
A	F	B	27° 44′ 15″
B	A	C	152° 36′ 25″
C	B	D	67° 20′ 30″
D	C	E	200° 38′ 45″
E	D	F	128° 27′ 30″
F	E	A	143° 12′ 35″

The bearing of AB is S 2° 22′ 15″ E. Compute the azimuths and bearings of the remaining sides of the traverse.

8-2. Given the same angles as in Problem 8-1 and the bearing of DE as S 15° 25′ 00″ W, compute the azimuths and bearings of the remaining sides of the traverse.

8-3. Given the same angles as in Problem 8-1 and the bearing of CD as due east, compute the azimuths and bearings of the remaining sides.

8-4. Given the same angles as in Problem 8-1 and the bearing of EF as N 10° 22′ 15″ W, compute the azimuths and bearings of the remaining sides.

8-5. The following deflection angles were measured in a traverse that originates at station A with a backsight on station R and closes on station G with a foresight on station S.

STATION	FROM	TO	MEASURED DEFLECTION ANGLE
A	R	B	88° 16′ R
B	A	C	26° 50′ R
C	B	D	46° 32′ L
D	C	E	61° 25′ R
E	D	F	8° 16′ R
F	E	G	61 18′ L
G	F	S	62° 48′ L

The fixed azimuth of AR is 97° 15′ 30″ and the fixed azimuth of GS is 291° 21′ 00″. What are the adjusted azimuths and bearings of the intermediate traverse sides?

8-6. Given the same angles as in Problem 8-5, the fixed azimuth of AR is 282° 50′ 00″ and the fixed azimuth of GS is 117° 06′ 00″. Compute the adjusted azimuths and bearings of the intermediate traverse sides.

8-7. Given the same angles as in Problem 8-5, the fixed azimuth of AR is 186° 18′ 20″ and the fixed azimuth of GS is 20° 26′ 10″. Compute the adjusted azimuths and bearings of the intermediate traverse sides.

8-8. Given the same angles as in Problem 8-5, the fixed azimuth of AR is 8° 57′ 30″ and the fixed azimuth of GS is 203° 10′ 00″. Compute the adjusted azimuths and bearings of the intermediate traverse sides.

8-9. The following angles to the right were measured with a 1″ theodolite in a traverse that originates at point K with a backsight on point J and closes on point O with a foresight to point P.

STATION	FROM	TO	MEASURED ANGLE TO RIGHT
K	J	L	212° 21′ 16.2″
L	K	M	161° 14′ 58.6″
M	L	N	215° 30′ 08.4″
N	M	O	90° 50′ 01.4″
O	N	P	148° 47′ 23.0″

The fixed azimuth of JK is 179° 52′ 55.5″ and the fixed azimuth of OP is 108° 36′ 53.6″. Compute the adjusted azimuths of the intermediate traverse sides.

8-10. Given the same angles as in Problem 8-9, the fixed azimuth of JK is 17° 15′ 52.5″ and the fixed azimuth of OP is 305° 59′ 50.1″. Compute the adjusted azimuths of the intermediate traverse sides.

8-11. Given the same angles as in Problem 8-9, the fixed azimuth of JK is 246° 14′ 18″ and the fixed azimuth of OP is 174° 58′ 04.1″. Compute the adjusted azimuths and bearings of the intermediate traverse sides.

8-12. Given the same angles as in Problem 8-9, the fixed azimuth of JK is 350° 40′ 18.8″ and the fixed azimuth of OP is 279° 24′ 09.9″. Compute the adjusted azimuths and bearings of the intermediate traverse sides.

8-13. The lengths in feet and the adjusted bearings of the sides of a closed traverse follow:

SIDE	LENGTH (ft)	BEARING	SIDE	LENGTH (ft)	BEARING
AB	802.60	N 89° 00′ 10″ E	DE	881.93	S 43° 34′ 30″ W
BC	1095.72	N 24° 20′ 30″ E	EF	852.28	S 58° 19′ 20″ E
CD	1464.05	N 65° 52′ 00″ W	FA	525.38	S 3° 53′ 20″ W

Compute the latitudes and departures of the sides.

8-14. Balance the traverse in Problem 8-13 by the compass rule. Then compute the coordinates of the stations, taking $X_A = 25,000.00$ ft and Y_A as 18,400.00 ft.

8-15. Using the results of Problem 8-14, compute the area bounded by the traverse in Problem 8-13 by Eq. (8-25).

8-16. The length in feet and the adjusted azimuths of the sides of a closed traverse are as follows:

SIDE	LENGTH	AZIMUTH	SIDE	LENGTH	AZIMUTH
JK	1866.05	130° 15′ 20″	MN	2050.49	264° 49′ 40″
KL	2547.55	65° 35′ 40″	NO	1981.55	162° 55′ 50″
LM	3403.92	335° 23′ 10″	OJ	1221.51	225° 08′ 30″

Compute the latitudes and departures of the sides.

8-17. Balance the traverse in Problem 8-16 by the compass rule. Then compute the coordinates of the stations, taking $X_J = 120,000.00$ ft and $Y_J = 82,000.00$ ft.

8-18. Using the results of Problem 8-17, compute the area bounded by the traverse in Problem 8-16 by Eq. (8-25).

8-19. The lengths in metres and the adjusted azimuths of the sides of a closed traverse are as follows:

SIDE	LENGTH	AZIMUTH	SIDE	LENGTH	AZIMUTH
QR	1248.04	180° 45′ 50″	TU	1371.40	315° 20′ 10″
RS	1703.84	116° 06′ 10″	UV	1325.30	213° 26′ 20″
ST	2276.60	25° 53′ 40″	VQ	816.97	275° 39′ 00″

Compute the latitudes and departures of the sides.

8-20. Balance the traverse in Problem 8-19 by the compass rule. Then compute the coordinates of the stations, taking $X_Q = 24,558.54$ m and $Y_Q = 30,880.64$ m.

8-21. Compute the area, in square metres, of the area bounded by the traverse of Problem 8-19 based on the results of Problem 8-20 using Eq. (8-25).

8-22. Compute the length and bearing of line CF in Fig. 8-13 by using the adjusted coordinates of the traverse given in Section 8-18.

8-23. Compute the length and bearing of the line CE in Fig. 8-13 by using the adjusted coordinates of the traverse given in Section 8-18.

8-24. Compute the length and bearing of the line DF in Fig. 8-13 by using the adjusted coordinates of the traverse given in Section 8-18.

8-25. Compute the length and bearing of the line BF in Fig. 8-13 by using the adjusted coordinates of the traverse given in Section 8-18.

8-26. A line passing through point A in Fig. 8-13 bears S 26° 16′ E. At what distance from point C does this line intersect line CD?

8-27. A line passing through point E of Fig. 8-13 bears S 62° 13′ E. At what distance from point C does this line intersect line CD?

8-28. A line passing through point C of Fig. 8-13 bears N 50° 50′ W. At what distance from point A does this line intersect line AB?

8-29. A line passing through point F of Fig. 8-13 bears N 88° 32′ E. At what distance from point B does this line intesect line BC?

8-30. Compute the coordinates of the intersection of lines AE and CF in Fig. 8-13. How far from A is the point of intersection?

8-31. Compute the coordinates of the intersection of lines DE and FA of Fig. 8-13.

8-32. Compute the coordinates of the intersection of lines AB and EF of Fig. 8-13.

8-33. What is the perpendicular distance between point E and line BC of Fig. 8-13. Compute the distance by putting line BC in the normal form.

8-34. Compute the perpendicular distance from point A to line BC of Fig. 8-13 by putting line BC in the normal form.

8-35. Compute the perpendicular distance between point C and line AB of Fig. 8-13 by putting line AB in the normal form.

8-36. Given the following coordinates of points in two coordinate systems

	SYSTEM 1		SYSTEM 2	
POINT	X	Y	X	Y
A	14,385.26	8,790.20	28,601.84	44,952.62
B	15,588.22	8,305.98	28,836.43	46,228.27
C			27,638.80	45,515.42
D			29,402.27	45,597.40
E			28,166.64	46,207.15

Transform the coordinates of points C, D, and E into system 1.

8-37. In Fig. 8-26, the azimuth of line AP is $176°\ 10'\ 22''$. The following directions and distances were measured at A using a total station instrument.

POINT	DIRECTION	DISTANCE (m)
B	$120°\ 44'\ 16''$	462.88
C	$145°\ 33'\ 08''$	634.10
D	$163°\ 17'\ 20''$	742.25
E	$190°\ 10'\ 04''$	521.72
P	$211°\ 59'\ 51''$	—
F	$222°\ 08'\ 11''$	579.90
G	$232°\ 15'\ 43''$	342.20
H	$277°\ 18'\ 05''$	331.28

The coordinates of A are: $X_A = 25,452.84$; $Y_A = 16,780.92$. Compute the coordinates of the other points in the closed figure.

8-38. Compute the area of the closed figure of Problem 8-37 using Eq. (8-25).

8-39. Compute the area of the closed figure of Problem 8-37 using Eq. (29) in Table C of Appendix C.

8-40. The following perpendicular offsets in feet are measured from a straight line to an irregular boundary at regular intervals of 20 ft.

$$
\begin{array}{lll}
h_1 = 24.8 & h_6 = 40.8 & h_{11} = 60.2 \\
h_2 = 41.4 & h_7 = 45.6 & h_{12} = 47.7 \\
h_3 = 36.6 & h_8 = 50.3 & h_{13} = 36.6 \\
h_4 = 32.6 & h_9 = 44.7 & h_{14} = 36.0 \\
h_5 = 36.7 & h_{10} = 51.9 &
\end{array}
$$

Compute the area lying between the straight line and the irregular boundary by the trapezoid rule.

8-41. Solve Problem 8-40 using Eq. (8-25).

8-42. Compute the area in Problem 8-40 by Simpson's one-third rule: (a) using h_1 as the first offset, and (b) using h_{14} as the first offset.

8-43. The following perpendicular offsets, in metres, are measured from a straight line to an irregular boundary at equal intervals of 10 m:

$$h_1 = 42.23 \qquad h_6 = 75.82 \qquad h_{11} = 92.88$$
$$h_2 = 41.04 \qquad h_7 = 69.07 \qquad h_{12} = 64.04$$
$$h_3 = 33.26 \qquad h_8 = 95.47 \qquad h_{13} = 43.36$$
$$h_4 = 36.83 \qquad h_9 = 93.00 \qquad h_{14} = 28.51$$
$$h_5 = 53.30 \qquad h_{10} = 93.31 \qquad h_{15} = 17.84$$

Compute the area, in hectares (1 hectare = 10,000 m²), lying between the straight line and the irregular boundary by the trapezoid rule.

8-44. Compute the area in Problem 8-43 by Simpson's one-third rule.

8-45. In the traverse of Fig. 8-29(b) the following lengths and bearings were measured:

SIDE	LENGTH (ft)	BEARING
AB	300.11	N 27° 53′ W
BC	412.54	N 56° 13′ E
CD	369.12	S 51° 45′ E
DE	405.27	—
EA	—	S 85° 49′ W

Compute the missing length and bearing.

8-46. In the traverse of Fig. 8-29(b) the following lengths and bearings were measured:

SIDE	LENGTH (ft)	BEARING
AB	424.36	N 32° 58′ W
BC	583.33	N 51° 08′ E
CD	521.93	S 56° 50′ E
DE	573.05	—
EA	—	S 80° 44′ W

Compute the missing length and bearing.

8-47. In the traverse of Fig. 8-29(b) the following lengths and bearings were measured:

SIDE	LENGTH (m)	BEARING
AB	133.25	N 38° 42′ W
BC	183.17	N 45° 24′ E
CD	163.89	S 62° 34′ E
DE	179.94	—
EA	—	S 75° 00′ W

Compute the missing length and bearing.

8-48. In the traverse of Fig. 8-29(c) the following lengths and bearings were measured:

SIDE	LENGTH (ft)	BEARING	SIDE	LENGTH (ft)	BEARING
AB	—	N 23° 06′ E	*DE*	—	S 29° 52′ E
BC	317.95	N 89° 30′ E	*EF*	356.35	S 83° 26′ W
CD	420.19	S 6° 28′ W	*FA*	261.64	N 28° 26′ W

Compute the missing sides.

8-49. In the traverse of Fig. 8-29(c) the following lengths and bearings were measured:

SIDE	LENGTH (ft)	BEARING	SIDE	LENGTH (ft)	BEARING
AB	—	N 28° 18′ E	*DE*	—	S 24° 40′ E
BC	529.70	S 85° 18′ E	*EF*	593.68	S 88° 38′ W
CD	700.04	S 11° 40′ W	*FA*	435.89	N 23° 14′ W

Compute the missing sides.

8-50. In the traverse of Fig. 8-29(c) the following lengths and bearings were measured:

SIDE	LENGTH (m)	BEARING	SIDE	LENGTH (m)	BEARING
AB	—	N 33° 20′ E	*DE*	—	S 19° 38′ E
BC	141.172	S 80° 16′ E	*EF*	158.224	N 86° 20′ W
CD	186.563	S 16° 42′ W	*FA*	116.176	N 18° 12′ W

Compute the missing sides.

8-51. Part off the southwesterly one-quarter portion of the area enclosed by the traverse of Fig. 8-13 by a line passing through point *E*. Give the position of the partition point along line *CD* with respect to point *C*.

8-52. Part off the easterly one-third of the area enclosed by the traverse of Fig. 8-13 by a line parallel to *BC*. Give the positions of the partition points with respect to points *A* and *D*.

8-53. Plot the traverse *ABCD* of Section 8-21 by the method of distance and bearings outlined in Section 8-31. Use a plotting scale 1 in. = 50 ft.

8-54. Plot the boundary *APRS* of Section 8-21 by coordinates using a scale 1 in. = 50 ft with 100-ft grid spacings.

BIBLIOGRAPHY

Graff, D. R. 1973. Survey coordinate systems. *Proc. Fall Conv. Amer. Cong. Surv. & Mapp.* p. 205.

Harris, L. W. 1970. A practical system for adjusting a network of survey traverses. *Proc. 30th Ann. Meet. Amer. Cong. Surv. & Mapp.* p. 635.

Hillman, H. F. 1978. Further uses and refinements of analytic geometry in surveying computations. *Proc. Ann. Meet. Amer. Cong. Surv. & Mapp.* p. 322.

Mikhail, E. M., and Gracie, G. 1981. *Analysis and Adjustment of Survey Measurements.* New York: Van Nostrand Reinhold.

Richardus, P. 1966. *Project Surveying.* New York: Wiley.

Smith, R. J., and Toler, J. P. 1978. Mapping control with the simultaneous double traverse system using radial side shots. *Proc. Fall Conv. Amer. Cong. Surv. & Mapp.* p. 345.

Staughton, H. W. 1975. Computing missing elements in a polygon. *Surveying and Mapping* 35:217.

Wolf, P. R. 1969. Horizontal position adjustment. *Surveying and Mapping* 29:635.

Chapter 9
Traversing with Inertial Surveying Systems

9-1. INTRODUCTION A traverse as discussed in Chapter 8 is surveyed by measuring distances between the traverse stations using a tape or an EDM, and by measuring the angles between the traverse sides using a transit or theodolite. Traverse calculations are performed by: (1) computing adjusted bearings or azimuths of the traverse lines; (2) computing the latitudes and departures of the lines; (3) adjusting the latitudes and departures between fixed or known points; and (4) computing plane rectangular X- and Y-coordinates of the intermediate stations.

In recent years, refinements in inertial navigation guidance systems have permitted traverse surveys to be performed by revolutionary methods based on a computer-controlled gyrostabilized inertial platform and three orthogonally oriented accelerometers attached to the platform. The gyroscopes in effect take the place of angle measuring instruments while the accelerometers perform the work of the distance measuring instruments.

The system is mounted in either a land vehicle or a helicopter as shown in Fig. 9-1. Beginning at a fixed point, the system is aligned so that the platform Y-axis points north, the X-axis points east, and the Z-axis points vertically upward, coinciding with the direction of gravity. After alignment, the X- and Y-coordinates (normally the latitude and longitude) and the

Figure 9-1. Inertial survey system mounted in truck and helicopter. Note the EDM mounted atop the inertial measurement unit. Courtesy of SPAN International.

elevation of the fixed point are entered into an onboard computer, together with other factors needed to control the system. The vehicle proceeds in the direction of another fixed point, stopping periodically to update the system and to mark intermediate traverse stations along the route. The accelerometer outputs are used to compute the latitude and longitude displacements (analogous to latitudes and departures in a plane coordinate system) as well as the vertical displacements from one marked point to the next. These increments are successively added to the initial values in order to compute the latitude, longitude, and elevation of each marked point along the traverse. These values are stored on magnetic tape for processing.

When the second fixed point is reached, the measured values of the latitude, longitude, and elevation are compared with the fixed values, and the discrepancies form the basis for an adjustment back through the line. The system is thus seen to be an interpolative process between fixed points of known coordinates and elevations.

9-2. COMPONENTS OF AN INERTIAL SURVEY SYSTEM The heart of the inertial survey system is the *inertial measurement unit*. This unit consists of an assembly of either two or three precision gyroscopes; a triad of accelerometers mounted so as to measure north-south, east-west, and vertical accelerations; the necessary sensing or pick-off elements needed to sense the orientations of the gyroscopes and displacements of the accelerometers; and torquers which provide rotations of the elements to bring them into the desired orientations. All of the elements are mounted together as a single unit in a case which is perfectly balanced and suspended in a gimbal support. This assembly constitutes the *stable platform*. The gimbal system isolates the inertial platform from the rotational motions of the vehicle.

The platform is highly sensitive to temperature changes. The gyros, accelerometers, sensor-torquers, and other elements continuously input a great amount of heat into the system. This requires that the heat equilibrium of the system be maintained by a very effective heat control (sensor-exchanger) mechanism.

A *data processing unit* is an on-board computer whose function is to execute the initial alignment of the platform, and then to compute the velocity changes of the system over very short time intervals on the order of 15–20 ms, to integrate these velocity changes in order to determine instantaneous velocities and distances traveled in the three coordinate directions, to compute instantaneous coordinates of the current position of the system, to correct for system errors, to control the orientation of the stable platform as a function of time and displacement, and to adjust the positions of the marked traverse points.

A *control and display unit* shown in Fig. 9-2 provides the interaction between the operator and the system. It displays the coordinates of the current position of the system computed by the computer. The operator enters the coordinates of fixed points in the traverse into the control unit in order to perform traverse adjustment. All control commands are entered into the on-board computer through the control unit. Auxiliary data needed to compute the positions of points off the traverse are also entered into the computer through the control unit. The control unit tells the operator the condition of any part of the system, or the contents of any storage location in the computer. It also warns the operator when critical failures have occurred or are about to occur.

Figure 9-2. Control and display unit in SPAN truck. The unit is set to display longitude which is 111°55′ 20.744″ W. Courtesy of SPAN International.

The *data storage unit* contains tape cassettes on which all of the survey and instrumental data are stored for subsequent use by the computer.

A *power supply unit* supplies power to the system at a constant voltage. The power is drawn from the generator-battery system of the land vehicle or helicopter. The system contains a separate standby battery which will operate the system for 5 to 10 min in case the engine or the alternator fails to function, in order to salvage the current data contained in the computer.

Offset arms or lever arms are calibrated distances from the center of the stable element to some point on the vehicle which is plumbed over a fixed station or a station to be marked. The calibrated values for a particular lever arm are entered into the control unit at the beginning of the survey, and any subsequent measurement at a traverse stop is transferred from the coordinate position of the stable element to that of the marked point or vice versa. As shown in Fig. 9-3, the azimuth Az of the vehicle axis is determined by the computer from the orientation of the stable platform. The known angle between the vehicle axis and the lever arm, measured after the system is initially installed in the vehicle, is the angle θ. Therefore, the azimuth of the lever arm from the center of the stable element to the reference point mounted on the outside of the driver's door can be computed in the computer. The length d of the lever arm together with its azimuth is then used to compute the latitude and longitude offsets to the point directly beneath the reference point. The calibrated vertical distance h between the stable element and the ground is the difference in elevation from the center line of the stable element to the ground point.

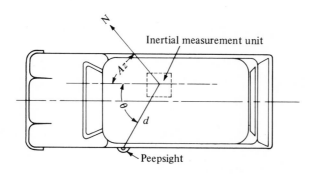

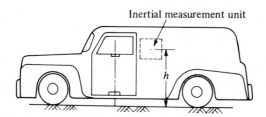

Figure 9-3. Offset or lever arm.

Some inertial survey systems contain an EDM which is used to measure distances to points off the main traverse. This is required when the vehicle cannot occupy a point whose position is to be determined or whose fixed position is needed as input to a traverse adjustment.

9-3. ACCELEROMETERS An accelerometer is an instrument which measures the acceleration of a vehicle along a straight line caused by application of a force applied to the vehicle along this straight line. A person sitting in an automobile at rest senses the vertical acceleration of gravity as an upward reaction force exerted on him by the seat. As the vehicle begins to accelerate forward in a straight line, he then experiences the reaction force of the back of the seat. When the automobile reaches a constant speed or velocity, he no longer feels the reaction force of the back of the seat because the acceleration has been reduced to zero. However, he still feels the reaction from the seat due to the force of gravity. If the automobile decelerates (negative acceleration), particularly if the braking is severe, and if the person has his seat belt fastened, he will feel the restraining or reaction force of the seat belt.

A simple spring-type accelerometer is shown in Fig. 9-4. As shown in view (a), a mass is held in a null position by a pair of restraining springs

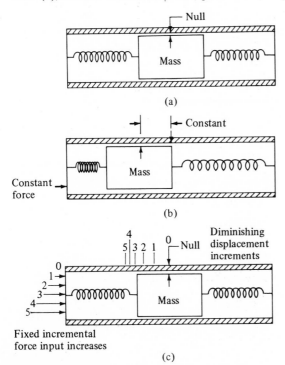

Figure 9-4. Spring-type accelerometer. *Source* Litton Aero Products Publication TT104, October 1975.

when the accelerometer is at rest. An applied force along the *sensitive axis* to the right as shown in view (b) causes the mass to be displaced to the left, compressing one spring and elongating the other. If the external force is constant, the acceleration is constant, and the mass will be displaced a constant amount. The system could be calibrated to measure the acceleration in terms of the displacement of the mass. However, as shown in Fig. 9-4(c), if the external force, and thus the acceleration, is increased linearly, the displacement of the mass is not linear due to the mechanics of the spring in compression. Thus, the spring accelerometer is not suitable for measuring accelerations.

In Fig. 9-5(a) a pendulous mass is suspended at a null position in a case at rest. Application of an external force to the right along the sensitive axis causes the mass to swing to the left through angle γ as shown in view (b). If the external force is constant, the mass will maintain the constant angle γ. However, as shown in view (c), if the force is increased linearly, the

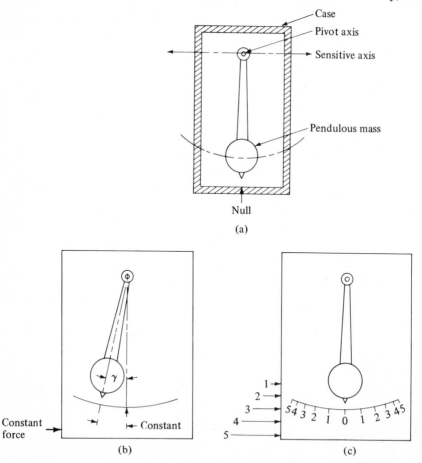

Figure 9-5. Pendulous-type accelerometer. *Source* Litton Aero Products Publication TT104, October 1975.

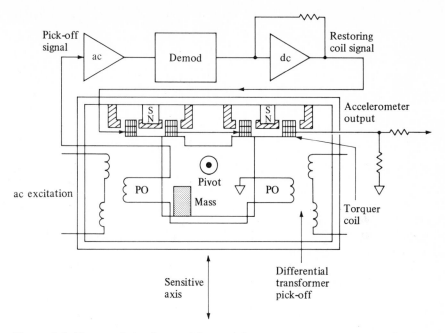

Figure 9-6. Torque-rebalancing pendulous accelerometer. *Source* Litton Aero Products Publication TT104, October 1975.

resulting angular displacement is not linear since the sine of angle γ is not linear. Furthermore the sensitive axis will shift from its original position.

Figure 9-6 is a simplified schematic diagram of an accelerometer in which coils attached to the pendulous mass interact with external coils attached to the case to produce an electrical signal which is amplified and sent to a torquer circuit. As the mass begins to displace, the pick-off coils sense the movement, and the current generated in the torquer circuit acts to return the mass to its null position. Since the mass is constrained by the torquer, no actual rotation takes place about the pivot axis. The amount of restoring current is a linear measure of acceleration. This is called a *torque-rebalanced* pendulous accelerometer and is the type used in inertial survey systems.

A triad of accelerometers is shown in Fig. 9-7(a). Each accelerometer must be carefully aligned so that its sensitive axis is in line with the corresponding

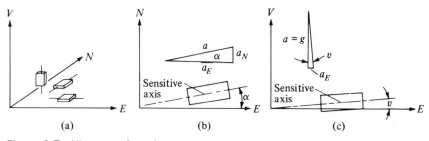

Figure 9-7. Alignment of accelerometers.

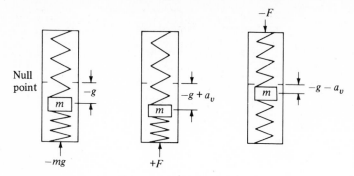

Figure 9-8. Accelerations in vertical direction.

coordinate axis in order to measure the acceleration of the system along that axis. A misalignment of the east accelerometer, for example, by an angle α shown in Fig. 9-7(b) will measure an acceleration $a_E = a \cos \alpha$ in the east-west direction. It will also measure an acceleration $a_N = a \sin \alpha$ in the north-south direction.

In Fig. 9-7(c), the east accelerometer is out of level, by angle v. Even if the system is at rest, the accelerometer would measure a component $a_E = a \sin v$ due to the acceleration caused by the force of gravity.

The vertical accelerometer measures two accelerations, that due to gravity and that due to vertical accelerations of the system. Using the spring-type accelerometer for illustration, Fig. 9-8(a) shows the accelerometer at rest. View (b) shows an upward acceleration, and view (c) shows a downward acceleration. (The principle shown in Fig. 9-8 is experienced while riding in an elevator.) The gravity acceleration must therefore be subtracted from the output of the vertical accelerometer before differences in elevation can be computed. This is performed by the computer.

9-4. DISPLACEMENTS FROM ACCELEROMETER OUTPUTS

Acceleration along a straight line is the second derivative of displacement along the line. Conversely, displacement is obtained by doubly integrating acceleration. This is shown in Fig. 9-9. The upper curve shows the accelerometer output in ft/s^2 over a 60-s period of time. A positive constant acceleration of 10 ft/s^2 during the first 10 s causes a linear increase in velocity from 0 to 100 ft/s. The resulting displacement in this 10-s interval is 550 ft. In the next 10-s interval, there is no acceleration, and the velocity is a constant 100 ft/s. The displacement in this interval is 1000 ft. In the third 10-s interval, a deceleration of 4 ft/s^2 causes a reduction of 40 ft/s in velocity to a value of 60 ft/s. The displacement in this interval is 830 ft. In the fourth 10-s interval, the acceleration is zero, and the velocity remains at 60 ft/s, giving a displacement of 600 ft. The acceleration is increasing uniformly from zero to 10 ft/s^2 in the fifth 10-s interval causing a parabolic increase in velocity from 60

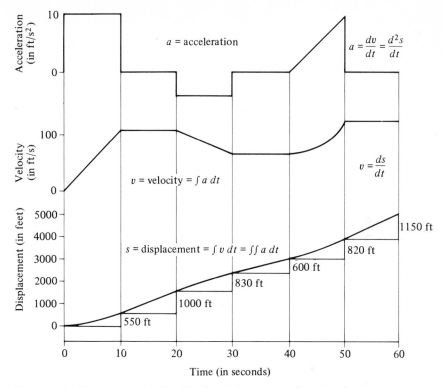

Figure 9-9. Displacement obtained by double integration of acceleration.

to 115 ft/s. The displacement in this interval is 820 ft. In the last interval, the acceleration is zero, and the velocity remains constant at 115 ft/s, resulting in a displacement of 1150 ft. The total displacement summed up in this 60-s interval is 4910 ft.

The double integration of acceleration in the inertial system's computer is performed at short intervals (every 17 ms in one current operating system), and the displacements determined during these short intervals are summed up to give the total displacement along a particular axis from the point of beginning. The horizontal linear displacements are converted to changes in latitude and longitude in the computer, based on the radii of curvature of the earth in the vicinity and on the dimensions of the ellipsoid used to compute the survey (see Section 10-26). Displacement in the vertical direction is the difference in elevation between successive traverse stations along the reference spheroid.

9-5. PLATFORM STABILIZATION In order to keep the accelerometer sensitive axes aligned in the correct direction, the platform on which they are mounted must be stabilized and oriented with respect to the survey north, east, and vertical directions at all times. This is accomplished by means of gyroscopes, gyro pick-offs, gyro torquers, gimbal torquers, and the necessary computations made by the computer. A system so oriented is called a north-oriented, locally leveled system. The gyroscope was briefly described in Section 7-16 as it applies to gyrocompassing. It contains a spin axis about which the wheel spins, an input axis about which disturbing torques are sensed, and a precession or *output* axis about which the gyroscope precesses. Two methods are used to stabilize a platform. In the first method, three single-degree-of-freedom gyroscopes are mounted on the stable platform with their input axes aligned respectively in the directions of the three axes. A single-degree-of-freedom axis gyroscope is shown in Fig. 9-10(a). The spin axis is free to rotate in a gimbal which in turn is free to rotate about an axis the bearings of which are attached to the platform. A disturbing torque about the input axis would cause the gyro to precess, thus rotating the gimbal about its axis. This disturbing torque about this output axis is sensed by pick-off coils and is used to control the orientation of the platform.

The second method, which is used in currently operating inertial survey systems, employs two two-degrees-of-freedom gyroscopes. This type is shown in Fig. 9-10(b). The gyroscope is free to rotate about two mutually perpendicular axes. Consequently, there are two input axes IA_1 and IA_2 and two output axes OA_1 and OA_2 with this arrangement. One of the input axes of one gyroscope is aligned in the north-south direction, and that of the second is aligned in the east-west direction. There are thus two input axes available in the vertical direction only one of which is used for orientation.

Although gyroscopes represent the highest form of mechanical perfection, there are numerous disturbing torques which causes the gyroscope to drift

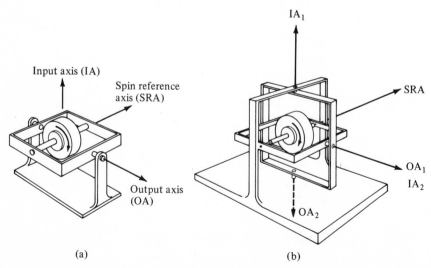

(a)   (b)

Figure 9-10. Gyroscopes. (a) Single-degree-of-freedom. (b) Two-degrees-of-freedom. *Source* Litton Aero Products Publication TT104, October 1975.

slowly over a long period of time. These drift rates can be determined quite accurately, and can then be compensated for by applying small torques to one or the other input axis causing the gyroscope to precess in a direction opposite to the drift. This stabilizes the gyroscope in inertial space. The applied countertorques are referred to as *gyro bias*.

Figure 9-11 is a schematic diagram of a stable platform which contains two two-degrees-of-freedom gyroscopes together with the three accelerometers. The torquers and pick-off coils are not shown in order to keep the diagram simple. The actual arrangement of the gyros and accelerometers are quite different than that shown. All of the elements are actually housed in a single case which is perfectly balanced about its geometric center, otherwise the platform would act like a pendulum.

Let it be imagined that the platform is perfectly leveled in both the X- and Y-directions with the sensitive axes of both the X- and Y-accelerometers also level. If the platform is stationary, the X- and Y-accelerometers would have zero output and the Z-accelerometer output would detect the acceleration due to gravity. Now if the platform is tilted in the Y-direction causing a rotation about the X-axis (and consequently about the X-gyro input axis), the Y-accelerometer would output a slight acceleration due to gravity, and the X-gyro would precess about its output axis. The amount of precession is measured by pick-off coils on the X-gyro output axis and generates a signal which is sent to a servomotor. The motor then reorients the platform

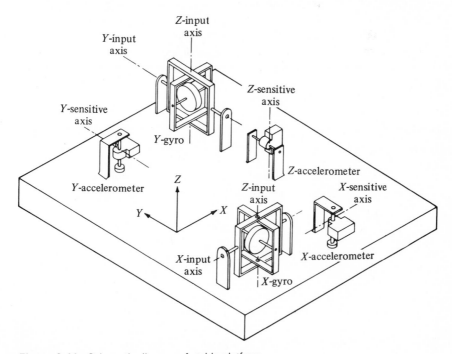

Figure 9-11. Schematic diagram of stable platform.

until the Y-accelerator output decreases to zero. This relevels the platform about the X-axis. Similarly, the platform is leveled about the Y-axis.

In order to isolate the stable platform from the rotational motions of the vehicle, the platform is suspended in a three- or four-gimbal suspension system as shown in Fig. 9-12. Any torques set up due to friction in the gimbal bearings are sensed by the gyroscopes and are counteracted by the gyro torquers, thus keeping the platform stable within the gimbal mount.

Three major rotations of the stable platform must be made while the inertial survey system is in operation. These rotations are controlled by the computer, and are made in order to keep the platform locally leveled and the Y-axis pointing north at all times throughout the survey.

In Fig. 9-13, imagine looking down on the equator from a point in space directly above the North Pole. The stable platform is initially leveled on a point at A on the equator as shown in view (a). If the system remains on the point, the earth rotation carries it from west to east to position B. However, since the platform is inertially stable, it will be out of level at position B. By the time position C is reached, the platform would be vertical instead of level. In order to compensate for the earth rotation rate at the equator, the system must be torqued or rotated about the north-south (Y) axis at the rate of earth rotation which is approximately 15.04° per hour. This compensation is shown in Fig. 9-13(b).

Figure 9-14 shows that as the system is positioned north of the equator, the necessary rotation about the platform Y-axis to compensate for earth rate rotation decreases with an increase in latitude. However, a counter-rotation about the platform X-axis is also necessary in order to keep the platform locally leveled. Furthermore, a rotation about the Z- or vertical axis is also required in order to keep the Y-axis pointing north. Corrections

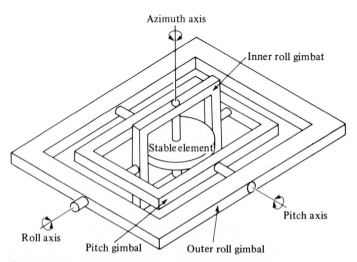

Figure 9-12. Schematic diagram of four-gimbal suspension mount. *Source* Litton Aero Products Publication TT104, October 1975.

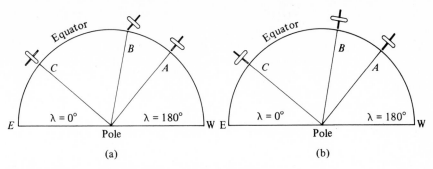

Figure 9-13. Correction for earth rate at the equator.

for earth rate rotation are functions of the rotational rate and the elevation and latitude of the point, and are accomplished by torquing the appropriate gyro input axes to cause the gyroscopes to precess at the necessary rates to overcome the undesired local orientation of the stable platform.

Figures 9-13 and 9-14 can also be used to illustrate another required correction which allow for a change in the position of the inertial system as the traverse survey proceeds. Referring again to Fig. 9-13, imagine the system to be transported along the equator instantaneously (to eliminate consideration of earth rotation) from position A to position C. If no compensation is made, the platform will be out of level at position C as shown in view (a). Compensation by a rotation about the Y-axis is shown in view (b). Figure 9-14 illustrates the system transport situation at different latitudes and longitudes. The computer determines the necessary rotations about the three axes as functions of the earth's local radius and of the change in both

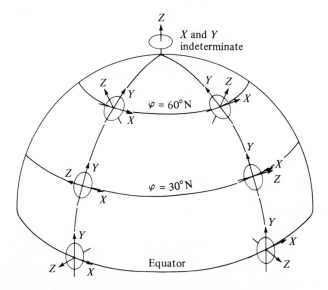

Figure 9-14. Earth rate compensation at different latitudes.

latitude and longitude as the survey progresses. The counter-rotations are made by torquing the appropriate gyro input axes.

Although earth rate and transport rate rotations are the major corrections to be made to the inertial survey system to keep it north-oriented and locally level, other minor corrections must also be made. These are discussed in some of the references given at the end of this chapter.

9-6. INITIAL ALIGNMENT AND SELF-CALIBRATION Before a traverse survey can be run between fixed points, the inertial survey system must be aligned to north and locally leveled, and it must be calibrated in order to determine any residual gyro and accelerometer biases existing in the system. A prior factory calibration determines the gyro drift rates, accelerometer biases, accelerometer scale factors, and heading corrections. Misalignment angles of the accelerometer and gyro-sensitive axes are also determined. These values are stored in the on-board computer for real-time compensations.

With the vehicle occupying a point of known or approximately known coordinates and elevation, the system is turned on, and the latitude and elevation are entered into the computer via the control unit. These coordinates must be known in order to determine earth and transport rate rotations and to compensate for the acceleration due to gravity in the vertical accelerometer at the starting point. The inertial platform is oriented to true north by gyrocompassing. The X- and Y-accelerometer outputs are examined by the system. If the outputs are not zero after earth rate torquing, this means that the platform is not level or that the wrong coordinates have been entered. Signals are sent to the gyro torquers which causes the gyros to precess the correct amount necessary to level the platform and to drive both the X- and Y-accelerometer outputs to zero.

The system must be completely at rest during self-calibration and alignment. This requires that the vehicle be shielded from external vibrations or wind effects, and that the vehicle wheels rest on a nonyielding surface. The alignment process requires 30 minutes to an hour, but it is totally under computer control. Thus the system can be left unattended during this process. If for any reason the power to the system is interrupted during a survey, the alignment process must be repeated before the survey can continue. In helicopter operation, a fuel stop for example may require that the engine be turned off, interrupting the power supply. Thus, a complete realignment would be necessary. The system should not be run for more than 4 hr before a new calibration and alignment is performed.

9-7. RUNNING THE TRAVERSE SURVEY As soon as alignment is completed, the earth rate compensation keeps the system in proper alignment with the survey coordinate axes. If the point of alignment is a fixed control point, the latitude, longitude, and elevation are again entered into the computer. If the initial point is not a fixed control point, the vehicle is then driven or flown to the control point and centered over the point using the peep sight. The known coordinates and elevations are then entered into the system. This process is called a position and elevation update.

As the vehicle is driven or flown to subsequent points in the traverse, errors in the system will accumulate, causing errors in the positions of the traverse stations along the way. Some of the errors accumulate with time while others accumulate with distance. Time-related errors are system noise effects, and changes in azimuth misalignment, accelerometer scale factor errors, and accelerometer misalignment. Distance-related errors are initial azimuth misalignment, accelerometer scale factor errors, and accelerometer misalignment errors. These accumulated errors are reduced in the computer by an error controller called a *Kalman filter.*

The Kalman filter is a series of mathematical models which statistically relate the different sources of error to one another. If the vehicle is stopped, and the accelerometer outputs are examined, the accelerations should all read zero, and the first integration of these accelerations, which are velocities, should also be zero since the system is not moving. However, because of system errors mentioned above, and because of change in the direction of the local vertical, the accelerometer outputs will have generated false velocities. These measured velocity errors are input into the Kalman filter which relates the velocity errors to the system errors, and makes a correction to the measured latitude, longitude, and elevation of the point. Stopping to measure the false velocities is performed every 3 to 4 min. The operation is called a *zero velocity update* (ZUPT). Each time a ZUPT is performed, the Kalman filter calculates a new set of relations between the system and velocity errors and recorrects all previous positions. This recursive updating of all of the positions back to the starting point based on measured velocity errors during each ZUPT takes place throughout the traverse until the next fixed control point is reached. During each ZUPT, the platform is leveled by zeroing out the X- and Y-accelerometers, and the effects of gravity changes are computed. A ZUPT must be performed following an EDM measurement to a point off the main traverse because of the time delay in making the EDM measurement.

If a helicopter is used, it should be landed to perform ZUPT's. However, on some survey projects, particularly when elevations are of secondary importance, the update is performed while the helicopter is hovering over a point on the ground. Several passes are made over the point from which computed velocities are determined. These are compared with the velocities measured by the system, and the differences are used in the Kalman filter to correct the positions of the points back through the system.

The Kalman filter computes the most likely estimates of the corrections in real time. It is a recursive process, that is, in every new estimate, the characteristics of all previous measurements are also included. The filter weighs every new observed data according to some *a priori* knowledge about the nature and usual magnitude of the system errors. However, if some evidence shows that some systematic changes have occurred, the *a priori* knowledge and the estimates based upon it are recursively updated. When many error sources are involved, then the filter "budgets" its activity, that is, it distributes the error according to its updated knowledge about the nature and likely magnitude of the system errors.

When the system stops at a point whose position and elevation are to be determined, a ZUPT is performed, after which the point identification number is entered into the control unit. This number, together with the coordinates and elevation of the point, after correction in the Kalman filter, are stored on magnetic tape in the data storage unit for further processing. This is called *marking* a position.

After all of the desired traverse stations are marked, the vehicle then proceeds to a fixed control station where a position and/or elevation update are performed. After correction in the Kalman filter, the measured coordinates and elevations are compared with their fixed value to give the error of closure in latitude, longitude, and/or elevation. The Kalman error controller then uses these error closures, the data from the previous ZUPT's, and known system errors to perform an adjustment back through the traverse to the point of beginning. This process is called *smoothing*. If the traverse is run on in fairly uniform manner between the fixed end points, the results of smoothing will give positions and elevations with accuracies to submetre values. Under ideal conditions, decimetre accuracies are obtained on linear traverses.

Improvement in the accuracy of smoothed coordinate and elevation values can be obtained by running the survey in the opposite direction and adopting a mean of both sets of smoothed values.

9-8. FINAL COORDINATES AND ELEVATIONS After a traverse has been completed, the smoothed latitudes, longitudes, and elevations, together with their identification numbers are stored on cassettes in the data storage unit. In inertial survey traversing, a network of interconnecting traverse lines is usually surveyed if maximum accuracy is desired. This network is then adjusted in a separate computer by reading the data tapes into the computer and entering appropriate weights of the various traverse lines. The adjustment is performed by the method of least squares (see Appendix A). The adjusted latitudes and longitudes are then converted to plane coordinates, if desired (see Chapter 11), and a final listing of coordinates and elevations is printed out for use by the engineer or surveyor.

9-9. DESIGN OF AN INERTIAL SURVEY A well-designed inertial control survey should conform to the following design criteria.

1. All fixed control points should be well identified and access provided prior to the survey in order to avoid time delays.

2. The positions of all intermediate traverse stations to be marked should be monumented and located for convenient access by the land vehicle or the helicopter. In the latter case, landing sites should be prepared at the stations.

3. The positions of the intermediate traverse stations should be determined as closely as possible with reference to existing maps. The inertial survey system has the ability to navigate to a point by distance and direction if the coordinates of the point have been entered. The display unit displays a continuous value of distance and angle that the vehicle must travel in order to reach the point. When it is within 100 m of the point, the display unit then displays the distance forward and the distance right or left that the system must travel in order to reach the point. This "steering" ability reduces search time, speeds up the survey operation, and permits setting of stations at desired coordinates.

4. Traversing should proceed at a uniform time rate in order to optimize the Kalman filtering.

5. Zero velocity updates should be performed at uniform intervals, no more than 4 min. The system's clock can be set to give an audible signal at any given interval of time telling the operator when a ZUPT is to be performed. The next time interval begins at the end of the ZUPT which takes about 40 s to perform.

6. Rapid accelerations, abrupt changes in direction, and bumpy roads are to be avoided.

Departure from any or all of the above criteria will degrade the accuracy of the survey. Present inertial survey systems are costly to operate. Advance planning is of the utmost importance, and improvisations which have to be made during the course of the survey are bound to result in higher costs and lower accuracy.

9-10. FLEXIBILITY OF INERTIAL SURVEYS The greatest advantage in traversing by inertial survey systems is the speed with which surveys can be performed. The system does not depend on any external source for alignment or positioning (except at the initial point), and is virtually independent of weather conditions. Surveys can be performed in remote areas without having to clear brush or construct towers. Human mistakes are greatly reduced both in the field operation and in the data reduction because the system is fully computerized.

Rapid profiling is performed by commanding the system to record data points either at regular time intervals or at regular distance intervals. There is no need to stop and record each data point as the profiling advances, except to stop periodically for ZUPT's. With the proper vehicle, the system can be run on railroad rails in order to determine the rail profiles.

The steering capability of the inertial system is used to recover monuments whose positions are known or to establish points at desired coordinates. This is advantageous in searching for "lost" corners (see Chapter 18).

If the system is mounted in a helicopter, an irregular line such as a meandering stream can be followed. The positions of the meander line can be stored on the magnetic tape at closely spaced time intervals, resulting in a faithful reproduction of the meander line in terms of a string of data points.

Using the EDM mounted on the inertial measuring unit, inaccessible points off to either side of a traverse can be surveyed quite rapidly and automatically tied to the traverse.

Ground elevations and profiles can be measured from a helicopter. As the terrain or the line is traversed, the helicopter hovers over the ground at regular intervals. A weighted wire is dropped to the ground, and a reel-type mechanism is used to measure the vertical distance from the system to the ground. This value is subtracted from the current elevation of the stable element and is stored on tape along with the latitude and longitude of the point. This technique is of particular advantage where terrain and ground cover conditions prevents the helicopter from landing or a land vehicle from navigating over the area.

Very few inertial survey systems are in operation at the time of this writing. Those which are in use are intensively employed in a variety of applications. It is not practical to attempt to cover all of the aspects of this revolutionary survey system in a single chapter. The student who is interested in pursuing the subject of inertial surveying in greater depth is urged to study the references listed in the Bibliography.

PROBLEMS

9-1. In Fig. 9-3, the azimuth of the axis of the vehicle is $(360° - Az) = 317° 22' 35''$. The angle θ is $65° 14' 20''$ and the length d of the lever arm is 1.37 m. The plane rectangular coordinates of the point over which the peepsight is located are: $X = 124,622.64$ m, and $Y = 78,990.16$ m. What are the coordinates of the inertial measurement unit?

9-2. If the east accelerometer sensitive axis is out of alignment by $0° 00' 22''$ counter clockwise from the X-coordinate axis, what is the effect on the X- and Y-coordinates of a point which lies 10 km from the initial point on an azimuth of $72° 25' 50''$ if the final point was reached by a straight line from the initial point?

9-3. Do a numerical double integration of the following accelerations: 0–5 s, $+4 \text{ ft/s}^2$; 5–10 s, zero acceleration; 10–15 s, -2 ft/s^2; 15–20 s, zero acceleration; 20–25 s, $+6 \text{ ft/s}.^2$ Plot an abscissa scale of 1 in. $= 5$ s, ordinate scales of 1 in. $= 10 \text{ ft/s}^2$, 1 in. $= 50 \text{ ft/s}$, and 1 in. $= 200 \text{ ft}$. Plot the acceleration, velocity, and displacement curves in this 25-s time interval.

BIBLIOGRAPHY

Ball, W. E., Jr., and Ives, J. C. 1975. Testing an airborne inertial survey system for BLM cadastral survey applications in Alaska. *Proc. Ann. Meet. Amer. Cong. Surv. & Mapp.* p. 107.

Ball, W. E., Jr. 1976. Performance characteristics of an airborne inertial survey system. *Proc. Fall Conv. Amer. Cong. Surv. & Mapp.* p. 133.

Ball, W. E., Jr. 1978. Adjustment of inertial survey system errors. *Proc. Ann. Meet. Amer. Cong. Surv. & Mapp.* p. 198.

Chapman, W. H., and Starr, L. E. 1979. Surveying from the air using inertial technology. *Proc. Ann. Meet. Amer. Cong. Surv. & Mapp.* p. 352.

Fishel, N., and Roof, E. 1977. Results of tests using an inertial rapid geodetic survey system. *Proc. Ann. Meet. Amer. Cong. Surv. & Mapp.* p. 100.

——. 1975. *Fundamentals of Inertial Navigation.* TT104 Litton Aero Products, Canoga Park, CA.

Gregerson, L. F. 1975. Inertial geodesy in Canada. *Proc. Amer. Geophysical Union.*

Gregerson, L. F., and Carrier, R. C. 1975. Inertial surveying system experiments in Canada. *Proc. IAG Congress.* Grenoble, France.

Griffin, A. 1977. Auto-surveyor cadastral resurveys in the California desert. *Proc. Fall Conv. Amer. Cong. Surv. & Mapp.* p. 71.

Huddle, J., and Mancini, A. 1975. Gravimetric and position determination using land-based inertial systems. *Proc. Ann. Meet. Amer. Cong. Surv. & Mapp.* p. 93.

Proceedings First Int. Symp. on Inertial Technology for Surveying and Geodesy. 1977. Canadian Institute of Surveying, Ottawa, Canada.

Chapter 10
Horizontal Control
Networks

10-1. INTRODUCTION The methods of measuring distances presented in Chapter 2 and those of measuring angles presented in Chapter 4 were combined in the form of traverse discussed in Chapter 8. The use of inertial surveying systems was described in Chapter 9. The objective of traverse is to establish the horizontal positions of points in a survey, these positions being expressed as X and Y coordinates on a plane coordinate system. The positions are determined by measurement, computations, and adjustments.

The horizontal positions of points in a network developed to provide accurate control for subsidiary surveys can be obtained in a number of different ways in addition to traversing. These methods, called *triangulation, trilateration, intersection, resection,* and *Doppler satellite positioning,* are the subject of the present chapter.

The method of surveying called *triangulation* is based on the trigonometric proposition that if one side and the three angles of a triangle are known, the remaining sides can be computed. Furthermore, if the direction of one side is known, the directions of the remaining sides can be determined. In Fig. 10-1(a) the side BA of triangle CBA is known, having been measured directly with a tape or EDM, or else computed from a previous triangle side. The three angles have been measured in the field. The triangle can be solved by the law of sines to give the lengths of AC and CB.

If the azimuth of AB, and thus of BA, is known, then the azimuths of AC and BC can be determined from the measured angles. Also, since this is a closed figure and since the three measured angles should add up to $180°$ plus any spherical excess (see Section 10-21), the internal consistency of the angle measurements can be determined. The triangle is now like a closed traverse because the directions and lengths of all the lines are known. If the coordinates of either A or B are known, the coordinates of the other two points can be computed by application of principles developed in Chapter 8.

The methods of triangulation are quite demanding. They require a considerable amount of precise angle measurement with a minimum amount of distance measurement. The triangles are developed into a net of interconnected figures, and certain lines, called *baselines*, must be measured in order to compute the lengths of the other lines in the net. The baselines must be measured with extreme precision, since errors propagated through the system originate with these measured lines.

In order to eliminate the effects of random errors as much as possible, triangulation systems always include more than the minimum number of measurements necessary to fix the positions of the points in the triangulation net. These extra, or redundant, measurements provide the data necessary for the adjustment of the net, usually by the method of least squares.

Since the advent of the long-range EDM's, a method of surveying, called *trilateration*, has been adopted to combine with triangulation in order to extend horizontal control. Trilateration is based on the trigonometric proposition that if the three sides (laterals) of a triangle are known, the three angles can be computed. In Fig. 10-1(b) all three sides of the triangle FED have been measured with EDM. Thus the three angles can be computed by formulas given in Table C-3 of Appendix C. This accomplishes the same thing as triangulation, namely, the determination of all the sides and all the angles in the triangle. Directions of the lines and positions of the points in the triangle can then be computed as in traverse and triangulation.

Frequently in a triangulation system an occasion arises in which a point, whose position is to be determined, is not occupied—for example, a prominent church spire. Point J in Fig. 10-1(c) is such a point. This point can be determined by measuring the two angles G and H at the two ends of a known

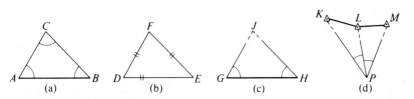

Figure 10-1. Basic horizontal-control configurations. (a) Triangulation; (b) trilateration; (c) intersection; (d) resection.

line and determining the third angle at J by computation since $G + H + J = 180°$ plus spherical excess. The angle J is called a *concluded angle*. The method is referred to as *intersection* because J lies at the intersection of the two lines from G and H.

The intersection shown in Fig. 10-1(c) has no geometric check because angle J is concluded. This is the major difference between the triangulation solution shown in view (a) and the intersection solution shown in view (c). As a practical matter, an intersection station such as J is observed from three or more regular triangulation stations in the net. This provides the necessary checks on the accuracy and reliability of the intersection station.

The method of locating a point P in Fig. 10-1(d) from at least three known points K, L, and M without having occupied the known points is called *resection*. In this system only the point to be determined is occupied, and the angles between the known stations are measured. A minimum of three known points is required to determine the position of an unknown point by the method of resection. The situation shown in Fig. 10-1(d) thus has no check because it just meets the minimum conditions. In order to provide the check necessary to test the reliability of the computed position of the unknown point, four or more known stations in the net are observed.

In satellite Doppler positioning, a ground receiver at the point to be located receives signals which are continuously transmitted from a passing satellite. These signals are obtained by modulating the phase of electromagnetic carrier waves generated at frequencies of approximately 150 MHz and 400 MHz. Included in these signals are the precise time of transmission and the orbital parameters of the satellite reference coordinate system. The ground receiver compares the incoming carrier frequencies transmitted by the satellite with reference frequencies generated by the ground receiver itself. A change in the incoming frequencies at regular time intervals is used to determine the position of the point occupied by the receiver antenna relative to the satellite position. The satellite referenced position of the ground point is then transformed into an earth-centered coordinate system from which latitude, longitude, and elevations of the ground point are computed.

10-2. CLASSES OF HORIZONTAL CONTROL Control surveys are executed under varying field standards of accuracy. The standards for a particular survey depend on the purpose of the survey. In the United States these standards have been prepared by the Federal Geodetic Control Committee and have been reviewed by the American Society of Civil Engineers, the American Congress on Surveying and Mapping, and the American Geophysical Union, as well as by other public and private organizations and individuals engaged in control surveys throughout the country.

There are three orders of accuracies, namely, first, second, and third. First order is the highest accuracy and is required for developing the national network of horizontal control, for the study of small crustal movements in

TABLE 10-1. Classification, Standards of Accuracy, and General Specifications for Horizontal Control

TRIANGULATION

CLASSIFICATION	FIRST ORDER	SECOND ORDER CLASS I	SECOND ORDER CLASS II	THIRD ORDER CLASS I	THIRD ORDER CLASS II
RECOMMENDED SPACING OF PRINCIPAL STATIONS	Network stations seldom less than 15km. Metropolitan surveys 3 to 8 km and others as required.	Principal stations seldom less than 10km. Other surveys 1 to 3 km or as required.	Principal stations seldom less than 5km or as required	As required	As required
STRENGTH OF FIGURE					
R_1 between bases		1			
Desirable limit	20	60	80	100	125
Maximum limit	25	80	120	130	175
Single figure Desirable limit					
R_1	5	10	15	25	25
R_2	10	30	70	80	120
Maximum limit					
R_1	10	25	25	40	50
R_2	15	60	100	120	170
BASE MEASUREMENT					
Standard error	1 part in 1,000,000	1 part in 900,000	1 part in 800,000	1 part in 500,000	1 part in 250,000
HORIZONTAL DIRECTIONS					
Instrument	0".2	0".2	0".2 or {1".0	1".0	1".0
Number of positions	16	16	8 or {12	4	2
Rejection limit from mean	4"	4"	5" or {5"	5"	5"

[*continues*]

363

TABLE 10-1. *(continued)*

CLASSIFICATION	TRIANGULATION				
			SECOND ORDER		THIRD ORDER
	FIRST ORDER	CLASS I	CLASS II	CLASS I	CLASS II
TRIANGLE CLOSURE					
Average not to exceed	1".0	1".2	2".0	3".0	5".0
Maximum seldom to exceed	3".0	3".0	5".0	5".0	10".0
SIDE CHECKS					
In side equation test, average correction to direction not to exceed	0".3	0".4	0".6	0".8	2"
ASTRO AZIMUTHS					
Spacing-figures	6–8	6–10	8–10	10–12	12–15
Number of observations/night	16	16	16	8	4
Number of nights	2	2	1	1	1
Standard error	0".45	0".45	0".6	0".8	3".0
VERTICAL ANGLE OBSERVATIONS					
Number of and spread between observations	3D/R–10"	3D/R–10"	2D/R–10"	2D/R–10"	2D/R–20"
Number of figures between known elevations	4–6	6–8	8–10	10–15	15–20
CLOSURE IN LENGTH					
(Also position when applicable) after angle and side conditions have been satisfied, should not exceed	1 part in 100,000	1 part in 50,000	1 part in 20,000	1 part in 10,000	1 part in 5,000

	Network stations seldom less than 10 km. Other surveys seldom less than 3 km.	Principal stations seldom less than 10 km. Other surveys seldom less than 1 km.	Principal stations seldom less than 5 km. For some surveys a spacing of 0.5 km between stations may be satisfactory.	Principal stations seldom less than 0.5 km.	Principal stations seldom less than 0.25 km.
RECOMMENDED SPACING OF PRINCIPAL STATIONS					
GEOMETRIC CONFIGURATION					
Minimum angle contained within, not less than	25°	25°	20°	20°	15°
LENGTH MEASUREMENT					
Standard error	1 part in 1,000,000	1 part in 750,000	1 part in 450,000	1 part in 250,000	1 part in 150,000
VERTICAL ANGLE OBSERVATIONS					
Number of and spread between observations	3D/R–10″	3D/R–10″	2D/R–10″	2D/R–10″	2D/R–20″
Number of figures between known elevations	4–6	6–8	8–10	10–15	15–20
ASTRO AZIMUTHS					
Spacing-figures	6–8	6–10	8–10	10–12	12–15
Number of observations/night	16	16	16	8	4
Number of nights	2	2	1	1	1
Standard error	0″.45	0″.45	0″.6	0″.8	3″.0
CLOSURE IN POSITION					
After geometric conditions have been satisfied should not exceed	1 part in 100,000	1 part in 50,000	1 part in 20,000	1 part in 10,000	1 part in 5,000

areas of seismic activity, and for large metropolitan control expansion. First-order accuracy is also used for measuring the performance of space vehicles, ballistic systems, electronic-measuring systems, precision cameras, and other engineering and scientific developments requiring high precision over moderately large distances.

Second-order control is subdivided into Class I and Class II. Class I applies to area networks between the first-order arcs and detailed surveys in very high-value land areas. After adjustment to the national network, it forms a part of it.

Second-order, Class II control is used to establish control along the coastline, inland waterways, and interstate highways. It is recommended for controlling extensive land subdivision and construction.

Third-order control is subdivided into Class I and Class II and is used for local horizontal control. Surveys of this order are used to establish control for local improvements and developments, topographic and hydrographic surveys, and for other such projects for which they provide sufficient accuracy. Class I is very important to the engineer, as it is the lowest classification permissible for designating points on the state plane coordinate systems (see Chapter 11).

The requirements for the different orders of accuracy for triangulation and trilateration are given in Table 10-1. The significance of strength of figures indicated in the table is presented in Section 10-6. The significance of the side check and that of the closure in length is given in Sections 10-22 and 10-23. The requirements for the different orders of accuracy for traverse are given in Table 10-10.

10-3. TRIANGULATION Triangulation nets are of two types. One is the arc or chain type, which is shown in Fig. 10-2. This is the type used for the main horizontal control net of the nation. The other is the area triangulation net, which is shown in Fig. 19-1, especially in the southeasterly portion. An area net is usually developed for county and municipal surveys and is invariably tied to arc triangulation to establish the overall position and orientation of the net.

Triangulation as a form of horizontal control is applied when a large area is to be surveyed and where the methods of traversing would not be expected to maintain a uniformly high accuracy over the entire area. Thus the National Geodetic Survey employs triangulation to establish a basic network of high-order control throughout the country. Also, the United States Geological Survey uses triangulation to form the control necessary for its national mapping activities.

Triangulation is employed in every sizable city to form a network of consistently accurate control monuments from which the city and private engineers can work in locating streets and utilities. Triangulation is necessary to control the locations of large bridge structures, state and federal highways,

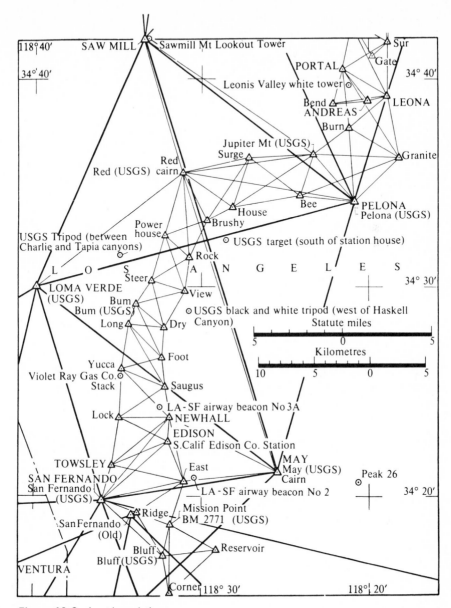

Figure 10-2. Arc triangulation.

dams, canals, and other engineering works of a massive nature. A large project, such as boundary location, power development, water-resources development, flood control, irrigation, or reclamation, requires triangulation in order to maintain the necessary accuracy throughout the system.

Once a triangulation system, whether large or small, has been developed, measured, and adjusted, the points in the system then furnish control for subsequent traversing, minor triangulation, trilateration, intersection, and resection needed for day-to-day engineering operations.

10-4. RECONNAISSANCE The success of any triangulation depends to a great extent on the reconnaissance, which is the most difficult and exacting task of an extensive survey. Although no two regions call for exactly the same treatment, certain general principles apply to all such work. The reconnaissance preliminary to minor triangulation for some isolated project may be very simple. Reconnaissance for triangulation of the largest size is a matter of much complexity and demands skill, experience, and judgment.

During the reconnaissance the sites for the future stations are selected and all information that will be valuable in the operations of the building and observing parties is collected. Heavily wooded country is the most difficult in which to carry on triangulation. A valley of proper width, with peaks on either side, is the most favorable. The stations are located on the higher points, provided their locations will give the best-shaped triangles. The ideal location is one in which low towers can be used and where there is little or no clearing to be done. A most important and difficult part of the reconnaissance is the determination of the heights of the towers necessary to make the line of sight between any two of them clear of all obstructions. A mistake on the part of the person making the reconnaissance may delay the large observing party an entire day while the line is being cleared or higher towers are being erected. This will be particularly true on first-order and second-order work, as practically all the observing is done after dark.

10-5. TYPES OF FIGURES IN TRIANGULATION Although triangles are the basic figures in a triangulation net, they are not allowed to exist alone because they do not provide sufficient checks on the measurements and subsequent computations. Triangles are combined to form other figures which are considered as geometric entities. The most common figure used is the braced quadrilateral. The quadrilateral, shown in Fig. 10-3(a), is comprised of four triangulation stations in which there are eight measured angles. The term *braced* comes from the fact that the two diagonal lines are observed, thus bracing the four-sided outer figure. A discussion of the

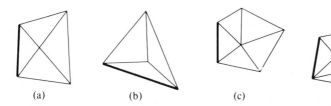

(a)   (b)   (c)   (d)

Figure 10-3. Types of triangulation figures.

geometric conditions existing in a fully observed or braced quadrilateral is given in Appendix A.

The figure shown in view (b) is a center-point triangle, which is actually a quadrilateral with one of its corners lying inside the other three. If the center point is moved downward outside the triangle, for example, a braced quadrilateral results.

A center-point pentagon is shown in view (c) and a quadrilateral containing a center point is shown in view (d). Other types of figures can be identified in Figs. 10-2 and 19-1. These figures are combined either into chains or into area-wide nets as stated in Section 10-2.

10-6. STRENGTH OF FIGURE

Since computed lengths are likely to be uncertain when the sines of small angles are involved, no very small angles should be included in a triangulation scheme if those angles are to be used in the computations. The National Geodetic Survey employs a method of testing the precision that may be expected in any given case. This method is based on the standard error of a computed length. The square of the standard error E of the logarithm of a side of a figure is

$$\sigma^2 \propto \tfrac{4}{3}(d^2)\frac{D-C}{D}\sum(\delta_A^2 + \delta_A\delta_B + \delta_B^2) \qquad (10\text{-}1)$$

in which d is the standard error of an observed direction; D is the number of directions observed in the figure; C is the number of conditions to be satisfied in the figure; and δ_A, δ_B are the respective logarithmic differences of the sines.

The subscripts A and B in Eq. (10-1) refer to the two *distance angles* in each of the triangles under consideration. In Fig. 10-4 the side QR is assumed to be known, and the side TU is to be determined. In triangle SQR, the side

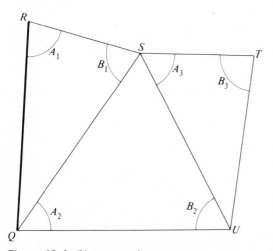

Figure 10-4. Distance angles.

SQ is computed from QR by the law of sines using the two angles opposite the known side and the side to be computed. These two angles are labeled A_1 and B_1. Thus

$$\frac{SQ}{\sin A_1} = \frac{QR}{\sin B_1}$$

The side QS is now used in the second triangle UQS to determine the forward side SU, using the two angles labeled A_2 and B_2. Thus

$$\frac{SU}{\sin A_2} = \frac{QS}{\sin B_2}$$

Finally, side US is used in the third triangle TUS to determine the side TU, using the two angles labeled A_3 and B_3. Thus

$$\frac{TU}{\sin A_3} = \frac{US}{\sin B_3}$$

Note that the A angle is opposite the side to be computed, whereas the B angle is opposite the known side. The third angle in each of the triangles is referred to as the *azimuth angle*. The strength of the solution of each triangle does not depend on the size of the azimuth angle, only on the A and B angles.

The respective logarithmic differences of the sines corresponding to a change of $1''$ in the angles A and B used in the computation of the triangle are expressed in units of the sixth decimal place. The summation $\sum$ is to be taken for the triangles used in computing the value of the side in question from the side supposed to be absolutely known.

In determining the value of D for any figure, the starting line is supposed to be completely fixed, and hence the directions observed along that line are not included. A direction is a pointing or sighting made from one station to another station with the theodolite. When all stations are occupied and all lines are observed, there will be two directions for all lines except the starting line. In Fig. 10-3 the heavier lines are the starting lines. The value of D for the different types of figures shown are as follows: in (a), $D = 10$; in (b), $D = 10$; in (c), $D = 18$; in (d), $D = 16$.

The number of conditions C to be satisfied in any figure can be computed from the following relationship:

$$C = (n' - s' + 1) + (n - 2s + 3) \tag{10-2}$$

in which n is the total number of lines, n' is the number of lines observed in both directions, s is the total number of stations, and s' is the number of stations occupied. Thus for a triangle with all stations occupied, $C = (3 - 3 + 1) + (3 - 6 + 3) = 1$, as is to be expected, since the only condition involved in a triangle is that the sum of the three angles should equal 180°. For a quadrilateral, with all stations occupied, $C=(6-4+1)+(6-8+3)=4$ (see Section 10-22 and Appendix A).

In Eq. (10-1) the values of the two terms $(D-C)/D$ and $\sum(\delta_A^2+\delta_A\delta_B+\delta_B^2)$ depend entirely on the figure chosen and are independent of the accuracy with which the angles are measured. The product of these two terms is therefore a measure of the strength of the figure with respect to length, insofar as the strength depends on the selection of stations and of lines over which observations are made. Hence, the strength of a figure is

$$R = \frac{D-C}{D}\sum(\delta_A^2 + \delta_A\delta_B + \delta_B^2) \tag{10-3}$$

When a required distance can be computed through two or more chains of triangles, the strengths of the figures are designated by R_1 and R_2 for the best and second-best chains, respectively.

In Table 10-2 the values $(\delta_A^2 + \delta_A\delta_B + \delta_B^2)$ are tabulated. The two arguments of the table are the distance angles in degrees, the smaller distance angle being given at the top of the table.

If in the quadrilateral shown in Fig. 10-5 the side AB is the known length, the side CD might be computed in four different ways. One solution would

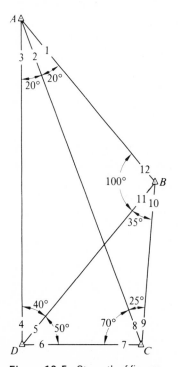

Figure 10-5. Strength of figures.

TABLE 10-2. Factors for Determining Strength of Figure

°	10°	12°	14°	16°	18°	20°	22°	24°	26°	28°	30°	35°	40°	45°	50°	55°	60°	65°	70°	75°	80°	85°	90°
10	428	359																					
12	359	295	253																				
14	315	253	214	187																			
16	284	225	187	162	143																		
18	262	204	168	143	126	113																	
20	245	189	153	130	113	100	91																
22	232	177	142	119	103	91	81	74															
24	221	167	134	111	95	83	74	67	61														
26	213	160	126	104	89	77	68	61	56	51													
28	206	153	120	99	83	72	63	57	51	47	43												
30	199	148	115	94	79	68	59	53	48	43	40	33											
35	188	137	106	85	71	60	52	46	41	37	33	27	23										
40	179	129	99	79	65	54	47	41	36	32	29	23	19	16									
45	172	124	93	74	60	50	43	37	32	28	25	20	16	13	11								
50	167	119	89	70	57	47	39	34	29	26	23	18	14	11	9	8							
55	162	115	86	67	54	44	37	32	27	24	21	16	12	10	8	7	5						
60	159	112	83	64	51	42	35	30	25	22	19	14	11	9	7	5	4	4					
65	155	109	80	62	49	40	33	28	24	21	18	13	10	7	6	5	4	3	2				
70	152	106	78	60	48	38	32	27	23	19	17	12	9	7	5	4	3	2	2	1			
75	150	104	76	58	46	37	30	25	21	18	16	11	8	6	4	3	2	2	1	1	1		
80	147	102	74	57	45	36	29	24	20	17	15	10	7	5	4	3	2	1	1	1	0	0	
85	145	100	73	55	43	34	28	23	19	16	14	10	7	5	3	2	2	1	1	0	0	0	0

	90	95	100	105	110	115	120	125	130	135	140	145	150	152	154	156	158	160	162	164	166	168	170
90	0																						
95	0	0																					
100	0	0	0																				
105	0	0	0	0																			
110	1	0	0	0	1																		
115	1	1	1	1	1	1																	
120	1	1	1	1	1	1	1																
125	2	2	2	2	2	2	2	2															
130	3	3	3	2	2	2	2	3	3														
135	4	4	4	4	3	3	3	4	4	4													
140	6	6	6	5	5	5	5	5	5	5	6												
145	9	9	8	8	7	7	7	7	7	7	8	9											
150	13	13	12	12	11	11	10	10	10	10	10	11	13										
152	16	15	14	14	13	13	12	12	12	12	12	13	15	16									
154	19	18	17	17	16	15	15	14	14	14	14	15	16	17	19								
156	22	22	21	20	19	19	18	18	17	17	17	17	18	19	21	22							
158	27	26	25	25	24	23	22	22	21	21	20	21	21	22	23	25	27						
160	33	32	31	30	30	29	28	27	26	26	25	25	26	26	27	28	30	33					
162	42	41	40	39	38	37	36	35	34	33	32	32	32	32	33	34	35	35	42				
164	54	53	51	50	49	48	46	45	44	43	42	41	40	40	41	42	43	45	48	54			
166	71	70	68	67	65	64	62	61	59	58	56	55	54	53	53	54	54	56	59	63	71		
168	98	96	95	93	91	89	88	86	84	82	80	77	75	75	74	74	74	74	76	79	86	98	
170	143	140	138	136	134	132	129	127	125	122	119	116	112	111	110	105	107	107	107	109	113	122	143

TABLE 10-3. Computations for Strength of Figure

TRIANGLE	KNOWN SIDE	COMPUTED SIDE	DISTANCE ANGLES		$\delta_{A^2}^2 + \delta_A\delta_B + \delta_{B^2}^2$
ABD	AB	BD	40°	40°	19
BCD	BD	CD	95°	35°	9
					$\overline{28} \times 0.6 = 17 = R_1$
ABC	AB	AC	25°	135°	15
ACD	AC	CD	90°	20°	33
					$\overline{48} \times 0.6 = 29$
ABD	AB	AD	40°	100°	6
ACD	AD	CD	70°	20°	38
					$\overline{44} \times 0.6 = 26 = R_2$
ABC	AB	BC	25°	20°	80
BCD	BC	CD	50°	35°	18
					$\overline{98} \times 0.6 = 59$

be first to find the side BD in the triangle ABD by using the distance angles 40° and 40°. The required side CD could then be found from the triangle BCD by using the angles 95° and 35°. In the quadrilateral there are a total of 12 directions, indicated by the numbered lines. Since the line AB is assumed fixed in direction and length, $D = 10$, $C = 4$, and $(D - C)/D = 0.60$. For convenience in illustrating the use of Table 10-2, the angles here given are in whole degrees. The strength of this figure is $0.60 \times (19 + 9) = 17$. Table 10-3 shows the strength of figure for all four sets of computations.

The values of R_1 and R_2 are 17 and 26, respectively. The value for R_1 is above the desirable limit for a single figure in second-order, Class I or II triangulation, as shown in Table 10-1, but is well below the maximum limit for both classes. This figure could be strengthened by shortening the line AD and thus increasing each angle at station A.

10-7. LENGTHS OF LINES In the earlier days when it was necessary to extend the first-order triangulation across the country rapidly, very long lines were used whenever possible, these lines in many cases being over 100 miles in length. Since one of the objectives of present-day work is to establish points for the use of the local surveyor and engineer, much shorter lines are used. Most of them range from 8 to 15 miles, with an occasional maximum of about 40 miles.

The average length of sight is usually determined by the nature of the country. With very short lines, extremely accurate centering of signals and instrument is required, many more stations must be occupied to accomplish the same linear advance, the area covered becomes a comparatively narrow strip, and the computations are increased. If the objective of the work is to distribute useful points with specified frequency for the purpose of local survey, this consideration may govern the lengths of the lines used in the main figures. It should be noted, however, that all the stations required probably do not need to be main stations. Satisfactory intersected points that will meet the specifications usually can be distributed among the main stations. If long sights are desired, the stations selected must be at commanding elevations so as to overlook the intervening country. Visibility is better in some regions than in others. Since it varies with the seasons, the prevalent atmospheric conditions should be introduced as a factor in considering the lengths of lines to be used.

10-8. STATION MARKS Except where the triangulation is of a temporary nature, the stations should be permanently marked and referenced. Bronze or copper markers cemented into solid ledge rock make the best station marks. In earth, a concrete monument makes an excellent permanent mark. It should be set deep enough in the ground to prevent movement by frost action.

Each station must have an azimuth mark at a distance of not less than $\frac{1}{4}$ mile from the station and must also have at least two reference marks. An azimuth mark or a reference mark consists of a metal tablet similar to one used for a station mark and bears an arrow pointing to the station. When a mark is set in a concrete post, the post may be smaller in diameter and shorter than one for the station mark.

In order that a station may be of future use to other engineers and surveyors, a very complete description of the station and its location should be prepared. This description should include the type of monument and references used and its location. The description of the location begins with the state and county, and the distance and direction from the nearest town; and it also includes the position in a particular quarter section (see Chapter 18) and the position with respect to the reference marks and nearby topographic features.

10-9. SIGNALS The type of signal used will depend on the length of the line and the accuracy required. As experience has shown that the air is steadier and that the lateral refraction is a minimum between dark and daylight, most precise work is done during those hours. The signal used for such work is practically an automobile headlight, the current for the light being supplied by dry batteries. A simple rheostat, operated by a light tender, is placed in the circuit to control the intensity of the light.

For third-order triangulation work, a small pole signal that is 5 or 6 ft high and is braced or guyed with wire is satisfactory for a sight under 3 or 4 miles. The signal can be found and identified more easily if a flag is attached and if the pole is painted black (or red) and white on alternate sections. For a greater distance, the pole should carry two cross targets made of cloth stretched on wooden frames and set at right angles to each other. Targets 3 ft square can usually be seen without difficulty at a distance up to 8 or 10 miles under average conditions. Tall signals, made of poles guyed with wire, may be necessary to project above obstructions, such as trees or buildings.

A tripod or quadripod surmounted by a pole with a flag or with cross targets is more substantial, and a signal of this kind should be built if the work is of such a nature that the station is to be used frequently for a considerable period, as in the case of a long tunnel or an important bridge. The best design is that in which the center pole does not come all the way to the ground, but is elevated sufficiently to allow space and headroom to set up the instrument over the station mark under the pole. The legs should be well anchored to prevent the structure from overturning in the wind.

Church spires, windmills, water tanks, and other prominent objects can be conveniently located by intersection.

The use of signals subject to phase should be avoided. If a signal is so situated that one side is illuminated by the sun while the other is in shadow, it is likely that a distant observer will see only the illuminated or the shaded side and his readings will be affected accordingly. A solid square or cylindrical object that is light in color is particularly bad in this respect. A signal consisting of cross targets set on a pole is nearly free from phase if the targets are set about a foot apart so that the upper one will not cast a shadow on the lower one. The color should be such that the target can be seen easily against the background.

10-10. TOWERS The most favorable country for triangulation is one with numerous high points on which the stations can be located. Where the ground is flat or the timber is dense, the theodolite and the signals must be elevated sufficiently to provide clear sights. The amount of this elevation may vary from a few feet to more than 100 ft in particularly difficult country or where extremely long sights are taken.

The National Geodetic Survey has developed a steel tower that can be easily erected and dismantled. It is built of light sections similar to those used on windmills. This tower consists of two independent structures. The instrument rests on the inner one, which is undisturbed by the movements of the observer on the outer one. When the instrument need be raised only a few feet above the ground, a substantial tripod of 2 by 4 in. lumber is built, and the observer stands on a platform which is entirely separated from this tripod.

On third-order triangulation the intervisibility of stations should be checked in the field, and for reasons of economy the use of towers and elevated signals should be kept to a minimum. When the lines extend across a flat terrain or across a large body of water, the curvature of the earth and the effects of refraction are definite factors. In Section 3-2 it was seen that the combined effect of these two factors can be approximated from the relationship

$$h = 0.574K^2$$

$$h = 0.0785K_M^2$$

(10-4)

in which h is the required height of eye, in feet or metres, K is the length of sight, in miles, and K_M is the length of sight, in kilometres. Thus if it is desired to see a point 10 miles away on the surface of a body of water, it would be necessary to elevate the eye of the observer 57.4 ft. If two towers of equal height are built, the line of sight would be tangent to the earth's surface at a distance of 5 miles, and two towers 14.4 ft high would replace the 57.4-ft tower. On account of refraction, it is desirable that the line of sight clear intervening objects by at least 10 ft. Hence towers about 24 ft high would be required.

10-11. BASELINES Since the computed sides of a triangulation system can be no more accurate than the baselines, every precaution to insure accuracy is taken in measuring these lines. The length of any base is determined primarily by the desirability of securing strong figures in the base net. Ordinarily, the longer the base, the easier it will be found to secure strong figures.

The base is connected to the triangulation system through a base net. This connection may be made through a simple figure, as shown in Fig. 10-6, or through a much more complicated figure. The net shown is a strong one, as it is expanded from the baseline AB to the side CD, which is a side of a main triangle of the triangulation system.

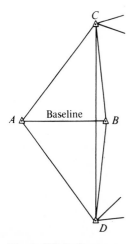

Figure 10-6. Base net.

Because of errors in the angular measurements, the accuracy of the computed lengths will decrease as the distance from the baseline increases. On extensive triangulation the accuracy is maintained by measuring additional baselines. The required frequency of the bases is dependent on the strength of the figures, a new base being measured whenever the sum of the values of R_1 in the triangles through which the computations are carried reaches a certain limit. These limits for first-order, second-order, and third-order work are shown in Table 10-1. They will be found to correspond to a chain of from 5 to 30 triangles, the number depending on the strengths of the figures involved.

Baselines are measured either by using invar tapes and related apparatus or by EDM's discussed in Chapter 2. The EDM shown in Fig. 10-7 is a long-range laser instrument suitable for the measurement of triangulation baselines. When a long-range EDM is employed, the development of a base net is not necessary since the side of a main triangle can be measured directly. On lower-order triangulation work the measurement of a baseline can frequently be eliminated by connecting the work to a triangulation system of a higher order.

When a baseline is to be measured by taping, the ground is prepared for the measurement by first clearing away obstructions, in order that the tape may hang freely when under tension. Stakes are then set on the line at distances apart equal to the length of the tape to be used in measuring the base. Copper strips are fastened to the tops of these stakes for marking the tape lengths. The tape is supported at the middle point as well as at the ends. For the intermediate support, stakes are set with their edges on line, and nails are driven in the sides to support the middle of the tape on line with the tops of the adjacent stakes. Where the grade of the base is uniform, the stakes need not project more than $1\frac{1}{2}$ to 2 ft above the ground surface, since the lower height makes them more rigid.

For first-order baselines, 4 by 4 in. stakes are used at the tape ends, with 2 by 4 in. stakes at the middle points, the stakes being lined in with a theodolite. The accuracy of the alignment should be such that when a 50-m tape is to be used, no stake is more than 6 in. off the line between the terminal stations and no marking strip is more than 1 in. off the line joining the strips on the two adjacent stakes.

For baselines of less precision, 2 by 4 in. and 1 by 4 in. stakes can be used. For second-order precision and when a 50-m tape is to be used for measuring, no part of the measured line should be more than 6 in. off the straight line between the terminal stations, nor should any one marking strip on a stake be more than 2 in. off the line between the strips on the two adjacent stakes. For third-order precision, the stakes should be so located that the error in the length of the base caused by poor alignment will not exceed one part in 150,000.

Levels must be run along the baseline to determine the differences of elevation between the tops of the stakes. These differences of elevation

Figure 10-7. Model 8 Geodimeter suitable for measuring baselines. Courtesy of AGA Geodimeter, Inc.

are used to reduce the slope measurements to the horizontal. The accuracy with which the differences of elevation must be obtained depends on the grade and the distances between the stakes as discussed in Section 2-12.

The baselines of the National Geodetic Survey are measured with 50-m invar tapes. They have a very unstable molecular arrangement and are easily kinked unless handled very carefully. They should not be reeled or unreeled rapidly or under a heavy tension, or wound on a reel having a small diameter, or dragged over the ground, or shaken violently, or subjected to sudden large changes in temperature. When the tape is not on a reel, a slight tension must be maintained constantly to prevent it from kinking.

These tapes are standardized by the Bureau of Standards, a tension of 15 kg being used. The following data are supplied by the Bureau: the weight of the tape in grams per meter; the coefficient of expansion per degree Celsius; the length at a specified temperature when supported at the 0-, 25-, and 50-m points; length at a specified temperature when supported at the 0-, 12.5-, 25-, 37.5-, and 50-m points; length at a specified temperature when supported throughout.

The standardized length of the tape supported throughout presupposes a frictionless surface as a support for the tape. Second-order bases are sometimes measured along railroad rails. In such cases the rail should be dry and care should be taken in lowering the tape to the rail. On first-order bases it is not desirable to have the tape supported throughout, because the error due to friction will vary with the surface conditions of the rail.

When a baseline must be measured with steel tapes, the measurements should be made either on a cloudy day or early in the morning or late in the afternoon, when the temperature of the tape will be the same as that of the surrounding air. Since the coefficient of expansion of steel is about $0.0000065/°F$ or $1.15 \times 10^{-5}/°C$, the temperature must be known more accurately than when invar tapes are used if equal precision is to be attained. The steel tapes should be standardized before any important base is measured.

The field temperature for each tape length is determined by two thermometers. Those used by the National Geodetic Survey are supported in channel-bar holders, which are fastened to the tape with narrow bands of adhesive tape. The thermometers are attached at the same points at which they are placed during the standardization of the tape, namely, at points 1 m toward the center from the terminal marks, the distances being measured from the marks to the nearer ends of the thermometer. The thermometers used on first-order and second-order bases are correct to within 0.3°C.

The tension in the tape is maintained by a spring balance. This balance should be sufficiently sensitive to permit maintaining a tension within 100 g, or about $\frac{1}{4}$ lb, of the standard tension. The balance should be tested before and after each day's work, also at midday if practicable, and oftener if it is suspected that the position of the pointer has changed.

Steel tubes pointed at the ends or ordinary steel line rods can be used in applying tension to the tape. A leather loop attached to the rear end of the tape and one attached to the spring balance at the forward end of the tape slip over the rods. These loops can be slipped up and down on the rods to correspond to the heights of the stakes.

The field party for measuring a baseline consists of six men, namely, front and rear tension men, front and rear contact men, a middle man, and a recorder. Usually the chief of the party makes the forward contact, as in that position he can best surpervise the manipulation of the tape and can set the pace of measurement. If any of the men are inexperienced, it is better to measure a practice section of several hundred metres before the recorded measurements are begun, each man being drilled in his position by an experienced man.

10-12. MEASURING BASELINE In the actual measurement of the baseline the line is broken up into sections about 1 km in length, and at least two measurements of a section are made with different tapes before proceeding to the next section. When three tapes are available, the total length is divided into three divisions of approximately equal length, and a different pairing of the tapes is used for the measurement of each division. The permissible standard errors of the measurements for various classes of triangulation are given in Table 10-1. These are given as the ratio of standard error to length of baseline. The standard errors reflect the accuracy of the length after corrections have been applied to eliminate systematic errors.

10-13. CORRECTIONS TO BASELINE MEASUREMENTS The principal corrections applied to the field measurements of a baseline are for incorrect length of tape, for temperature, for slope, and for reduction of the length to sea-level length. The corrections for sag and tension are usually avoided by having the tape standardized under the conditions to which it will be subjected in the field.

When the triangulation system is extensive, all linear distances are reduced to their equivalent sea-level lengths. In Fig. 10-8, A and B are two points on the earth's surface at an average elevation h above sea level. Since the lengths of arcs are proportional to their radii, $(D - C)/R = D/(R + h)$. This equation may be expanded into the form

$$\frac{D}{R} - \frac{C}{R} = \frac{D}{R} - \frac{h}{DR^2} + D\frac{h^2}{R^3} - D\frac{h^3}{R^4} + \cdots$$

where

$$C = D\frac{h}{R} - D\frac{h^2}{R^2} + D\frac{h^3}{R^3} - \cdots \qquad (10\text{-}5)$$

in which C is the correction to be subtracted from the measured length D, at an average elevation h, and R is the radius of curvature of the earth. For most work, the first term is sufficient and an average value of R can be used. When h is in feet, $R = 20,906,000$ ft; for h in metres, $R = 6,372,200$ m. The more exact values of R are dependent on the mean latitude and the azimuth of the base.

For a baseline at an average elevation of 6100 ft and a measured length of 4324.186 ft, the correction is

$$C = \frac{4324 \times 6100}{20,906,000} = 1.262 \text{ ft}$$

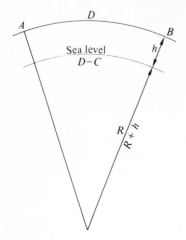

Figure 10-8. Reduction to sea level.

The corrections for incorrect length of tape, temperature, slope, sag, and other conditions have been discussed in Chapter 2.

10-14. BASELINE MEASUREMENT BY USE OF EDM   The use of EDM's eliminates the need for elaborate baseline preparation. The general requirement for a baseline that is to be measured by means of one of these instruments is a clear line of sight between the two ends of the base. A baseline may be chosen between two prominent triangulation stations since the measurement with an EDM is virtually independent of the character of the intervening terrain.

The time required to make the necessary instrument readings during the actual measuring operations is independent of the length of the baseline. When an EDM is used under normal conditions, the instruments can be set up and aligned and all the readings can be made in less than an hour. There is a tremendous saving in time, especially when baselines 20 to 30 miles in length are considered.

Because of the ease with which bases can be measured by using the EDM, check bases can be selected at more frequent intervals than when baselines must be measured with tapes. Check bases tend to strengthen the entire triangulation network.

The corrections for meteorological conditions and instrumental errors discussed in Sections 2-27 and 2-28 are applied to the measurements to obtain the corrected slope distance. In order to reduce the slope distance to sea level as required, either the reciprocal vertical angles are measured at the two ends of the line or else the elevations of the two ends of the line must be known. The methods of making the slope reductions are given in Sections 10-29 and 10-30.

10-15. MEASURING ANGLES IN TRIANGULATION The instruments to be used in measuring the angles in the triangulation network depend on the desired accuracy of the positions of the triangulation stations. For first-order work, direction theodolites which can be read directly to 0.2″ should be used. Examples of such an instrument are shown in Figs. 4-26 and 10-9. For second-order work the instrument should be capable of reading directly to 1″. Examples of second-order direction theodolites are shown in Chapter 4. For third-order triangulation an engineer's transit or theodolite reading to 30″ or 20″ can be employed if advantage is taken of the added precision gained by repeating angles.

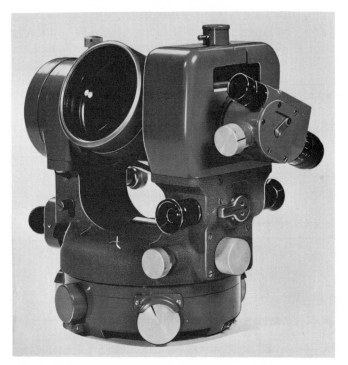

Figure 10-9. First-order direction theodolite. Courtesy of Kern Instruments, Inc.

Before the angles at a triangulation station which are to be used in the computations are measured accurately, a preliminary list of directions should be prepared by occupying the station and measuring either the angles or the directions to other stations in the network. This list eliminates much unnecessary delay in measuring the final angles because it allows the instrument to be pointed almost directly to the sighted stations by means of precomputed circle settings. Thus a distant station that may not be readily seen with the naked eye can be picked up in the field of the telescope after the circle setting is made.

10-16. TRIANGULATION ANGLES BY REPETITION

The repetition method of measuring triangulation angles is the same as that described in Section 4-17. In Fig. 10-10 each one of the angles *a* through *e* is measured by repetition, either three times with the telescope direct and three times with it reversed, or six times direct and six times reversed. The number of turnings depends on the required accuracy and on the number of sets of repetitions to be taken. When all the angles about the station have been measured by using the required number of repetitions, the complete procedure constitutes one *set* of observations. For third-order triangulation, six sets of 6 repetitions or three sets of 12 repetitions will suffice when a 30″ transit is used. The mean value of each measured angle is used to find the sum of the angles about the station, and then each angle receives an equal correction. A list of directions based on the adjusted angles is then prepared for subsequent use in computations, the initial direction being taken as 0° 00′ 00.0″.

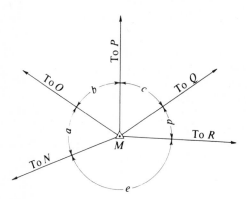

Figure 10-10. Angles about a station.

Example 10-1

As shown in Fig. 10-10, five angles were measured about station *M*. Each set of measurements consisted of six repetitions with the telescope direct and six with the telescope reversed, and three sets of measurements were taken. The mean values of the angles are shown in the second column of Table 10-4. Adjust the mean angles and prepare an abstract of directions to the observed stations. Take the direction to station *N* as 0° 00′ 00.0″.

Solution: Since the sum of the measured angles is 359° 59′ 56.0″, the angles are adjusted by adding 0.8″ to each measured value. The seconds in the adjusted angles are entered in the fourth column of the tabulation.

TABLE 10-4. Station Adjustment and Abstract of Directions

ANGLE	MEAN VALUE	CORRECTION	ADJUSTED VALUE	TO STATION	ABSTRACTED DIRECTION
a	54° 02′ 15.0″	+0.8″	15.8″	N	0° 00′ 00.0″
b	57° 33′ 12.7″	+0.8″	13.5″	O	54° 02′ 15.8″
c	55° 42′ 42.5″	+0.8″	43.3″	P	111° 35′ 29.3″
d	36° 57′ 22.5″	+0.8″	23.3″	Q	167° 18′ 12.6″
e	155° 44′ 23.3″	+0.8″	24.1″	R	204° 15′ 35.9″
	359° 59′ 56.0″		00.0″	N	360° 00′ 00.0″
	closure = −4.0″				

The directions in the last column are obtained by adding each adjusted angle in succession to the preceding direction. As a check the angle e is added to the direction of R to see if the sum is exactly 360°.

10-17. TRIANGULATION ANGLES BY DIRECTION METHOD
As indicated in Section 4-21, the direction instrument does not have a lower clamp. So it cannot be used to repeat angles. However, provision is made to advance the position of the horizontal circle relative to the reading microscope without rotating the telescope in azimuth. Angles in almost all first-order and second-order triangulations are measured by using an optical reading direction instrument as described in Chapter 4.

With the direction instrument set up over a selected triangulation station and the telescope direct, the line of sight is directed at each adjacent station in turn, the telescope being rotated in a clockwise direction, and the circle is read at each pointing by means of the optical micrometer. The telescope is then reversed, and another round of directions is observed. This entire set of readings constitutes one *position*. For first-order work a total of between 8 and 16 positions are observed, the number depending on the precision of the instrument. For second-order work, a total of between 4 and 8 positions are observed. The circle reading should be advanced by $180°/n$ for each new position, in which n is the number of positions to be observed. This interval tends to distribute the readings over the entire circle and eliminates the errors of graduation of the circle.

Each position gives a set of angles, as shown by way of illustration in Section 4-22. If four positions have been observed, then each angle will have four values. The mean value of each angle must be obtained, and an abstract is then prepared.

Example 10-2

The angles a through e in Fig. 10-10 have been measured by observing the directions from M to N, O, P, Q, and R with four positions of a direction theodolite, and the results are as shown in the accompanying tabulation. Taking the direction to station N as $0° 00' 00.00''$, prepare an abstract of directions to the observed stations.

ANGLE	POSITION 1	POSITION 2	POSITION 3	POSITION 4	MEAN ANGLE
a	$54° 02' 13.5''$	$54° 02' 15.0''$	$54° 02' 18.4''$	$54° 02' 12.0''$	$54° 02' 14.72''$
b	$57° 33' 14.0''$	$57° 33' 14.0''$	$57° 33' 10.2''$	$57° 33' 15.6''$	$57° 33' 13.45''$
c	$55° 42' 46.1''$	$55° 42' 44.3''$	$55° 42' 42.4''$	$55° 42' 46.3''$	$55° 42' 44.77''$
d	$36° 57' 21.0''$	$36° 57' 18.8''$	$36° 57' 24.1''$	$36° 57' 21.8''$	$36° 57' 21.43''$
e	$155° 44' 25.4''$	$155° 44' 27.9''$	$155° 44' 24.9''$	$155° 44' 24.3''$	$155° 44' 25.63''$
	$360° 00' 00.0''$	$360° 00' 00.0''$	$360° 00' 00.0''$	$360° 00' 00.0''$	$360° 00' 00.00''$

Solution: The angles resulting from each position are added to verify that their sum is $360°$. If it is not, a mistake has been made in obtaining the angles. The average of the four values of each angle is then computed. Finally, the direction abstract is obtained as in Example 10-1.

TO STATION	ABSTRACTED DIRECTION
N	$00° 00' 00.00''$
O	$54° 02' 14.72''$
P	$111° 35' 28.17''$
Q	$167° 18' 12.94''$
R	$204° 15' 34.37''$
N	$360° 00' 00.00''$

As a check, the angle e is added to the direction of R, as was done in Example 10-1.

10-18. NATURE OF TRIANGULATION COMPUTATIONS The purpose of triangulation is to compute some inaccessible distance or to obtain the coordinates of triangulation stations as a basis for further surveys. Before such computations are made, the errors in the field measurements must be eliminated or distributed. This distribution can be made by approximate methods or by the more exact method of least squares.

In a triangulation net that contains fixed control points, lines of fixed azimuth, and more than one measured base, five kinds of discrepancies are encountered. (1) The sum of the three angles in each triangle will seldom equal exactly $180°$ plus spherical excess (see Section 10-21). (2) In a quadrilateral the length of any unknown side will have two different values when computed through two different sets of triangles. (3) The measured lengths of the additional baselines will not agree with their computed lengths as obtained from the preceding base and the measured angles. (4) The azimuths of fixed lines will not agree with the azimuths computed from the measured angles. (5) The fixed position (coordinates) of a point will disagree with its position as computed through the triangle nets. Consequently, before triangulation computations can be made, the measured values must be adjusted in order to distribute the random errors in such fashion that the sum of the squares of the corrections to the measured angles will be a minimum.

In the adjustment procedure the five kinds of inconsistencies cited above give rise to five conditions that must be satisfied. These are referred to, respectively, as follows: (1) angle condition, (2) side condition, (3) length condition, (4) azimuth condition, and (5) position condition. The least-squares adjustment is made to satisfy all these conditions.

The order of the calculations is as follows: (1) reduction to center when the instrument has not been placed exactly over a given station; (2) correction for eccentricity of signal where the signal has not corresponded exactly with the station; (3) spherical excess; (4) approximate adjustment, or least-squares adjustment; (5) computation of the lengths of the sides; (6) computation of coordinates. A form of least-squares adjustment called *variation of coordinates* accomplishes steps (4) and (6) at the same time. If lengths of sides are needed, they can be obtained by inversing the line using Eq. (8-12). The method of least squares employing variation of coordinates is given in Appendix A.

10-19. REDUCTION TO CENTER   In the measurement of triangulation angles it often is not possible to occupy a station directly below an excellent target, such as a church spire or a lighthouse. Also, targets are frequently blown out of position and the angles read on them have to be corrected to the true position of the triangulation station. There are thus two types of problems: (1) when the instrument's point is not the true station and the measured angles must be corrected to what they would be at the station, and (2) when the target is out of position. In the example shown in Fig. 10-11 the angles were measured at an eccentric station A' instead of the true station A.

In computing the corrections to be applied to the measured angles or directions, the simplest method is to calculate first the directions of the lines with respect to the line between the eccentric station and the true station as a meridian. The corrections to refer these directions to the true station are obtained by solving the triangles such as ABA', ACA', and ADA' for the angles at the stations B, C, D, E, and F. In the triangle ABA'

$$\sin e = \frac{l}{L} \sin a$$

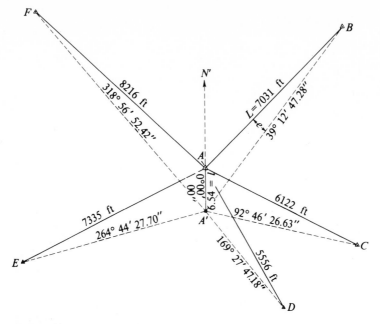

Figure 10-11. Reduction to center.

in which e is the angle $A'BA$; l equals AA' or the distance between the true station and the eccentric station; L equals AB or the distance from the true station to B; a is the direction of $A'B$ measured at the eccentric station with respect to the meridian through A' and A.

The small angle e is then added to the direction $A'B$ to give the corrected direction AB. Thus if $e = 0°\ 02'\ 01.30''$, the corrected direction $AB = 39°\ 12'\ 47.28'' + 0°\ 02'\ 01.30'' = 39°\ 14'\ 48.58''$.

The value of L is not known but its approximate value can be taken as $A'B$ which is obtained by a preliminary computation of the triangles using A' as the triangulation station.

Since the angles at B, C, D, E, and F will be very small, their values in seconds can be obtained by dividing both sides of the preceding equation by sin $1''$. Thus if e'' denotes the angle in seconds,

$$e'' = \frac{l}{\sin 1''} \times \frac{\sin a}{L} \tag{10-6}$$

For any given case, $l/\sin 1''$ will be a constant for solving all the triangles that are involved. In Table 10-5 are shown the computations for reducing to the true station A (Fig. 10-11), the directions measured from the eccentric station A'. The distance l from A to A' is 6.54 ft and $(l/\sin 1'')$ is 1.3490×10^6. The signs of the corrections are the same as the signs of sin a.

TABLE 10-5. Reduction to Center

STATIONS	B	C	D	E	F
Distance	7031′	6122′	5556′	7335′	8216′
a	39° 12′ 47.28″	92° 46′ 26.63″	169° 27′ 47.18″	264° 44′ 27.70′	318° 56′ 52.42″
sin a	+0.63221	+0.99883	+0.18287	−0.99579	−0.65675
l/sin 1″	1.3490×10^6	1.3490×10^6	1.3490×10^6	1.3490×10^6	1.3490×10^6
$e″$	+121.30″	+220.10″	+44.40″	−183.14″	−107.83″

The angles at station A can be obtained from the corrected directions. If the values of L used in the computations for $e″$ differ considerably from the more exact values, it may be necessary to make other computations in which the corrected lengths are used. Since the eccentric distance l will be, in general, small compared with the lengths of the sides, this recalculation will seldom be required.

10-20. CORRECTION FOR ECCENTRIC SIGNAL   If in Fig. 10-11 A' represents the station and A the eccentric signal, the corrections to be applied to the directions measured to the eccentric signal from stations $B, C, D, E,$ and F will be the same as those computed in the preceding section.

10-21. SPHERICAL EXCESS   When a theodolite is set up and leveled at a triangulation station, the vertical axis of the instrument assumes a plumb position and the line of sight sweeps through a vertical plane. The vertical plane containing the line of sight to a distant triangulation station intersects the sea level surface along an arc, which for the sake of this analysis is like the arc of a great circle. Thus in Fig. 10-12 a vertical plane containing

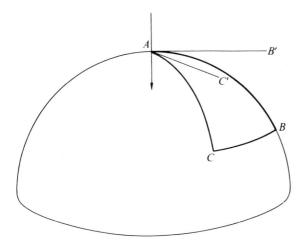

Figure 10-12. Spherical angles.

the line of sight from A to B intersects the sphere (sea level) along arc AB. The direction reading on the theodolite's horizontal circle is represented by the line AB'. Also, a vertical plane containing the line of sight from A to C intersects the sphere along the arc AC. The direction reading on the theodolite's horizontal circle is represented by the line AC'. The horizontal angle $B'AC'$ obtained from the two direction readings is equal to the spherical angle BAC. The same situation would exist at stations B and C. Thus the theodolite measures the spherical angles at the triangulation stations forming the spherical triangle ABC.

The amount by which the sum of the three angles of a spherical triangle exceeds 180° is called the *spherical excess*. Before the closing error in the angles of a large triangle can be ascertained, the spherical excess must be known. The amount of this excess is dependent on the area of the triangle and on the latitudes of the vertexes. It is approximately equal to 1″ for every 75.5 mile² of area. A more exact value for the excess in seconds is

$$E'' = \frac{bc \sin A}{2R^2 \sin 1''} = mbc \sin A \qquad (10\text{-}7)$$

in which b, c, and A are the two sides and the included angle of the triangle, and R is the radius of curvature of the earth. Most tables intended for geodetic computations contain values of m for various latitudes. When the triangles are small, as will be those encountered in third-order triangulation work, and even those in much first-order and second-order triangulation, the mean radius of the earth can be used. For a triangle in latitude 45°, $m = 2.355 \times 10^{-10}$ when the dimensions are in feet and $m = 2.535 \times 10^{-9}$ when the dimensions are in metres. The excess is divided equally between the three angles of the triangle, the corrections being subtracted from the observed values.

10-22. TRIANGULATION ADJUSTMENT   When a system of points is to be connected by a triangulation system, the connection can be made either by a chain of triangles, as shown in Fig. 10-13(a), by means of connected quadrilaterals, as shown in view (b) or by a combination of different kinds of figures. When triangles are used, the only adjustment to be made is to satisfy the condition that the sums of the angles of the individual triangles shall equal 180°. This adjustment is made ordinarily by correcting each angle by one-third of the closure in that triangle. Although this method involves the reading of fewer angles and simplifies the computations, the only check on the computed lengths of the sides is afforded by the measurement of a second baseline, such as CD in Fig. 10-13(a).

When the connection is made by means of quadrilaterals or more complex figures, additional checks on the angular measurements are provided by the added geometrical conditions in each figure. In a quadrilateral, for example,

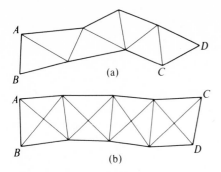

Figure 10-13. Triangulation nets.

checks on the computed lengths are afforded, since any length can be obtained through the use of four different combinations of triangles in each quadrilateral. Unless the measured angles have been adjusted, it is unlikely that the four computed values will agree exactly. The purpose of the adjustment is to correct the observed angles so that the various conditions relative to the sums of the angles will be satisfied, and in addition so that the computed length of a side will have the same numerical value, regardless of the triangles used in the computation.

In the quadrilateral shown in Fig. 10-14, the sum of the eight angles should equal 360° and the sums of the angles of any triangle should equal 180°. In addition, $b + c = f + g$ and $h + a = d + e$. The side condition can be developed as follows: In the triangle ABC,

$$BC = \frac{AB \sin b}{\sin e}$$

In the triangle BCD,

$$CD = \frac{BC \sin d}{\sin g} = \frac{AB \sin b \sin d}{\sin e \sin g}$$

In the triangle CDA,

$$DA = \frac{CD \sin f}{\sin a} = \frac{AB \sin b \sin d \sin f}{\sin a \sin e \sin g}$$

In the triangle DAB,

$$AB = \frac{DA \sin h}{\sin c} = \frac{AB \sin b \sin d \sin f \sin h}{\sin a \sin c \sin e \sin g}$$

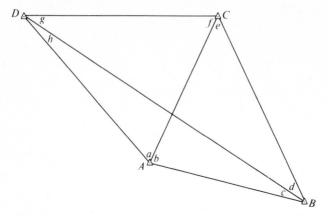

Figure 10-14. Triangulation adjustment in quadrilateral.

Hence

$$1 = \frac{\sin b \sin d \sin f \sin h}{\sin a \sin c \sin e \sin g}$$

If the logarithms of both members of the equation are considered,

$$(\log \sin b + \log \sin d + \log \sin f + \log \sin h)$$
$$- (\log \sin a + \log \sin c + \log \sin e + \log \sin g) = 0$$

Unless this condition is satisfied, the value of a computed length will depend on the triangles used in the computation.

The most accurate method of making the adjustments is according to the theory of least squares, as the most probable values for the measured angles and the computed lengths are then obtained. For the application of the least-squares method, refer to the adjustment of a quadrilateral given in Appendix A.

If the sides are less than a mile in length, the computations can be made without an adjustment when the angles have been quite accurately measured and when extreme precision in the computed lengths is unnecessary. When greater precision is required, approximate methods of adjusting will give computed lengths that will agree very closely with the least-squares values.

10-23. APPROXIMATE ADJUSTMENT OF A QUADRILATERAL

The first step in the adjustment of a quadrilateral like that in Fig. 10-14 by the approximate method is to correct each of the eight angles so that their sum will be exactly 360°. In the numerical example for which the computations are shown in Table 10-6, a total correction of $+1.71''$ is necessary. Hence five of the angles are increased by $0.21''$ and three of them are increased

TABLE 10-6. Computations for Adjustment of Quadrilateral by Approximate Method

OBSERVED ANGLES	FOR 360°	FOR OPPOSITE ANGLES	LOG SINES OF ADJUSTED ANGLES*	DIFFERENCES FOR 1″	CORRECTED ANGLES	CORRECTED LOG SINES
$a = 63°17'26.18″$	26.39″	26.97″	9.950 9970	10.6	63°17'28.12″	9.950 9982
$b = 84°18'22.69″$	22.91″	21.52″	9.997 8518	2.1	84°18'20.37″	9.997 8516
$c = 17°52'26.72″$	26.93″	25.54″	9.487 0263	65.3	17°52'26.69″	9.487 0338
$d = 30°41'18.49″$	18.71″	18.13″	9.707 8838	35.5	30°41'16.98″	9.707 8797
$e = 47°07'55.18″$	55.39″	54.81″	9.865 0576	19.5	47°07'55.96″	9.865 0598
$f = 66°35'54.84″$	55.06″	56.45″	9.962 7234	9.1	66°35'55.30″	9.962 7224
$g = 35°34'49.01″$	49.22″	50.61″	9.764 8106	29.4	35°34'51.76″	9.764 8140
$h = 14°31'45.18″$	45.39″	45.97″	9.399 4615	81.2	14°31'44.82″	9.399 4522
359°59'58.29″	00.00″	00.00″	39.067 9205	252.7	360°00'00.00″	39.067 9058
8)1.71			8915			39.067 9059
+0.21			290			

$$a + h = 77°49'11.78″ \qquad b + c = 102°10'49.84″$$
$$d + e = 77°49'14.10″ \qquad f + g = 102°10'44.28″$$
$$4)2.32″ \qquad\qquad\quad 4)5.56″$$
$$\;\;0.58″ \qquad\qquad\qquad\;\; 1.39″$$

$$\frac{290}{252.7} = 1.15″$$

* When using the hand calculator, the logarithm of the sine is always displayed as a negative number (log sine of 90° = 0); for example, the log sine of 63° 17′ 26.97″ is displayed as −0.049 0030. The sum of the left column of log sines is thus −0.932 1085, and the sum of the right column is −0.932 0795 which is 290 larger in the 7th place as shown in the table. The difference for 1 second is obtained for angle a, for example, by entering 63° 17′ 27.97″, obtaining the log sine (in the 8th place), then entering 63° 17′ 26.97″, obtaining the log sine, and finally subtracting to give +0.000 00106 or +10.6 in the 7th place as shown.

by 0.22″. The sums of the opposite angles are next made to agree by correcting each angle by one-fourth of the error in the sums. Thus a and h are increased by $\frac{1}{4} \times 2.32″ = 0.58″$ and d and e are decreased by equal amounts. In a like manner b and c are diminished and f and g are increased by 1.39″.

The logarithmic sines and the differences in the logarithms for 1″ of angle are tabulated next, the corrected values just obtained being used for the angles. Note that seven-place logarithms are used in the adjustment. The logarithmic sines and the difference for 1″ can be obtained on most handheld calculators. The sums of the alternate logarithmic sines fail to agree by 290, and hence the angles must be corrected by amounts that will eliminate this difference. The change must be effected by correcting each angle by an equal amount in order to keep their sum at 360° and to keep the sums of opposite pairs equal. The value of the correction in seconds is found by dividing the total change in logarithmic sines, or 290, by the total difference in logarithmic sines for 1″ of angle, or 252.7. The quotient 1.15″ is the correction applied to each angle, being added to the four angles whose logarithmic sines are to be increased and subtracted from the other four. If the corrections are accurate and are properly applied, the sums of the logarithmic sines of the corrected angles will agree within one or two in the last decimal place.

10-24. COMPUTATION OF LENGTHS

Two sides of each triangle are computed by using the law of sines, since one side of the triangle is always known and the three angles have been measured and adjusted. In computing the sides of the triangles in a quadrilateral, such as that shown in Fig. 10-14, the solution of two triangles is sufficient to compute the positions of the forward triangulation stations. The two triangles chosen must be the strongest route through the quadrilateral. In Fig. 10-14 the side AB is the known or measured line of the quadrilateral, and the strongest route is obtained by considering triangle CBA and triangle CDA in that order. To check the accuracy of the field work or the consistency of the figure after adjustment, the two triangles DBA and CBD can be computed in that order. This computation gives a check on the length of the side CD, which is the forward side of the quadrilateral. However, only the result for the strongest route will be used in further computations.

The tabulation in Table 10-7 shows the computations of determining the lengths of the sides of the quadrilateral in Fig. 10-14 by using the corrected angles found in Table 10-6. The stations in the triangle CBA are listed in the order determined by beginning with the unknown point C and going around the triangle in a clockwise direction. The corrected angles are filled in as shown. With the side BA known, the sides AC and CB are computed by using the laws of sines.

$$\frac{AC}{\sin(c + d)} = \frac{BA}{\sin e} \quad \text{or} \quad AC = \frac{BA \sin(c + d)}{\sin e} \tag{10-8}$$

TABLE 10-7. Computation of Lengths in Triangles

SIDES	STATIONS	ANGLES	ADJUSTED ANGLES	DISTANCES
BA				2899.06
	C	e	47° 07′ 55.96″	0.732 92548
	B	c + d	48° 33′ 43.67″	0.749 67381
	A	b	84° 18′ 20.37″	0.995 06538
AC			180° 00′ 00.00″	2965.31
CB				3935.94
CA				2965.31
	D	g + h	50° 06′ 36.58″	0.767 27890
	C	f	66° 35′ 55.30″	0.917 74558
	A	a	63° 17′ 28.12″	0.893 30193
AD			180° 00′ 00.00″	3546.82
DC				3452.35
BA				2899.06
	D	h	14° 31′ 44.82″	0.250 87197
	B	c	17° 52′ 26.69″	0.306 92610
	A	a + b	147° 35′ 48.49″	0.535 87391
AD			180° 00′ 00.00″	3546.82
DB				6192.52
BD				6192.52
	C	e + f	113° 43′ 51.26″	0.915 44565
	B	d	30° 41′ 16.98″	0.510 36357
	D	g	35° 34′ 51.76″	0.581 85393
DC			180° 00′ 00.00″	3452.35
CB				3935.95

Also,

$$\frac{CB}{\sin b} = \frac{BA}{\sin e} \quad \text{or} \quad CB = \frac{BA \sin b}{\sin e} \tag{10-9}$$

After the triangles are laid out as shown in Table 10-7, the adjusted angles are added to check that their sum is 180°. The length of the starting side BA of the first triangle is entered in the last column followed by the sines of the angles. The lengths of AC and CB are then computed by Eqs. (10-8) and (10-9) and shown in the table. The line $CA = AC$ is then used as the starting side of triangle DCA.

The sines of the angles are shown in Table 10-7 in order to be able to follow the steps involved in the triangle solutions. These need not be entered if the calculator generates the sines internally as most calculators do.

The two triangles DBA and CBD in the second route are solved by starting from side BA and following a similar procedure. This route provides a check on the side DC and also shows the consistency of the sides AD and CB.

10-25. COMPUTATION OF PLANE COORDINATES The objective of triangulation is to establish the horizontal positions of the triangulation stations in the network relative to one another and with respect to a horizontal datum. Chapter 11 contains a description of the state plane coordinate systems that should be the basis of any triangulation to be established on a plane coordinate system. The triangulation network must include at least one point, and preferably should include two or more points, the horizontal position of which is known with respect to the state plane coordinate system. In the network of Fig. 10-13(b), if A and B are points with known coordinates, the length and azimuth of the line AB can be computed by using Eqs. (8-12) and (8-13). If point C is also a point with known coordinates, then a check on the entire network is provided, because the discrepancy between the fixed position of C and its position as computed through the quadrilaterals can be determined.

If the coordinates of stations A and B in Fig. 10-14 are known, the length of line BA computed by Eq. (8-12) is used as the starting line in the computations for the lengths of the sides of the quadrilateral shown in Table 10-7. The azimuth from north of the line AB is determined by Eq. (8-13). The azimuth of the line AC is equal to the azimuth of AB minus angle b. The azimuth of BC is equal to the azimuth of BA plus angle $(c + d)$. Furthermore, when the azimuths of AC and BC have been computed, the difference between the two must be equal to angle e. If the figure has been adjusted, any discrepancy represents a mistake. These computations are shown in Table 10-8 for the two triangles CBA and DCA which will be used to compute the positions of C and D.

After the lengths of the lines AC and BC have been computed by solving the triangles and the azimuths of the lines AC and BC have been determined, the latitudes and departures of these two lines are computed by Eqs. (8-3) and (8-4). Applying the latitudes and departures of AC and BC to the coordinates of A and B, respectively, in accordance with Eqs. (8-7) and (8-8),

TABLE 10-8. Computation of Azimuths of Lines in Triangles

azimuth AB =	104° 42′ 56.60″		azimuth BA	=	284° 42′ 56.60″
−angle b = −	84° 18′ 20.37″		+angle $(c + d)$ = +		48° 33′ 43.67″
azimuth AC =	20° 24′ 36.23″		azimuth BC	=	333° 16′ 40.27″
	+360°				
	380° 24′ 36.23″				
	− 333° 16′ 40.27″				
angle e =	47° 07′ 55.96″ (check)				
azimuth AC =	20° 24′ 36.23″		azimuth CA	=	200° 24′ 36.23″
−angle a = −	63° 17′ 28.12″		+angle f	= +	66° 35′ 55.30″
azimuth AD =	317° 07′ 08.11″		azimuth CD	=	267° 00′ 31.53″
	− 267° 00′ 31.53″				
angle $(g + h)$ =	50° 06′ 36.58″ (check)				

TABLE 10-9. Position Computation

STATION	LENGTH	COS AZIMUTH	SIN AZIMUTH	Y	X
A				622,516.21	1,442,416.25
	2965.31	+0.937 22075	+0.348 73667	+2,779.15	+ 1,034.11
C				625,295.36	1,443,450.36
	3935.94	−0.893 19764	+0.449 66429	−3,515.57	+ 1,769.85
B				621,779.79	1,445,220.21
			Fixed	621,779.78	1,445,220.22
A				622,516.21	1,442,416.25
	3546.82	+0.732 76763	−0.680 47894	+2,598.99	− 2,413.54
D				625,115.20	1,440,002.71
	3452.35	+0.052 18331	+0.998 63752	+ 180.16	+ 3,447.65
C				625,295.36	1,443,450.36
			Fixed	625,295.36	1,443,450.36

gives two sets of values of the coordinates of C. These two sets of values should not differ by more than 0.01 or 0.02 ft. The slight discrepancy is due either to rounding-off errors or to slight inconsistencies in the triangles.

The computation just discussed is known as a *position computation*. The tabulation in Table 10-9 shows the position computation for the quadrilateral in Fig. 10-14 when the angles determined in Table 10-6 and the lengths determined in Table 10-7 are used. Each triangle is solved as if it were a two-sided traverse beginning on a known point and closing on a known point. The slight discrepancies are caused by rounding off errors.

The coordinates of all the stations throughout the triangulation system are computed by using the strongest route of triangles in the network. When a station is reached whose coordinates are fixed, the positions of the intermediate stations can then be adjusted. If the triangulation system is of great extent and high precision, this adjustment should be made by an application of the least-squares principle. If the system is moderate in extent, an application of the compass rule described in Section 8-15 will give highly satisfactory results. In this case a traverse extending from one fixed point to another fixed point and including all the intermediate triangulation stations is selected in as direct a line as possible. This traverse is then adjusted by the compass rule.

10-26. GEODETIC POSITIONS When the triangulation system extends over a large area, as does the work of the National Geodetic Survey, the coordinates of the stations are given as the latitudes and longitudes on an adopted reference ellipsoid. In computing these geodetic positions, the latitude ϕ and the longitude λ of one station either must be known or must be determined by astronomical observations. The coordinates ϕ' and λ' of any adjacent station can be computed when the azimuth a (reckoned from the south point) and the length s of the line connecting the two stations are known. The difference in latitude $\Delta\phi$ and the difference in longitude $\Delta\lambda$, in seconds of arc, which correspond with the latitude and departure of a side of a plane traverse, are calculated from the equations that follow:

$$-\Delta\phi = s \cos a \cdot B + s^2 \sin^2 a \cdot C + (\delta\phi)^2 D - hs^2 \sin^2 a \cdot E \quad (10\text{-}10)$$

$$\Delta\lambda = s \sin a \sec \phi' \cdot A' + \text{arc-sine corr.} \quad (10\text{-}11)$$

$$\phi' = \phi + \Delta\phi \quad (10\text{-}12)$$

$$\lambda' = \lambda + \Delta\lambda \quad (10\text{-}13)$$

$$-\delta\phi = s \cos a \cdot B + s^2 \sin^2 a \cdot C - hs^2 \sin^2 a \cdot E \quad (10\text{-}14)$$

$$h = s \cos a \cdot B \quad (10\text{-}15)$$

Because of the convergence of meridians, the difference between the forward azimuth and the back azimuth of a line will not exactly be 180°. The amount of this convergence, Δa, can be found from the relationship

$$-\Delta a = \Delta\lambda \sin \tfrac{1}{2}(\phi + \phi') \sec \tfrac{1}{2}(\Delta\phi) + (\Delta\lambda)^3 F \quad (10\text{-}16)$$

and the back azimuth is

$$a' = a + \Delta a + 180° \quad (10\text{-}17)$$

The constants B through F in the above equations are functions of the latitude of the known point and the specific ellipsoid (see Section 1-2) on which the calculations are made. A' is a function of the latitude of the adjacent station and the specified ellipsoid. In the United States at the present time, this ellipsoid is the Clarke Spheroid of 1866. Tables of these values are given in publications of the National Geodetic Survey and in textbooks on geodesy.

When the length of a side is less than 15 miles, the computations can be shortened by omitting the term involving E in $\Delta\phi$, as well as the factor $\sec \tfrac{1}{2}(\Delta\phi)$ and the term involving F in Δa.

By means of the preceding equations, the reverse problem can also be solved; that is, when the geographic positions of two points are known, the azimuth and length of the line connecting them can be computed.

Equations (10-10) through (10-17) are accurate for lines up to about 80 km or 50 miles. If the triangulation figures are extremely large, then more complex equations are required in order to maintain the necessary computational accuracy. These computations are more properly studied in the field of geodesy.

10-27. PRECISE TRAVERSE Under certain conditions a traverse of high precision may be substituted for a triangulation network. In flat, heavily wooded country the construction of towers and signals is a large item of expense in establishing a triangulation system. When such an area is crossed by railroads and paved highways, traverses may replace the triangulation. In establishing control for a city survey, where many control points are needed, the traverse can frequently be used to advantage.

In deciding between triangulation and traverse, the following considerations should be kept in mind. Traverse stations will usually be more available for the control of local surveys than will triangulation stations, since the latter are generally located in places more difficult of access. Because of this inaccessibility, the transportation of materials for the erection of towers and signals may be impossible and thus preclude the use of triangulation. On the other hand, triangulation stations are usually on high ground, and therefore are visible from much larger areas than are traverse stations. Besides, an arc of triangulation covers a belt of country at least 10 or 15 miles wide whereas the traverse is just a single line.

The electronic distance-measuring systems are successfully employed in measuring the lengths of lines in precise traverses. Their use, however, is restricted to traverses composed of lines at least 500 ft in length. The amount of time needed to measure a line 500 ft long with the EDM is about the same as that required to measure the same line by using a tape. On shorter lines, taping the lines will prove more economical. Also, the inherent instrument errors, representing only a very small fraction of the length of a long line, are too large for lines under 500 ft in length.

As in the case of triangulation and trilateration, traverses are classified as first order, second order, or third order, according to the degree of precision attained. The requirements for the various orders are shown in Table 10-10.

No check on the precision of the traverse is available until the traverse has been closed and the latitudes and departures and the closures have been computed. The discrepancy is sometimes of such size that it is difficult to tell whether it is caused by a blunder or by an accumulation of small errors in the traverse. When a blunder has been made, it is often difficult to locate the incorrect value and it may be necessary to rerun a large part or all of the traverse line. In triangulation as each quadrilateral is completed in the field, there are checks on both the angles and the lengths. Furthermore, the office computations for the adjustment of triangulation have automatic mathematical checks that are lacking in the computations for a traverse.

The lengths of lines used on traverses will, as a rule, be shorter than those for triangulation of the same order of precision, 5 miles being about the maximum length. The angles in a traverse are measured in the same manner as triangulation angles, that is, either by the direction method or by repetition. The measurement of the sides is conducted in the same manner as for base lines.

TABLE 10-10. Classification of Traverses

CLASSIFICATION	FIRST ORDER	SECOND ORDER		THIRD ORDER	
		CLASS I	CLASS II	CLASS I	CLASS II
RECOMMENDED SPACING OF PRINCIPAL STATIONS	Network stations 10–15 km. Other surveys seldom less than 3 km.	Principal stations seldom less than 4 km except in metropolitan area surveys where the limitation is 0.3 km.	Principal stations seldom less than 2 km except in metropolitan area surveys where the limitation is 0.2 km.	Seldom less than 0.1 km in tertiary surveys in metropolitan area surveys. As required for other surveys.	
HORIZONTAL DIRECTIONS OR ANGLES					
Instrument	$0''.2$	$0''.2 \big\}$ or $\big\{ 1''.0$	$0''.2 \big\}$ or $\big\{ 1''.0$	$1''.0$	$1''.0$
Number of observations	16	8 } { 12*	6 } { 8*	4	2
Rejection limit from mean	$4''$	$4''$ $5''$	$4''$ $5''$	$5''$	$5''$
		*May be reduced to 8 and 4, respectively, in metropolitan areas.			
LENGTH MEASUREMENTS					
Standard error	1 part in 600,000	1 part in 300,000	1 part in 120,000	1 part in 60,000	1 part in 30,000

RECIPROCAL VERTICAL ANGLE OBSERVATIONS

	3D/R−10″	3D/R−10″	2D/R−10″	2D/R−10″	2D/R−20″
Number of and spread between observations	3D/R−10″	3D/R−10″	2D/R−10″	2D/R−10″	2D/R−20″
Number of stations between known elevations	4–6	6–8	8–10	10–15	15–20
ASTRO AZIMUTHS					
Number of courses between azimuth checks	5–6	10–12	15–20	20–25	30–40
Number of observations/night	16	16	12	8	4
Number of nights	2	2	1	1	1
Standard error	$0″.45$	$0″.45$	$1″.5$	$3″.0$	$8″.0$
Azimuth closure at azimuth check point not to exceed	$1″.0$ per station or $2″ \sqrt{N}$	$1″.5$ per station or $3″ \sqrt{N}$. Metropolitan area surveys seldom to exceed $2″.0$ per station or $3″ \sqrt{N}$	$2″.0$ per station or $6″ \sqrt{N}$. Metropolitan area surveys seldom to exceed $4″.0$ per station or $8″ \sqrt{N}$	$3″.0$ per station or $10″ \sqrt{N}$. Metropolitan area surveys seldom to exceed $6″$ per station or $15″ \sqrt{N}$	$8″$ per station or $30″ \sqrt{N}$
POSITION CLOSURE					
After azimuth adjustment	0.04 m $\sqrt{K}$ or 1:100.000	$0.08 \sqrt{K}$ or 1:50,000	0.2 m $\sqrt{K}$ or 1:20,000	0.4 m $\sqrt{K}$ or 1:10,000	0.8 m $\sqrt{K}$ or 1:5,000

10-28. TRILATERATION A triangulation system can be converted to a pure trilateration system by measuring the lengths of the lines directly, using the EDM's, without measuring any horizontal angles in the network. The net can then be computed and adjusted in order to obtain the coordinates of the stations. However, in order to maintain the accuracy of the azimuths of the lines in the trilateration net, astronomical observations are made at selected stations. These measured azimuths impose conditions that must be satisfied in the adjustment process.

Trilateration is frequently combined with triangulation in order to strengthen a net that may have serious deficiencies in geometric conditioning; or trilaterated lines are used as check bases in a chain of figures in order to meet the criteria for strength of figures. It is, or course, possible to observe a net completely as both triangulation and trilateration. Such a hybrid network imposes several rigid geometric conditions that must be met in the adjustment of the net. However, in certain instances this rigidity is necessary. For example, the measurement of very small displacements due to earthquake fault movement must be made over fairly large areas. These measurements must be duplicated periodically with the same degree of reliability in order to reflect earth movement and to avoid measuring errors. The internal accuracy and reliability of such a network is greatly enhanced by a hybrid triangulation-trilateration system.

Before the lengths of the lines in the trilateration net can be used in any subsequent computations, their slope lengths determined by the instrument and corrected for atmospheric conditions must be reduced to the corresponding sea-level distances. Just as the slope measurements discussed in Sections 2-11 and 2-12 require auxiliary measurements to determine slope corrections, so do the lines in the trilateration net. The auxiliary data are either reciprocal vertical angles measured at the two ends of each line or the elevations of the two ends of each line.

10-29. REDUCTION OF SLOPE DISTANCE BY VERTICAL ANGLES In Fig. 10-15, the positive angle α and the negative vertical angle β are measured from stations A and B, respectively. Vertical distances and angles have been greatly exaggerated in the diagram. The angle e is the refraction angle of the line of sight, assumed to be the same at both stations. It is further assumed that the height of the instrument above the ground and the height of the signal above the ground in each instance are the same, or that the two angles have been reduced to an equivalent situation (see Section 2-30).

The angle subtended at the center of the earth between the vertical lines through A and B is θ. In triangle OVB the angle at V is $90° - \theta$, as shown, because the angle at B is $90°$.

The distance G is measured along the refracted line from A to B by the EDM. This distance, within the range of the instrument, can be taken as the

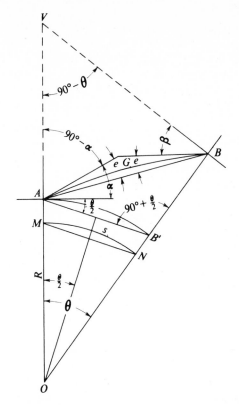

Figure 10-15. Reduction to sea-level distance by vertical angles.

straight-line slope distance *AB*. The difference between the two distances in 100 miles is less than a foot, decreasing to a negligible amount in the range of the instruments.

In the triangle *AVB*

$$(\beta + e) + (90° - \theta) + [(90° - \alpha) + e] = 180°$$

or

$$2e = \alpha + \theta - \beta$$

and

$$e = \frac{\alpha}{2} - \frac{\beta}{2} + \frac{\theta}{2}$$

In the triangle ABB'

$$\angle BAB' = \alpha - e + \frac{\theta}{2} = \alpha - \left(\frac{\alpha}{2} - \frac{\beta}{2} + \frac{\theta}{2}\right) + \frac{\theta}{2}$$

or

$$\angle BAB' = \frac{\alpha + \beta}{2} \tag{10-18}$$

Also the distance AB' is very nearly

$$AB' = AB \cos BAB' \quad \text{(approx)} \tag{10-19}$$

With the approximate value of AB', the value of θ can be obtained with sufficient accuracy by the relationship

$$\sin \frac{\theta}{2} = \frac{AB'}{2R} \tag{10-20}$$

in which R is the radius of the earth for the latitude of the area in the direction of the line.

In the triangle ABB'

$$\angle ABB' = 90° - \frac{\theta}{2} - \angle BAB' \tag{10-21}$$

The triangle ABB' can now be solved by the law of sines to give a more exact value of the distance AB'. Thus

$$AB' = \frac{AB \sin ABB'}{\sin[90° + (\theta/2)]} \tag{10-22}$$

The distance MN is the length of a chord connecting the sea-level positions of stations A and B, and is less than the length of the chord AB'. The amount to be subtracted from AB' is determined by an analysis of the sea-level correction given in Section 10-13. If C_1 denotes the amount by which the chord AB' must be reduced to obtain the chord MN,

$$C_1 = AB' \frac{h_A}{R} \tag{10-23}$$

in which h_A is the elevation of station A, and R is the radius of the earth for the latitude of the area in the direction of the measured line.

The sea-level length s is greater than the corresponding chord length MN. The amount to be added to MN to obtain s is found as follows:

$$MN = 2R \sin \frac{\theta}{2} = 2R\left(\frac{\theta}{2} - \frac{\theta^3}{48} + \frac{\theta^5}{3840} - \cdots\right)$$

$$s = R\theta = 2R\frac{\theta}{2}$$

$$s - MN = 2R\frac{\theta}{2} - 2R\frac{\theta}{2} + 2R\frac{\theta^3}{48} - 2R\frac{\theta^5}{3840}$$

If the last term is neglected, the result is

$$C_2 = R\frac{\theta^3}{24} = \frac{\overline{MN}^3}{24R^2} \tag{10-24}$$

in which C_2 is the amount to be added to the chord length to obtain the sea-level length of the line, R is the radius of the earth expressed in the same units as the measured distance, and θ is the angle subtended by the line at the center of the earth, expressed in radians.

The value of C_2 is about 14 ft in a distance of 100 miles and decreases to 1 ppm of the measured length at about 20 miles. Thus for distances less than 20 miles, the chord-to-arc correction given by Eq. (10-24) can be neglected.

The presentation given above for reducing slope distances by reciprocal vertical angles can be simplified by computing the effect of earth curvature as an angular value per 1000 ft of distance and then using this small angle in the reduction computations. In Fig. 10-16 the effect of curvature in 1 mile is shown to be 0.667 ft, in accordance with Eq. (3-1). The value 0.667 is computed for an average radius of curvature. The actual value is dependent on the radius of curvature of the earth along the line in question and will therefore vary a small amount from place to place.

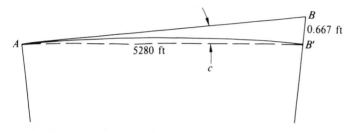

Figure 10-16. Angle of curvature.

The linear effect of curvature, BB' in Fig. 10-16, is converted to an angular value c by dividing BB' by the distance AB, giving

$$c'' = \frac{0.667}{5280 \text{ arc } 1''} = 26.06''/\text{mile}$$

or

$$c'' = 4.935''/1000 \text{ ft of distance} \qquad (10\text{-}25a)$$

In the metric system

$$c'' = \frac{0.0785}{1000 \text{ arc } 1''} = 16.19''/\text{km} \qquad (10\text{-}25b)$$

In the triangle ABB' of Fig. 10-15 the angle c is the same as $\theta/2$. By Eq. (10-18), $\angle BAB'$ is the mean of the reciprocal vertical angles and the angle at B' is $90° + c$. Thus the angle at B is $90° - c - \angle BAB'$, which corresponds to Eq. (10-21).

The triangle is solved by the law of sines to give the chord distance AB', which is then reduced to sea-level chord by Eq. (10-23), and to the sea-level length by Eq. (10-24).

Example 10-3

The slope distance AB in Fig. 10-15, corrected for meteorological conditions and instrument constants is 76,963.54 ft. The vertical angle at A is $+3°$ $02'$ $05''$, and at B is $-3°$ $12'$ $55''$. The elevation of A is 1056.67 ft. What is the sea-level length MN? (Note that these are the same data used in Example 3-1.)

Solution: The angle of curvature is, by Eq. (10-25a), $4.935 \times 76.963 = 379.8''$ or $6'$ $20''$. Angle $BAB' = (3° 02' 05'' + 3° 12' 55'')/2 = 3° 07' 30''$. The angle at $B' = 90° + 06' 20'' = 90° 06' 20''$. The angle at B is $90° - (3° 07' 30'' + 0° 06' 20'') = 86° 46' 10''$. The distance AB' is by the law of sines

$$AB' = \frac{76,963.54 \sin 86° 46' 10''}{\sin 90° 06' 20''} = 76,841.36 \text{ ft}$$

Assuming a mean radius of curvature of 20,906,000 ft, the reduction of the chord AB to the chord MN is, by Eq. (10-23), $76,841.36 \times 1056.67/20,906,000 = 3.88$ ft. Thus $MN = 76,841.36 - 3.88 = 76,837.48$ ft. The corresponding sea-level distance is obtained by adding the value of C_2 given by Eq. (10-24) to MN. Thus

$$C_2 = \frac{76,837^3}{24 \times 20,906,000^2} = 0.04 \text{ ft}$$

which is quite negligible. The sea level length is then

$$76,837.48 + 0.04 = 76,837.52 \text{ ft}$$

If reciprocal vertical angles have not been measured, then a value for atmospheric refraction must be assumed. If this is taken to be about 14% of curvature, then the combined curvature and refraction angle can be taken as $(c + r) = 4.935 \times 0.86 = 4.24''/1000$ ft, or in the metric system $16.19 \times 0.86 = 13.92''/\text{km}$. Assuming the vertical angle at the lower station to have been measured, then angle $BAB' = \alpha + (c + r)$. If the vertical angle were measured at the upper station, then angle $BAB' = |\beta + (c + r)|$. The vertical angle β is always negative.

10-30. REDUCTION OF SLOPE DISTANCE BY STATION ELEVATIONS
In Fig. 10-17, the elevations of A and B are shown. The distance AB is again assumed to be the straight-line distance measured with the EDM. This distance must be reduced to the chord distance AB'. The vertical distance BB' is the difference in elevation between A and B. Since the slopes are usually not great when trilateration is suitable, Eq. (2-6) can be used to determine the difference between the slope distance AB and the distance AC. Since $AB = AD$, then $AB - AC = CD$. By Eq. (2-6),

$$CD = \frac{\overline{BB'}^2}{2AB} = \frac{(\Delta h)^2}{2G} \tag{10-26}$$

in which Δh is the difference in elevation, in feet or metres, between the two ends of the line, and G is the measured slope distance, also in feet or metres. If a line is 5 miles long and the difference in elevation between the two ends is 1000 ft, the last term in Eq. (2-5) amounts to about 0.07 ft. So, for shorter lines and large differences of elevation, this term should be evaluated.

The angle $\theta/2$ in Fig. 10-17 may be obtained by Eq. (10-20). But the value of AB' is not known. Its approximate value, however, is AC, which can be found by subtracting the distance CD computed by Eq. (10-26) from the measured distance G; and the distance AC can be used in Eq. (10-20) instead of AB'. Also

$$\sin \frac{\theta}{2} = \frac{B'C}{BB'} = \frac{B'C}{\Delta h}$$

and

$$B'C = \Delta h \sin \frac{\theta}{2} \tag{10-27}$$

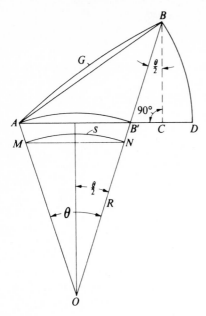

Figure 10-17. Reduction to sea-level distance by station elevations.

Then the distance $AB' = AB - (B'C + CD)$. Finally, the distance AB' is reduced to the sea-level distance by applying Eqs. (10-23) and (10-24).

The reduction using difference in elevation can be simplified by assuming that the angle $\theta/2$ in Fig. 10-17 is the curvature angle c'' given by Eq. (10-25) as was done in Section 10-29. Then the approximate value of AB' is

$$AB' = \sqrt{G^2 - \overline{BB'}^2} - BB' \sin c'' \qquad (10\text{-}28)$$

Since the value of c'' in the second part of the right side of Eq. (10-28) assumes that AB equals AB', the solution may require an iteration if the difference in elevation is substantial.

Example 10-4

The slope distance between two points A and B, corrected for meteorological and instrumental effects is 97,542.20 ft. The elevation of A is 4225.5 ft, and the elevation of B is 13,724.0 ft. What is the reduced sea-level distance?

Solution: The difference in elevation between the two points, BB of Fig. 10-17, is 9498.5 ft. The curvature angle c'' is $4.935 \times 97.54 = 481'' = 0° 08' 01''$. The approximate value of AB' is, by Eq. (10-28),

$$AB' = \sqrt{97,542.20^2 - 9,498.5^2} - 9498.5 \sin 0° 08' 01''$$

or

$$AB' = 97,078.62 - 22.15 = 97,056.47 \text{ ft}$$

The approximate value of AB' is now used to obtain a second value of c''. Thus, $c'' = 4.935 \times 97.06 = 479'' = 0° 07' 59''$. Then

$$AB' = 97,078.62 - 9498.5 \sin 0° 07' 59'' = 97,056.56 \text{ ft}$$

The difference between the first and second value of AB' is 0.09 ft, which is less than 1 ppm of the measured distance.

The reduction to sea level is, by Eq. (10-23), $97056.56 \times 4225.5/20,906,000 = 19.62$ ft, and the reduced chord MN is $97056.56 - 19.62 = 97036.94$ ft. Since the distance is less than 20 miles, the chord-to-arc correction given by Eq. (10-24) can be neglected. The sea-level distance is thus 97,036.94 ft.

10-31. ADJUSTMENT OF TRILATERATION Although there are several ways in which a quadrilateral or a network of figures in a trilateration system can be adjusted, one that can be used when extreme accuracy is not required is as follows: The angles in each of the triangles are computed from the lengths of the sides, and then these angles are adjusted by methods given in Section 10-23. Formulas for the solution of oblique triangles where all the sides are known are given in Table C-3 of Appendix C. As a check on the computed angles, the three angles in any triangle must add up to 180° exactly, whether or not spherical excess exists. If the spherical angles in a large triangle are needed for computation of spherical coordinates, one-third of the spherical excess is added to each computed angle.

Any approximate adjustment of trilateration will produce inconsistencies in subsequent computations. Therefore if the work is on a large project in which high accuracy is to be obtained, then a least-squares method of adjustment must be investigated. One such method is given in Appendix A for the adjustment of a small net. This method, called *variation of coordinates*, involves the prior determination of a set of approximate coordinates of all the points to be adjusted. Thus a preliminary computation of the angles in a limited number of the triangles is required so that preliminary position computations can be made. These computations follow the procedure outlined in Section 10-25 if the net is on a plane coordinate system.

One difficult problem in adjusting a trilateration net is that of assigning weights to the measured lengths. The standard errors for relatively short lines may be just as great as those for long lines because of the inherent errors in the measuring system. The order of accuracy of a long line is higher than that of a short line because these inherent errors are rather constant. For example, suppose that an inherent error of 2 in. exists in the EDM. Then the accuracy ratio for a line 1 mile long, is one part in 31,680, whereas that for a line 20 miles long, is one part in 633,600. If the standard error of each line can be determined by repeated measurements of each line, then the weights can be assumed to be inversely proportional to the squares of the standard errors in accordance with Eq. (5-28). For a limited net with relatively few lines, all of which have lengths of the same order of magnitude, a unit weight can be given to each line.

10-32. COMPUTATION OF LENGTHS OF SIDES If an approximate method of adjustment of the computed angles is used first, the lengths of the sides must be recomputed to be consistent with the adjusted angles. Before this recalculation can be performed, the strongest triangle in each of the quadrilaterals must be selected, and one side of this triangle must be held fixed and equal to the reduced sea-level distance. If, however, the trilateration begins from a line joining two points with fixed positions, then the fixed length of that line will be used to compute the sides.

If the trilateration has been adjusted by the method of variation of coordinates, the solution gives the adjusted coordinates of all the points. If the adjusted lengths are desired, they can be obtained by inversing each line by Eq. (8-12). The adjusted azimuths are obtained by Eq. (8-13).

10-33. COMPUTATION OF PLANE COORDINATES IN TRILATERATION If an approximate adjustment is used, the positions of the stations in the trilateration system may be computed by applying the procedures described in Section 10-25.

10-34. INTERSECTION The method of intersection mentioned in Section 10-1 is applicable to a situation in which a point is to be established in the network of a triangulation system, but in which the point is not occupied. The simplest case is that in which two angles are measured, one from each end of a known line. In Fig. 10-18 point P is observed from the two

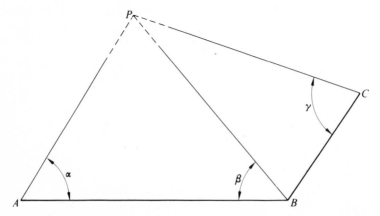

Figure 10-18. Intersection.

known points A and B where the angles α and β are measured. The azimuth of AP is then the azimuth of AB minus angle α. The azimuth of BP is the azimuth of BA plus angle β. The coordinates of P are then obtained by putting the two lines AP and BP in the form of Eq. (8-16) and solving the resulting pair of equations to obtain X_P and Y_P.

Example 10-5

The coordinates of A in Fig. 10-18 are $X_A = 1,455,460.24$ and $Y_A = 640,276.08$; and the coordinates of B are $X_B = 1,463,876.62$ and $Y_B = 637,406.51$. The angle $\alpha = 74° 29' 16''$, and angle $\beta = 37° 46' 28''$. Compute the coordinates of P.

Solution: By Eq. (8-13),

$$\tan A_{AB} = \frac{X_B - X_A}{Y_B - Y_A}$$

$$X_B = 1,463,876.62 \qquad\qquad Y_B = 637,406.51$$
$$X_A = 1,455,460.24 \qquad\qquad Y_A = 640,276.08$$

$$X_B - X_A = \quad 8,416.38 \qquad\quad Y_B - Y_A = -2,869.57$$

and

$$\tan A_{AB} = \frac{8,416.38}{-2,869.57} = -2.93297602$$

$$A_{AB} = \quad 108° 49' 36'' \qquad A_{BA} = \quad 288° 49' 36''$$
$$-\alpha = -74° 29' 16'' \qquad +\beta = +37° 46' 28''$$

$$A_{AP} = \quad 34° 20' 20'' \qquad A_{BP} = \quad 326° 36' 04''$$

Reducing X coordinates by 1,455,000 and Y coordinates by 637,000 for convenience gives $X'_A = 460.24$, $Y'_A = 3,276.08$, $X'_B = 8,876.62$, and $Y'_B = 406.51$. Then, by Eq. (8-16),

$$Y'_P - 3,276.08 = \cot 34° 20' 20'' \, (X'_P - 460.24) \tag{1a}$$

$$Y'_P - 406.51 = \cot 326° 36' 04'' \, (X'_P - 8,876.62) \tag{2a}$$

which reduce to

$$1.463810 \, X'_P - Y'_P = - \; 2,602.38 \tag{1b}$$
$$-1.516644 \, X'_P - Y'_P = -13,869.18 \tag{2b}$$

$$2.980454 \, X'_P \qquad\quad = \quad 11,266.80$$

$$X'_P = 3,780.23$$

then

$$Y'_P = 8,135.92$$

and

$$X_P = 1,455,000 + 3,780.23 = 1,458,780.23$$

$$Y_P = \quad 637,000 + 8,135.92 = \quad 645,135.92$$

Since there is no check on the position of the intersected point from observations only at A and B, a third point C is occupied, and the angle γ is measured. An intersection of lines AP and CP then gives an independent check on the computed position of the intersection point. If the intersection point is observed from more than the minimum of two control points, then its most probable position can be determined by the method of least squares. The variation of coordinates method is a convenient solution to a conditioned intersection. This is explained in Section A-9 of Appendix A.

10-35. RESECTION The method of resection is a convenient way of determining the position of an unknown point by occupying the point and measuring the horizontal angles between at least three, and preferably more, control points. The fact that the control points are not occupied, except perhaps to erect signals, makes the determination of the point somewhat risky, even when more than the minimum of three control stations are observed.

There are numerous approaches to the solution of the resection problem, and the interested reader should consult the readings at the end of this chapter for further study on the subject. The method presented here is but one of the available methods. The solution is set up for the hand or desk computer, and is shown in Table 10-11.

In Fig. 10-19, A, B, and C are points whose positions (coordinates) are known and are visible from the unknown station P. If station P is occupied and the angles p'' and p' are measured, then these angles and the coordinates of the control points are sufficient to determine the position of P. The lengths and azimuths of the lines AB, BC, and CA are computed from the coordinates of these three points. The angle at A from C to B may be computed from the azimuths of the lines AC and AB. Let the angle at B from A to P be denoted by x and the angle at C from P to A be denoted by y, as shown in Fig. 10-19. Then, since the sum of the interior angles in a four-sided figure is 360°,

$$x + y + A + p'' + p' = 360°$$

Therefore

$$x + y = 360° - (A + p'' + p')$$

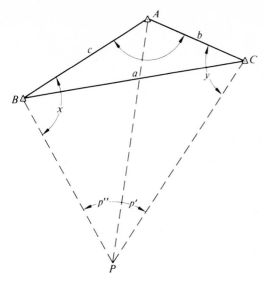

Figure 10-19. Three-point resection.

and

$$\tfrac{1}{2}(x + y) = 180° - \tfrac{1}{2}(A + p'' + p') \tag{10-29}$$

Furthermore, $\sin x/AP = \sin p''/c$ and $\sin y/AP = \sin p'/b$. So

$$AP = \frac{c \sin x}{\sin p''} = \frac{b \sin y}{\sin p'}$$

Let $\sin y/\sin x = \tan Z$. Then

$$\tan Z = \frac{c \sin p'}{b \sin p''} \tag{10-30}$$

When one is added to each side of the equation $\sin y/\sin x = \tan Z$, the result is

$$\frac{\sin y}{\sin x} + 1 = \tan Z + 1$$

or

$$\frac{\sin x + \sin y}{\sin x} = \tan Z + \tan 45° \tag{10-31}$$

When each side of the same equation is subtracted from one, the result is

$$1 - \frac{\sin y}{\sin x} = 1 - \tan Z$$

or

$$\frac{\sin x - \sin y}{\sin x} = 1 - \tan Z \tan 45° \qquad (10\text{-}32)$$

Dividing Eq. (10-32) by Eq. (10-31) gives

$$\frac{\sin x - \sin y}{\sin x + \sin y} = \frac{1 - \tan Z \tan 45°}{\tan Z + \tan 45°}$$

$$= \frac{1}{\tan(Z + 45°)} = \cot(Z + 45°)$$

The left-hand side of Eq. (10-32) may be expressed in terms of the product of two functions, rather than in terms of the sum of two functions. Thus

$$\frac{2 \cos \frac{1}{2}(x + y) \sin \frac{1}{2}(x - y)}{2 \sin \frac{1}{2}(x + y) \cos \frac{1}{2}(x - y)} = \cot(Z + 45°)$$

or

$$\cot \tfrac{1}{2}(x + y) \tan \tfrac{1}{2}(x - y) = \cot(Z + 45°)$$

Hence

$$\tan \tfrac{1}{2}(x - y) = \cot(Z + 45°) \tan \tfrac{1}{2}(x + y) \qquad (10\text{-}33)$$

When the triangle ABC has been solved for the angle at A and the lengths b and c, the angle $\frac{1}{2}(x + y)$ is computed by Eq. (10-29). Then $\tan Z$ is computed by Eq. (10-30), and the angle $\frac{1}{2}(x - y)$ is computed by Eq. (10-33). Adding $\frac{1}{2}(x + y)$ and $\frac{1}{2}(x - y)$ gives x; subtracting $\frac{1}{2}(x - y)$ from $\frac{1}{2}(x + y)$ gives y. These two angles complete the data necessary to solve the triangles PBA and PAC, the length of the common side PA affording a check on the computations. The azimuths of the three lines BP, AP, and CP are determined from the known angles and azimuths. Their lengths are determined from the triangle solutions. The latitude and departure of any one of the lines are computed to determine the coordinates of P. The latitude and departure of another of the lines will afford a check on the coordinates of P.

A numerical example, indicating the manner in which the computations for solving a three-point resection may be tabulated, is shown in Table 10-11.

TABLE 10-11. Computation of Three-Point Resection

DETERMINATION OF ANGLES

$$c = 1642.83 \text{ ft} \qquad p' = 38°\,12'\,20'' \qquad \sin p' = 0.618\ 4846$$
$$b = 1076.44 \text{ ft} \qquad p'' = 49°\,36'\,10'' \qquad \sin p'' = 0.761\ 5697$$
$$A = 141°\,28'\,30''$$
$$\text{sum} = \overline{229°\,17'\,00''}$$
$$\tfrac{1}{2}\,\text{sum} = 114°\,38'\,30''$$

$$\tfrac{1}{2}(x+y) = 180° - \tfrac{1}{2}\,\text{sum} = 65°\,21'\,30'' \qquad \tan\tfrac{1}{2}(x+y) = 2.179\ 9995$$

$$\tan Z = \frac{c \sin p'}{b \sin p''} = \frac{1642.83 \times 0.618\ 4846}{1076.44 \times 0.761\ 5697} = 1.239\ 4301$$

$$Z = 51°\,06'\,09'' \qquad Z + 45° = 96°\,06'\,09'' \qquad \cot(Z+45°) = -0.106\ 9133$$

$$\tan\tfrac{1}{2}(x-y) = \cot(Z+45°)\tan\tfrac{1}{2}(x+y) = -0.106\ 9133 \times 2.179\ 9995 = -0.233\ 0709$$

$$\tfrac{1}{2}(x+y) = 65°\,21'\,30''$$
$$\tfrac{1}{2}(x-y) = -13°\,07'\,11''$$
$$x = \overline{52°\,14'\,19''}$$
$$y = 78°\,28'\,41''$$

SOLUTION OF TRIANGLES (see Table 10-7)

TRIANGLE *PBA*

SIDES	STATIONS	ANGLES	DISTANCES
$c = BA$			1642.83
	$P = p''$	49° 36' 10"	0.761 5697
	$B = x$	52° 14' 19"	0.790 5679
	A	(78° 09' 31")	0.978 7194
AP		$\overline{180°\ 00'\ 00''}$	1705.38
PB			2111.26

TRIANGLE *PAC*

SIDES	STATIONS	ANGLES	DISTANCES
$b = AC$			1076.44
	$P = p'$	38° 12' 20"	0.618 4846
	A	(63° 18' 59")	0.893 4999
	$C = y$	78° 28' 41"	0.979 8483
CP		$\overline{180°\ 00'\ 00''}$	1555.09
PA			1705.38 Checks

Due regard must be given to the algebraic sign of $\tan \frac{1}{2}(x + y)$ and also of $\cot(Z + 45°)$ in the solution of Eq. (10-33). This determines whether $\frac{1}{2}(x - y)$ is positive or negative. If $\frac{1}{2}(x + y)$ is 90°, then the position of the unknown point is indeterminate since it lies on the circle passing through the three control stations.

In Fig. 10-20(a), $\frac{1}{2}(x + y) = 180° - \frac{1}{2}(A + p'' + p')$, as in Fig. 10-19. However, in Fig. 10-20(b), $\frac{1}{2}(x + y) = \frac{1}{2}(A - p'' - p')$. Otherwise the solution for x and y is the same for the conditions in Fig. 10-20(b) as for these in Figs. 10-19 and 10-20(a).

A three-point resection can always be made to check mathematically, but *the mathematical checks do not provide a check on the field work or on the control points used.* The field angles may be checked by measuring the angles p'', p', and the exterior angle at P, since the sum of these three angles should be 360°.

The value of a three-point resection may be increased considerably by sighting on a *fourth* control point. The result is a conditioned three-point resection, since there are two independent solutions that should give the same position for the unknown point. If the discrepancy between the two computed positions of the point based on a conditioned resection is unreasonably large, then the cause may be that the observer has sighted on an erroneously identified control point, the published position of a control point was wrong, a mistake was made in measuring the angles, or a mistake was made in the computations. If the control points are correct and the discrepancy is within reason, then its amount will give an indication of the accuracy of the field work. The least-squares adjustment of an overdetermined resection is explained in Section A-10 of Appendix A.

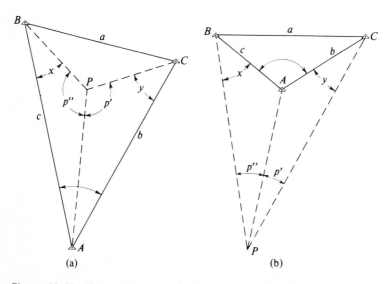

Figure 10-20. Three-point resection.

10-36. SATELLITE DOPPLER POSITIONING The three-dimensional position of a survey point can be determined by observations of electromagnetic signals transmitted by the Navy Navigation Satellite System known as Transit. It is based on the Doppler effect created by the change in frequency of the transmitted signals received at the survey point caused by a change in the position of the satellite with respect to the survey point.

The Doppler effect will be explained first by reference to sound waves, and will then be applied to electromagnetic waves. Sound travels in air at approximately 331.3 m/s. The pitch of the sound increases with the frequency of the sound waves. In Fig. 10-21(a), a source at S is emitting sound waves at a frequency F_T of 300 Hz. The waves are received at a receiver fixed at R. As long as the source is stationary, the frequency F_R received at R is the same as the frequency emitted at S. That is, $F_R = F_T$. If the source is moved toward the receiver at a velocity V_S of say 20 m/s as shown in Fig. 10-21(b), the frequency at which the waves are received at R will be higher than that at which they are emitted at the moving source. The sound wavelength is $331.3/300 = 1.104$ m. Therefore, the source moves through $V_S/\lambda_T = 20/1.104$ wavelengths per second. This increases the frequency at which the sound is being received at R by $20/1.104 = 18.12$ Hz, causing the pitch to be higher than when the source was stationary. The frequency received at R is thus $F_R = F_T + V_S/\lambda_T$ or 318.12 Hz.

In Fig. 10-21(c), the source is moving away from R at a velocity of 20 m/s; the frequency at which the waves are received at R will be lower than that at which they are emitted by the moving source. The change in frequency is again 18.12 Hz. Thus $F_R = F_T - V_S/\lambda_T$ or 281.88 Hz. The situation in Fig. 10-21(b) and (c) is diagrammed in Fig. 10-22. This is the phenomenon experienced by listening to the horn of a passing train. As the train approaches the observer at a constant velocity, the pitch is higher than if it were stationary. As the train passes the observer, the pitch suddenly drops. This is the well-known Doppler principle. The change in frequency is called the *Doppler shift*.

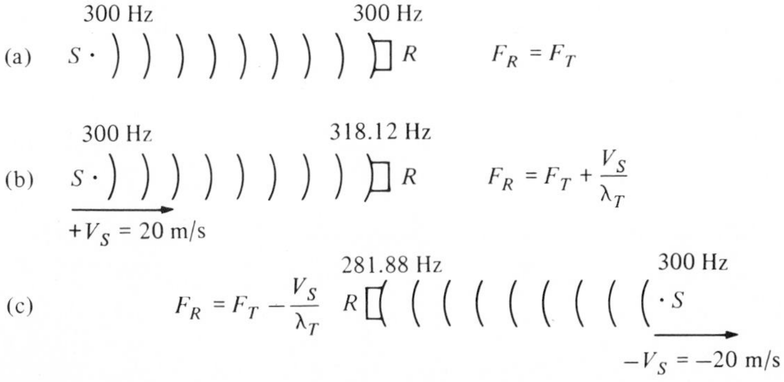

Figure 10-21. Doppler effect using sound waves.

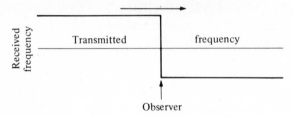

Figure 10-22. Change in received frequency due to Doppler effect.

In the above example, the Doppler shift is $+18.12$ Hz as the source approaches the observer, and is -18.12 Hz as the source moves away from the observer.

Recognizing that $F_T = V_T/\lambda_T$ and the Doppler shift F_D is V_S/λ_T, then multiplying the numerator and the denominator of each by an arbitrary time interval Δt gives

$$F_T = \frac{V_T \, \Delta t}{\lambda_T \, \Delta t} \quad \text{and} \quad F_D = \frac{V_S \, \Delta t}{\lambda_T \, \Delta t} \tag{10-34}$$

The frequency F_R received at the fixed point of observation is then

$$F_R = F_T + F_D = \frac{V_T \, \Delta t}{\lambda_T \, \Delta t} + \frac{V_S \, \Delta t}{\lambda_T \, \Delta t}$$

or

$$F_R = \left(\frac{V_T \, \Delta t/\lambda_T}{\Delta t} \right) + \left(\frac{V_S \, \Delta t/\lambda_T}{\Delta t} \right) \tag{10-35}$$

In Eq. (10-35), $V_T \, \Delta t$ is the distance that the transmitted wave travels in the time interval Δt; $V_S \, \Delta t$ is the distance covered by the traveling source along the line toward the observer during the same time interval. Dividing each of these by the wavelength λ_T gives the number of waves which reach the observer during this same time interval. Dividing by Δt gives the value of the frequency F_R received by the observer. The Doppler shift is then the second part of Eq. (10-35). That is,

$$F_D = F_R - F_T = \frac{V_S \, \Delta t/\lambda_T}{\Delta t} \tag{10-36}$$

In Fig. 10-23, the change in range Δr from the position at time t_1 to the position at time t_2 is the velocity of the source times the time interval Δt, that is, $\Delta r = V_S \, \Delta t$, or

$$V_S = \frac{\Delta r}{\Delta t} \tag{10-37}$$

Figure 10-23. Change in distance in time interval $t_1 - t_2$.

From the basic wave equation, the wavelength λ_T is

$$\lambda_T = \frac{V_T}{F_T} \tag{10-38}$$

Substituting Eqs. (10-37) and (10-38) into Eq. (10-36) and rearranging gives

$$F_D\,\Delta t = \frac{\Delta r\,\Delta t\,F_T}{V_T\,\Delta t}$$

and

$$\Delta r = \frac{V_T}{F_T}\,F_D\,\Delta t \tag{10-39}$$

If the source is not traveling directly toward (or away from) the observer, and if the time interval Δt is taken small enough, then Δr is as shown in Fig. 10-24. If the two positions X_1, Y_1, and X_2, Y_2 of the source of transmission are known at times t_1 and t_2, the value of Δr can be expressed in two dimensions as the difference of the two distances r_1 and r_2. Thus, by Eq. (8-12),

$$\Delta r = [(X_1 - X_R)^2 + (Y_1 - Y_R)^2]^{1/2} - [(X_2 - X_R)^2 + (Y_2 - Y_R)^2]^{1/2} \tag{10-40}$$

Substituting Eq. (10-40) into Eq. (10-39) gives

$$F_D\,\Delta t = \frac{F_T}{V_T}\,\{[(X_1 - X_R)^2 + (Y_1 - Y_R)^2]^{1/2}$$

$$- [(X_2 - X_R)^2 + (Y_1 - Y_R)^2]^{1/2}\} \tag{10-41}$$

This equation in the two unknowns X_R and Y_R is a function of the Doppler shift, the time interval, the transmitted frequency, the velocity of propagation of the transmitted wave, and the known positions $X_1 Y_1$ and $X_2 Y_2$ of the source of transmission. If additional Doppler shift measurements are made

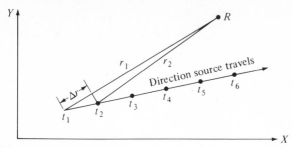

Figure 10-24. Change in distance at an angle with direction source moving.

as the source moves along its path, the position of the unknown point R can be determined by the method of least squares.

The principles discussed above are applied to the determination of the *three-dimensional* position of a ground survey station by means of signals received from passing satellites.

10-37. BEATS If a transmitter generates electromagnetic waves at a frequency F_T, a receiver receives these waves at a frequency F_R. If the receiver can generate a reference frequency F_G, the Doppler shift caused by the movement of the transmitting source can be determined by superimposing the incoming wave frequency F_R onto the reference frequency F_G. This superimposition creates a *beat* frequency as illustrated in Fig. 10-25. The frequency generated by the receiver is, as a numerical example, a constant 12 Hz, while that of the incoming wave is 15 Hz. This sets up a beat whose frequency F_B is $15 - 12 = 3$ Hz. The wave generated at the beat frequency is complex but can be represented by the dashed sine wave shown in Fig. 10-25. The beat phenomenon can be observed by striking two keys of a piano simultaneously and listening to the pulsating variation in loudness.

The beat frequency which is equal to the Doppler shift is detected in an electromagnetic wave receiver by a periodic increase or decrease in voltage. The number of beats in a given time interval Δt is called the *Doppler count N* for that interval.

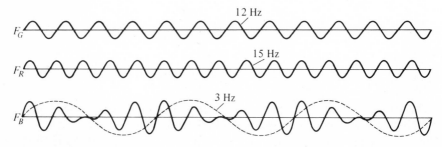

Figure 10-25. Beat frequency.

10-38. POSITIONING BY SATELLITES The Transit system mentioned in Section 10-36 is composed of five satellites which travel in polar orbits at an altitude of about 1075 km. The period of orbit is approximately 107 min. In the Arctic regions from 40 to 50 usable passes of the satellites can be obtained in a day, whereas only between 8 and 12 passes can be observed in the equatorial areas. Each satellite transmits electromagnetic waves at carrier frequencies of slightly less than 150 and 400 MHz. These carriers are phase modulated to send a stream of data in the form of coded messages to a receiver antenna set up at a station whose position is to be determined. Two such receivers in worldwide use for surveying applications using satellite signals are shown in Figs. 10-26 and 10-27. Every 2 min, the satellite sends messages which include all of the satellite orbital parameters necessary to determine the satellite's position in an earth centered three-dimensional coordinate system at any instant of time. A precise clock contained in the ground receiver is used to time these messages. The messages sent by the satellite are contained in the *broadcast ephemeris* whose predicted orbital parameters are updated every 12 hr based on observations made at four fixed ground stations, and transmitted to the satellite.

Figure 10-26. Magnavox MX1502 receiver. Courtesy of Magnavox Co.

Figure 10-27. JMR-1 receiver. Courtesy of JMR Instruments, Inc.

As the satellite passes over the ground station, the frequency F_R received at the station is compared with a constant reference frequency F_G generated by the receiver to determine the Doppler shift. This is shown schematically in Fig. 10-28. The reference frequency is high enough to prevent the Doppler shift from changing algebraic sign. The Doppler counts N_i, represented in Fig. 10-28, are obtained by summing up the beats which occur during each time interval. The receivers employed for surveying use a time interval dt of nominally 4.6 s which is controlled by the precise clock.

In Fig. 10-28, t_1 is the time that the signal is transmitted when the satellite is at position 1. Then $t_1 + r_1/c$ is the time at which the signal is received at the ground station, in which c is the velocity of the electromagnetic waves. Then t_2 and $t_2 + r_2/c$ have the same meaning for position 2 of the satellite. Summing over the time interval dt gives

$$N_1 = \int_{t_1+r_1/c}^{t_2+r_2/c} (F_G - F_R)\, dt$$

$$= \int_{t_1+r_1/c}^{t_2+r_2/c} F_G\, dt - \int_{t_1+r_1/c}^{t_2+r_2/c} F_R\, dt \qquad (10\text{-}42)$$

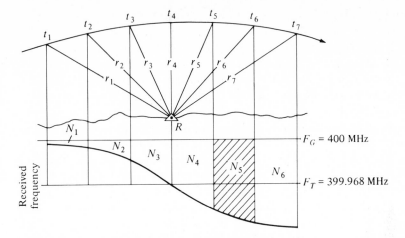

Figure 10-28. Doppler shift caused by passing satellite.

By the principle of *conservation of cycles*, the number of cycles received in the time interval from $t_1 + r_1/c$ to $t_2 + r_2/c$ must be the same as the number of cycles transmitted by the satellite in the time interval from t_1 to t_2. Equation (10-42) can therefore be written as

$$N_1 = \int_{t_1 + r_1/c}^{t_2 + r_2/c} F_G \, dt - \int_{t_1}^{t_2} F_T \, dt \qquad (10\text{-}43)$$

Assuming that F_G and F_T are constant while the satellite passes over the station, the integration in Eq. (10-43) gives

$$N_1 = F_G \left[(t_2 - t_1) + \frac{1}{c}(r_2 - r_1) \right] - F_T(t_2 - t_1)$$

or

$$N_1 = (F_G - F_T) \Delta t + \frac{F_G}{c} \Delta r \qquad (10\text{-}44)$$

Let X_1, Y_1, Z_1 and X_2, Y_2, Z_2 denote the known spatial coordinates of the satellite obtained from the data contained in the broadcast ephemeris, and let X_R, Y_R, Z_R denote the spatial coordinates of the receiver antenna. Then Eq. (10-44) can be written as

$$N_1 = \frac{F_G}{c} \{ [(X_1 - X_R)^2 + (Y_1 - Y_R)^2 + (Z_1 - Z_R)^2]^{1/2}$$

$$- [(X_2 - X_R)^2 + (Y_2 - Y_R)^2 + (Z_2 - Z_R)^2]^{1/2} \} + (F_G - F_T) \Delta t$$

$$(10\text{-}45)$$

Equation (10-45) in three dimensions is similar to Eq. (10-41) in two dimensions except for the term $(F_G - F_T) \Delta t$ which is known.

A series of Doppler counts $N_1, N_2, \ldots, N_n$ during one pass of a satellite is used to compute the values of X_R, Y_R, Z_R by the method of least squares. In the overall computations, the ionospheric refraction effects are eliminated by using both the 150 and the 400 MHz frequencies. The tropospheric refraction effects are reduced by making corrections for temperature, atmospheric pressure, and relative humidity measured at the ground station.

The position of the satellite at any instant of time obtained by the broadcast ephemeris can be in error by several metres in all three directions. Thus, the reduction of data from a single pass of the satellite would not be sufficient for survey accuracy. If enough passes of the satellites are observed, the results would tend to converge to an acceptable accuracy. If 100 passes were observed at a ground station, the standard deviation of the calculated position of the station would be on the order of 2 to 3 m. In midlatitude locations, this would require occupying the station for as long as 8 days.

Two of the satellites in the Transit system are observed from 20 stations in a global tracking network over 48-hr intervals. Their position at any instant of time, derived from these observations, are contained in the *precise ephemeris*, computed by the Naval Surface Weapons Center. The precise ephemeris is not, however, immediately available to the engineer or surveyor and thus the position of the ground point cannot be determined until a later date (as long as six months). Since the precise ephemeris is based on ground measurements, the position of the satellite is known to within about 2 to 3 m. Thus, better results would be obtained with fewer passes than when using the broadcast ephemeris. However, since only two (and sometimes only one) satellites are tracked, the field time would be increased in order to obtain a given number of passes.

Different observation and data reduction techniques can be employed to reduce the uncertainties in the position of the satellite, the effects of refraction, and other errors of the system. The method discussed in this section is called *point positioning*. In this technique, a single receiver occupies the ground station, and the ground coordinates are derived from several passes using either the broadcast or the precise ephemeris. Two other methods are employed to obtain better results in a shorter period of time. These are the *translocation* method and the *short-arc geodetic adjustment*.

10-39. TRANSLOCATION If two receivers are used, one can be placed at a base station whose position has been determined by triangulation or trilateration. A second receiver is located at a point whose position is to be determined. This is shown in Fig. 10-29 in which A is the known point and B is the point whose position is to be determined. Simultaneous observations are then made on the satellites. The coordinates of both points are determined by the point positioning method. However, since the system errors affect both points by the same amount (or nearly so), the *relative*

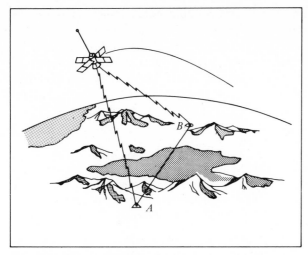

Figure 10-29. Translocation.

positions of the two points are quite accurate, even when using the broadcast ephemeris. The values of ΔX, ΔY, and ΔZ obtained between the two points are then added to the coordinates of the known station to obtain the position of the unknown station. Accuracies on the order of 3 m can be obtained with as few as 25 satellite passes using the broadcast ephemeris. If more than two receivers are used simultaneously, then more points can be located at the same time.

10-40. SHORT-ARC GEODETIC ADJUSTMENT

In the short arc geodetic adjustment, an overall simultaneous adjustment is made of all observed stations in a net together with the six orbital parameters contained in the ephemeris, and five error coefficients representing zero set, timing bias, frequency offset, frequency drift, and coefficient of refraction. In a typical network of ground points, several thousand normal equations (see Appendix A) must be solved in order to determine corrections to the orbital parameters and the error coefficients for each satellite pass and to compute the most probable position of each point. This large computer effort, however, results in the determination of the positions of the points to within less than a metre.

10-41. REDUCTION TO DESIRED COORDINATE SYSTEM

The X, Y, Z coordinates of a ground station determined from a single satellite by any of the methods previously discussed are with respect to an earth-centered Cartesian coordinate system defined by the plane of the satellite's orbit and its point of perigee. A transformation must then be made to place the station in an earth-centered coordinate system with the Z-axis in the direction of the mean polar axis and the X-axis passing through the Greenwich meridian. These axes are the basis for the ellipsoid defining the

World Geodetic System of 1972 (WGS-72). Another transformation must be made to bring the station into the desired reference ellipsoid used in the area of interest. This transformation establishes the elevation of the station. Finally, the plane rectangular coordinates of the station can be transformed into any desired system such as the transverse Mercator system discussed in Chapter 11.

PROBLEMS

10.1. Triangulation towers of equal heights are to be erected on opposite shores of a body of water 14 miles wide. The line of sight is to clear the water by 30 ft. The theodolite is presumed to sit directly atop each tower, and so no allowance need be made for the height of a tripod. Scaffolding is constructed around each tower but independent of the tower, so that the observer may move without disturbing the instrument. How high must the tower be?

10-2. Solve Problem 10-1 if the body of water is 18 miles wide.

10-3. Solve Problem 10-1 if the body of water is 10 miles wide.

10-4. A signal is to be erected across a broad, flat valley 12 miles from a triangulation station. If the theodolite stands on its tripod 5 ft above the triangulation station, how high must the signal be raised in order that the line of sight from the theodolite to the signal will just clear the intervening ground?

10-5. Solve Problem 10-4 if the signal is 8 miles from the triangulation station.

10-6. Solve Problem 10-4 if the signal is 16 miles from the triangulation station.

10-7. Determine R_1 and R_2 for each of the quadrilaterals shown in the accompanying illustration. The starting side is AB.

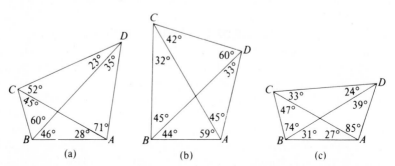

(a)   (b)   (c)

10-8. Determine the value of D and C for each of the figures shown in the illustration. The heavy line is the starting line. In (a) the center point is occupied. A dashed line going into a station means that a sight was made on that station but that the line was not sighted from that station. If all lines going into a station are dashed, the station is presumed not to have been occupied.

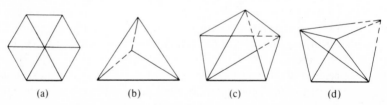

(a)   (b)   (c)   (d)

10-9. The coefficient of thermal expansion of a 50-m invar baseline tape is $3.60 \times 10^{-7}/°C$. What is the effect on a 12-km baseline measured with the tape if the temperature differs 12°C from standard? Express the answer to the nearest 0.001 m.

10-10. A 50-m invar tape is standardized under a 15-kg tension when supported at the 0–25–50 m points. What effect does a -0.8-kg systematic error in the applied tension have on a baseline that measures 15,000 m? $E = 2.0 \times 10^6$ kg/cm^2; $A = 0.040$ cm^2; $w = 25$ g/m.

10-11. The average elevation of a baseline is 1045 m and the length of the base is 17,875.224 m. What is the length reduced to sea level?

10-12. What is the sea-level length of a baseline that measures 12,848.653 m at an average elevation of 776 m?

10-13. A baseline between two baseline stations is measured in two sections along two tangents, one of which deflects 3° 22′ 12″ from the other. The length of the first section is 8762.333 m, and the length of the second section is 2979.550 m. What is the length of the line joining the two baseline stations?

10-14. In Problem 10-13 if the deflection is 2° 25′ 40″ and the two sections are 4,558.71 m and 18,584.06 m, what is the length of the baseline?

10-15. A direction theodolite is set up at E in the accompanying illustration. The following directions were observed:

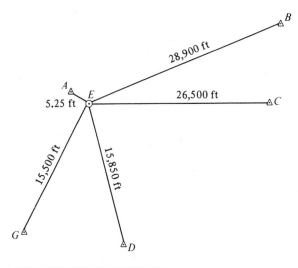

TO STATION	MEAN DIRECTION
A	254° 16′ 14.6″
B	26° 42′ 28.9″
C	47° 29′ 02.0″
D	123° 06′ 58.8″
G	166° 52′ 33.1″

Compute the directions from A to B, C, D, and G. Prepare an abstract of these directions, taking the direction from A to B as 0° 00′ 00.0″.

10-16. The hypotenuse of a 30°–60°–90° triangle is 25 miles long. The triangle is assumed to lie in latitude 45°. Compute the spherical excess to the nearest hundredth of a second of arc.

10-17. What is the spherical excess to the nearest 0.01″ of a 45°–45°–90° triangle if the hypotenuse is 20 km long? The triangle is assumed to be in latitude 45°.

10-18. In the quadrilateral represented in the accompanying illustration, all angles were measured with the same degree of precision, with the following results:

ANGLE	VALUE	ANGLE	VALUE
1	53° 54′ 53.5″	5	48° 41′ 49.8″
2	46° 38′ 32.8″	6	51° 51′ 34.4″
3	43° 39′ 49.8″	7	34° 22′ 24.8″
4	35° 46′ 38.0″	8	45° 03′ 52.8″

Adjust the quadrilateral by the approximate method of Section 10-23.

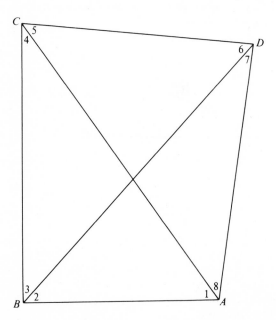

10-19. Adjust the quadrilateral given in Problem 10-18 by the method of condition equations given in Appendix A.

10-20. In the quadrilateral in Problem 10-18 the coordinates of A are $X_A = 1,495,056.52$ ft and $Y_A = 504,574.66$ ft, and the coordinates of B are $X_B = 1,497,509.00$ ft and $Y_B = 501,524.50$ ft. Compute the length and azimuth of AB. Then solve the triangle CAB. Finally, compute the coordinates of C and show a check. (Use the angles computed in Problem 10-18.)

10-21. Using the length of AC computed in Problem 10-20, solve the triangle DAC, and compute the coordinates of D. Show a check on the arithmetic.

10-22. In the quadrilateral represented in the accompanying diagram all angles were measured with the same degree of precision with the following results:

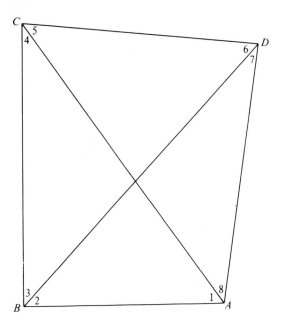

ANGLE	VALUE	ANGLE	VALUE
1	28° 46′ 16.2″	5	25° 15′ 07.2″
2	95° 49′ 21.2″	6	99° 20′ 33.9″
3	47° 40′ 30.6″	7	25° 36′ 34.7″
4	7° 43′ 50.4″	8	29° 47′ 46.0″

Adjust the quadrilateral by the approximate method of Section 10-23.

10-23. Adjust the quadrilateral given in Problem 10-22 by the method of condition equations given in Appendix A.

10-24. In the quadrilateral given in Problem 10-22 the coordinates of A are $X_A = 1,518,664.75$ ft and $Y_A = 473,340.01$ ft, and the coordinates of B are $X_B = 1,517,319.18$ ft and $Y_B = 474,602.18$ ft. Compute the length and azimuth of AB. Then solve the triangle DAB. Finally, compute the coordinates of D and show a check. (Use the angles computed in Problem 10-22.)

10-25. Using the length of BD computed in Problem 10-24, solve the triangle CDB, compute the coordinates of C, and show a check.

10-26. In the quadrilateral represented in the accompanying figure all angles were measured with the same degree of precision, with the following results:

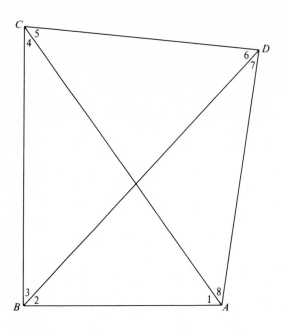

ANGLE	VALUE	ANGLE	VALUE
1	46° 32′ 02.1″	5	30° 43′ 17.1″
2	59° 29′ 00.7″	6	75° 17′ 40.5″
3	44° 57′ 36.8″	7	46° 07′ 47.4″
4	29° 01′ 22.5″	8	27° 51′ 13.0″

Adjust the quadrilateral by the approximate method of Section 10-23.

10-27. Adjust the quadrilateral given in Problem 10-26 by the method of condition equations given in Appendix A.

10-28. In the quadrilateral of Problem 10-26 the coordinates of A are $X_A = 1,522.680.20$ ft and $Y_A = 478,750.96$ ft, and the coordinates of B are $X_B = 1,519,708.08$ ft and $Y_B = 470,470.55$ ft. Compute the length and azimuth of AB. Then solve the triangle DAB. Finally, compute the coordinates of D and show a check. (Use angles computed in Problem 10-26.)

10-29. Using the length of BD computed in Problem 10-28, solve the triangle CDB, compute the coordinates of C, and show a check.

10-30. The distance AB in Fig. 10-15 is determined by EDM measurement to be 28,565.50 ft. Angle α is $+2°$ 12′ 22″, angle β is $-2°$ 16′ 06″. The elevation of A is 1733.50 ft. Compute the sea-level length MN.

10-31. If only angle α had been measured in Problem 10-30, compute the sea-level length MN.

10-32. The distance AB of Fig. 10-15 is measured by EDM as 14,853.55 m. Angle α is $+1°$ 49′ 18″, and angle β is $-1°$ 55′ 46″. The elevation of A is 842.55 m. Compute the sea-level length MN.

10-33. If only angle β had been measured in Problem 10-32, compute the sea-level length MN.

10-34. Trilateration station A is at an elevation of 576.055 m above sea level, and station B is at an elevation of 1693.740 m. The slope distance from A to B as measured by EDM is 20,454.50 m. Reduce this distance to sea level.

10-35. Trilateration station A is at an elevation of 1065.80 ft above sea level, and station B is 3450.22 ft above sea level. The slope distance between the two stations measured by EDM is 9753.53 ft. Reduce this distance to sea level.

10-36. In the quadrilateral diagram accompanying Problem 10-18, assume that only the sides are measured, using EDM, and reduced to the correct distances as follows:

LINE	REDUCED LENGTH (ft)
AC	6763.88
BC	5466.05
AD	5092.33
BD	6917.71
CD	4798.51

The fixed line AB is 3954.31 ft long. Using the law of cosines, compute angles 1 and 4 from triangle CAB, angles 2 and 7 from triangle DAB, angles 3 and 6 from triangle DBC, and angles 5 and 8 from triangle CDA. Then adjust the angles by the approximate method given in Section 10-23.

10-37. Solve Problem 10-36 with the following data. The fixed length of AB is 5414.17 ft.

LINE	REDUCED LENGTH (ft)
AC	10806.95
BC	8099.80
AD	6470.23
BD	7233.10
CD	5916.98

10-38. The coordinates of two control points A and B are: $X_A = 602,105.32$ m and $Y_A = 126,118.90$ m; $X_B = 601,048.82$ m and $Y_B = 125,613.48$ m. The clockwise angle at A from B to an unknown point P is $52°$ 18′ 46.2″; the counterclockwise angle at B from A to P is $37°$ 28′ 16.0″. Compute the coordinates of P by intersection.

10-39. The coordinates of two control points A and B are; $X_A = 473,261.25$ m and $Y_A = 220,535.70$ m; $X_B = 473,224.68$ m and $Y_B = 221,381.86$ m. The counterclockwise angle at A from B to P is $119°\ 52'\ 20.4''$; the clockwise angle at B from A to P is $23°\ 44'\ 50.0''$. Compute the coordinates of P by intersection.

10-40. The coordinates in feet of three control points are as follows:

POINT	Y	X
A	98,202.66	38,762.50
B	110,002.65	61,252.84
C	89,102.32	78,565.12

(a) From an unknown point P_1 lying southerly from A, the clockwise angle from A to B is measured as $22°\ 30'\ 14.6''$ and that from B to C as $29°\ 48'\ 50.6''$. Compute the coordinates of P_1.

(b) From a point P_2 lying easterly from A and inside the triangle formed by the control points, the clockwise angle from A to B is measured as $102°\ 28'\ 29.7''$ and that from B to C as $86°\ 20'\ 04.4''$. Compute the coordinates of P_2.

(c) From a point P_3 lying northerly from B, the clockwise angle from C to B is measured as $12°\ 52'\ 22.6''$ and that from B to A as $53°\ 28'\ 33.4''$. Compute the coordinates of P_3.

BIBLIOGRAPHY

Adler, R. K., and Shmutter, B. 1971. Precise traverse in major geodetic networks. *Canadian Surveyor* 25:389.

Ball, W. E., Jr. 1980. Bureau of Land Management satellite Doppler positioning techniques. *Proc. Ann. Meet. Amer. Cong. Surv. & Mapp.* p. 411.

Bomford, Brigadier G. 1962. *Geodesy.* 3rd ed. London: Oxford Univ. Press.

Brown, D. C. A primer on satellite Doppler surveying. DBA Technical Note 75-002, DBA Systems, Inc. Melbourne, Fla. 1975.

Burke, K. F. 1971. Why compare triangulation and trilateration? *Proc. Ann. Meet. Amer. Cong. Surv. & Mapp.* p. 244.

Colcord, J. E. 1966. Geodetic sections. *Surveying and Mapping* 26:455.

Danial, N. F. 1978. Another solution of the three-point problem. *Surveying and Mapping* 38:329.

Dracup, J. F. 1970. Standards and specifications for supplemental horizontal control surveys. *Proc. 30th Ann. Meet. Amer. Cong. Surv. & Mapp.* p. 509.

—— 1969. Trilateration—a preliminary evaluation. *Proc. Fall Conv. Amer. Cong. Surv. & Mapp.* p. 95.

Dracup, J. F. 1976. Tests for evaluating trilateration surveys. *Proc. Fall Conv. Amer. Cong. Surv. & Mapp.* p. 96.

Dracup, J. F., and Swift, P. H. 1970. Tellurometer traverse across St. Elias Mountain. *Surveying and Mapping* 30:43.

Gale, P. M. 1969. Control surveys for the city of Houston. *Surveying and Mapping* 29:669.

Harris, H. C. 1978. Some results of Doppler receiver testing. *Proc. Ann. Meet. Amer. Cong. Surv. & Mapp.* p. 252.

Haug, M. D., Moffitt, F. H., and Anderson, J. M. 1980. A simplified explanation of Doppler satellite positioning. *Surveying and Mapping* 40:29.

Hothem, L. D., and Strange, W. E. 1976. The use of Doppler satellite positioning for extension of offshore geodetic control. *Proc. Fall Conv. Amer. Cong. Surv. & Mapp.* p. 295.

Hothem, L. D., Strange, W. E., and White, M. Doppler satellite surveying system. *Journal of Surv. & Mapp. Div. ASCE*, No. SU1, Proc. Paper 14132, Nov. 1978.

Kesler, J. M. 1973. EDM slope reduction and trigonometric leveling. *Surveying and Mapping* 33:61.

Krakiwsky, E. J., Wells, D. E., and Thompson, D. B. 1972. Geodetic control for Doppler satellite observations for lines under 200 km. *The Canadian Surveyor* 32:141.

Laird, M. O. 1970. Small figure triangulation. *Proc. 30th Ann. Meet. Amer. Cong. Surv. & Mapp.* p. 44.

Lambert, A. F. 1974. A new instrument tower for angular survey measurements. *Proc. Ann. Meet. Amer. Cong. Surv. & Mapp.* p. 561.

——. *Manual of Second- and Third-Order Triangulation and Traverse.* Special Publication No. 145, U. S. Coast and Geodetic Survey. U. S. Government Printing Office.

Meade, B. K. 1967. High-precision transcontinental traverse surveys. *Surveying and Mapping* 28:41.

Parkin, E. J. 1965. Geodetic surveys for earth movement studies along the California aqueduct. *Surveying and Mapping* 25:561.

Reynolds, W. F. *Manual of Triangulation Computations and Adjustments.* Special Publication No. 138, U. S. Coast and Geodetic Survey. U. S. Government Printing Office.

Richardus, P. 1966. *Project Surveying.* New York: Wiley.

Simmons, L. G. 1949. *Natural Tables for the Computation of Geodetic Positions: Clarke Spheroid of 1866.* Spec. Pub. No. 241, U. S. Dept. of Commerce, Washington, D.C.

Small, J. B., and Parkin, E. J. 1967. Alaskan surveys to determine crustal movement. *Surveying and Mapping* 27:413.

Stansell, T. 1978. *The Transit Navigation Satellite System.* Magnavox Government and Industrial Electronics Co.

Stipp, D. W. 1962. Trilateration adjustment. *Surveying and Mapping* 22:575.

Stoughton, H. W. 1977. An algorithm to adjust a quadrilateral. *Surveying and Mapping* 37:201.

Turpin, R. D. 1966. Use of the resection principle to determine local survey coordinates. *Surveying and Mapping* 26:73.

U. S. Coast and Geodetic Survey. *Formulas and Tables for the Computation of Geodetic Positions.* Special Publication No. 8, U. S. Government Printing Office.

Walker, J. W. 1975. Establishment of mapping control by Doppler satellite point positioning. *Proc. Fall Conv. Amer. Cong. Surv. & Mapp.* p. 285.

Walker, J. W. 1977. Transformation of Doppler geodetic point position to the local datum. *Proc. Ann. Meet. Amer. Cong. Surv. & Mapp.* p. 349.

Wheeler-Holohan, P.. 1977. Satellite positioning—a modern survey tool. *Proc. Ann. Meet. Amer. Cong. Surv. & Mapp.* p. 166.

Chapter 11
State Plane Coordinate and Universal Transverse Mercator Systems

11-1. PURPOSE In Chapter 8 the use of plane rectangular coordinates was discussed as applied to plane surveying, not only for defining the positions of survey stations, but also for solving such diverse problems as determining areas, locating the intersections of lines and curves, parting off land, and computing lengths and azimuths. No mention was made, however, of the extent of any given plane coordinate system. The methods of plane surveying are based on the assumption that all distances and directions are projected onto a horizontal plane surface which is tangent to the surface of the earth at one point within the area of the survey.

If two surveys are made independently of each other, then the measurements in each survey are referred to two different horizontal planes, which, of course, do not coincide. Furthermore, if the Y axis of a plane coordinate system for each of the two surveys is assumed to be parallel to the true meridian at one station of the survey, then even the Y axes of the two systems are not parallel with one another because of the convergence of the meridians. In a given system the farther the survey departs from the point of tangency, the more will the distances and angles as measured on the ground differ from the corresponding distances and angles as projected onto the horizontal plane. When this discrepancy becomes intolerable, then the limits of plane surveying and its inherent simplicity have been reached.

In Chapter 10 are discussed briefly the methods of geodetic surveying wherein all distances are reduced to a common reference surface conforming closely to the sea-level surface. Angles in triangles are considered as spherical (sometimes spheroidal) angles. Coordinates of points are computed with reference to parallels of latitude and meridians of longitude by using angles computed near the center of the earth rather than distances. Geodetic surveying is employed so that precise surveys may be extended over great distances in any direction without suffering the limitations of plane-surveying methods. Geodetic-surveying methods are more complex and more expensive, they involve more difficult computations, and they require specialized personnel in their execution.

For more than a century, large surveying and mapping organizations of the federal government, notably the National Geodetic Survey, have established horizontal control monuments in the form of triangulation, traverse, and intersection stations. These stations have been located by the methods of geodetic surveying. The network formed by these control points is being constantly filled in and added to by the same organizations. All the control points throughout the country bear a definite relationship, one to another, being referred to one common spheroidal surface.

The state plane coordinate systems have been devised to allow methods of plane surveying to be used over great distances in any direction. At the same time, a precision approaching that of geodetic surveying is maintained. As a result surveyors and engineers can incorporate the network of control established by geodetic surveying into their own surveys for purposes of coordination, checking, and reestablishing lost points.

If a land surveyor runs a traverse from one of these stations to the nearest corner of the land that he is to survey, he can calculate readily the state-coordinate position for each corner of the land and has in effect keyed his survey to all the stations of the National Geodetic Survey. As a result all stations of the National Geodetic Survey become witnesses to the positions of the land corners whose coordinates on a state system are known. The material marks of the land corners—such as trees, stones, fence posts, or other monuments—may be destroyed, and yet the positions they occupied on the ground can be closely reproduced from any recoverable stations of the national survey that are within practical distances of these corners. In this manner a survey station or land corner which is described in terms of a state coordinate system is practically indestructible.

The system is of great value to highway engineers and others whose work covers very large areas. Many surveys, which in the past would have been open traverse with few checks on their accuracy, can now start from one known point and end on another, these points providing a closure for the traverse.

The state coordinate systems are shown on many federal maps, particularly on topographic maps. These maps thus become useful to engineers who desire data for reducing their surveys to a state grid. At the same time their own surveys become available for transfer to the map.

11-2. LIMITS OF STATE PLANE COORDINATE SYSTEMS

Just as a map of a considerable portion of the earth's surface is a compromise when compiled on a plane surface, namely, the map sheet, so must a state plane coordinate system be a compromise since it represents a spheroidal surface projected onto a curved surface which may be developed into a plane surface. However, this compromise does not prove to be at all serious for two reasons: (1) Any distortion suffered through the transformation of the spheroidal surface to the plane surface may be allowed for by simple arithmetic. (2) If the extent of the earth's surface to be represented by a plane surface is limited, then the distortion may be neglected entirely except for a special survey requiring great accuracy; in such a case, the distortion is eliminated by simple computation.

If a state is not too large, the entire state can be embraced in one projection. If the state is of such size that the distortions in one projection would be too great to be ignored, then the state is divided into *zones* of appropriate size. These zones are broken along county lines to permit an entire county to be included in the same zone.

Two types of projections are employed in developing the state plane coordinate systems, namely, the *transverse Mercator projection* and the *Lambert conformal projection*. The transverse Mercator projection employs a cylindrical surface, the axis of which is normal to the earth's axis of rotation and which intersects the surface of the earth along two ellipses equidistant from a meridian plane through the center of the area to be projected (see Fig. 11-1). The distortions occur in the east-west direction, and therefore the projection

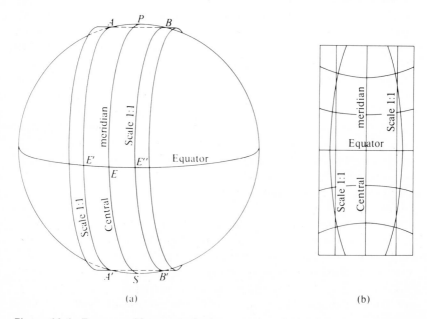

(a)   (b)

Figure 11-1. Transverse Mercator projection.

is used for states or zones that have relatively short east-west dimensions. An example is Vermont.

The Lambert conformal projection employs a conical surface, the axis of which coincides with the earth's axis of rotation and which intersects the surface of the earth along two parallels of latitude that are approximately equidistant from a parallel lying in the center of the area to be projected (see Fig. 11-5). The distortions occur in a north-south direction, and therefore the projection is used for states or zones with relatively short north-south dimensions. An example is Tennessee.

If the distance in an east-west direction in a transverse Mercator projection is limited to about 158 miles or if the distance in a north-south direction in a Lambert projection is limited to about 158 miles, then the distortions will be such that a distance at sea level will not differ from the corresponding projected distance by any more than one part in 10,000. Thus it becomes necessary to assign more than one projection to some individual states. California, for example, has seven Lambert zones; Nevada has three transverse Mercator zones; New York has three transverse Mercator zones to cover all but Long Island, which is covered by one Lambert zone. Florida has two transverse Mercator zones covering the peninsula proper and one Lambert zone which covers the western portion of the state. Table 11-1 shows the number and type of projections covering the various states.

11-3. GEODETIC AND GRID AZIMUTHS

11-3. GEODETIC AND GRID AZIMUTHS In placing a survey on a system of plane rectangular coordinates, all bearings or azimuths are referred to the same meridian. Since the system is rectangular, all grid north-and-south lines are parallel to a *central meridian*. In general there is a difference between the grid azimuth or bearing of a line and its geodetic azimuth or bearing. This difference is caused by the convergence of meridians, the amount increasing with the latitude and the distance from the central meridian. When the grid coordinates of the two ends of a starting line are known, its azimuth may be computed by either Eq. (8-13) or Eq. (8-14) and its length by Eq. (8-12).

The geodetic azimuth of a line is a computed azimuth based on triangulation and traverse observations, computations, and adjustments resulting from geodetic surveying. It is always reckoned from the south part of the meridian. The difference between true or astronomical azimuth as defined in Section 7-6 and geodetic azimuth is relatively small and caused by the fact that the plumb line at a station is deflected from a true normal to the spheroid of computation for geodetic positions. The geodetic azimuth of a line is that published by the National Geodetic Survey along with the geodetic positions in triangulation and traverse networks.

It is frequently important for the surveyor to know the difference between the geodetic and grid azimuths of a survey line for purposes of obtaining a check on a computed value or to provide a starting azimuth for a survey. On the Lambert grid, the angle between the geodetic and grid meridians is

TABLE 11-1. State Plane Coordinate Systems

STATE AND ZONE	GRID	STATE AND ZONE	GRID
Alabama		Illinois	
East	Transverse Mercator	East	Transverse Mercator
West	Transverse Mercator	West	Transverse Mercator
Alaska[a]		Indiana	
		East	Transverse Mercator
Arizona		West	Transverse Mercator
East	Transverse Mercator		
Central	Transverse Mercator	Iowa	
West	Transverse Mercator	North	Lambert
		South	Lambert
Arkansas			
North	Lambert	Kansas	
South	Lambert	North	Lambert
		South	Lambert
California			
Zone 1	Lambert	Kentucky	
Zone 2	Lambert	North	Lambert
Zone 3	Lambert	South	Lambert
Zone 4	Lambert		
Zone 5	Lambert	Louisiana	
Zone 6	Lambert	North	Lambert
Zone 7	Lambert	South	Lambert
Colorado		Maine	
North	Lambert	East	Transverse Mercator
Central	Lambert	West	Transverse Mercator
South	Lambert		
		Maryland	Lambert
Connecticut	Lambert		
		Massachusetts	
Delaware	Transverse Mercator	Mainland	Lambert
		Island	Lambert
Florida			
East	Transverse Mercator	Michigan[b]	
West	Transverse Mercator	East	Transverse Mercator
North	Lambert	Central	Transverse Mercator
		West	Transverse Mercator
Georgia			
East	Transverse Mercator	Minnesota	
West	Transverse Mercator	North	Lambert
		Central	Lambert
Hawaii		South	Lambert
Zone 1	Transverse Mercator		
Zone 2	Transverse Mercator	Mississippi	
Zone 3	Transverse Mercator	East	Transverse Mercator
Zone 4	Transverse Mercator	West	Transverse Mercator
Zone 5	Transverse Mercator		
		Missouri	
Idaho		East	Transverse Mercator
East	Transverse Mercator	Central	Transverse Mercator
Central	Transverse Mercator	West	Transverse Mercator
West	Transverse Mercator		

TABLE 11-1. *(continued)*

STATE AND ZONE	GRID	STATE AND ZONE	GRID
Montana		Rhode Island	Transverse Mercator
North	Lambert		
Central	Lambert	South Carolina	
South	Lambert	North	Lambert
		South	Lambert
Nebraska			
North	Lambert	South Dakota	
South	Lambert	North	Lambert
		South	Lambert
Nevada			
East	Transverse Mercator	Tennessee	Lambert
Central	Transverse Mercator		
West	Transverse Mercator	Texas	
		North	Lambert
New Hampshire	Transverse Mercator	North Central	Lambert
		Central	Lambert
New Jersey	Transverse Mercator	South Central	Lambert
		South	Lambert
New Mexico			
East	Transverse Mercator	Utah	
Central	Transverse Mercator	North	Lambert
West	Transverse Mercator	Central	Lambert
		South	Lambert
New York			
Long Island	Lambert	Vermont	Transverse Mercator
East	Transverse Mercator		
Central	Transverse Mercator	Virginia	
West	Transverse Mercator	North	Lambert
		South	Lambert
North Carolina	Lambert		
		Washington	
North Dakota		North	Lambert
North	Lambert	South	Lambert
South	Lambert		
		West Virginia	
Ohio		North	Lambert
North	Lambert	South	Lambert
South	Lambert		
		Wisconsin	
Oklahoma		North	Lambert
North	Lambert	Central	Lambert
South	Lambert	South	Lambert
Oregon			
North	Lambert	Wyoming	
South	Lambert	Zone I	Transverse Mercator
		Zone II	Transverse Mercator
Pennsylvania		Zone III	Transverse Mercator
North	Lambert	Zone IV	Transverse Mercator
South	Lambert		

[a] Alaska is divided into 10 zones although projection tables are not available. Zone 1 is an oblique transverse Mercator projection; Zones 2 through 9 are Mercator transverse projections; and Zone 10 is a Lambert projection.

[b] Michigan is also divided into 3 Lambert zones, the North, Central and South. These projections were computed for an elevation of 800 ft rather than for sea level.

designated by θ and is listed in the state projection tables. On the transverse Mercator grid the angle is $\Delta\alpha$, employed in geodetic computations to represent the convergence of the meridians. In current publications of state-coordinate position data, the appropriate value of θ or $\Delta\alpha$ is listed for each station.

A line on the surface of the spheroid representing a line of sight between two points or, more exactly, representing the most direct path between the two points, can be thought of as a "great circle," as would be the case if the sea-level surface were a true sphere. This line is referred to as a *geodetic line*. The geodetic line projected onto the Lambert grid or the transverse Mercator would in general show a slight curvature. Thus if long lines are involved and utmost accuracy is to be obtained, the difference between geodetic and grid azimuths is slightly different from the value of θ and $\Delta\alpha$ (see Sections 11-5 and 11-9).

11-4. TRANSVERSE MERCATOR PROJECTION

The transverse Mercator projection is represented diagrammatically in Fig. 11-1. View (a) shows the sea-level surface of the earth intersected by a cylindrical surface along the two ellipses $AE'A'$ and $BE''B'$, which are equidistant from a *central meridian PES*. This intersection along the two ellipses is effected by scaling down from a cylinder which would otherwise be tangent to the earth along the central meridian. View (b) shows a portion of the cylindrical surface developed into a plane surface on which the meridians and parallels of the earth's surface have been projected mathematically. On the development each ellipse appears as a straight line parallel to and equidistant from the central meridian. Along either of these two lines, a distance on the projection is the same as the corresponding distance on the sea-level surface. Between the two lines, a distance is smaller than the corresponding distance on the sea-level surface. Outside the two lines, a distance on the projection is larger than the corresponding distance on the sea-level surface. The discrepancy between these corresponding distances depends on the position of the line being considered with respect to the central meridian. It is seen that the scale of a line running in an east-west direction varies from point to point. It is also seen, however, that a line on the projection parallel to the central meridian has a constant scale throughout its length, whether this scale be larger than, equal to, or less than that on the corresponding sea-level line.

To apply the transverse Mercator projection to a state or a zone, the width of the projection is limited to 158 miles, and the lines along which the scale of the projection is exact are separated by about two-thirds this distance. At no point within these limits will the discrepancy between a sea-level distance and the projected distance, called the *grid distance*, be greater than one part in 10,000.

11-5. GRID AZIMUTH ON TRANSVERSE MERCATOR PROJECTION

With but two exceptions, a line run on the surface of the earth as a straight line does not have a constant azimuth with respect to the true

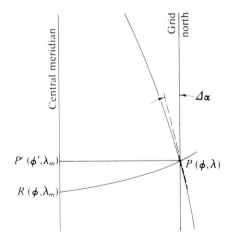

Figure 11-2. Relationship between geodetic and grid azimuths on transverse Mercator projection.

meridians because of the convergence of these meridians. The exceptions are a straight line run along the true meridian from which the azimuth angle is measured and a line run along the equator. For example, a secant line run to establish a parallel of latitude, as described in Section 18-8, extends due east or due west at one point only. Immediately on leaving this point, the secant runs in a southeasterly or southwesterly direction.

In the transverse Mercator system of plane coordinates, the azimuth of any line is referred to the central meridian. This azimuth is also referred to any line parallel to the central meridian. The direction of each of this series of parallel lines is *grid north*. Thus a straight line on the transverse Mercator projection has a constant grid azimuth throughout its length. The angle between the grid north and the true north at a point on this projection depends on two things: (1) the distance from the central meridian to the point, and (2) the latitude of the point.

In Fig. 11-2, P is a station whose latitude and longitude are, respectively, ϕ and λ. The line parallel to the central meridian passing through P is grid north. The curved line making the angle $\Delta\alpha$ with grid north at P is the true meridian through P. The straight line PP', making an angle of 90° with the central meridian, is the projection of the arc of a great circle (nearly) passing through P. It intersects the central meridian at P', whose coordinates are ϕ' and λ_m. The parallel of latitude through P is a curved line intersecting the central meridian at R. The latitude of R is ϕ. The value of $\Delta\alpha$ is the same as that given in Chapter 10 for computing geodetic positions. Thus

$$-\Delta\alpha = \Delta\lambda \sin \tfrac{1}{2}(\phi + \phi') \sec \tfrac{1}{2}(\phi - \phi') + (\Delta\lambda)^3 F \qquad (11\text{-}1)$$

Since the value of sec $\frac{1}{2}(\phi - \phi')$ is practically unity, and since the term containing F is small except for very exact determinations, the equation for $\Delta\alpha$ may be reduced to

$$-\Delta\alpha = \Delta\lambda \sin \tfrac{1}{2}(\phi + \phi') \qquad (11\text{-}1a)$$

in which $\Delta\lambda$ is the difference between the longitude of the station and the longitude of the central meridian for the given projection, and ϕ' is the latitude of the foot of the perpendicular from the point to the central meridian (P' in Fig. 11-2). The value of $\Delta\lambda$ is found from the relationship

$$\Delta\lambda = \lambda_m - \lambda \qquad (11\text{-}2)$$

in which λ_m is the longitude of the central meridian, and λ is the longitude of the station.

If the station lies east of the central meridian, then $\Delta\lambda$ is positive $(+)$. If the station lies west of the central meridian, $\Delta\lambda$ is negative $(-)$. Because the direction in which $\Delta\lambda$ is computed by Eq. (11-2) is opposite to that in which it is reckoned in Chapter 10, the algebraic sign of $\Delta\alpha$ as computed by Eq. (11-1a) is the same as the algebraic sign of $\Delta\lambda$. Thus if the station lies east of the central meridian, $\Delta\alpha$ is positive, and if to the west, $\Delta\alpha$ is negative. The value of ϕ' is found by solving Eqs. (11-5) and (11-6).

The relationship between the true azimuth and the grid azimuth of a line originating in P is given by the expression

$$\text{geodetic azimuth} - \text{grid azimuth} = \Delta\alpha \qquad (11\text{-}3)$$

The difference between the true and geodetic azimuths is explained in Section 11-3.

In Fig. 11-3 lines AB and CD lie to the west of the central meridian and EF and GH lie to the east of the central meridian. The grid lines joining the end points are straight lines in every case. The ground line, or geodetic line, joining any pair of end points, however, is curved when projected onto the grid. Furthermore, this curvature, which is greatly exaggerated in Fig. 11-3, increases outward from the central meridian. To allow for this curvature in going from the geodetic azimuth to the grid azimuth, or vice versa, when a line longer than 5 miles is involved, Eq. (11-3) must be modified by a small correction term. Thus

$$\text{geodetic azimuth} - \text{grid azimuth} = \Delta\alpha + \frac{(Y_2 - Y_1)(2X'_1 + X'_2)}{(6\rho_0^2 \sin 1'')_g} \qquad (11\text{-}3a)$$

in which X'_1 and X'_2 are the X coordinates of the origin and terminal of the line, respectively, *referred to the central meridian* (see Eq. 11-8); Y_1 and Y_2 are the Y coordinates; and ρ_0 is the mean radius of the earth in the vicinity of the projection. The subscript g indicates that the value of ρ_0 has been

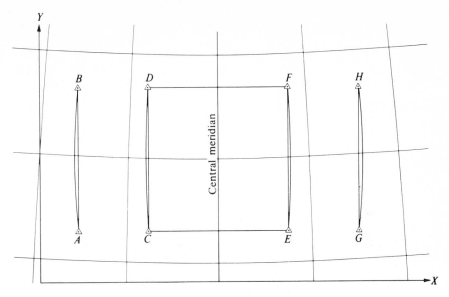

Figure 11-3. Curvature of geodetic lines on transverse Mercator projection.

reduced by the scale factor mentioned in Section 11-4. Due regard must be given to the sign of the X' coordinates and of the value $Y_2 - Y_1$. Notice that if the line is grid east-west, as is the line DF or CE in Fig. 11-3, no correction need be applied to $\Delta\alpha$ since $Y_2 - Y_1 = 0$.

11-6. GEOGRAPHIC TO PLANE COORDINATES ON TRANS-VERSE MERCATOR PROJECTION
The origin of coordinates on the transverse Mercator projection is taken far enough south and west of the zone covered to render all values of coordinates positive. To make all the X coordinates positive, the central meridian, which is the Y axis, is assigned a large X value denoted by X_0; usually $X_0 = 500,000$ ft. The X axis is a line perpendicular to the central meridian where $Y = 0$.

The Y coordinate of a point on the central meridian is equal to the sea-level distance along the central meridian measured from the intersection of the X axis and the central meridian to the point and adjusted for the scale along the central meridian. The X coordinate of a point on the central meridian is, of course, the same as the X value of the central meridian.

Before the Y and X coordinates of any point whose latitude and longitude are known can be determined, the sea-level length of a line passing through the point and perpendicular to the central meridian must be computed. The latitude of the foot of this perpendicular on the central meridian must then be determined. In Fig. 11-4, P is a point whose latitude and longitude are ϕ and λ. The line $P'P$ is the sea-level line passing through P and perpendicular to the central meridian at P'. The azimuth (from south) of the line from P'

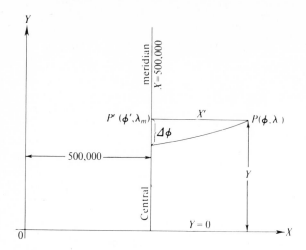

Figure 11-4. Rectangular coordinates on transverse Mercator projection.

to P is therefore 270°. The sea-level length of this line may be determined by the equation for $\Delta\lambda$ in the computation for geodetic positions (see Chapter 10 and Special Publication 241, U. S. Coast and Geodetic Survey). This length is given by Eq. (10-11).

$$\Delta\lambda = s \sin\alpha \sec\phi \cdot A + \text{arc-sine correction}$$

in which $\Delta\lambda$ is the difference in longitude between the point and the central meridian given by Eq. (11-2), s is the sea-level length of the line $P'P$, in metres, α is the azimuth of $P'P$, A is the factor A for the known point P, and ϕ is the latitude of point P. Since the azimuth of $P'P$ is 270°, $\sin\alpha = -1$. Transposing terms then gives the following equation for the distance s, in metres:

$$s = \frac{\Delta\lambda \cos\phi}{A} - \text{arc-sine correction} \tag{11-4}$$

In Eq. (11-4), $\Delta\lambda$ is positive if the point is east of the central meridian and is negative if the point is west of the central meridian.

The latitude of P' is determined from the equation for $-\Delta\phi$ in the computation for geodetic position, which is

$$-\Delta\phi = Bs \cos\alpha + Cs^2 \sin^2\alpha + (\delta\phi)^2 D - Ehs^2 \sin^2\alpha$$

in which $-\Delta\phi$ is the minus difference in latitude going from P' to P, C is the factor C for the unknown point P', and α is the azimuth of the line $P'P$, which is 270°. Since $\cos\alpha = 0$, the first term drops out. Also, because the difference in latitude between P' and P can never be very large, the terms containing D and E are both neglected. So

$$-\Delta\phi = Cs^2 \tag{11-5}$$

The distance s in metres is obtained from Eq. (11-4). The value of C should be obtained for the latitude of P'. However, since this is as yet unknown, the preliminary value of C for the known latitude of the point P is used. Then when the preliminary value of ϕ' is determined, a better value of C is obtained to give a corrected value of $-\Delta\phi$. It should be noted that the numerical value of $\Delta\phi$ is always added to the latitude ϕ of point P to obtain ϕ'. Therefore

$$\phi' = \phi + \Delta\phi \tag{11-6}$$

Once the value of ϕ' has been determined, the Y coordinate of P', and hence of P, is obtained by interpolation in the tables prepared for the particular projection by the National Geodetic Survey. This coordinate is a true length along the central meridian reduced for the scale along the central meridian. Therefore it may be tabulated for each latitude.

The length of the line $P'P$ on the plane projection is obtained by the equation

$$X' = s_g + \frac{s_g^3}{(6\rho_0^2)_g} \tag{11-7}$$

in which X' is the grid length of $P'P$ corresponding to s (the sea-level length of $P'P$), s_g is the distance s converted to feet and reduced by the overall scale reduction of the projection, and ρ_0 is the mean radius of curvature of the spheroid, in feet, in the vicinity of the projection. This radius is assumed to be constant for any given projection and has also been reduced by the overall scale reduction.

The X coordinate of the point P may now be determined by adding the value of X' to the X coordinate of the central meridian. Thus

$$X = X_0 + X' \tag{11-8}$$

If $\Delta\lambda$ is negative, then s and consequently X' will also be minus, and the value of X will be less than X_0.

To summarize, the equations are repeated here in their order of solution as follows:

$$\Delta\lambda = \lambda_m - \lambda \tag{11-2}$$

$$s = \frac{\Delta\lambda \cos\phi}{A} - \text{arc-sine correction} \tag{11-4}$$

$$-\Delta\phi = Cs^2 \tag{11-5}$$

$$\phi' = \phi + \Delta\phi \tag{11-6}$$

The value Y is found by interpolating in an appropriate table with ϕ' as argument.

$$X' = s_g + \frac{s_g^3}{(6\rho_0^2)_g} \qquad [11\text{-}7]$$

$$X = X_0 + X' \qquad [11\text{-}8]$$

11-7. TABLES FOR TRANSVERSE MERCATOR PROJECTION

The National Geodetic Survey has prepared and published tables for each projection in each state and makes them available through the U. S. Government Printing Office. These tables are used in place of Eqs. (11-1) through (11-8) and are convenient for use with desk or hand calculators.

Table 11-2 shows a portion of the projection tables for Arizona. The function Y_0 is the Y coordinate of the intersection of the given parallel of latitude with the central meridian. The remainder of the quantities are used in the following equations, taken from Special Publication No. 257 for Arizona.

$$X = X' + 500{,}000 \qquad (11\text{-}9)$$

which is the same as Eq. (11-8) in which $X_0 = 500{,}000$.

$$X' = H \cdot \Delta\lambda'' \pm ab \qquad (11\text{-}10)$$

which performs the same function as Eq. (11-7).

$$Y = Y_0 + V\left(\frac{\Delta\lambda''}{100}\right)^2 \pm c \qquad (11\text{-}11)$$

in which the middle term of the right-hand side is essentially the same as Eq. (11-5).

$$\Delta\alpha'' = \Delta\lambda'' \sin\phi + g \qquad (11\text{-}12)$$

A comparison of Eq. (11-12) with Eq. (11-1a) shows that $\sin\phi$ is the same as $\sin\frac{1}{2}(\phi + \phi')$. This is permissible since $\Delta\alpha$ is small, and $\Delta\phi$ given by the middle term of Eq. (11-11) is also small. The g term is similar to the F term of Eq. (11-1) and is usually quite small.

Note that the functions Y_0, H, V, and a are based on the latitude of the point in question, whereas b, c, and g are based on the longitude.

The value of $\Delta\lambda$ needed for use in the tables is given by Eq. (11-2). The second term correction to $\Delta\alpha$ is given in Eq. (11-3a).

Example 11-1.

The latitude and longitude of a control point in Arizona are $\phi = 33°\,22'\,14.656''$ and $\lambda = 109°\,52'\,43.255''$. Compute the coordinates of the control point on the transverse Mercator east zone for Arizona. Compute $\Delta\alpha$, neglecting the correction term.

Solution: The difference in longitude is first computed by Eq. (11-2). Thus

$$\Delta\lambda = 110°\,10'\,00.000'' - 109°\,52'\,43.255'' = +0°\,17'\,16.745'' = +1036.745''$$

The work then proceeds as follows.

$(\Delta\lambda''/100)^2 = 107.484$

$H =$	84.806357	(from Table 11-2)
$V =$	1.130978	(from Table 11-2)
$a =$	-0.820	(from Table 11-2)
$b =$	$+3.656$	(from Table 11-2)
$H \cdot \Delta\lambda =$	$+87,922.57$	
$ab =$	-3.00	

$X' = +87,919.57$ (by Eq. 11-10)

$c = -0.02$ (from Table 11-2)

$V(\Delta\lambda''/100)^2 \pm c = 121.54$

$Y_0\text{(minutes)} = 860,907.68$ (from Table 11-2)

$Y_0\text{(seconds)} = +1,481.21$ (interpolation in Table 11-2)

$Y_0 = 862,388.89$ (Y value on central meridian for given latitude)

$X = 587,919.57$ (by Eq. 11-9)

$Y = 862,510.43$ (by Eq. 11-11)

$\Delta\alpha'' = +1036.745 \sin 33°\,22'\,15'' + 0 = +570.27'' = +0°\,09'\,30.27''$
                 (by Eq. 11-12)

The values of the remaining functions given in the tables are needed to compute latitude and longitude from X and Y coordinates according to the following equations.

$$Y_0 = Y - P\left(\frac{X'}{10,000}\right)^2 - d \tag{11-13}$$

in which X', P, and d are obtained from the given coordinates. The value of X' is given by Eq. (11-9).

TABLE 11-2. Transverse Mercator Projection for Arizona East and Central Zones

LATITUDE	Y_0 (ft)	ΔY_0 (per second)	a	H	ΔH (per second)	V	ΔV (per second)
33° 20′	848 779.94	101.064 33	−0.823	84.842 642	269.39	1.130 341	4.73
33° 21′	854 843.80	101.064 67	−0.821	84.826 479	269.51	1.130 625	4.73
33° 22′	860 907.68	101.065 00	−0.820	84.810 309	269.64	1.130 909	4.73
33° 23′	866 971.58	101.065 17	−0.819	84.794 131	269.74	1.131 193	4.72
33° 24′	873 035.49	101.065 33	−0.817	84.777 947	269.87	1.131 476	4.72

$\Delta\lambda''$	b	Δb	c
1000	+3.540	+0.315	−0.022
1100	+3.855	+0.305	−0.027
1200	+4.160	+0.294	−0.032
1300	+4.454	+0.282	−0.038
1400	+4.736	+0.268	−0.043

Table for g $\Delta\alpha'' = \sin\phi(\Delta\lambda'') + g$

LATITUDE	0″	1000″	2000″	3000″	4000″	5000″	6000″
			$\Delta\lambda''$				
31°	0.00	0.00	0.02	0.08	0.19	0.37	0.64
32°	0	0	0.02	0.08	0.19	0.38	0.65
33°	0	0	0.02	0.08	0.19	0.38	0.65
34°	0	0	0.02	0.08	0.19	0.38	0.65
35°	0	0	0.02	0.08	0.19	0.38	0.65

Y	P	ΔP
800,000	1.56187	1638
900,000	1.57825	1649

X'	d
100,000	+0.02
150,000	+0.04

Table for e

Y \ X'	200,000	300,000	400,000	500,000
0	0	0.1	0.3	0.6
1,000,000	0.0	0.1	0.4	0.8
2,000,000	0.0	0.2	0.5	1.0

	EAST AND CENTRAL ZONES	
Y	M	ΔM
500,000	0.006 2436	664
600,000	0.006 3100	667
700,000	0.006 3767	672
800,000	0.006 4439	676
900,000	0.006 5115	680

Constants for East Zone: Central meridian 110° 10′ 00.000″. Scale reduction (central meridian) 1 : 10,000. $(1/6\rho_0^2 \sin 1'')_q = 0.7872 \times 10^{-10}$.

The latitude is obtained by interpolating in the tables using the value of Y_0 from Eq. (11-13). Then the value of H for this latitude can be interpolated.

$$\Delta\lambda''(\text{approx.}) = \frac{X'}{H} \qquad (11\text{-}14)$$

Then a is obtained from the known latitude and b from the approximate value of $\Delta\lambda''$. The final value for $\Delta\lambda''$ is then given by

$$\Delta\lambda'' = \frac{X' \mp ab}{H} \qquad (11\text{-}15)$$

and the value of $\Delta\alpha''$ is

$$\Delta\alpha'' = MX' - e \qquad (11\text{-}16)$$

Note that P and M are functions of the Y coordinate of the point, d is a function of the X coordinate of the point, and e depends on both X and Y.

Example 11-2.

The state plane coordinates of a point on the Arizona east system are $X = 626,545.10$ ft and $Y = 868,452.25$ ft. Determine the latitude and longitude of the point. Determine the value of $\Delta\alpha''$, neglecting the small correction term.

Solution: The value of X' is $626,545.10 - 500,000.00 = +126,545.10$. Then from Table 11-2, $P = 1.57308$ and $d = +0.02$. By Eq. (11-13),

$$Y_0 = 868,452.25 - 1.57308(12.65451)^2 - 0.02 = 868,200.32$$

Interpolating in the tables for ϕ gives

$$868,200.32 = Y_0$$
$$\underline{-866,971.58 \ (\text{for } \phi = 33° 23')}$$

$$1,228.74 \ (\text{difference from tabular value})$$

$1,228.74/101.06517 = 12.158''$

$$\phi = 33° 23' 12.158''$$

$H = 84.790852$ \hfill (from Table 11-2)

$\Delta\lambda''(\text{approx.}) = 126,545.10/84.790852 = +1492''$ \hfill (by Eq. 11-14)

$a = -0.819$ and $b = +4.984$; \quad $ab = -4.082$

$\Delta\lambda'' = (126,545.10 + 4.082)/84.790852 = +1492.486''$

\qquad $= +0° 24' 52.486''$ \hfill (by Eq. 11-15)

$\lambda = 109° 45' 07.514''$ \hfill (by Eq. 11-2)

$M = 0.0064902$ \quad and \quad $e = 0$

$\Delta\alpha'' = 0.0064902 \times 126,545.10 = +821.30'' = +0° 13' 41.30''$

TABLE 11-3. Scale Factor for Arizona East and Central Zones

X' (ft)	SCALE FACTOR	X' (ft)	SCALE FACTOR	X' (ft)	SCALE FACTOR
0	0.999 9000	290 000	0.999 9963	500 000	1.000 1862
5000	0.999 9000	295 000	0.999 9996	505 000	1.000 1919
10000	0.999 9001	300 000	1.000 0030	510 000	1.000 1977
15000	0.999 9003	305 000	1.000 0065	515 000	1.000 2036
20000	0.999 9005	310 000	1.000 0100	520 000	1.000 2095

The projection tables for Arizona as well as for the other states using the transverse Mercator projection also include a section for use in reducing sea-level distances to grid distances. Table 11-3 shows portions of the table containing the reduction factors included in the Arizona Tables for the east and central zones. Note that the scale factor varies as the distance east or west from the central meridian. The scale factor for the central meridian where $X' = 0$ is 0.999 9000, and changes from less than unity to greater than unity at $X' \cong \pm 295,832$ ft.

11-8. LAMBERT CONFORMAL PROJECTION The Lambert projection is represented diagrammatically in Fig. 11-5. In view (a) is shown the sea-level surface of the earth intersected by a cone along two parallels of latitude AL_1B and CL_2D. These are known as the *standard parallels* of the projection. View (b) shows a portion of the conical surface developed into a plane surface on which the meridians and parallels of the earth's surface have been projected mathematically. The developments of the two parallels appear as arcs of concentric circles with the center at the apex of the developed cone (not shown on the illustration). Along these two parallels the distances on the projection are the same as corresponding distances on the sea-level surface. Between the two parallels, a distance on the projection is smaller than the corresponding distance on the sea-level surface. Outside the two parallels a projected distance is larger. The discrepancy between these corresponding distances depends on the position of the line being considered with respect to the two standard parallels. It is seen that the scale of a line running in a north-south direction varies from point to point. It is also seen, however, that a due east-west line has a constant scale throughout its length, whether this scale be larger than, equal to, or less than that on the corresponding sea-level line.

To apply the Lambert conformal projection to a state or a zone the width of the projection in a north-south direction is limited to 158 miles, and the standard parallels are separated by about two-thirds this distance. At no point within these limits will the discrepancy between a sea-level distance and the grid distance be greater than one part in 10,000.

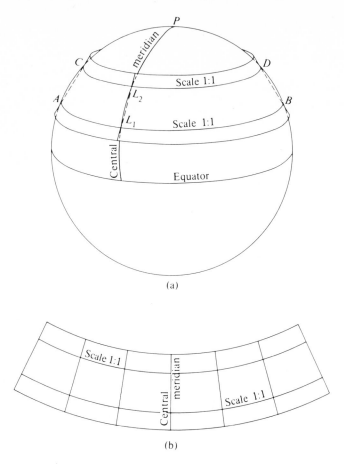

Figure 11-5. Lambert conformal projection.

11-9. GRID AZIMUTH ON LAMBERT CONFORMAL PRO-
JECTION   The central meridian and all lines parallel thereto on the Lambert projection are grid-north lines. Therefore the azimuth of any line on the plane projection is referred to any of these grid-north lines. The relationship between geodetic azimuth and grid azimuth is indicated in Fig. 11-6. The point P has geographic coordinates ϕ and λ. Two true meridians are shown. Line AB is the central meridian of the projection, and PB is the meridian through P. The line through P, parallel to the central meridian, is the grid meridian or grid-north line at P. The point B is the apex of the developed cone. The angle at B between the meridian through P and the central meridian, which is designated θ, is known as the *mapping angle* for the point.

To compute the mapping angle for a point, first determine the difference

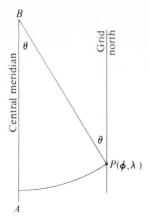

Figure 11-6. Relationship between geodetic and grid azimuths on Lambert projection.

in longitude $\Delta\lambda$ between the central meridian and the point by the equation

$$\Delta\lambda = \lambda_m - \lambda \qquad (11\text{-}17)$$

Then multiply $\Delta\lambda$ by a constant of the given projection which gives the relationship between $\Delta\lambda$ and θ. That is

$$\theta = l\,\Delta\lambda \qquad (11\text{-}18)$$

where l is the constant. If the point is east of the central meridian, then $\Delta\lambda$ and θ are positive. If the point is west of the central meridian, $\Delta\lambda$ and θ are negative.

As seen from Fig. 11-6, the relationship between the geodetic azimuth and the grid azimuth at a point is

$$\text{geodetic azimuth} - \text{grid azimuth} = +\theta \qquad (11\text{-}19)$$

Due regard must be paid to the algebraic sign of θ. Tables giving the value of θ for every minute of longitude are available for the various Lambert projections. These tables are discussed in Section 11-11.

In Fig. 11-7 the arc designated by ϕ_0 is a certain parallel of latitude lying slightly north of an arc midway between the standard parallels. The Y coordinate of the intersection of this parallel with the central meridian is designated by Y_0, as shown. The line AB lies north of this parallel. The grid line joining the two points A and B is a straight line. The geodetic line joining the two points, when projected onto the grid, is seen to curve outward away from the parallel designated by ϕ_0. The geodetic line CD is also seen to curve outward away from this parallel, whereas the grid line between C and D is straight. To allow for this curvature in going from the geodetic azimuth to

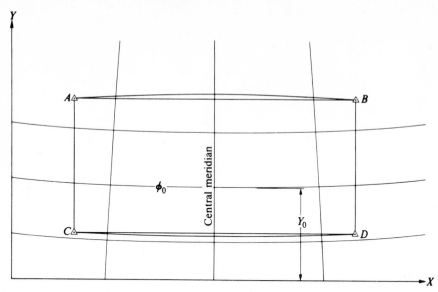

Figure 11-7. Curvature of geodetic lines on Lambert projection.

the grid azimuth, or vice versa, when a long line is involved, Eq. (11-19) must be modified by a small correction term. Thus

$$\text{geodetic azimuth} - \text{grid azimuth} = +\theta - \frac{X_2 - X_1}{2\rho_0^2 \sin 1''}\left(Y_1 - Y_0 + \frac{Y_2 - Y_1}{3}\right)$$

$$(11\text{-}20)$$

in which X_1 and Y_1 and X_2 and Y_2 are the coordinates of the ends of the line, Y_0 is as defined above, and ρ_0 is the mean radius of curvature of the earth at latitude ϕ_0. The value of $1/(2\rho_0^2 \sin 1'')$ and the value of Y_0 are given in the projection tables for each zone.

Notice that if a line runs grid north or grid south, as does the line AC or BD in Fig. 11-7, there is no correction term to θ because $X_2 - X_1 = 0$. Also if a line other than a grid north-south line crosses the parallel designated by ϕ_0, its projection onto the grid will have a reverse curvature.

11-10. GEOGRAPHIC TO PLANE COORDINATES ON LAMBERT CONFORMAL PROJECTION   On a Lambert projection the central meridian is the Y axis, and the X axis is perpendicular to the central meridian. The origin of coordinates is far enough south and west to render all the co-ordinate values positive. The Y axis is assigned a large X value. It is usually, but not always, 2,000,000 ft. This large value is designated by X_0; that is, $X_0 = 2,000,000$ ft. Except for three projections, the Y value of the X axis is 0 ft.

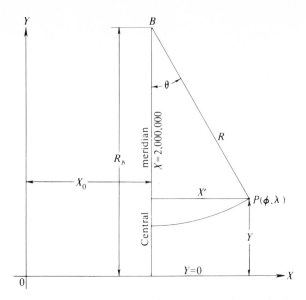

Figure 11-8. Rectangular coordinates on Lambert projection.

In Fig. 11-8, P is a point whose geographic coordinates are ϕ and λ. The parallel of latitude through P is a circle whose radius is R. As the latitude increases, the corresponding value of R decreases. The value of R for any latitude may be obtained from the tables to be discussed in the next section. The mapping angle θ for point P may be determined by Eq. (11-18), or it may be obtained by interpolation in the proper table for the given projection. The distance along the central meridian from the X axis to point B is designated by R_b. It is constant for a given projection. Thus the X axis is at a distance R_b south of B.

If the latitude of point P in Fig. 11-8 is known, R can be obtained. Also, if the longitude of the point P is known, θ can be obtained. Then the Y and X coordinates of P can be computed by means of the following equations:

$$Y = R_b - R \cos \theta \qquad (11\text{-}21)$$

$$X = X_0 + R \sin \theta \qquad (11\text{-}22)$$

Because of the large numbers involved, the calculator employed to solve Eqs. (11-21) and (11-22) must be able to generate sines and cosines to 10 places after the decimal.

11-11. TABLES FOR LAMBERT PROJECTIONS The National Geodetic Survey has prepared tables for each projection in each state and makes them available through the U. S. Government Printing Office. Table 11-4 shows a portion of Tables I and II for the north zone of the State of Ohio.

TABLE 11-4. Lambert Projection for Ohio–North

TABLE I

LATITUDE	R (ft)	Y' Y VALUE ON CENTRAL MERIDIAN (ft)	TABULAR DIFFERENCE FOR 1″ OF LATITUDE (ft)	SCALE EXPRESSED AS RATIO
40° 51′	24,128,050.71	431,107.76	101.19850	0.9999463
40° 52′	24,121,978.80	437,179.67	101.19883	0.9999452
40° 53′	24,115,906.87	443,251.60	101.19883	0.9999443
40° 54′	24,109,834.94	449,323.53	101.19917	0.9999434
40° 55′	24,103,762.99	455,395.48	101.19950	0.9999426

TABLE II

1″ OF LONGITUDE = 0.65695032″ OF θ

LONGITUDE	θ
84° 46′	$-1° 29′ 20.71461″$
84° 47′	$-1° 30′ 00.13163″$
84° 48′	$-1° 30′ 39.54865″$
84° 49′	$-1° 31′ 18.96567″$
84° 50′	$-1° 31′ 58.38269″$

Constants

$$R_b = 24,559,158.47 \text{ ft}$$
$$X_0 = C = 2,000,000 \text{ ft}$$
$$Y_0 = 510,419.96 \text{ ft}$$
$$l = 0.65695032$$
$$\lambda_m \text{ (central meridian)} = 82° 30′ 00.000″$$
$$1/(2 \, \rho_0^2 \sin 1″) = 2.3574 \times 10^{-10}$$

Table I gives for every minute of latitude the value of R, the Y coordinate of the intersection of the parallel of latitude with the central meridian, and the tabular difference for 1″ of latitude for both R and Y'. The values of R decrease, while the values of Y' increase. Table I also shows the amount by which a sea-level length must be increased or decreased to obtain the corresponding grid length. The values in the last column of Table I are the scale factors by which the sea-level lengths are multiplied to give the corresponding grid lengths. Note that the scale varies with the latitude.

Table II gives the value of θ for each full minute of longitude. To interpolate, the number of seconds of longitude is multiplied by the constant appearing in the heading, and the product is added to or subtracted from the tabular value. Table II is not really necessary, since $\Delta\lambda$ is readily obtained by Eq. (11-17). The value of θ can then be found by Eq. (11-18), in which l is the value given at the head of Table II or is included among the constants of the projection.

The constants following Table II are those discussed in Sections 11-8, 11-9, and 11-10.

Example 11-3.

The latitude and longitude of a control point in Ohio are $\phi = 40° 52' 44.602''$ and $\lambda = 84° 48' 16.225''$. Compute the plane coordinates of the control point on the projection for the Ohio–north zone.

Solution: The value of R for the point is obtained from Table I by interpolation, as follows:

$$R \text{ for } 40° 52' = 24{,}121{,}978.80 \text{ ft}$$
$$101.19883 \times 44.602'' = \underline{\quad -4{,}513.67}$$
$$R \text{ for } 40° 52' 44.602'' = 24{,}117{,}465.13 \text{ ft}$$

The value of θ is obtained from Table II by interpolation.

$$\theta \text{ for } 84° 48' = -1° 30' 39.54865''$$
$$0.656\ 95032 \times 16.225'' = \underline{\quad -10.65902''}$$
$$\theta \text{ for } 84° 48' 16.225'' = -1° 30' 50.20767''$$

Then the plane coordinates may be computed by Eqs. (11-21) and (11-22)

$$R_b = 24{,}559{,}158.47 \qquad\qquad X_0 = 2{,}000{,}000.00$$
$$R \cos \theta = \underline{24{,}109{,}046.29} \qquad R \sin \theta = \underline{-637{,}190.01}$$
$$Y = \quad 450{,}112.18 \text{ ft} \qquad\qquad X = \quad 1{,}362{,}809.99 \text{ ft}$$

Example 11-4.

A line joins the control point in Example 11-3 with a point whose coordinates are $X = 1{,}414{,}516.20$ ft and $Y = 459{,}168.24$ ft. Compute the geodetic azimuth of this line in the direction of the second point.

Solution: The grid azimuth is found by Eq. (8-14).

$$X_1 = 1{,}362{,}809.99 \qquad Y_1 = 450{,}112.18$$
$$X_2 = 1{,}414{,}516.20 \qquad Y_2 = 459{,}168.24$$
$$\Delta X = -51{,}706.21 \qquad \Delta Y = -9{,}056.06$$

$$\tan \alpha_s = \frac{-51{,}706.21}{-9{,}056.06}$$
$$= +5.709\ 57016$$
$$\text{grid } \alpha_s = 260° 03' 56.59''$$

The grid azimuth is taken from the south in this example because a geodetic azimuth is always reckoned from the south meridian.

The value of θ for the first point was determined in Example 11-3 by using the longitude of the point. However, in this example we will assume that only the plane coordinates of the ends of the line are known. It can be seen from Fig. 11-8 that

$$\tan \theta = \frac{X'}{R_b - Y} = \frac{X - X_0}{R_b - Y} \tag{11-23}$$

The computations for determining θ follow:

$$X = 1,362,809.99 \qquad R_b = 24,559,158.47 \qquad \tan \theta = \frac{-637,190.01}{24,109,046.29}$$

$$X_0 = 2,000,000.00 \qquad Y = 450,112.18 \qquad \tan \theta = -0.02642\ 94988$$

$$X' = -637,190.01 \qquad R_b - Y = 24,109,046.29 \qquad \theta = -1°\ 30'\ 50.21''$$

The correction term in Eq. (11-20) is now computed

$$X_2 - X_1 = +51,706.21 \qquad\qquad Y_1 = 450,112.18$$
$$Y_2 - Y_1 = +9,056.06 \qquad\qquad Y_0 = 510,419.96$$

$$\qquad\qquad\qquad\qquad\qquad Y_1 - Y_0 = -60,307.78$$

$$\frac{1}{2\rho_0^2 \sin 1''} = 2.357 \times 10^{-10} \qquad (Y_2 - Y_1)/3 = +3,018.69$$

$$\qquad\qquad\qquad\qquad\qquad\qquad\qquad -57,289.09$$

$$\text{correction term} = 5.171 \times 10^4 \times 2.357 \times 10^{-10} \times -0.573 \times 10^5 = -0.70''$$

Finally, the geodetic azimuth is determined by applying Eq. (11-20).

$$\text{grid azimuth} = 260°\ 03'\ 56.59''$$
$$+\theta = -1°\ 30'\ 50.21''$$
$$\overline{258°\ 33'\ 06.38''}$$
$$-\text{ correction term} = +0.70''$$
$$\text{geodetic azimuth} = 258°\ 33'\ 07.08''$$

Example 11-5.

The state plane coordinates of a point in the Ohio–north zone are $X = 2,097,462.33$ ft and $Y = 440,028.92$ ft. Determine the latitude and longitude of the point.

Solution: The value of θ is obtained by Eq. (11-23) given in Example 11-4, as follows:

$$X = 2,097,462.33 \qquad R_b = 24,559,158.47 \qquad \tan \theta = \frac{+97,462.33}{24,119,129.55}$$

$$X_0 = 2,000,000.00 \qquad Y = 440,028.92 \qquad \tan \theta = +0.00404\,08726$$

$$\underline{X' = +97,462.33} \qquad \underline{R_b - Y = 24,119,129.55} \qquad \theta = +0° \, 13' \, 53.4853''$$

$$\theta = +833.4853''$$

By Eq. (11-18),

$$\Delta\lambda = \frac{+833.4853}{0.656\,95032} = +1268.719'' = +0° \, 21' \, 08.719''$$

By Eq. (11-17), the longitude is

$$\lambda = 82° \, 30' \, 00.000'' - 0° \, 21' \, 08.719''$$

$$\lambda = 82° \, 08' \, 51.281''$$

The value of R can be obtained by a rearrangement of Eq. (11-21). Thus

$$R = \frac{R_b - Y}{\cos \theta} = \frac{24,119,129.55}{0.99999\,18355} = 24,119,326.47$$

Finally, the latitude is obtained from the value of R by interpolating in Table I of the projection tables.

$$R \text{ for } 40° \, 52' = 24,121,978.80$$

$$\underline{R \text{ for point} = 24,119,326.47}$$

$$\text{difference for seconds} = 2,652.33$$

$$\text{number of seconds} = \frac{2,652.33}{101.19883} = 26.209''$$

$$\phi = 40° \, 52' \, 26.209''$$

11-12. STATE PLANE COORDINATES BY AUTOMATIC DATA PROCESSING If the engineer or surveyor is faced with the task of converting several points from geographic to plane coordinates, or vice versa, on frequent occasions, it becomes desirable to program the solution for high-speed data processing or programmable calculators. In anticipation of this move to automatic data processing, the National Geodetic Survey has developed basic formulas for the different types of projections covering the states and the U. S. possession. These formulas, published in Coast and Geodetic Survey Publication 62-4, cover the projections given in Table 11-1 and, in addition, the 10 Alaska zones, the Lambert projections covering Puerto Rico, Virgin Islands, St. Croix, and American Samoa, and an approximate azimuthal equidistant projection covering the island of Guam.

Once the engineer completes the programming of the formulas for a Lambert projection, it becomes an easy matter to convert from one zone to another, or from one state to another using a Lambert grid. This is done simply by changing the list of constants that are given in Publication 62-4 for each projection. Since the formulas are basic, there is no table look-up involved. Trigonometric and other variable functions are developed by the computer.

11-13. APPLICATION OF THE STATE COORDINATE SYSTEMS

As mentioned in Section 11-1, the national network of horizontal control monuments is available to any engineer or surveyor who desires to connect his work with the federal system by methods of plane surveying. It is made usable through the state coordinate systems computed for each state, because the positions of these monuments are published both in geographic and state plane coordinates. If a point to be used is one whose Y and X coordinates are not known, they may be computed from the latitude and longitude of the point by the equations discussed in Sections 11-6 and 11-10.

When a federal monument has been established, it is the practice to set also a second monument, called an *azimuth mark*, which is far enough away to give an accurate point for backsighting. In addition to the azimuth marks so set, other control monuments in the vicinity can be used as backsights. The grid azimuths, as well as the geodetic azimuths, of these lines to azimuth marks are usually published along with the position data. If, however, no grid azimuth is available for a particular line whose geodetic azimuth is known, then the angle $\Delta\alpha$ or the angle θ is computed as explained in Section 11-5 or Section 11-9 and the grid azimuth may be obtained by Eq. (11-3) or (11-3a) or by Eq. (11-19) or (11-20). If the plane coordinates of the two ends of the line are known, then the grid azimuth may be determined by Eq. (8-13) or (8-14) of Section 8-20.

When a traverse survey is to be related to the state plane coordinate system, the first consideration is the necessary accuracy of the results. If the relative accuracy that is expected is no better than, say, one part in 5000, then the traverse may be related to the state coordinate system simply by beginning at a point whose state coordinates are known, carrying the traverse on grid azimuths, closing on another point whose state coordinates are known, and then adjusting the traverse to fit between these two points. This type of traverse is described in Section 8-19.

Since the state plane coordinate tables first became available to engineers and surveyors, a great advance has been made in surveying instrumentation, most notably the adoption of the glass circle direction theodolites described in Chapter 4 and the development of EDM's described in Chapter 2. Survey accuracy has been greatly enhanced with very little additional effort on the part of the surveyor. Thus it is not unusual nor is it difficult to obtain relative accuracies on the order of one part in 25,000 or better. Also, it is to be realized that land values constantly rise, a fact that requires upgrading of survey operations. These factors must be borne in mind when referring to some of the existing older publications describing the use of the state coordinate systems.

If the expected relative accuracy of a traverse survey is better than one part in 10,000, a few refinements must be made to the measurements. Since the transverse Mercator projections and the Lambert conformal projections are computed for the sea-level surface,† the distances measured in the traverse must be reduced to their corresponding sea-level length. The method for making this reduction for baseline and trilateration measurements using EDM was presented in Chapter 10.

The sea-level reduction can also be expressed as a reduction factor derived from Eq. (10-5). Taking only the first term on the right-hand side, the sea-level length can be expressed as

$$s = D - C = D - D\frac{h}{R} = D\left(1 - \frac{h}{R}\right)$$

and the sea-level reduction factor r_{SL} is given by

$$r_{SL} = \left(1 - \frac{h}{R}\right) \tag{11-24}$$

A horizontal distance D at elevation h is then reduced to its sea-level length s by multiplying by the sea-level reduction factor.

Table 11-5 gives the values of r_{SL} for an average radius of the sea-level surface of 20,906,000 ft when the elevation of the line is given in feet. If the elevation is expressed in metres, it must be converted to feet by multiplying by 39.37/12 before entering the table. If the radius of the earth in the project vicinity is substantially different from the average given above, the values given in Table 11-5 may not be sufficiently accurate. The reduction is then made using Eq. (11-24).

Example 11-6.

A line measures 22,456.72 ft at elevation 3260 ft. The average radius of curvature in the area is 20,847,000. Compute the sea-level distance using Table 11-5 and also by Eq. (11-24).

---

† Michigan's Lambert zones are computed for the 800-ft elevation rather than for sea level.

TABLE 11-5. Sea-Level Reduction Factors

ELEVATION h (ft)	SEA-LEVEL REDUCTION FACTOR	ELEVATION h (ft)	SEA-LEVEL REDUCTION FACTOR
500	0.9999 761	5,500	0.9997 369
1,000	0.9999 522	6,000	0.9997 130
1,500	0.9999 283	6,500	0.9996 891
2,000	0.9999 043	7,000	0.9996 652
2,500	0.9998 804	7,500	0.9996 413
3,000	0.9998 565	8,000	0.9996 173
3,500	0.9998 326	8,500	0.9995 934
4,000	0.9998 087	9,000	0.9995 695
4,500	0.9997 848	9,500	0.9995 456
5,000	0.9997 608	10,000	0.9995 217

Solution: By interpolation in Table 11-5,

$$r_{SL} = 0.9998\ 565 - 239 \times \frac{260}{500} = 0.9998\ 441$$

and

$$s = 22{,}456.72 \times 0.9998\ 441 = 22{,}453.22 \text{ ft}$$

By Eq. (11-24),

$$r_{SL} = 1 - \frac{3260}{20{,}847{,}000} = 0.9998\ 436$$

and

$$s = 22{,}456.72 \times 0.9998\ 436 = 22{,}453.21 \text{ ft}$$

The difference between the two values in this example is insignificant.

The sea-level lengths of the lines in the survey must be further reduced to their corresponding grid lengths by application of the grid scale factors given in the projection tables. When working on the transverse Mercator projection, the grid scale factor is a function of the distance east or west of the central meridian as shown in Table 3 for Arizona. Note that if the area is less than about 295,000 ft east or west of the central meridian ($X' = \pm 295{,}000$ ft), the grid lengths will be less than the sea-level lengths; whereas beyond 295,000 ft, the grid lengths will be greater than the sea-level lengths.

When working on the Lambert conformal projection, the grid scale factor is a function of the latitude of the area as shown in Table I for Ohio–north

(Table 11-4). The factor is thus constant for east-west lines but changes in the north-south direction.

The value of X' needed for determining the grid scale factor on the transverse Mercator projection is obtained by Eq. (11-8). When using the Lambert conformal projection, the latitude of the area must be known in order to determine the scale factor. This can be obtained either by the method shown in Example 11-5 or else by scaling it from a map of the area.

It is often desirable and expedient to assign a combination factor for sea-level and grid lengths to an area whose elevation does not fluctuate too much. For example, if the average elevation in a given area is 2600 ft above sea level, the sea-level reduction factor is 0.9998756. If the area is in such a latitude on a Lambert projection that the grid scale factor is 0.9999000, the combination factor in this area would be $(0.9998756)(0.9999000) = 0.9997756$. A ground length of 10,000 ft in this area would therefore correspond to a grid length of 9997.756 ft.

Figure 11-9 illustrates the application of the sea-level reduction and grid scale corrections to the length of a line. The line AB lying at an elevation h is projected to the sea-level length $A'B'$ by the sea-level reduction factor. The trace of the grid projection surface is shown at five different positions. If the grid surface lies below sea level as shown at position 1, then the grid scale factor is less than unity and so is the combination factor. If the grid surface generally coincides with sea level as shown at position 2, the grid scale factor is unity and the combination factor is less than unity.

If the grid surface lies above the sea-level surface as at positions 3, 4, and 5, the grid scale factor is greater than unity. At position 3 the combination factor is less than unity. At position 4 the grid surface is coincident with the ground surface and hence the combination factor is unity. At position 5 the grid surface lies above the ground surface and the combination factor is greater than unity.

The lengths of traverse lines are usually short enough to allow the methods given in Sections 8-4, 8-5, and 8-6 to be used for computing azimuths from

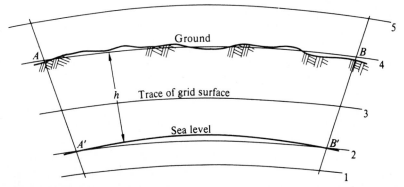

Figure 11-9. Relationship between sea-level length, grid length, and lengths of lines at different elevations.

measured angles. After the measured lengths have been reduced to their appropriate grid lengths and the azimuths have been computed and adjusted, the traverse is then computed and adjusted on the state coordinate system by the methods presented in Chapter 8. If, however, the lines in the traverse are long, as is quite feasible using EDM, then the azimuths of the lines must be modified by the small correction term given in either Eq. (11-3a) or (11-20).

In Fig. 11-3 assume that the grid azimuth of CD is known. This is the direction of the straight line from C to D. If a theodolite is set up at C and a backsight is made to D, the grid projection of the line of sight is the curved line CD. If the theodolite is now sighted on E, the measured angle would be larger than the plane angle DCE. If this measured angle were to be added to the grid azimuth of CD to obtain the grid azimuth of CE, the resulting value would be too large by the small angle between the straight line CD and the curved line CD. The small angle is the correction term of Eq. (11-3a). The same analysis can be made of the angle at A from B to C shown in Fig. 11-7. Thus when the traverse contains very long lines, a preliminary solution for the coordinates of the points will allow an evaluation of the correction term at each end of each line. The corrected grid azimuths are then used to compute the final coordinates of the traverse stations.

Triangulation of moderate extent may be executed and computed by using Y and X coordinates on the state coordinate systems. A line makes an ideal baseline when the coordinates of both ends are known beforehand, since both the grid azimuth and the grid distance may be computed from the coordinates by inversing the line. Plane angles are used to solve the triangles and to carry grid azimuths through the triangulation system.

If the system is fairly extensive with long lines, the plane angles must be determined after correcting the directions of the lines for the small azimuth correction term. This requires that a preliminary triangulation computation be performed as outlined in Chapter 10 in order to obtain preliminary coordinates with which to compute the correction terms. Since the triangulation will ultimately be adjusted to some framework of control points, the variation of coordinates adjustment method also requires a preliminary set of coordinates. It therefore requires only a slight amount of additional effort to evaluate the azimuth correction terms at each end of each line.

In performing trilateration on the state coordinate system, the measured lengths are all reduced to sea level by the methods outlined in Sections 10-29 and 10-30. They are then reduced to grid lengths for use in computing the coordinates of the trilateration stations by applying the appropriate grid scale factors. If an approximate adjustment is to be made as explained in Section 10-31, the computed angles are plane angles. These plane angles will not give exactly the azimuths that would be obtained by applying the small azimuth correction term to each line. However, the difference is quite negligible considering that the adjustment is approximate. Therefore no azimuth correction need be applied when calculating the azimuths of the trilateration lines. The only occasion for this correction will arise in trilateration when the azimuth of one of the lines has been determined by astronomical observations as a check on the orientation of the trilateration net. This must be assumed to be the geodetic azimuth of the line unless the deflection of the vertical line from a normal to the spheroid surface at the point of observation is known (see Section 1-2). Correction for the deflection of the vertical is outside the scope of this discussion. The grid azimuth is then computed from the geodetic azimuth by Eq. (11-3a) or (11-20) and compared with its value obtained from the trilateration computations.

If a least-squares adjustment of a large trilateration net is made on a state coordinate system by the method of variation of coordinates, no additional refinement need be made to the lines other than their reduction to grid length. If the trilateration is to be combined with triangulation, then the observation equations (see Appendix A) which involve measured angles must be modified to take into account the small azimuth correction terms.

Because of the inferior accuracy and reliability in the determination of the position of a point by the methods of intersection and resection, there is little or no justification for applying any azimuth correction to the lines. If the positions of the known points are on the state coordinate system, then the computations are straightforward as outlined in Sections 10-34 and 10-35, and the new point is automatically on the state coordinate system.

A problem frequently arises in which a large survey extends from one coordinate-projection zone into an adjacent zone. The solution requires a conversion from plane coordinates of one zone to geodetic coordinates, as illustrated in Examples 11-2 and 11-5, and then a conversion from geodetic coordinates to plane coordinates of the next zone, as illustrated in Examples 11-1 and 11-3. The transition usually requires that at least two points in each zone near the common boundary of the two zones be converted, in order that a line of known grid azimuth may be established for each zone.

11-14. LAND AREAS FROM STATE COORDINATES
If the state coordinates of the corners of a parcel of land are known, the grid area encompassed within the boundary may be computed by Eq. (8-25) or (8-26) of Section 8-20. Since this grid area does not, as a rule, correspond to the actual ground area, it should be *divided* by the *square* of the combination factor used to determine grid lengths from ground lengths. For example, if the grid area of a parcel of land lying in an area whose combination factor is 0.9997750 was computed by Eq. (8-25) of Section 8-20 to be 45,675 ft^2, then the actual ground area is $(45,675)/(0.9997750)^2 = 45,696$ ft^2.

11-15. PROJECT LENGTHS FROM STATE COORDINATES
Engineering projects require the field location of construction lines and survey points. Since most large projects will be established with reference to control

points on the state coordinate system, the construction plans and drawings will contain state coordinates and grid lines. If the state coordinates of two ends of a construction line are used to inverse the line, the grid length and grid bearing or azimuth will result. The proper length to be laid off, however, is the actual ground length. The difference between the grid length and the ground length can be taken into account by one of two methods. The first method, which is quite practical, is to ignore the difference. This method assumes that the errors in layout measurements are larger than the errors introduced by neglecting the difference between the grid and the ground distance. For example, in an area whose elevation is 2500 ft, the sea-level reduction factor is 0.9998804. Suppose the grid scale factor is 0.9999000 or one part in 10,000. The combination factor is then 0.9997804. This means that the difference between the ground length and the grid length is only 0.02 ft/100 ft or 0.22 ft/1000 ft. This small discrepancy, amounting to about one part in 5000, could be ignored on most construction projects. In high country the sea-level reduction factor becomes significant, particularly if the project lies in an area of the state projection where the grid scale factor is less than unity.

The second method is to compute the combination sea-level grid factor, and then to *divide* all grid distances indicated on, or derived from, the plans in order to obtain the correct ground or project lengths. In order to accomplish this without misunderstanding between the surveyor, engineer, and contractor, a note should accompany each construction drawing giving explicit instructions to the user. The note could possibly read, "All distances shown on this set of plans (this drawing) or derived from plane coordinates shown on the plans (drawing) are grid distances on the—Coordinate System, Zone—. To obtain ground distances for laying out construction lines, divide grid distances by 0.9998940."

11-16. LOST AND OBLITERATED CORNERS When the coordinates of a land corner on a state system are known and the corner cannot be found by the usual methods, its ground position can be determined by locating the point corresponding to those coordinates. This is accomplished by running a traverse from other stations, for which the state-coordinate data are available, to a new point in the immediate vicinity of the point sought for. The coordinates of the new point and the sought-for point are compared, and the azimuth and distance from the new point to the old one are laid out on the ground. The point thus established is the ground position corresponding to the coordinates of the required corner. If any remnants of the original corner remain in position, they may be found by a careful and detailed examination of the ground within a short distance of the restored position.

Marks that are otherwise difficult to find, and which if found are apt to be considered as of doubtful origin, may be recovered and proved authentic by the use of the state coordinate system. A fragment of stone or a piece of decayed wood may constitute a satisfactory recovery if found in the ground position corresponding to the coordinates of the landmark. If no marks are found in the position indicated by the coordinates, it may safely be concluded that the monument has been destroyed, and the coordinate position may be accepted as primary evidence of the ground location.

As the accuracy of the restored position reflects the errors of both the original survey and the restoration survey, care must be taken to maintain a high order of accuracy in the restoration survey, especially when the restoration survey is longer than the original one. Although it is not necessary to employ the same stations in the resurvey that were used in the original survey, it is desirable to use stations that were in the same traverse or survey net. A station on some other traverse may not be so well related in position to the lost mark because of the methods that were used in the previous adjustment.

11-17. DESCRIPTIONS BY COORDINATES A legal description of a parcel of land identifies the land for title purposes and provides information necessary for locating the land on the ground. A description that is satisfactory for title purposes may be wholly inadequate for a field location of the property. One of the simplest forms of land description for title purposes is afforded by the public-land method of surveying, in which a description by reference to township, range, and section can apply to one and only one parcel of land. Thus "the northwest quarter of the southeast quarter ($NW\frac{1}{4}SE\frac{1}{4}$) of Section Ten (10), Township Two South (T2S), Range Six East (R6E) of the Meridian of Michigan" is satisfactory for title purposes. It is also satisfactory for survey purposes as long as the controlling monuments are in existence. For the sake of perpetuating the corners, it would be helpful to add the coordinates of the center of the section, its south and east quarter-corners, and its southeast corner. If a corner monument is later obliterated, these coordinates will provide an effective means of recovering and identifying any part of the monument that may remain in position. If the original monument is entirely destroyed, the coordinates become strong evidence of where it stood and make possible an accurate replacement.

When the description is by metes and bounds, it may be simplified by reference to a plat which is made a part of the conveyance. The azimuths and lengths of the sides of the parcel, as well as the coordinates of the place of beginning, can be shown on this plat.

Where no plat is available, the following form taken from the Second Progress Report of the Joint Committee of the Real Property Division, American Bar Association, and the Surveying and Mapping Division, American Society of Civil Engineers, Proceedings, A.S.C.E., June, 1941, may be used:

———— situated in the Town of ————, County of ————, State of ————, and bounded as follows:

"Beginning at a drill hole in a stone mound which is set in the corner of a stone wall on the north line of Farm Road at the southwest corner of land of Peter L.

Prince and at the southeast corner of land hereby conveyed, the coordinates of which monument referred to the Massachusetts Coordinate System, Mainland Zone, are: $x = 417{,}603.29$, $y = 316{,}042.17$. Thence, on an azimuth of $81° 39' 30''$, 123.39 feet along the northerly line of Farm Road to an iron pin at the southwest corner of the tract hereby conveyed; thence, on an azimuth of $181° 47' 30''$, 145.82 feet along the easterly line of land of Arthur C. Hicks to an iron pin in a stone wall at the northwest corner of the tract hereby conveyed; thence, on an azimuth of $276° 32' 00''$, 62.04 feet along a stone wall on the southerly line of Peter L. Prince to a drill hole in a stone bound in the wall at the northeast corner of land hereby conveyed; thence, on an azimuth of $335° 10' 20''$, 133.09 feet along a stone wall on the westerly line of land of said Prince, to the place of beginning."

Zero azimuth is grid south in the Massachusetts Coordinate System, Mainland Zone.

This description was written June 1, 1939, from data secured by survey made by Fred L. Connor, Civil Engineer, in March and April, 1939.

11-18. ESTABLISHMENT OF STATE SYSTEMS

The legislatures of the different states have enacted laws establishing the legal status of the state coordinate systems and have defined its use and limitations. Assembly Bill No. 546 of the California legislature which was passed by the State Assembly on May 13, 1947, and by the State Senate on June 16, 1947, is typical of the legislation giving legal status to the state coordinate systems. In 1953 this bill was incorporated into the State of California Public Resources Code under Division 8, Surveying and Mapping, Chapter 1, and reads as follows.

CHAPTER 1 CALIFORNIA COORDINATE SYSTEM

§8801. **California coordinate system:** The system of plane coordinates which has been established by the United States Coast and Geodetic Survey for redefining and stating the positions or locations of points on the surface of the earth within the State of California is the "California Coordinate System." (Added Stats. 1953, c. 108, p. 832,§ 1.)

§8802. **Division of state into zones; area included in each of the zones designated:** For the purpose of the use of the California Coordinate System the State is divided into seven zones.

The area in the following counties constitutes Zone 1: Del Norte, Humboldt, Lassen, Modoc, Plumas, Shasta, Siskiyou, Tehama, and Trinity.

The area included in the following counties constitutes Zone 2: Alpine, Amador, Butte, Colusa, El Dorado, Glenn, Lake, Mendocino, Napa, Nevada, Placer, Sacramento, Sierra, Solano, Sonoma, Sutter, Yolo, and Yuba.

The areas included in the following counties constitutes Zone 3: Alameda, Calaveras, Contra Costa, Madera, Marin, Mariposa, Merced, Mono, San Francisco, San Joaquin, San Mateo, Santa Clara, Santa Cruz, Stanislaus, and Tuolumne.

The area included in the following counties constitutes Zone 4: Fresno, Inyo, Kings, Monterey, San Benito, and Tulare.

The area included in the following counties constitutes Zone 5: Kern, San Bernardino, San Louis Obispo, Santa Barbara, and Ventura.

The area included in the following counties constitutes Zone 6: Imperial, Orange, Riverside, and San Diego.

The area included in Los Angeles County constitutes Zone 7. (Added Stats. 1953, c. 108, p. 833, § 1.)

§8803. California coordinate system, zone 1: As established for use in Zone 1, the California Coordinate System shall be named, and on any map on which it is used it shall be designated, the "California Coordinate System, Zone 1."

The California Coordinate System, Zone 1, is a Lambert conformal projection of the Clarke spheroid of 1866, having standard parallels at north latitudes 40 degrees 00 minutes and 41 degrees 40 minutes, along which parallels the scale shall be exact. The point of control of coordinates is at the intersection of the meridian 122 degrees 00 minutes west longitude with the parallel 39 degrees 20 minutes north latitude, such point of control being given the coordinates: x (east) = 2,000,000 feet and y (north) = 0 feet. (Added Stats. 1953, c. 108, p. 833, § 1.)

§8804. California coordinate system, zone 2: As established for use in Zone 2, the California Coordinate System shall be named, and on any map on which it is used it shall be designated, the "California Coordinate System, Zone 2."

The California Coordinate System, Zone 2, is a Lambert conformal projection of the Clarke spheroid of 1866, having standard parallels at north latitudes 38 degrees 20 minutes and 39 degrees 50 minutes, along which parallels the scale shall be exact. The point of control of coordinates is at the intersection of the meridian 122 degrees 00 minutes west longitude with the parallel 37 degrees 40 minutes north latitude, such point of control being given the coordinates: x (east) = 2,000,000 feet and y (north) = 0 feet. (Added Stats. 1953, c. 108, p. 833, § 1.)

§8805. California coordinate system, zone 3: As established for use in Zone 3, the California Coordinate System shall be named, and on any map on which it is used shall be designated, the "California Coordinate System, Zone 3."

The California Coordinate System, Zone 3, is a Lambert conformal projection of the Clarke spheroid of 1866, having standard parallels at north latitudes 37 degrees 04 minutes and 38 degrees 26 minutes, along which parallels the scale shall be exact. The point of control of coordinates is at the intersection of the meridian 120 degrees 30 minutes west longitude with the parallel 36 degrees 30 minutes north latitude, such point of control being given the coordinates: x (east) = 2,000,000 feet and y (north) = 0 feet. (Added Stats. 1953, c. 108, p. 833, § 1.)

§8806. California coordinate system, zone 4: As established for use in Zone 4, the California Coordinate System shall be named, and on any map on which it is used it shall be designated, the "California Coordinate System, Zone 4."

The California Coordinate System, Zone 4, is a Lambert conformal projection of the Clarke spheroid of 1866, having standard parallels at north latitudes 36 degrees 00 minutes and 37 degrees 15 minutes, along which parallels the scale shall be exact. The point of control of coordinates is at the intersection of the meridian 119 degrees 00 minutes west longitude with the parallel 35 degrees 20 minutes north latitude, such point of control being given the coordinates: x (east) = 2,000,000 feet and y (north) = 0 feet. (Added Stats. 1953, c. 108, p. 834, § 1.)

§8807. California coordinate system, zone 5: As established for use in Zone 5, the California Coordinate System shall be named and on any map on which it is used shall be designated, the "California Coordinate System, Zone 5."

The California Coordinate System, Zone 5, is a Lambert conformal projection of the Clarke spheroid of 1866, having standard parallels at north latitudes 34 degrees 02 minutes and 35 degrees 28 minutes, along which parallels the scale shall be exact. The point of control of coordinates is at the intersection of the meridian 118 degrees 00 minutes west longitude with the parallel 33 degrees 30 minutes north latitude, such point of control being given the coordinates: x (east) = 2,000,000 feet and y (north) = 0 feet. (Added Stats. 1953, c. 108, p. 834, § 1.)

§8808. California coordinate system, zone 6: As established for use in Zone 6, the California Coordinate System shall be named, and on any map on which it is used it shall be designated, the "California Coordinate System, Zone 6."

The California Coordinate System, Zone 6, is a Lambert conformal projection of the Clarke spheroid of 1866, having standard parallels at north latitudes 32 degrees 47 minutes and 33 degrees 53 minutes, along which parallels the scale shall be exact. The point of control of coordinates is at the intersection of the meridian 116 degrees 15 minutes west longitude with the parallel 32 degrees 10 minutes north latitude, such point of control being given the coordinates: x (east) = 2,000,000 feet and y (north) = 0 feet. (Added Stats. 1953, c. 108, p. 834, § 1.)

§8809. California coordinate system, zone 7: As established for use in Zone 7, the California Coordinate System shall be named, and on any map on which it is used it shall be designated, the "California Coordinate System, Zone 7."

The California Coordinate System, Zone 7, is a Lambert conformal projection of the Clarke spheroid of 1866, having standard parallels at north latitudes 33 degrees 52 minutes and 34 degrees 25 minutes, along which parallels the scale shall be exact. The point of control of coordinates is at the intersection of the meridian 118 degrees 20 minutes west longitude with the parallel 34 degrees 08 minutes north latitude, such point of control being given the coordinates: x (east) = 4,186,692.58 feet and y (north) = 4,160,926.74 feet. (Added Stats. 1953, c. 108, p. 834, § 1.)

§8810. What plane coordinates consist of: The plane coordinates of a point on the earth's surface, to be used in expressing the position or location of the point in the appropriate zone of this system, shall consist of two distances, expressed in feet and decimals of a foot. One of these distances, to be known as the "x-coordinate," shall give the position in an east-and-west direction from the Y axis; the other, to be known as the "y-coordinate," shall give the position in a north-and-south direction from the X axis. The Y axis of any zone shall be parallel with the meridian passing through the point of control of that zone. The X axis of any zone shall be at right angles to the meridian which passes through the point of control of that zone. (Added Stats. 1953, c. 108, p. 835, § 1.)

§8811. Depending upon and conformity with plane coordinates of triangulation and traverse stations of United States Coast and Geodetic Survey within state: The plane coordinates of a point in any zone shall be made to depend upon and conform with the plane coordinates of the triangulation and traverse stations of the United States Coast and Geodetic Survey within the State of California, as those coordinates shall be determined to conform with the provisions of this chapter. (Added Stats. 1953, c. 108, p. 835, § 1.)

§8812. Survey extending from one zone into another; positions delineated; naming zone used: If the survey of any parcel of land extends from one coordinate zone into another, the positions of all points delineated upon the map thereof may be referred to either of these zones. The zone which is used shall be specifically named in the title upon the map. (Added Stats. 1953, c. 108, p. 835, § 1.)

§8813. Manner of marking position of California coordinate system; requirements of survey or map: The position of the California Coordinate System shall be as marked on the ground by triangulation or traverse stations whose geodetic positions have been rigidly adjusted on the North American datum of 1927 and established in conformity with the standards adopted by the United States Coast and Geodetic Survey for first-order and second-order work, and whose coordinates have been computed on the system defined. Any survey or map purported to be based on the California Coordinate System shall have established connections to at least two of such stations. (Added Stats. 1953, c. 108, p. 835, § 1, as amended Stats. 1953, c. 853, p. 2182, § 2.)

HISTORICAL NOTE

The 1953 amendment relocated in the first sentence the phrase "whose geodetic positions have been rigidly adjusted on the North American datum of 1927." Previously the phrase followed the reference to standards for first-order and second-order work.

The 1953 amendatory act, in section 1 thereof, made a similar amendment to Stats. 1947, c. 1307, p. 2845, § 5.

Section 3 of the 1953 amendatory act read: "Section 2 of this act becomes operative only if Division 8 of the Public Resources Code is enacted by the Legislature at its 1953 Regular Session, and in such case at the same time as said Division 8 takes effect, at which time Section 5 of Chapter 1307 of the Statutes of 1947, as amended by this act, is repealed."

§8814. Requirements of map, survey, conveyance or other instrument to constitute, when recorded, constructive notice, reference to recorded data controlling in case of conflict: Any map, survey, conveyance or other instrument delineating, or affecting the title to, real property which delineates, describes or refers to such property, or any part thereof, by reference to coordinates based on the California Coordinate System shall, in order to constitute, when recorded, constructive notice thereof under the recording laws, also delineate, describe or refer to such property, or such part thereof, by reference to data appearing of record in any office the records of which constitute constructive notice under the recording laws, sufficient to identify it without recourse to such coordinates; and in case of conflict between them the references to such recorded data shall be controlling. (Added Stats. 1953, c. 1195, p. 2710, § 2.)

HISTORICAL NOTE

Former section 8814, added by Stats. 1953, c. 108, p. 835, § 1, repealed by Stats. 1953, c. 1195, p. 2710, § 1, read: "If any instrument affecting the title to real property contains a reference to a point or line established in accordance with the California

Coordinate System of Surveying and also a reference to a point or line established with reference to data appearing of record in the county recorder's office, in a manner which will cause a conflict in position, the reference to the point or line established with reference to recorded data is controlling."

§8815. Use of term limited to coordinates based on California coordinate system: The use of the term "California Coordinate System" on any map or in any field notes referring to the system shall be limited to coordinates based on the California Coordinate System as defined in this chapter. (Added Stats. 1953, c. 108, p. 836, § 1.)

§8816. Use of system optional. The use of the California Coordinate System by any person, corporation, or governmental agency engaged in land surveying or mapping is optional. (Added Stats. 1953, c. 108, p. 836, § 1.)

11-19. REMARKS ON THE STATE COORDINATE SYSTEMS

The various state coordinate systems are made possible by virtue of the highly precise geodetic-surveying methods used in establishing the initial framework of horizontal control that defines these systems. Each traverse or local triangulation network which has been executed by methods to give position closures of 1 in 10,000 or better is an addition to the state system and helps to define it more completely. The National Geodetic Survey is continuously breaking the larger triangulation networks down to more usable control for local surveyors and engineers. In turn, the highway departments, utilities, railroads, and county and city engineers are filling in between the federally established control. The fact that these surveys are all related to the state coordinate systems means that they are all related to one another. This is the most singular advantage of the state coordinate systems.

Other advantages of using the state coordinate system are as follows: (1) A traverse of relatively low accuracy run between a pair of control points is actually raised in accuracy after an adjustment between the control points is made. (2) The use of well-established control points in a traverse eliminates serious mistakes in measuring both distances and angles. (3) A point whose Y and X coordinates have been determined can, if lost, always be replaced with the degree of precision with which it was originally established. (4) Maps which have been controlled by coordinated points will always conform when joined, no matter how unrelated were the projects that necessitated the maps. (5) The use of a common reference system for surveys reduces or eliminates costly duplication in the way of numerous control surveys over the same area by various engineers and surveyors. (6) The use of the state coordinate system permits surveys to be carried over state-wide distances by using plane-surveying methods with results that approach those obtained by geodetic methods. (7) Photogrammetric mapping can be conducted at much less expense when all control points in the area to be mapped are on the same system.

Many additional advantages and applications of the state-coordinate systems could be listed. The reader is urged to take this subject under serious consideration when practicing the civil engineering profession. The use of the state systems continues to increase, and many economies are being realized by their application.

11-20. UNIVERSAL TRANSVERSE MERCATOR (UTM) SYSTEM

The universal transverse Mercator system shown in Fig. 11-10 employs 60 zones numbered 1 through 60 commencing at $\lambda = 180°$, each of which embraces 6° of longitude, and lies between 80° north latitude and 80° south latitude. The central meridian for each zone lies at the center of that zone. Coordinates are expressed in metres. The geometry of each zone of the UTM is identical with that of the state transverse Mercator zone described in Sections 11-4 through 11-6. However, since it is a world-wide system, plane coordinates are computed from different reference ellipsoids for different countries. At present, the North American continent uses the Clarke spheroid of 1866; all but the northern countries of Africa use the Clarke spheroid of 1880; India and surrounding countries extending into the Middle East and Southeast Asia employ the Everest ellipsoid; Japan and Manchuria use the

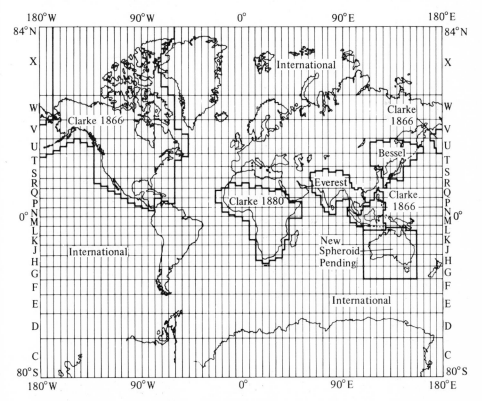

Figure 11-10. Universal transverse Mercator (UTM) system. The zones are numbered from 1 to 60 eastward from 180° West Longitude.

Bessel ellipsoid; and the remainder of the countries are on the International ellipsoid.

The zone number of any point can be determined from the following equations

$$N = (180° + \lambda_E)/6$$
$$N = (180° - \lambda_W)/6$$

$$(11\text{-}25)$$

in which N is the zone number and λ_E and λ_W denote east and west longitude. If the values in parentheses are not exactly divisible by 6, then N is the next higher zone.

Example 11-7.

Compute the zone number for a point P whose longitude is $25° 42' 16''$ E. Compute the zone number for a point Q whose longitude is $123° 45' 36''$ W.

Solution: By Eq. (11-25)

$$N_P = (180° + 25° 42' 16'')/6 = 34 + 1 = 35$$
$$N_Q = (180° - 123° 45' 36'')/6 = 9 + 1 = 10$$

The origin of coordinates for each zone is the intersection of its central meridian and the equator. In the northern hemisphere, the coordinates of the origin in the northern hemisphere are $X_0 = 500,000$ m; $Y_0 = 0$ (see Section 11-6). In the southern hemisphere $X_0 = 500,000$ m; $Y_0 = 10,000,000$ m. The scale factor along the central meridian is 0.9996. The ellipses of intersection of the enveloping cylinder (see Fig. 11-1) with the ellipsoid along which the scale is unity lie 180,000 m east and 180,000 m west of the central meridian. The scale at the outer east-west limits of each zone is about 1.0010 at the equator.

Formulas for the direct conversion of latitude and longitude into plane coordinates on the UTM grid and for converting to grid azimuths using a computer or programmable calculator are given in the Department of the Army Technical Manual TM5-241-4/1. TM5-241-4/2 is used for conversion from plane coordinates to latitude and longitude. Both publications also contain tables for use in making these conversions when it is required to make the conversions only occasionally.

The UTM system is employed by the U. S. Department of Defense in its world-wide mapping programs. Surveys of many military and defense-oriented installations are often required to be expressed in UTM grid co-ordinates. Because the UTM grid embraces a much wider area than the transverse Mercator or Lambert grids used to define the state plane co-ordinate systems, it is found quite useful to those responsible for planning over large areas embracing one or more states.

PROBLEMS

11-1. The latitude and longitude of a point on the Arizona east projection are, respectively, 33° 21′ 46.864″ N and 110° 32′ 14.755″ W. Compute the state plane coordinates of the point.

11-2. Solve Problem 11-1 for a point in latitude 33° 22′ 20.565″ N and longitude 110° 28′ 16.226″ W.

11-3. Solve Problem 11-1 for a point in latitude 33° 23′ 42.580″ N and longitude 109° 52′ 15.588″ W.

11-4. The plane coordinates of a point on the Arizona east projection are $X = 392,546.50$ ft and $Y = 862,545.90$ ft. Compute the latitude and longitude of the point.

11-5. Solve Problem 11-4 for a point whose coordinates are $X = 396,842.50$ ft and $Y = 870,216.44$ ft.

11-6. Solve Problem 11-4 for a point whose coordinates are $X = 612,006.70$ ft and $Y = 856,454.86$ ft.

11-7. The plane coordinates of a point on the Arizona east projection are $X = 203,146.22$ ft and $Y = 516,422.87$ ft. Compute the scale factor at the point.

11-8. Solve Problem 11-7 for a point whose coordinates are $X = 517,865.14$ ft and $Y = 632,767.16$ ft.

11-9. Solve Problem 11-7 for a point whose coordinates are $X = 506,815.84$ ft and $\dot{Y} = 425,550.70$ ft.

11-10. Compute the angle $\Delta\alpha$ at the point given in Problem 11-1.

11-11. Compute the angle $\Delta\alpha$ at the point given in Problem 11-2.

11-12. Compute the angle $\Delta\alpha$ at the point given in Problem 11-3.

11-13. Three points on the Arizona east projection have the following coordinates:

POINT	Y	X
A	1,424,516.50	177,548.82
B	1,247,030.76	156,210.08
C	1,250,385.26	402,381.80

Compute the plane angle at B from A to C. Compute the spherical angle at B from A to C (see Fig. 11-3).

11-14. Given the plane coordinates of Problem 11-13, compute the plane and spherical angle at A from C to B.

11-15. Given the plane coordinates of Problem 11-13, compute the plane and spherical angle at C from B to A.

11-16. The latitude and longitude of a point on the Ohio–north projection are, respectively, 40° 51′ 50.632″ N and 84° 47′ 16.600″ W. Compute the state plane coordinates of the point.

11-17. Solve Problem 11-16 for a point in latitude 40° 53′ 16.622″ N and longitude 84° 46′ 22.300″ W.

11-18. Solve Problem 11-16 for a point in latitude 40° 54′ 50.255″ N and longitude 82° 10′ 10.560″ W.

11-19. The plane coordinates of a point on the Ohio–north projection are $X_A = 1,989,503.70$ ft and $Y_A = 446,716.09$ ft. Compute the latitude and longitude of the point.

11-20. Solve Problem 11-19 for a point with coordinates $X_A = 2{,}045{,}662.22$ ft and $Y_A = 445{,}162.34$ ft.

11-21. Solve Problem 11-19 for a point with coordinates $X_A = 2{,}562{,}716.82$ ft and $Y_A = 432{,}868.85$ ft.

11-22. Compute the mapping angle at the point given in Problem 11-19.

11-23. Compute the mapping angle at the point given in Problem 11-20.

11-24. Compute the mapping angle at the point given in Problem 11-21.

11-25. The coordinates of two points A and B on the Ohio–north projection are $X_A = 1{,}989{,}476.22$ ft and $Y_A = 444{,}526.24$ ft; $X_B = 2{,}008{,}196.21$ ft and $Y_B = 439{,}986.22$ ft. Compute the geodetic azimuth of AB and BA, both measured from south. Use Eq. (11-20).

11-26. Three points on the Ohio–north projection have the following coordinates:

POINT	Y	X
P	447,196.20	1,858,433.60
Q	186,540.02	1,854,100.36
R	181,102.35	2,002,216.80

Compute the plane angle and the spherical angle at Q from P to R.

11-27. Given the plane coordinates of Problem 11-26, compute the plane and spherical angles at P from R to Q.

11-28. Given the plane coordinates of Problem 11-26, compute the plane and spherical angles at R from Q to P.

BIBLIOGRAPHY

Berry, R. M. 1972. Simple algorithms for the calculation of scale factors for plane coordinate systems. *Proc. Ann. Meet. Amer. Cong. Surv. & Mapp.* p. 260.

Burkholder, E. F. 1980. The Michigan scale factor. *Proc. Ann. Meet. Amer. Cong. Surv. & Mapp.* p. 180.

Department of the Army. 1958. *Universal Transverse Mercator Grid.* Technical Manual TM 5-241-8, Dept. of Defense, Washington, D.C.

Department of the Army. 1958. *Universal Transverse Mercator Grid Tables for Latitude 0°–80°: Clarke 1866 Spheroid (Metres): Vol. II: Transformation from Geographic to Grid.* Technical Manual TM5-241-4/1, Dept. of Defense, Washington, D.C.

Department of the Army. 1958. *Universal Transverse Mercator Grid Tables for Latitude 0°–80°: Clarke 1866 Spheroid (Metres): Vol. II: Transformation from Grid to Geographic.* Technical Manual TM5-241-4/2, Dept. of Defense, Washington, D.C.

Gale, P. 1971. Practical uses of coordinates-or, control surveys in operation. *Proc. Fall Conv. Amer. Cong. Surv. & Mapp.* p. 45.

Graff, D. R. 1973. Survey coordinate systems. *Proc. Fall Conv. Amer. Cong. Surv. & Mapp.* p. 205.

Hillman, H. 1977. Conversion between latitude and longitude and state plane, and other coordinate systems with modern portable calculators. *Proc. Fall Conv. Amer. Cong. Surv. & Mapp.* p. 345.

Hillman, H. 1980. Power series algorithm for conversion between latitude and longitude and the plane coordinate systems. *Proc. Ann. Meet. Amer. Cong. Surv. & Mapp.* p. 168.

Laird, M. O. 1978. Design criteria for a cohesive North American plane coordinate system. *Surveying and Mapping* 38:125.

Mitchell, H. C., and Simmons, L. G. *The State Coordinate Systems (a Manual for Surveyors)*. Special Publication No. 235, U. S. Coast and Geodetic Survey. U.S. Government Printing Office.

Stephenson, R. B. 1978. Using your state coordinate system. *Proc. Fall Conv. Amer. Cong. Surv. & Mapp.* p. 378.

Chapter 12
Practical Astronomy

12-1. USE OF ASTRONOMY The engineer or surveyor needs a knowledge of practical astronomy in order to make intelligently the observations and computations necessary for the determination of latitude, time, longitude, and azimuth. The purpose of the present chapter is not to supplant a text on astronomy, but rather to present, in condensed form, methods of observation and computation which can be used when the survey is being conducted with an ordinary engineer's transit or a theodolite. For the development of the formulas used, and for more exact methods, reference can be made to the readings at the end of the chapter.

The various heavenly bodies are considered to be located on the surface of a sphere of infinite radius. The center of this sphere is the center of the earth. The field observations usually consist of the measurement of horizontal and vertical angles to some heavenly body. The computations involve the solution of the spherical triangle formed by the observer's zenith, the celestial north pole, and the heavenly body.

Where only moderate precision is required, the field observations can be made either with a sextant or with a transit that is equipped preferably with a full vertical circle and a prismatic eyepiece. A good watch is needed for determining the instant at which the observation is made. An ephemeris is required for the computations. The *American Ephemeris and Nautical Almanac*, which is printed each year by the U.S. Government Printing Office, contains tables that furnish the positions of the principal heavenly bodies. The Nautical Almanac Office of the U.S. Naval Observatory annually prepares the ephemeris for the U.S. Department of Interior's Bureau of Land Management (see Chapter 18) for use in surveying the public lands. This publication is also available through the U.S. Government Printing Office. The various instrument makers publish small pocket editions containing essential data for solar and Polaris calculations.

Although the engineer is ordinarily interested primarily in the azimuth of a certain line, most azimuth methods in use require that the instant at which the observation is made be known within a very few minutes, and that the latitude also be known. It is thus frequently necessary to determine the time and the latitude, by observation, before the azimuth can be ascertained.

12-2. DEFINITIONS In Fig. 12-1 are shown the *astronomical* or *celestial triangle PZS* and the principal points and planes on the celestial sphere. The *celestial poles P* and *P'* are the points of intersection, with the celestial sphere, of the axis of rotation of the earth, if that axis be extended indefinitely. The *celestial equator E'S"WVE* is the intersection of the celestial sphere with the

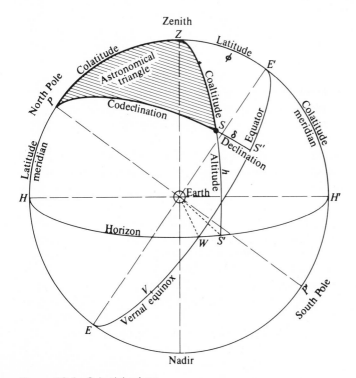

Figure 12-1. Celestial sphere.

plane passing through the center of the earth and perpendicular to the axis of rotation.

If a vertical line at any point on the earth's surface is extended, it intersects the upper portion of the celestial sphere at the *zenith Z*, and the lower portion of the sphere at the *nadir*. A *vertical circle* is a great circle passing through the zenith and the nadir; ZSS' is an arc of a vertical circle. The *meridian plane HPZE'H'P'E* is a vertical circle passing through the celestial poles, the *meridian* being the intersection of this plane with the celestial sphere. An *hour circle* is a great circle passing through the north and south celestial poles; PSS'' is an arc of an hour circle.

The *horizon* is the intersection of the celestial sphere with the plane that passes through the center of the earth and is perpendicular to a vertical line; $HWS'H'$ is half of the horizon.

The *azimuth* of a heavenly body S is the angular distance $H'S'$, measured along the horizon from the south point of the meridian to the vertical circle through the heavenly body. Azimuth is considered positive when measured clockwise. In Fig. 12-1 the azimuth $H'S'$, measured from the south, equals 180° minus the angle Z in the astronomical triangle. Many engineers prefer to measure azimuth from the north, particularly when observations are made on stars near the north pole.

The *altitude h* of a heavenly body S is the angular distance $S'S$, measured upward from the horizon along a vertical circle. The *coaltitude*, or *zenith distance*, ZS equals 90° minus the altitude. The altitude corresponds to the vertical angle to the body. The zenith distance corresponds to the zenith angle to the body (see Section 4-18).

The *latitude ϕ* of a point on the earth's surface is the angular distance $E'Z$ measured along the meridian from the equator to the zenith of that point. It is numerically equal to the altitude HP of the pole measured at the point. Latitudes north of the equator are considered positive; those south of the equator, negative. The *colatitude PZ* equals 90° minus the latitude.

The *vernal equinox V* is that intersection of the plane of the earth's orbit with the equator at which the sun appears to cross the equator from south to north. This occurs about March 21. The *right ascension* of a heavenly body S is the angular distance VS'' measured eastward along the equator from the vernal equinox to the hour circle through the body. The *declination δ* of a heavenly body S is the angular distance $S''S$ measured from the equator to the body along the hour circle through the body. If the body is north of the equator, the declination is considered positive; if south, negative. The *codeclination*, or *polar distance*, PS equals 90° minus the declination.

The *hour angle t* of a heavenly body S is the angular distance $E'S''$ measured westward along the equator from the meridian to the hour circle through the heavenly body. Hour angles measured to the west are positive; to the east, negative. Three hour angles are of particular importance. The *local hour angle* (LHA) is the hour angle of a heavenly body with respect to a local meridian. The *Greenwich hour angle* (GHA) is the hour angle with respect to the meridian of Greenwich. The *sidereal hour angle* (SHA) of a heavenly body is its angular distance, measured westward along the equator, from the vernal equinox to the hour circle through the heavenly body.

If the equatorial plane of Fig. 12-1 is viewed along the line PP', the hour circles become lines radiating outward from P as shown in Fig. 12-2. The line PG is the hour circle passing through Greenwich Observatory. The line PL is the local meridian. The longitude of the local meridian is thus the angle GPL or arc GL. The line PSS'' is the hour circle passing through the heavenly body S and corresponds to the hour circle PSS'' of Fig. 12-1. Point V is the position of the vernal equinox which lies on the equator. The local hour angle of the heavenly body is $\angle LPS''$ or arc LS''; the Greenwich hour angle of the body is $\angle GPS''$ or arc GLS''; the sidereal hour angle is the arc $VGLS''$ measured westward; and the right ascension of the body is the arc VS'' measured eastward. Thus it is seen that the sidereal hour angle of a body is equal to 360° minus the right ascension of the body.

The *transit* of a heavenly body is its passage over the observer's meridian. The *upper branch* of the meridian is the half of the meridian, from the north celestial pole to the south celestial pole, that contains the observer's zenith, as arc $PZE'H'P'$ in Fig. 12-1. The *lower branch* of the meridian is the half containing the nadir. An *upper transit* of a body occurs at the instant at which the body is on the upper branch of the meridian. A *lower transit* occurs at the instant at which the body is on the lower branch of the meridian.

The *north meridian* is the arc of the meridian from the observer's zenith to the nadir and containing the north celestial pole. This is the arc $ZPHE$-nadir in Fig. 12-1. The *south meridian* contains the south celestial pole.

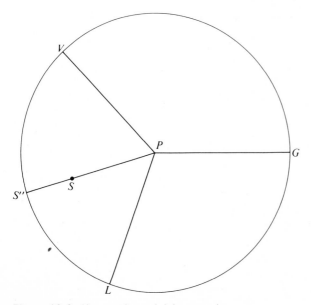

Figure 12-2. Hour angles and right ascension.

TABLE 12-1. Mean Refraction[a]

APPARENT ALTITUDE	REFRACTION	APPARENT ALTITUDE	REFRACTION	APPARENT ALTITUDE	REFRACTION	APPARENT ALTITUDE	REFRACTION
0°00'	34'54"	16°00'	3'19"	32°	1'32"	64°	0'28"
30'	29'04"	30'	3'13"	33°	1'29"	65°	27"
1°00'	24'25"	17°00'	3'07"	34°	1'25"	66°	26"
30'	20'51"	30'	3'02"	35°	1'22"	67°	25"
2°00'	18'09"	18°00'	2'56"	36°	1'19"	68°	0'23"
30'	16'01"	30'	2'51"	37°	1'17"	69°	22"
3°00'	14'15"	19°00'	2'46"	38°	1'14"	70°	21"
30'	12'48"	30'	2'42"	39°	1'11"	71°	20"
4°00'	11'39"	20°00'	2'37"	40°	1'09"	72°	0'19"
30'	10'40"	30'	2'33"	41°	1'06"	73°	18"
5°00'	9'47"	21°00'	2'29"	42°	1'04"	74°	17"
30'	9'02"	30'	2'26"	43°	1'02"	75°	16"
6°00'	8'23"	22°00'	2'22"	44°	1'00"	76°	0'14"
30'	7'50"	30'	2'18"	45°	0'58"	77°	13"
7°00'	7'20"	23°00'	2'15"	46°	0'56"	78°	12"
30'	6'53"	30'	2'12"	47°	0'54"	79°	11"

Altitude	Corr.	Altitude	Corr.	Altitude	Corr.	Altitude	Corr.
8° 00'	6' 30"	24° 00'	2' 09"	48°	0' 52"	80°	0' 10"
30'	6' 08"	30'	2' 06"	49°	50"	81°	09"
9° 00'	5' 49"	25° 00'	2' 03"	50°	48"	82°	08"
30'	5' 32"	30'	2' 00"	51°	47"	83°	07"
10° 00'	5' 16"	26° 00'	1' 58"	52°	0' 45"	84°	0' 06"
30'	5' 02"	30'	1' 55"	53°	44"	85°	05"
11° 00'	4' 49"	27° 00'	1' 53"	54°	42"	86°	04"
30'	4' 36"	30'	1' 50"	55°	40"	87°	03"
12° 00'	4' 25"	28° 00'	1' 48"	56°	0' 39"	88°	0' 02"
30'	4' 15"	30'	1' 46"	57°	38"	89°	01"
13° 00'	4' 05"	29° 00'	1' 44"	58°	36"	90°	00"
30'	3' 56"	30'	1' 42"	59°	35"		
14° 00'	3' 47"	30° 00'	1' 40"	60°	0' 33"		
30'	3' 39"	30'	1' 38"	61°	32"		
15° 00'	3' 32"	31° 00'	1' 36"	62°	31"		
30'	3' 25"	30'	1' 34"	63°	29"		
16° 00'	3' 19"	32° 00'	1' 32"	64°	28"		

[a] To be subtracted from observed altitudes.

TABLE 12-2. Corrections to Mean Refraction[a]

BAROMETRIC PRESSURE OR ELEVATION

BAROMETER (in.)	ELEVATION (ft)	MULTIPLIER	BAROMETER (in.)	ELEVATION (ft)	MULTIPLIER	BAROMETER (in.)	ELEVATION (ft)	MULTIPLIER
30.5	−451	1.03	27.2	2670	0.92	23.6	6538	0.80
30.2	−181	1.02	26.9	2972	0.91	23.3	6887	0.79
30.0	00	1.01	26.6	3277	0.90	23.0	7239	0.78
29.9	+91	1.01	26.3	3586	0.89	22.7	7597	0.77
29.6	366	1.00	26.0	3899	0.88	22.4	7960	0.76
29.3	643	0.99	25.7	4215	0.87	22.1	8327	0.75
29.0	924	0.98	25.4	4535	0.86	21.8	8700	0.74
28.7	1207	0.97	25.1	4859	0.85	21.5	9077	0.73
28.4	1493	0.96	24.8	5186	0.84	21.2	9460	0.72
28.1	1783	0.95	24.5	5518	0.83	20.9	9848	0.71
27.8	2075	0.94	24.2	5854	0.82	20.6	10242	0.70
27.5	2371	0.93	23.9	6194	0.81	20.3	10642	0.69

TEMPERATURE

TEMPERATURE (°F)	MULTIPLIER	TEMPERATURE (°F)	MULTIPLIER	TEMPERATURE (°F)	MULTIPLIER
−20	1.16	+30	1.04	+80	0.94
−10	1.13	+40	1.02	+90	0.93
0	1.11	+50	1.00	+100	0.91
+10	1.08	+60	0.98	+110	0.90
+20	1.06	+70	0.96	+120	0.88

[a]Multiply mean refraction by tabular values.

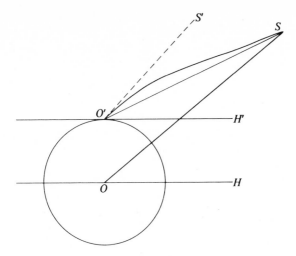

Figure 12-3. Refraction and parallax.

The *refraction* of a body is the angular increase in its apparent altitude due to the refraction of light. In Fig. 12-3 the angle $S'O'H'$ is the measured altitude and angle $S'O'S$ is the refraction. The *correction* for refraction is always negative, as seen from the figure. It is a maximum when the altitude is small and decreases to zero when the altitude is 90°. Values of the correction for refraction are given in Table 12-1. Owing to the uncertainty in the correction when the altitude is small, observations on heavenly bodies within 20° or 30° of the horizon should be avoided if possible.

The values of refraction given in Table 12-1 are based on a barometric pressure of 29.6 in. Hg and a temperature of 50°F. Barometric pressure changes with the weather conditions and also the elevation. Since atmospheric refraction is a function of temperature and pressure, the mean values given in Table 12-1 should be modified according to the multipliers given in Table 12-2.

Example 12-1.

The observed altitude to a star is 33° 15′ when the temperature is 72°F and the barometer reads 28.0 in. What is the refraction correction?

Solution: The mean refraction for an altitude of 33° 15′, from Table 12-1, is $-1' 28'' = -88''$. From Table 12-2, the pressure multiplier is 0.95 and the temperature multiplier is 0.96. The refraction correction is therefore $-88'' \times 0.95 \times 0.96 = -80'' = -1' 20''$.

Example 12-2.

The observed altitude to the sun's limb is 27° 30′ when the temperature is 48°F, the elevation of the point of observation is 4850 ft, and the barometer reading is unknown. What is the refraction correction?

Solution: Since the barometric pressure is not known, the standard sea-level pressure of 30.0 in. is assumed. From Table 12-1, the mean refraction for an altitude of $27° 30'$ is $-1' 50'' = -110''$. From Table 12-2, the elevation multiplier is 0.85, and the temperature multiplier is 1.00. The refraction correction is therefore $-110'' \times 0.85 \times 1.00 = -94'' = -1' 34''$.

The *correction for parallax* (in altitude) is the small angular increase in the apparent altitude of a heavenly body to allow for the fact that the observations are made at the surface of the earth instead of at its center. In Fig. 12-3 the angle $SO'H'$ is the angle at the surface corrected for refraction. The angle SOH is the true altitude, or the altitude that would be measured at the center of the earth. Observations on the sun, the moon, and the planets must be corrected for parallax. Observations on the stars need no parallax correction because any star is virtually at an infinite distance from the earth. Parallax correction is always added.

12-3. TIME   There are three kinds of time that may be involved in astronomical observations, namely, *sidereal time*, *apparent time* or *true solar time*, and *mean solar time*. A *sidereal day* is the interval of time between two successive upper transits of the vernal equinox over the same meridian. The sidereal day therefore begins at the instant at which the vernal equinox is on the upper branch of the meridian. The sidereal time at any instant, referred to a particular meridian, is the hour angle of the vernal equinox referred to the same meridian. The sidereal time for the local meridian shown in Fig. 12-2 is the arc LV.

An *apparent solar day* is the interval of time between two successive lower transits of the sun's center over the same meridian. Because of the obliquity of the plane of the earth's orbit (which is about $23° 26.5'$) and because of the variation in the earth's angular velocity about the sun in its elliptical orbit, the interval between two successive lower transits of the sun's center over the same meridian varies from day to day. Since it would not be feasible to keep time on clocks and watches if the apparent solar day were used, the mean solar day is used.

A *mean solar day* is the interval of time between two successive lower transits of the mean sun over the same meridian. The mean sun is a fictitious body which travels at a uniform rate along the equator. The mean sun is sometimes ahead of and sometimes behind the apparent, or true, sun. There are 24 hr in a day. The mean solar time at any instant for a given meridian is the hour angle of the mean sun referred to the same meridian plus 12 hr. At midnight, the hour angle of the mean sun is 12 hr. At mean solar noon, the hour angle of the mean sun is 0 hr.

The difference between apparent solar time and mean solar time is called the *equation of time*, or apparent − mean = equation of time. Its value for each day of the year is given in the ephemeris.

A *solar year* contains 365.2422 solar days. Because the earth rotates on its axis each day and orbits around the sun once a year, any point on the earth faces the sun one time less than it faces any of the stars and the vernal equinox in a period of one year. Therefore the solar year contains 366.2422 sidereal days. The mean solar day is longer than the sidereal day by $3^m\ 55.909^s$ of mean solar time or $3^m\ 56.555^s$ of sidereal time.

A solar interval of time can be converted to a corresponding sidereal interval by adding 9.856^s for each hour in the solar interval. A sidereal interval of time can be converted to a corresponding solar interval by subtracting 9.830^s for each hour in the sidereal interval. The application of these conversions are shown in examples in Sections 12-11, 12-12, and 12-16.

12-4. STANDARD TIME *Standard time* is the mean solar time referred to meridians 15° apart, measured from Greenwich which lies at 0° longitude. Since there are 24 hr in the mean solar day, then $24^h = 360°$ and $1^h = 15°$. Therefore the standard meridians are 1 hr apart. Eastern standard time (EST) is the time of the 75th meridian west of Greenwich; Central standard time (CST), of the 90th meridian; Mountain standard time (MST), of the 105th meridian; and Pacific standard time (PST), of the 120th meridian.

Conversion from degrees, minutes, and seconds, to hours, minutes, and seconds, or the reverse, is based on the following relationships:

$$360° = 24^h \qquad 24^h = 360°$$
$$1° = 4^m \qquad 1^h = 15°$$
$$1' = 4^s \qquad 1^m = 15'$$
$$1'' = 0.067^s \qquad 1^s = 15''$$

In Fig. 12-4 the mean sun, which travels westward along the equator, is shown to about 10° 30′ or 42^m beyond the 90th meridian. Central standard

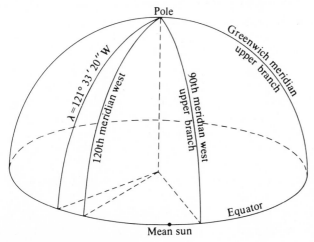

Figure 12-4. Time (mean).

time is therefore 12 : 42 P.M. At the Greenwich meridian at the same instant, the time is 6^h later. So the Greenwich standard time is 6 : 42 P.M. The standard time referred to the Greenwich meridian is called Greenwich civil time (GCT), Greenwich mean time (GMT), or universal time. The mean sun has not yet reached the 120th meridian, and the Pacific standard time is therefore before noon. Since the difference between CST and PST is 2^h, the Pacific standard time, or time based on the 120th meridian, is 10^h 42^m A.M.

12-5. LOCAL TIME

Local time is the time based on the observer's meridian. *Local civil time* (LCT), which is based on mean solar time, is the hour angle of the mean sun measured from the local meridian plus 12 hr. *Local apparent time* (LAT), which is based on the true sun, is the hour angle of the sun's center plus 12 hr. For example, when the actual sun's center is on the observer's meridian, the time is local apparent noon. *Local sidereal time* is the hour angle of the vernal equinox measured from the observer's meridian.

At the meridian in longitude 121° 33′ 20″, standard time is based on the 120th meridian. For the instant shown in Fig. 12-4, the time is 10 : 42 A.M., PST. Local civil time is earlier by an amount corresponding to 1° 33′ 20″, or 6^m 13.3^s. The local civil time at longitude 121° 33′ 20″ is therefore 10 : 35 : 46.7 A.M. The local civil time at longitude 72° 03′ 14″ for the instant shown in Fig. 12-4 is found by adding to the Central standard time the amount corresponding to 90° − 72° 03′ 14″. The time interval for 90° − 72° 03′ 14″, or 17° 56′ 46″, is 1^h 11^m 47.1^s, and the LCT is 1^h 53^m 47.1^s P.M.

12-6. TIME BY RADIO

Where a shortwave radio receiver is available, time may be determined several times daily by picking up the signals sent out by the U.S. Naval Observatory. These signals are also sent out by some broadcasting stations at noon and at 10 P.M., EST. The signals are supposed to be accurate within 0.1″. Each week the U.S. Naval Observatory distributes, on request, a list of corrections.

Time signals are also sent out by radio station WWV of the Bureau of Standards at Washington, D.C., on frequencies of 2.5, 5, 10, and 15 megacycles. These signals are sent out continuously as the ticks of a grandfather clock, superimposed on a 440-cycle hum. There is no tick at the 59th second of each minute. Also the musical tone is cut off at the beginning of each minute that is a multiple of 5 min and is resumed 1 min later. There is a voice announcement on the hour and the half hour.

12-7. OBSERVING THE SUN WITH THE THEODOLITE

Observations on the sun are made to determine the altitude or the azimuth of the sun's center or both. Altitude or zenith distance is read on the vertical circle of a theodolite, whereas azimuth is related to horizontal-circle readings.

Because it is quite difficult to set the intersection of the cross hairs directly on the center of the sun, the cross hairs are usually made to become tangent to the edges of the sun, and then the necessary amounts are added to or subtracted from the circle readings to reduce the observation to the sun's center.

In Fig. 12-5 is shown the design of a glass reticule which can be used to great advantage when a survey requires many solar observations. The usual horizontal and vertical cross hairs and the stadia hairs are augmented by a pair of vertical tick marks having the same spacing as the stadia hairs. Inside the stadia hairs and these two tick marks is inscribed a solar circle with an angular radius of $15' \, 45''$. This is slightly smaller than the *semidiameter* of the sun, which is approximately $0° \, 16'$. The sun's image can thus be centered on this circle with a very high degree of accuracy.

Although the sun can be observed directly through the telescope if the appropriate kind of dark filter is attached to the eyepiece, the usual practice is to view the image of the sun projected onto a white card held behind the eyepiece of the theodolite, as shown in Fig. 12-6. Two methods will be discussed in this section, each using the white-card technique. The first method is referred to as the center-tangent or disappearing-segment method. The second is known as the quadrant-tangent method.

When the center-tangent method is used, the observation is always made on the trailing edge, or *limb*, of the sun. During the morning hours, the sun's altitude is increasing, and the lower limb is observed for altitude (vertical-circle reading). During the afternoon hours, the sun's altitude is decreasing, and the upper limb is observed for altitude. Throughout all hours of the day, the sun moves from east to west, and the east limb is observed for azimuth (horizontal-circle reading).

When the theodolite is to be leveled over a point for an astronomical observation to measure the altitude or zenith distance of a heavenly body, the bubble of the telescope level must be used as described in Section 4-18. This procedure brings the vertical axis of the instrument more perfectly

Figure 12-5. Reticle for sighting sun's center.

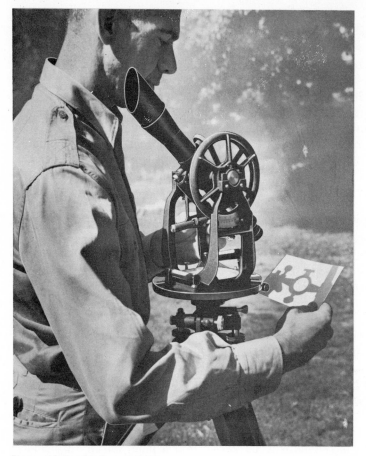

Figure 12-6. Viewing sun's image.

vertical than would be possible by using the plate levels. Vertical circle compensators eliminate this problem.

With the theodolite set up and leveled, the telescope is pointed in the direction of the sun, and the telescope bubble is centered. The vertical circle is read to determine the index error (see Section 4-18). If the theodolite is equipped with a pendulum compensator, there will be no index error. The telescope is raised and pointed as nearly as possible toward the sun without actually looking through the telescope. A white card is held from 3 to 4 in. behind the eyepiece, and the telescope is rotated both in altitude and azimuth until it casts a circular shadow on the card (see Fig. 12-6). The cross hairs are then focused on the card. The telescope is now focused to make the image of the sun clear and sharp on the card. The image of the sun produced by an erecting telescope will be upside down when the observer looks directly at the card. Therefore the lower edge of the sun's image is actually the upper limb of the real sun, and the upper edge of the image is the lower limb of the real sun. The right edge of the sun's image, as seen on the card, is the western limb of the real sun; the left edge of the image is the eastern limb of the real sun.

In the morning, when it is desired to make the horizontal cross hair tangent to the lower limb of the sun, the vertical-motion tangent screw is turned so that the horizontal cross hair cuts the sun's image in the position indicated by the dashed lines in Fig. 12-7(a). The altitude of this cross hair now remains stationary. As the sun's image moves downward and to the right, as indicated by the arrow, the small segment becomes smaller. The observer moves the vertical cross hair by means of the upper horizontal-motion tangent screw so that this cross hair continuously bisects the disappearing segment. He stops at the instant the edge of the sun's image becomes tangent to the horizontal cross hair. At this point the vertical cross hair passes through the center of the sun and the horizontal-circle reading needs no correction. However, the horizontal cross hair is tangent to the lower limb, and the vertical angle must be increased by one-half the angle subtended by the sun. This is the sun's semidiameter and is about $0°\,16'$. Its value is given for different times of the year in the ephemeris. The vertical angle is corrected for index error, parallax, and refraction before the semidiameter is added. The observation is repeated with the telescope reversed as explained in Section 12-20.

In the afternoon, when it is desired to make the horizontal cross hair tangent to the sun's upper limb, the horizontal cross hair is first set as shown in Fig. 12-7(c). Also, the semidiameter must be subtracted from the vertical-circle reading after corrections for index error, parallax, and refraction have been applied.

If it is desired to make the vertical cross hair tangent to the eastern limb of the sun, either in the morning or in the afternoon, the vertical cross hair should be set to cut the sun's image in the position indicated by the dashed lines in Fig. 12-7(b) or (d). The horizontal cross hair is then moved so as to bisect the disappearing segment continuously until tangency is reached. At this point the horizontal cross hair passes through the center of the sun and the vertical-circle reading need be corrected only for index error, parallax, and refraction. The horizontal-circle reading, however, is too small by an

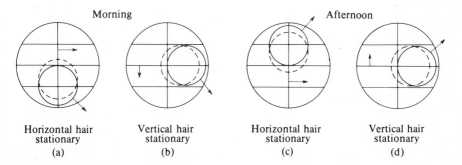

Morning		Afternoon	
Horizontal hair stationary	Vertical hair stationary	Horizontal hair stationary	Vertical hair stationary
(a)	(b)	(c)	(d)

Figure 12-7. Image of sun as seen on card held behind eyepiece of an erecting telescope.

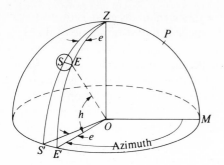

Figure 12-8. Correction to horizontal-circle reading for sun's semidiameter.

amount found by multiplying the semidiameter by sec h, where h is the altitude or vertical angle. This amount is added to the horizontal-circle reading.

In Fig. 12-8, Z is the observer's zenith, P is the pole, arc ZPM is part of the observer's meridian, S is the sun's center, and angle MOS' is the azimuth of the sun's center. The arc ZEE' is part of the vertical circle tangent to the sun's eastern limb. This arc lies in the vertical plane containing the line of sight of the theodolite when it is brought tangent to the eastern limb. Thus the azimuth of the sun's eastern limb is the angle MOE', which is smaller than the angle MOS' by the small angle e. In the triangle ZSE the angle at Z is e, the angle at E is $90°$, side ZS is the coaltitude of the sun, or $90° - h$, and side SE is the sun's semidiameter d. By the law of sines of spherical trigonometry for the triangle ZSE,

$$\frac{\sin Z}{\sin SE} = \frac{\sin E}{\sin ZS} \quad \text{or} \quad \frac{\sin e}{\sin d} = \frac{\sin 90°}{\sin(90° - h)} = \frac{1}{\cos h} = \sec h$$

Since the semidiameter d is small, it may be assumed that

$$e = d \sec h \tag{12-1}$$

in which e is the amount by which the horizontal-circle reading to the sun's eastern limb must be increased. Both d and e are expressed in minutes of arc.

Example 12-3.

On the morning of October 16, 1979, the horizontal cross hair of a theodolite was brought tangent to the lower limb of the sun, as indicated in Fig. 12-7(a), and the vertical circle read $+40° 22'$. The vertical circle read $+0° 02'$ when the telescope bubble was centered. If the mean refraction can be assumed, what was the true altitude of the sun's center?

Solution: The computations may be arranged as follows:

vertical-circle reading	$= +40° 22'$	
index correction	$= - \ 0° 02'$	
	$+40° 20'$	
mean refraction	$= - \quad 1' 09''$	(Table 12-1)
	$40° 18' 51''$	
sun's parallax	$= + \qquad 7''$	(from ephemeris)
	$40° 18' 58''$	
semidiameter for October 16	$= + \quad 16' 04''$	(from ephemeris)
true altitude of sun's center	$= \quad 40° 35' 02''$	

Example 12-4.

On the afternoon of May 21, 1979, the vertical cross hair of a theodolite was brought tangent to the eastern limb of the sun, as indicated in Fig. 12-7(d). The clockwise horizontal-circle reading was 322° 44′ 30″ and the vertical circle read +27° 52′. The index error of the vertical circle was determined to be zero. What was the corrected horizontal-circle reading, and what was the altitude of the sun's center with mean refraction?

Solution: The calculations follow:

horizontal-circle reading =	$322° 44' 30''$
semidiameter × sec h = $+$	$17' 54''$
corrected horizontal-	
circle reading =	$323° 02' 24''$

vertical-circle reading =	$+27° 52'$
refraction = $-$	$1' 48''$
	$+27° 50' 12''$
parallax = $+$	$8''$
true altitude of sun =	$+27° 50' 20''$

When the position of the sun is observed in the manner just described, several observations should be taken in rapid succession, and the mean values of the corrected horizontal-circle readings and the vertical angles should be used. Each of the observed values should be corrected for index error, refraction and parallax, semidiameter, and d sec h. The watch time of each observation should be noted to the nearest second, whether or not the times are exactly correct. Then after the several horizontal-circle readings and vertical angles have been corrected, both the values of the corrected horizontal-circle readings and the values of the corrected vertical angles should be plotted versus the time. Each plot should be very nearly a straight line. Values that are not close to the adopted straight line should be rejected before the mean values for the group of observations are computed (see Section 12-20).

When the sun is observed by the quadrant-tangent method, the white card is used in the manner previously described. In Fig. 12-9 is shown the motion of the sun's image as it would be projected through an erecting telescope onto the card. In the morning the first part of the observation is made as indicated in Fig. 12-9(a). The vertical cross hair is always kept tangent to the sun's western, or leading, limb by means of the upper horizontal tangent screw. The vertical motion of the theodolite is not disturbed, and thus the horizontal cross hair remains at the same altitude. At the instant at which the sun's lower limb (the upper edge of the sun's image) becomes tangent to the horizontal cross hair, as shown by the solid circle, horizontal motion of the theodolite is stopped. At the same instant, the watch time is noted. Next, the vertical cross hair is held at the same azimuth, as shown in Fig. 12-9(b), while the horizontal cross hair is kept tangent to the sun's upper limb (the lower edge of the sun's image) by means of the vertical slow-motion screw. At the instant at which the eastern limb becomes tangent to the vertical cross hair, as shown by the solid circle, the theodolite motion is stopped and the watch time is noted.

In the afternoon the sun's upper limb is first made tangent, as indicated in Fig. 12-9(c), and then the eastern limb is made tangent as shown in Fig. 12-9(d).

When the quadrant-tangent method is used, the observer must watch for tangency at two points of the sun's disk, instead of at only one point in the center-tangent method. It is therefore more difficult to make the observation. Presumably, a pair of observations, one carried out as in Fig. 12-9(a) and

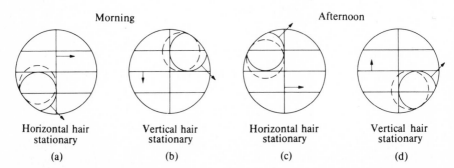

	Morning		Afternoon	
Horizontal hair stationary	Vertical hair stationary	Horizontal hair stationary	Vertical hair stationary
(a)	(b)	(c)	(d)

Figure 12-9. Image of sun as seen on card held behind eyepiece of an erecting telescope (quadrant-tangent method).

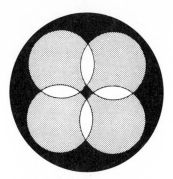

Figure 12-10. Pattern of sun's image produced by Roelof's solar prism.

the other as in (b), could be made in quick succession, and the readings could be averaged to eliminate the semidiameter corrections to both the horizontal-circle and vertical-circle readings. Moreover, if another pair of observations is made with the telescope reversed, the index error of the vertical circle could also be eliminated. However, a set of solar observations should never be used for computation until all the corrections have been made to the circle readings and the resulting angles have been plotted as a function of time. This plot uncovers any possible mistakes which would otherwise escape notice if all observations are presumed valid and the results are averaged. In each instance shown in Fig. 12-9, the corrections would consist of index correction, refraction and parallax, semidiameter, and $d \sec h$. Section 12-20 includes a discussion of mistakes in observations and the ways in which defective observed values may be rejected on a valid basis.

An ingenious device, which is known as *Roelof's solar prism* and which can be fitted to the objective of the telescope, produces four overlapping images of the sun to give the solar-disk pattern shown in Fig. 12-10. By direct viewing through the theodolite the small diamond-shaped area in the center of the pattern can be bisected simultaneously by the horizontal and vertical cross hairs with a high degree of accuracy. Because of the pattern symmetry, such a pointing is made directly on the sun's center, and the semidiameter corrections are thus eliminated. Because of the direct-viewing feature, the observations can be made quite rapidly. Of course, corrections for index error, refraction, and parallax must be applied to the vertical-circle reading. There is, however, no correction to the horizontal-circle reading.

12-8. OBSERVING A STAR WITH THE THEODOLITE Unlike the sun, a star appears as a point of light in the heavens when viewed through a theodolite telescope. It is therefore quite easy to observe the instant at which the star's image crosses either the horizontal cross hair when the star's altitude is being measured or the vertical cross hair when a horizontal angle to the star is being measured. Because observations are taken on stars under

such conditions that the amount of natural light coming through the telescope usually is not sufficient to allow the observer to see the cross hairs, artificial light must be provided. A good method is to shine a flashlight obliquely across the objective lens at such an angle that there is sufficient light to make the cross hairs visible, but not enough to blot out the image of the star. Some instruments have internal illumination, the brightness of which can be controlled by a rheostat.

If a star is to be observed for determining the azimuth of a line or a back-sight has to be made on a ground station during the observation for any other reason, then the ground station itself must be illuminated. Suitable light can usually be provided by holding a flashlight behind the mark when the station is being observed.

Preparations for star observations should be made in daylight or during twilight hours, if possible, simply for convenience. Some stars may actually be observed in the twilight hours. This condition is, of course, most desirable, because the observer and the notekeeper then have at least some natural light in which to work.

12-9. TIME BY TRANSIT OF THE SUN As previously stated, the primary purpose of most astronomical observations on many surveys is the determination of the azimuth of a line, but most methods of observation require that the time be known within a few seconds or a few minutes. Where standard time cannot be obtained, field observations for time may be necessary. Time at an established meridian may be determined by noting the instant of the passage of the sun's center or of a star across that meridian. Where a good map is available, the longitude of the place of observation can be scaled with sufficient accuracy from the map.

When the sun is used, the watch correction is computed by noting the instant at which the sun's center crosses the meridian. This instant is observed by first setting the telescope in the meridian and cutting a small segment from either the upper or lower limb of the sun with the horizontal cross hair by using the vertical-motion tangent screw. The sun's image is then tracked by the vertical-motion tangent screw until the small segment is bisected by the vertical cross hair. The instant of watch time of crossing is noted. This instant is local apparent noon.

The Greenwich apparent time of observation is obtained by adding the longitude (presumed west) in time units to the local apparent time of 12^h. The equation of time, obtained from the ephemeris, is subtracted algebraically to obtain the Greenwich civil time of observation. Finally, the longitude of the standard meridian in time units is subtracted from the Greenwich civil time to give the standard time on which the watch is based. From the correct standard time, the watch error may be computed. Typical computations illustrating the procedure for determining the standard time of the sun's transit are shown in Table 12-3.

TABLE 12-3. Computations for Time from Sun's Transit

Date: October 14, 1973. Longitude: $113° 42' 45''$ or $7^h 34^m 51^s$. Watch time of transit of sun: $12^h 20^m 04^s$ P.M., MST.

local apparent time	=	$12^h 00^m 00^s$
longitude west of Greenwich	=	$+ 7^h 34^m 51^s$
Greenwich apparent time	=	$19^h 34^m 51^s$
equation of time at 0^h GCT	=	$+00^h 13^m 51^s$
change for $19^h 34^m 51^s$	=	$+00^h 00^m 11^s$

(from ephemeris of sun)

equation of time at instant of observation	=	$+00^h 14^m 02^s$
Greenwich civil time (GAT—equation of time)	=	$19^h 20^m 49^s$
longitude to MST	=	$- 7^h 00^m 00^s$
Mountain standard time	=	$12^h 20^m 49^s$
watch time	=	$-12^h 20^m 04^s$
watch slow		45^s

12-10. TIME BY ALTITUDE OF THE SUN Time may be determined by measuring the altitude of the sun's center at any instant, and the watch error may be found by noting the watch time at the instant the observation is made. Since it is very difficult to locate the sun's center accurately by eye, the altitude is measured to either the lower limb or the upper limb. This observed angle is corrected for the index error of the vertical circle, refraction, parallax, and semidiameter of the sun. Better results can be obtained if several altitudes are measured in quick succession, the corresponding instants of time being noted. The observed limb depends on whether the observation is made in the morning or in the afternoon. See Figs. 12-7(a) and (c). Corrections for index error, parallax, refraction, and semidiameter are applied to each vertical angle, and the means of the altitudes and the times are used in computing the hour angle by one of the following formulas:

$$\sin \tfrac{1}{2}t = \sqrt{\frac{\cos s \sin(s - h)}{\cos \phi \sin p}} \tag{12-2}$$

$$\tan \tfrac{1}{2}t = \sqrt{\frac{\cos s \sin(s - h)}{\cos(s - p) \sin(s - \phi)}} \tag{12-3}$$

$$\cos t = \frac{\sin h - \sin \phi \sin \delta}{\cos \phi \cos \delta} \tag{12-4}$$

in which t is the hour angle, s is $\tfrac{1}{2}(\phi + h + p)$, ϕ is the latitude obtained from other observations or scaled from a map, δ is the declination obtained from the ephemeris, h is the corrected altitude, and p is the polar distance, or 90° minus the declination.

The hour angle t can be converted into time by applying the following relations: 15° of arc = 1^h, $15' = 1^m$, and $15'' = 1^s$. If the observation is made in the morning, the result is the time interval before local apparent noon. If it is an afternoon observation, the result is the time since local apparent noon. The approximate time must be known for obtaining the declination from the ephemeris. When the time is first computed, it may be found that the estimated time was too much in error, and a second computation must then be made with a corrected declination. The resulting local apparent time can be converted to any standard time by applying the equation of time and the longitude correction, as was done in the example in the preceding section.

The results obtained by this method will usually be unsatisfactory if the observations are made when the sun is close to the horizon or close to the meridian. When the observed altitude is less than 20° to 30°, the refraction correction is likely to be uncertain. For observations made within an hour of local apparent noon, a small error in the observed vertical angle will seriously affect the hour angle.

12-11. TIME BY ALTITUDE OF A STAR   The preceding method may also be used for observations made on a star instead of the sun. For accurate results two stars should be used, one nearly directly east and the other nearly directly west of the place of observation. Both stars should be at least 30° above the horizon. No correction for parallax need be made in the case of a star. Errors of observation will be minimized if several altitudes are measured in quick succession, half with the telescope normal and half with it inverted. By using one star east and one west of the meridian, instrumental and refraction-correction errors are largely eliminated from the mean of the results.

In Fig. 12-1 the arc $E'S''$ along the equator is the hour angle of the star, and the arc VS'' is the right ascension of the star. It is thus seen that the local sidereal time equals the local hour angle of the star plus the right ascension of the star. The local sidereal time is reduced to Greenwich sidereal time by adding the longitude (presumed west) in time units. The ephemeris gives the sidereal time corresponding to 0^h GCT. Since the difference between the sidereal time at 0^h GCT and the Greenwich sidereal time of observation is a sidereal interval of time, the corresponding solar interval of time is obtained by subtracting from the sidereal interval 9.83^s for each sidereal hour in the interval. The amount to be subtracted is tabulated in the ephemeris. The solar interval of time since 0^h GCT is the Greenwich civil time of the observation. The standard time of observation is obtained by subtracting the longitude of the standard meridian in time units; so the watch error can be easily determined. The computations for a single observation made on a star are given in Table 12-4.

Some ephemerides give the sidereal hour angle of selected stars in angular units, but not the right ascension. Thus in the computation shown in Table 12-4, the SHA of Arcturus would be $24^h - 14^h 14^m 30.4^s = 9^h 45^m 29.6^s$ or $146° 22' 24''$.

TABLE 12-4. Computations for Time from Altitude of Star

Date: Aug. 28, 1974. Latitude: 43° 17' 13''. Longitude: 110° 40' 45''. Star (west): α Bootis (Arcturus). Observed vertical angle: 39° 17' 30''. Watch time: $7^h 44^m 25^s$ P.M., MST

$$\sin \tfrac{1}{2}t = \sqrt{\frac{\cos s \sin(s - h)}{\cos \phi \sin p}}$$

observed vertical angle	=	39° 17' 30''
refraction	=	−1' 10''
h	=	39° 16' 20''
ϕ	=	43° 17' 13''
δ (from ephemeris) = 19° 18' 54'' 90° − δ = p =		70° 41' 06''
$(h + \phi + p)$	=	152° 74' 39''
$s = \tfrac{1}{2}(h + \phi + p)$	=	76° 37' 20''
$s - h$	=	37° 21' 00''

$\cos s = 0.231\ 3706$

$\sin(s - h) = 0.606\ 6824$
$\cos \phi = 0.727\ 9290$
$\sin p = 0.943\ 7144$

$$\sin \tfrac{1}{2}t = \sqrt{\frac{0.231\ 3706 \times 0.606\ 6824}{0.727\ 9290 \times 0.943\ 7144}}$$

$\sin \tfrac{1}{2}t = 0.452\ 0318$

$\tfrac{1}{2}t = 26° 52' 27''$
$t = 53° 44' 54''$
$= 3^h 34^m 59.6^s$

right ascension star (from ephemeris)	=	14 14 30.4
local sidereal time	=	17 49 30.0
longitude (time)	=	7 22 43
Greenwich sidereal time	=	24 71 73.0
sidereal time, Greenwich 0^h, Aug. 29 (from ephemeris)	=	22 27 16.8
sidereal interval since 0^h	=	2 44 56.2
correction to solar interval (2.75^h × 9.830)	=	−27.0
Greenwich civil time, Aug. 29	=	2 44 29.2
or Greenwich civil time, Aug. 28	=	26 44 29.2
longitude to Mountain standard time	=	7 00 00
Mountain standard time	=	19 44 29.2
or Mountain standard time	=	7 44 29.2 P.M.
watch time	=	7 44 25
watch correction	=	+4.2^s

12-12. TIME BY THE TRANSIT OF A STAR

The most accurate method of determining time is by noting the instant of the passage of a star across the meridian. The stars chosen for this work are those near the equator, as their apparent motions are more rapid. Usually several stars are observed and the mean of the results is used. The stars can be identified by computing beforehand their approximate times of transit and their altitudes. The altitude will equal the colatitude of the observer plus the declination of the star. Unless the theodolite is equipped with a prismatic eyepiece, the altitudes should not exceed 55° or 60°. An inspection of Fig. 12-1 will show that the right ascension of a star will be the local sidereal time for the instant at which the star crosses the meridian. The sidereal time is changed to standard time as shown in Table 12-5. Here, again the SHA of Enif would be given in some ephemerides as 34° 15′ 30″ and the right ascension could be obtained from that SHA.

TABLE 12-5. Computations for Time from Transit of Star

Date: Nov. 2, 1974. Longitude: 110° 30′ 45″ W. Star: Enif. Time of upper transit: 7^h 17^m 46^s P.M., MST.

local sidereal time	=	21^h 42^m 58^s	= RA (from ephemeris)	
longitude west of Greenwich	=	+7 22 03		
Greenwich sidereal time	=	29 05 10		
Greenwich sidereal time	=	5 05 01		
sidereal time Greenwich 0^h	=	2 47 29	(from ephemeris for Nov. 3)	
sidereal interval since 0^h	=	2 17 32		
correction to solar interval		−23	(2.29^h × 9.830)	
solar interval since 0^h Nov. 3	=	2 17 09		
solar interval since 0^h Nov. 2	=	26 17 09		
longitude to MST	=	−7 00 00		
Mountain standard time	=	19 17 09		
Mountain standard time	=	7 17 09	P.M.	
watch time	=	7 17 46		
watch correction	=	−37^s		

12-13. LATITUDE

The latitude of a point on the earth's surface can be obtained by scaling it from a map, by measuring the altitude of the sun at local apparent noon, or by measuring the altitude of a circumpolar star. Where the latitude is to be used for the computation of time or of azimuth, it is generally accurate enough to scale the latitude from a good map. The final calculated results will probably be as good as can be obtained with the engineer's transit, for they will not be seriously affected by a slight error in the latitude used in the computations.

12-14. LATITUDE BY SUN'S ALTITUDE AT NOON

Where a meridian has been determined, the latitude can be obtained by measuring the

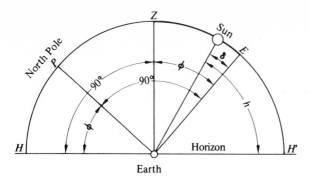

Figure 12-11. Latitude by sun's altitude at noon.

altitude of the sun's lower or upper limb at the instant the sun crosses the meridian. Where the meridian is unknown, this altitude may be measured when the sun reaches the highest point in its path. The observed angle is corrected for the index error of the vertical circle, for refraction and parallax, and for the semidiameter of the sun. The sun's declination at the instant of apparent noon and the semidiameter of the sun can be obtained from the ephemeris. From Fig. 12-11, which is a section taken in the plane of the meridian, it is evident that the latitude, which is measured by the arc ZE, will be

$$\phi = 90° - (h - \delta) \tag{12-5}$$

North declination is considered positive and south declination is negative. Instrumental errors and the correction for semidiameter can be eliminated if one vertical angle is measured to the sun's lower limb with the telescope normal and another is measured immediately afterward to the upper limb with the telescope inverted. If the second observation is made within 1 or 2 min of apparent noon, the sun's altitude will not have changed appreciably during that interval.

12-15. LATITUDE BY OBSERVATION ON CIRCUMPOLAR STAR

A circumpolar star is one that never goes below the observer's horizon. Thus, as shown in Fig. 12-11, the codeclination or polar distance PH must be less than the latitude of the place. Also from Fig. 12-11 it will be noted that the altitude of the pole is also a measure of the latitude of the observer. Latitude can be determined by measuring the vertical angle to a circumpolar star when it is on the meridian. The star most frequently used by engineers is Polaris (α Ursae Minoris), because it is quite close to the pole and it can be readily identified. The brighter stars about the North Pole are shown in Fig. 12-12. Polaris can usually be identified without difficulty by first locating

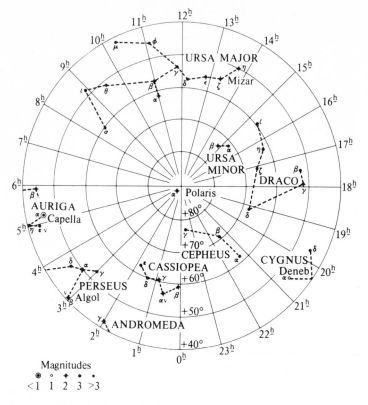

Figure 12-12. Stars about the North Pole.

the Great Dipper (Ursa Major). The edge of the dipper that is formed by the stars α and β points toward Polaris.

If the meridian is unknown, the observation can be made when the star reaches either the highest or the lowest point in its path. These positions are known as *upper culmination* and *lower culmination*. Upper culmination occurs when the star is on the upper branch of the meridian, whereas lower culmination occurs when the star is on the lower branch of the meridian. The time of culmination, which occurs approximately $3^m 56^s$ earlier each day, may be found with the aid of special tables in the ephemeris or may be computed as follows: At the instant of upper culmination, the local sidereal time equals the right ascension of the star. At lower culmination, it is the right ascension plus 12 hr. The corresponding standard time may be computed from the sidereal time by proceeding as shown in Section 12-12.

A special table for determining the time of culmination is provided in the ephemeris. Table 12-6 gives the local civil time of upper culmination of Polaris for the meridian of Greenwich for intervals of 10 days during the year 1979. A similar table for the current year is included in the ephemeris. To find the

TABLE 12-6. Local Civil Time of Upper Culmination of Polaris for the Meridian of Greenwich—1979

DATE	CIVIL TIME OF UPPER CULMINATION		VARIATION PER DAY	DATE	CIVIL TIME OF UPPER CULMINATION		VARIATION PER DAY
1979	h	m	m	1979	h	m	m
Jan. 1	19	28.7	3.96	July 10	7	01.0	3.91
11	18	49.2	3.96	20	6	21.9	3.91
21	18	09.6	3.96	30	5	42.8	3.91
31	17	30.0	3.96	Aug. 9	5	03.8	3.91
Feb. 10	16	50.5	3.96	19	4	24.7	3.91
20	16	10.9	3.95	29	3	45.6	3.91
Mar. 2	15	31.4	3.95	Sept. 8	3	06.5	3.91
12	14	51.9	3.95	18	2	27.4	3.92
22	14	12.4	3.94	28	1	48.2	3.92
Apr. 1	13	33.0	3.94	Oct. 8	1	09.0	3.92
11	12	53.6	3.93	18	0	29.8	3.92
21	12	14.3	3.93	27	23	50.6	3.93
May 1	11	35.0	3.93	Nov. 6	23	11.3	3.93
11	10	55.7	3.92	16	22	31.9	3.94
21	10	16.5	3.92	26	21	52.5	3.94
31	9	37.4	3.91	Dec. 6	21	13.1	3.94
June 10	8	58.3	3.91	16	20	33.6	3.95
20	8	19.1	3.91	26	19	54.1	3.95
30	7	40.1	3.91	1980			
				Jan. 5	19	14.6	3.95

local time of upper culmination at Greenwich for any date not listed in this table, subtract from the time for the preceding tabular date the product of the variation per day and the number of days elapsed. To find the local time of upper culmination at any other place, subtract 10^s for each 15° the place of observation is west of Greenwich or add 10^s for each 15° the place is east of Greenwich. If standard time is desired, add or subtract 4 min for every degree of longitude by which the place is west or east of the standard time meridian (60°, 75°, 90°, 105°, etc.). To determine the time of lower culmination, add $11^h 58.0^m$ to the time of upper culmination for the same date if the time of upper culmination is less than $11^h 58.0^m$; or subtract $11^h 58.0^m$ if the time of upper culmination is greater than $11^h 58.0^m$.

Example 12-5.

What is the Pacific standard time of upper culmination of Polaris on Dec. 12, 1979, at a place in longitude 121° 14′ 56″ W? What is the Pacific standard time of lower culmination on the same date?

Solution: The computations may be arranged as follows:

upper culmination for meridian of Greenwich, Dec. 6

$$\text{(Table 12-6)} = 21^h 13.1^m$$

correction to Dec. 12, 6 × 3.94	$= -23.6$
upper culmination for meridian of Greenwich, Dec. 12	$= \overline{20^h 49.5^m}$
correction to longitude of place, $(121.25/15) \times 10^s = 81^s$	$= -1.4^m$
upper culmination for meridian of place of observation	$= \overline{20^h 48.1^m}$
reduction to PST, $1° 14' 56''$ converted to time	$= +5.0^m$
PST of upper culmination at place of observation	$= \overline{20^h 53.1^m}$
	$= 8^h 53.1^m \text{P.M.}$
PST of upper culmination, Dec. 12	$= 20^h 53.1^m$
interval between upper culmination and lower culmination	$= 11^h 58.0^m$
PST of lower culmination, Dec. 12	$= \overline{8^h 55.1^m \text{A.M.}}$

Since the altitude of Polaris changes very slowly, a more accurate value of the latitude will be obtained if a number of vertical angles are measured in rapid succession, half of them with the telescope normal and the others with it inverted. The mean of the vertical angle is corrected for refraction. The declination of the star is obtained from the ephemeris. If the observation is made at lower culmination, the latitude is found from the relationship

$$\phi = h + (90° - \delta) \tag{12-6}$$

If the observation is made at upper culmination, the relationship is

$$\phi = h - (90° - \delta) \tag{12-7}$$

Example 12-6.

The altitude of Polaris was measured as $38° 14' 40''$ at upper culmination at the place and date given in Example 12-5. The temperature was $42°F$; the barometric pressure was 29.2 in.; and there was no index error. What is the latitude of the place of observation?

Solution: The declination of Polaris for Dec. 12, 1979, is obtained from the ephemeris as $89° 10' 31''$. Thus by Eq. (12-7),

measured h	$=$	$38° 14' 40''$
refraction correction (Tables 12-1 and 12-2)	$=$	$1' 14''$
h	$=$	$\overline{38° 13' 26''}$
$-(90° - \delta)$	$=$	$-0° 49' 29''$
$\phi =$		$\overline{37° 23' 57''}$

12-16. LATITUDE FROM POLARIS AT ANY HOUR ANGLE

During part of the summer, both upper and lower culmination of Polaris may occur during daylight hours. If the time is known within a few minutes, the altitude may be observed whenever the star is visible and the latitude may be computed without serious error by the relationship

$$\phi = h - p\cos t + \tfrac{1}{2}p^2\sin^2 t\tan h\sin 1'' \qquad (12\text{-}8)$$

in which h is the true altitude, obtained by correcting the observed altitude for index error and refraction; t is the hour angle, obtained by subtracting the star's right ascension from the sidereal time at the instant of observation; and p is the polar distance, expressed in seconds.

The nearer Polaris is to culmination, the smaller will be the error due to errors in time. For hour angles of 3, 9, 15, and 21^h, an error of 1^m in time causes an error of about $12''$ in the computed latitude. For hour angles of 6 and 18 hr, an error of 1 min in time causes an error of about $17''$ in the computed latitude.

The method of calculating the latitude will be apparent from the example in Table 12-7.

TABLE 12-7. Computations for Latitude from Polaris at Any Hour Angle

Date: Aug. 22, 1974. Longitude: $110°\,37'\,01''$ W. Time: $7^h\,22^m\,20^s$ P.M., MST. Observed vertical angle on Polaris: $42°\,37'\,30''$.

Mountain standard time	=	$19^h\,22^m\,20^s$
longitude west from Greenwich	=	$+\ 7\quad 00\quad 00$
Greenwich civil time, Aug. 22	=	$26\quad 22\quad 20$
mean solar interval since 0^h, Aug. 23	=	$2\quad 22\quad 20$
correction to sidereal interval	=	$+23.4\ (2.37^h \times 9.856)$
sidereal interval since 0^h, Aug. 23	=	$2\quad 22\quad 43.4$
sidereal time, Greenwich 0^h, Aug. 23	=	$22\quad 03\quad 37.6$ (from ephemeris of sun)
Greenwich sidereal time of observation	=	$24\quad 26\quad 21.0$
longitude to place of observation	=	$-\ 7\quad 22\quad 43$
local sidereal time of observation	=	$17\quad 03\quad 38.0$
right ascension of Polaris	=	$-\ 2\quad 08\quad 11$ (from ephemeris)
hour angle of Polaris	=	$14\quad 55\quad 27$
		$\times 15$
hour angle of Polaris, t		$223°\,51'\,45''$

$$\phi = h - p\cos t + \tfrac{1}{2}p^2\sin^2 t\tan h\sin 1''$$

observed altitude = $42°\,37'\,30''$		$90°\,00'\,00''$
refraction $\quad = \quad -01'\,02''$		$\delta = 89°\,08'\,41''$ (from ephemeris)
$h = 42°\,36'\,28''$		$p = \ 0°\,51'\,19'' = 3079''$

$$p\cos t = 3079'' \times (-0.721\ 0048) = -2220'' = -0°\,37'\,00''$$

$$\tfrac{1}{2}p^2\sin^2 t\tan h\sin 1'' = \tfrac{1}{2} \times 3079^2 \times 0.692\ 9301^2 \times 0.919\ 7977 \times 4.848 \times 10^{-6} = 10''$$

$$\phi = 42°\,36'\,28'' - (-0°\,37'\,00'') + 0°\,00'\,10'' = 43°\,13'\,38''$$

12-17. LONGITUDE The longitude of a point can be scaled from a map, or it can be calculated by determining the difference in time between the unknown point and some point whose longitude is known. The radio time signals sent out from the Naval Observatory at Washington, D.C., provide eastern standard, or 75th-meridian, time. If local mean time is determined by any of the methods of Sections 12-9 to 12-12, the difference between this local time and 75th-meridian time is the amount the unknown point is east or west of the 75th meridian, expressed in hours, minutes, and seconds. This longitude difference can be changed to degrees, minutes, and seconds of arc by multiplying by 15, and the unknown longitude can be readily determined. Compared with Eastern standard time, the local time will be slow if the point is west of the 75th meridian, and it will be fast if the point is east of this meridian.

12-18. AZIMUTH FROM OBSERVATIONS ON POLARIS AT ELONGATION The two most common methods of determining the azimuth of a line are by observations on circumpolar stars and by solar observations. The simplest method is to observe Polaris either at the apparently most westerly point in its path (western elongation) or at the apparently most easterly point (eastern elongation). The azimuth of Polaris at elongation may be taken from the ephemeris, or it may be computed from the following relationship:

$$\sin Z = \sin p \sec \phi \qquad (12\text{-}9)$$

in which Z is the azimuth measured from the north; p is the polar distance, or $90°$ minus the declination; and ϕ is the latitude.

The time of elongation can be determined from the time of upper culmination by adding or subtracting the proper interval given in Table 12-8. Also, it can be calculated in the following manner: The hour angle can be computed from the relationship

$$\cos t = \tan \phi \cot \delta \qquad (12\text{-}10)$$

The sidereal time at elongation will equal the right ascension of the star plus its hour angle. The sidereal time can be changed to standard time by the method of Section 12-12.

If the exact time is unknown, the star can be followed with the vertical cross hair in the theodolite. When it appears to move vertically on the cross hair, it is at elongation. The horizontal angle is then measured to some fixed point, or the line of sight is brought to the ground, a point on the line is established, and the angle is measured the next morning. If Polaris was at eastern elongation, its azimuth is added to the clockwise-measured angle to

TABLE 12-8. Mean Time Interval between Upper Culmination and Elongation[a]

LATITUDE	TIME INTERVAL	LATITUDE	TIME INTERVAL	LATITUDE	TIME INTERVAL
10°	$5^h\ 58.3^m$	40°	$5^h\ 55.7^m$	52°	$5^h\ 53.9^m$
15°	$5^h\ 58.0^m$	42°	$5^h\ 55.4^m$	54°	$5^h\ 53.6^m$
20°	$5^h\ 57.6^m$	44°	$5^h\ 55.2^m$	56°	$5^h\ 53.1^m$
25°	$5^h\ 57.2^m$	46°	$5^h\ 54.9^m$	58°	$5^h\ 52.7^m$
30°	$5^h\ 56.7^m$	48°	$5^h\ 54.6^m$	60°	$5^h\ 52.1^m$
35°	$5^h\ 56.2^m$	50°	$5^h\ 54.3^m$	62°	$5^h\ 51.6^m$

[a] Eastern elongation precedes and western elongation follows upper culmination.

obtain the required azimuth of the line. If the star was at western elongation, its azimuth is subtracted from the clockwise-measured angle.

12-19. AZIMUTH FROM OBSERVATION ON POLARIS AT ANY HOUR ANGLE
If the time is accurately known, the observations on Polaris can be made whenever the star is visible, and the azimuth can be calculated by either of the two following relationships:

$$\sin Z = -\sin t \cos \delta \sec h \qquad (12\text{-}11)$$

$$\tan Z = \frac{\sin t}{\sin \phi \cos t - \cos \phi \tan \delta} \qquad (12\text{-}12)$$

in which Z is the azimuth of Polaris, measured from the north; t is the hour angle, obtained from the time as in Section 12-16; δ is the declination, taken from the ephemeris; h is the altitude, obtained by correcting the observed altitude for index error and refraction; and ϕ is the latitude.

If the latitude is unknown, the altitude is measured and the first equation is used. When the altitude is measured, the same observations can be used for the determination of both azimuth and latitude. The signs of the trigonometric functions must be taken into account in using these equations. When the star is near elongation, the effect of errors in time on the computed azimuth will be negligible. When the star is near culmination, an error of 5^m in the time in a latitude of about 40° will cause an error of about $0°\ 02'$ in the computed azimuth.

12-20. SOLAR OBSERVATION FOR AZIMUTH
If the time is known to within 1 or 2 min, the azimuth of a line can be determined by measuring the horizontal angle from the line to the sun's center and measuring the altitude of the sun's center. A total of eight measurements, four with the

telescope direct and four with it reversed and taken in quick succession, will give results reliable to within 15^s when a 30" transit is used.

As shown in Fig. 12-13 the instrument is set up and leveled over a selected point, as station TT41, and the line of sight is directed toward the sun. The telescope bubble is centered and the vertical circle is read to determine the index error, which is recorded. The horizontal circle is set to read 0° 00′ 00″, and a backsight is taken on another selected point as station TT42. The upper clamp is loosened, the telescope is pointed at the sun, and the image is obtained on a white card as described in Section 12-7. Four observations are taken with the image located as indicated in Fig. 12-7 or 12-9. In the morning, these positions would correspond to those in views (a), (b), (a), and (b). In the afternoon the positions would be as in views (c), (d), (c), and (d). For each observation the times to the nearest second at the instant of tangency, the horizontal-circle reading, and the vertical angle are recorded.

The telescope is then reversed and is again pointed at the sun and adjusted to give an image on the card. This pointing is made with the upper clamp of the theodolite loosened. Four observations are taken with the telescope reversed, the positions of the image being the same as those with the telescope direct, and the corresponding readings are recorded.

Finally, the upper clamp is loosened and a sight is taken back to station TT42. The horizontal-circle reading, which should be 180° 00′ 00″, is recorded.

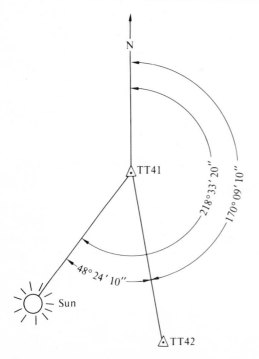

Figure 12-13. Azimuth of line by solar observation.

The index error should be determined after all the observations are made to see whether or not it has changed.

The measured horizontal and vertical angles are corrected to give the true horizontal and vertical angles to the sun's center, as explained in Section 12-7 and as demonstrated in Examples 12-3 and 12-4. In order to determine the validity of the observations, a plot is prepared showing corrected horizontal angles and corrected vertical angles as functions of time. If the observations have been made within a period of about 10 min or less, the two functions should plot as straight lines.

In Fig. 12-14 eight observations are plotted for an afternoon observation. The horizontal-angle function is shown as a straight line, but the vertical-angle function is shown as two separate straight lines that are parallel. The offset e between these two lines is analyzed as twice the unaccounted-for systematic error of the instrument. If each vertical angle read with the telescope direct were to be reduced by $e/2$ and each vertical angle read with the telescope reversed were increased by $e/2$, the resulting eight values would plot on one straight line. This, however, is not necessary, since the mean of the eight measured vertical angles is free from the systematic error.

In Fig. 12-15 eight morning observations have been plotted. The second vertical angle measured with the telescope reversed is seen to be wrong by the amount e. This error is obviously caused by a mistake in reading the vertical circle. Before the means of the observed values are found, both the horizontal angle and the vertical angle for the second reverse readings and the horizontal and vertical angles for the corresponding symmetrical direct readings would be rejected. These latter angles would be those for the third direct readings. This rejection is necessary in order to eliminate systematic

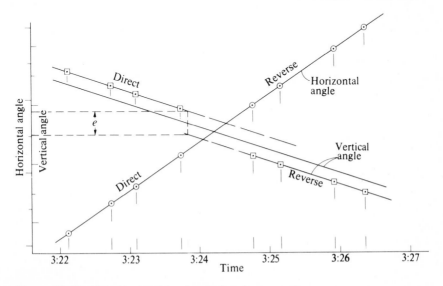

Figure 12-14. Plot of horizontal and vertical angles versus time.

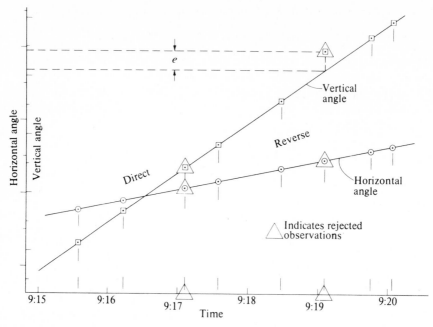

Figure 12-15. Rejection of observations by analysis of angle-time plot.

instrument errors to the fullest possible extent. Note that the corresponding two instants of time would also be rejected. The means of the remaining six values of horizontal angles, vertical angles, and times would be used in the computations for azimuth.

Occasionally the plotted results for a series of solar observations will leave some doubt in regard to the validity of the corrected vertical angles. Since the sun's altitude is a function of time, the change of vertical angle with respect to time can be determined by expressing the altitude h in terms of the sun's hour angle t by solving Eq. (12-4) for sin h. If the resulting equation is differentiated for h with respect to t, the slope of the curve for vertical angle versus time can be expressed as follows:

$$\frac{dh}{dt} = -\frac{\cos \phi \cos \delta \sin t}{\cos h} \tag{12-13}$$

The mean corrected vertical angle for the set of observations gives h. Angle t is obtained by converting the mean watch time of observation first to Greenwich civil time, then to Greenwich apparent time by applying the equation of time, and, finally, to local apparent time by subtracting the west longitude of the place of observation (or adding if the longitude is east). The result is the hour angle of the sun in hours, minutes, and seconds. This angle is converted to degrees, minutes, and seconds for use in Eq. (12-13). In the morning the hour angle is considered negative; in the afternoon, positive.

TABLE 12-9. Computations for Azimuth from Solar Observation

Date: Oct. 15, 1973. Transit at Station TT41. Latitude: 37° 52′ 30″ N.

mean corrected horizontal angle TT42 to sun	=	48° 24′ 10″
mean corrected vertical angle to sun = h	=	35° 07′ 35″
mean instant of observation	=	$2^h 57^m 02^s$ P.M., PDST
	=	$1^h 57^m 02^s$ P.M., PST
longitude of standard meridian	=	$8^h 00^m 00^s$
		$\overline{9^h 57^m 02^s}$
		$+12^h$
Greenwich civil time of observation	=	$\overline{21^h 57^m 02^s}$
declination of sun at 0^h, GCT, Oct. 15, 1973	=	$-8° 22.8′$
change in declination in $21.95^h = 21.95$		
$\times (-0.93′)$	=	$-20.4′$
declination at instant of observation	=	$\delta = -8° 43.2′$

$$\tan \tfrac{1}{2} Z = \sqrt{\frac{\sin(s - \phi) \sin(s - h)}{\cos s \cos(s - p)}}$$

$p =$ 98° 43′ 12″
$\phi =$ 37° 52′ 30″
$h =$ 35° 07′ 35″
2 $\overline{)171° 43′ 17″}$
$s =$ 85° 51′ 39″
$s - h =$ 50° 44′ 04″
$s - \phi =$ 47° 59′ 09″
$s - p =$ $-12° 51′ 33″$

$$\tan \tfrac{1}{2} Z = \sqrt{\frac{0.742\,9794 \times 0.774\,2208}{0.072\,1793 \times 0.974\,9201}}$$

$\tan \tfrac{1}{2} Z = 2.859\,1040$
$\tfrac{1}{2} Z = 70° 43′ 20″$
$Z = 141° 26′ 40″$ (counterclockwise from north)
$Z =$ 218° 33′ 20″   (clockwise from north)
horizontal angle $= -48° 24′ 10″$
azimuth TT41 to TT42 $= \overline{170° 09′ 10″}$

$$\cos Z = \frac{\sin \delta - \sin \phi \sin h}{\cos \phi \cos h}$$

$$= \frac{-0.151\,6059 - 0.613\,9408 \times 0.575\,3820}{0.789\,3520 \times 0.817\,8848}$$

$$\cos Z = -0.781\,9969$$

$Z =$ 141° 26′ 38″   (counterclockwise from north)
$Z =$ 218° 33′ 22″   (clockwise from north)
horizontal angle $= -48° 24′ 10″$
azimuth TT41 to TT42 $= \overline{170° 09′ 12″}$

If dh/dt is to be expressed as angular minutes of altitude per minute of time, then Eq. (12-13) must be modified to give the following relationship:

$$\frac{dh}{dt} = -\frac{15 \cos \phi \cos \delta \sin t}{\cos h} \tag{12-14}$$

This slope can then be plotted on the diagram for vertical angle versus time, and a line having this slope should be parallel with the line passing through all valid vertical angles. Observations can then be rejected on this basis. If an observation for a vertical angle is rejected, it is necessary to reject also the corresponding horizontal angle. Furthermore, a symmetrical set of observations must also be rejected, as previously mentioned.

The azimuth of the sun is computed from either of the following relationships:

$$\tan \tfrac{1}{2}Z = \sqrt{\frac{\sin(s - \phi) \sin(s - h)}{\cos s \cos(s - p)}} \tag{12-15}$$

$$\cos Z = \frac{\sin \delta - \sin \phi \sin h}{\cos \phi \cos h} \tag{12-16}$$

in which Z is the azimuth of the sun, measured from north clockwise in the morning and counterclockwise in the afternoon; s is $\frac{1}{2}(\phi + h + p)$; ϕ is the latitude; h is the true altitude, obtained from the mean of the corrected vertical angles; p is the polar distance, or $(90° - \delta)$; and δ is the declination, obtained for the mean instant of time from the ephemeris. The best results will be obtained if the observations are made when the local mean time is between 8 and 10 A.M. or between 2 and 4 P.M. Except when the observation is made early in the morning or late in the afternoon, Eq. (12-15) should be used to compute the azimuth of the sun.

The horizontal angle between the backsight and the position of the sun represented by the mean of the corrected angles is used with the azimuth of the sun to obtain the azimuth of the line. Computations showing the use of both formulas to determine the azimuth of the line from station TT41 to station TT42 in Fig. 12-13 are given in Table 12-9.

12-21. AZIMUTH BY SOLAR ATTACHMENT The solar attachment is a device fastened to a transit for rapidly determining the direction of a meridian. The three types commonly encountered are the Burt, the Saegmuller and the Smith attachments. Although these three attachments differ greatly in appearance, the principles involved in their use are the same.

The Saegmuller attachment, shown in Fig. 12-16, consists of an auxiliary telescope that is equipped with a bubble and has two motions at right angles

Figure 12-16. Transit equipped with Saegmuller solar attachment.
Courtesy of Keuffel & Esser Co.

to each other. If the main telescope is put in the plane of the meridian and the line of sight is directed to the celestial equator, as shown in Fig. 12-17, then the vertical axis of the attachment also lies in the meridian and points to the North Pole. If the line of sight of the auxiliary telescope is at right angles to the polar axis, it will generate an equatorial plane. When the line of sight is inclined to this plane by an amount equal to, and in the direction of, the sun's declination and the telescope is revolved on its polar axis, the line

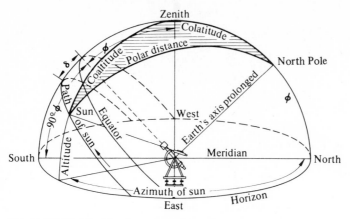

Figure 12-17. Meridian by solar attachment.

of sight will follow the sun's path in the heavens for the given day, the declination being assumed to remain constant.

Thus when the main telescope is elevated through a vertical angle equal to the colatitude, and the angle between the planes of the horizontal cross hairs of the two telescopes is made equal to the declination of the sun, the main telescope must be in the meridian when the auxiliary telescope is directed toward the sun. Only one position for the main telescope can be found that will permit the smaller telescope to be directed toward the sun. When this position is found, the main telescope will be in the meridian.

In setting off the declination angle, the two telescopes are brought into the same plane by directing the lines of sight to the same distant object. The main telescope is then depressed or elevated, according as the declination is north or south, and the declination angle is set on the vertical circle. The auxiliary telescope is revolved about its horizontal axis until its bubble comes to the center of its tube. The angle between the two telescopic lines of sight will be the declination angle. The main telescope is then elevated until its latitude is equal to the colatitude of the place, and it is clamped in this position. With the vertical motions of the two telescopes clamped, the horizontal clamps are released and the telescopes are turned until the small telescope is directed toward the sun. The main telescope is then in the meridian.

Because of refraction, the sun will appear higher in the sky than it really is. Hence the declination angle used in making the observations is obtained by adding a refraction correction to the declination of the sun as obtained from the ephemeris. The refraction correction is dependent on the declination, hour angle, and latitude. Tables giving the values of this correction are found in the abridged editions of the ephemeris published by the various instrument makers.

Although the solar attachment has been used in the subdivision of much of the public lands, the results obtained may be considerably in error, unless the instrument is in very good adjustment.

12-22. AZIMUTH BY EQUAL ALTITUDES The meridian can be found by means of two equal altitudes of a heavenly body, one east and the other west of the meridian. If the declination of the sun is considered to be constant, its path, as shown in Fig. 12-17, is symmetrical with respect to the meridian. The meridian is therefore midway between any two positions of the sun that are at equal distances above the horizon.

To use this method a horizontal angle is measured between a fixed line of reference and the sun's center when it is at any convenient altitude in the morning and the horizontal cross hair is located as in Fig. 12-7(a). The horizontal circle and vertical circle are read. The index error is determined and the vertical angle to the sun's lower limb is obtained, uncorrected for parallax and refraction. For the afternoon observation the vertical angle is increased by the diameter of the sun, and the circle reading is set off on the transit to take into account the index error of this pointing. The sun is tracked until the image reaches the position indicated in Fig. 12-7(c), at which instant the horizontal cross hair is tangent to the sun's upper limb. Consequently, the altitude of the sun is identical with its altitude in the morning.

Since the sun's declination is not constant, the mean between the two positions of the sun will not be the exact south point. The correction for the change in declination is $C = \frac{1}{2}d/(\cos \phi \sin t)$, in which d is the change in declination during the elapsed time, ϕ is the latitude, and t is the hour angle of the sun, or approximately half the time interval between the two observations. If the sun is moving north, the variation per day of the declination is positive, and the mean of the two positions of the sun lies to the west of the south point. When the correction is added to 180°, due heed being paid to

TABLE 12-10. Computations of Azimuth by Equal Altitudes of Sun

Date: Aug. 17, 1974. Transit at station 1. Latitude: 43° 27′ N.

MST	Vernier A	Vernier B	Vertical Angle	To station
A.M.	0° 00′ 00″	180° 00′ 00″		2
10ʰ 02ᵐ 43ˢ	34° 24′ 30″	214° 24′ 30″	+46° 38′	☉ lower limb
P.M.				
2ʰ 50ᵐ 52ˢ	148° 18′ 30″	328° 18′ 30″	+47° 10′	☉ upper limb
	0° 00′ 00″	180° 00′ 00″		2

elapsed time $= 4^h\ 48^m\ 09^s$

$t \qquad = 2^h\ 24^m\ 05^s = 36°\ 01′ \qquad \frac{1}{2}d = (2.4)(-47.5″) = -114″$

azimuth of sun at

$\cos 43°\ 17′ = 0.727972$ middle position $\quad = 179°\ 55′\ 34″$

$\sin 36°\ 01′ = 0.588021$ mean horizontal angle $= \ \ 91°\ 21′\ 30″$

       azimuth of line 1–2 $\quad = \ \ 88°\ 34′\ 04″$

$$C = \frac{-114}{0.727\ 972 \times 0.588021}$$

$$= -266″ = -0°\ 04′\ 26″$$

the sign of the variation per day, the result is the azimuth, measured from the north, of the sun at the middle position.

When the observations are made at night, a series of observations on different stars can be made. No corrections are needed in this case, since any change in the declination of a star is negligible. An example of the computations for an observation on the sun is shown in Table 12-10.

12-23. ORDER OF MAKING OBSERVATIONS In most azimuth observations the time and the latitude will be known with sufficient accuracy to permit the use of solar observations, or of observations on Polaris at any hour angle. The time is obtained from a good watch, or by comparing a fair one with the time furnished by radio time signals, a Western Union Telegraph clock, or time service by telephone. The latitude can usually be scaled with sufficient accuracy from an existing map.

If it is necessary to make observations where no preliminary information is available, an approximate latitude can be obtained by measuring the maximum altitude of the sun. The approximate local apparent time would be obtained at the same observation, since the maximum altitude will occur at the instant of local apparent noon. A more exact latitude could be obtained by measuring the maximum or the minimum altitude of Polaris or any other convenient star.

From the observed latitude, a very close approximation to the true meridian could be established by observations on Polaris at elongation. Local time can be obtained by an observation on the sun at local apparent noon, the previously located meridian being used, or by measuring the altitude of the sun in the manner described in Section 12-7.

By repeating the observations and using the data obtained from each preceding observation, the location of the meridian, the local time, and the latitude can eventually be determined with considerable exactness.

The easiest way of determining the longitude is by comparing the local mean time with some standard time, sent out by some radio transmitter. With the small compact short wave receivers now available, there should be few times when this method cannot be used. Books on astronomy contain methods by which the approximate longitude can be obtained by observations on the moon.

PROBLEMS†

12-1. Determine the sun's declination and the local apparent time for the instant of 10:42:50 A.M., CST, on Oct. 12 at a place whose longitude is 89° 25′ 25″ W.

12-2. What is the PST of local apparent noon on Dec. 12 at a place whose longitude is 121° 35′ W? What is the sun's declination at this instant?

† The ephemeris for the current year is to be used in the problems for Chapter 12.

12-3. What is the sun's hour angle on Aug. 12 at a place whose longitude is 76° 22′ W at the instant of 5:15:20 A.M., PST?

12-4. When the altitude of the sun's lower limb is measured at local apparent noon on Sept. 5, the vertical-circle reading is +44° 25′. The index error is +0° 03′. What is the latitude of the place of observation if the longitude is 85° 20′ W?

12-5. Determine the PST of eastern and western elongation of Polaris on April 5 at a place whose longitude is 122° 40′ W and whose latitude is 42° 32′ N.

12-6. Polaris was observed at upper culmination in longitude 105° 04′ 30″ W on the night of Dec. 19, and the vertical angle was +38° 30′. What was the Mountain standard time? What is the latitude of the place of observation?

12-7. What is the azimuth of Polaris at 8:55:10 P.M., EST on Feb. 10, at a place whose latitude is 44° 20′ and whose longitude is 74° 21′ 30″ W?

12-8. From the following observations made on Polaris at longitude 7^h 22^m 43^s west, determine the latitude of the place and the azimuth of the line AB.

Date: Sept. 16. Transit at station A at elevation of 6110 ft. Temperature: 75°F.

MST P.M.	VERNIER A	VERNIER B	VERTICAL ANGLE	TELESCOPE	POINTING
	0° 00′ 00″	180° 00′ 00″		Direct	NW to station B
7^h 12^m 16^s	54° 24′ 00″	234° 24′ 00″	+42° 58′ 30″	Direct	N to Polaris
7 14 42	234° 24′ 30″	54° 25′ 00″	42° 59′ 00″	Reversed	N to Polaris
	180° 00′ 00″	0° 00′ 30″		Reversed	NW to station B

12-9. The following observations were made on the sun in latitude 42° 16′ 30″ N.

Date: Aug. 12. Transit at station C at elevation of 905 ft. Temperature: 82°F.

EST A.M.	VERNIER A	VERNIER B	VERTICAL ANGLE	TELESCOPE	POINTING
	0° 00′ 00″	180° 00′ 00″		Direct	SW to station D
10^h 24^m 32^s	268° 48′ 30″	88° 49′ 00″	+49° 57′	Direct	Sun's lower limb
25 20	268° 38′ 00″	88° 38′ 30″	50° 20′	Direct	Sun's eastern limb
26 04	269° 16′ 00″	89° 16′ 00″	50° 11′	Direct	Sun's lower limb
28 12	89° 28′ 00″	269° 27′ 30″	50° 48′	Reversed	Sun's eastern limb
28 58	90° 07′ 00″	270° 07′ 30″	50° 39′	Reversed	Sun's lower limb
29 41	89° 54′ 00″	269° 54′ 30″	51° 02′	Reversed	Sun's eastern limb
	180° 00′ 00″	0° 00′ 30″		Reversed	SW to station D

The longitude of the place of observation is 82° 02′ 20″ W. Correct the horizontal and vertical angles to the sun's center, and plot the corrected values versus the corresponding times of observation. Compute the azimuth of the line CD and the watch correction.

BIBLIOGRAPHY

American Ephemeris and Nautical Almanac. U.S. Government Printing Office.

Buckner, R. B. 1975. Reasons and methods for accurate direction in land surveys. *Surveying and Mapping* 35:305.

Chait, B. 1970. Azimuth and longitude from near meridian star observations by theodolite. *The Canadian Surveyor* 24:215.

Chauvenet, W. 1891. *Spherical and Practical Astronomy*. Philadelphia: Lippincott.

Hoskinson, A. J. and Duerksen, J. A. 1947. *Manual of Geodetic Astronomy*. Special Publication No. 237, U. S. Coast and Geodetic Survey. U.S. Government Printing Office.

Hosmer, G. L. and Robbins, J. M. 1948. *Practical Astronomy*. New York: Wiley.

Nassau, J. J. 1948. *Text Book on Practical Astronomy*. New York: McGraw-Hill.

Chapter 13
Horizontal and Vertical Curves

13-1. REMARKS The horizontal alignment of a highway, railroad, or canal consists of a series of straight lines and curves. The straight portions are called *tangents*. As shown in Fig. 13-1 successive tangents change direction by deflection angles designated Δ_i. The change in direction at each intersection is distributed along a curve or series of curves in order to make a smooth transition. The curves shown in Fig. 13-1 are simple circular with radii designated R_i and central angles Δ_i. Horizontal transition curves can also be made up of multiple circular curves of different radii called *compound curves* or of circular curves joined to the tangents by spirals of third degree.

The grade line on a profile of any length is likewise made up of straight lines and curves as indicated in Fig. 13-2. Intersections of the grade lines are designated V_i, grades are designated g_i, and the lengths of the curves joining the grade lines are designated L_i. The curves joining the grade lines are called *vertical curves* and are generally parabolas of second degree. Their function is to make a smooth transition from one grade to another—in other words, to spread out the change in grade over a distance L.

This present chapter is designed to give the reader an understanding of the basic geometry of horizontal and vertical curves and how they are located or

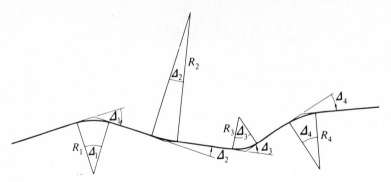

Figure 13-1. Horizontal alignment consisting of tangents joined by simple circular curves.

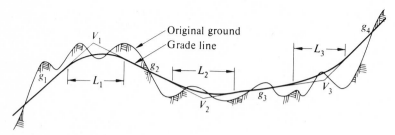

Figure 13-2. Grade lines on a profile joined by vertical curves. The vertical scale is exaggerated.

constructed in the field. Those interested in a more thorough treatment of the subject should consult texts on route location.

The treatment of curves given in this chapter is expressed in the foot unit. However, all of the equations to be developed are just as valid when using the metric system as long as a station is understood to be 100 m. Thus, the designation m can be used wherever the designation ft is used except where noted.

13-2. NOTATION FOR CIRCULAR CURVES   The notation commonly used on circular curves is shown in Fig. 13-3. The point at which the two tangents to the curve intersect is called the *vertex*, which is designated V, or the *point of intersection* (abbreviated PI). The deflection angle between the tangents, which is equal to the angle at the center of the curve, is denoted by Δ. If the survey (shown in Fig. 13-3) is progressing to the right, the straight line to the left of the PI is the *back tangent* and the one to the right is the *forward tangent*. The beginning point of the curve is called the *point of curvature* (abbreviated PC). This is sometimes referred to as the *tangent-to-curve point* (abbreviated TC) or the *beginning of curve* (abbreviated BC). The end of the curve is the *point of tangency* (or the PT). This is also referred to as the *curve-to-tangent point* (the CT), or the *end of curve* (the EC).

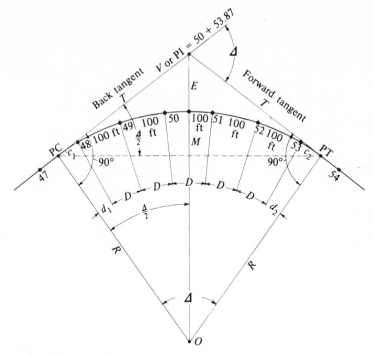

Figure 13-3. Circular curve.

The distance from the PI to the PC or the PT is the *tangent distance T.* The distance from the PI to the middle point of the curve, measured along the bisector of the central angle, is the *external distance E.* The distance from the middle point of the curve to the middle point of the chord joining the PC and the PT is the *middle ordinate M.* The distance along the line joining the PC with the PT is the *long chord C* (see Fig. 13-6). The angle subtended at the center of the curve by a 100-ft chord is the *degree of curve D,* by *chord* definition. The angle subtended at the center of the curve by a 100-ft arc is the *degree of curve D,* by *arc* definition. The *radius* of the curve is designated by *R.*

As the station method is commonly used in recording the horizontal distances, and as the PC of a curve will rarely be at a full station, the distance c_1 from the PC to the first full station on the curve will be, in general, less than 100 ft. The central angle subtended by this distance is d_1. The distance from the last full station on the curve to the PT is c_2, and the corresponding central angle is d_2.

A curve can be designated by either the radius or the degree of curve. The designation by the radius is finding widespread use in highway practice. Usually some integral multiple of 50 ft is used as the radius of the curve.

The *length of a curve* is the difference in stationing between the PC and the PT of the curve. By arc definition this corresponds to the length of the curve measured along the actual arc. By chord definition it does not correspond either to the length measured along the arc or to the length measured along a series of chords unless the curve is, by chance composed of full 100-ft chords only. The length L of a curve herein defined is a mathematical value used to compute other elements used in laying out the curve in the field.

If the physical length of a curve measured along the actual arc is needed for property description, but the curve is defined by the chord definition, its length is obtained by $R\Delta$, in which Δ is the central angle in radians obtained with the aid of Table A in Appendix C.

13-3. RADIUS AND DEGREE OF CURVE The relationship between the radius and the degree of curve by chord definition is shown in Fig. 13-4. In either of the two right-angled triangles formed by bisecting the central angle D,

$$\sin \tfrac{1}{2}D = \frac{50}{R} \tag{13-1}$$

from which

$$R = \frac{50}{\sin \tfrac{1}{2}D} \tag{13-2}$$

The relationship between the radius and the degree of curve by arc definition is shown in Fig. 13-5. When D is expressed in degrees,

$$\frac{100}{D} = \frac{2\pi R}{360°}$$

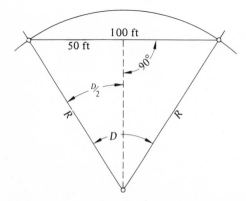

Figure 13-4. Relation between R and D by chord definition.

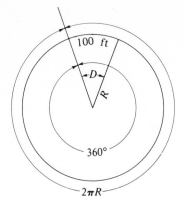

Figure 13-5. Relation between R and D by arc definition.

from which

$$R = \frac{5729.578}{D} \qquad (13\text{-}3)$$

or

$$D = \frac{5729.578}{R} \qquad (13\text{-}4)$$

From Eqs. (13-2) and (13-3) if the degree of curve is $1°$, the radius of the chord definition curve would be 5729.651 ft, whereas the radius of the arc definition curve would be 5729.578 ft. The difference between the two values would be of no significance in laying out the curves. However, for sharper curves with more curvature and shorter radii, the difference between the two definitions is quite obvious. Suppose that $D = 20°$. Then by chord definition the radius is 287.94 ft, whereas by arc definition it is 286.48 ft.

13-4. EQUATIONS FOR CIRCULAR CURVES The relationships involving the radius R of a circular curve, the deflection angle Δ between the tangents, and other elements of the circular curve are shown in Fig. 13-6. The following equations apply to both the chord definition and the arc definition:

$$T = R \tan \tfrac{1}{2}\Delta \qquad (13\text{-}5)$$

$$E = R \sec \tfrac{1}{2}\Delta - R = R(\sec \tfrac{1}{2}\Delta - 1)$$

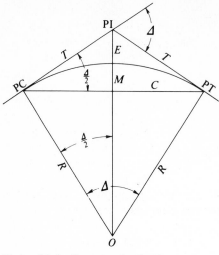

Figure 13-6. Elements of a circular curve.

or

$$E = R \text{ exsec } \tfrac{1}{2}\Delta \tag{13-6}$$

$$M = R - R \cos \tfrac{1}{2}\Delta = R(1 - \cos \tfrac{1}{2}\Delta)$$

or

$$M = R \text{ vers } \tfrac{1}{2}\Delta \tag{13-7}$$

$$C = 2R \sin \tfrac{1}{2}\Delta \tag{13-8}$$

The length of the curve, or the difference in stationing between the PC and PT, is computed by the relationship

$$L = \frac{\Delta}{D} \text{ (expressed in stations)}$$

$$L = 100 \frac{\Delta}{D} \text{ (expressed in feet or metres)} \tag{13-9}$$

13-5. SELECTION OF CURVE Any two given tangents can be connected by an infinite number of circular arcs. The curve to be used in a particular case is determined by assuming D, R, T, E, or L. Since all these quantities are independent, only one can be assumed. The other values are calculated from the relationships developed in the preceding section.

Field conditions frequently decide which quantity should be assumed. Thus when the survey is following the bank of a stream, the external distance may be the limiting factor. On a winding road the tangent lengths may be restricted. On high-speed modern pavements and railroads, an attempt is made to keep the degree of curve below a given maximum. Wherever possible a radius of more than 1000 ft (300 m) is adopted. This radius corresponds to a degree of curve of about $5° 44'$.

13-6. STATIONS OF PI, PC, AND PT   When a road or railroad is planned, its location is first made on a map sheet to fit the existing conditions of man-made and natural features, culture, and topography. This is referred to as a *paper location*. The coordinates of the PI's are carefully scaled from the map (see Section 8-33). The coordinates can be used in one of two different ways. First, each tangent can be inversed using Eqs. (8-11) and (8-12), from which all the tangent lengths and deflection angles are determined. The line is then staked out and stationed by the methods discussed in Sections 6-5, 6-6, and 6-7.

The alternative way is to run a random control traverse through the project area and then compute the lengths and bearings of tie lines by inversing between the points on the control traverse and the PI's, as discussed in Section 8-23. The tie lines are run in the field to locate the PI's. The tangents are then stationed by surveying between the PI's.

In the latter method for locating the PI's and stationing the tangents there are inevitable errors in the location survey. In order to ensure that the curves will fit the tangents as located, the deflection angles at each PI between the back tangent and the forward tangent must be measured by double centering.

The net result of the field location of the tangents is a set of lines stationed from the point of beginning, throughout the project, together with the deflection angles between the lines. A section of such a field location is shown in Fig. 13-7(a). The distance to the first PI is 814.72 ft, to the second PI is

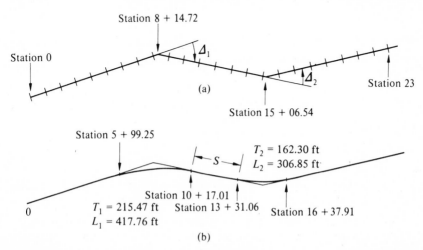

Figure 13-7. "Old" stationing along straight lines and "new" stations along tangents and curves.

1506.54 ft, and so on. The station of the first PI is thus 8 + 14.72 and the station of the second PI is 15 + 06.54.

In order to make the stationing continuous along the tangents and the curves throughout the project, the line must be restationed to reflect the difference between the straight-line distance of Fig. 13-7(a) and the combination straight-line and curve distances of Fig. 13-7(b). In this restationing, the new stations of all the PC's and PT's are required.

The station of the first PC is obtained by the following

$$PC = PI - T \qquad (13\text{-}10)$$

in which T is the tangent distance of the first curve. In the first curve of Fig. 13-7(b), the PC station is

$$
\begin{array}{rl}
PI = & 8 + 14.72 \\
T = & -2 + 15.47 \\
\hline
PC = & 5 + 99.25
\end{array}
$$

The PT station is then computed by the following

$$PT = PC + L \qquad (13\text{-}11)$$

in which L is the length of the curve computed by Eq. (13-9). The station of the PT for the first curve of Fig. 13-7(b) is

$$
\begin{array}{rl}
PC = & 5 + 99.25 \\
L = & 4 + 17.76 \\
\hline
PT = & 10 + 17.01
\end{array}
$$

The second and subsequent PC stations are determined as follows. The distance S of Fig. 13-7(b) is the difference between the length of the line joining the two PI's and the sum of the T_1 and T_2. Then

$$PC_2 = PT_1 + S_{1-2} \qquad (13\text{-}12)$$

In this example the length of the line between the two PI's is $1506.54 - 814.72 = 691.82$, and $S = 691.82 - (215.47 + 162.30) = 314.05$. Then by Eq. (13-12)

$$
\begin{array}{rl}
PT_1 = & 10 + 17.01 \\
S_{1-2} = & 3 + 14.05 \\
\hline
PC_2 = & 13 + 31.06
\end{array}
$$

The station of the PT for the second curve is obtained by Eq. (13-11), which is 16 + 37.91. The remainder of the curves in the project are treated in the same manner as the second curve.

The difference between the "old" stationing along the straight lines and the "new" stationing along the tangents and curves is accounted for where required by means of an equation. For example, a point on the located line may be defined as old station 56 + 42.70 = new station 51 + 17.66. The meaning of this equation should be obvious.

13-7. CENTRAL ANGLE AND CHORD TO FIRST CURVE STATION   In Fig. 13-8, the PC station is 20 + 54.00, the first curve station is 21 + 00, and the second curve station is 22 + 00. By either the chord or the arc definition, the angle at the center subtending a full station is equal to the degree of curve D. Let c_1' equal the *difference in stationing* between the PC and the first station on the curve expressed in feet. Then by direct proportion

$$\frac{d_1}{D} = \frac{c_1'}{100}$$

or

$$d_1 = \frac{c_1' D}{100} \qquad\qquad (13\text{-}13)$$

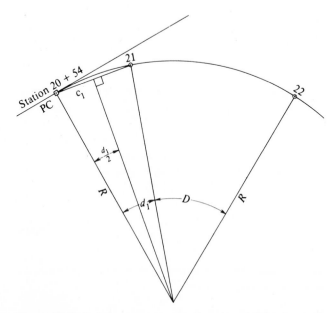

Figure 13-8. Relation between central angle and deflection angle.

In Fig. 13-8 let $D = 15°$. Then since $c_1' = 46.00$ ft, $d_1 = 46.00 \times 15/100 = 6.90° = 6° 54'$.

The *actual* chord length between the PC and station $21 + 00$, designated c_1 in Fig. 13-8, is used to lay out the point on the ground. It is found from the relationship

$$\frac{\frac{1}{2}c_1}{R} = \sin \tfrac{1}{2}d_1$$

from which

$$c_1 = 2R \sin \tfrac{1}{2}d_1 \tag{13-14}$$

The value of c_1 in Fig. 13-8 will depend on whether the degree of curve is determined by the chord or arc definition, since the value of R and consequently T will be different for the two definitions for a given D.

13-8. CENTRAL ANGLE AND CHORD FROM LAST CURVE STATION TO PT

If c_2' denotes the difference in stationing between the last station on the curve and the PT, then the central angle d_2 between these two points is

$$d_2 = \frac{c_2' D}{100} \tag{13-15}$$

The chord length c_2 between the last curve station and the PT is

$$c_2 = 2R \sin \tfrac{1}{2}d_2 \tag{13-16}$$

13-9. CENTRAL ANGLE AND CHORD BETWEEN ANY TWO CURVE POINTS

From Sections 13-7 and 13-8 it is obvious that the angle at the center subtended between any two points on the curve is proportional to the difference in stationing between the two points. Let c' be the difference in stationing between any two points on the curve, d be the central angle between the two points, and D be the degree of curve, either by chord or arc definition. Then

$$d = \frac{c'D}{100} \tag{13-17}$$

The chord distance c between the same two points, either by chord or arc definition, is given by the relationship

$$c = 2R \sin \tfrac{1}{2}d \tag{13-18}$$

Example 13-1.

The PC of an arc definition curve is at station $17 + 73.35$; the length of the curve, by Eq. (13-9), is 788.96 ft; the degree of curve D is $7° 30'$. What is the value of d_1 and d_2? What are the chord lengths c_1 between the PC and the first curve station, and c_2 between the last curve station and the PT? What is the chord length between station $19 + 00$ and station $23 + 50$?

Solution: By Eq. (13-11) the PT station is

$$1773.35 + 788.96 = 25 + 62.31$$

Then

$$c_1' = 1800.00 - 1773.35 = 26.65 \text{ ft}$$

$$c_2' = 2562.31 - 2500.00 = 62.31 \text{ ft}$$

By Eq. (13-13)

$$d_1 = 26.65 \times \frac{7.5}{100} = 1.99875° = 1° 59' 56''$$

By Eq. (13-15)

$$d_2 = 62.31 \times \frac{7.5}{100} = 4.67325° = 4° 40' 24''$$

By Eq. (13-3)

$$R = \frac{5729.58}{7.5} = 763.95 \text{ ft}$$

Then by Eq. (13-14)

$$c_1 = 2 \times 763.94 \sin 0° 59' 58'' = 26.65 \text{ ft}$$

Also, by Eq. (13-16)

$$c_2 = 2 \times 763.94 \sin 2° 20' 12'' = 62.29 \text{ ft}$$

Note the c_1 and c_1' are equal to one another, and c_2 differs from c_2' by only 0.02 ft.

By Eq. (13-17) the central angle between station 19 and station 23 + 50 is

$$d = 450.00 \times \frac{7.5}{100} = 33.75° = 33° 45'$$

By Eq. (13-18) the chord joining these two points on the curve is

$$2 \times 763.94 \sin 16° 52' 30'' = 443.52 \text{ ft}$$

Example 13-2.

What is the chord length between any two successive stations of the curve of Example 13-1? Between any two half stations? Between any two quarter stations?

Solution: The central angle between any two successive stations is D; between any two half stations it is $D/2$; and between any two quarter stations it is $D/4$. Then by Eq. (13-18)

$$c_{100} = 2R \sin \frac{D}{2} \qquad (13\text{-}19)$$

$$c_{50} = 2R \sin \frac{D}{4} \qquad (13\text{-}20)$$

$$c_{25} = 2R \sin \frac{D}{8} \qquad (13\text{-}21)$$

in which c_{100}, c_{50}, and c_{25} are the respective chord lengths desired. Thus

$$c_{100} = 2 \times 763.94 \sin 3° 45' \qquad = 99.94 \text{ ft}$$
$$c_{50} = 2 \times 763.94 \sin 1° 52' 30'' = 49.99 \text{ ft}$$
$$c_{25} = 2 \times 763.94 \sin 0° 56' 15'' = 25.00 \text{ ft}$$

Example 13-2 shows that the chord lengths c_{100}, c_{50}, and c_{25} are all less than, if not equal to, the corresponding difference in stationing when using the arc definition. However, using a chord definition curve, $c_{100} = 100$ ft, by definition; but c_{50} is greater than 50 ft and c_{25} is greater than 25 ft. This can be seen by an inspection of Fig. 13-9. The chord between station 23 and station 24 is 100 ft, by definition. The chord between station 23 and station 23 + 50 is longer than 50 ft because it is the hypotenuse of a right triangle

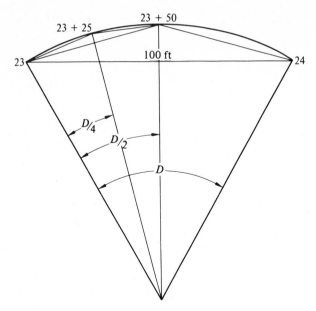

Figure 13-9. Chord lengths on chord-definition curve.

one side of which is 50 ft. Also, the chord between station 23 and station 23 + 25 is longer than 25 ft.

13-10. DEFLECTION ANGLES TO POINTS ON CURVE The angle formed between the back tangent and a line from the PC to a point on the curve is the *deflection angle* to the point. This deflection angle, measured at the PC between the tangent and the line to the point, is one-half the central angle subtended between the PC and the point. This relationship comes directly from the geometry of a circle. An angle between a tangent and a chord is measured by one-half the intercepted arc, whereas the central angle is measured by the whole arc. Thus in Fig. 13-10 the central angle d_1 between the PC at station 12 + 62.50 and the first full station on the curve is 3°, and the deflection angle $\frac{1}{2}d_1$ between the back tangent and the line directed to station 13 + 00 is 1° 30′. Similarly, the central angle between the PC and the second full station on the curve is $d_1 + D = 3° + 8° = 11°$ and the deflection angle, which is $\frac{1}{2}d_1 + \frac{1}{2}D$, is 5° 30′. If each half station were to be located, then the deflection angle to station 13 + 50 in Fig. 13-10 would be $\frac{1}{2}d_1 + \frac{1}{4}D = 3° 30′$.

The deflection angle to each succeeding full station is obtained by adding the value $\frac{1}{2}D$ to the preceding deflection angle. To locate half stations, $\frac{1}{4}D$ is added; to locate quarter stations, $\frac{1}{8}D$ is added. The last deflection is obtained by adding $d_2/2$ to the next to last calculation. This is the deflection angle to the PT and must equal $\Delta/2$ as shown in Fig. 13-3. This provides a check on the calculations.

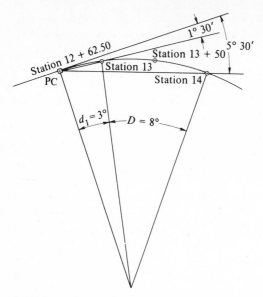

Figure 13-10. Deflection angles to points on curve.

Example 13-3.

Compute the deflection angles to each half station of the curve given in Example 13-1.

Solution: The value of c'_2 in this case is $2562.31 - 2550.00 = 12.31$ ft, and d_2 is $(12.31 \times 7.5)/100 = 0° 55' 24''$. The value of Δ is, by Eq. (13-9), $59° 10' 20''$. The calculations for the deflection angles are as follows:

STATION	DEFLECTION ANGLE
PC = Station 17 + 73.35	0° 00'
$+ d_1/2$	$+0° 59' 58''$
18 + 00	0° 59' 58''
$+ D/4$	$+1° 52' 30''$
18 + 50	2° 52' 28''
$+ D/4$	$+1° 52' 30''$
19 + 00	4° 44' 58''
$+ D/4$	$+1° 52' 30''$
19 + 50	6° 37' 28''
$\vdots$	$\vdots$
25 + 00	27° 14' 58''
$+ D/4$	$+1° 52' 30''$
25 + 50	29° 07' 28''
$+ d_2/2$	$+0° 27' 42''$
PT = Station 25 + 62.31	29° 35' 10'' = $\Delta/2$

13-11. CURVE LOCATION To locate the position of a simple curve in the field, the intersection angle Δ is measured at the PI with the theodolite. Based on an assumed value for one of the elements of the curve, the other elements and the stationing of the PC and the PT are computed. A table of deflection angles is prepared for locating full stations, half stations, or quarter stations. With the theodolite still at the PI, the PC and the PT are located by laying off the tangent distance T.

The instrument is next set up at the PC and a backsight is taken along the back tangent toward the PI. The first deflection angle $\frac{1}{2}d_1$ is turned off, and the actual chord distance c_1 is laid off on the line of sight to locate the first curve station. When full stations are located, the second deflection angle $\frac{1}{2}d_1 + \frac{1}{2}D$ is turned off by setting the horizontal circle to this reading. The chord distance between the first and second curve stations is laid off by measuring from the first curve station, and the forward end of the chord is brought on the line of sight. This procedure establishes the second curve station. Each station is located by a line of sight from the PC and a chord distance from the preceding station, until the end of the curve is reached, or until a setup on the curve becomes necessary. The previously located position of the PT provides a check on the accuracy of the field work.

The complete computations for the curve shown in Fig. 13-3 are given in Table 13-1. The order in which these computations are made is indicated by

TABLE 13-1. Computations for Circular Curve Full Stations, Chord Definition

$$\Delta = 23°\,18' \qquad \tfrac{1}{2}\Delta = 11°\,39' \qquad D \text{ (assumed)} = 4°\,00' \text{ chord definition}$$

PI = station $50 + 53.87$	$c_1' = 41.52$ $\qquad(6)$
$R = 50/\sin 2°\,00' = 1432.69$ $\qquad(1)$	$\tfrac{1}{2}d_1 = \dfrac{41.52 \times 4°}{200} = 0°\,49'\,49''$ $\qquad(7)$
$T = 1432.69 \tan 11°\,39' = 295.39$ $\qquad(2)$	
$L = 100 \times 23.3/4 = 582.50$ $\qquad(3)$	$c_1 = 2 \times 1432.69 \times \sin 0°\,49'\,49'' = 41.52$ $\qquad(8)$
$\qquad$ PI $= 50 + 53.87$	$c_2' = 40.98$ $\qquad(9)$
$\qquad$ $T = \;\;2 + 95.39$	
$\qquad$ PC $= 47 + 58.48$ $\qquad(4)$	$\tfrac{1}{2}d_2 = \dfrac{40.98 \times 4}{200} = 0°\,49'\,11''$ $\qquad(10)$
$\qquad$ $L = \;\;5 + 82.50$	$c_2 = 2 \times 1432.69 \times \sin 0°\,49'\,11'' = 40.99$ $\qquad(11)$
$\qquad$ PT $= 53 + 40.98$ $\qquad(5)$	

Station	Deflection angle	(12)
PC $= 47 + 58.48$	$0°\,00'\,00''$	
48	$0°\,49'\,49''$	
49	$2°\,49'\,49''$	
50	$4°\,49'\,49''$	
51	$6°\,49'\,49''$	
52	$8°\,49'\,49''$	
53	$10°\,49'\,49''$	
PT $= 53 + 40.98$	$11°\,39'\,00''$	
	(checks $\tfrac{1}{2}\Delta$)	

TABLE 13-2. Computations for Half Stations, Arc Definition

$\Delta = 28°\,46'$ $\frac{1}{2}\Delta = 14°\,23'$	R (assumed) = 850 ft arc definition

$D = 5729.58/850 = 6.74068°$	(1)	$D/4 = 1°\,41'\,07''$		(1)
$T = 850 \tan 14°\,23' = 850 \times 0.5645$		$c_1' = 47.43$		(6)
$\qquad = 217.98$ ft	(2)	$c_2' = 29.33$		(7)

$$L = \frac{100 \times 28.76667}{6.74068} = 426.76 \qquad (3)$$

Station	Deflection angle (13)
PC = 24 + 02.57	0° 00′ 00″
24 + 50	1° 35′ 54″
25	3° 17′ 01″
25 + 50	4° 58′ 08″
26	6° 39′ 15″
26 + 50	8° 20′ 22″
27	10° 01′ 29″
27 + 50	11° 42′ 36″
28	13° 23′ 43″
PT = 28 + 29.33	14° 23′ 01″
	(checks $\frac{1}{2}\Delta$)

$$\begin{aligned}
PI &= 26 + 20.55 \\
T &= 2 + 17.98 \\
\overline{PC} &= \overline{24 + 02.57} \qquad (4)\\
L &= 4 + 26.76 \\
\overline{PT} &= \overline{28 + 29.33} \qquad (5)
\end{aligned}$$

$$\tfrac{1}{2}d_1 = \frac{47.43 \times 6.74068}{200} = 1°\,35'\,54'' \qquad (8)$$

$$\tfrac{1}{2}d_2 = \frac{29.33 \times 6.74068}{200} = 0°\,59'\,18'' \qquad (9)$$

$$\begin{aligned}
c_1 &= 2 \times 850 \sin 1°\,35'\,54'' = 47.42 & (10)\\
c_2 &= 2 \times 850 \sin 0°\,59'\,18'' = 29.32 & (11)\\
c_{50} &= 2 \times 850 \sin (6.74068/4)° = 49.99 & (12)
\end{aligned}$$

the numbers in parentheses. The deflection angles are computed more exactly than the angles can be turned off in the field. This is done to check the value of the final deflection angle, since it must agree with $\frac{1}{2}\Delta$. The curve is based on the chord definition for degree of curve.

The complete computations for an arc definition curve are shown in Table 13-2. Every half station is to be located by chords and deflection angles. Since the radius is assumed to be an integral number of feet, the degree of curve must be computed to at least five places after the decimal as shown in the table. The 1″ discrepancy in the check on deflection angles is due to rounding-off errors and is of no consequence.

13-12. MOVING UP ON CURVE It frequently happens that because of the length of the curve or because of obstacles on the line, the entire curve cannot be staked out with the transit at the PC of the curve. If the transit is properly manipulated, it can be moved forward along the curve as many times as may be necessary, and the deflection angles previously computed can be used in staking out the remaining portions of the curve.

A portion of the curve for which the computations are tabulated in Table 13-1 is shown in Fig. 13-11 to a distorted scale. Let it be assumed that the theodolite is moved to station 50, which has been located by turning off a

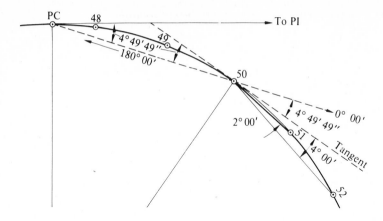

Figure 13-11. Moving up on a curve.

deflection angle of 4° 49′ 49″ from the tangent at the PC. If a backsight is taken on the PC with the circle reading 180°, it is apparent that the circle will read 4° 49′ 49″ when the line of sight is directed forward along the tangent at station 50. Since the angle between this tangent and the chord to station 51 is $\frac{1}{2}D$, or 2° 00′, the circle will read 6° 49′ 49″ when the line of sight is directed along the chord to station 51. This reading of 6° 49′ 49″ is the previously computed deflection angle for station 51. The reading of 8° 49′ 49″, previously computed for station 52, will be the reading when the line of sight is directed along the chord from station 50 to station 52.

Thus if the circle is set at 180° and a backsight is taken on the PC, the previously computed deflection angles can be used in locating stations beyond the one occupied by the theodolite. If the instrument is in good adjustment, the circle can be set to 0° 00′ and the backsight made with the telescope upside down. Then when the telescope is plunged to its direct position, the line of sight is oriented along the line marked 0° 00′ in Fig. 13-11.

If a second or subsequent curve setup is required, the backsight circle reading then corresponds to the computed deflection angle for the station on which the backsight is made. For example, if a setup is made at station 52 of Fig. 13-11 and a backsight is taken to station 50, the circle reading should read 180° + 4° 49′ 49″ when the backsight is made with the telescope direct or 4° 49′ 49″ when the backsight is made with the telescope upside down. The remainder of the curve stations are then set using the deflection angles as computed.

13-13. INTERSECTION OF CURVE AND STRAIGHT LINE One of the common problems in right-of-way surveys is the intersection of a straight line with a curve. Thus in Fig. 13-12 *DG* represents a property line intersected by the curve *ABC*. When the curve is flat, the intersection *B* can be located on the ground by setting points at *E* and *F* on the curve on both

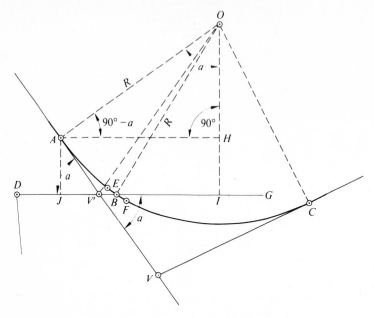

Figure 13-12. Intersection of curve and straight line.

sides of the intersection and close enough together so that no appreciable error will be introduced by considering the curve to be straight line between these two points. The point B is found at the intersection of the chord EF and the property line DG.

The intersection can be obtained mathematically if the distance AV' and the angle $GV'V$ are measured on the ground. The figure $V'BOA$ can be considered as a traverse with two missing quantities, the length $V'B$ and the direction of the line BO. These values can be calculated by the methods of Section 8-28. From the computed direction, the central angle AOB and the station of B on the curve can be calculated.

A second mathematical solution will be apparent from the following relationships: In Fig. 13-12,

$$\text{angle } GV'V = \text{angle } DV'A = \text{angle } IOA = a$$

In triangle AOH,

$$AH = R \sin a \quad \text{and} \quad OH = R \cos a$$

In triangle $AV'J$,

$$V'J = AV' \cos a \quad \text{and} \quad AJ = AV' \sin a$$

Also,

$$V'I = AH - V'J$$

$$OI = OH + AJ$$

$$\cos IOB = \frac{OI}{R}$$

$$\tan IOV' = \frac{V'I}{OI}$$

$$BI = R \sin IOB$$

$$V'B = V'I - BI$$

and

$$\text{angle } AOB = \text{angle } AOI - \text{angle } IOB$$

When these computations have been completed, B may be located by measuring the computed distance $V'B$ along $V'G$ or by means of a deflection angle and a chord measured from the PC of the curve. The length of the chord AB is $2R \sin \frac{1}{2}AOB$.

The problem of intersecting a line with a curve can be solved by the methods given in Section 8-20 if the line and the curve are both on a co-ordinate system.

13-14. VERTICAL CURVES When the grade line of a highway or a railroad changes grade, provision must be made for a vehicle to negotiate this transition smoothly and to provide vision over the crest of a hill far enough ahead to give the operator of the vehicle ample time to react to a dangerous situation. The parabola is most commonly used for connecting two different grades in order to provide for this transition. It is easy to compute elevations on a parabola, and such a curve also provides a constant rate of change of grade. The results of the vertical-curve computations are the grade elevations at selected points along a route from the beginning of the curve to its end. These elevations are used, in turn, to control grading operations when the roadbed is to be brought to the desired grade by excavation and the construction of embankments.

The length of a vertical curve in 100-ft stations is designated as L and is measured along the horizontal. The two grades in the direction of stationing are g_1 and g_2. The total change in grade is $(g_2 - g_1)$. The rate of change of grade per station, designated as r, is found by dividing the total change in grade by the length of the curve in stations. Thus

$$r = \frac{g_2 - g_1}{L} \tag{13-22}$$

in which r is the rate of change of grade per station; g_1 is the initial grade, in percent; g_2 is the final grade, in percent; and L is the length of the curve, in stations. When r is specified, the required length L is found by the relationship

$$L = \frac{g_2 - g_1}{r} \qquad (13\text{-}23)$$

If two given grades are to be connected by a vertical curve, then either r or L must be assumed and the other value is computed. The sharpness of a vertical curve in railroad location is usually defined by the allowable rate of change of grade.

When a vertical curve is laid out so that the intersection of the grade lines, called the *point of grade intersection* and designated as V, lies midway between the two ends of the curve measured horizontally, then the vertical curve is called an *equal-tangent* parabolic vertical curve.

Two methods are available for computing the elements of a vertical curve. The first method, presented in Section 13-15, treats the curve analytically. By this method problems involving high and low points on the curve, vertical clearance, and curve intersections can be dealt with most efficiently. Also, the analytic method can be adapted to computer programming very easily. The second method, presented in Section 13-16, is somewhat more easily applied for simple vertical-curve problems. It takes advantage of the geometric properties of the parabola. The solution of a complicated curve problem, however, is more difficult by the geometric method.

13-15. VERTICAL CURVES BY EQUATION OF PARABOLA

As shown in Fig. 13-13, the beginning of a vertical curve is abbreviated as BVC, the intersection of the tangents as V, and the end of the vertical curve as EVC. The initial grade is g_1 and the final grade is g_2. The y values are elevations in feet, and the x values are stations beyond the BVC. The y axis is passed through the BVC, and the x axis lies on the datum. The equation of a parabola with the axis in the y direction is

$$y = ax^2 + bx + c \qquad (13\text{-}24)$$

When $x = 0$, y is the elevation of the BVC. Therefore c is the elevation of the BVC, and Eq. (13-24) becomes

$$y = ax^2 + bx + (\text{elevation of BVC}) \qquad (13\text{-}25)$$

The first derivative of y with respect to x from Eq. (13-24) is

$$\frac{dy}{dx} = 2ax + b \qquad (13\text{-}26)$$

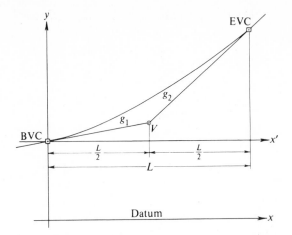

Figure 13-13. Parabola with rectangular coordinate axes.

When $x = 0$, the slope of the curve is g_1. Since this slope equals the first derivative of the curve at $x = 0$, it follows that $b = g_1$. Equation (13-24) then becomes

$$y = ax^2 + g_1x + \text{(elevation of BVC)} \qquad (13\text{-}27)$$

The second derivative is the rate of change of slope or grade of the curve. So $2a = r$, and Eq. (13-24) becomes

$$y = \frac{r}{2}x^2 + g_1x + \text{(elevation of BVC)} \qquad (13\text{-}28)$$

which is the equation of the equal-tangent parabolic vertical curve used to connect two grades. In Eq. (13-28), y is the elevation of a point on the curve, and x is the distance in stations between the BVC and the point. If the elevations of the points above or below the BVC are desired, the x axis becomes the x' axis, as shown in Fig. 13-13, and c becomes 0. For these conditions,

$$y' = \frac{r}{2}x^2 + g_1x \qquad (13\text{-}29)$$

The value of r must be assigned its proper algebraic sign. Equation (13-22) gives the algebraic sign directly. In Fig. 13-14 it is seen that if the vertical curve opens upward, r is plus; and if it opens downward, r is minus.

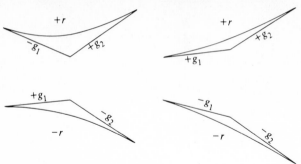

Figure 13-14. Algebraic sign of r.

Example 13-4.

Two grades, for which $g_1 = +1.25\%$ and $g_2 = -2.75\%$, intersect at station $18 + 00$, and the elevation of the intersection is 886.10 ft. If the length of the curve is to be 600 ft, what are the elevations of the BVC, the EVC, and all full stations on the curve?

Solution: See Fig. 13-15. The BVC is at station 15, and the EVC is at station 21. The elevation of the BVC is obtained by going backward along grade g_1 from the point of intersection for a distance of 300 ft. The elevation of the EVC is obtained by going forward from the point of intersection along grade g_2 for 300 ft. The computations follow:

$$\text{elevation BVC} = 886.10 - 3 \times 1.25 = 882.35 \text{ ft}$$

$$\text{elevation EVC} = 886.10 - 3 \times 2.75 = 877.85 \text{ ft}$$

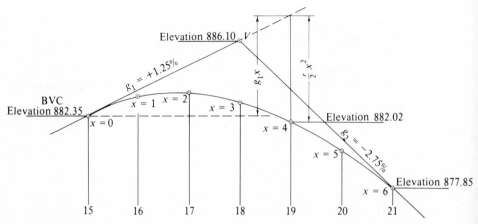

Figure 13-15. Parabolic vertical curve.

TABLE 13-3. Computations for Vertical Curve by Equation of Parabola

STATION	x	x^2	$\dfrac{r}{2}x^2$	$g_1 x$	ELEVATION BVC	ELEVATION CURVE
BVC = 15	0	0	0	0	882.35	882.35
16	1	1	− 0.33	+1.25	882.35	883.27
17	2	4	− 1.33	+2.50	882.35	883.52
18	3	9	− 3.00	+3.75	882.35	883.10
19	4	16	− 5.33	+5.00	882.35	882.02
20	5	25	− 8.33	+6.25	882.35	880.27
EVC = 21	6	36	− 12.00	+7.50	882.35	877.85

By Eq. (13-22) the change of grade per station is

$$r = \frac{-2.75 - 1.25}{6} = -0.667\% \text{ per station}$$

The equation of the curve is $y = -0.333x^2 + 1.25x + 882.35$.

The elevations of the points on the curve are computed by preparing the accompanying table. Note that the elevation of a point is the sum of the values in the fourth, fifth, and sixth columns of Table 13-3.

13-16. VERTICAL CURVES BY TANGENT OFFSETS FROM GRADE LINES Three properties of an equal-tangent vertical curve are as follows: (1) The offsets from the tangent to the curve at a point are proportional to the squares of the horizontal distances from the point; (2) offsets from the two grade lines are symmetrical with respect to the point of intersection of the two grade lines; (3) the curve lies midway between the point

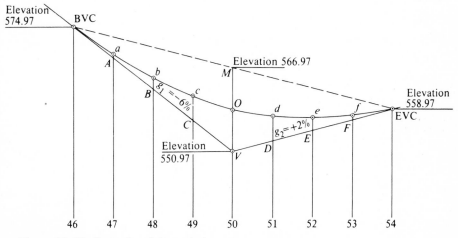

Figure 13-16. Properties of the vertical curve.

of intersection of the grade lines and the middle point of the chord joining the BVC and the EVC. Proof of these properties is left to the reader.

In Fig. 13-16, by the first property, $bB = 4aA$; $cC = 9aA$; $OV = 16aA$; $eE = 4fF$; $dD = 9fF$; $OV = 16fF$. By the second property, $aA = fF$; $bB = eE$; $cC = dD$. By the third property, $MO = OV$, or $OV = \frac{1}{2}MV$. The distance MV is obtained by subtracting the elevation of V from the elevation of M. The elevation of M is the mean of the elevations of the BVC and the EVC, since it lies at the middle point of the chord joining the two points. The elevations of points A, B, C, D, E, and F, the BVC, and the EVC are computed from the assigned grades of the two tangents. By employing the properties of the curve, the elevations of the points on the curve can then be computed.

Example 13-5.

For the two grade lines of Fig. 13-16, $g_1 = -6\%$ and $g_2 = +2\%$. The elevation of the intersection of the grade lines at station 50 is 550.97 ft. The length of the curve is 800 ft. Compute the stations and elevations of the BVC and the EVC and the elevations at all other stations on the curve.

Solution: The computations may be arranged as follows:

station of BVC $= 50 - 4 =$ station 46

elevation of BVC $= 550.97 + 4 \times 6 = 574.97$ ft

station of EVC $= 50 + 4 =$ station 54

elevation of EVC $= 550.97 + 4 \times 2 = 558.97$ ft

$$\text{elevation of middle point of chord} = \frac{574.97 + 558.97}{2} = 566.97 \text{ ft}$$

$$\text{offset to curve at intersection} = \frac{566.97 - 550.97}{2} = 8.00 \text{ ft} = VO$$

offset at A and $F = (\frac{1}{4})^2 \times 8.00 = 0.50$ ft

offset at B and $E = (\frac{2}{4})^2 \times 8.00 = 2.00$ ft

offset at C and $D = (\frac{3}{4})^2 \times 8.00 = 4.50$ ft

elevation of $A = 574.97 - 6.00 = 568.97$ ft

elevation of $B = 568.97 - 6.00 = 562.97$ ft

elevation of $C = 562.97 - 6.00 = 556.97$ ft

elevation of $V = 556.97 - 6.00 = 550.97$ ft

elevation of $F = 558.97 - 2.00 = 556.97$ ft

elevation of $E = 556.97 - 2.00 = 554.97$ ft

elevation of $D = 554.97 - 2.00 = 552.97$ ft

elevation of $V = 552.97 - 2.00 = 550.97$ ft (check)

elevation of a = station 47 = 568.97 + 0.50 = 569.47 ft

elevation of b = station 48 = 562.97 + 2.00 = 564.97 ft

elevation of c = station 49 = 556.97 + 4.50 = 561.47 ft

elevation of o = station 50 = 550.97 + 8.00 = 558.97 ft

elevation of d = station 51 = 552.97 + 4.50 = 557.47 ft

elevation of e = station 52 = 554.97 + 2.00 = 556.97 ft

elevation of f = station 53 = 556.97 + 0.50 = 557.47 ft

The foregoing computations should be arranged as shown in Table 13-4.

TABLE 13-4. Computations for Vertical Curve by Offsets from Both Tangents

STATION	TANGENT ELEVATION	OFFSET FROM TANGENT	ELEVATION CURVE
BVC = 46	574.97	0	574.97
47	568.97	+0.50	569.47
48	562.97	+2.00	564.97
49	556.97	+4.50	561.47
50	550.97	+8.00	558.97
51	552.97	+4.50	557.47
52	554.97	+2.00	556.97
53	556.97	+0.50	557.47
EVC = 54	558.97	0	558.97

13-17. INTERMEDIATE POINTS ON VERTICAL CURVES

Occasions will frequently arise where the elevations of points on vertical curves must be computed at intervals of 50, 25, or even 10 ft. Also the elevations of random points on the curve are quite frequently necessary. The most direct way of computing the elevations of these intermediate points is by use of Eq. (13-28) where x is the station or plus beyond the BVC and y is the elevation of the point in feet. In Example 13-4 if the elevation of station 17 + 22.33 is desired, then x = 2.2233 stations beyond the BVC. The elevation of station 17 + 22.33 is therefore

$$y = (-0.333)(2.2233)^2 + (1.25)(2.2233) + 882.35 = 883.48 \text{ ft}$$

The elevation of an intermediate point can be computed by the tangent offset method, although not quite so readily as by using the equation of the curve. In Example 13-5, the elevation of station 49 + 52 is obtained by first computing the tangent elevation at the station and then by computing the tangent offset at the station. The tangent elevation is $574.97 - (3.52)(6) = 553.85$ ft. The tangent offset is $(3.52/4)^2 \times 8.00 = 6.20$ ft. Therefore the elevation of the curve at station 49 + 52 is 553.85 + 6.20 = 560.05 ft.

13-18. LOCATION OF HIGHEST OR LOWEST POINT When
g_1 and g_2 have opposite algebraic signs, either a high point or a low point will occur between the BVC and the EVC. Furthermore, this point may not fall on a full station or on a previously selected point. Sometimes it is necessary to determine the station and elevation of the high point or the low point in order to locate a clearance point, a drainage structure, or some other feature. The tangent to the curve at this point will be a horizontal line; that is, the slope of the curve will be zero. The position of the point can be determined, then, by equating to zero the first derivative of Eq. (13-28). For the high or low point,

$$\frac{dy}{dx} = rx + g_1 = 0$$

or

$$x = -\frac{g_1}{r} \tag{13-30}$$

The value of x in Eq. (13-30) is the distance in stations from the BVC to the high or low point. The elevation of the point is found by substituting the value of x obtained from Eq. (13-30) in Eq. (13-28), and solving for y.

The high point of the curve of Example 13-4 occurs at $x = 1.25/0.666 = 1.875$ stations beyond the BVC. Thus the summit is at station $16 + 87.5$. The elevation at the point is 883.52 ft.

In Example 13-5 the value of r found by Eq. (13-22) is $[2 - (-6)]/8 = +1.00$, and the low point occurs at $x = 6/1.00 = 6$ stations beyond the BVC. It is at station 52 where the elevation was previously computed.

13-19. MINIMUM LENGTH OF VERTICAL CURVE The length
of a vertical curve on a highway should be ample to provide a clear sight that is sufficiently long to prevent accidents. The American Association of State Highway and Transportation Officials (AASHTO) has developed criteria for the distance required to pass another vehicle traveling in the same direction on a vertical curve and also for the distance required to stop a vehicle in an emergency. The former distance, called the *safe passing sight distance* and designated as S_{sp}, is based on the assumptions that the eyes of the driver of a vehicle are about 3.75 ft above the pavement and the top of an on-coming vehicle is about 4.50 ft above the pavement. The latter required distance, called the *safe stopping sight distance* and designated as S_{np}, is based on the assumption that an obstruction ahead of the vehicle is 0.50 ft above the pavement. The values of S_{sp} and S_{np} are given in Table 13-5.

If it is assumed that the safe passing sight distance is less than the length L of the curve, then

$$L = \frac{S_{sp}^2(g_1 - g_2)}{33.0} \tag{13-31}$$

TABLE 13-5. AASHTO Sight-Distance Recommendations

DESIGN SPEED (mph)	S_{sp} (ft)	S_{np} (ft)
30	1100	200
40	1500	275
50	1800	350
60	2100	475
70	2500	600
80	2700	750

in which distances are in feet and g_1 and g_2 are expressed as ratios. If it is assumed that S_{sp} is longer than the curve, then

$$L = 2S_{sp} - \frac{33.0}{g_1 - g_2} \qquad (13\text{-}32)$$

If the safe stopping sight distance is less than the length of the curve, then

$$L = \frac{S_{np}^2(g_1 - g_2)}{14.0} \qquad (13\text{-}33)$$

If S_{np} is longer than the curve, then

$$L = 2S_{np} - \frac{14.0}{g_1 - g_2} \qquad (13\text{-}34)$$

On a highway with four or more traffic lanes, the safe stopping sight distance can be used to determine the required length of a vertical curve, because there is little probability of meeting oncoming vehicles in the passing lane. On a two-lane highway, however, the safe passing sight distance must be used if a vehicle is permitted to pass another one traveling in the same direction on the vertical curve. Use of the safe passing sight distance results in excessive lengths of vertical curves, and in most instances causes excessive excavation (see Chapter 17). Two methods are employed on two-lane roads to allow the safe stopping sight distance to be used for computing L. One method is to prohibit passing on crests and to indicate the restriction by appropriate center-line marking. The second method is to widen the pavement at the crest to permit two lanes in both directions for a sufficient distance.

Example 13-6.

The grades at a crest are $g_1 = +2\%$ and $g_2 = -3\%$, and the design speed is 60 mph. Compute the lengths of the vertical curves required for the safe passing sight distance and the safe stopping sight distance recommended by the AASHTO.

Solution: By Eq. (13-31) and Table 13-5

$$L = \frac{2100^2 \times 0.05}{33} = 6682\,\text{ft}$$

Also, by Eq. (13-32),

$$L = 2 \times 2100 - \frac{33}{0.05} = 3540\,\text{ft}$$

The length required for safe passing is therefore 6682 ft, since S_{sp} is less than L.

By Eq. (13-33)

$$L = \frac{475^2 \times 0.05}{14} = 806\,\text{ft}$$

Also, by Eq. (13-34),

$$L = 2 \times 475 - \frac{14}{0.05} = 670\,\text{ft}$$

The length required for safe stopping is thus 806 ft, since S_{np} is less than L.

In this example it would not be necessary to apply Eq. (13-32) or (13-34) after it is found by Eq. (13-31) or (13-33) that the required sight distance is less than L.

13-20. COMPOUND CURVES A compound curve consists of two or more consecutive circular arcs, the PT of one curve being the PC of the next and the centers of the curves being on the same side of the curve. Such a curve

TABLE 13-6. Basis of Equations for Compound Curves

SIDE	AZIMUTH	LENGTH	DEPARTURES E	DEPARTURES W	LATITUDES N	LATITUDES S
1–2	$0°$	R_s	0		R_s	
2–3	$90°$	T_s	T_s			0
3–4	$90° + \Delta$	T_l	$T_l \cos \Delta$			$T_l \sin \Delta$
4–5	$180° + \Delta$	R_l		$R_l \sin \Delta$		$R_l \cos \Delta$
5–1	Δ_s	$R_l - R_s$	$(R_l - R_s) \sin \Delta_s$		$(R_l - R_s) \cos \Delta_s$	

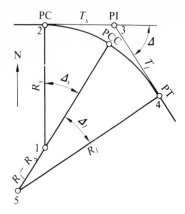

Figure 13-17. Compound curve.

is shown in Fig. 13-17. Although it is beyond the scope of this text to go deeply into the problems of compound curves, it may be pointed out that many of the compound-curve equations can be developed by considering the polygon 1-2-3-4-5 in Fig. 13-17 as a five-sided traverse.

The values for this traverse are shown in Table 13-6.

Since the traverse is a closed one, the algebraic sums of the latitudes and departures must equal zero. From the departures,

$$T_s + T_l \cos \Delta - R_l \sin \Delta + (R_l - R_s) \sin \Delta_s = 0 \qquad (13\text{-}35)$$

From the latitudes, north latitudes being considered negative for convenience, $-R_s + T_l \sin \Delta + R_l \cos \Delta - (R_l - R_s) \cos \Delta_s = 0$. If R_l is added and subtracted, this equation can be written

$$(R_l - R_s) - (R_l - R_s) \cos \Delta_s + T_l \sin \Delta - (R_l - R_l \cos \Delta) = 0 \quad (13\text{-}36)$$

From Eqs. (13-35) and (13-36) and the relation $\Delta = \Delta_s + \Delta_l$, the values of T_s, Δ_l, and Δ_s can be found when Δ, T_l, R_s, and R_l are known. For a complete discussion of compound curves, a text on route surveying should be consulted.

13-21. REVERSED CURVES A reversed curve is composed of two simple curves turning in opposite directions, as shown in Fig. 13-18. The point of reverse curve, PRC, is the PT of the first curve and the PC of the second one.

If the angles Δ_1 and Δ_2 and the distance between intersection points PI_1 and PI_2 have been measured in the field, one radius, or one degree of curve,

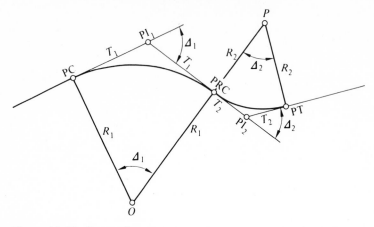

Figure 13-18. Reversed curve.

can be assumed and the other calculated. If R_1 is assumed, the first tangent distance T_1 is computed from the relationship

$$T_1 = R_1 \tan \tfrac{1}{2}\Delta_1$$

This length subtracted from the total distance between the intersection points gives the value of T_2. Then

$$R_2 = T_2 \cot \tfrac{1}{2}\Delta_2$$

The use of reversed curves on railroads is limited to sidings and crossovers. The necessity of elevating the outer rail on a railroad, and the outer edge of a highway-prevents the use of reversed curves except where very low speeds are encountered.

13-22. EASEMENT CURVES The amount of superelevation, in feet per foot of width, on a curve can be determined from the equation

$$e = \frac{v^2}{32.2R} = 0.067\frac{V^2}{R} \tag{13-37}$$

in which v is the velocity, in feet per second; V is the velocity, in miles per hour; and R is the radius of the curve, in feet.

On a straight track or pavement the two edges are at the same elevation. On a circular arc the outer edge is elevated the proper amount for the radius of the curve and for the speed expected. These two requirements lead to an impossible condition at the PC of a simple curve, since the PC is both on the curve and on the tangent and at the same point there should be super-elevation for the curve and none for the tangent.

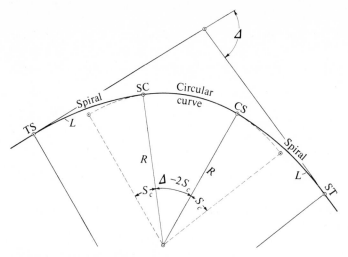

Figure 13-19. Spiral easement curves.

The introduction of a spiral easement, or transition, curve between the tangent and the circular arc, as indicated in Fig. 13-19, permits the gradual elevation of the outer edge. There are many curves that can be used as easement curves. However, the one recommended by the American Railway Engineering Association (AREA) is probably as simple as any. With the aid of tables found in railroad engineering handbooks, the computations can be made almost as quickly as the computations for a simple curve.

The AREA spiral is based on the assumption that the superelevation is to change at a uniform rate along the spiral. The equation of this curve is

$$y = \frac{l^3}{6RL}$$

in which y is the tangent offset at any point whose distance from the beginning point TS (tangent to spiral) of the spiral is l; R is the radius of the circular arc; and L is the length of the spiral. The radius of this curve decreased from infinity at the TS (tangent to spiral) to R at the SC (spiral to curve). Between the circular arc and the forward tangent is a second spiral whose ends are the CS (curve to spiral) and the ST (spiral to tangent).

The principal differences between the circular curve and the spiral are as follows: The length of the spiral is twice the length of a circular arc with radius R and central angle S_c. In the case of the usual spiral it is accurate enough for practical purposes to assume that the deflection angle from the tangent at the TS to any point on the spiral is one-third (instead of one-half) the central angle subtended by the chord from the TS to the point. The deflection angles are proportional to the squares of the distances along the spiral (rather than to the distances themselves, as in the case of the circular curve).

For the development of spirals and their applications, the reader should refer to Meyer and Gibson's *Route Surveying and Design* or Skelton's *Route Surveys*. These texts contain tables that simplify the field computations.

PROBLEMS

13-1. The degree of curve is 6° 00′. Compute the radius of the curve on the basis of (a) chord definition and (b) arc definition.

13-2. The degree of curve is 2° 30′. Compute the radius of the curve on the basis of (a) chord definition and (b) arc definition.

13-3. The degree of curve is 16° 00′. Compute the radius of the curve on the basis of (a) chord definition and (b) arc definition.

13-4. What is the degree of curve, arc definition of a 1900-ft radius curve?

13-5. What is the degree of curve, arc definition of a 300-ft radius curve?

13-6. A circular curve is to connect two tangents that intersect at an angle Δ of 23° 26′ at station 73 + 54.58. The tangent must be less than 500 ft. Using the arc definition, determine the degree (to the nearest 30′) of the flattest possible curve that will satisfy the conditions. Compute the deflection angles to each full curve station.

13-7. In Problem 13-6 compute the value of c_1 and c_2 given by Eqs. (13-14) and (13-16).

13-8. The deflection angle between two tangents is 30° 14′. The maximum permissible tangent distance is 300 ft. Using the arc definition, determine the degree of flattest possible curve (nearest 30′) that will fit the conditions.

13-9. Station of the PI of Problem 13-8 is at 20 + 52.50. Compute the deflection angles to the curve stations.

13-10. Solve Problems 13-8 and 13-9 on the basis of chord definition.

13-11. A circular curve is to join two tangents that intersect at an angle Δ of 26° 54′ 30″ at station 36 + 90.45. Using the arc definition, compute the radius (to a full 100 ft) of the flattest possible curve for which the arc length of the curve do not exceed 600 ft. Compute the deflection angles to each half curve station. Compute the values of c_1, c_2, and c_{50}.

13-12. The conditions are as stated in Problem 13-11 except that the maximum length of curve is 1000 ft. Perform the same computations.

13-13. The external distance of a curve joining two tangents is not to be less than 35 ft. The deflection angle Δ is 52° 22′ at station 148 + 52.50. On the basis of arc definition, compute the degree of curve (to the nearest 30′) of the sharpest curve that will satisfy the limitation. Compute R, T, M, and C. Compute c_{100} for this curve.

13-14. The conditions are as stated in Problem 13-13 except that the external distance is not to be less than 20 ft. Perform similar computations.

13-15. Compute the chord length to the nearest 0.001 ft joining two adjacent full stations on a circular curve (a) when the degree by arc definition is 15°; (b) when it is 10°; (c) when it is 5°.

13-16. Compute the chord length to the nearest 0.001 ft joining two adjacent quarter stations on a circular curve when the degree by arc definition is (a) 20°; (b) 16°; (c) 8°.

13-17. Solve Problem 13-16 on the basis of chord definition.

13-18. The PC of a circular curve is located at station 56 + 12.20. The degree of curve is 3° 00′. The curve is to be located by deflection angles. Stations 57, 58, and 59 can be located with the transit set up at the PC. Because of an obstruction beyond station 59, the transit must be moved up to occupy station 59 in order to locate the remainder of the curve. If a backsight of 0° 00′ is taken on the PC with the telescope inverted and the telescope is then plunged back to normal, what deflection angles should be turned to locate the remainder of the stations to the PT at station 64 + 37.76?

13-19. What is the central angle of the curve of Problem 13-18?

13-20. The PC of a circular curve is located at station 72 + 95.60. The degree of curve is 4° 30′. The curve is to be located by deflection angles. Station 73 through 76 can be located from the PC transit setup. The transit is advanced to station 76 and a backsight of 0° 00′ is taken to the PC with the telescope inverted. If the telescope is then plunged back to normal, what deflection angles should be turned to locate the remainder of the stations to the PT at station 85 + 52.20?

13-21. What is the central angle of the curve of Problem 13-20?

13-22. A series of horizontal circular curves are to join a series of tangents that have been located and stationed in the field from a beginning point designated station 0 + 00. The following data are given for each point of intersection.

PI NUMBER	ORIGINAL PI STATION	DEFLECTION ANGLE	DEGREE OF CURVE (Arc Definition)
1	9 + 14.56	8° 33′ L	2° 00′
2	19 + 25.37	21° 20′ R	3° 30′
3	38 + 37.80	14° 43′ L	3° 00′
4	60 + 02.20	19° 36′ R	2° 00′
5	77 + 54.86	End of line	

Compute the "new" stationing of the PC's, PT's and the end of the line.

13-23. Compute the "new" stationing in Problem 13-22 on the basis of the chord definition.

13-24. In Fig. 13-12, $\Delta = 44°\ 15′$, $D = 4°\ 30′$ (chord definition), the PC at A is at station 52 + 25.30, $AV' = 216.85$ ft, and angle $GV'\ V$ is 55° 22′. Compute $V'B$ and the stationing of B on the curve.

13-25. In Fig. 13-12, $\Delta = 50°\ 20′$, $D = 5°\ 00′$ (arc definition), the PC at A is at station 39 + 18.70, $AV' = 242.66$ ft, and angle $GV'V = 37°\ 55′$. Compute $V'B$ and the stationing of B on the curve.

13-26. A vertical curve joining two grade lines is 1200 ft long. Also, $g_1 = +2.4\%$ and $g_2 = -3.4\%$. The intersection of the grade lines at station 76 is at elevation 475.50 ft. Compute the elevation at each half station on the curve.

13-27. Compute the station and elevation of the summit of the curve of Problem 13-26.

13-28. A 1400-ft vertical curve joins two grades in which $g_1 = +2.14\%$ and $g_2 = -1.50\%$. The intersection of the two grades at station 80 + 00 is at elevation 1562.50 ft. Compute the elevations of the full curve stations.

13-29. Compute the station and elevation of the summit of the curve in Problem 13-28.

13-30. A 1500-ft vertical curve is to join two grades for which $g_1 = -3.0\%$ and $g_2 = +2.7\%$. The intersection is at station $166 + 00$ and at elevation 2565.60 ft. Compute the elevations of full curve stations.

13-31. Compute the station and elevation of the low point of the curve of Problem 13-30.

13-32. At what station or stations on the curve in Problem 13-30 is the elevation 2580.00 ft?

13-33. A grade g_1 of -2.2% passes station 60 at an elevation of 1350.65 ft, and a grade g_2 of $+2.0\%$ passes station 85 at an elevation of 1348.02 ft. Compute the station and elevation of the point of intersection of these two grades. (The point-slope form of the equation of a line can be used.) Compute the elevations of the BVC, the EVC, and each full station of a 1000-ft vertical curve joining these two grades.

13-34. If g_1 is $+3.50\%$ and g_2 is -2.00%, compute the minimum length of a vertical curve necessary to provide safe passing sight distance for a design speed of 50 mph. Compute the length necessary to provide the safe stopping sight distance for a design speed of 70 mph.

13-35. If $g_1 = +2.80\%$ and $g_2 = -5.00\%$, compute the minimum length of a vertical curve necessary to provide safe passing sight distance for a design speed of 40 mph. Compute the length necessary to provide the safe stopping sight distance for a design speed of 60 mph.

13-36. In Fig. 13-17, $T_s = 300$ ft, $R_s = 430$ ft, $\Delta_s = 18° 15'$, $\Delta = 42° 10'$, and the PI is at station $53 + 19.70$. Using the arc definition for degree of curve, compute T_l, R_l, Δ_l, and the stationing of the PCC and the PT.

13-37. In Fig. 13-17, $T_s = 350$ ft, $R_s = 500$ ft, $\Delta_s = 24° 22'$, $\Delta = 51° 50'$, and the PI is at station $101 + 42.55$. Using the chord definition for degree of curve, compute T_l, R_l, and Δ_l and the stationing of the PCC and the PT.

13-38. In Fig. 13-18, the distance from PI_1 to PI_2 is 1074.65 ft. The intersection angles are $\Delta_1 = 58° 20'$ and $\Delta_2 = 40° 50'$. The degree of the first curve, by chord definition, is $D_1 = 5° 00'$. Compute R_2 and D_2 necessary to fit the given conditions.

13-39. A pavement 54-ft wide is to be superelevated to allow safe negotiation of a $2° 00'$ (arc definition) circular curve at a design speed of 70 mph. What will be the theoretical difference in elevation between opposite edges of the pavement?

13-40. If the gauge width between rails of railroad track is 4 ft, $8\frac{1}{2}$ in., what is the superelevation of the outer rail for a design speed of 45 mph and a degree of curve (chord definition) of $3° 30'$?

BIBLIOGRAPHY

Myers, C. F. and Gibson, D. W. 1980. *Route Surveying and Design*. New York: Harper & Row.

Pryor, W. T. 1975. Metrication. *Surveying and Mapping* 35:229.

Skelton, R. R. 1949. *Route Surveys*. New York: McGraw-Hill.

Chapter 14
Tacheometry

14-1. USES OF TACHEOMETRY The term *tacheometry* in surveying
is used to denote the procedures for obtaining horizontal distances and
differences in elevation by rapid indirect methods, which are based on the
optical geometry of the instruments employed. The procedure is sometimes
referred to as stadia, optical distance measurement or telemetry. The instruments employed are the engineer's transit or theodolite and the leveling rod
or stadia rod, the telescopic alidade and the leveling rod or stadia rod, the
theodolite and the subtense bar, and the self-reducing theodolite and the
leveling rod. Horizontal distances are obtained with each of these combinations of instruments without resorting to direct taping. Differences in
elevation can be determined indirectly with most of the combinations.

Tacheometry is used to measure the lengths of traverse sides, to check the
more accurate taped distances in order to uncover gross errors or mistakes,
to determine differences of elevation between points, and to carry lines of
levels where a relatively low order of accuracy is permissible. Its most general
use is found in the compilation of planimetric and topographic maps by
field methods alone, by which distances, elevations, and directions to points
are to be determined from field control points whose positions have been
established by a higher order of accuracy.

The principles of stadia measurement by use of the transit or theodolite and the leveling rod or stadia will be developed thoroughly in this chapter. The other methods of tacheometry will then be discussed. The similarity and the difference in the principles of the various techniques can thus be better understood.

14-2. PRINCIPLE OF STADIA MEASUREMENTS For the measurement of stadia distances, the telescope of the transit or theodolite is equipped with three horizontal cross hairs, the upper and lower hairs being called the *stadia hairs*. The actual vertical separation of the stadia hairs in the reticle is designated as *i*. In Fig. 14-1, points *a* and *b* represent the positions of the upper and lower stadia hairs with a spacing $ab = i$. By the laws of optics a ray of light that is parallel with the optical axis of a lens will pass through the *principal focus* of the lens on the opposite side. In Fig. 14-1, a transit with a movable objective lens is used to illustrate the principles of stadia. The stadia equations developed from Fig. 14-1 will be modified later on to take into account the internally focused telescope described in Chapter 3 which has all but replaced the externally focusing telescopes found on older surveying instruments. Referring to Fig. 14-1, point *F* is the principal focus of the objective lens, and it is located on the optical axis in front of the movable objective lens at a distance *f* from the center of the lens. The distance *f* is known as the *focal length* of the lens, and it has a fixed value for a given lens.

Imagine a ray of light to be directed along the line *aa'*, which is parallel with the optical axis of the telescope. This ray of light will be bent by the lens and will continue in the direction *a'FB*. Similarly, a ray along *bb'* will travel in the direction *b'FA* on the other side of the lens. The vertical distance *AB* can be measured on a graduated rod. It is obtained by subtracting the stadia-hair reading at *B* from the stadia-hair reading at *A*. The distance *AB*, designated as *s*, is called the *stadia interval*.

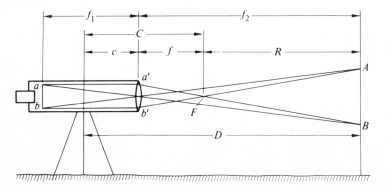

Figure 14-1. Principle of transit stadia measurement.

- external focusing instrument

The distance $a'b'$ is equal to ab and is therefore equal to i. By similar triangles, $a'b'/f = AB/R$, in which R is the distance from the principal focus in front of the lens to the graduated rod. So the distance R may be found from the equation

$$R = \frac{f}{i} s \qquad (14\text{-}1)$$

in which f is the focal length, which is fixed for a given telescope; i is the spacing of the stadia hairs, which is also fixed for a given telescope; and s is the vertical distance on the rod between the upper stadia-hair reading and the lower stadia-hair reading. The ratio f/i is called the *stadia interval factor* and is designated as K. Equation (14-1) thus becomes

$$R = Ks \qquad (14\text{-}1a)$$

The manufacturer can space the stadia hairs with relation to the focal length so as to obtain any convenient value of K desired. The most common value of f/i, or K, is 100, and is assumed so in this textbook.

The horizontal distance (D in Fig. 14-1) between the center of the instrument and the point at which the rod is held is $R + f + c$, in which c is the distance from the center of the instrument to the lens. This small distance c varies slightly when focusing the objective lens. Its exact value depends on the distance from the instrument to the rod, but the variation is not enough to affect the accuracy of stadia measurements. If C is substituted for $f + c$, then the horizontal distance D from the center of the instrument to the rod is given by the equation

$$D = Ks + C \qquad (14\text{-}2)$$

The distance C is called the *stadia constant*. Equation (14-2) is the stadia equation for a horizontal line of sight using a movable objective telescope. It is applicable for lines of sight inclined by as much as 3° where the distance to be determined is not critical. For this reason when a stadia interval is measured, it is permissible and desirable to set the lower cross hair on the nearest footmark to facilitate reading the interval, even though the line of sight is not quite horizontal.

The value of C will vary from about 0.6 to 1.4 ft in the older transits found in everyday use. These instruments, which contain the movable objective lens, are being replaced by transits with the internal-focusing lens. The geometry of the optics of the newer telescopes is such that the value of the stadia interval factor changes slightly when focusing the internal lens. This change is almost compensated for throughout the range of focus by a corresponding change in the value of C. This fact allows the constant C to be neglected; that is, $C = 0$. The value of the stadia interval factor is then assumed to be constant throughout the focusing range.

If the older instrument is used, the value of C can be assumed to be 1 ft. When K is 100, or nearly so, Eq. (14-2) can be written as follows:

$$D = K(s + 0.01) \tag{14-2a}$$

in which the interval s is simply increased by 0.01 ft.

For long sights when C can be entirely neglected and when the internal-focusing instrument is employed, Eq. (14-2a) can be reduced to the simple form

$$D = Ks \tag{14-2b}$$

14-3. DETERMINATION OF STADIA CONSTANT
If the instrument is externally focusing, the stadia constant is readily determined by setting up the transit, focusing on a distant point at least 1000 ft away, and measuring the distance between the objective lens and the cross-hair ring. This is the focal length f. The telescope is then focused on an object between 200 and 300 ft away. The distance between the objective lens and the center of the instrument is then measured. This is the distance c in Fig. 14-1. The constant C is the sum of f and c.

If the instrument contains the fixed-objective lens, then the value of C can be assumed to be zero.

14-4. DETERMINATION OF STADIA INTERVAL FACTOR
If it is desired to determine the stadia interval factor $f/i = K$, a straight line from 400 to 800 ft in length is run on ground that is as nearly level as practicable. The instrument is set on this line. If the instrument is externally focusing, a point P is located on the line at a distance $f + c = C$ from the center of, and in front of, the instrument. If the instrument is internally focusing, then the position of P is directly under the instrument.

Points spaced about 50 ft apart are set on line, and the distances $R_1, R_2,$ $R_3,$ and so on are measured from point P to the successive points on line. A leveling rod equipped with two targets is held at each point in succession, and the intervals $s_1, s_2, s_3,$ and so on are determined to the nearest 0.001 ft. Successive values of K are determined by the relationships $K_1 = R_1/s_1,$ $K_2 = R_2/s_2, K_3 = R_3/s_3,$ and so on, and the mean of the values of K thus obtained is taken as the stadia interval factor of the instrument.

14-5. INCLINED STADIA MEASUREMENTS
Inclined stadia measurements are more frequent than are horizontal measurements. Each inclined measurement is reduced to give the horizontal distance from the instrument to the rod and the difference in elevation between the telescope

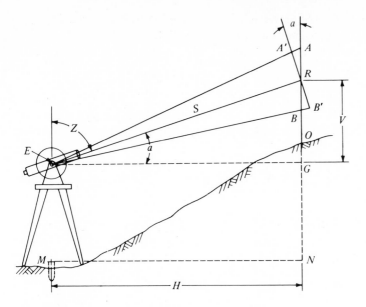

Figure 14-2. Inclined stadia measurement ($C = 0$).

axis at the center of the instrument and the point at which the middle cross hair strikes the rod. This point on the rod is referred to as the middle cross-hair reading. In Fig. 14-2, an internal-focusing instrument in which $C = 0$ is located at M. The apex of the angle subtended by the upper and lower stadia hairs is assumed to fall at the middle of the telescope. The rod is located at O, the horizontal distance between the instrument and the rod is $MN = EG = H$, and the difference in elevation between the telescope axis at E and the middle cross-hair reading at R is $RG = V$. Thus, V represents the total rise (or fall in the case of a minus vertical angle) of the line of sight from the instrument to the rod with the telescope inclined by the vertical angle a. The angle AEB is only about $0° 34'$, and therefore the angles $RA'E$ and $RB'E$ are considered right angles. Therefore $A'B' = AB \cos a = s'$. The slope distance S is ER. By the ratio $f/i = K$, remembering that AB is the stadia interval s,

$$S = Ks' = Ks \cos a$$

The horizontal distance $H = S \cos a$. Thus

$$H = Ks \cos a \cos a$$

or

$$H = Ks \cos^2 a \qquad (14\text{-}3a)$$

in which K is the stadia interval factor f/i, s is the stadia interval, and a is the vertical angle of the line of sight read on the vertical circle of the theodolite. If zenith angles are observed, the horizontal distance is given by

$$H = Ks \sin^2 Z \qquad (14\text{-}3b)$$

The distance RG, which equals $S \sin a$, is the vertical distance V between the telescope axis and the middle cross-hair reading. Thus V is given by the equation

$$V = Ks \sin a \cos a \qquad (14\text{-}4)$$

or

$$V = \tfrac{1}{2}Ks \sin 2a = \tfrac{1}{2}Ks \sin 2Z \qquad (14\text{-}5)$$

Example 14-1.

Stadia readings were made from an instrument set up to three different points. The value of K is 100. The recorded values are $s_1 = 0.848$ m and $a_1 = +3° 44'$; $s_2 = 1.455$ m and $a_2 = +0° 20'$; $s_3 = 1.820$ m and $a_3 = -12° 20'$. Compute the horizontal distance and the rise or fall of the line of sight.

Solution: By Eq. (14-3a),

$$H_1 = 100 \times 0.848 \cos^2 3° 44' = 84.44 \text{ m}$$
$$H_2 = 100 \times 1.455 \cos^2 0° 20' = 145.50 \text{ m}$$
$$H_3 = 100 \times 1.820 \cos^2 12° 20' = 173.70 \text{ m}$$

By Eq. (14-5),

$$V_1 = 50 \times 0.848 \sin 7° 28' = +5.51 \text{ m}$$
$$V_2 = 50 \times 1.455 \sin 0° 40' = +0.85 \text{ m}$$
$$V_3 = 50 \times 1.820 \sin(-24° 40') = -37.98 \text{ m}$$

14-6. READING A STADIA INTERVAL When using a foot rod, the most convenient method of measuring a stadia interval is to first read the vertical angle and the middle cross hair and then set the lower cross hair on the nearest whole foot mark. Then read the upper stadia hair and get the difference which is the stadia interval. When using a metric rod, the lower stadia hair is set to the nearest decimetre mark before reading the upper stadia hair.

14-7. DIFFERENCE IN ELEVATION BETWEEN TWO POINTS

When stadia methods are used to determine differences in elevation between points, three situations will prevail. In the first situation, which is associated with traversing and mapping details about a point, the difference in elevation between the point on the ground over which the instrument is set up and the point on the ground at which the rod is held is the desired value. In Fig. 14-3 the instrument occupies point M, and the rod is held at point O. A leveling rod or stadia rod is held alongside the instrument, and the height of the telescope axis above point M is measured. This height is recorded in the notes as the HI which for this situation is understood to be the height of the instrument *above the ground*. The line of sight is directed at the rod held at O, and the stadia interval is read and recorded. Then the reading of the middle cross hair at R that corresponds to a vertical angle a is read and recorded, along with the vertical angle. The difference in elevation between points M and O is equal to $+ME + GR - RO$. But the distance ME equals the HI, and the distance GR is the vertical distance V computed by Eq. (14-5). RO is the middle cross-hair reading, which will be referred to simply as the *rod reading*. Therefore the difference in elevation between M and O is $(+\text{HI} + V - \text{rod reading})$ if the vertical angle is positive, or $(+\text{HI} - V - \text{rod reading})$ if the vertical angle is negative.

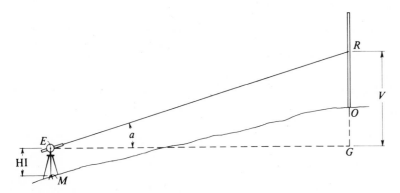

Figure 14-3. Difference in elevation between instrument ground station and rod station.

Example 14-2.

The HI of the instrument of Example 14-1 is measured as 1.52 m. The middle cross-hair readings for the three observations are: $R_1 = 1.42$ m; $R_2 = 0.44$ m; and $R_3 = 2.00$ m. Compute the difference in elevation between the point over which the instrument was set up and the points where the rod was held.

Solution: $DE_1 = +1.52 + 5.51 - 1.42 = + 5.61 \text{ m}$

$DE_2 = +1.52 + 0.85 - 0.44 = + 1.93 \text{ m}$

$DE_3 = +1.52 - 37.98 - 2.00 = - 38.46 \text{ m}$

It is convenient whenever possible to read the vertical angle when the rod reading equals the HI. The difference in elevation is then $+ \text{HI} \pm V - \text{HI} = \pm V$ directly. This procedure, however, is not always possible on account of obstructions on the line of sight.

In the second situation, which is sometimes associated with mapping details about a point, the difference in elevation between the telescope axis and the point on the ground at which the rod is held is the desired value. Again refer to Fig. 14-3, and let it be assumed that the elevation of point M has been established. The distance ME added to the elevation of M gives the elevation of the telescope axis *above the datum,* or the HI as understood in differential leveling and defined in Section 3-24. Then if the elevation of point O is desired, the difference in elevation between points E and O is added to the HI. In this instance the difference in elevation between E and O is ($+ V$ − rod reading). If the vertical angle is negative, the desired difference is ($- V$ − rod reading).

In the third situation, which is associated with running a line of levels by stadia, the difference in elevation between two rod stations is desired. The elevation of the ground at the instrument setup is of no consequence. In Fig. 14-4 the difference in elevation between the backsight station B and the foresight station F is desired. The instrument is set up at A, and a stadia interval, vertical angle a_B, and rod reading R_B are observed on point B and recorded. Then an interval, vertical angle a_F, and rod reading R_F are observed on point F and recorded. The vertical distances $O_B R_B = V_B$ and $O_F R_F = V_F$ are computed by Eq. (14-5). The difference in elevation in this instance is ($+$ backsight rod reading $- V_B + V_F -$ foresight rod reading). Sketches

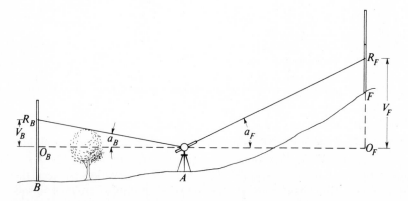

Figure 14-4. Difference in elevation between two rod stations.

should be drawn of other situations of this nature, in which the backsight and foresight vertical angles are, respectively, + and −, + and +, − and −, and − and +. The values to be added and subtracted to arrive at the difference in elevation between the two rod stations will become apparent from the sketches.

Example 14-3.

In Fig. 14-4, $s_B = 1.144$ m, $a_B = +0° 18'$, $R_B = 2.14$ m; $s_F = 1.458$ m, $a_F = +3° 16'$, and $R_F = 2.05$ m. What is the difference in elevation from B to F?

Solution: By Eq. (14-5),

$$V_B = 50 \times 1.144 \sin 0° 36' = 0.60 \text{ m}$$

$$V_F = 50 \times 1.458 \sin 6° 32' = 8.29 \text{ m}$$

The difference in elevation from B to F is then

$$DE = +2.14 - 0.60 + 8.29 - 2.05 = +7.78 \text{ m}$$

14-8. STADIA TRAVERSE A stadia traverse is sometimes run to obtain supplementary control based on existing higher-order control. This supplementary control is used to furnish instrument stations for the compilation of field data needed for the production of planimetric and topographic maps, or to provide picture points for controlling photogrammetric mapping. The traverse work may be performed concurrently with the gathering of data, or it may be performed separately from the map-compilation phase.

Three quantities are obtained from the results of the stadia traverse: (1) the length of each traverse side, (2) the azimuth or bearing of each traverse side, and (3) the difference in elevation between the ground stations at the ends of each traverse side. The lengths of the sides are obtained by the stadia reduction equations for horizontal distance given in Sections 14-2 and 14-5. The azimuths of the sides are most conveniently determined by carrying a traverse by azimuth as described in Section 8-7. Differences in elevation are obtained by the method discussed as the first situation of Section 14-7.

The most satisfactory results are obtained if observations for each line of the traverse are taken in both directions. This procedure reveals mistakes, eliminates certain index errors in the vertical circle of the theodolite, and gives a better value for the horizontal distance and difference in elevation, each of these distances being taken as the mean of two values.

In the stadia traverse of Fig. 14-5, a traverse begins and closes on station A. The azimuth of AD has been previously established as $218° 00'$. The

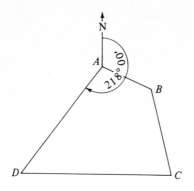

STA.	INT.	HOR. DIST.	AZIMUTH	VERT Δ	D.E.
⊼ @ A;	H.I.=1.52				
D	1.770	176.99	218°00'	+0°21'	+1.08
B	0.709	70.52	114°16'	−4°12'	−5.18
⊼ @ B;	H.I.=1.55				
A	0.706	70.22	294°16'	+4°12'	+5.16
C	1.148	114.73	166°53'	+1°25'	+2.84
⊼ @ C;	H.I.=1.55				
B	1.147	114.63	346°53'	−1°24'	−2.80
D	1.989	198.85	270°14'	+0°55'	+3.18
⊼ @ D;	H.I.=1.53				
C	1.988	198.77	90°14'	−0°44' on 2.12	−3.16
A	1.770	176.99	38°08'	−0°22'	−1.13
				e=+8'	

Figure 14-5. Stadia traverse with corresponding notes.

elevation of A is 104.56 m. The HI at A is 1.52 m, having been measured by holding the rod alongside the transit, and is recorded as shown in the notes. The value of the azimuth of AD is set on the clockwise circle, and a backsight is taken on D, using the lower motion. This orients the theodolite circle at A. The middle cross hair is set on 1.52 on the rod held at D, and a vertical angle of $+0°21'$ is read and recorded. The stadia interval of 1.770 m is now read and recorded (see Section 14-6).

Station B is now sighted using the upper motion, and the clockwise circle reading of $114°16'$ is recorded as the azimuth of AB. The middle cross hair is set on 1.52 m and a vertical angle of $-4°12'$ is read and recorded. A stadia interval of 0.709 m is read and recorded. This completes the stadia observations at A.

The instrument is next set up at station B, the HI above the ground is measured, and the back azimuth of the line AB is set on the horizontal circle. This is $294°16'$ and is recorded as shown. The theodolite is oriented by backsighting on station A using the lower motion. The stadia interval is read with

TABLE 14-1. Results of Measurements in Stadia Traversing

LINE	DISTANCE	ADJUSTED AZIMUTHS	AVERAGE DE's +	AVERAGE DE's −	ADJUSTED DE's +	ADJUSTED DE's −
AB	70.37	$114°\,16' - 02' = 114°\,14'$		5.17		5.13
BC	114.68	$166°\,53' - 04' = 166°\,49'$	2.82		2.88	
CD	198.81	$270°\,14' - 06' = 270°\,08'$	3.17		3.27	
DA	176.99	$38°\,08' - 08' = 38°\,00'$		1.11		1.02
	560.85		5.99	6.28	6.15	6.15
				-5.99		
				$e = -0.29$		

the rod held on A and is recorded as 0.706 m. When the middle cross hair is set at the HI, the vertical angle is $+4°\,12'$. Station C is next sighted by using an upper motion, and the azimuth of BC, which is read on the horizontal circle, is recorded as $166°\,53'$. A stadia interval and a vertical angle to the HI are read and recorded.

The procedure just described is repeated at each station, and a final sight is taken on A from D in order to close the traverse. Notice that when the instrument was set up at station D and the rod was held at C, the HI could not be sighted with the middle cross hair. Therefore the rod reading of 2.12 m is recorded, together with the corresponding vertical angle of $-0°\,44'$.

All the distances and differences in elevation (DE's) are computed using Eqs. (14-3) and (14-5) and are entered in the notes as shown.

The azimuth of line DA should be $218°\,00' - 180° = 38°\,00'$, but the field results show $38°\,08'$. Thus, the azimuth error is $+8'$. This is distributed throughout the traverse by reducing each *forward* azimuth in proportion to the number of instrument setups. Thus the measured azimuth of AB receives $-02'$ correction, BC receives $-04'$, CD receives $-06'$, and DA receives $-08'$ correction. The average distances, the adjusted azimuths, and the average differences in elevation for each line are shown in Table 14-1.

The sum of the plus DE's is less than the sum of the minus DE's as shown in the table. These are adjusted in proportion to the length of each line. Line AB gets a correction of $0.29 \times 70.37/560.85 = 0.04$ m; line BC gets corrected $0.29 \times 114.68/560.85 = 0.06$ m, and so on. The adjusted DE's shown in the table are then used to compute the elevations of B, C, and D.

Latitudes and departures are computed for each line and the traverse is adjusted by the method described in Chapter 8.

14-9. DETAILS ABOUT A POINT   When the transit stadia method is used to locate details about a point for the purpose of plotting these details, the theodolite is set up over the point and the HI above the ground is determined by the method discussed in the first situation of Section 14-7. The

POINT	INTERVAL	HOR. DIST.	AZIMUTH	VERT. ∠	ROD READ.	D.E.	ELEV.
Ⅱ @ STA. G		B.S. 43°22' on STA.F			H.I.= 5.2 ft.		491.0
1	3.28	328	37°10'	−3°18'	7.4	−21.1	469.9
2	3.32	333	39°15'	−3°16'	8.1	−21.8	469.2
3	4.06	407	152°50'	—	8.8	−3.6	487.4
4	4.51	452	158°05'	−1°06'	5.2	−8.7	482.3
5	1.29	129	292°22'	+3°45'	5.2	+8.5	499.5
6	1.20	111	316°30'	+16°20'	8.2	+29.7	520.7
7	1.31	124	316°40'	+14°10'	5.2	+31.3	522.3

Figure 14-6. Stadia detail notes.

horizontal circle is oriented by setting the known azimuth of a line on the circle and backsighting along this line by using the lower motion. The upper motion is then used for all subsequent pointings from the station, so that the azimuth of each line directed to a detail point may be determined. As stadia readings are taken by the instrumentman, the note keeper should keep a detailed sketch up to date in the field notebook, numbering the detail points to correspond with the numbers assigned in the notes. Word descriptions of detail points—such as "fence corner," "3' oak," and "road int's'n,"—are used to advantage in documenting the stadia observations.

A form of notes for gathering details about a point is shown in Fig. 14-6. The instrument is set up on station G and oriented by backsighting on station F with the azimuth of the line GF on the horizontal circle. The value of the HI is the height of the telescope axis above the ground, and all differences in elevation are applied to the elevation of the ground point. For example, on the sight to point 1, the value of the vertical distance by Eq. (14-5) = -18.9 ft. The difference in elevation is then $+5.2 - 18.9 - 7.4 = -21.1$ ft, and the elevation of point 1 is $491.0 - 21.1 = 469.9$ ft. On the sight taken to point 3, the telescope bubble was centered when the rod reading was made, and since there was no vertical angle, the difference in elevation is $+5.2 - 8.8 = -3.6$ ft. On the sights taken to points 4, 5, and 7, the middle cross hair was sighted on the HI on the rod, and the difference in elevation in each instance is simply the vertical distance determined by Eq. (14-5).

An alternative method of determining elevations is discussed in the second situation of Section 14-7. The HI above the vertical datum at the instrument setup is determined, and then the difference in elevation between the telescope axis and the ground point is obtained as described in Section 14-7.

14-10. STADIA LEVELING The operations in stadia leveling are quite similar to those in differential leveling, but in stadia leveling the line of sight does not have to be horizontal. This feature is advantageous in hilly country where a relatively low order of accuracy is acceptable because the lines of sight can be quite long.

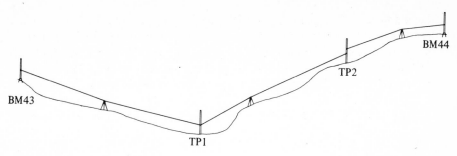

Figure 14-7. Stadia leveling.

In Fig. 14-7 a line of stadia levels is run from BM43 to BM44. The instrument is set up, and backsight readings are taken on BM43. These consist of an interval, a rod reading, and a corresponding vertical angle. Foresight readings consisting of an interval, a rod reading, and a corresponding vertical angle are then taken on TP1. This procedure is repeated at the second and third setups.

The following procedure is used to detect mistakes in reading the vertical circle and the interval, and in computing the values for the backsights and the foresights. The line of sight is directed at the rod, and the lower stadia hair is set on some footmark (or decimetre mark) as shown in Fig. 14-8(a). The readings of the lower, middle, and upper cross hairs are observed and recorded. The vertical circle is then read, and the angle is recorded opposite

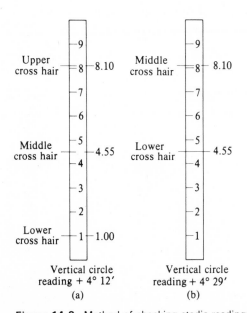

Figure 14-8. Method of checking stadia readings.

the middle cross-hair reading. Now the line of sight is either raised or lowered by means of the vertical tangent screw until the middle cross hair is set on the original reading of the upper or lower cross hair. In Fig. 14-8(b) the middle cross hair is set on the original reading of the upper cross hair. The vertical circle is again read, and the angle is recorded opposite this cross-hair reading.

If the value of K for the transit is 100 or nearly so, then the angle between the upper and middle cross hairs or between the middle and lower cross hairs is $\tan^{-1}(1/200) = 0°17'$, approximately. Therefore the two values of the vertical angle should differ by about $0°17'$. If they do not, then a mistake has been made in either a cross-hair or vertical-circle reading and the readings can be checked before leaving the setup. The computation of the backsight and foresight values can be checked, because there will be two values for each distance.

A form of notes for stadia leveling is shown in Fig. 14-9. The readings shown are those for the three setups of Fig. 14-7. The stadia interval for the first backsight taken on BM43 is $9.80 - 3.00 = 6.80$. Assuming the stadia constant to be zero, the value of V for $2°13'$ is 26.3 ft by Eq. (14-5). The value of the backsight is $+6.40 - 26.3 = -19.9$ ft. The value of V for $2°30'$ is 29.6. The value of the backsight using this angle is then $+9.80 - 29.6 = -19.8$ ft. The mean, rounded to 0.1 ft, is -19.9 ft as shown in the notes.

The stadia interval for the foresight taken on TP1 is $8.25 - 1.00 = 7.25$. The value of V for $1°41'$ is 21.3. The value of the foresight is then $-21.3 - 4.60 = -25.9$ ft. Using $1°23'$, $V = 17.5$. The value of the foresight is then $-17.5 - 8.25 = -25.75$. The mean, rounded to 0.1 ft, is -25.8 ft as shown in the notes.

After all the values of the backsights and foresights between benchmarks have been computed, the difference in elevation between the points is then

STATION	ROD	VERT.∠	B.S.	ROD	VERT.∠	F.S.
	9.80	+2°30'				
B.M. 43	6.40	+2°13'	−19.9			
	3.00					
	7.70	−3°08'		8.25	−1°23'	
T.P I	5.85	−3°25'	+27.9	4.60	−1°41'	−25.8
	4.00			1.00		
	6.20	−2°21'		8.00	+3°20'	
T.P 2	4.10	−2°38'	+23.4	4.50	+3°03'	+32.6
	2.00			1.00		
				5.00	+0°27'	
B.M. 44				3.50	+0°10'	−2.6
				2.00		

Figure 14-9. Stadia leveling notes.

the algebraic sum of the backsights and the foresights. In the example given here, DE $= -19.9 - 25.8 + 27.9 + 32.6 + 23.4 - 2.6 = +35.6$ ft.

14-11. ERRORS IN STADIA MEASUREMENTS

The accuracy of stadia measurements is largely dependent on the instrument and the rod used, and on atmospheric conditions. For sights up to 400 ft it is usually possible to obtain the intercepted distance on the rod to 0.01 ft. At 100 m, the intercept can be measured with an accuracy of 3 mm. Consequently, if the stadia interval factor and the rod used are correct, the resulting horizontal distance should be within 1 ft or 3 dm of the correct distance. For longer sights the error depends on the magnifying power of the telescope, the coarseness of the stadia hairs, the type of rod used, and the weather conditions. Under reasonably favorable conditions, the error should not exceed 10 ft in a 1000-ft sight or 3 m in a 300-m sight.

Errors due to imperfections in the graduation of the rod can be kept to a negligible quantity by standardizing the rod. For most stadia work some type of pattern rod, with coarser graduations than are found on leveling rods, is used. Unless the painting of the rod has been carelessly done, no correction should be necessary for work of moderate precision.

If the rod is not held plumb, the resulting error in horizontal distance will be small when the vertical angle is small, but may be considerable when large vertical angles are encountered. The effect on an observed distance of 1000 ft, determined by reading a 12-ft rod the top of which is 0.5 ft out of plumb, for vertical angles of 0° 30′, 10°, and 25°, is given in Table 14-2.

Errors from this source can be eliminated by using a rod equipped with a plumbing level. In computing the values just given, the constant C has been assumed to be zero.

When a cross hair appears to cover an appreciable space on the rod, the observations can be made best by using the tops of the hairs, rather than by attempting to use their centers.

Errors in observing the intercepted distance on the rod will be random in character. On important sights all three hairs can be read. If the hairs are equally spaced, the middle-hair reading should be the mean of the upper and

TABLE 14-2. Error in Horizontal Distance Due to Rod Being Out of Plumb

		CORRECTED HORIZONTAL DISTANCES		
VERTICAL ANGLE	OBSERVED DISTANCE	ROD PLUMB	ROD LEANING TOWARD	ROD LEANING AWAY
+ 0° 30′	1000	999.9	999.4	998.7
+10° 00′	1000	969.8	976.1	961.9
+25° 00′	1000	821.4	836.7	804.7

lower readings. When sights are taken between instrument stations, observations should be made from both ends of the line.

14-12. ERRORS IN STADIA ELEVATIONS The errors in elevations obtained from stadia measurements are caused by errors in measuring the vertical angles and in reading the rod, and by not keeping the rod plumb.

The effect of an error of 1′ in a vertical angle is practically independent of the size of the angle, but is proportional to the length of the sight. For a distance of 100 ft, the error in elevation is about 0.03 ft.

The effect of an error of 1 ft in an observed inclined distance increases with the size of the vertical angle. The error in elevation is 0.05 ft for a 3° angle, about 0.1 ft for a 5° angle, about 0.2 ft for a 10° angle, and about 0.4 ft for a 25° angle. It is necessary therefore that intervals be observed more carefully when the vertical angles are large.

The error in elevation caused by the rod not being plumb is considerable on a long sight when the vertical angle is large. The effect of the top of a 12-ft rod being 0.5 ft out of plumb when the observed distance is 1000 ft is as shown in Table 14-3.

TABLE 14-3. Error in Vertical Distance Due to Rod Being Out of Plumb

		DIFFERENCES OF ELEVATION		
VERTICAL ANGLE	OBSERVED DISTANCE	ROD PLUMB	ROD LEANING TOWARD	ROD LEANING AWAY
+ 0° 30′	1000	8.7	8.7	8.7
+10° 00′	1000	171.0	172.1	169.6
+25° 00′	1000	383.0	390.1	375.2

14-13. SELF-REDUCING STADIA INSTRUMENTS When a major portion of the surveying to be performed consists of stadia measurements, the self-reducing stadia instrument is of particular value. The theodolite shown in Fig. 14-10 is a self-reducing tacheometer in which the stadia hairs are replaced with three curved lines, as shown in Fig. 14-11. The lower curve is referred to as the zero curve. It is placed on a convenient full graduation, which is 1.000 m in Fig. 14-11. The upper curve then determines the horizontal distance interval. In (a) this interval is 1.572 − 1.000 = 0.572 m. Multiplying by 100 gives the horizontal distance as 57.2 m. In (b) the interval is 0.485, giving a horizontal distance of 100 × 0.485 = 48.5 m. The middle curve determines the vertical-distance interval, together with the factor for the part of the curve that is being used. In (a) the vertical-distance interval is 0.401 m, or 40.1 cm. The factor +0.2 is to be applied to the interval in centimetres, or the factor to be applied to the interval in metres is +20. The

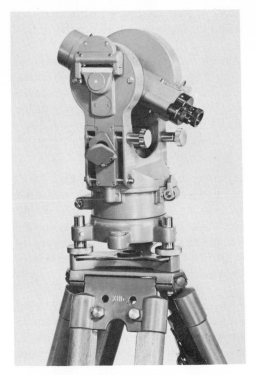

Figure 14-10. Automatic stadia-reduction instrument. Courtesy of Wild Heerbrugg Instruments, Inc.

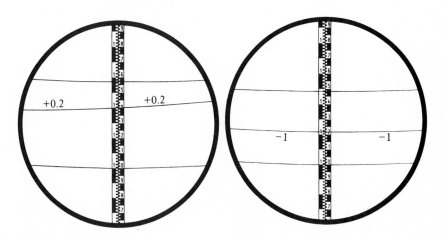

Distance : 57.2 m   Distance: 48.5 m

Difference of elevation: +0.2·40.1 = 8.02 m   Difference of elevation : −1·21.7 = −21.7 m

Figure 14-11. Stadia intercept curves in field of view of reduction tacheometer.

vertical distance is therefore $0.401 \times 20 = 40.1 \times 0.2 = +8.02$ m. In (b) the vertical-distance interval is 21.7 cm and the factor is -1. Hence the vertical distance is $21.7 \times -1 = -21.7$ m.

As the line of sight is inclined, the curves approach or recede from the zero curve to give a continuous solution of Eqs. (14-3) and (14-5). The vertical-distance curve is interrupted to give a new curve as the line of sight becomes more and more inclined. Each new curve has a different multiplying factor, which is analogous to a different stadia interval factor in the transit. There is only one horizontal-distance curve, however, and its factor is always 100.

14-14. SUBTENSE BAR The subtense bar, shown in Fig. 14-12, establishes a short but very precise baseline at one end of a line to be measured. It contains a target at each end, together with a sighting target at its middle point. The separation of the end targets is controlled by invar wires under a slight but firm spring tension. Although temperature fluctuations cause the bar to expand and contract, the low thermal expansion of the invar holds the targets at practically a fixed distance apart. Slack is increased or taken up by the springs.

The subtense bar is leveled by means of a bull's-eye level and leveling screws, just as is a transit or theodolite. The bar is thus brought into a horizontal plane. Also, the bar is brought perpendicular to the line to be measured by taking a sight to the far end of the line through a small low-power telescope located at the midpoint of the bar.

Figure 14-13 shows a plan view of a subtense measurement. The instrument used to measure the horizontal angle γ between the two targets T and T' is located at A. The horizontal distance AB is given by the relationship

$$AB = \frac{TT'}{2} \cos \frac{\gamma}{2}$$

or

$$D = \frac{b}{2} \cot \frac{\gamma}{2} \tag{14-6}$$

Figure 14-12. Subtense bar. Courtesy of Wild Heerbrugg Instruments, Inc.

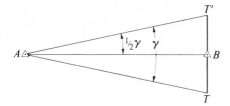

Figure 14-13. Subtense-bar principle.

in which D is the horizontal distance between the ends of the line; b is the length of the subtense bar, or the distance between the targets; and γ is the subtended horizontal angle. Both D and b are, of course, expressed in the same units. The subtense bar is usually 2 m long. The length of the line in metres is then cot $\gamma/2$. A table is supplied with the subtense bar which gives the distances in metres corresponding to values of γ.

The instrument used to measure the subtended angle must be capable of measuring to 1″ of angle or less. This accuracy can be obtained by measuring the angle with a 1″ theodolite in several positions of the circle. There is no need to reverse the telescope in these measurements because both targets are at the same vertical angle and at the same distance from the theodolite. Also, the angle is obtained by the difference between the two directions to the targets. Thus an instrumental error for one pointing equals that for the other pointing. These errors are eliminated by the subtraction of one direction from the other.

The accuracy of subtense measurement is a function of the subtended angle and the length of the bar. Differentiating Eq. (14-6) with respect to γ gives

$$dD_\gamma = -\frac{D^2}{b}\,d\gamma$$

and differentiation with respect to b gives

$$dD_b = -\frac{D}{b}\,db$$

If it is assumed that the uncertainty in $d\gamma$ is 1″, or 0.00000485, and the uncertainty in db is 0.2 mm, then the uncertainty dD in length for various distances is given in Table 14-4, together with the ratio of dD to the distance D.

The precision of a distance measured by use of the subtense bar is seen to fall off quite rapidly as the distance increases. This change is caused by the effect of the angular error. As would be expected, an error in the length of the bar produces an error of constant percent in the distance because the ratio of dD_b to D is constant. If the accuracy of the angular measurement is increased by using multiple pointings, so that the uncertainty is reduced to

TABLE 14-4. Accuracy of Subtense Measurements

D (m)	dD_y (m)	RATIO $dD_y : D$	dD_b (m)	RATIO $dD_b : D$
50	0.006	1 : 8330	0.0050	1 : 10,000
75	0.014	1 : 5350	0.0075	1 : 10,000
100	0.024	1 : 4170	0.010	1 : 10,000
150	0.055	1 : 2720	0.015	1 : 10,000
200	0.097	1 : 2060	0.020	1 : 10,000
250	0.152	1 : 1640	0.025	1 : 10,000
300	0.218	1 : 1380	0.030	1 : 10,000
400	0.388	1 : 1030	0.040	1 : 10,000
500	0.606	1 : 820	0.050	1 : 10,000
600	0.874	1 : 690	0.060	1 : 10,000

$\pm 0.5''$, each ratio in the second column is halved. Thus up to a distance of about 150 m or about 500 ft, the precision can be held to about one part in 5000.

14-15. TELESCOPIC ALIDADE The telescopic alidade, one type of which is shown in Fig. 14-14, is a tacheometric instrument which consists of a telescope similar in all details to that of a transit; an upright post, which supports the standards of the horizontal axis of the telescope; and a straight-edge, called the *blade*, whose edges are essentially in the same direction as the line of sight. The blade contains a circular level, a compass trough with provision for raising or lowering the compass needle, and a knob at each end to facilitate moving the blade. The eyepiece of the telescope is sometimes fitted

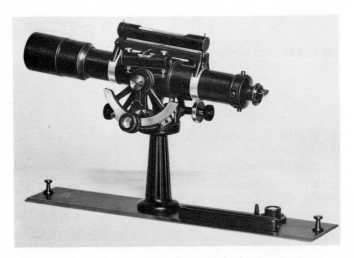

Figure 14-14. Telescopic alidade. Courtesy of W. & L. E. Gurley Co.

with a prism attachment to allow the observer to view downward rather than directly along the line of sight. The telescope can be rotated 180° about its line of sight in a sleeve by loosening the sleeve ring, and can be rotated about its horizontal axis through an elevation angle or depression angle of about 30°. A vertical clamp and tangent screw control the vertical movement of the telescope, just as on a transit.

The reticle of the telescope contains an additional tick mark in addition to the standard stadia cross hairs. This allows the observer to read a quarter interval which, when multiplied by four, is entered as a full interval in the notes. This is frequently necessitated when brush or vegetation obscures a full or half interval. Additionally, some reticles contain vertical tick marks like those shown in Fig. 12-5 that allow a full or half interval to be read when the rod is held horizontally instead of vertically.

The vertical circle has three sets of graduations. When the observer faces the circle, the graduations to the left are in degrees and half degrees and there is a vernier to read directly to the nearest minute. A horizontal line of sight gives a vertical-circle reading of 30°, the circle reading increasing from 30° for an elevation angle and decreasing from 30° for a depression angle. The vertical angle, together with its proper algebraic sign, is therefore equal to the circle reading minus 30°.

The graduations to the right are the H scale and V scale of a Beaman arc. The Beaman arc is a circular scale which solves the stadia equations automatically. The H-scale reading gives the number of feet or metres which must be subtracted for each 100 ft or metres of stadia distance or for each foot or metre of stadia interval. Thus, in Fig. 14-15, the H-scale reads 0.50. For a stadia interval of 1.00, the horizontal distance is $1.00(100 - 0.50) = 99.50$. If $s = 2.00$, then the horizontal distance is $2.00(100 - 0.50) = 199.00$, and so on. The stadia interval of 3.75 gives a horizontal distance of $3.75(100 - 0.50) = 373.13$. The expression for the horizontal distance H is thus

$$H = s(100 - H\text{-scale reading}) \tag{14-7}$$

This is the same as the value of H in Eq. (14-3) when using vertical angles.

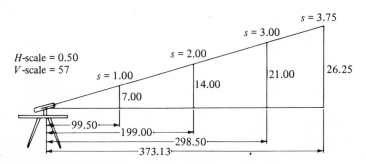

Figure 14-15. Reduction of Beaman arc readings.

The V-scale of the Beaman arc of the alidade shown in Fig. 14-14 reads 50 when the telescope is level and the control bubble is centered. The reading increases with an upward slope and decreases downward. Before making a V-scale reading, the observer raises or lowers the line of sight with the vertical motion tangent screw in order to make the V-scale read a whole number, following which he reads the H-scale. Subtracting 50 from the V-scale reading $(57 - 50 = +7$ in Fig. 14-15) gives the total rise $(+)$ or fall $(-)$ of the line of sight for every foot or metre of stadia interval. Thus, the value of V can be expressed as

$$V = s(V\text{-scale reading} - 50) \qquad (14\text{-}8)$$

This value, referred to as the *product*, is the same as the value of V in Eq. (14-5) when using vertical angles. In Fig. 14-15, the line of sight is seen to rise 7 ft or metres if the value of s is 1.00, and it rises 26.25 if the interval is 3.75. Most V-scales on modern alidades read 0 when the telescope is level and the control bubble is centered. Thus, Eq. (14-8) is modified to give the product

$$V = s \times V\text{-scale reading} \qquad (14\text{-}9)$$

The alidade is used in the field in conjunction with the plane table described in Section 14-16. The plane table is, by its nature, not as stable as a transit or theodolite and it can never be assumed to be horizontal no matter how carefully it is leveled when first set up. Thus as the alidade is moved over the surface of the plane table and pointed in different directions, the vertical-circle vernier and the Beaman arc indices invariably contain sizable index errors of a nature discussed in Section 4-18.

In order to eliminate the index error, the vertical circle is equipped with a level bubble which, when centered, brings the vertical-circle vernier and the index marks of the H and V scales in their proper position independent of the rest of the alidade. This level bubble is called the *control bubble*, the *vernier bubble*, or the *index bubble*. Together with the vernier and index marks, it is moved by a separate tangent screw. The control bubble *must always be centered* before a vertical-circle reading is made or before the V-scale setting is made.

Some kinds of telescopic alidades contain automatic pendulum-type compensators which serve to bring the vertical-circle vernier and the Beaman arc indices into their proper position with respect to the vertical even though the plane table is not horizontal. This eliminates the need for the control bubble.

The telescope shown in Fig. 14-14 is made horizontal, when desired, by means of a rather sensitive striding level that sits on a pair of collars on the telescope and is secured by a small knob on the top of the telescope tube. If the striding level is in adjustment and it is centered by means of the vertical clamp and tangent screw, the V-scale reading should read 50 when the control

Figure 14-16. Telescopic microptic alidade. Courtesy of A. Lietz Co.

bubble is brought to center. The adjustment of the bubble is described in Appendix B.

The microptic alidade, shown in Fig. 14-16, contains a vertical circle graduated on glass, and the circle is viewed through the eyepiece directly above the telescope eyepiece. The control bubble is viewed from the eye position by means of a mirror mounted over the bubble. The vertical circle contains the three sets of graduations just mentioned, namely, the V scale, the H scale, and the vertical-angle scale. The degrees and minutes are read in an upper window, the V scale is read in a middle window, and the H scale is seen in a lower window, as shown in Fig. 14-17. The V scale on the microptic

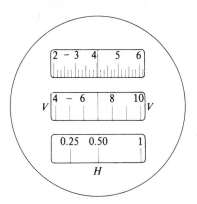

Figure 14-17. Circle readings with microptic alidade. (Vertical circle = $-4°\,01'$; V scale = -7; H scale = 0.50).

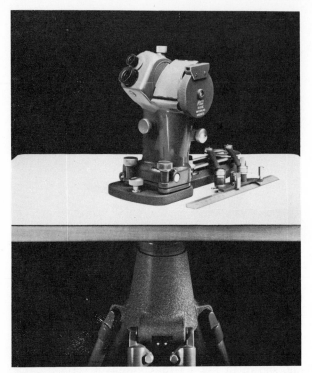

Figure 14-18. Self-reducing alidade. Courtesy of Kern Instruments, Inc.

alidade reads 0 for a horizontal line of sight. For the first eight graduations on either side of zero, a plus sign or a minus sign appears on the scale. Beyond that, the observer must logically deduce the proper algebraic sign.

Example notes for Beaman arc readings discussed above are given in Sections 14-17 and 14-18.

The self-reducing alidade shown in Fig. 14-18 contains a fixed telescope with a tilting objective prism which is used to raise or lower the line of sight. The stadia curves are shown in Fig. 14-19. As the tilting prism is turned, the curves sweep past the field of view and their spacings are continuously changing so as to solve the stadia equations automatically. The outer curves are for horizontal distance, the factor always being 100. There are three sets of inner curves representing vertical distances. Each of these curves has a different factor. It is 20 for slopes of 0° to 12°, 50 for slopes of 12° to 27°, and 100 for slopes of 27° to 40°. Beyond 40°, the stadia equation is used. In Fig. 14-19(a) there is a small vertical tick mark on each side. This marking indicates a vertical-distance factor of 100. In Fig. 14-19(b) there are two such marks on each side to indicate that the factor is 20. The horizontal-distance interval in (b) is $1.717 - 1.500 = 0.217$ m and the horizontal distance H is $0.217 \times 100 = 21.7$ m. The vertical-distance interval is $1.662 - 1.551 =$

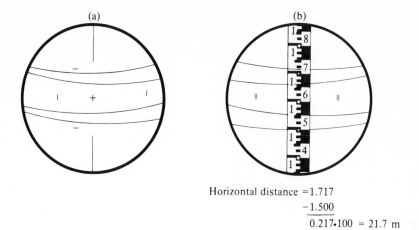

$$\text{Horizontal distance} = 1.717$$
$$\underline{-1.500}$$
$$\overline{0.217 \cdot 100} = \underline{21.7 \text{ m}}$$

Figure 14-19. Stadia curves of self-reducing alidade. (a) Field of view of telescope without rod; (b) field of view of telescope with rod.

0.111 m and the vertical distance V is $0.111 \times 20 = 2.22$ m. When two sets of five tick marks are visible, the vertical-distance factor is 50.

14-16. PLANE TABLE   Although the plane table shown in Fig. 14-20 is not a tacheometric instrument, it is used in conjunction with the telescopic alidade described in Section 14-15. The plane table–alidade combination is an extremely useful and versatile instrument as can be appreciated by reference to the remainder of this chapter and to practically all of Chapter 15. It consists of a drawing board which is mounted on a tripod in such a manner that the board can be leveled and rotated in azimuth without disturbing the tripod. This condition is realized by a tripod head, known as the *Johnson head*, which is shown in Fig. 14-21. One wing nut beneath the head controls the leveling, and the other nut controls the rotation of the board. Standard size boards are 18 by 18 in., 18 by 24 in., and 24 by 31 in.

The plane table itself provides the lower motion for the alidade, and movement of the alidade on the board can be considered as the upper motion, these motions being similar to those of the theodolite. The rotation of the board is used in backsighting, whereas the rotation of the alidade over the face of the board is used in foresighting.

The primary use of the combination of the plane table and alidade is in field compilation of maps. For this purpose it is much more versatile than is the transit. The drawing paper used for plane-table work must be of high quality, must be well seasoned to prevent undue expansion and contraction, must contain a surface with a reasonable amount of tooth or roughness to take pencil lines without undue grooving of the paper, and must be tough enough to stand erasures. For high accuracy, plane-table sheets containing thin aluminum sheets laminated with the paper are used. Celluloid sheets are

Figure 14-20. Plane table and alidade.

Figure 14-21. Johnson head. Courtesy of W. & L. E. Gurley Co.

sometimes used where there is likely to be an accumulation of moisture on the sheets.

In addition to the board and alidade, such accessories as stadia rods, a scale, triangles, plotting needles, pencils, and an eraser are needed. When a map is being compiled, the map is kept clean by first covering it with a piece of low-grade paper, this paper being torn away to expose the map sheet. On threatening days, a cover of plastic or other waterproof material should be provided to protect the map sheet from a sudden shower.

14-17. PLANE-TABLE TRAVERSE In compiling a map it frequently becomes necessary to establish the control by the plane table itself, rather than by means of a more-precise theodolite traverse. If high-quality, durable paper is used, and if the plane-table man exercises care in taking the stadia readings and plotting the points, highly satisfactory traversing can be accomplished by using the plane table. Two general techniques are in use for running a plane-table traverse. In the first method, which is applicable to large-scale mapping, each selected traverse point is occupied by the plane table and the board is oriented by backsighting on the previous point. When this method is used, each line in the traverse can be observed from both directions so that checks are provided on the values of the distances and differences in elevation between successive traverse stations, as discussed in Section 14-8. In the second method, which is applicable to small-scale mapping, every other point in the traverse is occupied by the plane table, and the board is oriented with respect to the magnetic meridian by means of the compass needle on the blade of the alidade. In either method the traverse may be run concurrently with the map compilation, or traversing may precede the map compilation. Running the traverse separately ahead of map compilation affords the advantage of permitting adjustment of the positions of the traverse stations and eliminates much erasing and map revision in the event there is a relatively large closure.

Method 1: Orienting by Backsighting. In Fig. 14-22 the points A, B, C, D, and E are selected as traverse stations because these points are located advantageously with respect to the topography that is to be mapped. They will be occupied for the purpose of map compilation after the traverse has been plotted on the plane-table sheet.

The plane table is set up over point A and leveled by unclamping both wing nuts of the Johnson head, bringing the circular bubble on the blade of the alidade to the center by tipping and tilting the board, and finally clamping the upper wing nut. During this operation, the alidade must be held on the board, preferably over the center, and must not be allowed to slip off the board. The board is then so oriented in azimuth that the entire traverse will fall on the plane-table sheet. The HI of the alidade above the ground is now measured. Point a, which represents the map position of A, is arbitrarily

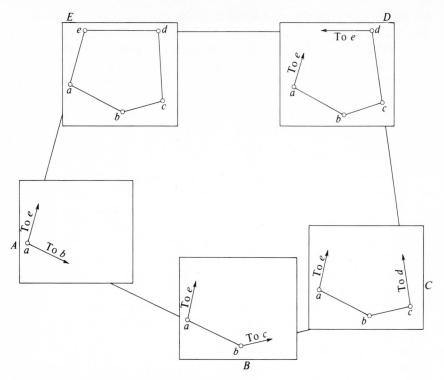

Figure 14-22. Plane-table traversing by backsighting.

plotted on the sheet. Point *a* being used as the pivot for the alidade blade, station *E* is sighted and a line is drawn on the sheet along the blade in the direction of *E*. The stadia interval is read on a rod held at *E*, the *V* scale of the Beaman arc is set at the nearest graduation mark, the *H*-scale is read, and the rod reading of the middle hair is taken and recorded. The form of notes is shown in Fig. 14-23. The distance *AE* and the difference in elevation between *A* and *E* can be computed (see Section 14-7 and Fig. 14-3). Next, the blade is pivoted about *a* until the line of sight is directed toward point *B* and the line *ab* is drawn on the sheet along the blade. The stadia readings are made and from them the distance and difference in elevation between *A* and *B* are computed.

STATION	INT.	H-SCALE	HORIZ. DIST.	V-SCALE	PROD. V	ROD READING	D.E.
		兀 @ A	H.I.= 4./'				
E	3.96	1.7	389.3	−/3	−5/.5	5.5	−52.9
B	4.02	0.25	40/.0	+5	+20./	4.6	+/9.6
		兀 @ B	H.I.= 3.8'				
A	4.0/	0.25	400.0	−5	−20./	3./	−/9.4
C	2./6	0	2/6.0	0	0	6.8	−3.0

Figure 14-23. Beaman arc notes for plane-table traversing.

The plane table is next set up over B, the blade is aligned along the line ba, and the board is rotated until the line of sight is directed at A. The board is now oriented, and it is clamped in this position. After the clamp is tightened, the line of sight should be checked to see whether or not the act of clamping affected the orientation. Stadia readings are made on station A, and the average of the two computed distances between A and B is plotted to the selected scale from a to define the plane-table position b of B. The blade is next pivoted about b until the line of sight is directed to station C, and the line in the direction of bc is drawn. Stadia readings are made on C to determine the first value of the length of the line BC and the difference in elevation.

The foregoing procedure is repeated at each station. Note that the position of the forward point is not plotted until two values of the length of the line are determined. At station E a backsight along the line through e and d orients the board, and stadia readings to D determine the second value of the length of the line DE. The position of e can then be plotted. Stadia readings on A from E give sufficient data to plot the final position of a along the line ea, and the closure can be adjusted by a graphical application of the compass rule. This adjustment is shown in Fig. 14-24. The point a' is the final position of a. Lines through points b, c, d, and e are drawn parallel with aa'. The distances bb', cc', dd', and ee' are in proportion to the distances from point a to the successive points. The dashed lines are the lines of the adjusted traverse.

The closure in elevations is found in the manner described in Section 14-8. Note that this type of plane-table traverse is similar in all respects to a stadia traverse performed with a transit or theodolite. Much of plane-table traversing, however, is graphical work.

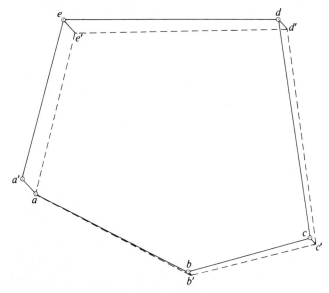

Figure 14-24. Graphical adjustment of traverse.

Method 2: Orienting by Compass Needle. In Fig. 14-25 the points A through F are traverse stations whose plane-table positions are to be obtained. Station A is occupied and the board is so oriented that the entire traverse will fall on the map sheet. The compass needle is unclamped, and the alidade is rotated in azimuth until the compass needle points to the north graduation on the end of the compass trough. A line representing the magnetic meridian is drawn the full length of the blade. The alidade is then pivoted in turn to F and B, and the rays af and ab are drawn. The necessary stadia readings are taken to determine the lengths of the lines AF and AB, from which points f and b are plotted by scaling the distances af and ab. Differences in elevation are computed for determining the elevations of F and B.

The plane table is next set up at station C and leveled. The compass needle is released and the blade is aligned with the line representing the magnetic meridian. The board is rotated in azimuth until the compass needle points to the north graduations on the end of the compass trough. The board is now oriented. With point b as the pivot, the alidade is rotated until the line of sight is directed to B. The ray bc is drawn back toward c. Stadia readings give the length of the line BC, and the scaled distance bc locates point c. The elevation of station C is also determined from the stadia readings. The alidade is now rotated about point c until station D is sighted, and the ray cd is drawn on the map sheet. Stadia readings give the distance CD, from

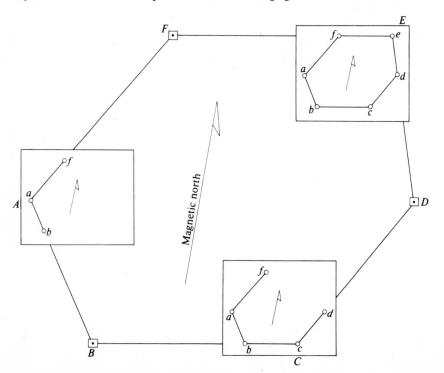

Figure 14-25. Plane-table traverse by compass needle.

which point d is plotted. The elevation of station D is also determined from the stadia readings.

At the last setup on station E the procedure is the same as that described for the setup at C. The closure at station F is determined from this setup, and a graphical adjustment of the traverse is made. The elevations are adjusted by the method described in Section 14-8.

14-18. METHOD OF RADIATION   When either a point and the magnetic meridian or two intervisible points have been properly located on a plane-table sheet, either by plane-table traverse or by plotting the computed positions from a transit traverse, the details about the point can be located on the plane-table sheet by the method of radiation. The plane table is set up and leveled over the point. The board is oriented either by using the compass needle or by backsighting along a plotted line. If the elevation of the ground point is known, the HI above the datum is determined by adding the distance from the ground to the telescope axis of the alidade to the elevation of the ground point.

All points whose positions and elevations are to be determined are sighted by using the map position of the occupied point as the pivot for the alidade. Rays are drawn in the directions of the successive points, and the stadia readings on the points give the lengths of the successive lines, from which the points can be plotted by scaling. This procedure is illustrated in Fig. 14-26. The backsight point should be checked frequently to detect any slipping of the board. The elevations of the points are determined as described in Sections 14-7 and 14-9, and as illustrated in the notes in Fig. 14-27.

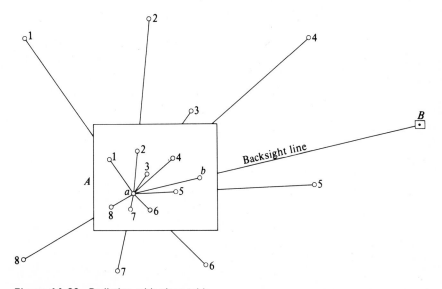

Figure 14-26. Radiation with plane table.

POINT	INT.	H-SCALE	HORIZ. DIST.	V-SCALE	PROD. V	ROD READING	D.E.	H.I.	ELEV.
			𝜆 @ A					415.82	414.60
1	1.422	0.35	141.70	-5	-7.11	2.14	-9.25		406.57
2	1.385	0.60	137.67	-8	-11.08	1.86	-12.94		402.88
3	0.616	2.00	60.37	+14	+8.62	1.45	+7.17		422.99
4	1.698	0	169.80	+2	+3.40	2.62	+0.78		416.60

Figure 14-27. Beaman arc notes using method of radiation.

14-19. PLANE-TABLE LEVELING

The plane-table method of leveling is excellent for getting rapid results of relatively low accuracy. It is in all respects the same as stadia leveling. The plane table is set up at an arbitrary point, as in differential leveling, and a backsight is taken on a point of known elevation. The backsight values consist of the stadia interval, the V-scale reading, and rod reading from which the HI above the datum is determined.

In Fig. 14-28 an interval of 9.80 is read on the rod held at BM50 whose elevation is 362.2 ft. The V-scale reading is 47 and the rod reading is 7.7 ft. The product is $(47 - 50) \times 9.81 = 29.4$. This is the vertical distance by which the line of sight falls from the alidade to the rod reading. The difference in elevation between the foot of the rod and the alidade is $+7.7 + 29.4 = +37.1$ ft, and the HI is $362.2 + 37.1 = 399.3$ ft. When a foresight is taken on the rod held at TP1, the interval is 6.64, the V-scale reading is 52, and the rod reading is 6.2 ft. The product is $(52 - 50) \times 6.64 = +13.3$ ft, and the difference in elevation between the alidade and the foot of the rod at TP1 is $+13.3 - 6.2 = +7.1$ ft. The elevation of TP1 is therefore $399.3 + 7.1 = 406.4$ ft.

A form of notes for plane-table leveling is given in Fig. 14-29. For the backsight taken on TP2 and for the foresight taken on BM51, the telescope is level, giving a V-scale reading of 50 which is not recorded. The interval is recorded on level sights to provide the distances used in adjusting the level line.

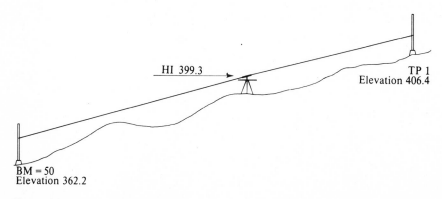

HI 399.3

TP 1
Elevation 406.4

BM = 50
Elevation 362.2

Figure 14-28. Plane-table leveling.

STA.	INT.	V–SCALE	PROD. V	ROD	B.S.	INT.	V–SCALE	PROD. V	ROD	F.S.
B.M.50	9.80	47	−29.4	7.7	+37.1					
T.P.1	3.22	40	−32.2	6.0	+38.2	6.64	52	+13.3	6.2	+7.1
T.P.2	4.85	—	—	9.7	+9.7	5.08	51	+5.1	4.7	+0.4
T.P.3	5.50	55	+27.5	3.4	−24.1	7.35	49	−7.4	8.8	−16.2
B.M.51						4.08	—	—	6.1	−6.1

Figure 14-29. Beaman arc notes for plane-table leveling.

14-20. PLANE-TABLE INTERSECTION   One of the decided advantages of the plane table is the ease with which a point can be located by intersection. The procedure is shown in Fig. 14-30 and is described in Section 6-13. With the board oriented at station A and by backsighting on station B, the alidade blade is pivoted at a and a ray of indefinite length is drawn toward the point P to be located. The plane table is then set up at station B and the board is oriented by backsighting on A. With the alidade pivoted at b, a ray toward P is drawn on the plane-table sheet. The intersection of these two rays defines the map position p of P. The difference in elevation between A and P and that between B and P can be obtained if the vertical angle to P has been measured from each setup. The product of the distance AP, scaled from the map, and the tangent of the vertical angle at A is the difference in elevation between the alidade at A and the point P. The difference in elevation

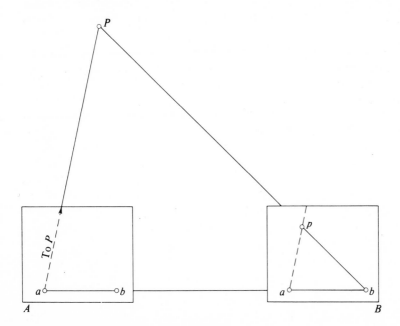

Figure 14-30. Intersection with plane table.

between the alidade at B and the point P can be determined in the same manner.

Suppose that the HI at A is 714.2 ft, the distance from A to P scaled on the map is 1380 ft, and the vertical angle to P is $-0°40'$. The elevation of P is $714.2 - 1380 \tan 0°40' = 698.1$ ft. If the HI at B is 708.0 ft, the distance BP scales 1970 ft, and the vertical angle from B to P is $-0°18'$, then the elevation of P is $708.0 - 1970 \tan 0°18' = 697.7$ ft. Thus a check can be provided against a gross error or a mistake.

14-21. ERRORS IN PLANE-TABLE SURVEYS Errors in plane-table surveys are caused by instrumental errors, by errors in drafting, and by the instability of the board.

The instrumental errors are practically the same as those for transit stadia work. By adjusting and properly manipulating the alidade, errors from this source can be kept to a minimum. Except for large-scale maps the error caused by not placing the station on the sheet directly over the station on the ground will cause no perceptible error.

In the case of a small-scale map the width of a pencil line may represent many feet on the ground and considerable care must be exercised in drawing the rays on such a map. An extremely hard, well-pointed pencil should be used for this purpose. A needle should be used in plotting station points.

Weather conditions may greatly affect the precision of a plane-table map. Stretching and shrinking of the sheet will be considerable when the weather is changeable. Errors from this source can be practically eliminated by using paper sheets that are mounted on thin sheets of aluminum. If mounted sheets are not used, all horizontal control that is to appear on a sheet should be plotted at one time. When the change in the paper is considerable, allowance for it must be made in plotting points. Frequently the change in the two dimensions will not be the same.

On windy days the measurement of stadia distances may be next to impossible. If a sheltered station can be occupied, sights from the more exposed one should be left for calmer weather. Vertical angles are affected by the board not being perfectly horizontal. As many boards soon become warped, the index correction should be determined for all important pointings by means of the striding level; otherwise the control bubble must be centered. The topographer must be careful not to lean on the board while sighting or plotting. To guard against possible movement of the board, the orientation should be checked at frequent intervals, particularly when new stations are being located.

PROBLEMS

14-1. Determine the stadia interval factor K of an external-focusing transit from the accompanying data. The taped distances are from the point of focus of the telescope, which is 1.1 ft in front of the center of the instrument.

TAPED DISTANCE (ft)	STADIA INTERVAL (ft)	TAPED DISTANCE (ft)	STADIA INTERVAL (ft)
50.0	0.501	300.0	2.997
100.0	1.001	400.0	3.988
150.0	1.502	500.0	4.985
200.0	2.000	600.0	5.982

14-2. The following stadia notes were recorded for a closed traverse:

STATION	INTER-VAL	HORIZONTAL DISTANCE	AZIMUTH	VERTICAL ANGLE	DE	ELEV-ATION	ADJUSTED ELEVATION
1						250.0	250.0
	3.69	368.9	120° 32′	+0° 56′	+6.0		
2							
	5.70	570.0	138° 33′	−0° 16′	−2.7		
3							
	4.39		180° 19′	−7° 06′			
4							
	4.87		141° 12′	−1° 05′			
5							
	4.63		209° 45′	−7° 29′			
6							
	5.63		231° 03′	−3° 26′			
7							
	4.92		316° 32′	+0° 07′			
8							
	4.39		320° 23′	+0° 14′			
9							
	5.02		285° 06′	−0° 18′			
10							
	5.78		277° 16′	−0° 32′			
11							
	4.35		62° 38′	+1° 06′			
12							
	4.53		83° 38′	+0° 50′			
13							
	5.92		46° 46′	+2° 33′			
14							
	6.50		7° 05′	+10° 50′			
1							

It is to be noted that the observations were made in the forward direction only, and that the middle cross hair was set on the HI to obtain the vertical angle. The stadia interval factor K is 100, and the constant C is zero.

(a) Compute the horizontal distance and the difference in elevation for each line to the nearest tenth of a foot.

(b) Compute the elevations of the traverse stations. Adjust the elevations in accordance with the principles of Section 14-8.

— compute EOC (−2.8′) = Σ DE

— adjust ELEV. by proportioning sides

$1 \text{ in} = 400 \text{ ft}.$

(c) Plot the stadia traverse to a scale of 1 in. = 200 ft by the method shown in Fig. 8-32. Using a sheet of drafting paper 18 by 24 in., lay off a $1\frac{1}{2}$-in. border on all sides. With the long dimension in the east-west direction, locate station 1 in the center of that dimension and 1 in. down from the top border. This position will allow the traverse to be conveniently centered on the map sheet.

(d) Distribute the error of closure in the position of station 1 by the method shown in Fig. 14-24.

14-3. The following stadia obervations were made in a closed traverse, using an internal-focusing transit in which $K = 100$.

OCCUPIED STATION	OBSERVED STATION	STADIA INTERVAL	AZIMUTH	VERTICAL ANGLE
A	G	4.71	180° 00′	+5° 22′
	B	′7.68	38° 13′	+1° 49′
B	A	7.66	218° 13′	−1° 49′
	C	6.00	126° 54′	+0° 50′
C	B	6.00	306° 54′	−0° 50′
	D	6.38	67° 51′	−1° 57′
D	C	6.40	247° 51′	+1° 57′
	E	11.30	192° 32′	+9° 29′
E	D	11.33	2° 32′	−9° 31′
	F	9.70	263° 07′	−4° 52′
F	E	9.71	83° 07′	+4° 52′
	G	4.39	302° 44′	−9° 02′
G	F	4.39	122° 44′	+9° 02′
	A	4.70	0° 07′	−5° 22′

The center cross hair was set on the HI in each instance when the vertical angle was read.

(a) Compute the average horizontal distance and the average difference in elevation for each line.

(b) The given elevation of station A is 763.2 ft. Compute the adjusted elevations of the remaining stations in accordance with the principles of Section 14-8.

(c) After having adjusted the azimuths of the lines, plot the traverse to a scale of 1 in. = 100 ft by the method shown in Fig. 8-32. Using a sheet of drafting paper 18 by 24 in., lay off a 1-in. border on all sides. With the long dimension in the east-west direction, locate station A $3\frac{1}{2}$ in. in from the west border and $7\frac{1}{2}$ in. down from the north border of the sheet. This position will allow the traverse to be conveniently centered on the map sheet.

(d) Distribute the error of closure in the position of station A by the method shown in Fig. 14-32.

14-4. With a theodolite set 1.58 m above station A, a sight is taken on a rod held at station B. The interval is 1.420 m; the rod reading of the middle cross hair is

3.54 m; the vertical angle is $-5°$ 13'. With the instrument set 1.55 m above station B, a sight is taken on the rod held at station A. The interval is 1.430 m; the rod reading of the middle cross hair is 2.35 m; the vertical angle is $+6°$ 00' The instrument is internal focusing and $K = 101$. What is the average length of the line AB? What is the average difference in elevation between the two points?

14-5. With a transit set 5.0 ft above station R, a sight is taken on a rod held at station S. The interval is 5.85 ft; the rod reading of the middle hair is 9.9 ft; the vertical angle is $+4°$ 12'. With the transit set 5.2 ft above station S, a sight is taken on the rod held at R. The interval is 5.83 ft; the rod reading of the middle hair is 12.6 ft; the vertical angle is $-3°$ 02'. The transit is internally focusing and has a K value of 100. What is the average length of the line RS? What is the average difference in elevation between the two points?

14-6. With a transit set 5.2 ft above station C, a sight is taken on a rod held at station D. The interval is 3.82 ft; the rod reading of the middle hair is 12.8; the vertical angle is $+0°$ 04'. With the transit set 5.3 ft above station D, a sight is taken on the rod held at C. The interval is 3.85 ft; the middle cross-hair reading is 9.8 ft; the vertical angle is $+1°$ 42'. The transit is internally focusing and has a K value of 100. What is the average length of the line CD? What is the average difference in elevation between the two points?

14-7. The following notes were recorded when carrying stadia levels between two temporary benchmarks. The value of C is 0 and $K = 100$. The elevation of TBM5 is 1322.2 ft. Compute the elevation of TBM6.

STATION	BACKSIGHT			FORESIGHT		
	INTERVAL	ROD	VERTICAL ANGLE	INTERVAL	ROD	VERTICAL ANGLE
TBM5	2.14	8.6	$-0°$ 25'			
TP1	5.32	4.5	$-3°$ 15'	5.62	6.3	$+1°$ 52'
TP2	4.88	10.0	$-1°$ 02'	2.04	2.2	$+6°$ 20'
TP3	5.08	8.0	$+0°$ 36'	7.06	5.0	$+0°$ 13'
TP4	2.96	5.5	$+5°$ 52'	4.41	8.5	$-4°$ 00'
TBM6				5.02	12.2	$-1°$ 13'

14-8. The following notes were recorded when carrying stadia levels between two temporary benchmarks. The telescope is internally focusing with a value of $K = 100$. The elevation of TBM12 is 875.52 m. Compute the elevation of TBM13.

STATION	BACKSIGHT			FORESIGHT		
	INTERVAL	ROD	VERTICAL ANGLE	INTERVAL	ROD	VERTICAL ANGLE
TBM12	1.298	2.38	$-3°$ 13'			
TP1	1.183	2.68	$-0°$ 50'	1.225	1.89	$+2°$ 16'
TP2	1.798	1.28	$+4°$ 11'	1.695	3.11	$-0°$ 10'
TP3	1.774	2.07	$-0°$ 25'	1.969	2.41	$0°$ 00'
TP4	1.798	2.93	$-3°$ 18'	1.466	3.29	$+6°$ 24'
TBM13				1.189	2.01	$+1°$ 00'

14-9. A plane table is set up over a control station whose elevation is 772.2 ft. The distance from the ground to the alidade, measured with the stadia rod, is 4.3 ft. Assume that $K = 100$ and $C = 0$. A V-scale reading of 50 on the Beaman arc indicates a level sight. The following notes were recorded from observations made on detail points.

POINT	INTERVAL	H	HORIZONTAL DISTANCE	V	PRODUCT $(V - 50) \times s$	ROD READING	DE	ELEVATION
17	0.55	8		78		5.6		
18	1.82	3		67		7.2		
19	1.74	2		38		12.6		
20	3.98	—		55		4.5		
21	6.62	—		49		5.8		
22	4.85	—		50		3.2		

Compute the distance from the plane table to each point and the elevations of the points.

14-10. A plane table is set up over a control station whose elevation is 484.8 ft. The HI above the ground is 4.1 ft. The value of K is 100 and $C = 0$. The V-scale reading is zero when the line of sight is level. Complete the following notes recorded at the station.

POINT	INTERVAL	H	HORIZONTAL DISTANCE	V	PRODUCT $V \times s$	ROD READING	DE	ELEVATION
30	2.03	—		+ 4		5.1		
31	1.55	2		− 14		7.8		
32	1.88	4		− 20		4.2		
33	3.74	—		+ 6		5.3		
34	2.16	7		+ 26		8.8		
35	3.10	3		− 18		12.0		

14-11. The plane-table operator sets over an unknown ground point and measures a distance of 1.31 m from the ground up to the alidade. The rodman holds the rod on a point whose elevation is 485.4 m. The plane-table operator reads a stadia interval of 1.664 m, a V-scale reading of +8, and a center cross hair reading of 1.89 m on the rod. Compute the elevation of the unknown ground point.

14-12. The mean angle measured between the two targets of a 2-m subtense bar is $0° 30' 14.22''$. The standard error of the angle is $\pm 0.37''$. Assume that the standard error of the distance between the targets is ± 0.25 mm. Compute the distance, in metres between the theodolite and the subtense bar. Compute the standard error of this distance, in metres.

14-13. In Problem 14-12 if the subtended angle is $0° 42' 14.22'' \pm 0.25''$, what is the distance, in feet, between the theodolite and the subtense bar? What is the standard error of this distance, in feet?

14-14. In Problem 14-12 if the subtended angle is $1° 55' 40.5'' \pm 0.75''$, what is the distance, in metres, between the theodolite and the subtense bar? What is the standard error of this distance, in metres?

BIBLIOGRAPHY

Low, J. W. 1952. *Plane Table Mapping.* New York: Harper & Row.

Chapter 15
Topographic Surveys

15-1. GENERAL PROCEDURES Topographic surveying is the process of determining the positions, on the earth's surface, of the natural and artificial features of a given locality and of determining the configuration of the terrain. The location of the features is referred to as *planimetry*, and the configuration of the ground is referred to as *topography* or *hypsography*. The purpose of the survey is to gather data necessary for the construction of a graphical portrayal of planimetric and topographic features. This graphical portrayal is a topographic map. Such a map shows both the horizontal distances between the features and their elevations above a given datum. On some maps the character of the vegetation is shown by means of conventional signs.

Topographic surveying or mapping is accomplished by ground methods requiring the use of the transit, plane table and alidade, level, hand level, tape, and leveling rod in various combinations. Total station EDM's are used to advantage in topographic surveying. The vast majority of topographic mapping is accomplished by aerial photogrammetric methods, as described in Chapter 16. In the photogrammetric methods, however, a certain amount of field completion and field editing must be done by ground methods described in this chapter.

The preparation of a topographic map, including the necessary control surveys, is usually the first step in the planning and designing of an engineering project. Such a map is essential in the layout of an industrial plant, the location of a railway or highway, the design of an irrigation or drainage system, the development of hydroelectric power, city planning and engineering, and landscape architecture. In time of war, topographic maps are essential to persons directing military operations.

15-2. SCALES AND ACCURACY Since a topographic map is a representation, on a comparatively small plane area, of a portion of the surface of the earth, the distance between any two points shown on the map must have a known definite ratio to the distance between the corresponding two points on the ground. This ratio is known as the scale of the map. As stated in Section 8-33, this scale can be expressed in terms of the distance on the map, in inches, corresponding to a certain distance on the ground, in feet. For example, a scale may be expressed as 1 in. = 200 ft. The scale can be expressed also as a ratio, such as 1 : 6000, or as a fraction, as 1/6000. In either of these last two cases, 1 unit on the map corresponds to 6000 units on the ground. A fraction indicating a scale is referred to as the *representative fraction*. It gives the ratio of a unit of measurement on the map to the corresponding number of the same units on the ground.

The scale to which a map is plotted depends primarily on the purpose of the map, that is, the necessary accuracy with which distances must be measured or scaled on the map. The scale of the map must be known before the field work is begun, since the field methods to be employed are determined largely by the scale to which the map is to be drawn. When the scale is to be 1 in. = 50 ft, distances can be plotted to the nearest $\frac{1}{2}$ or 1 ft, whereas if the scale is 1 in. = 1000 ft, the plotting will be to the nearest 10 or 20 ft and the field measurements can be correspondingly less precise.

15-3. METHODS OF REPRESENTING TOPOGRAPHY Topography may be represented on a map by hachures or hill shading, by contour lines, by form lines, or by tinting. *Hachures* are a series of short lines drawn in the direction of the slope. For a steep slope the lines are heavy and closely spaced. For a gentle slope they are fine and widely spaced. Hachures are used to give a general impression of the configuration of the ground, but they do not give the actual elevations of the ground surface.

A *contour line*, or *contour*, is a line that passes through points having the same elevation. It is the line formed by the intersection of a level surface with the surface of the ground. A contour is represented in nature by the shoreline of a body of still water. The *contour interval* for a series of contour lines is the constant vertical distance between adjacent contour lines. Since the contour lines on a map are drawn in their true horizontal positions with respect to the ground surface, a topographic map containing contour lines shows not only the elevations of points on the ground, but also the shapes of the various topographic features, such as hills, valleys, escarpments, and ridges.

The classical illustration used to show the relationship between the configuration of the ground and the corresponding contour lines is shown in Fig. 15-1. This illustration used to be printed on the backs of the U.S. Geological Survey quadrangle maps along with an explanation of the topographic map and how it is interpreted and used. Unfortunately the U.S. Geological Survey discontinued this feature of its quadrangle series in the early 1950s. The upper part of the illustration shows a stream lying in a valley between a cliff on the left and a rounded hill on the right. The stream is seen to empty into the ocean in a small bay protected by a sand hook. Other features such as terraces, gulleys, and a gentle slope behind the cliff can be identified. An abrupt cliff to the right of the sand hook plunges almost vertically to the ocean. The lower part of the figure is the contour line or topographic map representation of this terrain. The contour interval of this map is 20 units and could represent either 20 ft or 20 m.

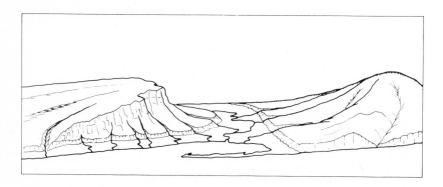

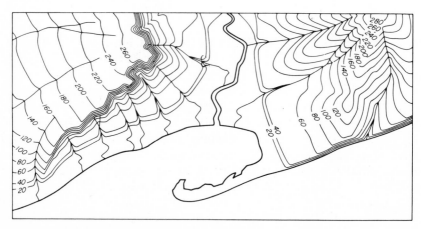

Figure 15-1. Contour line representation of terrain. By permission of U.S. Geological Survey.

On maps intended for purposes of navigation, peaks and hilltops along the coast are sometimes shown by means of *form lines*. Such lines resemble contours, but are not drawn with the same degree of accuracy. All points on a form line are supposed to have the same elevation, but not enough points are actually located to conform to the standard of accuracy required for contour lines.

On aeronautical charts and on maps intended for special purposes, such as those that may accompany reports on some engineering projects, elevations may be indicated by tinting. The area lying between two selected contours is colored one tint, the area between two other contours another tint, and so on. The areas to be flooded by the construction of dams of different heights, for example, might be shown in different tints.

15-4. CONTOUR LINES   The configuration of the ground and the elevations of points are most commonly represented by means of contour lines, because contours give a maximum amount of information without obscuring other essential detail portrayed on the map. Some of the principles of contours are represented in Fig. 15-2. Four different contour intervals are shown in views (a), (b), (c), and (d). The steepness of the slopes can be determined from the contour interval and the horizontal spacing of the contours. If all four of these sketches are drawn to the same scale, the ground slopes are the steepest in (d), where the contour interval is 20 ft, and are the flattest in (c), where the interval is 1 ft.

The elevation of any point not falling on a contour line can be determined by interpolating between the two contour lines that bracket the point. Quite

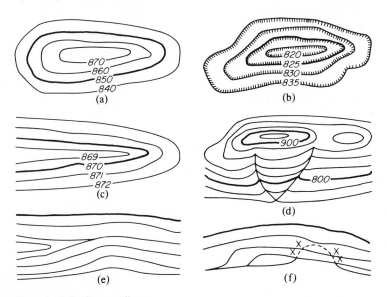

Figure 15-2. Contour lines.

often when the scale of the map is large and the terrain is flat, the successive contours are spaced so far apart horizontally that interpolation between adjacent contours does not have much significance. Therefore, in such an instance, the accuracy and utility of the map is greatly increased by showing the elevations of points at regular intervals in some form of a grid pattern. Elevations between these points are then determined by interpolation. Spot elevations are shown on Fig. 15-3.

A contour cannot have an end within the map. It must either close on itself, or commence and end at the edges of the map. A series of closed contours represents either a hill or a depression. From the elevations of the contour lines shown, a hill is represented in Fig. 15-2(a) and a depression in view (b). As indicated, a depression contour is identified by short hachures on the downhill side of the contour. A ravine is indicated by the contours in Fig. 15-2(c). If the elevations were reversed, the same contours would represent a ridge. View (e) is incorrect, as two contours are shown meeting and continuing as a single line; this would represent a knife-edged ridge or ravine, something not found in nature. View (f), if not incorrect, is at least unusual. Several contours are shown merging and continuing as a single line. This would be correct only in the case of a vertical slope or a retaining wall.

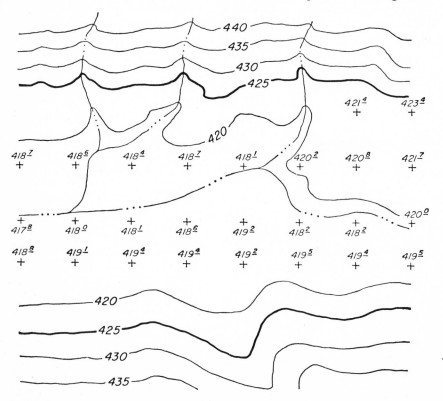

Figure 15-3. Spot elevations to supplement contour lines.

Also, one contour is shown to cross two others. Thus each point marked x has two elevations, a condition found only at a cave or an overhanging cliff.

A series of equally spaced contour lines represents a constant slope along a line normal to the contours. A series of straight, parallel, equally spaced contours represents man-made excavations or embankments. The steepest direction from any point on a topographic map is that which runs normal to successive contour lines near the point.

The drainage of the terrain is the primary agent in shaping the topography. Its influence on the shape of the contour lines can be seen in Figs. 15-1, 15-2 (d), and 15-3. Note that as contour lines cross gulleys or streams or other drainage features, the contour lines form modified V's pointing upstream. The form of the V's is determined by the type of underlying soil or rock. In general, if the underlying material is fine grained like a clay soil, the V will be smooth and rounded, and if the material is coarse and granular, the V will be quite sharp.

As a convenience in scaling elevations from a topographic map, each fifth contour is drawn as a heavier line. This is called an *index contour*. When the interval is 1 ft, contours whose elevations are multiples of 5 ft are shown heavy. When the interval is 10 ft, the heavy contours have elevations that are multiples of 50 ft. Enough contours should be numbered to prevent any uncertainty regarding the elevation of a particular contour. Where the contours are fairly regular and closely spaced, only the heavy contours need be numbered.

15-5. FIELD METHODS   Among the factors that influence the field method to be employed in the compilation of a topographic map are the scale of the map, the contour interval, the type of terrain, the nature of the project, the equipment available, the required accuracy, the type of existing control, and the extent of the area to be mapped. The area to be mapped for highway or railroad location and design takes the form of a strip with a width varying from 100 ft to perhaps more than 1000 ft (30 to 300 m). The control lines are the sides of a traverse which have been established by a preliminary survey and which have been stationed and profiled as outlined in Chapter 3. The method of locating topography most commonly employed for this purpose is the *cross-section* method.

To make an engineering study involving drainage, irrigation, or water impounding or to prepare an accurate map of an area having little relief, each contour line must be carefully located in its correct horizontal position on the map by following it along the ground. This is the *trace contour* method.

When an area of limited extent is moderately rolling and has many constant slopes, points forming a grid are located on the ground and the elevations of the grid points are determined. This is the *grid* method of obtaining topography.

If the area to be mapped is rather extensive, the contour lines are located by determining the elevations of well-chosen points from which the positions of points on the contours are determined by interpolation. The topographer determines the shapes of the contours by experience and judgment. This is known as the *controlling-point* method.

15-6. CROSS-SECTION METHOD The cross-section method of obtaining topography can be performed by transit and tape, by transit stadia, by level and tape, by hand level and tape, by plane table and tape, or by plane-table stadia. Horizontal control is established by a theodolite-tape traverse, by an EDM traverse, or by a stadia traverse between fixed control points. Stakes are set every 50 or 100 ft or at other pertinent intervals, the spacing depending on the terrain. Vertical control is obtained by profile leveling which may be performed either before the topography is taken or concurrently with the mapping.

When the theodolite and tape are used, the instrumentman occupies each station or plus station on the line. He determines the HI by holding the leveling rod alongside the instrument. A right angle is turned off the line, and the rodman, holding one end of the tape, proceeds along this crossline until a break in the slope occurs. If possible, the instrumentation takes a level sight on the rod and records the reading and the distance to the point. If the rod cannot be sighted with the telescope level, a vertical angle is read and the slope distance is recorded. The rise or fall of the line of sight equals the slope distance times the sine of the vertical angle as discussed in Section 3-3. The horizontal distance to the point is obtained by Eq. (2-1). The rod reading must be taken into account in determining the elevation of the point, as discussed in Section 14-7.

The rodman proceeds to the next break in the topography, and the process is repeated. If the width of strip on each side of the line is greater than the length of the tape, a third man is necessary to hold the tape. As the distance out becomes greater, the slope distance cannot be measured directly in case vertical angles must be read, and the true rise or fall of the line of sight will be somewhere between the measured distance times the sine of the vertical angle and the measured distance times the tangent of the angle. Where the accuracy will deteriorate because of this difficulty, the distance can be measured with an EDM attached to the theodolite. Most of the modern EDM's display both horizontal and vertical distances from the EDM to the reflector. When used for cross sectioning, the reflector is held on a staff or rod which contains a level bubble to bring the rod vertical. The height of the reflector above ground together with the HI of the EDM are then used in determining the DE to the point.

Table 15-1 shows the data recorded for nine regular cross sections spaced at 50-ft intervals, and additional ones at station 22 + 28 and station 23 + 70. The center-line elevations are obtained either from prior profile leveling or else by leveling carried along as the cross-section measurements are made. If by the latter method, then profile-level notes would be recorded on the left-hand side of the field book in combination with the cross-section notes on the right-hand side. The cross-section notes consist of the station number, the horizontal distance to right or left of center line to the point as the denominator, and the elevation of the point as the numerator. Note that the stationing increases from the bottom upward. A glance at the right-hand side shows that when one views the notes, he can imagine looking in the forward direction of the line, with the values to the left of him entered as such and those to the right of him entered as such.

TABLE 15-1. Recorded Data for Nine Cross Sections

		CROSS SECTIONS						
STATION	CENTER-LINE ELEVATION	L			C	R		
26 + 00	550.0			550.2/93	550.0/0	547.6/23	539.0/61	532.4/106
25 + 50	534.5			545.0/107	534.5/0	532.1/59	527.7/109	
25 + 00	539.3		551.3/103	549.1/31	539.3/0	537.3/10	521.7/100	
24 + 50	544.2		551.0/94	555.1/52	544.2/0	532.3/52	524.5/106	
24 + 00	551.0		550.0/101	556.3/65	551.0/0	529.1/100		
23 + 70	545.0	548.0/100	557.3/67	553.9/45	545.0/0	539.0/27	528.6/100	
23 + 50	546.6		547.4/102	557.8/61	546.6/0	525.8/94		
23 + 00	550.0		552.0/108	559.8/69	550.0/0	547.6/20	532.0/109	
22 + 50	542.7		554.8/89	549.2/47	542.7/0	538.0/46	535.0/102	
22 + 28	539.5			550.0/92	539.5/0	530.0/62	530.7/96	
22 + 00	539.3	546.6/92	545.5/61	541.4/19	539.3/0	539.7/32	528.6/102	

Instead of the tape or EDM, stadia can be used to obtain the horizontal distances to and the elevations of points on the cross section, as outlined in Chapter 14. When stadia is used only to obtain elevations, two people handling the tape can measure the distance to the right or left of the line. Perpendicularity to the line is estimated by the rodman. In this way the theodolite need not be set up at each station. The station number, the distance to the right or left of the line, and the elevation of the point are recorded in a systematic manner. It is the best practice to determine the elevation of each point as soon as it is observed in order to preclude any mistakes or possible misunderstandings at a later date.

When the level and tape or EDM are used, the HI is determined by backsighting on a point of known elevation. The distances to all points observed are recorded along with their elevations. When the leveling rod is out of sight, a new setup must be made and the HI must be determined either by backsighting on another point of known elevation or by establishing a turning point and backsighting on the turning point.

In any of the foregoing methods the located points are plotted either in the field or in the office. The positions of the contour lines are obtained by interpolating between the elevations of the plotted points as discussed below in Section 15-7.

The use of the plane table and alidade for cross-section compilation is the same as the use of the theodolite, except that the topographer does the sketching directly on the plane-table sheet. The sheet is prepared beforehand by plotting the control line to the desired map scale, marking the positions of the stations, and entering the elevations obtained from profile leveling directly on the plane-table sheet. In the field the board is oriented by backsighting along the line on the ground with the blade of the alidade aligned along the plotted position of the line. As points on the crosslines are observed, their elevations are plotted, the necessary interpolations are made, and the contour lines are sketched in.

In compiling topography by the cross-section method, the positions of all planimetric features, such as buildings, fences, streams, and property lines, must be located with respect to the control line and plotted on the topographic map. The positions of the features can be located by transit-stadia or plane-table methods. The interval and azimuth to various points are read when the transit is used, or the interval is read and the direction is plotted on the plane-table sheet when the plane table and alidade are used. Other methods of locating the positions of points with respect to a traverse line are discussed in Section 6-13.

15-7. METHODS OF INTERPOLATING   In locating contours on a map by interpolation, the positions of points on the contours can be determined either mathematically or mechanically. In Fig. 15-4 it is required to locate the 5-ft contours on the line connecting points a and b, whose elevations

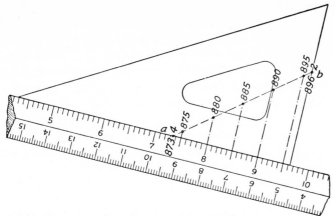

Figure 15-4. Interpolating with triangle and scale.

are 873.4 and 896.2, respectively. It is evident that contours at elevations 875, 880, 885, 890, and 895 will cross this line.

The distances on the map from the point a to the contours can be calculated by proportion. The horizontal distance between a and b on the map is scaled and found to be 2.78 in. The corresponding vertical distance on the ground is $896.2 - 873.4 = 22.8$ ft. The vertical distance from a to the 875-ft contour is $875.0 - 873.4 = 1.6$ ft. The horizontal distance from a to the 875-ft contour is $(2.78/22.8)\ 1.6 = 0.17$ in. The distance between two adjacent 5-ft contours is $(2.78/22.8)\ 5 = 0.61$ in. The horizontal distances can also be expressed in terms of the distances on the ground.

In Fig. 15-4 is illustrated a method by which the points on the contours can be located mechanically, a triangular engineer's scale and a small celluloid triangle being used. The method is an application of the geometric method of dividing a line into any number of equal parts. The 7.34-in. mark on the scale is pivoted on a, whose elevation is 873.4 ft; the corner of the triangle is placed at the 9.62-in. mark; and both the scale and the triangle are turned until the edge of the triangle passes through b, whose elevation is 896.2 ft. The scale is then held in place while the corner of the triangle is moved successively to the 9.50-, 9.00-, 8.50-, 8.00-, and 7.50-in. points on the scale. Where the edge of the triangle crosses the line ab in the various positions are the corresponding contour points.

This method is very rapid and accurate and entirely eliminates mathematical computations. Any edge of the triangular scale can be used, provided the length on the scale corresponding to the difference in elevation between the two plotted points is shorter than the length of the straight line between the points on the map. When the difference in elevation is considerable and the map distance is short, it may be necessary to let one division on the scale represent several feet in elevation. Thus the smallest division on the scale may correspond to a difference in elevation of 10 ft, instead of 1 ft as in Fig. 15-4. By changing the value of a division, some side of the scale can always be used, regardless of the difference in elevation of the length of the line on the map.

Instead of the triangle and scale, a piece of tracing cloth, on which equally spaced horizontal lines have been ruled, can be used in exactly the same manner. The tracing cloth is turned until lines corresponding to the given elevations pass through the two points on the map. The contour points are then pricked through the tracing cloth to the map beneath.

In most cases the positions of the contour crossings can be estimated with sufficient precision, and exact interpolation is unnecessary. When the positions of the contour lines have been plotted, the contour lines are sketched in freehand by joining points at the same elevation.

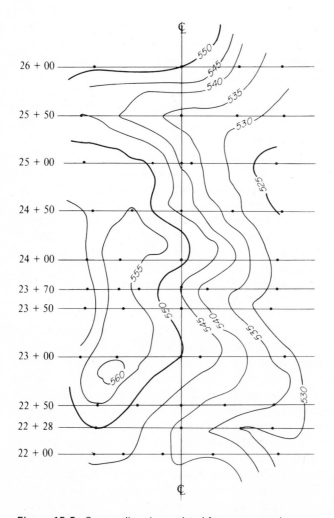

Figure 15-5. Contour lines interpolated from cross-section notes.

The cross-section notes shown in Table 15-1 are used to plot the contour lines shown in Fig. 15-5. The center line is first laid out in the proper position on the map sheet, usually by the coordinates of its points of intersection. The center-line tangent is then stationed, and cross-section lines are drawn through the station marks. Distances are laid out to the points where elevations were determined by reference to the cross-section notes. These points are represented by small dots in the illustration. Elevations are entered at the points. The positions of the contour-line crossings are then determined; and, finally, the contour lines are sketched in and numbered. Notice that an extra cross section was needed at station 22 + 28 in order to pick up the stream bed, and at station 23 + 70 in order to locate the swale in the side of the hill.

15-8. CONTOUR LOCATION WITH HAND LEVEL When the hand level and tape are used, a 5-ft stick is generally used as a support for the hand level. As the rod viewed through the level is not magnified, the length of sight is limited to the visibility of rod readings with the naked eye. A rod with coarse graduations, such as a stadia rod, or a rod equipped with a target is commonly used.

The notes shown in Fig. 15-6 are for the topography between stations 105 and 109. The contours for a 5-ft contour interval are located directly on the ground. On the left-hand page of the notebook are entered the station numbers and the elevations of the ground surface at the stations, as furnished by the profile leveling party. On the right-hand page are sketches showing the locations of the contours to scale. The heavy line at the center of the page is red in the notebook and represents the center line as staked on the ground. The small spaces are $\frac{1}{4}$ in. square and the scale used for the sketch is 1 in. = 100 ft. The notes begin at the bottom of the page so that when the topographer faces in the direction in which the station numbers increase, objects on either side of the line can be sketched in their natural positions with reference to the center line.

In Fig. 15-7 is represented the cross section at station 105. To take measurements at this station, the leveler first holds the 5-ft stick, with the hand level resting on it, on the ground at the station, which is marked A in the figure. Since the elevation of the ground at A is 958.7 ft, the hand level is at an elevation of 958.7 + 5 = 963.7, and to locate contour 955 the reading on the rod must be 963.7 − 955 = 8.7 ft. The rodman moves the leveling rod along the downhill slope on a line at right angles to the center line until the line of sight of the hand level cuts the 8.7-ft mark on the rod. The rod is then on the 955-ft contour, or at the point marked B in the figure. This point is located by measuring the distance from A, which is found to be 138 ft. The topographer plots the 955-ft contour on the cross-ruled, right-hand page of the notebook, as shown in Fig. 15-6, and marks the distance from the center line to the contour.

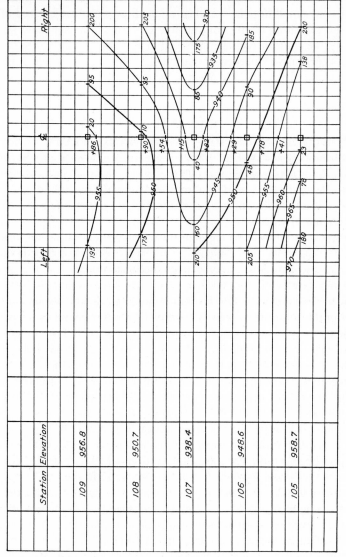

Figure 15-6. Topography with hand level.

604

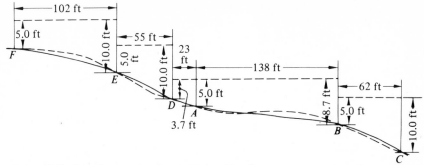

Figure 15-7. Locating contour lines with hand level.

The leveler now moves to point B, Fig. 15-7, and holds the hand level on the 5-ft stick at this point. The elevation of the level is $955 + 5 = 960$ ft, and to locate the 950-ft contour the rodman moves down the slope until a point is found where the rod reading is $960 - 950 = 10$ ft. This point is marked C in Fig. 15-7. The measured distance from C to C is 62 ft, and the total distance from the center line to C is 200 ft. The topographer plots this point, marking its distance from the center line. If the cross sections are to be carried about 200 ft on either side of the line, no further points are necessary on this side.

The contour points on the left-hand side of the station are next determined. The leveler again holds the hand level on the 5-ft stick at the station. Since the elevation of the hand level is 963.7 ft, the rodman moves up the slope at right angles to the center line until the point D is found where the rod reading is 3.7 ft, and thus locates the 960-ft contour. The distance to D from A is found to be 23 ft. This point is plotted on the left-hand side of the center line on the cross-ruled page of the notebook, as shown in Fig. 15-6. Since the ground slopes up from the center line on this side, the leveler goes ahead with the hand level and the rodman holds the leveling rod at point D, Fig. 15-7. The leveler now moves up the slope to a point where, with the hand level held on the 5-ft stick, he makes a reading on the leveling rod of 10 ft. This point, which is marked E, is on the 965-ft contour. The distance from D to E measures 55 ft, and the topographer plots the point, at a distance of $55 + 23 = 78$ ft from the survey line, on the cross-ruled page of the notebook. In like manner the 970-ft contour is located and its location is plotted. When the ground surface is unobstructed by vegetation, points on the uphill side can be located by the topographer standing at the contour just below the one required. The next higher contour will be wherever his line of sight with the hand level on the 5-ft stick strikes the ground.

This procedure is repeated at every station along the line. Points where the contours cross the survey line are also located by plus distances from the preceding stations. In case the ground along a portion of the cross section is so flat that the 5-ft contours are too far apart for a sight, the elevations of points at regular intervals, such as every 50 ft, should be determined, until the next contour or the required width of the section is reached.

The contour lines are drawn through the points of equal elevation, as shown in Fig. 15-6. As soon as a point is plotted in the notebook, the contour line on which it lies is started and the elevation of the contour is marked on the line. When a point at a succeeding station is located in the sketch, the contour having the same elevation is immediately extended to pass through that point, the contour lines being adjusted and smoothed out as the plotting progresses. With the ground before him, an experienced topographer can usually draw the contours quite readily in this manner. When there is any doubt concerning the path of a contour, it may be desirable to plot a few points before the contour lines are drawn. In this case the elevations of the plotted points, as well as their distances from the traverse line, should be recorded.

15-9. TRACE-CONTOUR METHOD The trace-contour method of locating contour lines is most effectively performed by using the plane table and alidade, although the transit-stadia method can be employed. Control is provided by a transit-tape, transit-EDM, transit-stadia, or plane-table traverse. The traverse is computed, adjusted, and plotted on the plane-table sheet. The elevations of the control points are entered directly on the sheet.

In Fig. 15-8 the plane table is set up at station S, the plotted position of which is s, and is oriented by backsighting on station R. The elevation of station S is 338.1 ft. The rod is held alongside the alidade, and the alidade is found to be 4.1 ft above the station. The HI is therefore 342.2 ft. In order to locate the position of the 336-ft contour, the rodman backs off until the topographer reads $342.2 - 336 = 6.2$ ft on the rod with the telescope leveled. At this point the foot of the rod is on the 336-ft contour line. The topographer reads the interval, draws a ray from the plotted position of station S in the direction of the rod, and scales the distance along this ray to plot the point. The rodman walks along the contour, and the topographer reads intervals to successive points along the contour line.

When the 336-ft contour line is traced out as far as practicable from this setup, the topographer locates the 334-ft contour line by obtaining a rod reading of 8.2 ft with the telescope level. All points on the same contour are joined by sketching on the sheet. When the ground is beyond the limit of the rod, the topographer runs a short plane-table traverse to a new setup, from which additional contours are located. This method of locating contour lines on a topographic map is the most accurate and also the costliest and most time consuming. However, for certain purposes it is the only suitable method to give the required accuracy.

The high-water line of a proposed reservoir may be obtained by a combination of plane-table traversing and trace contouring. The desired contour is at the elevation of the top of the spillway of the reservoir dam. The topographer, in effect, traverses along the water line and determines the position of the high-water line as he proceeds.

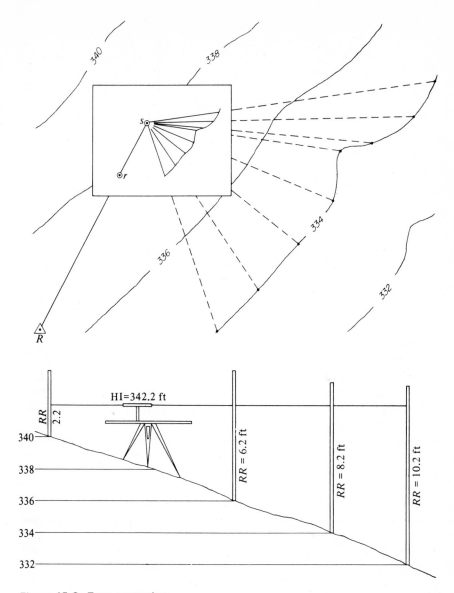

Figure 15-8. Trace contouring.

When the theodolite is used for the trace-contour method, the instrumentman sets up on a control point and determines the HI by using the leveling rod. He sets the azimuth of the backsight line on the horizontal circle and backsights on another control station by using the lower motion. This orients the horizontal circle. All subsequent pointing is done by using the upper motion. With the telescope leveled, he directs the rodman to a contour line by getting the proper reading of the middle cross hair, reads the interval and azimuth to the point, and records these data along with the rod reading.

The sketching may be done in the field or in the office. A 360° protractor is used to plot azimuths. The center of the protractor is set over the point representing the occupied point, and the protractor is rotated so that the reading at the backsight line coincides with its azimuth. This orients the protractor, and any other azimuth may be plotted opposite the corresponding graduation.

Short stadia traverses must be run off the main traverse, as is the case when the plane table is used. These traverses are plotted with respect to the main traverse by using the protractor and scale.

All planimetric features such as fences, roads, streams, and buildings are located by the methods discussed in Sections 14-9 and 14-18.

15-10. GRID METHOD The grid method of obtaining topography may be used in areas of limited extent where the topography is fairly regular. A level is usually used for determining elevations of the grid points, although a theodolite can be used by bringing the telescope horizontal for each sighting.

If the boundary of the area to be mapped has not been previously surveyed, the first step is to run a traverse around the area and establish the corners. Then, in order to determine the topography, the area is usually divided, as far as possible, into squares or rectangles of uniform size. The dimensions of these divisions depend on the required accuracy and the regularity of the topography but are usually between 25 and 100 ft. The form chosen for the divisions will depend somewhat on the shape of the area. The size of the divisions should be such that, for the most part, the ground slopes can be considered uniform between the grid points at the corners of the divisions. The grid points either are defined by stakes or are located by ranging out and measuring from stakes already set.

Figure 15-9 represents a tract of land of which a topographic map is to be prepared. It is assumed that a traverse survey locating the boundaries has already been made. The tract is to be divided by means of lines running in two perpendicular directions and spaced 100 ft apart. When a tract is divided in this manner, it is customary to designate by letters the lines that extend in one direction, and by figures the lines at right angles to that direction. The point of intersection of any two lines is then designated by the letter and figure of the respective intersecting lines. The intersections of the dividing lines with the exterior boundary lines on the far sides of the tract are designated by the proper letters or numbers affected with an accent or a subscript, as A', B', $2'$, $3'$.

Different methods may be followed in laying out the tract. Any method is satisfactory if it accurately defines the positions of the points of intersection so that they can be readily located when the levels are being taken. For this purpose it is not usually necessary to mark all points of intersection, but

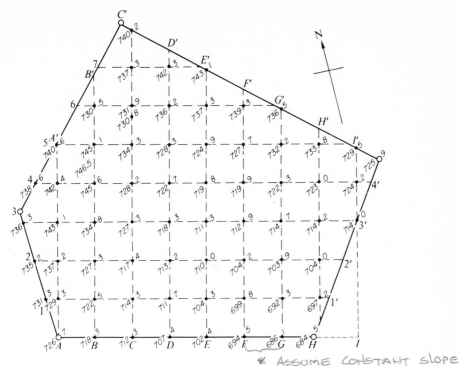

Figure 15-9. Topography by grid method.

※ ASSUME CONSTANT SLOPE
BETWEEN POINTS

enough points should be marked by stakes to permit the remaining points to be located easily and quickly by merely ranging them in from the points that are marked. The rougher and more irregular the surface of the tract, the more stakes must be set. Also, a tract of irregular form usually requires a comparatively greater number of stakes than a tract of rectangular form.

After the required stakes have been set, levels are taken over the tract in order to determine the elevations at all points of intersection and also at any intermediate points where the slope changes abruptly. Such an intermediate point is generally located in a direct line between two intersections by its distance from the intersection having the lower letter or number. This distance is measured with a tape, approximated by pacing, or merely estimated by the eye, according to the conditions and to the degree of accuracy required, and is recorded as a plus. Thus on line CC' there is a low point 80 ft beyond stake $C5$ and its elevation is 730.8; this point would be designated by $C5 + 80$. The high point whose elevation is 746.5, and which is situated between lines 4 and 5 and between lines A and B, would be designated in the notes as $A + 80, 4 + 45$. The levels should be taken in the order that is most advantageous for the nature of the ground. The object is to take rod readings at each of the intersections and at other points with as few settings of the level as possible. To be sure that rod readings are taken at all the intersections, those taken from each setting are checked off on the sketch, or sketches, in which they are all shown.

An EDM mounted on a theodolite or a total-station instrument can be used in place of the level. A point that commands a view of a substantial portion of the area to be mapped is located with reference to the control that is used to establish the grid intersections. Its elevation is also determined. After the grid has been staked out, the point is occupied by the EDM instrument, and the reflector is held at each grid intersection. The slope distance to the grid point is measured along with the vertical angle from which the elevation of the grid point can be determined. In order to locate planimetry, the theodolite is oriented in azimuth by backsighting on a control point, and then the upper motion is used to determine the direction to each planimetric feature. The EDM is used to measure the distance to the point.

After the field work has been completed, the control points, tract boundary, and the grid are plotted to the desired scale. The values of the elevations of the grid intersections are then written at the corresponding map positions of the intersections. The positions of the contour lines are located and sketched by interpolation between the grid intersections. Finally, the planimetric features are plotted using a 360° protractor as discussed in Section 15-9.

15-11. CONTROLLING-POINT METHOD The compilation of a topographic map by determining the positions and elevations of carefully selected controlling points is applicable to nearly every condition encountered in mapping. It is the method used most extensively in mapping a large area to a relatively small scale because of the economy realized. The method can be applied, instead of the cross-section method, to the mapping of a strip of terrain for route-location studies.

The accuracy of the map, the speed of progress, and the faithful delineation of the true shapes of the contour lines all depend on the experience and judgment of the topographer. The method is the most difficult to master, but it is also the most valuable because of its universal application. Although the accuracy of the map depends on the accuracy of the technical operations of making the observations on the points, the largest contributing factors in the success of mapping by this method are the topographer's knowledge of land shapes, slopes, and stream gradients, his facility for making maximum use of the surveying equipment, and his ability to decide where to select points so that he takes neither too many nor too few observations.

The plane table with telescopic alidade, because of its versatility and because it provides its own drawing board, is the most desirable equipment for this method of compilation. The theodolite can be used to good advantage, although its use requires that a draftsman having many qualifications of a topographer accompany the field party in order to obtain accuracy, completeness, and true expression of land forms in the map. Only the use of the plane table and alidade will be considered here.

Control. When a given area is to be mapped, the horizontal control may be established by making a simple transit-tape, transit-EDM, or transit-stadia traverse, which is computed, adjusted, and plotted on the plane-table sheet. The traverse measurements may be performed by using the plane table itself, either prior to or concurrently with the mapping. Unless the traverse is relatively short, it should be executed and adjusted graphically before compilation of the topography begins. The control may consist of a network of interconnected traverses made with transit and tape, transit and EDM, transit and stadia, or the plane table. The major lines in the traverse should be adjusted first, and the adjustment of the shorter cross lines should follow. On a very extensive survey the primary control may be either a simple or a very elaborate triangulation system, and additional control may be provided by traverses connecting the triangulation stations.

Vertical control is established by direct differential leveling, stadia leveling, and plane-table leveling. Benchmarks are established by direct leveling, supplemented by less-precise stadia leveling or plane-table leveling.

The accuracy that must be obtained in the basic horizontal and vertical control will depend on the scale to which the map is to be compiled, the contour interval, and the required accuracy of the topography. In general, the horizontal positions of basic control points should be located to within 1/200 in. on the final map. Thus if the final map scale is 1 in. = 800 ft, the horizontal control should be accurate to within 4 ft. The basic vertical control should be established to within one-tenth of the contour interval.

Locating Contour Lines. With the control plotted on the plane-table sheet, the topographer sets the plane table over a horizontal control point and orients the board by backsighting on another plotted control point; or if the mapping is to a small scale, he may orient the board by means of the compass needle as described in Section 14-17. He determines the HI either by backsighting on a vertical control point or if the elevation of the point over which the plane table is set is known, by measuring the distance from the ground up to the alidade. If the contour interval is 5 ft or greater, the topographer estimates this distance by his experience. It is usually about 4 ft.

Points are selected along ridges, along draws, streams, and drainage channels, at the tops and bottoms of lines with constant grade, and at points between which the topographer can estimate the crossing of the contour lines. As far as the topographer is concerned, drainage has more influence on land form than any other feature. Consequently, the drainage lines should be located fairly accurately. The positions of the contour lines as they cross the drainage lines are obtained by interpolation and are sketched in. The elevations of the points at the changes in slope will allow the topographer to interpolate for the contour crossings. The positions and elevations

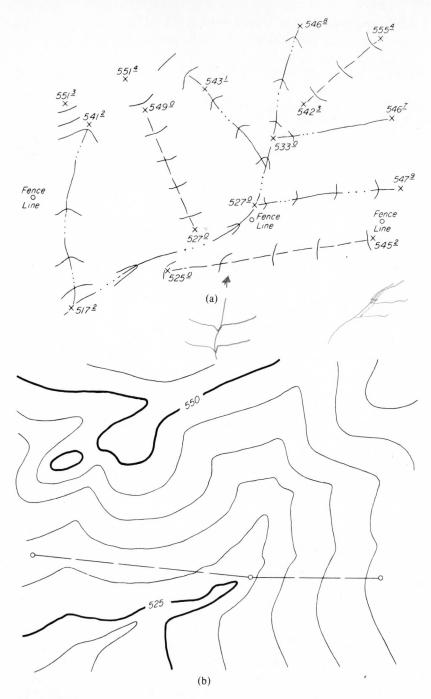

Figure 15-10. Controlling-point method of locating contour lines.

of the points located by the plane table are determined by the method discussed in Section 14-18.

In Fig. 15-10(a) the points marked with a cross (×) are controlling points whose plane-table positions have been determined. These points control the positions and gradients of the drainage lines and the tops and bottoms of lines with uniform slopes. From the positions and elevations of these points, the topographer sketches in the drainage lines and the lines along which contour crossings are to be interpolated. He then sketches in the contour crossings giving them proper forms by analyzing the shapes of the drainage channels and the configuration of the land. Three points on a fence line have been located as shown. The map after the contour sketching has been completed is shown in Fig. 15-10(b).

When the location of points in the area around a plane-table setup has been completed, the topographer moves to another control point. If the basic control is not sufficiently dense to allow all topography to be completed from the control points themselves, additional points must be located for plane-table setups. The supplementary setups may be established by traversing between control points or by plane-table intersection. These methods were discussed in Chapter 14.

PROBLEMS

15-1. A 100-ft grid is located over an area that measures 1500 ft in the east-west direction and 1500 ft in the north-south direction. The southwesterly corner is designated *A*-1 and the southeasterly corner is designated *P*-1. The northwesterly corner is designated *A*-16 and the northeasterly corner is designated *P*-16. Elevations to the nearest foot are obtained at the grid points by stadia leveling. These are tabulated as follows:

16	393	390	388	391	392	393	391	386	383	379	370	360	350	328	320	309
15	380	374	373	377	384	390	387	380	377	377	374	370	363	358	350	342
14	379	367	357	360	376	382	379	374	368	368	370	370	365	362	357	351
13	378	368	359	350	362	370	370	368	361	356	361	364	359	353	350	352
12	378	370	359	340	340	352	351	360	350	337	346	361	350	339	340	347
11	380	367	351	338	334	330	327	320	330	322	334	344	337	323	320	336
10	376	365	359	347	340	333	325	315	309	314	325	332	331	320	310	318
9	375	372	366	360	354	350	340	322	302	300	307	320	325	320	308	307
8	372	372	370	368	365	362	361	339	312	294	283	290	310	317	306	298
7	366	365	365	366	364	362	361	347	319	300	285	276	290	307	300	289
6	363	353	350	360	362	359	350	347	328	303	288	270	273	290	287	279
5	361	340	335	343	352	350	338	334	328	307	290	273	259	264	270	269
4	360	344	329	328	337	339	325	321	321	310	292	274	260	250	253	258
3	360	343	320	317	318	318	313	312	312	307	293	271	257	247	239	242
2	358	343	323	316	310	308	303	302	303	300	283	264	251	241	234	236
1	353	340	326	318	310	303	297	292	291	282	268	255	241	233	226	230
	A	*B*	*C*	*D*	*E*	*F*	*G*	*H*	*I*	*J*	*K*	*L*	*M*	*N*	*O*	*P*

Plot a grid to a scale of 1 in. = 200 ft and draw in 10-ft contour lines.

15-2. The cross-section notes on page 615 were taken for the purpose of plotting a strip of topography for road location. The positions of stream crossings are marked (*). Plot the tangent between stations 18 and 27 to a scale of 1 in. = 100 ft, and sketch in each 2-ft contour line.

15-3. The cross-section notes on page 616 were taken for the purpose of plotting contour lines of a strip of topography. Plot the tangent from station 0 to station 3, using a scale of 30 ft to the inch. Plot the notes on a sheet of $8\frac{1}{2} \times 11$-in. paper, and sketch the 5-ft contour lines. Show the location of the stream (*).

15-4. The elevations for the gridded area shown in Fig. 15-11 are as follows:

F	690	709	726	732	735	740	743	747
E	696	715	729	736	740	737	739	740
D	702	723	733	742	737	737	745	737
C	707	726	732	732	729	740	740	730
B	709	725	723	722	730	738	731	727
A	711	721	715	715	732	730	719	722
	1	2	3	4	5	6	7	8

$$P = 741 \qquad Q = 751$$

Construct a 1-in. grid on a sheet of $8\frac{1}{2} \times 11$-in. paper and sketch in the 5-ft contour lines. Label and broaden every fifth contour line.

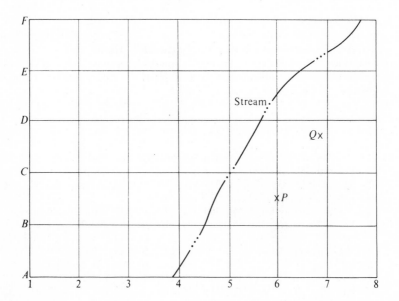

Figure 15-11

15-5. The controlling points shown in Fig. 15-12 on page 615 were obtained by plane table and alidade. Trace the figure and sketch the 10-ft contour lines. Number and broaden every fifth contour line including the 400-ft line.

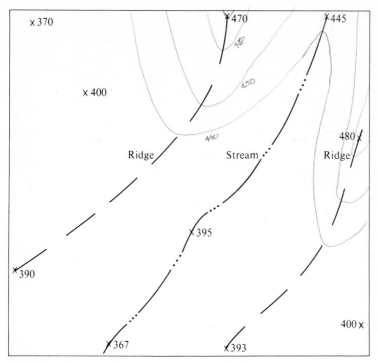

Figure 15-12

Cross-Section Notes

STATION	LEFT			CENTER LINE		RIGHT	
27 + 00	$\dfrac{473.1}{200}$	$\dfrac{461.4}{107}$	$\dfrac{450.8}{53}$	$\dfrac{450.4}{0}$	$\dfrac{439.4}{200}$		
26 + 50	$\dfrac{458.0}{200}$	$\dfrac{449.8}{97}$	$\dfrac{429.3}{17}$	$\dfrac{427.7}{0}$	$\dfrac{425.9}{116}$	$\dfrac{421.3^*}{200}$	
26 + 00		$\dfrac{445.2}{200}$	$\dfrac{428.2}{65}$	$\dfrac{425.7}{0}$	$\dfrac{425.3^*}{8}$	$\dfrac{423.6}{94}$	$\dfrac{421.9}{200}$
25 + 93				$\dfrac{425.5^*}{0}$			
25 + 50	$\dfrac{430.8}{176}$	$\dfrac{429.0}{140}$	$\dfrac{428.6}{0}$	$\dfrac{424.3}{154}$	$\dfrac{422.0}{200}$		
25 + 00	$\dfrac{436.1}{200}$	$\dfrac{430.8}{91}$	$\dfrac{431.0}{0}$	$\dfrac{430.3}{65}$	$\dfrac{423.2}{200}$		

Cross-Section Notes—*continued*

STATION	LEFT			CENTER LINE	RIGHT		
24 + 90				431.6/0	422.5*/200		
24 + 50			439.3/200	432.4/0	425.0*/136	424.8/200	
24 + 00		443.2/200	435.8/93	432.1/0	427.7*/94	428.1/115	425.8/200
23 + 76		444.9/200	433.8/73	431.4*/0	426.9/200		
23 + 15		438.4*/182	438.0/150	437.5/0	437.5/80	431.2/200	
23 + 00		440.0/200	438.1/87	439.2/0	438.8/113	432.1/200	
22 + 00		448.9/200	448.2/164	450.0/0	439.8/139	433.7/200	
21 + 00	459.9/200	460.0/153	457.2/88	446.8/0	433.4/102	424.6/200	
20 + 00			467.0/200	438.6/0	420.7/143	416.1/200	
19 + 00			479.2/200	443.5/0	422.4/111	414.0/200	
18 + 00	499.1/200	475.4/118	461.2/63	447.0/0	433.2/80	419.2/200	

Cross-Section Notes

STATION	LEFT					CENTER LINE	RIGHT				
3 + 00	405/50	400/43	395/32	390/17	385/3	384/0	380/18	376*/35			
2 + 50		400/46	395/35	390/25	385/9	382/0	380*/8	385/39	390/50		
2 + 00	400/47	395/35	390/27	385/20	384*/10	385/0	390/5	395/11	400/20	400/33	395/50

Cross-Section Notes—*Continued*

STATION				LEFT				CENTER LINE			RIGHT			
1 + 89								$\frac{390}{0}$						
1 + 78								$\frac{395}{0}$						
1 + 69								$\frac{400}{0}$						
1 + 62								$\frac{405}{0}$						
1 + 54								$\frac{410}{0}$						
1 + 50	$\frac{400}{50}$	$\frac{395}{42}$	$\frac{390^*}{32}$	$\frac{395}{19}$	$\frac{400}{13}$	$\frac{405}{9}$	$\frac{410}{4}$	$\frac{412}{0}$	$\frac{415}{6}$	$\frac{415}{12}$	$\frac{410}{18}$	$\frac{405}{27}$	$\frac{400}{33}$	$\frac{395}{50}$
1 + 42								$\frac{415}{0}$						
1 + 25								$\frac{417}{0}$						
1 + 00	$\frac{400}{50}$	$\frac{405}{44}$	$\frac{410}{36}$	$\frac{415}{24}$				$\frac{416}{0}$	$\frac{415}{4}$	$\frac{410}{10}$	$\frac{405}{21}$	$\frac{400}{30}$	$\frac{395}{48}$	
0 + 95								$\frac{415}{0}$						
0 + 84								$\frac{410}{0}$						
0 + 68								$\frac{405}{0}$						
0 + 50	$\frac{415}{50}$	$\frac{410}{33}$	$\frac{405}{14}$					$\frac{403}{0}$	$\frac{400}{11}$	$\frac{395}{27}$	$\frac{390}{50}$			
0 + 32								$\frac{400}{0}$						
0 + 00	$\frac{415}{52}$	$\frac{410}{44}$	$\frac{405}{33}$	$\frac{400}{14}$				$\frac{397}{0}$	$\frac{395}{7}$	$\frac{390}{28}$	$\frac{385}{48}$			

(15-6) Reduce the following transit-stadia notes and plot the positions of the points to a scale of 1 in. = 20 ft. Sketch 1-ft contour lines. Assume $K = 100$ and $C = 0$. Line AB is due north.

		TRANSIT AT B, ELEVATION $= 459.5$ ft BS $0° 00'$ ON A				
POINT	INTERVAL	HORIZONTAL DISTANCE	AZIMUTH	VERTICAL ANGLE	DE	ELEVATION
1	0.56		0° 00′	−1° 20′		
2	0.99		1° 00′	−0° 28′		
3	0.25		10° 15′	+4° 36′		
4	0.85		12° 15′	+1° 58′		
5	1.06		16° 45′	−0° 03′		
6	0.58		17° 45′	−2° 22′		
7	0.76		30° 00′	−1° 49′		
8	1.08		40° 45′	−2° 14′		
9	1.43		45° 00′	−2° 48′		
10	0.83		50° 00′	−3° 15′		
11	0.59		61° 45′	−2° 32′		
12	1.12		61° 45′	−4° 41′		
13	0.86		68° 45′	−4° 05′		
14	1.00		68° 45′	−4° 40′		
15	1.04		75° 15′	−6° 07′		
16	0.91		77° 00′	−6° 21′		
17	0.45		80° 30′	−3° 04′		
18	0.68		83° 30′	−3° 38′		
19	0.80		86° 30′	−6° 12′		
20	1.00		88° 45′	−9° 05′		

BIBLIOGRAPHY

Kellie, A. C. 1979. An evaluation of field techniques for topographic mapping. *Proc. Ann. Meet. Amer. Cong. Surv. & Mapp.* p. 471.

Chapter 16
Photogrammetry

16-1. SCOPE Photogrammetry is the science of making measurements on photographs. Terrestrial photogrammetry applies to the measurement of photographs that are taken from a ground station, the position of which usually is known or can be readily determined. Aerial photogrammetry applies to the measurement of photographs taken from the air. As the science of aerial photogrammetry has developed, it has come to include all operations, processes, and products involving the use of aerial photographs. Among these are included the measurement of horizontal distances, the determination of elevations, the compilation of planimetric and topographic maps, the preparation of mosaics and orthophotos, and the interpretation and analysis of aerial photographs for geological, agricultural, and engineering investigation, and for evaluating timber stands.

This chapter discusses the elementary principles of aerial photogrammetry as they apply to the work of the survey engineer. Some of the operations to be discussed are only approximations to more rigorous operations involved in photogrammetry, but which are most helpful to the surveyor. These operations involve the use of paper print aerial photographs which can be used for determining approximate distances and elevations, for planning a survey, and for the construction of mosaics.

More rigorous operations are involved in the compilation of very accurate topographic maps, in the production of orthophotos, and in the digitizing of points on the terrain to obtain digital terrain models. These operations require dimensionally stable material such as glass plates on which the photographs are printed, together with rather complex photogrammetric instrumentation.

The entire field of photogrammetry cannot be incorporated into a single chapter in a textbook. This chapter is intended as an introduction to photogrammetry in order that the student may gain some appreciation of the potential applications of the science to surveying. References at the end of the chapter should be studied by the student who is interested in pursuing this most fascinating field in depth.

16-2. AERIAL CAMERA The aerial camera is comparable to a surveying instrument in that it gathers data in the form of light rays with certain directions and records the data on a photographic negative. Figure 16-1 shows a schematic diagram of an aerial camera with its component parts. The optical axis usually is supposed to be essentially vertical when a photograph is taken.

The *lens assembly*, which includes the shutter and the diaphragm, forms the photographic image with the proper amount of light admitted. The lens itself is composed of several elements, as shown in Fig. 16-2. It must be highly corrected to satisfy the requirements of a rather wide angular coverage, a high resolution, and a very minor amount of lens distortion. Because of atmospheric haze which contains an overabundance of blue light, the lens is invariable fitted with a yellow, orange, or red filter to absorb some of the

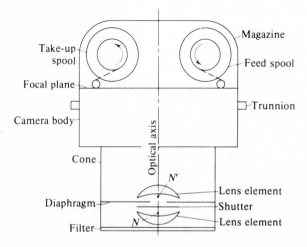

Figure 16-1. Component parts of aerial camera.

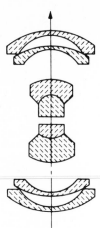

Figure 16-2. Wild Aviogon aerial camera lens.

blue light. The *cone* and the *camera body* hold the lens assembly in its proper position with respect to the *focal plane*.

The camera body houses the drive mechanism for the shutter assembly and the *magazine*. The upper surface of the camera body, in general, defines the focal plane of the camera. This surface contains four *fiducial marks*, located either with one in the middle of each side of the focal plane opening as seen in the aerial photograph of Fig. 16-4, or else in each corner as shown in Fig. 16-10. Aerial cameras used for mapping have focal-plane openings with dimensions of 9 × 9 in., (230 × 230 mm) and produce aerial photographs of corresponding sizes. The purpose of the fiducial marks is to define the point at which the optical axis of the lens intersects the focal plane. This point is called the *principal point* of the photograph.

The magazine is a light-tight container for the film. It contains a feed spool and a take-up spool, the mechanism for advancing the film after each exposure, and a device for holding the film flat in the focal plane at the instant the exposure is made.

The two points on the optical axis marked N and N' in Fig. 16-1 are the *front* and *rear nodal points* of the lens system. On striking the front nodal point any ray of light is so refracted by the lens that it appears to emerge from the rear nodal point parallel with its original direction.

The focal length of the lens is the distance between the rear nodal point of the lens and the focal plane. It is designated by the symbol f. The focal length is fixed for a given camera and its value is precisely determined by calibration. The nominal values of focal lengths used in mapping and in the construction of mosaics are generally 3.5, 6, 8.25, and 12 in. (88, 152, 210, and 305 mm).

The camera is suspended in a camera mount by means of a pair of trunnions on either side of the camera. It is free to rotate in the mount about all

Figure 16-3. Aerial camera installed in mount.

three axes. The mount is secured over an opening in the bottom of the airplane. Figure 16-3 shows an aerial camera installed in the mount. Some of the features just described can be seen in the picture.

16-3. TYPES OF AERIAL PHOTOGRAPHS An aerial photograph taken with the optical axis held essentially vertical is called a *vertical photograph*. Because of movement of the airplane at the instant of exposure, virtually all vertical photographs contain a certain amount of tilt. The optical axis may be accidentally inclined as much as 5° from the vertical. Tilt, however, does not present serious difficulties in mapping and introduces very little difficulty in constructing mosaics. In some of the discussions in this chapter a vertical photograph is one that is assumed to be truly vertical unless otherwise stated. A vertical photograph is shown in Fig. 16-4.

An aerial photograph taken with the optical axis purposely tilted by a sizable angle from the vertical is called an *oblique photograph*. A high oblique is a photograph on which the apparent horizon appears. A low oblique is a photograph taken with the optical axis purposely tilted from the vertical, but not enough to include the horizon. Oblique photographs are shown in Fig. 16-5(a) and (b).

Most planimetric and topographic mapping, mosaic construction, and orthophoto production are done by using vertical photographs, although high obliques are sometimes used in the preparation of small-scale planimetric maps and charts because of their greater ground coverage.

Figure 16-4. Vertical photograph showing fiducial marks at the middles of the four edges. Courtesy of Pacific Aerial Surveys, Oakland, California.

Figure 16-5. Oblique photographs. (a) High oblique; (b) low oblique. Courtesy of Pacific Aerial Surveys, Oakland, California.

16-4. PHOTOGRAPHIC SCALE Photographic scale is the ratio between a distance measured on a photograph and the corresponding ground distance. A vertical photograph resembles a planimetric map in that it shows the planimetric and cultural features of a portion of the ground in their relative positions. It is different from a planimetric map, however, in two respects: (1) The vertical photograph does not contain standard map symbols, which are essential to a map; (2) the planimetric map has a uniform scale throughout, whereas the scale of a vertical photograph varies in different portions of the photograph. The photograph is a perspective projection of the ground onto the focal plane of the camera. Consequently, points lying in a plane closer to the camera at the time of exposure will have larger images than those points lying in a plane farther from the camera. The scale will vary across the area of the photograph also because of tilt of the optical axis at the time of exposure.

In Fig. 16-6(a) points A, O, and B all lie at the same elevation. The horizontal distances AO and OB are equal to $A'O'$ and $O'B'$ on a reference datum.

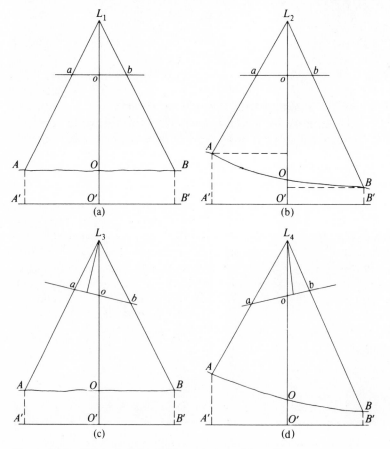

Figure 16-6. Effect of relief and tilt on photographic scale.

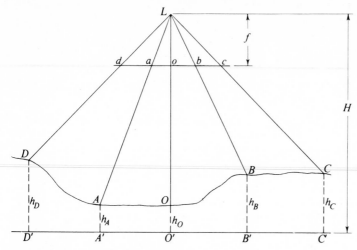

Figure 16-7. Photographic scale.

A truly vertical photograph taken with the camera at L_1 would show the positions of A, O, and B at a, o, and b. The ratio ao/AO equals the ratio ob/OB, and the scale of the photograph is uniform across the photograph.

In Fig. 16-6(b) points A, O, and B are at different elevations, and the horizontal distances between A and O and between O and B are $A'O'$ and $O'B'$, respectively. If a vertical photograph is taken with the camera at L_2, the points A, O, and B would appear at a, o, and b on the photograph. The ratio $ao/A'O'$ does not equal the ratio $ob/O'B'$, and the scale of the photograph is seen to vary from point to point because of the variation in the elevations of the ground points.

In Fig. 16-6(c) points A, O, and B all lie at the same elevation. A tilted photograph taken with the camera at L_3 shows the positions of the ground points at a, o, and b. The ratio ao/AO does not equal the ratio ob/OB, and consequently the scale varies across the photograph. In Fig. 16-6(d) the combined effect of both relief and tilt on the scale of a photograph is shown.

The scale of a vertical photograph at a point, along a line or in an area, can be determined from the relationship between the focal length of the camera, the flying height of the aircraft at the time of exposure, and the elevation of the point, the line, or the area. In Fig. 16-7 points O and A are at the same elevation, and so $AA' = OO'$ or $h_A = h_O$. Points B and C lie at elevation $h_B = h_C$, and D lies at elevation h_D. By similar triangles

$$\frac{ao}{AO} = \frac{Lo}{LO}$$

or

$$\frac{ao}{AO} = \frac{f}{H - h_A} = \frac{f}{H - h_O}$$

Similarly,

$$\frac{bc}{BC} = \frac{f}{H - h_B} = \frac{f}{H - h_C}$$

But the ratios ao/AO and bc/BC are the scales of the photograph along lines ao and bc, respectively. Therefore, in general,

$$S_E = \frac{f}{H - h} \tag{16-1}$$

in which S_E is the scale of a vertical photograph for a given elevation; f is the focal length, either in inches or in millimetres; H is the flying height above the datum, in feet or metres; and h is the elevation of the point, line, or area above the datum, in feet or metres. The focal length given for most modern aerial camera lenses is expressed in millimetres. Although a scale can be expressed in terms of the number of feet per millimetre, this is quite uncommon. When working with photo scales, it is more convenient to convert millimetres to either inches or feet by dividing by 25.4 or 304.8, respectively.

Example 16-1

In Fig. 16-7 the elevation of points O and A is 267 ft, that of points B and C is 524 ft, and that of D is 820 ft. The flying height above sea level is 1769 ft, and the focal length of the camera is 8.23 in. Determine the scale of the photograph along the line ao, along the line bc, and at point d, expressing each scale as a number of feet corresponding to 1 in.

Solution: The value of S_E for line ao is 8.23/(1769 − 267). Reducing the numerator to unity gives a scale of 1 in. = 182.5 ft. The value of S_E for line bc is 8.23/(1769 − 524), and the scale is 1 in. = 151.3 ft.

The value of S_E at point d is 8.23/(1769 − 820), and the scale is 1 in. = 115.3 ft. The corresponding representative fractions expressing the scale are 1/2190, 1/1816, and 1/1384, respectively.

Quite often the *average* scale of a single photograph or of a set of photographs is desired in order to be able to measure distances in any area of the photograph or photographs. If the average scale is known, it can be applied to a scaled distance to give a reasonable value of the corresponding ground length, provided that the relief is not extremely variable. The average scale is given by the relationship

$$S_A = \frac{f}{H - h_{av}} \tag{16-2}$$

in which S_A is the average scale of the photograph, which may be reduced to either an engineer's scale or a representative fraction; f is the focal length, in inches or millimetres; H is the flying height above the datum, usually sea level, in feet or metres; and h_{av} is the average elevation of the area covered by the photography, in feet or metres.

If the photographic scale is to be expressed as a representative fraction, the numerator of Eq. (16-1) or (16-2) must be expressed in feet or metres and then reduced to unity. In Example 16-1 the representative fraction form of the scale at point d is $(8.23/12)\,\text{ft}/(1769 - 820)\,\text{ft}$ or $0.686\,\text{ft}/(1769 - 820)\,\text{ft}$. If the numerator is reduced to unity, the representative fraction is $1/1384$.

The scale of a photograph can be determined by comparing a distance measured on the photograph with the corresponding known ground distance. The ground distance may have been measured directly, or it may be a distance of common knowledge as, for example, the length of a section line (see Chapter 18), a city block, or a stretch of a highway.

The scale of a photograph can be determined by comparing a distance measured on the photograph with the corresponding distance measured on a map of known scale. The photograph scale is then found by the following relationship:

$$\frac{\text{photo scale}}{\text{map scale}} = \frac{\text{photo distance}}{\text{map distance}} \qquad (16\text{-}3)$$

Both the photo scale and the map scale must be in the same units and must be expressed as a fraction. For example, if the distance between two road intersections is 4.34 in. on a photograph and 1.55 in. on a map drawn to a scale of 1 in. = 800 ft,

$$\frac{1\,\text{in.}/X\,\text{ft}}{1\,\text{in.}/800\,\text{ft}} = \frac{4.34\,\text{in.}}{1.55\,\text{in.}}$$

Hence $X = 285$ ft, and the photo scale is 1 in. = 285 ft. Note that the quantity X is the denominator of the photograph scale. Equation (16-3) can be expressed also as

$$D_P = \frac{M}{P} D_M \qquad (16\text{-}4)$$

in which D_P is the denominator of photograph scale; D_M is the denominator of map scale, in the same units as D_P; P is the photograph distance; and M is the map distance, in the same units as P.

Example 16-2

The distance between two points measures 24.62 mm on a map whose scale is 1/24,000. The distance between the same two points appearing on a vertical aerial photograph measures 32.05 mm. What is the scale of the photograph?

Solution: By Eq. (16-4),

$$D_P = \frac{24.62}{32.05} \times 24{,}000 = 18{,}436 \text{ or } 18{,}450 \text{ (approx.)}$$

The photo scale is thus 1/18,450.

Example 16-3

If the focal length of the camera lens of Example 16-2 is 152.4 mm, what is the flying height above the ground in metres?

Solution: Assuming a datum to lie at the average elevation of the area, then by Eq. (16-2),

$$\frac{1}{18{,}450} = \frac{152.4 \text{ mm}}{(H - 0) \text{ m}} = \frac{0.1524 \text{ m}}{(H - 0) \text{ m}}$$

giving $H = 2812$ m above the ground (approx.).

16-5. RELIEF DISPLACEMENT The term *relief displacement* is applied to the displacement of the image of a ground point on a photograph from the position the image would have if the point were on the datum. This displacement is due to the elevation of the ground point above or below the datum. In Fig. 16-8 ground points A and B lie at elevations h_A and h_B above the datum. Their images on a vertical photograph are at points a and b, respectively. The datum positions A' and B' would have images at a' and b' on the photograph. Point o is the principal point of the photograph. It is located by joining opposite fiducial marks at the corners or the edges of the photograph. Because of the elevations of A and B above datum, both a and b have been displaced outward along the radial lines oa and ob, respectively.

The amount of relief displacement for a point is given by

$$d = \frac{rh}{H} \tag{16-5}$$

in which d is the relief displacement ($a'a$ for point A and $b'b$ for point B in Fig. 16-8) expressed in inches or millimetres; r is the radial distance from the principal point to the image of the point in the same units as d; h is the eleva-

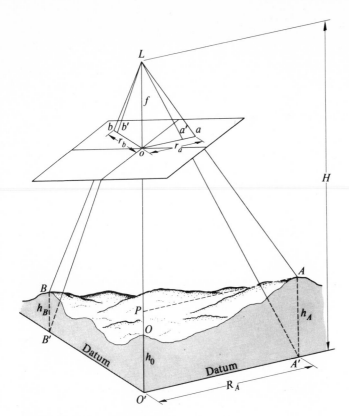

Figure 16-8. Relief displacement.

tion of the point above the datum, in feet or metres; and H is the flying height above the datum in the same units as h.

From Eq. (16-5) it is seen that the relief displacement of a point depends on the position of the point on the photograph. At the principal point, the displacement is zero. It increases outward toward the edges of the photograph. Also, the relief displacement increases as the elevation of the ground point increases, and decreases with an increase in flying height.

Assume that a photograph contains all the corners of a tract of land and that the elevations of these corners are known. Unless all the corners are at the same elevation, the sides of the tract cannot be scaled directly because of variation in scale discussed in Section 16-4. The relief displacement of each point can be computed, provided the flying height above sea level has been determined, by measuring the radial distance to each point and applying Eq. (16-5). The positions of the points are then corrected by the amount of relief displacement and the new positions are therefore at a common datum. In Fig. 16-9 points a, b, c, d, and e are the photographic images of the corners of a tract of land. The elevations of the points and the flying height above sea level are known. The positions of the five points, corrected to a common

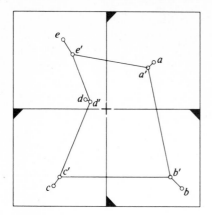

Figure 16-9. Points brought to common datum.

Figure 16-10. Effect of relief displacement on straight lines.

datum, are a', b', c', d', and e'. These are the positions of the five points on a map that is at the datum scale of f/H, in which H is the flying height used to compute the relief displacements. The corrected lines may be scaled; and the lengths of the lines, the angles between the lines, and the area of the tract can be determined to a fair approximation depending on how much the photograph was tilted at the instant of exposure.

If the flying height above the datum for a given photograph is not known, it can be determined with varying accuracy by measuring the length of a ground line, the end points of which lie at about the same elevation, and then scaling the corresponding photograph distance. The scale at the elevation of the line being known, H may be determined by Eq. (16-1). The photographic distance may also be compared with a corresponding map distance to give the scale by Eq. (16-3), and then H may be determined.

Figure 16-10 shows a vertical photograph taken from an altitude of approximately 2000 ft above sea level and covering an area at an average elevation of about 500 ft. The scale is, by Eq. (16-2), about 1/3000 or 1 in. = 250 ft. Lines ab and ac are actually straight fence lines, and the angle at a from b to c is approximately 90° if measured at ground point A from B to C. Because of relief displacement, however, it is obvious that the fence lines are displaced outward from the principal point as the lines go over ridges, and are displaced inward as the lines cross the gulleys. This relief displacement, which is nothing more than a manifestation of perspective in an aerial photograph, is of rather minor consequence in a photograph taken at a great altitude over ground with very little relief. However, if a photograph is to be used as a map substitute, then the effect of relief displacement and scale variation must be recognized, especially where the scale of the photograph is large and there is relatively large terrain relief.

16-6. PHOTOGRAPH OVERLAP When aerial photography is used for mapping, mosaic construction, or orthophotography, flight lines are laid out on a flight map with a spacing that will cause photographs to cover an overlapping strip of ground. This overlap between flight strips amounts to about 25% of the width of the photograph. The actual spacings on the flight map may be determined from the scale at which the photographs are to be made by solving Eq. (16-3). The photographic distance is 75% of the width of the photograph, provided that a 25% overlap between flight strips is to be maintained. This overlap is shown in Fig. 16-11.

Each photograph in the line of flight covers an area that overlaps the area covered by the previous photograph by about 60%. This is illustrated in Figs. 16-11 and 16-12. The large overlap between successive photographs serves three primary purposes. First, it provides coverage of the entire ground area from two viewpoints, such coverage being necessary for stereoscopic viewing and measuring. Second, it allows only the central portion of each photograph to be used in mosaic construction, eliminating to a great extent

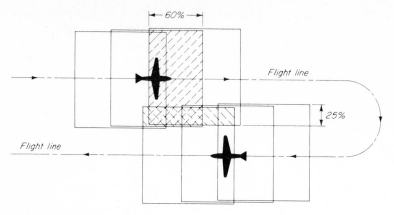

Figure 16-11. Photographic overlap.

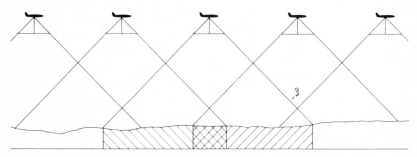

Figure 16-12. Overlap along flight line.

the effect of relief displacement. Third, the small overlap area between alternate photographs allows control to be extended along the strip by photogrammetric methods.

16-7. GROUND CONTROL FOR PHOTOGRAMMETRY In order that aerial photographs may be used for making simple measurements of distances and elevations, for constructing planimetric and topographic maps, and for constructing mosaics and orthophotos, a certain amount of ground control is necessary to fix the scale of the photographs, the map, or the mosaic, and to establish a vertical datum with which to establish contour lines on the map. This control may be obtained simply by measuring the ground distance between two identifiable points that appear on a photograph. Or an elaborate system of triangulation and level nets may be required to control the horizontal and vertical positions of identifiable points that appear on the photographs. In general, triangulation and traverse stations and benchmarks cannot be identified on the photographs unless they have been marked in the field before the pictures were taken. As a result, shorter traverse lines and level lines, usually of a lower order of accuracy than the main system, are run to points that can be identified. These points are called *picture points*, and the network of picture points in a group of photographs is called *photo control*.

The horizontal photo control may be obtained by running a traverse from a fixed point to the picture points, or it may be obtained by a simple triangulation system of the desired order of accuracy. A large amount of photo control work is conducted by a combination of a theodolite and an electronic distance-measuring system. Vertical photo control is obtained by running differential, stadia, or plane-table levels from benchmarks to the points, or, for large contour-interval mapping by using aneroid barometers.

The choice of the field method to be used in establishing photo control depends on the desired accuracy, which in turn depends on the scale of the map or mosaic and the contour interval to be used in compiling the map.

A picture point must be positively identifiable between the photograph and the ground. It must be sharp and well defined as seen on the photograph under magnification. The picture point must fall in the correct position on the photograph. A picture point falling, for example, near the very edge of the photograph may be entirely useless for the purpose for which it is intended. It should be reasonably accessible on the ground so that the expense of the photo-control survey can be kept to a minimum. Finally, the picture point must be well described and documented directly on the photograph used in the field when the photo-control survey is performed.

When a very high degree of accuracy is required in the photogrammetric process, natural features are generally not used as picture points. Instead, the field survey crew sets out panels, referred to as *premarked points* or panel points, just prior to the photography. The most common panel configuration is shown in Fig. 16-13. The size of the inner square is a function of the proposed scale of the photography and the size of the measuring mark in the instrument to be used for measuring the photographs (see Section 16-10).

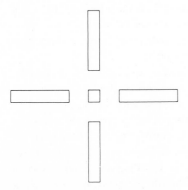

Figure 16-13. Panel configuration.

The outer four legs of the panel are used to find and identify the panel point on the photograph. The control survey is run to the center of the inner square to determine its position and elevation.

16-8. MOSAIC A mosaic is an assembly of a series of overlapping aerial photographs to form one continuous picture of the terrain. It may consist of a single strip of photographs, termed a *strip mosaic*, or it may contain many overlapping strips. When photographic film has been processed, each negative of a flight strip is numbered consecutively, and each flight strip bears a number that is also assigned to each photograph of that strip. These identifying numbers appear when the photographs are printed from the negatives. If a set of aerial photographs are laid down and stapled to a board in consecutive order in such a manner that the identifying numbers show on the finished assemblage, the result is a crude mosaic with very little accuracy but with high utility. It is called an *index mosaic* because it is a visual method of indexing each photograph in a set.

A mosaic that is constructed for its pictorial quality alone is usually not controlled to any extent by photo control. To prepare a mosaic for this purpose, all but the central portion of each photograph is trimmed away, leaving a small amount of overlap, and the edges are brought to a featheredge by sandpapering. The photographs are pasted onto a mounting board, masonite, for example, by some type of adhesive. Gum arabic is a common type of mosaic adhesive, because it is easy to work with and can be cleaned off readily with water. Each photograph is laid on the preceding one and is shifted on the board until the images in the overlap areas of the successive photographs match. It is then squeegeed into place.

A mosaic that is constructed to give both high pictorial quality and good accuracy must be controlled by picture points. The accuracy and density of the control is, of course, dependent on the accuracy desired in the finished mosaic. The control provided for mosaics can be obtained by ground surveys supplemented by graphically located points. These supplementary points are located by a method known as *radial-line plotting* (see References). The control can also be provided in the form of a planimetric map mounted directly on the mounting board to the desired scale of the mosaic. This type of control is used to lay the mosaic shown in Fig. 16-14.

Because of differences of flying heights between successive photographs, and because of displacements of images due to relief and photographic tilt, the photographs of a set must be brought to a common scale, and they must be corrected for tilt displacement and, to a limited extent, for relief displacement. This ratioing to a common scale, while eliminating the effect of tilt at the same time, is performed in an enlarger whose negative holder and easel can be tilted. Such an enlarger is called a *rectifier*. When each negative has been rectified so that it satisfies the positions of the plotted control points, a photograph is made that will be used in the subsequent mosaic construction.

Figure 16-14. Laying mosaic to control.

Photographs for a controlled mosaic are trimmed, sanded, and pasted essentially in the manner discussed previously. After the assembly is completed, the mosaic is photographed to the final desired scale by means of a copy camera, and reproductions of the copy negative are made for use.

Mosaics possess numerous advantages over maps prepared by conventional ground methods. The mosaic can be produced more rapidly, and the cost of duplicating by ground surveys the wealth of detail and completeness found on a mosaic would usually be prohibitive. An objection to the mosaic is that it is not a topographic map, and elevations cannot be obtained from it. In fact, the mosaic is not a map at all, because of inherent displacements of images due to relief and residual tilt effects; but it is an excellent map substitute.

16-9. STEREOSCOPY AND PARALLAX The word *stereoscopy* is defined as the viewing of an object or image in three dimensions and necessarily implies binocular, or two-eyed, vision. The optical axes of a person's eyes will converge when looking at an object, and the angle measured at the object between the two optical axes is called the *parallactic angle*. This angle will increase as the object is brought closer to the eyes, and will decrease as the object recedes from the eyes. Two objects at different distances will be interpreted by the eyes as forming two different parallactic angles, and the difference in the two angles is interpreted as depth.

In Fig. 16-15(a) two marks of the same size and shape are plotted on a sheet of paper separated by a distance of about $2\frac{1}{4}$ in. If each eye concentrates on but one mark, the image of the two marks will appear as one mark at a distance d_1, and the parallactic angle formed at this image is ϕ_1. In Fig. 16-15(b) an additional pair of marks is plotted with a slightly larger separation. The angle ϕ_2 is formed at the single image of this pair, the image appearing to lie at a distance d_2 from the eyes. Thus the four marks on a plane surface viewed stereoscopically will appear as two marks separated in the third dimension of depth by the amount $d_2 - d_1$.

In Fig. 16-16(a) a tower is photographed from two consecutive camera stations L and L' and forms images as shown. Notice the relief displacement of the top of the tower with respect to the bottom. The two photographs lined up side by side with the flight line parallel with the eye base will form a three-dimensional image of the tower, as shown in Fig. 16-16(b). Furthermore, all points in the area of overlap between the two photographs will be

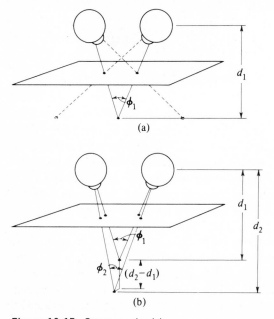

(a)

(b)

Figure 16-15. Stereoscopic vision.

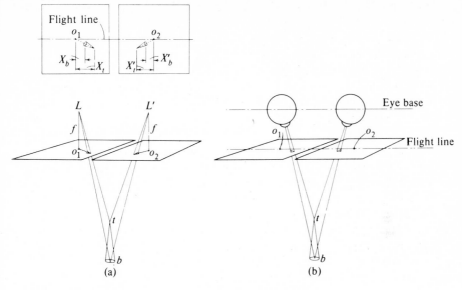

Figure 16-16. Images of tower on pair of aerial photographs.

seen stereoscopically, that is, in three dimensions. Because of the dimensions of the photograph and because of the eyes' resistance to focus closely and at the same time diverge abnormally, a stereoscope is used to form the stereoscopic image. The stereoscope can be of a simple lens type shown in Fig. 16-17(a) or of a mirror type shown in Fig. 16-17(b). When the photographs are laid down, the flight line of one photograph must line up with the same flight line of the other photograph before the stereoscope is placed over the pair. This adjustment is shown in Fig. 16-18.

In Fig. 16-16(a) it is noted that the image of the top of the tower has moved through a total algebraic distance of $x_t - x'_t$ between two successive exposures, and that the image of the bottom of the tower has moved through a total distance of $x_b - x'_b$ between exposures. This movement of an image over the focal plane between exposures is called the *parallax* of the point. Parallax increases as the point lies closer to the camera, that is, as the elevation of the point increases. It is the gradual difference in parallax between successive points as they are photographed from two successive camera stations that gives a gradual difference in the parallactic angles formed between the optical axes of the eyes. Thus a continuous three-dimensional image is formed.

Parallax is a direct indication of elevation and can be measured on a pair of overlapping photographs by means of a *parallax bar*. Such a measuring device is shown in use under a mirror stereoscope in Fig. 16-19. A schematic view of the parallax bar is shown in Fig. 16-20(a). Two measuring marks are located on the bottom surfaces of two pieces of glass which are attached to the bar. The left mark is fixed to the bar while the right mark is moved left

a

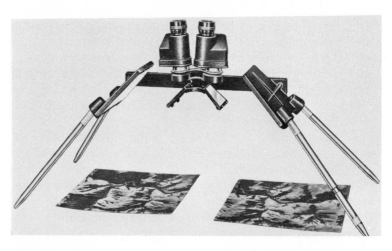

b

Figure 16-17. Stereoscopes. (a) Lens stereoscope; (b) mirror stereoscope.

or right by turning the micrometer screw. A movement to the left is an increase in parallax and to the right is a decrease. If the bar is placed under a mirror stereoscope, the two measuring marks will appear as a single mark (see Fig. 16-15). This is called the floating mark. By turning the micrometer screw one way or the other, this single mark appears to move up or down according to the increase or decrease in the parallax of the marks.

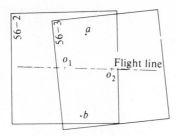

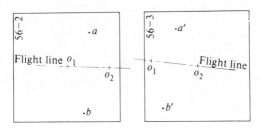

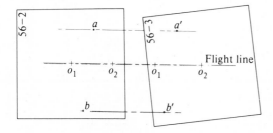

Figure 16-18. Adjusting photographs for stereoscopic viewing.

In Fig. 16-20(b), two photographs are oriented on a large sheet of paper such that their flight lines are carefully aligned as shown in Fig. 16-18. Prior to this alignment, the position of the left principal point o_1 must be transferred to its correct position on the right-hand photograph, and the right principal point o_2 must be transferred to the left-hand photograph. The separation S between any two *conjugate* points such as o_2 and o_2 is set to be equal to the spacing between the measuring marks when the movable mark is in its midposition. The photographs are now taped to the paper. The parallax bar is placed on the photographs as shown, and the mirror stereoscope is then placed over the photographs as shown in Fig. 16-19.

Figure 16-21 shows how the *difference* in parallax between two points can be measured. The parallax bar is moved over the photographs to point a, and the micrometer screw is turned until the floating mark appears to

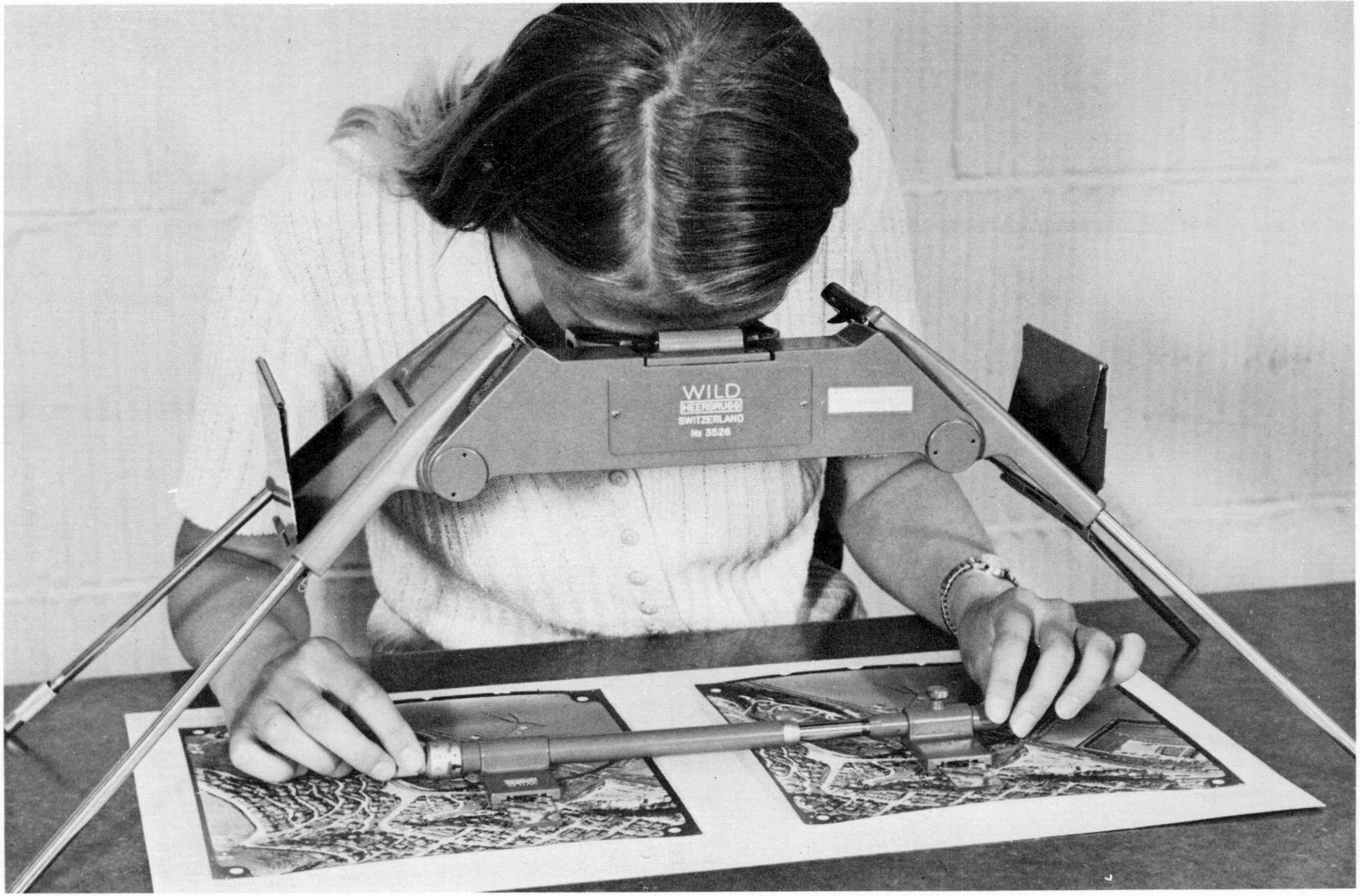

Figure 16-19. Parallax bar used under a mirror stereoscope.

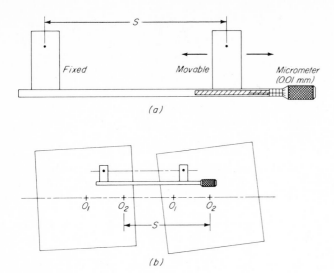

Figure 16-20. Diagram of parallax bar.

touch the stereoscopic image at *a*. The micrometer is read as 12.10 mm. The bar is moved to point *b*, the floating mark is again brought to the surface of the stereoscopic image at *b*. The micrometer is read as 9.65 mm. The difference in parallax, *dp*, from *a* to *b* is thus $9.65 - 12.10 = -2.45$ mm. The difference in parallax between any two points, such as the top and bottom of a tree or building, or between the top and bottom of a hill can be measured in this manner. The distance o_1-o_2 is measured with a scale on each photograph to give an average photograph base *b*. The difference in elevation *dh* between two points is then given by

$$dh = \frac{dp\,H}{dp + b} \tag{16-6}$$

in which *dh* is in feet or metres, *dp* and *b* are in millimetres, and *H* is the flying height *above the ground*, expressed in feet or metres. The value of *H* is obtained from Eq. (16-2) in which $h_{av} = 0$.

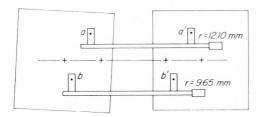

Figure 16-21. Micrometer readings of parallax bar.

Example 16-4

A pair of overlapping vertical photographs were taken with a lens having a 6-in. focal length. The average scale of the photography is 1 : 10,000. A vertical control point, whose elevation is 680 ft, appears in the overlap area. The average distance between principal points is 3.660 in., or 92.96 mm. Parallax-bar measurements give the measured differences of parallax between the known control point and five unknown points. These differences are shown in the accompanying tabulation. Compute the elevations of the five points.

POINT	dp (mm)	$dp + b$ (mm)	dh (ft)	ELEVATION (ft)
1	+4.62	97.58	+237	917
2	−0.88	92.08	− 48	632
3	+1.90	94.86	+100	780
4	+5.76	98.72	+292	972
5	−3.35	89.61	−187	493

Solution: We must assume that the elevation of the known point is nearly equal to the average elevation of the area shown on the photographs. Then, by Eq. (16-2), in which the average elevation is assumed to be zero,

$$S_A = \frac{6/12}{H} = \frac{1}{10,000} \quad \text{or} \quad H = 5,000 \, \text{ft}$$

With b given as 92.96 mm, the differences of elevation are computed by Eq. (16-6), as shown in the tabulation. Finally, these are used with the elevation of the known point to give the values in the last column.

The parallax bar is useful for obtaining rough profiles along any selected line drawn on one of the two photographs used under the stereoscope. Since the photos may be tilted, the vertical accuracy obtained by Eq. (16-6) will not give absolute elevations any better than about 1/300 times the flying height. However differences in elevation between two points located fairly close together are reliable up to about 1/1000 times the flying height.

16-10. STEREOSCOPIC PLOTTING INSTRUMENTS Thus far in this chapter, discussion has been confined to the use of paper prints in which the photographs were assumed to be vertical. These conditions would not be sufficient to obtain measurements necessary for the compilation of accurate maps. In order to take into account the tilts and difference of flying heights of consecutively exposed photographs, and to provide dimensional stability, glass plate photographs are measured in different kinds of stereoscopic plotting instruments. A stereoscopic plotting instrument, or plotter, is used to plot planimetric and topographic maps to a predetermined scale by using photographs that have been taken with a precision aerial camera. The elevations of selected ground points measured in a plotter can be determined with varying degrees of accuracy, the error depending on the precision of the plotter. This ranges from 1/1000 times the flying height to as little as 1/10,000 times the flying height. The error in contour lines drawn by means of the plotter will vary from 1/500 to 1/2500 times the flying height, the amount again depending on the precision of the plotter and also on the type of ground cover existing at the time the photographs were made.

In order to maintain accuracy, the photographs used in a plotting instrument are printed on glass plates. These glass plates are either the same size as the negatives or they are reduced in size by a carefully controlled ratio. For use in some plotters, the glass plates are made by projection through a lens which compensates within tolerance the distortions due to the aerial camera lens. In other plotters the camera-lens distortion is compensated for directly in the projection systems.

Every plotting instrument has four main features: a projection system, a viewing system, a measuring system, and a drawing system. These features will be described with reference to Fig. 16-22, in which shows a plotting instrument of simple design. This is the Multiplex plotter. Each of the two projectors contains a projector lens with a principal distance of 30 mm, a light source, and a condenser lens which concentrates the light so that it passes through the projector lens. A small reproduction on glass, called a *diapositive* and measuring 64 × 64 mm, is placed in each of the two projectors, corresponding to two overlapping photographs. A blue-green filter is placed below the light source in one projector, and a red filter is put in the other projector. When the lamps of the projectors are turned on, two cones of rays are projected downward toward the map table, each cone of rays having a different color.

Because of tilt in the photographs and because of unequal flying heights, a ray of light through a ground point and coming from one projector may fail to intersect the ray through the same point coming from the other projector; and, similarly, for other pairs of rays through other points. Each projector is free to rotate about each of the three axes and also to move along the three axes. By a systematic manipulation of these motions of each projector, the discrepancies existing between the rays are eliminated, and the two overlapping cones of rays form a stereoscopic image in the area above the map table. This process is referred to as *relative orientation*. It sets into the two projectors the proper amount of relative tilts that existed in the aerial camera at the time the two aerial photographs were taken along the flight line.

The stereoscopic image is viewed by the operator through a pair of spectacles, one lens of which is blue-green and the other red. In this way the cone

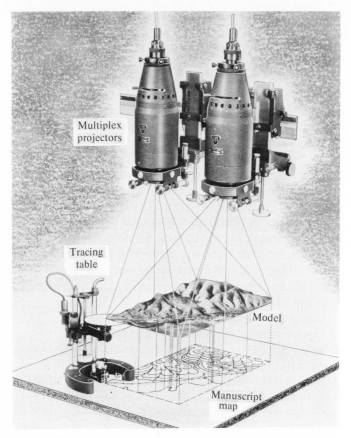

Figure 16-22. Stereoscopic plotting instrument. Courtesy of U.S.
Geological Survey.

of rays projected downward from the left projector can be seen with the left
eye only, whereas that coming from the right projector is seen with the right
eye only. The operator thus is able to see a three-dimensional image of the
overlap area covered by the two successive photographs. This is called the
anaglyphic method of viewing. The stereoscopic image seen by the operator
is called a *stereoscopic model*.

The small movable table that sits on the map table is called the *tracing
table*. The tracing table contains a white circular platen that moves up and
down by means of a knurled knob, the up and down movement being
recorded on a vernier or a counter. In some types of tracing tables this
movement is recorded in millimetres; in other types it is given directly in
feet or metres of elevation. At the center of the platen is a small hole, underneath which is a small light bulb. When the operator views the model
formed on the platen, he also sees a small pinpoint of light at the center of the
platen. This is the measuring mark, and its up-and-down motion is recorded
as just mentioned.

Directly beneath the measuring mark is the plotting pencil, which can be raised from or lowered to the map sheet. The position of the plotting pencil can be adjusted in both directions so that the pencil and the measuring mark form a line that is perpendicular to the surface of the map table.

Before the stereoscopic model can be measured and plotted in the form of a planimetric or topographic map, it must be brought to the correct scale and datum. The scale is controlled by the horizontal positions of at least two photo-control points appearing in the area of overlap, their positions having been plotted on the map sheet. By varying the spacing between the two projectors in the direction of the flight line, the scale is varied until the images of the control points received by the measuring mark are directly over the corresponding plotted positions of the control points. This operation is called *scaling* the model.

Three, four, or five vertical-control points appearing in the area of overlap are used to bring the model to the correct datum. The operator raises or lowers the measuring mark until it appears to be in contact with a vertical control point as seen in the model. He reads the elevation counter and determines the discrepancy between the known elevation and the measured elevation. Repeating this at each vertical control point, the operator then analyzes the discrepancy at each point to decide the amount by which the projectors as a unit must be tipped and tilted to make the measured elevations in the model agree with the known elevations. After the tips and tilts have been introduced, the elevations are again read to determine any minor amount of adjustment which must be made. This process is referred to as *leveling* the model.

Taken together, the process of scaling and leveling the stereoscopic image is referred to as *absolute orientation*. Thus the tilts and flying heights of the original photographs have been recreated in the instrument. The plotter thus creates a spatial model of a portion of the earth's surface. For the instrument shown in Fig. 16-22 the distance between the two projector lenses is the air base reduced to the scale of the map. The height of the projectors above the map table represents, to the scale of the map, the flying height above a level datum, although not necessarily the sea-level datum. Vertical distances measured above the map sheet to points on the stereoscopic model are elevations, to scale, of the points above the level datum. The stereoscopic model can be measured in all three dimensions by means of the measuring mark.

When plotting planimetric features on the map sheet, the operator follows the features with the measuring mark, raising it when the terrain rises and lowering it when the terrain is falling. The plotting pencil traces these features in their correct orthographic positions on the map sheet.

When plotting a contour line, the operator sets the measuring mark to correspond to the elevation of the contour line to be plotted by reference to the tracing table elevation counter. With the pencil raised from the map sheet, he moves the measuring mark until it appears to come in contact

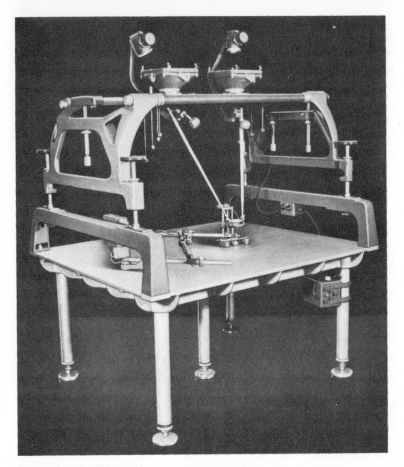

Figure 16-23. Early Kelsh plotter. Courtesy of Kelsh Instrument Co.

with the spatial model. The measuring mark is thus on the contour line. The operator lowers the pencil to the map sheet and moves the tracing table in such a manner that the measuring mark is at all times in apparent contact with the model. The pencil traces this movement as a contour line orthographically on the map sheet. To verify that he has not missed a contour line at high points, the operator usually measures the elevations of these points and records their values on the map sheet.

The Kelsh plotter, shown in Fig. 16-23, is used extensively for the compilation of medium-scale and large-scale engineering topographic maps used for design purposes. It is also used for relatively small-scale work. Operating principles are the same as those for the Multiplex plotters. The diapositive is the same size as the original negative, or about 9×9 in. The principal distance of the lens is nominally 6 in., but it can be varied a slight amount to accommodate small variations from the nominal 6-in. focal length of aerial-camera lenses. Illumination of small portions of the diapositives is accomplished by means of two lamps seen in Fig. 16-23. As the tracing table is moved over the map sheet, two guide rods guide the lamps around, illuminating only that part of the model being viewed on the tracing table. The scale of the Kelsh spatial model is approximately twice that of a Multiplex model from the same aerial photography.

The instrument shown in Fig. 16-24 performs the same function as the plotters previously discussed, but the design and operating principles are quite different. The projection takes place by means of two steel space rods machined to remarkable precision. These rods are analogous to two rays of light coming from two conjugate points on the diapositives and intersecting in space below the projectors. The two space rods can be seen in Fig. 16-24. The two diapositive holders can be rotated about three axes and can also be translated in a direction parallel with the flight line. Viewing of the three-dimensional image formed by two diapositives takes place by means of a pair of optical trains commencing on the undersides of the diapositives, continuing through a series of prisms and lenses, and ending at a pair of binocular eyepieces seen in Fig. 16-24.

Two measuring marks, one in each optical train, are superimposed onto the images of the diapositives, and they appear as one floating mark when the viewer observes the stereoscopic image. This feature is similar to the parallax bar principle mentioned in Section 16-9. The image of the mark is caused to move through the model in an X- and Y-direction by moving the intersection of the space rods using a pair of handwheels shown in Fig. 16-24. This movement of the floating mark enables the details seen in the model to be plotted on the map sheet directly in front of the observer. The floating mark is made to appear to move up or down by rotating the large foot disk seen in the illustration. The up-and-down motion is registered on a height

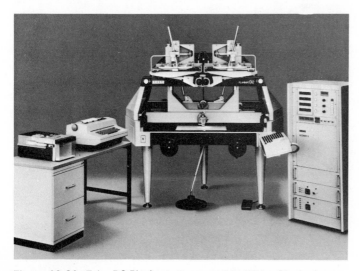

Figure 16-24. Zeiss D2 Planimat. Courtesy of Carl Zeiss, Oberkochen.

display and the mark can be set to a selected elevation for following a contour line.

There are no projector lenses in the instrument shown in Fig. 16-24. The perspective centers, replacing the lenses, are located at the pivot points of the space rods and are not seen in the illustration. By physically varying the vertical distance between these pivot points and the plane of the diapositives, all of the focal lengths given in Sec. 16-2 can be accommodated.

The instrument shown in Fig. 16-25 also uses space rods in place of optical rays to produce the stereoscopic model. Its design and operating principles are similar to those of the instrument shown in Fig. 16-24. This instrument includes the coordinatograph shown to the right. The movement of the measuring mark actuated by the two handwheels is translated to the coordinatograph pencil by means of electrical pulses. The map can be drawn at widely different scales by setting off the appropriate ratios between the model scale and the map scale.

The instrument shown in Fig. 16-26 is called an analytical plotter because all of the tilts and tips of the photographs are, in effect, generated by a computer based on the principles of spatial analytical geometry. Thus, the diapositives need not be rotated about the three axes in order to perform relative and absolute orientation. Small portions of each diapositive are greatly enlarged and projected onto one another through cross-polarized light on the face of the screen directly in front of the operator. The operator

Figure 16-25. Wild Aviomap stereo-plotter. Courtesy of Wild Heerbrugg Instruments, Inc.

Figure 16-26. Matra Optique Traster 77 Analytical plotter. Courtesy of Matra-Division Optique.

views these superimposed images through polarized glasses to get the stereoscopic effect. A measuring mark projected onto the screen allows the operator to measure and trace the features of the model, the movement being translated to the coordinatograph pencil. A television camera that moves with the pencil allows the progress of the drawing to be monitored via a small television screen.

The instruments briefly discussed above are representative of a wide variety of stereoscopic plotting instruments in use at the present time. More detailed descriptions are given in modern photogrammetry textbooks (see Moffitt and Mikhail, 1980, in the Bibliography).

16-11. ADVANTAGES AND DISADVANTAGES OF PHOTO-GRAMMETRIC MAPPING The main advantages of the compilation of topographic maps by using aerial photographs over ground methods are as follows: the speed of compilation, the reduction in the amount of control surveying required to control the mapping, the high accuracy of the locations of planimetric features, the faithful reproduction of the configuration of the ground by continuously traced contour lines, and the freedom from interference by adverse weather and inaccessible terrain. Also, by the proper selection of flying heights, focal lengths, and plotting instruments, and by proper placement of ground control, photogrammetric mapping can be designed for any map scale ranging from 1 in. = 20 ft down to 1 in. = 20,000 ft and smaller (1/250 to 1/250,000), and for a contour interval as small as $\frac{1}{2}$ ft (1 dm). Because of the wealth of detail that can be seen in a spatial model, the resultant photogrammetric map will be more complete than will a comparable map produced by ground methods.

Among the disadvantages of mapping by using aerial photographs are the following: the difficulty of plotting in areas containing heavy ground cover, such as high grass, timber, and underbrush; the high cost per acre of mapping areas 5 acres or less in extent; the difficulty of locating positions of contour lines in flat terrain; and the necessity for field editing and field completion. Field completion is required where the ground cannot be seen in the spatial model because of ground cover, where spot elevations must be measured in flat terrain, and where such planimetric features as overhead and underground utility lines must be located on the map. Editing is necessary to include road classification, boundary lines not showing on the photography, drainage classification, and names of places, roads, and other map features.

16-12. DIGITIZED STEREOSCOPIC MODEL Most modern stereoscopic plotting instruments can be digitized in all three directions, with the three-dimensional position of a point being read out to about the nearest 0.01 mm. The Zeiss Planimat shown in Fig. 16-24 is typical of such an instrument. The X and Y spindles actuated by the two handwheels can be shaft encoded to give the X- and Y-model coordinates of any point in digital form. The foot disk turns the Z spindle, which can also be shaft encoded to give the Z coordinate of the point in digital form. The X, Y, and Z coordinates are then recorded on cards, magnetic tape, or paper tape for direct input into a data processor.

All desired points in a model are digitized in the above fashion. If a property or cadastral survey is being executed by photogrammetric methods, then all property corners, which have been prepaneled, are measured, along with the horizontal and vertical control points. The entire stereoscopic model can then be transformed mathematically in three dimensions to fit the ground coordinates and elevations of the control points. This process is just like absolute orientation. Since the points in the cadastral survey have also been measured, they are also transformed into the ground coordinate system.

A stereoscopic model can be completely digitized on a regular grid array in the X and Y directions, the grid spacing depending on the roughness of the terrain and the purpose to which the *digitized terrain model* is to be put. Here again, the grid points are all transformed mathematically into the ground system by means of the control. The digitized terrain can then be analyzed in a data processor to determine optimum highway alignment and grade, high-water lines, earthwork quantities, and many other engineering studies limited only by the imagination of the engineer. Since much of the process of digitizing can be made automatic, this method lends itself to rapid computer location of contour lines by interpolation between grid

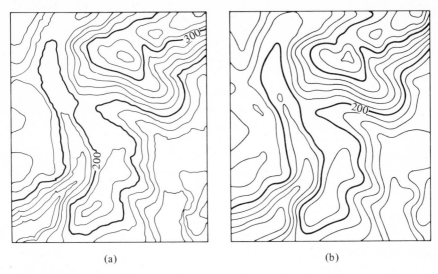

(a)   (b)

Figure 16-27. Contour lines plotted (a) manually, and (b) by computer control.

points. This is analogous to the grid method of contouring discussed in Section 15-10. The contour lines are then plotted from the digital data by means of a coordinatograph similar to that shown in Fig. 8-35.

Figure 16-27(a) shows a portion of a contour map as it was plotted manually in a stereoscopic plotting instrument. The same area was digitized on a regular grid array in the plotter. The resulting contour map generated in a computer and plotted on a coordinatograph driven by the computer output is shown in Fig. 16-27(b). The main difference between the two drawings is the smoothness of the contour lines generated by the computer.

The use of the stereoscopic model to digitize highway profiles and cross sections is discussed in Section 17-14. In this process the points to be digitized lie along predetermined lines in the model. These lines have been defined with reference to a completed topographic map at the same scale as the stereoscopic model.

16-13. ORTHOPHOTOS The effect of relief and tilt displacements on the positions of points on photographs and mosaics was discussed in previous sections. These effects are caused by the perspective nature of a photograph. A mosaic is actually a piecing together of a series of perspective projections. If it is a controlled mosaic, the overall accuracy will be good. Local scaling accuracy is not so good, however, on account of the effects of perspective. The greatest value of the mosaic is its wealth of photographic detail presented to the user in reasonably correct horizontal position.

On the other hand a pair of photographs properly oriented in a stereoscopic plotting instrument will present a theoretically perfect model of the terrain for measurement. The measurements are made orthogonally or orthographically to form a map underneath the model. This is imagined to be accomplished first by setting in the proper tilts in the projectors by the orientation process, and second by raising or lowering the tracing table when tracing the map in order to compensate exactly for scale changes due to relief. The features compiled on the map sheet are represented by lines and symbols, between which accurate scaling is possible.

In order to combine the accuracy of scaling a map with the pictorial representation of the mosaic, the images on the aerial photograph can be manipulated optically mechanically or optically electronically to remove the perspective aspect and the tilt effect. The resulting photo is called an *ortho-photo*, and the process is referred to as *orthophotography*. There are at present several different ingenious methods for preparing orthophotos. The easiest to comprehend is the *fixed-line-element* rectification of a photograph and will be presented here.

Referring to Fig. 16-28, a vertical-moving table containing a large film stage is located in the model area of an optical-projection type of stereo-plotter (see Fig. 16-22). During operation, which takes place in a darkened room with a red safelight, the film is covered by a large curtain on rollers to protect it from the light. This is shown in Fig. 16-29. The curtain contains an opening channel across its width, which is in turn covered by a ribbon curtain that moves at 90° with the larger curtain. The ribbon curtain contains a small slit surrounded by a white area, which constitutes the viewing platen. This platen is a strict analogy with the tracing table of the plotting instru-

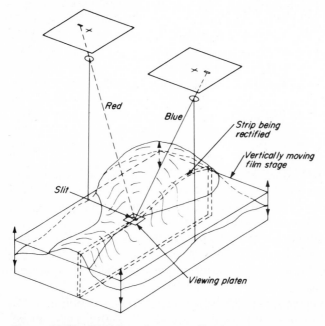

Figure 16-28. Fixed-line-element rectification.

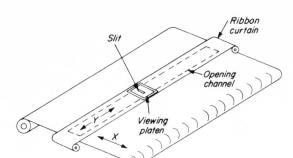

Figure 16-29. Curtain used to cover film during exposure in line-element rectification.

ment, and the small slit constitutes the floating mark. The slit lies perpendicular to the motion of the ribbon curtain.

After the stereoscopic model has been completely oriented, orthochromatic film, which is not sensitive to red light, is taped in place on the film stage and covered with the compound curtain, and is then placed in the model area. The projector lamps, which are filtered for anaglyphic viewing, are turned on to allow the operator to view a small portion of the model on the viewing platen. He can raise or lower the entire film support in order to keep the small slit in contact with the surface of the model. As the operator moves along a line in the model from front to back (or side to side), keeping the slit in constant contact with the model, an exposure of the image is "painted" onto the film in true orthographic projection. The film is exposed only by the blue light. When the model has been traversed completely along the strip, the slit is automatically stepped over a distance equal to the slit length, carrying the large curtain with it. The model is then traversed in the opposite direction. This back and forth motion exposes the film along contiguous strips over the entire model, producing a negative in orthographic projection. After the negative is developed, prints are made to produce the orthophoto at a known scale. The instrument shown in Fig. 16-30 produces orthophotos by the method just described.

The different types of a rectification employed to obtain orthophotos in a variety of instruments are shown in Fig. 16-31. The terrain and its corresponding topographic map is shown in view (a). The fixed line element method described above is shown in view (b). In view (c), the line elements are rotated to give better match between the adjacent strips. This is referred to as the *rotating-line* method. Some orthophoto rectifiers employ *plane area* rectification, diagrammed in Fig. 16-31(d) while others use the *curved area* method shown in view (e).

A series of orthophotos can be pieced together in the form of an ortho mosaic to cover a large area. Also, contour lines obtained in the stereo-plotting process can be combined with the imagery as shown in Fig. 16-32.

Figure 16-30. SFOM 693 Orthophotograph. Scanning is done in the *X*-direction. Courtesy of Matra-Division Optique.

The orthophoto lying to the left in Fig. 16-32 is the equivalent of a planimetric map in all respects except that it shows features photographically rather than symbolically. Careful examination of the orthophoto will reveal slight and unavoidable mismatching between scan lines, particularly in the building on the waterfront along the right edge of the illustration. This mismatching has virtually no effect on the overall accuracy with which the orthophoto can be scaled.

The utility of the orthophoto can be considerably enhanced by adding the contour lines as shown to the right in Fig. 16-32. If the tone of the orthophoto is generally light, then black contour lines are used. On the other hand, white contour lines would be more visible if the tone is fairly dark.

The difference between an aerial photograph and the corresponding orthophoto can be appreciated by a study of Fig. 16-33. A square grid has been ruled on a negative used to produce a diapositive from which the orthophoto was prepared. The left picture is the aerial photograph showing the grid. The right picture is the processed orthophoto. Notice how irregular the grid lines appear. This irregularity is a manifestation of relief displacement having been corrected by the orthophoto process. The instrument that produced this orthophoto uses the curved area method of rectification diagrammed in Fig. 16-31(e).

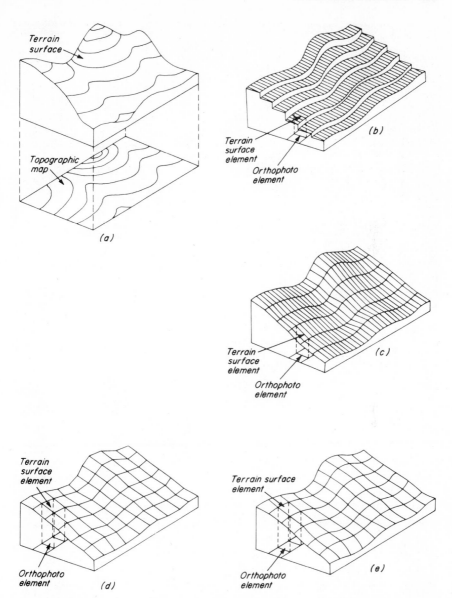

Figure 16-31. Methods of differential rectification used to produce orthophotos. (a) Terrain and corresponding contour map. (b) Fixed-line-element strip rectification. (c) Rotating line-element strip rectification. (d) Plane area element rectification. (e) Curved area element rectification. After Edmond, Bendix Technical Journal, Vol. 1, No. 2, 1968.

Figure 16-32. Orthophoto of southern approach to the Golden Gate Bridge. Courtesy of Pugh–Nolte & Associates.

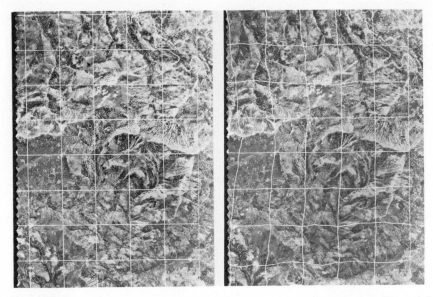

Figure 16-33. Comparison between aerial photograph and orthophotograph. The aerial photograph is to the left. Courtesy of Hobrough Ltd.

PROBLEMS

16-1. A camera with a focal length of 152.35 mm and a picture size of 230 × 230 mm is used to photograph an area from an altitude of 8000 ft above sea level. The average ground elevation is 1400 ft above sea level. What is the average scale of the photography: (a) as an engineer's scale 1 in. = so many ft; (b) as a representative fraction?

16-2. In Problem 16-1 what ground area is covered by a single photograph?

16-3. Overlap and sidelap of the photography in Problem 16-1 are 60% and 25%, respectively. What is the ground spacing, in feet, between exposures? What is the ground spacing, in feet, between flight lines.

16-4. The flight lines in Problem 16-3 are to be plotted on a map that is drawn to a scale of 1:25,000. What is the flight spacing on the map, in inches?

16-5. If the aircraft flies at ground speed of 150 mph along the flight line in Problem 16-3, what is the time interval, to the nearest 0.1 s, between exposures?

16-6. Two points lying at elevation 895 ft appear on a photograph taken as described in Problem 16-1. The distance between these points scales 22.78 mm on the photograph. What is the ground distance, in feet, between the two points?

16-7. A surveyor uses an aerial photograph in the field for general reconnaissance. In order to determine an approximate scale of the photograph, he scales 2.56 in. between two road intersection points. This same distance on a USGS quadrangle map measures 0.90 in. The scale of the quadrangle map is 1:62,500. What is the approximate scale of the photograph?

16-8. If the focal length of the aerial camera used to take the photograph described in Problem 16-7 is 210 mm, from what altitude was the photograph taken?

16-9. The image of a tall building with vertical sides appears on an aerial photograph taken with a lens having a focal length of 152.4 mm. The photo scale is 1/7500. The distance from the principal point to the image of the bottom of the building is measured as 74.16 mm, and that to the top scales as 81.54 mm. Compute the height of the building, to the nearest metre.

16-10. The photograph diagramed in Fig. 16-9 was taken with a 6-in. focal-length lens from a flying height of 6200 ft above sea level. The following data are given as an aid in solving this problem:

POINT	PHOTOGRAPH AZIMUTH	RADIAL DISTANCE (in.)	GROUND ELEVATION (ft)
a	46°	3.550	1465
b	135°	5.625	1602
c	216°	4.885	1090
d	291°	1.020	1355
e	326°	3.315	1820

The bearing of the line AB on the ground is S 20° 30′ W. Plot a 9 × 9-in. photograph, and plot the positions of the photograph axes and the positions of a, b, c, d, and e. Compute the relief displacement for each point, and plot the datum positions a', b', c', d', and e'. Make a graphical survey of the reduced polygon, determining ground lengths and bearings of the lines, and the area of the figure in square feet and in acres.

16-11. A 6-mile strip of terrain is to be photographed for highway mapping. The aerial camera contains a 6-in. focal-length lens and takes 9 × 9 in. photographs. The ground elevation varies from a low of 700 ft to a high of 1200 ft, and the average elevation is 950 ft. The average width of photographic coverage is to be 1800 ft.

(a) What flying height above sea level should be used for this flight?

(b) If the overlap between photos is 65% and the centers of the first and last photographs are to fall outside the limit of the strip, how many photographs will be required?

(c) What will be the largest and smallest photographic scales? Express each as an engineer's scale.

(d) What percent of the flying height with respect to the 950-ft level is represented by the total variation in the elevation of the terrain?

16-12. Panels are to be set out in an area to be photographed with an 88-mm focal-length lens for the purpose of making a control survey by photogrammetric methods. What are the theoretical dimensions, in centimetres, of the inner square shown in Fig. 16-13 if the photo image of this square is to be 80 × 80 μm and if the flying height is to be 3000 m above the ground?

16-13. Photographs are taken at a height of 1480 m above ground with a 152.46-mm focal-length lens. The photo base is 92.68 mm. The differences in parallax between a control point A whose elevation is 102.5 m and five other points are measured on an overlapping pair of these photographs. These differences are as follows:

POINT	DIFFERENCE IN PARALLAX (mm)
B	+ 3.58
C	− 3.02
D	+ 6.22
E	+ 0.25
F	− 1.08

Determine the elevations of the five points, to the nearest 0.1 m.

16-14. Aerial photographs are taken for use in the Kelsh plotter. After the projectors have been oriented to fit the plotted control at a scale of 1 in. = 50 ft, the distance between the projector lenses measures 18.08 in. With the measuring mark of the tracing table set at a point in the stereoscopic model whose ground elevation is 486 ft, the vertical distance between the tracing table platen and the projector lens measures 29.16 in. The platen is elevated 4.86 in. above the map table.

(a) What is the flying height above sea level of these photographs?

(b) What is the air base of the photographs, in feet?

(c) What vertical distance, in feet of ground elevation, is represented by a vertical movement of the measuring mark of 18.7 mm?

(d) What datum elevation is represented by the surface of the mapping table?

BIBLIOGRAPHY

American Society of Photogrammetry. 1981. *Manual of Photogrammetry*, 4th Edition.
———. 1968. *Canadian Surveyor* 22:entire issue.
Hallert, B. 1960. *Photogrammetry*. New York: McGraw-Hill.
Moffitt, F. H., and Mikhail, E. M. 1980. *Photogrammetry*, 3rd Ed. New York: Harper & Row.
Scher, M. B. 1969. Orthophoto maps for urban areas. *Surveying and Mapping* 29:431.
Schwidefsky, K. 1959. *An Outline of Photogrammetry*. New York:Pitman.
Wolf, P. R. 1974. *Elements of Photogrammetry*. New York:McGraw-Hill.
Zeller, Dr. M. 1952. *Text Book of Photogrammetry*. London: H. K. Lewis.

Chapter 17
Earthwork

17-1. REMARKS Earthwork operations involve the determination of the volumes of materials that must be excavated or embanked on an engineering project to bring the ground surface to a predetermined grade, and the setting of stakes to aid in carrying out the construction work according to the plans. Although the term earthwork is used, the principles involved in determining volumes apply equally well to volumes of concrete structures, to volumes of stock piles of crushed stone, gravel, sand, coal, and ore, and to volumes of reservoirs. The field work includes the measurements of the dimensions of the various geometrical solids that make up the volumes, the setting of grade stakes, and the keeping of the field notes. The office work involves the computations of the measured volumes and the determination of the most economical manner of performing the work.

The units used for measurement in earthwork computations are the foot, the square foot, and the cubic yard. In the metric system, the corresponding units are the metre, the square metre (m^2) and the cubic metre (m^3). This chapter employs the foot-yard system. However, metric units can be substituted with equal validity.

Because of the inherent accuracy of modern topographic maps of large scale produced by photogrammetric methods, much of the field work to be discussed in this chapter is eliminated, except for earthwork of limited extent. The measurements for the determination of volumes can be made directly from stereoscopic models or on topographic maps prepared by photogrammetric methods (see Sections 16-10, 17-13, and 17-14).

17-2. CROSS SECTIONS   A cross section is a section taken normal to the direction of the proposed center line of an engineering project, such as a highway, railroad, trench, earth dam, or canal. A simple cross section for a railroad embankment is shown in Fig. 17-1. The cross section for a highway or an earth dam would have similar characteristics. It is bounded by a base b, side slopes, and the natural terrain. The inclination of a side slope is defined by the horizontal distance s on the slope corresponding to a unit vertical distance. The slope may be a rise (in excavation) or a fall (in embankment). A side slope of $3\frac{1}{2}:1$, for example, means that for each $3\frac{1}{2}$ ft of horizontal distance the side slope rises or falls 1 ft. This can be designated as $3\frac{1}{2}:1$ or 1 on $3\frac{1}{2}$.

17-3. PRELIMINARY CROSS SECTION   In making a preliminary estimate and in determining the location of a facility, such as a highway or railroad, a preliminary line is located in the field as close to the final location of the facility as can be determined from a study of the terrain supplemented by maps or aerial photographs of the area. The preliminary line is stationed, and profile levels are taken. The configuration of the ground normal to the line is obtained by determining the elevations of points along sections at right angles to the line. This is identical to the process of obtaining elevations for topographic mapping described in Section 15-6.

The values of the elevations and the corresponding distances out to the right or left of the preliminary line can be plotted on specially printed cross-section paper, at a relatively large scale of from 1 in. = 5 ft to 1 in. = 20 ft (however, see Sections 17-13 and 17-14). When the location and grade of a trial line representing a tentative location of the center line of the facility have been established, the offset distance from the preliminary line to the

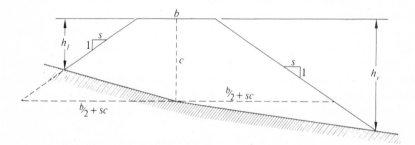

Figure 17-1. Cross-section for railroad embankment.

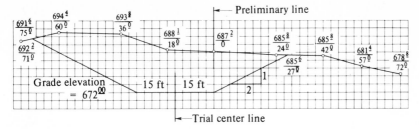

Figure 17-2. Preliminary cross section.

trial line is plotted, and the grade elevation of this trial line is plotted in relation to the terrain cross section. In Fig. 17-2 the elevations of, and distances to, the points plotted on the ground line were determined with reference to the preliminary center line. These are shown as fractions, with elevations as the numerators and distances from the preliminary line as the denominators. The offset distance of 15 ft from the preliminary line to the trial line is plotted, and the base of the roadbed is plotted at grade elevation 672.00 ft. At the edges of the roadbed, the side slopes of 2 : 1 are laid off and drawn to intersect the terrain line at the scaled distances and elevations shown as fractions lying under the terrain line. These points of intersection are called *catch points.*

The cross-sectional area bounded by the base, the side slopes, and the ground line of each trial cross section along the trial line is determined from the plotted cross section by using a planimeter or by computation based on the formulas for areas given in Section 8-20. This procedure is discussed in Section 17-8. The volumes of excavation and embankment for this trial line are computed from the successive areas and the distances between the areas by the methods described in Sections 17-9 to 17-11. The volumes for various trial lines are compared. The necessary changes in line and grade are then made to locate the final line and establish the final grade. This location will require a minimum of earthwork costs and, in the case of a highway project, for example, it will at the same time meet the criteria of curvature, maximum grade, and safe sight distances.

17-4. FINAL CROSS SECTIONS The line representing the adopted center line of a facility is staked out in the field and stationed. This line is located by computing and running tie lines from the preliminary line as discussed in Section 8-23. Deflection angles are measured between successive tangents, and horizontal curves are computed and staked out. Reference stakes are sometimes set opposite each station on both sides of the center line at distances of 25, 50, or 100 ft from the center line. These stakes are used to relocate the center line after grading operations are begun. Stakes at a distance on either side equal to half the base width are sometimes driven to facilitate taking final cross sections and setting construction or slope stakes. The center line and the reference lines are then profiled.

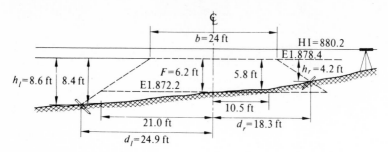

Figure 17-3. Slope staking.

When the final line has been located and profiled, a cross section is taken at each station to determine the area of the cross section and at the same time to locate the limits of excavation or embankment. These limits are defined on the ground by stakes. The process of setting these stakes is called *slope staking*.

In Fig. 17-3 the level is set up and a backsight is taken on the leveling rod held at a station on the center line whose elevation has been determined from the profile levels. The HI is established as 880.2 ft. The base width given on the plans is 24 ft, the side slopes are $1\frac{1}{2}:1$, and the grade elevation at the station is 878.4 ft. If the leveling rod were held so that the foot of the rod were at grade, the reading of the rod would be 1.8 ft. This is equal to the HI minus the grade elevation, and is called the *grade rod*. The grade rod can be plus or minus. The vertical distance from a point on the ground line to the grade line at any section is called the *fill* at the point if the section is in embankment, or the *cut* if the section is in excavation. It is seen that the fill at the left edge of the cross section of Fig. 17-3 is 8.6 ft, at the center line the fill is 6.2 ft, and at the right edge the fill is 4.2 ft.

The grade rod may be determined from the center-line cut or fill and the rod reading at the center line by the following relationship:

$$\text{grade rod} = \text{ground rod} + \text{center cut}$$

or   (17-1)

$$\text{grade rod} = \text{ground rod} - \text{center fill}$$

in which the ground rod is the rod reading at the center line. The ground rod in Fig. 17-3 is 8.0 ft. So the grade rod is $8.0 - 6.2 = 1.8$ ft, as determined before.

With the grade rod established for the cross section, the amount of cut or fill at any point in the section can be determined by reading the rod held at the point and applying the following relationship:

$$\text{cut or fill} = \text{grade rod} - \text{ground rod}$$ (17-2)

If the result is plus, the point is above grade, indicating cut ($+$); if the result is minus, the point is below grade, indicating fill ($-$).

The location, on the ground, of the slope stake is determined by trial. When the ground surface is horizontal, the position of the slope stake is at a distance from the center line equal to one-half the base width plus the product of the side-slope ratio and the center cut or fill. This is shown in the dashed-line portion of Fig. 17-1. If the ground in Fig. 17-3 were horizontal, each slope stake would be located at a distance from the center line equal to

$$12 + 1.5 \times 6.2 = 21.3 \text{ ft}$$

However, since the ground slopes transversely to the center line, it can be seen that the right slope stake will be less than 21.3 ft from the center line while the left slope stake will be greater than 21.3 ft from the center line. By trial, a point is found where 12 ft plus 1.5 times the fill computed by Eq. (17-2) equals the actual distance from the center line. With some experience the point can be found by one or two trials. If it is assumed that the ground appears to rise about 1.5 ft between the center line and the right-hand edge, the fill at the right-hand edge will be $6.2 - 1.5 = 4.7$ ft and the distance out should be

$$12.0 + 1.5 \times 4.7 = 19.0 \text{ ft}$$

On the basis of this estimate, the rod is held 19 ft from the center line and a rod reading is taken. If the actual reading at this point is found to be 5.8 ft, then by Eq. (17-2) the depth below the grade is

$$1.8 - 5.8 = -4.0 \text{ ft}$$

In order that the point will be at the intersection of the side slope with the ground surface, the distance from the center should be

$$12.0 + 1.5 \times 4.0 = 18.0 \text{ ft}$$

Since the actual distance was 19.0 ft, the trial point is incorrect. The rod is therefore moved nearer the center. Had the ground surface at the right-hand edge been horizontal, the slope stake could have been set at the 18.0-ft point. Since the ground is sloping downward toward the center line, the fill will be somewhat more than 4.0 ft, and the distance out will be greater than 18.0 ft. Consequently, the next trial is taken at 18.2 ft, where a rod reading of 6.0 ft is obtained. The depth below the grade is

$$1.8 - 6.0 = -4.2 \text{ ft}$$

and the distance from the center line should be 18.3 ft. This is within 0.1 ft of where the rod is being held, and so the stake is set at 18.3 ft from the center and the fill is recorded as 4.2 ft.

In the field notes these two dimensions are recorded in fractional form, the numerator representing the fill and the denominator representing the distance out from the center. To distinguish between cut and fill, either the letters C and F or the signs $+$ and $-$ are used to designate them. As the point is located with respect to the finished grade, a point below grade indicates a fill and is designated by a $-$ sign. The amount of cut or fill is marked on the side of the stake toward the center stake, and the distance out is marked on the opposite side. The stake is usually driven slantingly to distinguish it from a center-line stake and to prevent it from being disturbed during the grading operations.

In Fig. 17-3 there is a decided break in the ground surface between the center line and the right slope stake. This break is located by taking a rod reading there and measuring the distance from the center line. In a similar manner the break on the left side of the center and the left slope stake are located. The field notes for this particular station could be recorded as follows:

			CROSS SECTION				
STATION	GRADE ELEVATION	GROUND ELEVATION	L		C		R
15 + 00	878.4	872.2	−8.6	−8.4	−6.2	−5.8	−4.2
			24.9	21.0	0	10.5	18.3

17-5. CROSS SECTIONING BY SLOPE MEASUREMENT The
position of a slope stake can be determined by measuring the distance out from the center line along the slope and determining the difference in elevation between the center-line stake and the tentative position of the slope stake. Figure 17-4 shows a cross section in cut with a roadbed width b and side slope s: 1. The cut at the center line is designated c. A plus slope measurement

$$M'P' = t$$

is made from a point M' vertically above the center-line stake at M to a point P' vertically above the tentative position of the slope stake at P. The slope of $M'P'$ is the vertical angle α. In the figure

$$MM' = PP'$$

Then

$$V = t \sin \alpha \quad \text{and} \quad H = t \cos \alpha$$

The vertical distance V is the difference in elevation between M and P.

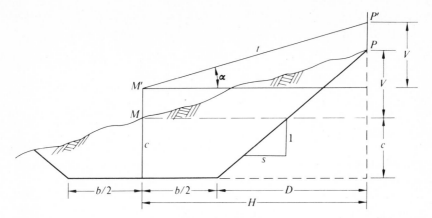

Figure 17-4. Slope staking by slope measurements.

If P is the correct position of the slope stake then the total rise of the side slope is $c + V$ (or $c - V$ on the downhill side). The side slope is in the ratio

$$\frac{1}{s} = \frac{c + V}{D} = \frac{c + V}{H - b/2}$$

Thus considering both an uphill and a downhill sight,

$$H = \frac{b}{2} + s(c \pm V) \tag{17-3}$$

If the computed value of $H = t \cos \alpha$ is the same as the value computed by Eq. (17-3), the slope stake is in the correct position. If

$$H = t \cos \alpha$$

is larger than that obtained by Eq. (17-3), the slope stake is too far out and must be moved in. Thus the final position of the slope stake is determined by trial.

If the cross section is a fill section, then

$$H = \frac{b}{2} + s(f \mp V) \tag{17-4}$$

in which f is the center fill, considered a positive quantity in this equation. If the sight is to the uphill side, the negative sign is used for V, and if downhill the plus sign is used.

Various methods of making the measurements can be used. If a theodolite is set up at M and MM' is measured, the vertical angle α is measured to a point P' on the leveling rod or staff such that

$$PP' = MM'$$

The slope distance is then measured with a steel or metallic tape.

The same procedure as above can be performed using the principle of transit stadia. The values of H and V are those discussed in Section 14-5. The leveling rod held at the slope-stake position should be carefully plumbed using a rod level. This method eliminates the use of the tape.

Several of the EDM's described in Chapter 2 are mounted directly onto transits or theodolites. Others are equipped with vertical circles. These instruments can be used for slope staking in a manner similar to the use of the transit and tape. The reflector is set on a staff at the same height as the horizontal axis of the transit or theodolite or of the instrument itself. The vertical angle α and the slope distance t are measured with the instrument combination. After a trial position of the slope stake has been made, the reflector is moved inward or outward along the cross section according to the difference between the two computed values of H, and a second determination is made. Thus the slope-stake position is fixed by trial and error.

Some EDM's have provisions for computing the values of H and V internally as discussed in Chapter 2. This eliminates having to do some of the computations in the field as the slope staking progresses.

17-6. LOCATION OF SLOPE STAKES BY INVERSING In modern transportation engineering, all of the surveys are placed on a coordinate system. The preliminary line from which preliminary cross sections are measured (however, see Sections 17-13 and 17-14) is surveyed on the coordinate system. The final alignment is laid out with respect to the control points as discussed in Section 8-23. The preliminary cross sections as shown in Fig. 17-2 are used to compute the positions of the catch points. The base width, side slopes, and grade elevation for each cross section are entered into a computer program. The computer then intersects the side slope lines with the appropriate lines on the preliminary cross section, using Eqs. (8-15) and (8-16). Output from the computer gives the distance to the left (d_L) and to the right (d_R) where slope stakes are to be located together with the cuts and fills at the catch points and the areas of the cross section. This is described in detail in Section 17-13.

If each station on the final located line is occupied with the theodolite or theodolite EDM combination, a right angle is laid off from the line, and the calculated distance left or right is measured off and the slope stake is set. This, however, requires an instrument setup at each station. If a control point with a commanding view of a substantial portion of the project is convenient, the

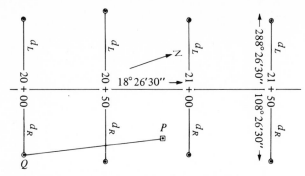

Figure 17-5. Location of slope stakes by inversing.

instrument can be set up at this control point from which angles and distances are measured off to locate several slope stakes.

In Fig. 17-5, the location center line is on a coordinate system. Thus, the coordinates of each center-line station is known, as are the coordinates of a control point P. The azimuth of the center line is shown as $18° 26' 30''$. The azimuth of each of the cross lines going to the left of the center line is $90°$ less, or $288° 26' 30''$. The azimuth of these lines going to the right is $90°$ more, or $108° 26' 30''$. The latitudes and departures can be computed for each cross line going to the left and to the right since the distances (the d_L's and d_R's) and the azimuths are known. The coordinates of the catch points where the slope stakes are to be set can then be computed based on the coordinates of the center-line stations.

The length and azimuth of any line from the control point P to a catch point such as point Q to the right of station $20 + 00$ are computed by Eqs. (8-12) and (8-13). The angles to be laid off at P from a backsight on another control station to each catch point can be computed from the known azimuths.

All of the above computations are carried out in one program in a data processor, and the output is given to the field engineer. The control point is occupied with a combination theodolite EDM or a total station instrument, all of the catch points that can be seen from the control point are located, and the slope stakes are set.

17-7. DISTANCE BETWEEN CROSS SECTIONS The horizontal distance between cross sections is dependent on the precision required, which in turn is dependent on the price per cubic yard paid for excavation. As the unit cost for highway or railroad grading is usually small, a station distance of 100 ft is generally sufficiently precise. For rock excavation and for work done under water, the cost per cubic yard mounts very rapidly and the distance between sections is often reduced to 10 ft.

In addition to the cross sections taken at regular intervals, other sections are taken at the point of curvature (PC) and the point of tangency (PT) of

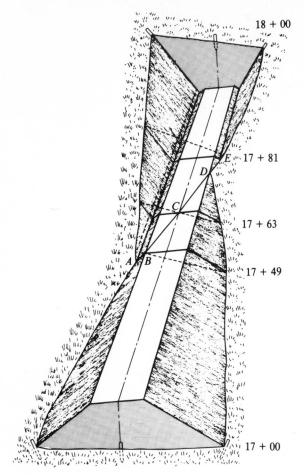

Figure 17-6. Transition from fill to cut.

each curve, at all breaks in the ground surface, and at all grade points. A *grade point* is a point where the ground elevation coincides with the grade elevation. In passing from cut to fill or from fill to cut, as many as five sections may be needed in computing the volume when the change occurs on a side hill. In Fig. 17-6 these sections are located at *A, B, C, D,* and *E*. Unless extreme precision is required, the sections at *A* and *D* are usually omitted. Although stakes are set only at the grade points *B, C,* and *E,* full cross sections are taken at the stations and the measurements are recorded in the field notes.

It will be noted in Fig. 17-6 a wider base of cross section is used when excavation is encountered. The additional width is needed to provide ditches for draining the cut.

17-8. CALCULATION OF AREAS The purpose of cross sectioning is to obtain the measurements necessary to compute the area of the plane of

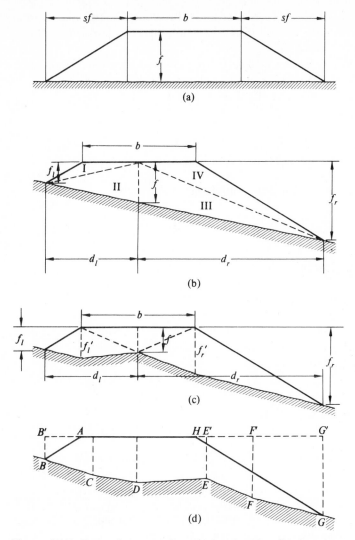

Figure 17-7. Types of cross sections. (a) Level section; (b) three-level section; (c) five-level section; (d) irregular section.

the cross section bounded by the ground, the side slopes, and the roadbed. The cross-section areas are then used to compute earthwork volumes, either by the average end-area method discussed in Section 17-9 or by the prismoidal formula given in Section 17-10.

Four classes of cross-section configurations are shown in Fig. 17-7. The first is the level section shown in view (a). The area of the cross section is given by

$$A = (b + fs)f \qquad (17\text{-}5)$$

in which b is the width of the roadbed, f is the center fill, and s is the side-slope ratio. If the cross section is in excavation, then c is the center cut and is substituted for f in Eq. (17-5).

Figure 17-7(b) shows a three-level section. The cross-section notes contain information only at the center line and at the two slope stakes. The area of the cross section is the sum of the areas of four triangles labeled I, II, III, and IV. The areas are

$$A_1 = \frac{1}{2} \times \frac{b}{2} \times f_l; \quad A_{II} = \frac{1}{2} f \times d_l; \quad A_{III} = \frac{1}{2} f \times d_r; \quad A_{IV} = \frac{1}{2} \times \frac{b}{2} \times f_r$$

Combining the areas of the four triangles gives

$$A = \frac{b}{4} (f_1 + f_r) + \frac{f}{2} (d_l + d_r) \qquad (17\text{-}6)$$

in which f_l and f_r are the fill at left and right slope stakes, respectively; d_l and d_r are the distances to left and right slope stakes, respectively; and f is the center fill. If the section is in cut, the symbol c is substituted for f.

A five-level section is shown in view (c). The fill is determined at points measured out a distance equal to one-half the roadbed width on both sides of the center line. The area of this class of section is

$$A = \tfrac{1}{2}(f'_l d_l + f b + f'_r d_r) \qquad (17\text{-}7)$$

in which f'_l and f'_r are the fill at the left and right edge of the roadbed, respectively; and d_l and d_r are the distances to left and right slope stakes, respectively.

The cross section shown in Fig. 17-7(d) is called an *irregular section* in which the ground surface is so irregular that a three- or five-level section would not give enough information to obtain an accurate determination of the area. The area of the section may be obtained by computing the areas of the trapezoids forming the figure $B'BCDEFGG'$ and subtracting from their sum the areas of the two triangles ABB' and GHG'.

Another method, which can also be applied to any of the preceding figures, is to consider the cross section as a traverse. The field notes provide the coordinates of the corners with respect to the finished grade and the center line as coordinate axes, the horizontal distances from the center line being the x coordinates with due regard for algebraic sign, and the vertical cuts or fills the y coordinates. For an eight-sided traverse for example, the area can be expressed by one of the following equations, which are like Eq. (8-25) or (8-26) in Section 8-20.

$$\text{area} = \tfrac{1}{2}[X_1(Y_2 - Y_8) + X_2(Y_3 - Y_1) + X_3(Y_4 - Y_2) + \cdots]$$
$$\text{area} = \tfrac{1}{2}[Y_1(X_2 - X_8) + Y_2(X_3 - X_1) + Y_3(X_4 - X_2) + \cdots]$$

TABLE 17-1. Computations for Area of Cross Section

A	B	C	D	E	F	G	H
F0.0	F5.2	F6.8	F7.2	F6.1	F7.4	F9.6	F0.0
-12.0	-19.8	-10.0	0	$+15.0$	$+20.0$	$+26.4$	$+12.0$
(1)	(2)	(3)	(4)	(5)	(6)	(7)	(8)

$$
\begin{array}{ll}
Y_i(X_{i+1} - X_{i-1}) & = \text{double area} \\
0.0(-19.8 - 12.0) & = \quad\ \ 0.0 \\
5.2(-10.0 + 12.0) & = +\ \ 10.4 \\
6.8(\quad 0.0 + 19.8) & = +134.6 \\
7.2(+15.0 + 10.0) & = +180.0 \\
6.1(+20.0 - \ \ 0.0) & = +122.0 \\
7.4(+26.4 - 15.0) & = +\ \ 84.4 \\
9.6(+12.0 - 20.0) & = -\ \ 76.8 \\
0.0(-12.0 - 26.4) & = +\ \ \ 0.0 \\
& \overline{2)454.6} \\
\text{area} = & \quad 227.3\ \text{ft}^2
\end{array}
$$

Since for a cross section in earthwork the y coordinates of two of the points are zero, the computations will be shortened if the second equation is used. An application of this equation to the area in Fig. 17-7 is given in Table 17-1 in which the roadbed width is 24 ft.

17-9. VOLUME BY AVERAGE END AREAS According to the end-area formula, the volume, in cubic feet, between two cross sections having areas A_0 and A_1 is

$$V_e = \tfrac{1}{2}(A_0 + A_1)L \tag{17-8}$$

in which L is the distance between the sections. The volume, in cubic yards, is

$$V_e = \frac{1}{2}(A_0 + A_1)\frac{L}{27} = \frac{L}{54}(A_0 + A_1) \tag{17-9}$$

Although this relationship is not an exact one when applied to many earthwork sections, it is the one most commonly used because of the ease of its application and because of the fact that the computed volumes are generally too great and thus the error is in the favor of the contractor.

An example of the calculation of earthwork volumes by the average end-area formula for cut sections is given in Table 17-2. Note that the section at station 36 + 00 is three-level, and at station 37 + 00 is five-level. The roadbed is 36 ft, and side slopes are 2:1.

TABLE 17-2. Computation of Volume of Earthwork by Average End-Area Formula

STATION	CROSS-SECTION NOTES	AREA
36 + 00	$\dfrac{C11.7 \quad C7.4 \quad C4.1}{41.4 \quad 0.0 \quad 26.2}$	$\dfrac{36}{4}(11.7 + 4.1) + \dfrac{7.4}{2}(41.4 + 26.2)$ $= 392.3 \text{ ft}^2$
37 + 00	$\dfrac{C12.4 \quad C9.1 \quad C7.6 \quad C6.3 \quad C2.9}{42.8 \quad 18.0 \quad 0.0 \quad 18.0 \quad 23.8}$	$\dfrac{1}{2}(9.1 \times 42.8 + 7.6 \times 36 + 6.3 \times 23.8)$ $= 406.5 \text{ ft}^2$

$$V = \frac{100}{54}(392.3 + 406.5) = 1479 \text{ yd}^3$$

17-10. VOLUME BY PRISMOIDAL FORMULA When the more exact volume must be known, it can be calculated by means of the prismoidal formula

$$V_p = \frac{L}{6}(A_0 + 4M + A_1) \tag{17-10}$$

in which M is the area of the middle section and V_p is the volume, in cubic feet. In general, M will *not* be the mean of the two end areas. It can be shown that this formula is correct for determining the volumes of prisms, pyramids, wedges, and prismoids that have triangular end sections and sides that are warped surfaces. Since the earthwork solids are included in this group, except for slight irregularities of the ground, the prismoidal formula gives very nearly the correct volume of earthwork. The error in the use of the end-area formula arises chiefly from the fact that in its application the volume of a pyramid is considered to be one-half the product of the base and the altitude, whereas the actual volume is one-third the product of those quantities.

The area of the middle section can be obtained by taking intermediate cross sections on the ground or when the same number of points have been taken on adjacent sections, by computing the area of a section that has dimensions equal to the means of the corresponding dimensions of the two end sections. An example of this method for three-level sections having bases of 24 ft and side slopes of $1\frac{1}{2}$: 1 is given in Table 17-3.

By the end-area formula,

$$V_e = (100/54)(215.04 + 82.20) = 550.4 \text{ yd}^3$$

Thus the error in applying the end-area formula to this solid is 11.3 yd^3, or about 2%.

TABLE 17-3. Computation of Volume of Earthwork by Prismoidal Formula

STATION	L	C	R	AREA
15 + 00	$\dfrac{-8.6}{24.9}$	$\dfrac{-6.4}{0.0}$	$\dfrac{-4.2}{18.3}$	$\dfrac{1}{2} \times 12 \times 12.8 + \dfrac{1}{2} \times 6.4 \times 43.2 = 215.04 \text{ ft}^2$
16 + 00	$\dfrac{-4.6}{18.9}$	$\dfrac{-2.8}{0.0}$	$\dfrac{-1.4}{14.1}$	$\dfrac{1}{2} \times 12 \times 6.0 + \dfrac{1}{2} \times 2.8 \times 33.0 = 82.20 \text{ ft}^2$
M	$\dfrac{-6.6}{21.9}$	$\dfrac{-4.6}{0.0}$	$\dfrac{-2.8}{16.2}$	$\dfrac{1}{2} \times 12 \times 9.4 + \dfrac{1}{2} \times 4.6 \times 38.1 = 144.03 \text{ ft}^2$

$$V_p = \frac{100}{6} \times (215.04 + 4 \times 144.03 + 82.2) = 14{,}556.0 \text{ ft}^3 = 539.1 \text{ yd}^3$$

17-11. VOLUME THROUGH TRANSITION   In passing through a transition from cut to fill, at least one of the cross sections will contain both cut and fill areas (the section at C in Fig. 17-6). Thus there will be both excavation volumes and embankment volumes between some of the cross sections. Let the cross sections of Fig. 17-6 be represented as shown in Fig. 17-8. The

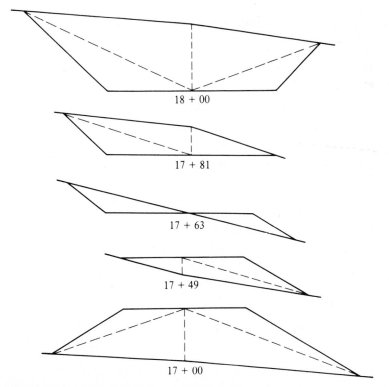

Figure 17-8. Cross sections in transition from fill to cut.

cross section at station $17 + 00$ is a complete three-level section in fill; that at station $17 + 49$ is a partial three-level section in fill in which the fill goes to zero at the left edge of the roadbed. The section at station $17 + 63$ is a side-hill section with the grade point at the center line. The left side of the section is in cut and the right side in fill.

The cross section at station $17 + 81$ is a partial three-level section in cut in which the cut begins from zero at the right edge of the roadbed. Finally, at station $18 + 00$ the section is a full three-level section in cut.

The volume between station $17 + 00$ and station $17 + 49$ is all embankment. Between station $17 + 49$ and station $17 + 63$, the left part of the volume is a prism in excavation and the right side is embankment. The same is true between station $17 + 63$ and station $17 + 81$. Finally, the volume lying between station $17 + 81$ and station $18 + 00$ is all excavation.

The volume of the prism of excavation between station $17 + 49$ and station $17 + 63$ and the prism of embankment between station $17 + 63$ and station $17 + 81$ should be computed as pyramids—that is, one-third the base times the altitude—since they are pyramidal in shape. The area of the partial cross section corresponds to the base of the pyramid, and the difference in stationing corresponds to the altitude of the pyramid.

17-12. VOLUME BY TRUNCATED PRISMS The volume of a borrow pit (see Section 17-19) or of the excavation for the foundation of a building is frequently found by dividing the surface area into squares or rectangles, according to the methods of Section 15-10. The original ground-surface elevations at the corners of these figures are usually determined by direct leveling. The final elevations are obtained either by repeating the leveling after the excavation has been completed or, in the case of building foundations, from the plans of the structure. The vertical depth at each corner is calculated by subtracting the final elevation from the original ground-surface elevation.

The volume of any prism is taken as the horizontal area of the prism multiplied by the average of the corner depths. Thus if a, b, c, and d represent the depths at the corners of a truncated square or rectangular prism, the volume is

$$V = A \times \frac{a + b + c + d}{4} \tag{17-11}$$

in which A is the horizontal area of the prism.

The computations can be shortened by noting that some corner heights are common to more than one prism. The volume of the assembled prisms is

$$V = A \times \frac{\sum h_1 + 2 \sum h_2 + 3 \sum h_3 + 4 \sum h_4}{4} \tag{17-12}$$

in which A is the area of one of the equal squares or rectangles, $\sum h_1$ is the sum of the vertical heights common to one prism, $\sum h_2$ is the sum of those heights common to two prisms, $\sum h_3$ is the sum of those common to three prisms, and $\sum h_4$ is the sum of those common to four prisms.

As an example of the application of Eq. (17-12), consider that the area of Fig. 15-9 bounded by points B-1, B-4, F-4, F-2, D-2, and D-1 is to be brought to a finished grade elevation of 700.00 ft. Assume that the sides of the excavation are vertical. The values of the corner heights are then as follows:

h_1	h_2	h_3	h_4
B-1 $=$ 22.5	B-2 $=$ 27.3	D-2 $=$ 13.2	C-3 $=$ 27.3
B-4 $=$ 45.6	B-3 $=$ 34.8	$\sum h_3 = \overline{13.2}$	D-3 $=$ 18.3
F-4 $=$ 19.9	C-4 $=$ 28.2		E-3 $=$ 11.3
F-2 $=$ 4.2	D-4 $=$ 22.7		C-2 $=$ 17.4
D-1 $=$ 11.7	E-4 $=$ 19.8		$\sum h_4 = \overline{74.3}$
$\sum h_1 = \overline{103.9}$	F-3 $=$ 12.9		
	E-2 $=$ 10.0		
	C-1 $=$ 14.3		
	$\sum h_2 = \overline{170.0}$		

If each square in Fig. 15-9 is, say, 25 ft on a side, then by Eq. (17-12) the volume in cubic feet is

$$V = 625 \times \frac{103.9 + 2 \times 170.0 + 3 \times 13.2 + 4 \times 74.3}{4} = 122{,}000 \text{ ft}^3$$

17-13. VOLUMES FROM TOPOGRAPHIC MAPS

In modern highway-location practice, accurate topographic maps with a large scale and a small contour interval are prepared by photogrammetric methods supplemented by field completion surveys. These maps are known as design maps. They are compiled typically at a scale of 1 in. $=$ 50 ft and with a 2-ft contour interval. The positions of all horizontal-control monuments are accurately plotted by means of their rectangular coordinates with respect to a grid which shows on the map.

The design map is used to study various possible positions for the location of the highway, to aid in establishing the various geometrical properties of the roadbed, to establish limits of the right of way, and to determine the quantities of earthwork for the various possible locations. These quantities are used for comparison of the different lines, for estimates in bidding, and in some instances for determining the actual pay quantities the contractor is held to.

The design map covers a strip of terrain that usually varies from about 600 to 2000 ft in width. When it has been compiled and verified in the field, the design engineer can then project trial lines onto the map with different grade lines. A study of the earthwork involved, together with other controlling factors, will establish a final line on the design map. The coordinates of the points of intersection are scaled from the grid lines, as described in Section 8-32, and tie lines from the control monuments are computed. Thus the final line may be located and staked out in the field by running these tie lines as outlined in Section 8-23.

The earthwork quantities for a given trial line are determined from the design map by the method that will now be described. The trial line is laid out on the map, and the curves joining the tangents at their intersections are drawn in. The line is marked from beginning to end at every 50 or 100 ft, the interval depending on the regularity of the terrain and the accuracy desired. These marks are stations and plusses.

A cross line is plotted normal to the trial line at each station and plus, and the elevations of, and distances right or left from the center line, to points representing breaks in the terrain are determined by interpolation between the contour lines and by scaling on the map. These elevations and distances are recorded opposite each station number on a set of notes, with the elevations as the numerators and the distances out from the center line as the denominators. The notes representing these cross sections are called *terrain notes*. The cross sections are taken far enough out to allow for deep cuts and fills, and to allow for line changes on successive trial lines. A cross section, with the terrain notes shown, is represented in Fig. 17-9. The actual terrain cross sections, however, are not plotted.

A grade line that is established by the design engineer is superimposed on a profile of the line plotted from elevations obtained from the design map. The lengths of the vertical curves are established, and the elevations of the stations on the vertical curves are computed. The grade elevation at each center-line station and plus is thus obtained.

Roadbed notes are prepared for each station and plus, taking into account the roadbed width, ditches in cuts, side slopes, roadbed crown, and super-elevation on curves. The roadbed notes for a section coinciding with the section of Fig. 17-9 are given in Fig. 17-10, the roadbed itself being shown graphically. The actual roadbed cross sections, however, are not plotted. Point a is at the center line. Points b and c are known as hinge points, because the roadbed cross section breaks at these points when the section is in cut. Points e and f are at the edges of the shoulder in cut or in fill.

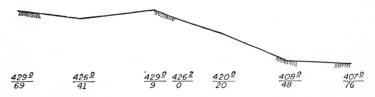

Figure 17-9. Terrain notes.

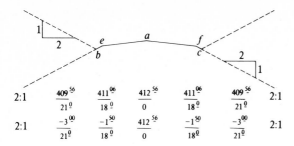

Figure 17-10. Roadbed notes.

The configuration of the roadbed will remain constant as long as the high-way is on a tangent. Since the notes will be used in subsequent computations, the values of the elevations at points b, e, f, and c can be tabulated as vertical distances from the center-line elevation, as shown in the lower notes of Fig. 17-10. Thus as long as the configuration of the roadbed does not change, the numerators of the lower set of fractions will not change, except for that representing the grade elevation at the center line. This elevation is shown as 412.56 in Fig. 17-10. In the process of computation these vertical offset distances will be converted to elevations. The roadbed notes will vary on curves because of superelevation, and will vary in different materials because of changes in the side slopes.

The notes for the terrain cross section and the roadbed notes are key-punched for input into a high-speed electronic computer and are brought together by a prepared computer program, the details of which are outside the scope of this book. The computer determines the distances to, and the elevations of, the catch points at each station and the area of each cross section bounded by the roadbed, the side slopes, and the terrain. The distance to the catch point (on a final line) is used in the field to locate the position of each slope stake. A graphical portrayal of the solution of Fig. 17-9 and Fig. 17-10 in the computer is shown in Fig. 17-11. The distances to the left and right slope stakes are, respectively, 58.6 and 33.0 ft. The area of the cross section is computed as that of a closed traverse in the computer, and the result is 949.3 ft^2.

Using the average-end-area method, the computer determines the volumes between successive cross sections from the computed areas and the spacing of the cross sections in stations and plusses. These values are part of the data recorded by the computer and furnished to the design engineer.

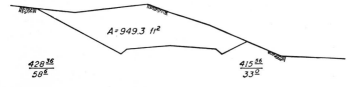

Figure 17-11. Terrain and roadbed notes combined.

When computations are to be made for a second trial line, the terrain notes need not be changed, provided they were taken from the map at sufficient distances from the center line. The roadbed notes need not be changed if the grade line has not been changed. However, the amount of line shift at each station must be introduced into the computer. For example, if the section of Fig. 17-11 is shifted to the right by 20 ft, this is portrayed in graphical form in Fig. 17-12. The resulting distances to the left and right slope stakes determined by the computer are, respectively, 55.6 and 23.0 ft. The area of the cross section is 667.6 ft^2.

If a grade change is to be made, the elevation of the roadbed center line at each section must be changed correspondingly before the data are introduced into the computer.

When the final line has been established by analysis of various trial lines, the computer output will give, in addition to the cross-sectional areas and earthwork volumes, the distances out from the center line to the left and right slope stakes. These listings are used by the field party to locate the slope stakes for construction.

A device used to scale distances automatically and record elevations along terrain cross sections is shown in Fig. 17-13. The round base plate is oriented on the topographic map in such fashion that the cylindrical movable arm is caused to move along a line perpendicular to the center line of a trial line. With the pointing device, located at the end of the arm, set at the center line, the elevation counter is set to read the ground elevation at the center line as interpolated between the contour lines on the map. The vertical counter is then set for the contour interval. As the arm is moved, say, to the left of the center line, the pointer is set on the first contour line encountered, and the operator depresses the register button. This operation records or card punches the distance to the left of the center line and the contour elevation of the point. When the movement of the arm is continued to the next contour line, the vertical counter automatically adds or subtracts the contour interval while the movement of the arm measures the distance, and the values are again registered. When the terrain cross section to the left has been completed, the pointer is again indexed on the center line. The measurements are then taken from the center line to the right. Information pertinent to the terrain cross section, such as the station number, is set into the control console shown in Fig. 17-13 to be registered automatically with the cross-section data.

The operator must depress a plus-minus button when going across ascending or descending contour lines in order that the elevation counter will

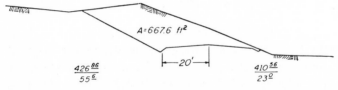

Figure 17-12. Cross section after line shift.

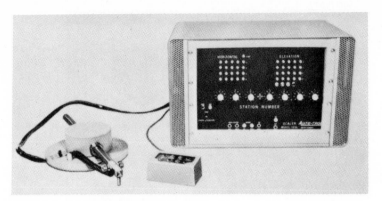

Figure 17-13. Automatic recording scaler. Courtesy of Autotrol Corp.

register the correct value of the contour line. If the elevation of a point between contour lines is to be recorded, the operator interpolates between the contour lines to obtain the value of the intermediate elevation. By depressing a button on the small control unit seen in Fig. 17-13, the small increments of elevation are added to or subtracted from the elevation contour until the intermediate elevation is registered on the console display. This value can be obtained by the operator, without having to observe the console display, simply by counting audible signals emitted by the counter. The register is then depressed to record or punch this intermediate elevation together with the corresponding distance from the center line.

The automatic scaling device is accurate to about 0.001 in. over the travel of the arm. It is certainly far more accurate than are the locations of the map features and the contour lines. The scaler eliminates mistakes and scaling errors. The scaling is much easier and far more efficient than that done by means of an ordinary engineer's scale. By automatically punching out the data necessary for machine computation, the scaler eliminates not only the data-recording time but also the card-punching operation, which is always subject to mistakes.

The foregoing method of determining earthwork quantities in highway location entirely eliminates the field surveys for determining preliminary and final cross sections. Furthermore, the planning can be conducted well in advance of all but the initial control survey, and time is provided for purchasing the necessary right of way without too much danger of unscrupulous land speculation. The success of the entire method, of course, depends on the accuracy and reliability of the design map.

17-14. EARTHWORK DATA FROM PHOTOGRAMMETRIC MODEL

The compilation of a design map by photogrammetric methods requires that a pair of overlapping aerial photographs be oriented in a stereoscopic plotting instrument and then that the resulting spatial model be

Figure 17-14. Automatic scaler used to measure cross sections directly in stereoscopic model.

oriented to control that has been plotted on the map sheet. When these conditions are satisfied, the operator of the stereoplotter then proceeds to compile the planimetric and topographic features, using the measuring mark of the tracing table to guide him. This process is discussed in Section 16-10. The resulting topographic map is a graphical representation of the more complete and more detailed spatial model from which it was derived. In order to increase the accuracy of earthwork quantities, it would be logical to measure the spatial model directly, thus eliminating the inaccuracy and the incompleteness of the map.

Figure 17-14 shows the scaling device discussed in Section 17-13 in which the pointer has been replaced by the pencil holder of the tracing table of a Kelsh plotter. The topographic map which was compiled from the aerial photography has been used to study possible trial center-line locations for a highway facility. After a trial center line has been drawn on the map, the map in turn is reoriented under the stereoscopic model (see Fig. 16-22).

The base of the scaler is now oriented so that the tracing table is constrained to move perpendicular to the center line drawn on the map. This condition is readily apparent from a study of Fig. 17-14. The stereoplotter operator then sets the tracing-table pencil on the center line at a desired station, sets the measuring mark on the surface of the stereo model, and indexes the elevation and distance counters. The elevation counter is actuated by the vertical movement of the tracing table, being sensed by an encoder, and it displays the elevation of the tracing table at all times.

After registering the center-line elevation, the operator then moves the tracing table, say, to the left of the center line until a significant break in the terrain appears in the stereoscopic model. After setting the measuring mark in apparent contact with the surface of the model at this point, he then depresses the register button, causing the elevation of the point and its distance from the center line to be automatically recorded or punched out. A repetition of this procedure thus provides terrain cross-section notes that can be used to compute earthwork quantities along this trial line.

17-15. EARTHWORK QUANTITIES BY GRADING CONTOURS

A grading contour is a line of constant elevation that is plotted on a topographic map to represent a true contour line after the proposed grading has been performed. Because grading operations produce smooth surfaces with regular slopes, grading contours are either a series of straight, equally spaced lines or a series of curved lines that are equally spaced. Simple landscape grading is shown in Fig. 17-15. The original contour lines are shown as solid lines with a contour interval of 1 ft. The grading contours are drawn on the map as straight dashed lines. The irregular dashed lines have been drawn

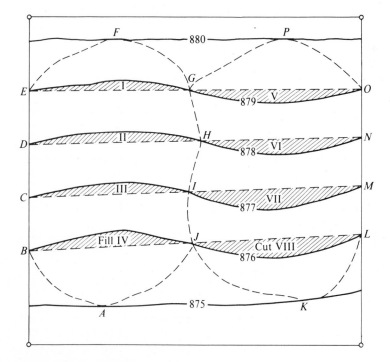

Figure 17-15. Landscape grading.

through the grade points to show the areas of cut and fill. The area marked I is enclosed by the original 879-ft contour line and the 879-ft grading contour. It is a horizontal surface at elevation 879 ft. Similarly, the area marked II is a horizontal surface at elevation 878 ft. The volume of fill necessary between these two surfaces is obtained by planimetering each area, determining the average of the two areas, and multiplying the average by the vertical separation or the contour interval which, in this instance, is 1 ft. The volume between areas II and III and the volume between areas III and IV are similarly obtained. The solid defined by the embankment between point F and area I is pyramidal in shape, as is also the solid defined by the embankment between point A and area IV. Each of these two volumes should be taken as $\frac{1}{3}bh$, in which b is the shaded area representing the base of the solid and h is the contour interval, or 1 ft. The volume of excavation between K and P may be determined in the same manner.

In Fig. 17-16 the location of a highway roadbed is plotted to scale on a design map on which the contour interval is 5 ft. The line is stationed by scaling 100-ft distances. The grade elevation at station $55 + 00$ is to be 465.00 ft and that at station $58 + 33$ is to be 462.00 ft. The side slopes are 2 : 1 in cut and are 3 : 1 in fill. The grade line along the center line intersects the ground surface at stations $53 + 62$ and $58 + 64$. These grade points are located by finding where the grade elevations coincide with the ground elevations determined by interpolation between contour lines.

The road is in cut between stations $53 + 62$ and $58 + 64$. Since the side slopes in cut are 2 : 1, a distance 10 ft out from either edge at station $55 + 00$ will be at elevation 470 ft after the slopes have been formed. Similarly, a point 16 ft out from an edge at station $58 + 33$ will also be at elevation 470 after the slopes have been formed.

A straight line shown dashed in the figure on either side of the road as a line joining two points at elevation 470 is a grading contour line. The 470-ft grading contours intersect the 470-ft contour lines on the ground at points a and b. Grading contours at elevation 475 will lie 10 ft farther out; those at 480, another 10 ft farther out, and so on. In Fig. 17-16 these points are plotted along lines normal to the center line at stations $55 + 00$ and $58 + 33$, and the elevations of the grading contours are shown at these stations. The top of the cut runs out at elevation 513 ft on one side of the road and at 497 ft on the other side.

A line joining the intersections of the grading contours with the corresponding ground contours represents the edge of the cut or the intersection of the ground by the side slopes. Any point on this line represents a slope-stake position. For example, the slope stakes at station $57 + 00$ would be located at points c and d.

Figure 17-17(a) shows the prismoid lying between the 475- and 480-ft contour lines in Fig. 17-16. To determine its volume, the area A_{475}, enclosed by the 475-ft ground contour and the 475-ft grading contour on the map is measured by using a planimeter. The area A_{480}, enclosed by the 480-ft

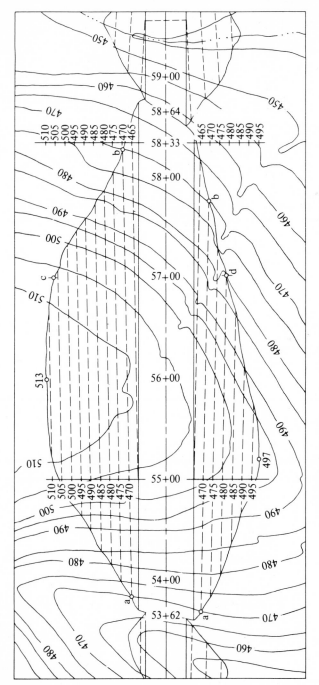

Figure 17-16. Design map showing grading contours for roadbed in cut.

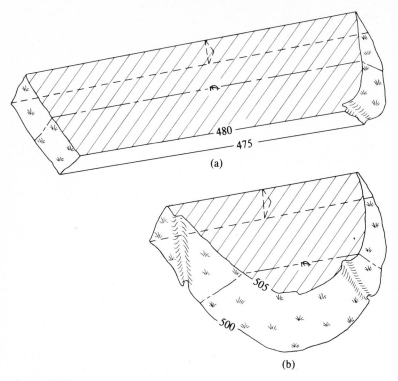

Figure 17-17. Volumes between successive contour surfaces.

contour lines, is also measured. The volume found by the average-end-area method is

$$V_{475-480} = 5 \times \frac{A_{475} + A_{480}}{2}$$

in which the volume is in cubic feet and the areas are in square feet, since the vertical distance between the two areas is in feet.

Figure 17-17(b) shows the prismoid between the 500- and 505-ft contour lines in Fig. 17-16, and its volume is determined in the same manner.

The volume in Fig. 17-16 lying between the 510-ft contour and the point at elevation 513 ft is computed by treating the earthwork as a pyramid. This volume is

$$V_{510-513} = \tfrac{1}{3} \times 3 \times A_{510}$$

Similarly, the volume lying between the 495-ft contour and the point at elevation 497 ft is

$$V_{495-497} = \tfrac{1}{3} \times 2 \times A_{495}$$

The method of procedure to be used in determining earthwork volumes from a topographic map by means of grading contours is summarized as follows:

1. Plot the proposed facility, such as a building, highway, or dam, to scale in the desired position. This can be done by referencing the facility to the horizontal control shown on the map, or by using coordinates. (A design map must always show coordinate grid lines or grid ticks.)
2. Establish all grade points from the grade elevations shown on the plans of the facility by interpolating between existing contour lines. Locating these points allows a visualization of areas of cut and fill.
3. From known grade elevations at certain points and the side-slope ratios given on the plans, plot the positions of points that will lie on the grading contours.
4. Draw the grading contours in their proper positions, and terminate them where they join the existing ground-contour lines.
5. Measure the areas enclosed by successive contours, and record the areas in a tabular form.
6. Compute the volumes lying between the successive contour surfaces by using the average-end-area method. Apply the pyramidal formula where the cut or fill runs out. In some instances the prismoidal formula may be applied. In these instances the vertical separation between surfaces will be twice the contour interval, since the area enclosed by the middle one of three consecutive contour lines will be the area M of the middle section.

The method of determining earthwork quantities by the grading-contour technique presupposes a reliable topographic map of reasonably large scale. The accuracy obtainable is equivalent to that obtained by field cross-sectioning methods and the work is performed more rapidly. The method is advantageously applied where the facility involves complicated warped grading surfaces, the dimensions of which are difficult to compute. Such would be the case in computing earthwork volumes at proposed highway interchanges.

17-16. RESERVOIR VOLUMES FROM CONTOUR MAPS   Two general methods are used in determining reservoir volumes from contour maps. One method is to planimeter the area enclosed by each contour. The volume is then computed by the average-end-area method, A_0 and A_1 being the planimetered areas of two adjacent contours and L the contour interval. The prismoidal formula can also be used where there is an odd number of contours, alternate contours being considered as middle sections and L as twice the contour interval.

A second method, which can be used when the reservoir is regular in shape, is to scale the dimensions of vertical cross sections from the contour map.

The volume is then calculated from the cross-sectional areas, as in the case of a route survey.

17-17. MASS DIAGRAM
Many of the problems connected with the handling of earthwork on highway and railroad construction can be solved by means of a mass diagram. It is usually plotted directly below the profile of the route, the ordinate at any station representing the algebraic sum of the volumes of cut and fill up to that station. From the diagram the most economical distribution of materials can be determined.

There are two common types of grading contracts. In one, the contractor bids a lump sum for the work, handling the materials as he sees fit. The mass diagram is of great value to him in determining the most economical manner of doing the work. In the other form of contract, payment is on the basis of the number of cubic yards of material handled. In this contract the successful bidder is paid a specified price per cubic yard for excavation. Embankment is taken care of by a clause stating that the excavation shall be deposited as directed by the engineer, who naturally directs that the excavated material be placed in embankments.

Since it would be unfair to expect the contractor to haul the materials taken from the excavations unlimited distances in forming the embankments, the contract will state the distance—500, 1000, or 1500 ft—that the excavation must be hauled without increased compensation. As soon as the contract limit has been exceeded, an additional sum per cubic yard is paid for each 100 ft of excess distance. The specified distance is the *limit of free haul*, and any excess is *overhaul*. The volumes and lengths involved in the calculation of overhaul payments can be taken from the mass diagram.

17-18. SHRINKAGE
Material taken from an excavation may occupy a greater or less volume when deposited in an embankment. As the material is first excavated and placed in a truck or other conveyance, it usually will occupy a greater volume than in its original position. Solid rock that must be broken up for handling may occupy about twice as much space as it did before being excavated. When earth is placed in an embankment, it will be compacted and may occupy less space, especially if it has been removed from the surface of a cultivated field and deposited at the bottom of a 15- or 20-ft embankment. The weight of a high fill may cause the original ground surface to settle or subside. When the fill is across marshy ground, this subsidence may be considerable. As the material is transported from excavation to embankment, some of it is generally lost in transit. Hence, for embankments composed of anything but rock, a volume somewhat greater than the calculated volume of the embankment will be required to make the actual embankment. An allowance is usually made for this shrinkage. The amount allowed may vary from 5% to 15%, the actual allowance depending on the

character of the material handled and the condition of the ground on which the embankment is placed.

17-19. COMPUTATION OF MASS DIAGRAM   In studies that are made to decide on the best line location and grade line, earthwork quantities are determined from preliminary cross sections taken in the field, from topographic maps, or from stereoscopic models. If the mass diagram is to be used in computing overhaul payments, the volumes used are usually taken from the final cross sections unless the contract specifies the use of large-scale maps for this purpose. In the latter event, the quantities computed for the final line discussed in Section 17-13 are used to compute the mass diagram.

Points beyond which it is not feasible to haul material define the limits of a mass diagram. A limit point may be the beginning of a project, the end of a project, a bank of a river, or an edge of a deep ravine where a bridge will be constructed.

The ordinates to the mass diagram are computed by adding the volumes between successive stations to the previous volumes, thus obtaining the accumulated volume up to any station. The volumes, and thus the ordinates, are in cubic yards. The abscissas are in stations. Excavation volumes are considered plus and embankment volumes are minus. Before embankment volumes are used to compute ordinates, they must be increased to allow for shrinkage. This allowance is usually expressed as a percent shrinkage or as a shrinkage factor. In the tabulation in Table 17-4, the shrinkage factor is 10%.

The initial ordinate in the tabulation is arbitrarily set at some large number so that all ordinates are positive. In Table 17-4 the initial ordinate is 10,000 yd^3. Between stations 7 and 9, stations 19 and 22, stations 33 and 35, stations 41 and 45, and stations 50 and 52, both excavation and embankment occur. These are portions of the route similar to that shown in Fig. 17-8. Since excavation will be carried from one side of the roadway to the other during grading, only the net amount of cut or fill is used to compute the mass-diagram ordinate. In Fig. 17-18 the mass diagram is shown plotted below the center-line profile.

An inspection of the mass diagram will show that a rising curve indicates excavation; a descending curve, embankment. Maximum and minimum points on the mass diagram occur at grade points on the profile. If a horizontal line is drawn to intersect the diagram at two points, excavation and embankment (adjusted for shrinkage) will be equal between the two stations represented by the points of intersection. Such a horizontal line is called a *balance line* because the excavation balances the embankment between the two points at its ends.

Since the ordinates to the diagram represent the algebraic sums of the volumes of excavation and embankment referred to the initial ordinate, the total volumes of excavation and embankment will be equal where the final

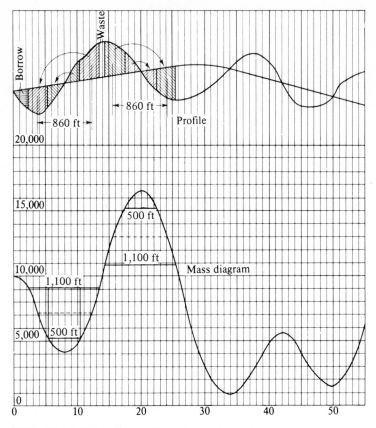

Figure 17-18. Center-line profile and mass diagram.

TABLE 17-4. Calculation of Ordinates to Mass Diagram

STATION	CUT VOLUME (yd^3)	FILL VOLUME (+10%)	MASS DIAGRAM	STATION	CUT VOLUME (yd^3)	FILL VOLUME (+10%)	MASS DIAGRAM
0			10,000.0	28			6,040.4
		183.5				1,611.2	
1			9,816.5	29			4,429.2
		622.2				1,338.2	
2			9,194.3	30			3,091.0
		1,034.7				1,002.9	
3			8,159.6	31			2,088.1
		1,268.2				652.7	
4			6,891.4	32			1,435.4
		1,231.5				357.2	

TABLE 17-4. (*continued*)

STATION	CUT VOLUME (yd^3)	FILL VOLUME (+10%)	MASS DIAGRAM	STATION	CUT VOLUME (yd^3)	FILL VOLUME (+10%)	MASS DIAGRAM
5			5,659.9	33			1,078.2
		919.1			39.2	150.0	
6			4,740.8	34			967.4
		502.8			235.8	51.8	
7			4,238.0	35			1,151.4
	20.6	163.6			465.2		
8			4,095.0	36			1,616.6
	190.2	12.0			711.5		
9			4,273.2	37			2,328.1
	615.7				904.4		
10			4,888.9	38			3,232.5
	941.5				904.4		
11			5,830.4	39			4,136.9
	1,150.0				757.0		
12			6,980.4	40			4,893.9
	1,500.0				516.5		
13			8,480.4	41			5,410.4
	1,772.7				280.2	90.4	
14			10,253.1	42			5,600.2
	1,754.6				126.7	316.1	
15			12,007.7	43			5,410.8
	1,540.4				98.4	640.0	
16			13,548.1	44			4,869.2
	1,262.3				20.0	770.6	
17			14,810.4	45			4,118.6
	931.8					789.0	
18			15,742.2	46			3,329.6
	546.5					728.0	
19			16,288.7	47			2,601.6
	203.1	30.6				577.2	
20			16,461.2	48			2,024.4
	101.0	278.5				355.7	
21			16,283.7	49			1,668.7
	18.0	586.2				115.9	
22			15,715.5	50			1,552.8
		1,005.3			407.0	100.0	
23			14,710.2	51			1,859.8
		1,377.4			735.5	26.6	
24			13,332.8	52			2,568.7
		1,676.0			983.3		
25			11,656.8	53			3,552.0
		1,860.3			1,285.9		
26			9,796.5	54			4,837.9
		1,917.1			1,522.4		
27			7,879.4	55			6,360.3
		1,839.0					

ordinate equals the initial ordinate. If the final ordinate is greater than the initial ordinate, there is an excess of excavation; if it is less than the initial ordinate, the volume of embankment is the greater and additional material must be obtained to complete the embankments.

In deciding on the grade line to be adopted, reasonable rates of grade should not be exceeded and an attempt should be made to balance the volumes of cut and fill over moderately short stretches of the line. If this is not done, some of the material may have to be hauled excessively long distances. Where it is impossible to avoid long hauls, it is often cheaper to waste material in one place and obtain the volumes necessary to complete the embankments from borrow pits located along the right of way.

17-20. CALCULATION OF OVERHAUL When the contract prices are known, the economical limit of haul can be determined. This limit is reached when the cost of haul equals the cost of excavation. For longer distances it will be cheaper to waste in one place and borrow in another. As an example, the economical limit of haul will be 1100 ft if the cost of excavation and hauling 500 ft or less is 24 cents per cubic yard, and if the cost of hauling each additional 100 ft is 4 cents. At this distance, the cost for 1 yd³ will be 24 cents for excavation and hauling 500 ft, plus 6 × 4 = 24 cents for overhaul. For longer distances the cost of excavating and hauling will be greater than 48 cents, or the cost of excavating twice, and it will be cheaper to borrow and waste. For distances under 1100 ft the cost of excavating and hauling will be less than 48 cents, and hauling will be the cheaper method. These figures are on the assumption that fill material is available within the limits of the right of way. If additional land for borrow pits must be purchased, the economical limit of haul would be greater.

To determine the cost of the grading shown in Fig. 17-18, two horizontal lines are drawn. The length of one is 500 ft, the limit of free haul; the length of the other is 1100 ft, the assumed economical limit of haul. Although excavation and embankment balance between station 0 + 00 and station 14 + 00 it would not be economical to haul material that distance. It will be cheaper to borrow 900 yd³ needed to make the fill from station 0 + 00 to station 2 + 30. The cost of this fill will be

$$900 \times \$0.24 = \$216.00$$

The 500-ft line intersects the mass diagram at station 5 + 40 and at station 10 + 40. When the excavation between the grade point at station 8 + 00 and station 10 + 40 is moved backward along the line, it will complete the fill from station 5 + 40 to station 8 + 00. The volume, as scaled from the diagram, is 1200 yd³. Since no part of it is hauled more than 500 ft, the cost of this grading will be

$$1200 \times \$0.24 = \$288.00$$

A horizontal line 1100 ft long extends from station 2 + 30 to station 13 + 30. The volume from station 10 + 40 to station 13 + 30 is 3800 yd³. This is the volume needed to complete the fill between station 2 + 30 and station 5 + 40. All of this material is hauled more than 500 ft. The positions of the centers of gravity of the masses of excavation and embankment can be found by drawing a horizontal line midway—vertically—between the 1100- and 500-ft horizontal lines. This line intersects the mass diagram at station 3 + 70, the center of gravity of the embankment, and at station 12 + 30, the center of gravity of the excavation. The average haul is the difference between these two stations, or 860 ft. Since the charge for overhaul is only for the distance in excess of 500 ft, the average overhaul distance is 860 − 500 = 360 ft. The cost of excavating and moving the material between station 10 + 40 and station 13 + 30 will be

excavation and 500-ft haul	3800 × $0.24 = $ 912.00
overhaul—3.6 stations	3800 × 3.6 × 0.04 = 547.20
	total cost = $1459.20

17-21. ECONOMICAL HANDLING OF MATERIAL
When there are frequent grade points on the profile, some study may be necessary to decide whether a given excavation should be moved forward or back along the line, or perhaps wasted. The mass diagram facilitates the choice of the plan requiring the least amount of work. Since the areas between horizontal balance lines and the mass diagram are the graphical products of volumes and distances, they are indicative of the amount of work involved in the grading. The smaller the areas, the less will be the work, and the cheaper the cost. Where the work may be done in two or more ways, the most economical way will be the one for which the areas between the mass diagram and the horizontal lines are a minimum.

These facts will be apparent from an investigation of the profile and mass diagram shown in Fig. 17-19. Although excavation and embankment balance between station 40 + 00 and station 70 + 00, there must be some waste and borrow if the economical limit of haul is not to be exceeded. If the horizontal

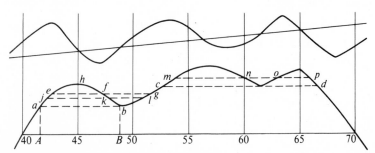

Figure 17-19. Economical hauling of materials.

line *ab* is lowered, some material would be hauled more than 2500 ft. Consequently, the line cannot be below *ab*. With the line in this position, there will be a volume of waste represented by the ordinate *Aa*. However, the total amount of waste will not be increased if the line *ab* is raised since there is to be additional waste beyond *b* and any increase to the left of *a* will be compensated for by an equal decrease beyond *b*. The most economical position for the line will be *efg*, where *ef* = *fg*. With the line in this position, there is no increase in the total amount of waste, but the labor involved is a minimum. The area *ehf* + *fgb* is a measure of the work required to make the corresponding excavations and embankments. Should the line be lowered to the position *jkl*, the corresponding area is *jhk* + *klb*. This area is greater than the former since the approximate trapezoid that has been added, or *efkj*, is greater than the one taken away, or *fglk*.

Similarly, if the line *efg* is raised, the area taken away from *ehf* will be less than the area added to *fgb*, and the resulting area will be greater. Consequently the labor involved will be a minimum when *ef* = *fg*.

Beyond station 50 + 00, the horizontal line cannot be below *cd* without exceeding the economical limit of haul. For this line the most economical position is *mnop*, where *mn* + *op* − *no* equals the economical limit of haul. By projecting these points onto the profile, it will be evident which parts of the excavations are to be moved forward and which backward in making the embankments.

PROBLEMS

17-1. Plot the following preliminary cross-section notes to a scale of 1 in. = 10 ft.

227.4	230.1	231.4	236.0	238.2	238.2	239.8	237.0
33.6	24.3	13.6	7.1	0	16.4	27.2	38.7

Assume a 28-ft roadbed at grade elevation of 226.00 ft and side slopes of $1\frac{1}{2}$:1. The roadbed center line coincides with the preliminary center line at this station. Plot the roadbed cross section, and scale the distances from the center line to the ⚹catch points and their elevations. Compute the cross-sectional area by the method of coordinates. ⚹ Point of intersection of proposed grade line w/ surface of the ground

17-2. Change the grade elevation in Problem 17-1 to 224.50 ft and shift the roadbed 8 ft to the left. Then recompute the cross-sectional area.

17-3. Plot the following preliminary cross-section notes to a scale of 1 in. = 10 ft.

537.2	541.1	545.4	546.3	546.4	544.0	554.0	556.2
58.8	30.5	18.7	0	3.6	20.0	36.2	45.8

Assume a 36-ft roadbed at grade elevation of 554.00 ft and side slopes of 2:1. The roadbed center line coincides with the preliminary center line at this station. Plot the roadbed cross section and scale the distances from the center line to the catch points and their elevations. Compute the cross-sectional area by the method of coordinates.

17-4. Change the grade elevation in Problem 17-3 to 556.50 ft and shift the roadbed 12 ft to the right. Then recompute the cross-sectional area.

17-5. Change the grade elevation in Problem 17-3 to 542.00 ft and the side slopes to 1:1. Assume that the roadbed center line coincides with the preliminary center line. Compute the cross-sectional area.

17-6. Change the grade elevation in Problem 17-3 to 538.00 ft and the side slopes to 1:1. Shift the roadbed 10 ft to the left. Compute the cross-sectional area.

17-7. The center cut at a cross section to be slope staked is 4.8 ft. The roadbed width is 50 ft and side slopes are 2:1. A vertical angle (refer to Fig. 17-4) of $+7° 12'$ is measured to a tentative position of the slope stake, and a slope distance of 42.5 ft is measured. Is the tentative position too far out or too far in, and approximately by how much?

17-8. Change the center cut in Problem 17-7 to 7.2 ft, and the vertical angle to $+6° 10'$, with a tentative slope distance of 54.7 ft.

17-9. Change the center cut in Problem 17-7 to 12.2 ft, and the vertical angle to $+4° 00'$, with a tentative slope distance of 50.0 ft.

17-10. From the accompanying final cross-section notes, compute the total volume of cut and the total volume of fill between station 43 + 00 and station 48 + 00 by the average-end-area method. The roadbed width is 24 ft in cut and 20 ft in fill; and the side slopes are 1:1 in cut and $1\frac{1}{2}$:1 in fill.

| | | | CROSS SECTION | | |
| | GROUND | GRADE | | | |
STATION	ELEVATION	ELEVATION	L	C	R
48 + 00	561.2	567.32	$\dfrac{F9.3}{24.0}$	$\dfrac{F6.1}{0}$	$\dfrac{F4.8}{17.2}$
47 + 00	565.9	568.32	$\dfrac{F4.8}{17.2}$	$\dfrac{F2.4}{0}$	$\dfrac{F2.0}{13.0}$
46 + 68	567.7	568.64	$\dfrac{F2.9}{14.5}$	$\dfrac{F0.9}{0}$	$\dfrac{F0.0}{10.0}$
46 + 00	569.3	569.32	$\dfrac{F1.8}{12.7}$	$\dfrac{C0.0}{0}$	$\dfrac{C2.7}{14.7}$
45 + 82	571.2	569.50	$\dfrac{C0.0}{12.0}$	$\dfrac{C2.0}{0}$	$\dfrac{C4.1}{16.1}$
45 + 00	573.8	570.32	$\dfrac{C2.2}{14.2}$	$\dfrac{C3.5}{0}$	$\dfrac{C6.2}{18.2}$
44 + 50	576.0	570.82	$\dfrac{C4.0}{16.0}$	$\dfrac{C5.2}{0}$	$\dfrac{C8.0}{20.0}$
44 + 00	576.6	571.32	$\dfrac{C3.5}{15.5}$	$\dfrac{C5.3}{0}$	$\dfrac{C7.8}{19.8}$
43 + 40	579.7	571.92	$\dfrac{C5.6}{17.6}$	$\dfrac{C7.8}{0}$	$\dfrac{C9.0}{21.0}$
43 + 00	580.4	572.32	$\dfrac{C5.8}{17.8}$	$\dfrac{C8.1}{0}$	$\dfrac{C11.6}{23.6}$

17-11. Compute the volume in cubic yards between stations 47 and 48 in Problem 17-10 by the prismoidal formula.

17-12. Compute the volume in cubic yards between stations 44 and 45 in Problem 17-10 by the prismoid formula, using the cross-sectional area at station 44 + 50 as the area of the midsection.

17-13. Assuming the unit prices used in Section 17-19, compute the cost of grading in Fig. 17-17.

17-14. Given the following five-level sections, compute the volume of earthwork lying between them by (a) the average-end-area method, and (b) by the prismoidal formula. Compute the side slopes.

Station 44 + 00	$\dfrac{C9.3}{42.6}$	$\dfrac{C10.2}{24.0}$	$\dfrac{C8.4}{0}$	$\dfrac{C7.8}{24.0}$	$\dfrac{C5.9}{35.8}$
Station 45 + 00	$\dfrac{C3.4}{30.8}$	$\dfrac{C3.0}{24.0}$	$\dfrac{C2.2}{0}$	$\dfrac{C1.6}{24.0}$	$\dfrac{C1.7}{27.4}$

17-15. Given the following notes for three irregular sections, compute the volume of earthwork lying between station 15 and station 16 by (a) the average-end-area method and (b) by the prismoidal formula. The roadbed width is 60 ft and the side slopes are $1\frac{1}{2}:1$. $V_e = 2320.2$ yd³ = $V_p = 2422.6$ yd³

Station 15 + 00	$\dfrac{F4.6}{36.9}$	$\dfrac{F7.0}{29.2}$	$\dfrac{F6.8}{18.8}$	$\dfrac{F9.4}{7.4}$	$\dfrac{F8.0}{0}$	$\dfrac{F7.8}{13.6}$	$\dfrac{F13.2}{42.2}$	$\dfrac{F20.0}{60.0}$	660.5
Station 15 + 50	$\dfrac{F5.6}{38.4}$	$\dfrac{F6.0}{30.8}$	$\dfrac{F14.0}{23.2}$	$\dfrac{F9.2}{0}$	$\dfrac{F9.2}{11.1}$	$\dfrac{F11.5}{29.7}$	$\dfrac{F5.4}{38.1}$		709.4
Station 16 + 00	$\dfrac{F1.8}{32.7}$	$\dfrac{F3.8}{28.8}$	$\dfrac{F2.3}{20.0}$	$\dfrac{F5.0}{6.1}$	$\dfrac{F5.2}{0}$	$\dfrac{F10.3}{11.7}$	$\dfrac{F7.4}{31.2}$	$\dfrac{F10.2}{45.3}$	426.5

17-16. In Fig. 15-9, the area bounded by points *B*-1, *B*-5, *H*-5, *H*-3, *F*-3, and *F*-1 is to be brought to a level grade of 695.00 ft. The sides of the squares are 50 ft. Assuming vertical faces for all sides of the excavation, compute the volume in cubic yards to be removed, using Eq. (17-12).

17-17. Change the grade elevation of Problem 17-16 to 688.00 ft and compute the volume in cubic yards to be removed.

17-18. In Fig. 15-9 the area bounded *A*-1, *A*-4, *D*-4, *D*-6, *F*-6, and *F*-1 is to be brought to a level grade of 697.00 ft. The sides of the squares are 50 ft. Assuming vertical faces for all sides of the excavation, compute the volume in cubic yards to be removed, using Eq. (17-12).

BIBLIOGRAPHY

Meyer, C. F., and Gibson, D. W. 1980. *Route Surveying and Design.* New York: Harper & Row.

Skelton, R. R. 1949. *Route Surveys.* New York: McGraw-Hill.

Stipp, D. W. 1968. Trigonometric construction staking. *Surveying and Mapping* 28:437.

Chapter 18
United States Public
Land Surveys

18-1. HISTORICAL The United States rectangular surveying system was devised with the object of marking on the ground and fixing for all time legal subdivisions for the purposes of description and disposal of the public domain under the general land laws of the United States. This system has been used in 30 states including Alaska. It has not been used in the older states along the Atlantic seaboard and in a few others where the lands were in private hands before the federal government came into being.

The rectangular system of survey of the public lands was inaugurated by a committee appointed by the Continental Congress. In 1784 this committee reported "An ordinance for ascertaining the mode of locating and disposing of lands in the western territory, and for other purposes therein mentioned." The ordinance, as finally passed on May 20, 1785, provided for townships 6 miles square, containing 36 sections 1 mile square. The first public surveys were made under the direction of the Geographer of the United States. The area surveyed now forms a part of the state of Ohio. In these initial surveys only the exterior lines of the townships were surveyed, but the plats were marked by subdivisions into sections 1 mile square, and mile corners were established on the township lines. The sections were numbered from 1 to 36, commencing with number 1 in the southeast corner of the township and closing with number 36 on the northwest corner thereof. By this method number 6 was in the northeast corner, and number 7 was west of, and adjacent to, number 1.

The act of congress approved May 18, 1796, provided for the appointment of a surveyor general and directed the survey of the lands northwest of the Ohio River and above the mouth of the Kentucky River. Under this law it was provided that "The sections shall be numbered, respectively, beginning with number one in the northeast section and proceeding west and east alternately through the township, with progressive numbers, till the thirty-sixth be completed." This method of numbering sections, shown in Fig. 18-1, is still in use.

Since that time the laws relating to the survey of the public lands have been amended several times. Until quite recently, it was the practice of the government to award contracts for the survey of certain portions of the public lands. These awards were frequently made, in payment of political obligations, to persons wholly unfit to be entrusted with such work. As a result, much of the work has been carelessly done. Fraudulent returns have been made to the advantage of speculators in timber and mineral lands. In more than one instance the government has been furnished with the field notes of a survey executed in the comparative comfort of a tent rather than upon the ground. In 1910, after most of the damage had been done, Congress abolished this contract system and authorized the interior department to employ a permanent body of surveyors, known as United States surveyors.

The interest of the present-day engineer or surveyor in the rectangular system is in the retracement of old lines and in the subdivision of the section into smaller units. For this reason it is imperative that he be familiar with the methods used in the original survey.

6	5	4	3	2	1
7	8	9	10	11	12
18	17	16	15	14	13
19	20	21	22	23	24
30	29	28	27	26	25
31	32	33	34	35	36

Figure 18-1. Method of numbering sections.

18-2. MANUAL OF SURVEYING INSTRUCTIONS Various regions of the United States have been surveyed under different sets of instructions issued at periods ranging from 1785 to the present time. The earliest instructions were issued to surveyors in manuscript or in printed circulars. Regulations more in detail, improving the system for greater accuracy, permanency, and uniformity, were issued in book form in editions of 1855, 1881, 1890, 1894, 1902, 1919, 1930, and 1947. The latest edition is the *Manual of Surveying Instructions, 1973*. The methods outlined in the following articles are taken from the edition of the *Manual of Instructions for the Survey of the Public Lands of the United States, of 1947*, prepared and published under the direction of the Director of the Bureau of Land Management, and printed by the U.S. Government Printing Office. These methods are the same as those outlined in the 1973 Manual.

Although the methods have, in general, been the same on all surveys, there have been important variations in detail. The local engineer or surveyor, before beginning the retracement of old land lines, should familiarize himself with the exact methods in use at the time the original survey was made.

18-3. GENERAL PROCEDURE The first step in the survey of an area is to divide it into tracts approximately 24 miles square by means of meridians and parallels of latitude. These 24-mile tracts are then divided into 16 townships, which are approximately 6 miles on a side. The last step, so far as the federal government is concerned, is to divide the township into 36 sections, each approximately 1 mile square. The subdivision of the sections into smaller units is the task of the local engineer and surveyor.

18-4. INITIAL POINT The place of beginning of the survey of any given region is called the *initial point*. Through this point is run a meridian, called the *principal meridian*, and a parallel of latitude, called the *baseline*. The initial point is selected with a view to its control of extensive agricultural areas within reasonable geographical limitations. On the establishment of an initial point, the position of the point in latitude and longitude is determined by accurate field astronomical methods.

Since surveys in widely separated sections of the country have been in progress simultaneously, a large number of initial points have been established. The positions of these points and the areas governed by them are given in Table 18-1. Since many of the points were established before present-day facilities for accurate field astronomical determinations were available, some of the values shown are only approximately correct. Present instructions call for the establishment of the initial point in such a manner as to make it as nearly permanent as possible.

TABLE 18-1. Meridians and Baselines of the United States Rectangular Surveys

MERIDIAN	GOVERNING SURVEYS (WHOLLY OR IN PART) IN THE STATES OF—	LONGITUDE OF PRINCIPAL MERIDIAN WEST FROM GREENWICH	LATITUDE OF BASELINE NORTH FROM EQUATOR
Black Hills	South Dakota	104° 03′ 16″	43° 59′ 44″
Boise	Idaho	116° 23′ 35″	43° 22′ 21″
Chicasaw	Mississippi	89° 14′ 47″	35° 01′ 58″
Choctaw	Mississippi	90° 14′ 41″	31° 52′ 32″
Cimarron	Oklahoma	103° 00′ 07″	36° 30′ 05″
Copper River	Alaska	145° 18′ 37″	61° 49′ 04″
Fairbanks	Alaska	147° 38′ 26″	64° 51′ 50″
Fifth Principal	Arkansas, Iowa, Minnesota, Missouri, North Dakota, and South Dakota	91° 03′ 07″	34° 38′ 45″
First Principal	Ohio and Indiana	84° 48′ 11″	40° 59′ 22″
Fourth Principal	Illinois	90° 27′ 11″	40° 00′ 50″
Fourth Principal	Minnesota and Wisconsin	90° 25′ 37″	42° 30′ 27″
Gila and Salt River	Arizona	112° 18′ 19″	33° 22′ 38″
Humboldt	California	124° 07′ 10″	40° 25′ 02″
Huntsville	Alabama and Mississippi	86° 34′ 16″	34° 59′ 27″
Indian	Oklahoma	97° 14′ 49″	34° 29′ 32″
Kateel River	Alaska	158° 45′ 31″	65° 26′ 17″
Louisiana	Louisiana	92° 24′ 55″	31° 00′ 31″
Michigan	Michigan and Ohio	84° 21′ 53″	42° 25′ 28″
Mount Diablo	California and Nevada	121° 54′ 47″	37° 52′ 54″
Navajo	Arizona	108° 31′ 59″	35° 44′ 56″
New Mexico Principal	Colorado and New Mexico	106° 53′ 12″	34° 15′ 35″
Principal	Montana	111° 39′ 33″	45° 47′ 13″
Salt Lake	Utah	111° 53′ 27″	40° 46′ 11″
San Bernardino	California	116° 55′ 48″	34° 07′ 13″
Second Principal	Illinois and Indiana	86° 27′ 21″	38° 28′ 14″
Seward	Alaska	149° 21′ 26″	60° 07′ 37″
Sixth Principal	Colorado, Kansas, Nebraska, South Dakota, and Wyoming	97° 22′ 08″	40° 00′ 07″
St. Helena	Louisiana	91° 09′ 36″	30° 59′ 56″
St. Stephens	Alabama and Mississippi	88° 01′ 20″	30° 59′ 51″
Tallahassee	Florida and Alabama	84° 16′ 38″	30° 26′ 03″
Third Principal	Illinois	89° 08′ 54″	38° 28′ 27″
Uintah	Utah	109° 56′ 06″	40° 25′ 59″
Umiat	Alaska	152° 00′ 05″	69° 23′ 30″
Ute	Colorado	108° 31′ 59″	39° 06′ 23″
Washington	Mississippi	91° 09′ 36″	30° 59′ 56″
Willamette	Oregon and Washington	122° 44′ 34″	45° 31′ 11″
Wind River	Wyoming	108° 48′ 49″	43° 00′ 41″

18-5. PRINCIPAL MERIDIAN The principal meridian is a true meridian that is astronomically determined and is extended from the initial point, either north or south, or in both directions, as the conditions may require, to the limits of the area being subdivided. Monuments are placed on this line at intervals of 40 chains ($\frac{1}{2}$ mile), and at its intersection with navigable bodies of water, streams 3 chains or more in width, and lakes of an area of 25 acres or more.

Two independent sets of measurements of this line are made. Should the difference between the two sets exceed 20 links (13.2 ft) per 80 chains, it is required that the line be remeasured to reduce the difference. If tests for alignment show the line to have deviated more than 3′ from the true cardinal course, the alignment must be corrected.

It should be borne in mind that the only equipment available for most of the early surveys was a compass and a chain. Therefore discrepancies far in excess of these limits will be found in rerunning many old lines. In other cases the agreement will be surprisingly close, when one considers that much of the country was often heavily wooded at the time the surveys were made.

18-6. BASELINE From the initial point the baseline is extended east and west on a true parallel of latitude to the limits of the area being surveyed. Monuments are placed on this line at intervals of 40 chains and at its intersection with all meanderable bodies of water. The manner of making the measurement of the baseline and the accuracy of both the alignment and the measurement must be the same as that required in the survey of the principal meridian.

Because of the convergence of meridians, a line making an angle of 90° with the principal meridian will, theoretically, be an east-and-west line for only an infinitesimal distance from the meridian. If extended, the resulting line is an arc of a great circle, and gradually departs southerly from the true parallel, which is a small circle. On this account, the baseline, being a parallel of latitude, is run as a curve with chords 40 chains in length. Three methods are employed in running the baseline—namely, the solar method, the tangent method, and the secant method.

The solar method is based on the fact that a line making a right angle with the meridian does not differ appreciably from the theoretical parallel of latitude in a distance of 40 chains. If a transit, equipped with a solar attachment (see Section 12-21) and in good adjustment, is employed, the true meridian may be determined by observation at the end of each half-mile interval. At each point an angle of 90° is turned from the new meridian, and this line is extended 40 chains to a point at which another meridian is determined. Thus the direction of the line beyond each half-mile monument is determined by turning off a right angle from the meridian at that monument.

18-7. BASELINE BY TANGENT METHOD
In the tangent method of locating a parallel of latitude, the direction of the tangent is determined by turning off a horizontal angle of 90° to the east or the west from the meridian. Points on the true parallel are established at half-mile intervals by offsets to the north from this tangent. The establishment of a baseline in latitude 42° 30′ N is illustrated in Fig. 18-2.

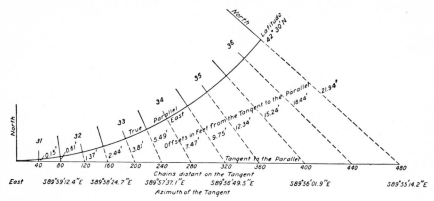

Figure 18-2. Baseline by offsets from tangent.

TABLE 18-2. Offsets (in Feet) From Tangent to Parallel

LATITUDE	1 MILE	2 MILES	3 MILES	4 MILES	5 MILES	6 MILES
30°	0.38	1.54	3.46	6.15	9.61	13.83
31°	0.40	1.60	3.60	6.40	10.00	14.40
32°	0.42	1.66	3.74	6.65	10.40	14.97
33°	0.43	1.73	3.89	6.91	10.80	15.56
34°	0.45	1.80	4.04	7.18	11.22	16.16
35°	0.47	1.86	4.19	7.45	11.65	16.77
36°	0.48	1.93	4.35	7.73	12.09	17.40
37°	0.50	2.01	4.51	8.02	12.53	18.05
38°	0.52	2.08	4.68	8.32	12.99	18.71
39°	0.54	2.15	4.85	8.62	13.47	19.39
40°	0.56	2.23	5.02	8.93	13.95	20.09
41°	0.58	2.31	5.20	9.25	14.46	20.82
42°	0.60	2.40	5.39	9.58	14.97	21.56
43°	0.62	2.48	5.58	9.92	15.50	22.33
44°	0.64	2.57	5.78	10.28	16.06	23.12
45°	0.66	2.66	5.98	10.64	16.62	23.94
46°	0.69	2.75	6.20	11.02	17.21	24.79
47°	0.71	2.85	6.42	11.41	17.83	25.67
48°	0.74	2.95	6.65	11.81	18.46	26.58
49°	0.76	3.06	6.88	12.24	19.12	27.53
50°	0.79	3.17	7.13	12.68	19.81	28.52

TABLE 18-3. Azimuths of the Tangent to the Parallel

LATITUDE	1 MILE	2 MILES	3 MILES	4 MILES	5 MILES	6 MILES
30°	89° 59' 30.0"	89° 59' 00.0"	89° 58' 29.9"	89° 57' 59.9"	89° 57' 29.9"	89° 56' 59.9"
31°	59' 28.8"	58' 57.5"	58' 26.3"	57' 55.0"	57' 23.8"	56' 52.6"
32°	59' 27.5"	58' 55.0"	58' 22.5"	57' 50.0"	57' 17.5"	56' 45.1"
33°	59' 26.2"	58' 52.5"	58' 18.7"	57' 45.0"	57' 11.2"	56' 37.4"
34°	59' 24.9"	58' 49.9"	58' 14.8"	57' 39.7"	57' 04.6"	56' 29.6"
35°	59' 23.6"	58' 47.2"	58' 10.8"	57' 34.4"	56' 58.0"	56' 21.6"
36°	59' 22.2"	58' 44.5"	58' 06.7"	57' 28.9"	56' 51.1"	56' 13.4"
37°	59' 20.8"	58' 41.7"	58' 02.5"	57' 23.3"	56' 44.1"	56' 05.0"
38°	59' 19.4"	58' 38.8"	57' 58.2"	57' 17.6"	56' 36.9"	55' 56.3"
39°	59' 17.9"	58' 35.8"	57' 53.7"	57' 11.6"	56' 29.5"	55' 47.5"
40°	59' 16.4"	58' 32.8"	57' 49.2"	57' 05.6"	56' 21.9"	55' 38.3"
41°	59' 14.8"	58' 29.6"	57' 44.5"	56' 59.3"	56' 14.1"	55' 28.9"
42°	59' 13.2"	58' 26.4"	57' 39.6"	56' 52.8"	56' 06.0"	55' 19.3"
43°	59' 11.5"	58' 23.1"	57' 34.6"	56' 46.2"	55' 57.7"	55' 09.2"
44°	59' 09.8"	58' 19.6"	57' 29.5"	56' 39.3"	55' 49.1"	54' 58.9"
45°	59' 08.0"	58' 16.1"	57' 24.1"	56' 32.2"	55' 40.2"	54' 48.2"
46°	59' 06.2"	58' 12.4"	57' 18.6"	56' 24.7"	55' 31.0"	54' 37.2"
47°	59' 04.3"	58' 08.6"	57' 12.9"	56' 17.2"	55' 21.4"	54' 25.7"
48°	59' 02.3"	58' 04.6"	57' 06.9"	56' 09.2"	55' 11.5"	54' 13.9"
49°	59' 00.2"	58' 00.5"	57' 00.7"	56' 01.0"	55' 01.2"	54' 01.4"
50°	58' 58.1"	57' 56.2"	56' 54.3"	55' 52.4"	54' 50.5"	53' 48.6"

The values of the offsets, which depend on the latitude and are proportional to the squares of the distances from the starting meridian, are shown in Table 18-2. Also the values of the azimuth of the tangent from the south point at the end of each mile are shown in Table 18-3.

As the field instructions require the location of all landmarks with respect to measurements along the true line and also require the blazing of trees along the true line, this method is not so convenient as the secant method, particularly when the country is so heavily wooded as to require the clearing of two lines, one for the tangent and another for the parallel.

18-8. BASELINE BY SECANT METHOD

As shown in Fig. 18-3 the secant used in locating a parallel of latitude is the one passing through the 1- and 5-mile points on the parallel. The secant is located by offsetting to the south from the initial point and turning off a known angle from the meridian at the offset point. The secant is extended 6 miles, points on the parallel being located at half-mile intervals by offsets from the secant. The offsets at the $\frac{1}{2}$- and $5\frac{1}{2}$-mile points are to the north. The 1- and 5-mile points are on the secant, and the remaining points are south of the secant. The offsets and the values of the azimuth of the secant from the north at half-mile intervals up to 3 miles are shown in Table 18-4. For the other 3 miles the offsets are the same as for points at corresponding distances from the 3-mile point; and the azimuths are numerically equal to those at corresponding distances from the 3-mile point but are measured from the south.

The advantage of the secant method lies in the fact that the offsets are much smaller than those for the tangent method. By clearing a line of moderate width, both the secant and the parallel will be contained in the same clearing. Measurements to landmarks will be substantially the same along both the secant and the parallel.

18-9. STANDARD PARALLELS

The next step in the subdivision of the district being surveyed is to run the *standard parallels* or *correction lines*. These lines are parallels of latitude that are established in exactly the same

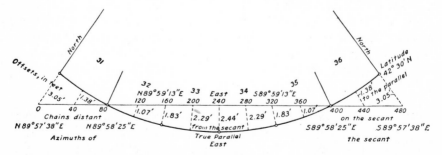

Figure 18-3. Baseline by offsets from secant.

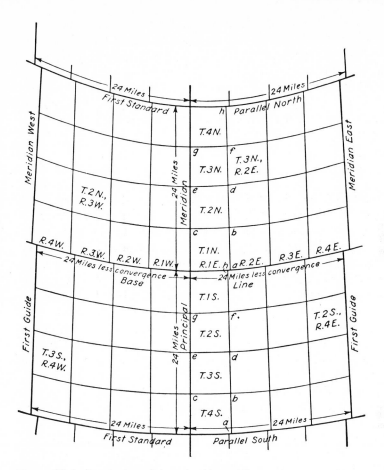

Figure 18-4. Standard parallels and guide meridians.

manner as the baseline. They are located at intervals of 24 miles north and south of the baseline, and extend to the limits of the district being surveyed. These standard parallels are numbered, as shown in Fig. 18-4, as the first (second, third, and so on) standard parallel north (or south). In the earlier surveys standard parallels have been placed at intervals of 30 or 36 miles north and south of the baseline. In connecting new work with old work, it has been necessary to establish intermediate correction lines which have been given local names, as "Fifth Auxiliary Standard Parallel North" or "Cedar Creek Correction Line."

18-10. GUIDE MERIDIANS The survey district is next divided into tracts approximately 24 miles square by means of guide meridians. These lines are true meridians which start at points on the baseline or standard parallels at intervals of 24 miles east and west of the principal meridian and

TABLE 18-4. Azimuths of the Secant, and Offsets (in Feet) to the Parallel

LATITUDE	0 MILES	½ MILE	1 MILE	1½ MILES	2 MILES	2½ MILES	3 MILES
30°	89° 58′ 30″ 1.92 N	89° 58′ 45″ 0.86 N	89° 59′ 00″ 0.00	89° 59′ 15″ 0.68 S	89° 59′ 30″ 1.16 S	89° 59′ 45″ 1.44 S	90° (E or W) 1.54 S
31°	89° 58′ 26″ 2.00 N	89° 58′ 42″ 0.90 N	89° 58′ 58″ 0.00	89° 59′ 13″ 0.70 S	89° 59′ 29″ 1.20 S	89° 59′ 44″ 1.50 S	90° (E or W) 1.60 S
32°	89° 58′ 23″ 2.08 N	89° 58′ 39″ 0.94 N	89° 58′ 55″ 0.00	89° 59′ 11″ 0.72 S	89° 59′ 28″ 1.24 S	89° 59′ 44″ 1.56 S	90° (E or W) 1.66 S
33°	89° 58′ 19″ 2.16 N	89° 58′ 36″ 0.97 N	89° 58′ 53″ 0.00	89° 59′ 09″ 0.76 S	89° 59′ 26″ 1.30 S	89° 59′ 43″ 1.62 S	90° (E or W) 1.73 S
34°	89° 58′ 15″ 2.24 N	89° 58′ 32″ 1.01 N	89° 58′ 50″ 0.00	89° 59′ 07″ 0.79 S	89° 59′ 25″ 1.35 S	89° 59′ 42″ 1.69 S	90° (E or W) 1.80 S
35°	89° 58′ 11″ 2.32 N	89° 58′ 29″ 1.05 N	89° 58′ 47″ 0.00	89° 59′ 05″ 0.81 S	89° 59′ 24″ 1.39 S	89° 59′ 42″ 1.74 S	90° (E or W) 1.86 S
36°	89° 58′ 07″ 2.42 N	89° 58′ 26″ 1.09 N	89° 58′ 45″ 0.00	89° 59′ 03″ 0.84 S	89° 59′ 22″ 1.45 S	89° 59′ 41″ 1.81 S	90° (E or W) 1.93 S
37°	89° 58′ 03″ 2.50 N	89° 58′ 22″ 1.13 N	89° 58′ 42″ 0.00	89° 59′ 01″ 0.88 S	89° 59′ 21″ 1.51 S	89° 59′ 40″ 1.88 S	90° (E or W) 2.01 S
38°	89° 57′ 58″ 2.60 N	89° 58′ 19″ 1.17 N	89° 58′ 39″ 0.00	89° 58′ 59″ 0.91 S	89° 59′ 19″ 1.56 S	89° 59′ 40″ 1.95 S	90° (E or W) 2.08 S
39°	89° 57′ 54″ 2.70 N	89° 58′ 15″ 1.21 N	89° 58′ 36″ 0.00	89° 58′ 57″ 0.94 S	89° 59′ 18″ 1.61 S	89° 59′ 39″ 2.02 S	90° (E or W) 2.15 S

40°	89° 57' 49" 2.79 N	89° 58' 11" 1.26 N	89° 58' 33" 0.00	89° 58' 55" 0.97 S	89° 59' 16" 1.67 S	89° 59' 38" 2.09 S	90° (E or W) 2.23 S
41°	89° 57' 45" 2.89 N	89° 58' 07" 1.30 N	89° 58' 30" 0.00	89° 58' 52" 1.01 S	89° 59' 15" 1.73 S	89° 59' 37" 2.17 S	90° (E or W) 2.31 S
42°	89° 57' 40" 2.99 N	89° 58' 03" 1.35 N	89° 58' 26" 0.00	89° 58' 50" 1.05 S	89° 59' 13" 1.80 S	89° 59' 37" 2.25 S	90° (E or W) 2.40 S
43°	89° 57' 35" 3.10 N	89° 57' 59" 1.40 N	89° 58' 23" 0.00	89° 58' 47" 1.08 S	89° 59' 12" 1.86 S	89° 59' 36" 2.32 S	90° (E or W) 2.48 S
44°	89° 57' 30" 3.21 N	89° 57' 55" 1.45 N	89° 58' 20" 0.00	89° 58' 45" 1.12 S	89° 59' 10" 1.93 S	89° 59' 35" 2.41 S	90° (E or W) 2.57 S
45°	89° 57' 24" 3.32 N	89° 57' 50" 1.50 N	89° 58' 16" 0.00	89° 58' 42" 1.16 S	89° 59' 08" 2.00 S	89° 59' 34" 2.49 S	90° (E or W) 2.66 S
46°	89° 57' 19" 3.45 N	89° 57' 46" 1.55 N	89° 58' 12" 0.00	89° 59' 39" 1.20 S	89° 59' 06" 2.06 S	89° 59' 33" 2.58 S	90° (E or W) 2.75 S
47°	89° 57' 13" 3.57 N	89° 57' 41" 1.61 N	89° 58' 09" 0.00	89° 58' 36" 1.24 S	89° 59' 04" 2.14 S	89° 59' 32" 2.67 S	90° (E or W) 2.85 S
48°	89° 57' 07" 3.70 N	89° 57' 36" 1.66 N	89° 58' 05" 0.00	89° 58' 33" 1.29 S	89° 59' 02" 2.21 S	89° 59' 31" 2.77 S	90° (E or W) 2.95 S
49°	89° 57' 01" 3.82 N	89° 57' 31" 1.72 N	89° 58' 01" 0.00	89° 58' 30" 1.34 S	89° 59' 00" 2.30 S	89° 59' 30" 2.87 S	90° (E or W) 3.06 S
50°	89° 56' 54" 3.96 N	89° 57' 25" 1.79 N	89° 57' 56" 0.00	89° 58' 27" 1.38 S	89° 58' 58" 2.38 S	89° 59' 29" 2.97 S	90° (E or W) 3.17 S

extend north to their intersection with the next standard parallel, as shown in Fig. 18-4. Because of the convergence of the meridians, the distance between these lines will be 24 miles only at the starting points. At all other points the distance between them will be less than 24 miles. Guide meridians are designated by number, as the first (second, third, and so on) guide meridian east (or west). As they extend from one standard parallel to the next, their lengths are limited to 24 miles (30 or 36 miles on older work). At their intersection with the standard parallel to the north a new monument is established. On the standard parallels, two sets of monuments are thus to be found. The monuments that were established as the parallel was first located are called *standard corners* and apply to the area north of the parallel. The second set, found at the intersection of the parallel with the meridians from the south, are called *closing corners* and govern the area to the south of the parallel. The distance between the closing corner and the nearest standard corner should be measured and recorded in the field notes.

The guide meridians are located with the same care for securing accuracy of alignment and measurement as is the principal meridian. The monuments are placed at half-mile intervals, except that all discrepancies in measurement are thrown into the last half mile. For this reason the distance from the closing corner to the first monument south of the standard parallel may be more or less than 40 chains.

18-11. TOWNSHIP EXTERIORS The regular order of establishing the township exteriors can be followed by reference to Fig. 18-4. Beginning at *a*, on the baseline (or standard parallel) and 6 miles east of the principal meridian (or guide meridian), a line is run due north for 6 miles, monuments being established at intervals of 40 chains. From the township corner *b*, a random line is run to the west to intersect the principal meridian (or a guide meridian) at *c*. As this line is run, temporary monuments are set at intervals of 40 chains. From the amount by which the random line fails to strike *c*, the direction of the true line between *c* and *b* can be calculated. The line is then run on this calculated course from *c* back to *b* and the temporary monuments are replaced by permanent ones in proper position. The true line is blazed through timber, and distances to important items of topography are adjusted to the correct true-line measurements.

On account of the convergence of meridians and errors in field work, the northern boundary of the township will not be exactly 6 miles in length. *All discrepancies due to convergence and to errors in measurement or alignment are thrown into the most westerly half mile, all other distances being made exactly 40 chains.* Although every attempt is made to keep lines within 0° 14' of the true courses, the boundaries of a township are not considered defective unless the error exceeds 0° 21'. If the random north boundary fails to close on the corner previously set by more than 3 chains for line or by more than 3 chains, minus the theoretical convergence, for distance, it is evident that the

0° 21′ limit has been exceeded, and the lines are retraced and the error is corrected.

The eastern boundary of the next township to the north is next run by extending the meridian from b to d, a distance of 6 miles, and locating permanent monuments at intervals of 40 chains. A random is then run westerly from d toward e and corrected back as the true line with monuments at intervals of 40 chains, measured from d, any discrepancy being thrown into the most westerly half mile. The next step is to run the line df, a random from f to g, and the true line from g to f. From f the meridian is extended to its intersection h with the baseline (or standard parallel), and a closing corner is established at the intersection. The distance from the closing corner to the nearest standard corner is measured and recorded in the field notes. Because of errors in measurement, the last half mile on the meridian may be more or less than 40 chains in length.

The two other meridians of the 24-mile block are run in a similar manner, random lines being run to the west at the township corners to connect with the corners previously set. From the township corners on the third meridian of the block, randoms are also run to the east to connect with the township corners on the guide meridian. As these lines are corrected back to the west, monuments are established at intervals of 40 chains, measured from the guide meridian, and the discrepancy is thus thrown into the most westerly half mile of the township, as before.

The instructions relative to the establishment of township exteriors in the case of irregular and partial surveys, as given in the *Manual of Surveying Instructions, 1947* are as follows:

As the remaining unsurveyed public lands are found to contain less and less extensive areas surveyable under the law, it becomes necessary to depart from the ideal procedure in order more directly to reach the areas authorized for survey. The many possible combinations are entirely too numerous to state in detail, but where an irregular order appears to be necessary such departure from the ideal order of survey will be specifically outlined in the written special instructions. Such departure should always be based on the principle of accomplishing, by whatever plan, the same relation of one township boundary to another as would have resulted from regular establishment under ideal conditions.

In authorizing surveys to be executed it will not usually be provided that exteriors are to be carried forward until the township is to be subdivided; thus, where causes operate to prevent the establishment of the boundaries in full, it is not imperative that the survey of the exterior lines be completed; under such conditions it may be found necessary to run section lines as offsets to township exteriors and such section lines will be run either on cardinal courses or parallel to the governing boundaries of such townships, or even established when subdividing, as existing conditions may require.

18-12. NUMBERING TOWNSHIPS The townships of a survey district are numbered meridionally into ranges and latitudinally into tiers with respect to the principal meridian and the baseline of the district. As illustrated in Fig. 18-4, the third township south of the baseline is in tier 3 south. Instead of tier, the word *township* is more frequently used; thus any township in this particular tier is designated as "township 3 south." The fourth township west of the principal meridian is in range 4 west. By this method of numbering, any township is located if its tier, range, and principal meridian are given, as township 3 south, range 4 west, of the fifth principal meridian. This is abbreviated T 3 S, R 4 W, 5th PM.

18-13. CONVERGENCE OF MERIDIANS The angular convergence between two meridians is dependent on the distance between the meridians, the latitude, and the dimensions of the earth. The convergence θ, in seconds of arc, can be expressed by the equation

$$\theta = \frac{s\sqrt{1 - e^2 \sin^2 \phi} \tan \phi}{a \text{ arc } 1''} \tag{18-1}$$

in which s is the distance between the meridians, e is the eccentricity of the spheroid, a is the semimajor axis of the spheroid, and ϕ is the latitude. When tables for computing geographic positions are available (see Section 10-26), the equation may be written as follows:

$$\theta = sA \tan \phi \tag{18-2}$$

TABLE 18-5. Convergence of Meridians 6 Miles Long and 6 Miles Apart

| | CONVERGENCE | | | CONVERGENCE | |
| | ON THE PARALLEL | | | ON THE PARALLEL | |
LATITUDE	(ft)	ANGLE	LATITUDE	(ft)	ANGLE
30°	27.67	03′ 00.1″	41°	41.63	04′ 31.0″
31°	28.79	03′ 07.5″	42°	43.12	04′ 40.7″
32°	29.94	03′ 14.9″	43°	44.65	04′ 50.7″
33°	31.11	03′ 22.6″	44°	46.24	05′ 01.1″
34°	32.32	03′ 30.4″	45°	47.88	05′ 11.7″
35°	33.55	03′ 38.4″	46°	49.58	05′ 22.8″
36°	34.80	03′ 46.6″	47°	51.34	05′ 34.3″
37°	36.10	03′ 55.0″	48°	53.17	05′ 46.2″
38°	37.42	04′ 03.7″	49°	55.07	05′ 58.5″
39°	38.79	04′ 12.5″	50°	57.04	06′ 11.4″
40°	40.19	04′ 21.7″	51°	59.11	06′ 24.8″

in which

$$A = \frac{\sqrt{1 - e^2 \sin^2 \phi}}{a \text{ arc } 1''}$$

The value of A, for s in metres, may be taken from the tables.

The linear convergence on the parallel, for two meridians of length l and at a distance s apart, is given by the equation

$$c = slA \tan \phi \sin 1'' \qquad (18\text{-}3)$$

The values of the convergence for two meridians 6 miles long and 6 miles apart, for latitudes between 30° and 51°, are given in Table 18-5.

18-14. SUBDIVISION OF TOWNSHIP In the work previously described the exterior lines of the township are generally marked by monuments at half-mile intervals. However, because of errors in field work and the convergence of meridians, the most westerly and the most northerly half miles of certain townships may be more or less than 40 chains in length. In subdividing the township into sections the aim is to secure as many sections as possible that will be 1 mile on a side. To accomplish this, the error due to convergence of meridians is thrown as far to the west as possible by running lines parallel to the east boundary of the township, rather than running them as true meridians. The earlier practice has been to run these lines as true meridians. Errors in linear measurements are thrown as far to the north as possible by locating monuments at intervals of 40 chains along the lines parallel to the east boundary of the township. Thus all the accumulated error falls in the most northerly half mile, which may be more or less than 40 chains in length.

The order in which the lines are run to subdivide the township into sections is indicated by the numbers on the lines of the township shown in Fig. 18-5. The instructions for monumenting these lines, as given in the *Manual of Surveying Instructions, 1947* are:

> The subdivisional survey will be commenced at the corner of sections 35 and 36, on the south boundary of the township, and the line between sections 35 and 36 will be run parallel to the east boundary of the township, or to the mean course thereof, if it is imperfect in alignment, but within limits, establishing the quarter-section corner at 40 chains, and at 80 chains, the corner of sections 25, 26, 35, and 36. From the last-named corner, a random line will be run eastward, without blazing, parallel to the south boundary of section 36, to its intersection with the east boundary of the township, placing at 40 chains from the point of beginning, a post for temporary quarter-section corner. If the random line intersects said township boundary exactly at the corner of sections 25 and 36, it will be blazed back and established as the true line, the permanent quarter-section corner being established thereon, midway

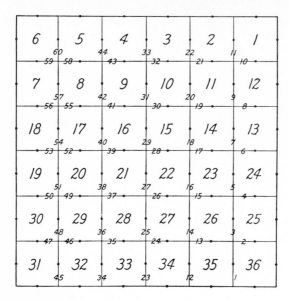

Figure 18-5. Subdivision of township.

between the initial and terminal section corners. If the random intersects said township boundary to the north or south of said corner, the falling will be carefully measured, and from the data thus obtained, the true return course will be calculated, and the true line blazed and established, and the position of the quarter-section corner determined, as directed above. The meridional section line will be continued on the same plan, likewise the successive latitudinal section lines except that each random will be run parallel to the true south boundary of the section to which it belongs. After having established the west and north boundaries of section 12, the line between sections 1 and 2 will be projected northward, on a random line, parallel to the east boundary of the township, or to its mean course, as the case may be, setting a post for temporary quarter-section corner at 40 chains, to its intersection with the north boundary of the township. If the random intersects said north boundary exactly at the corner of sections 1 and 2, it will be blazed back and established as the true line, the quarter-section corner being established permanently in its original temporary position, and the fractional measurement thrown into that portion of the line between the permanent quarter-section corner and the north boundary of the township. If, however, said random intersects the north boundary of the township to the east or west of the corner of sections 1 and 2, the falling will be carefully measured, and from the data thus obtained the true course will be calculated and the true line established, the permanent quarter-section corner being placed upon the same at 40 chains from the initial corner of the random line, thereby throwing the fractional measurement in that portion lying between the quarter-section corner and the north boundary of the township. When the north boundary of a township is a baseline or standard parallel, the line between sections 1 and 2 will be run as a true line parallel to the east boundary of the township, or to its mean course, as the case may be; the quarter-section corner will be placed at 40 chains, and a closing corner will be established at the point of intersection with such base or standard line;

and in such case, the distance from said closing corner, to the nearest standard corner on such base or standard line, will be carefully measured and noted.

The successive ranges of sections proceeding from east to west will be surveyed in the same manner; then after having established the west and north boundaries of section 32, a random line will be initiated at the corner of sections 29, 30, 31, and 32, which will be projected westward parallel to the south boundary of the township, setting a temporary quarter-section corner at 40 chains, to an intersection with the west boundary of the township, where the falling will be measured and the bearing of the true line calculated, whereupon the line between sections 30 and 31 will be permanently marked between the section corners, and the quarter-section corner thereon will be established at 40 chains from the east, thereby placing the fractional measurement in the west half mile as required by law. The survey of the west two ranges of sections will be continued on the same plan, and the random line between sections 6 and 7 will be run westward parallel to the true line between sections 7 and 18; the random will be corrected to a true line and the fractional measurement placed in the west half mile; finally the random line between sections 5 and 6 will be run northward parallel to the true line between sections 4 and 5; the random will be corrected to the true line and the fractional measurement placed in the north half mile.

In this manner the section lines are marked by monuments at half-mile intervals. The monuments on the east and west boundaries of a section are 40 chains apart, except for the fractional measurement placed in the most northerly half mile of the township. As the east and west boundaries are not exactly parallel and as there may be errors in chaining, the north and south boundaries of the section may not be exactly 80 chains. If the error does not exceed 50 links, the lines are allowed to stand and the quarter-section corner is located midway between the two section corners, except for the west range of sections, where the quarter-section corner is placed at a distance of 40 chains from the northeast corner of the section.

In addition to the section and quarter-section corners, meander corners are established at the intersection of the section lines with all navigable streams, with all streams 3 or more chains in width, and with lakes of the area of 25 acres and upward.

As noted in Section 18-2, some of the details of subdivision have been changed from time to time. According to the present instructions, double sets of corners are found only when the north boundary of the township is a baseline or a standard parallel. In some of the earlier surveys, closing corners are to be found on all four sides of a township. The first set of corners was established when the exterior boundaries of the township were located. Those monuments on the north boundary belong to sections lying to the north of the line. Those on the west boundary of the township belong to sections lying to the west of that line. The other set of corners was established at the time the township was subdivided. For instance, lines numbered 11, 22, 33, 44, and 60 in Fig. 18-5 were run to an intersection with the north boundary of the township, and closing corners were established at the intersections. In a similar manner, lines numbered 47, 50, 53, 56, and 59 were extended to an intersection with the west line of the township, and a closing corner was established at each intersection. The local surveyor should acquaint himself with the regulations in effect at the time any particular survey was made.

18-15. MEANDERING The purpose of meandering is to locate the mean high-water line of all important rivers and lakes, in order to determine the areas of land that may be sold for agricultural purposes. The United States Supreme Court has given the principles governing the use and purpose of meandering shores in its decision in a noted case (R. R. Co. vs. Schurmeir, 7 Wallace, 286-287) as follows:

> Meander lines are run in surveying fractional portions of the public lands bordering on navigable rivers, not as boundaries of the tract, but for the purpose of defining the sinuosities of the banks of the stream, and as the means of ascertaining the quantity of land in the fraction subject to sale, which is to be paid for by the purchaser. In preparing the official plat from the field notes, the meander line is represented as the borderline of the stream, and shows to a demonstration that the watercourse, and not the meander line as actually run on the land, is the boundary.

Meander corners are set at every point where standard, township, or section lines intersect the bank of a navigable stream, any river 3 chains or more in width, and any lake with an area of 25 acres or more. The meander corners are connected by traverses, which follow the bank or shoreline. The true bearings and the lengths of the sides of the traverse are measured. Since the meander lines are not strict boundaries, the work is simplified by adjusting the locations of the stations on the meander line to give bearings to the exact quarter degree and lengths of whole chains, or multiples of 10 links, with odd links only in the final course.

The latitudes and departures of the meander courses should be computed before leaving the vicinity, and if misclosure is found, indicating an error in the field measurements, the lines should be rerun.

In the case of lakes that are entirely within the boundaries of a section, a quarter-section line is run, if one crosses the lake, to the margin of the lake, where a *special meander corner* will be established. If the lake is entirely within a quarter section, an *auxiliary meander corner* will be established at some suitable point on its margin and connected with a regular corner on the section boundary.

18-16. MARKING LINES BETWEEN CORNERS The marking of the survey lines on the ground was supposed to be done in such a manner as to perpetuate the lines. This was done by monumenting the regular corners, by recording in the field notes all natural topographic features, and, whenever living timber was encountered, by means of blazing and hack marks on the trees.

A *blaze* is an axe mark that is made on a tree trunk at about breast height, so that a flat scar is left on the tree surface. The bark and a small amount of the live wood tissue are removed, and the smooth surface that is exposed forever brands the tree. A sufficient number of trees standing within 50 links of the line, on either side of it, are blazed on two sides quartering toward the line, in order to render the line conspicuous and readily traceable in either direction. The blazes are placed opposite each other and so that they coincide in direction with the line where the trees stand very near it, but approach nearer to each other and toward the line the farther the line passes from the blazed trees.

A *hack* is also an axe mark that is made on a tree trunk at about breast height, but in this case a horizontal notch is cut into the surface of the tree. The notch is made V shaped. Two hacks are cut to distinguish those made in the survey from accidental marks resulting from other causes. All trees intersected by the line will have two hacks or notches resembling a double-V ($\lessgtr$) cut on each of the sides facing the line.

Unfortunately for the present-day surveyor, the monuments used on the early surveys were not permanent in character and forest fires and settlers have destroyed most of the timber, so that the retracing of the old lines is often difficult and in many instances impossible.

18-17. OBJECTS NOTED IN THE SURVEY The field notes and plat of a survey furnish not only a technical record of the procedure, but also a report on the character of the land, soil, and timber traversed by the survey, and a detailed schedule of the topographical features along every line. It is necessary to include, in addition, accurate connections showing the relation of the rectangular surveys to other surveys, to natural objects, and to improvements. This information is often of considerable value in relocating obliterated lines.

According to the *Manual of Surveying Instructions, 1947*, the information that should appear in the field notes is as follows:

1. The precise course and length of every line run, noting all necessary offsets therefrom, with the reason for making them, and the method employed.
2. The kind and diameter of all bearing trees, with the course and distance of the same from their respective corners, and the markings; all bearing objects and marks thereon, if any; and the precise relative position of witness corners to the true corners.
3. The kind of material of which the corners are constructed, their dimensions and markings, depth set in the ground, and their accessories.
4. Trees on line. The name, diameter, and distance on line to all trees that intersects, and their markings.

5. Intersections by line of land objects. The distance at which the line intersects the boundary lines of every reservation, townsite, or private claim, noting the exact bearing of such boundary lines, and the precise distance to the nearest boundary corner; the center line of every railroad, canal, ditch, electric transmission line, or other right-of-way across public lands, noting the width of the right-of-way and the precise bearing of the center line; the change from one character of land to another, with the approximate bearing of the line of demarcation, and the estimated height in feet of the ascents and descents over the principal slopes typifying the topography of the country traversed, with the direction of said slopes; the distance to and the direction of the principal ridges, spurs, divides, rim rock, precipitous cliffs, and so on; the distance to where the line enters or leaves heavy or scattering timber, with the approximate bearing of the margin of all heavy timber, and the distance to where the line enters or leaves dense undergrowth.

6. Intersections by line of water objects. All unmeandered rivers, creeks, and smaller watercourses that the line crosses; the distance measured on the true line to the center of the same in the case of the smaller streams, and to both banks in the case of the larger streams, the course downstream at points of intersection, and their widths on line, if only the center is noted. All intermittent watercourses, such as ravines, gulches, arroyos, draws, dry drains, and so on.

7. The land's surface: whether level, rolling, broken, hilly or mountainous.

8. The soil: whether rocky, stony, gravelly, sandy, loam, clay, and so on, and also whether first, second, third, or fourth rate.

9. Timber: the several kinds of timber and undergrowth, in the order in which they predominate.

10. Bottom lands to be described as upland or swamp and overflowed, and the depth of overflow at seasonal periods to be noted. The segregation of lands fit for cultivation without artificial drainage, from the swamp and overflowed lands.

11. Springs of water, whether fresh, saline, or mineral, with the course of the stream flowing therefrom. The location of all streams, springs, or waterholes, which because of their environment may be deemed to be of value in connection with the utilization of public grazing lands.

12. Lakes and ponds, describing their banks, tributaries, and outlet, and whether the water is pure or stagnant, deep or shallow.

13. Improvements; towns and villages; post offices; Indian occupancy; houses or cabins, fields, or other improvements, with owner's name; mineral claims; mill sites; United States mineral monuments, and all other official monuments not belonging to the system of rectangular surveys will be located by bearing and distance or by intersecting bearings from given points.

14. Coal banks or beds, all ore bodies, with particular description of the same as to quality and extent; all mining surface improvements and underground workings; and salt licks.
15. Roads and trails, with their directions, whence and whither.
16. Rapids, cataracts, cascades, or falls of water, in their approximate position and estimated height of their fall in feet.
17. Stone quarries and ledges of rocks, with the kind of stone they afford.
18. Natural curiosities, petrifactions, fossils, organic remains, and so on, also all archaeological remains, such as cliff dwellings, mounds, fortifications, or objects of like nature.
19. The general average of the magnetic declination in the township, with the maximum known range of local attraction and other variations, will be stated in the general description, and the general average for the township, subject to local attraction, will be shown upon the plat.

In addition to the field notes the surveyors are required to prepare, as the work progresses, an outline diagram showing the course and length of all established lines with connections, and a topographical sketch embracing all features usually shown on the completed official township plat. These maps will be made to scale and will be kept up with the progress of the field work. The interiors of the sections will be fully completed, and the topographical features will be sketched with care while in the view of the surveyor. These maps will then form the basis of the official plat, the ultimate purpose of which is a true and complete graphic representation of the public lands surveyed.

18-18. SPECIMEN FIELD NOTES To give some idea of the manner in which the field notes are recorded, the notes in Table 18-6 for 1 mile of the subdivision of T 15 N, R 20 E of the principal base and meridian of Montana have been taken from the Manual of 1947. That portion of the notes covering the adjustment of the transit, and a comparison of the chain with the one used in the survey of the township exterior, has been omitted.

18-19. CORNER MONUMENTS The law provides that the original corners established during the progress of the survey shall forever remain fixed in position, and that even evident errors in the execution of the survey must be disregarded where these errors were undetected prior to the sale of the lands. The original monuments thus assume extreme importance in the location of land boundaries. Unfortunately, most of the public lands were surveyed before the present regulations relative to the character of monuments went into effect, and as a consequence most of the monuments used were of a very perishable nature. Their disappearance or destruction has rendered the relocation of old lines a very difficult task.

TABLE 18-6. T 15 N, R 20 E, Principal Base and Meridian of Montana[a]

Chains	
	Beginning the subdivisional survey at the cor. of secs. 1, 2, 35, and 36, on the S bdy. of the Tp., which is monumented with a sandstone 8 × 6 × 5 in. aboveground, firmly set, marked and witnessed as described in the official record. ,
	N 0° 01′ W, bet. secs. 35 and 36.
	Over level bottom land.
20.00	Enter scattering timber.
29.30	SE cor. of field; leave scattering timber.
31.50	A cabin bears W, 6.00 chs. dist.
39.50	Enter State Highway No. 25, bears N along section line, and E.
40.00	Point for the $\frac{1}{4}$ sec. cor. of secs. 35 and 36.
	Bury a granite stone, 12 × 12 × 12 in. mkd. *X*, 2 ft underground, from which
	An iron post, 30 in. long, 2 in. diam., set 24 in. in the ground for, a reference monument, with brass cap mkd. with an arrow pointing to the cor. and $\frac{1}{4}$ S 36 RM, bears E 46 lks. dist.
	An iron post 30 in. long, 2 in. diam., set 24 in. in the ground, for a reference monument with brass cap mkd. with an arrow pointing to the cor. and $\frac{1}{4}$ S 35 RM, bears W 46 lks. dist.
50.50	NE cor. of field.
51.50	Highway turns to N 70° W.
57.50	Enter heavy timber and dense undergrowth, edge bears N 54° E and S 54° W.
72.00	Leave undergrowth.
80.00	Point for the cor. of secs. 25, 26, 35, and 36.
	Set an iron post, 30 in. long, 2 in. diam., 24 in. in the ground with brass cap mkd.

T 15 N	R 20 E
S 26	S 25
S 35	S 36

1945

from which

    A green ash, 13 in. diam., bears N 22° E, 26 lks. dist. mkd. T 15 N R 20 E S 25 BT.

    A green ash, 23 in. diam., bears S $71\frac{1}{4}$° E, 37 lks. dist. mkd. T 15 N R 20 E S 36 BT.

    A green ash, 17 in. diam., bears S 64° W, 41 lks. dist. mkd. T 15 N R 20 E S 35 BT.

    A cottonwood, 13 in. diam., bears N $21\frac{1}{4}$° W, 36 lks. dist. mkd. T 15 N R 20 E S 26 BT.

    Land, level bottom; northern 20 chs. subject to overflow.

    Soil, alluvial, silt and loam.

    Timber, green ash and cottonwood; undergrowth, willow.

[a] *Manual of Instructions for the Survey of Public Lands of the United States of 1947.*

The present regulations call for the use of iron-pipe monuments, from 1 to 3 in. in diameter and about 3 ft long. One end of the pipe is split for a distance of about 4 or 5 in., and the two halves are spread to form flanges or foot plates. A brass cap is securely riveted to the opposite end of the pipe, and finally the pipe is filled with concrete. The 3-in. monuments are used for township corners, the 2-in. ones for section corners, and the 1-in. ones for quarter-section corners. The caps of the iron posts are suitably and plainly marked with steel dies at the time the posts are used. The posts are set with about three-fourths of their length in the ground, and earth and stone, if the latter is at hand, are tamped into the excavation to give the post a solid anchorage.

When the procedure has been duly authorized, the use of durable native stone may be substituted for the iron post. No stone can be used that is less than 20 in. in length, or less than 6 in. in either of its minor dimensions, or less than 1000 in.^{3} in volume. The stone is marked by letters chiseled on the faces. To aid in the identification of the sections to which it refers, grooves or notches are cut in the faces or edges. The number of these grooves or notches is a measure of the number of miles to the township line in that direction. Thus the corner common to sections 10, 11, 14, and 15 would have two notches on the east edge and four notches on the south edge.

Where a corner point falls upon solid surface rock, and excavation is prevented, a cross (**X**) is cut at the exact corner point; and if feasible, the monument is erected in the same position and is supported by a large stone mound of broad base, which is so well constructed that the monument will possess thorough stability.

Where the corner point falls exactly at the position occupied by a sound living tree, which is too large to be removed, the tree should be appropriately marked for the corner.

According to the earlier instructions, a corner monument could be a wooden post, or a mound of earth under which was buried a charred stake or a quart of charcoal, or some other fairly permanent and distinguishable mark when nothing more permanent was available.

18-20. WITNESSES The corner monuments are witnessed by measuring the bearing and distance to natural or artificial objects in the immediate vicinity. Bearing trees, or other natural objects, are selected for marking, when they are available within 5 chains of the corner monument. One tree, or object, is marked in each section cornering at the monument, when available, and the true course and horizontal distance from the exact corner point to the center of the tree at its root crown, or to the cross (**X**) upon a marked object, are carefully determined and recorded with the description of the tree, or object, and its marks.

Trees that are used as witnesses are blazed, and the blazes are scribed with letters and figures to aid in identifying the location. Thus a tree used as a

witness for a section corner might bear the inscription T 20 N   R 12 E
S 24 BT; this indicates that the tree stands in Section 24, T 20 N, R 12 E, BT
standing for bearing tree. A tree used as a witness for a quarter-section
corner is marked $\frac{1}{4}$ S 32 BT, when the tree is in Section 32.

A cross (**X**) and the letters "BO" and the section number are chiseled into
a bearing object if it is of rock formation, and the record is made sufficiently
complete to enable another surveyor to determine where the marks will be
found.

Where stone is available, and the surface of the ground is favorable, a
mound of stone is employed as a witness when a full quota of trees or other
bearing objects are not available. The mound consists of not less than five
stones and has a base not less than 2 ft wide and a minimum height of $1\frac{1}{2}$ ft.
In stony ground the size of the mound is sufficiently increased to make it
conspicuous. The nearest point on the base is about 6 in. distant from the
monument, and the size and position of the mound are recorded in the notes.

On open prairies where no trees, rock, or other natural objects are
available, the corners may be witnessed by digging pits 18 in. square and
12 in. deep, with the nearest side 3 ft distant from the corner monument.
The pits are oriented with a square side (and not a corner) toward the monument. The field notes should contain a description of the pits, including the
size and position, as the regulations relative to this form of witness have been
changed from those set forth in the earlier editions of the Manual.

When it is impossible to utilize any of the types of witnesses just mentioned,
a suitable memorial will be deposited at the base of the monument. A memorial may consist of any durable article that will serve to identify the
location of the corner in case the monument is destroyed. Such articles as
glassware, stoneware, a stone marked with a cross (**X**), a charred stake, a
quart of charcoal, or pieces of metal constitute a suitable memorial. A full
description of such articles is embodied in the field notes wherever they are
so employed.

18-21. FILING OF FIELD NOTES AND PLATS   Copies of the
field notes and plats of the public land surveys excepting those in Illinois,
Indiana, Iowa, Kansas, Missouri, and Ohio may be procured from the
Manager, Eastern States Land Office, Bureau of Land Management, 7981
Eastern Avenue, Silver Spring, Maryland 20910, and from state offices of the
bureau. The bureau's copy of public survey records of the excepted states
has been transferred to the National Archives and Records Service, General
Services Administration, Washington, D.C. 20408.

In those states where the public land surveys are considered as having been
completed, the field notes, plats, maps, and other papers relating to those
surveys have been transferred to an appropriate state office for safekeeping
as public records. No provision has been made for the transfer of the survey
records to the State of Oklahoma, but in the other states, the records are

filed in the following offices where they may be examined and copies made or requested:

Alabama: Secretary of State, Montgomery, Alabama 36104.

Arkansas: Department of State Lands, State Capitol, Little Rock, Arkansas 72201.

Florida: Board of Trustees of the Internal Improvement Trust Fund, Elliott Building, Tallahassee, Florida 32304.

Illinois: Illinois State Archives, Secretary of State, Springfield, Illinois 62706.

Indiana: Archivist, Indiana State Library, 140 North Senate Avenue, Indianapolis, Indiana 46204.

Iowa: Secretary of State, Des Moines, Iowa 50319.

Kansas: Auditor of State and Register of State Lands, Topeka, Kansas 66612.

Louisiana: Register, State Land Office, Baton Rouge, Louisiana 70804.

Michigan: Department of Treasury, Bureau of Local Government Services, Treasury Building, Lansing, Michigan 48922.

Minnesota: Department of Conservation, Division of Lands and Forestry, Centennial Office Building, Saint Paul, Minnesota 55101.

Mississippi: State Land Commissioner, P.O. Box 39, Jackson, Mississippi 39205.

Missouri: State Land Survey Authority, P.O. Box 1158, Rolla, Missouri 65401.

Nebraska: State Surveyor, State Capitol Building, P.O. Box 4663, Lincoln, Nebraska 68509.

North Dakota: State Water Conservation Commission, State Office Building, Bismarck, North Dakota 58501.

Ohio: Auditor of State, Columbus, Ohio 43215.

South Dakota: Commissioner of School and Public Lands, State Capitol, Pierre, South Dakota 57501.

Wisconsin: Department of Natural Resources, Box 450, Madison, Wisconsin 53701.

In the remainder of the states, the surveys are in progress, and the records are held by the state offices of the Bureau of Land Management, as follows:

Arizona: Phoenix	New Mexico: Santa Fe
California: Sacramento	Oregon: Portland
Colorado: Denver	Utah: Salt Lake City
Idaho: Boise	Washington: Portland, Oregon
Montana: Billings	Wyoming: Cheyenne
Nevada: Reno	Alaska: Anchorage

Copies of the field notes can be obtained from these offices.

18-22. SUBDIVISION OF SECTIONS The function of the United States surveyor ends with the establishment of monuments at half-mile intervals on the external lines of the sections and the filing of the detailed field notes and a plat. No monuments are set by him within the section, except in the case of a meanderable lake entirely within a section. As the rectangular system provides for the disposal of the public lands in units of quarter-quarter sections of 40 acres, the function of the local surveyor begins when he is employed as an expert to identify the lands that have passed into private ownership. This may be a simple task or a most complicated one, the difficulty depending on the condition of the original monuments. The work of the local surveyor usually includes the subdivision of the section into the smaller units shown on the official plat. It is apparent that he cannot properly serve his client unless he is familiar with the legal requirements concerning the subdivision of sections. If any of the original monuments are missing, it is necessary for him to know not only the methods used in the original survey, but also the principles that have been adopted by the courts, in order to be able to restore the missing corners legally.

After all the original monuments have been found or any missing ones have been replaced, the first step in the subdivision is the location of the center of the section. Regardless of the location of the section within the township, this point is always at the intersection of the line joining the east and west quarter-section corners, called the *east-and-west quarter line*; with the line joining the north and south quarter-section corners, called the *north-and-south quarter line*. By locating these lines on the ground, the section is divided into quarter sections containing approximately 160 acres each.

The method of dividing these quarter sections into the 40-acre parcels depends on the position of the section within the township. For any section except those along the north and west sides of the township, the subdivision is accomplished by bisecting each side of the quarter section and connecting the opposite points by straight lines. The intersection of these lines is the center of the quarter section.

The aim in subdividing the sections along the north and west sides of the township is to secure as many regular parcels as possible and to throw all irregularities caused by convergence or field work as far west and north as possible. The method of subdividing these sections is shown in Fig. 18-6. The corners on the north-and-south section lines are set at intervals of 20.00 chains, measured from the south, the discrepancy being thrown into the most northerly quarter mile. Similarly, the monuments on the east-and-west lines are set at intervals of 20.00 chains, measured from the east, the fractional measurement being thrown into the most westerly quarter mile. The fractional lots formed are numbered in a regular series progressively from east to west or from north to south, in each section. As section 6 borders on both the north and the west boundaries of the township, the fractional lots are numbered commencing with number 1 in the northeast, thence progres-

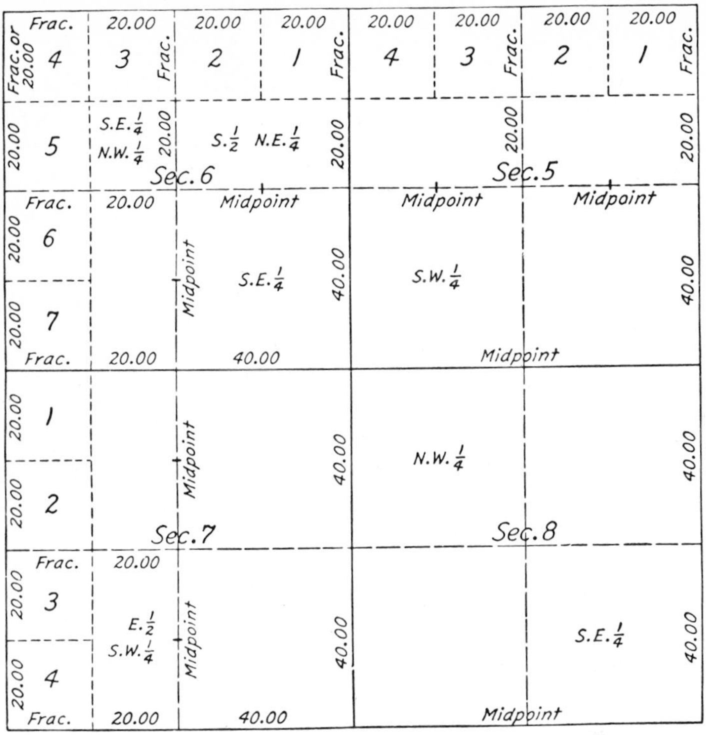

Figure 18-6. Subdivision of sections.

sively west to number 4 in the northwest, and south to number 7 in the southwest fractional quarter-quarter section.

It is the duty of the local surveyor to set these corners in the same positions they would have occupied had they been set in the original survey. For this reason, distances shown as 20.00 chains indicate distances of 20 chains according to the chain used by the original surveyor, and not 1320.00 ft according to the tape of the local surveyor. It is therefore necessary to establish these points by proportionate measurement, the local surveyor determining the length of the chain, in terms of his tape, by measuring the distance between the two existing monuments. Thus if the measured distance from the north quarter section corner of Section 6 to the northwest corner is 2637.24 ft, and if the field notes show this distance to be 39.82 chains, the local surveyor will measure $20.00 \times 2637.24/39.82 = 1324.58$ ft westward from the quarter-section corner in establishing the quarter-quarter-section corner.

Where some corners fall within meanderable bodies of water, every effort is made to run the necessary division lines in the same positions they would occupy in a regular section. Thus when a quarter-section corner for an interior section could not be set, the quarter-section line is run from the opposite existing corner on a bearing that is the mean of the bearings of the two adjacent section lines. This same principle holds in the establishment of quarter-quarter-section lines.

18-23. LEGAL DESCRIPTIONS

The rectangular system of subdivision provides a very convenient method of describing a piece of land that is to be conveyed by deed from one person to another. If the description is for a 40-acre parcel, the particular quarter of the quarter section is first given, then the quarter section in which the parcel is located, then the section number, followed by the township, range, and principal meridian. Thus the parcel of land adjacent to lots 3, 4, and 5 in Section 6, Fig. 18-6, can be described as the SE $\frac{1}{4}$ NW $\frac{1}{4}$, Section 6, T 8 N, R 12 E of the Third Principal Meridian. Lot 1 in Section 7 may also be described as the NW $\frac{1}{4}$ NW $\frac{1}{4}$, Section 7, T 8 N, R 12 E of the Third Principal Meridian. The legal descriptions of the larger parcels appear in Fig. 18-6.

18-24. RELOCATING LOST CORNERS

Nothing in the practice of surveying calls for more skill, persistence, and good judgment on the part of the surveyor than the relocation of a missing corner. In the original survey it was presumed that permanent monuments were being carefully established and witnessed, so that long lines of monuments would be perfect guides to the place of any one that chanced to be missing. Unfortunately, the "monuments" were often nothing but green sticks driven into the ground, lines were carelessly run, monuments were inaccurately placed, witnesses were wanting in permanence, and recorded courses and lengths were incorrect. As the early settlers made little effort to perpetuate either the corner monuments or the witnesses, the task of the present-day surveyor in reconciling much conflicting evidence (or in even finding any kind of evidence) is often an extremely difficult task.

A *lost corner* is a point of a survey whose position cannot be determined, beyond reasonable doubt, either from original traces or from other reliable evidence relating to the position of the original monument, and whose restoration on the earth's surface can be accomplished only by means of a suitable surveying process with reference to interdependent existent corners.

A thorough search will often show that an apparently lost corner is really not lost. Where wooden stakes were used, a careful slicing of the earth may disclose a discoloration in the soil caused by the decaying of the stake. Roots or depressions in the soil may give an indication of the locations of the original witness trees. Fence lines, the testimony of nearby residents, and the field notes of adjacent surveys may be of value in ascertaining the location of the corner. It must be remembered that no matter how erroneously the original work is done, lines and monuments that can be identified still govern the land boundaries. The surveyor has no power to "establish" corners. In any case of disputed lines, unless the parties concerned settle the controversy by agreement, the determination of the line is necessarily a judicial act, and it must proceed upon evidence. The surveyor serves as an expert witness in the court proceedings, presenting all evidence he has been able to discover relative to the proper location of the corner in question. As a disinterested surveyor of proper training should be much more competent than any court to decide on the proper location of any questionable line or corner, the surveyor should exert all his influence toward effecting a settlement by agreement, and thus save his client the expense of court proceedings. If this is impossible, he should insist that proper monuments, which are consistent with the court decree, be established, and if possible, be made a part of the decree.

When it is certain that a corner is lost, it is replaced in a manner that is in accord with the methods used in its original location. Lost corners are established by either single or double proportionate measurement. The method of double proportionate measurement is generally applicable to the restoration of lost corners of four townships and of lost interior corners of four sections. Its application is illustrated in Fig. 18-7. In the lower enlarged

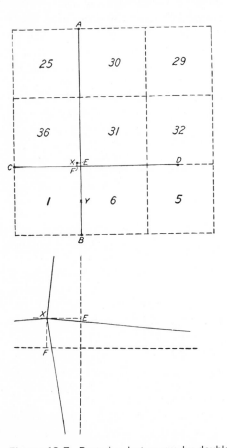

Figure 18-7. Restoring lost corner by double-proportionate measurement.

sketch, E represents the location of the corner if only the existing corners A and B are considered, the point being determined by proportionate measurement. If the known points C and D are used, a second location F is found. The restored corner is placed at X, the intersection of a west line through E with a north line through F. In this manner the distances from the replaced corner to the existing corners are consistent with the original field notes.

As a quarter-section corner was supposedly located on the straight line between the adjacent section corners, if there is no evidence on the ground to the contrary, a missing quarter-section corner is located on the straight line connecting the adjacent section corners. The distances from the section corners are determined by proportion, by comparing the present-day measurement between existing corners with the same distance as recorded in the field notes. Thus in Fig. 18-7 the quarter-section corner Y between Sections 1 and 6 would be located on the line between B and X by a single proportionate measurement.

Before attempting work of this sort, the detailed rules and examples shown in the *Manual of Instructions for the Survey of the Public Lands of the United States*, from which much of the material in this chapter has been taken, should be carefully examined.

BIBLIOGRAPHY

Brown, C. M. *Boundary Control and Legal Principles.* New York: Wiley, 1957.

Brown, C. M., and Eldridge, W. H. *Evidence and Procedures for Boundary Location.* New York: Wiley, 1962.

Bureau of Land Management. 1947 and 1973. *Manual of Instructions for the Survey of the Public Lands of the United States.* U.S. Government Printing Office.

Proportionate measurements—a panel presentation. *Surveying and Mapping* 28:621.

——. 1955. *Restoration of Lost or Obliterated Corners and Subdivision of Sections.* U.S. Government Printing Office.

Cole, G. M. 1978. Florida's restoration of corners program. *Proc. Ann. Meet. Amer. Cong. Surv. & Mapp.* p. 28.

Eickbush, F. D. 1975. Alaskan surveys based on protraction diagrams. *Proc. Ann. Meet. Amer. Cong. Surv. & Mapp.* p. 59.

Grimes, J. S. 1959. *Clark on Surveying and Boundaries.* Indianapolis: Bobbs-Merrill.

Lindsey, J. A. 1969. Relocating government corners on the state coordinate systems. *Surveying and Mapping* 29:401.

Mann, C. V. 1966. How shall we preserve the federal public land survey within Missouri. *Surveying and Mapping* 26:85.

Skelton, R. H. *Boundaries and Adjacent Properties.* Indianapolis: Bobbs-Merrill.

Wattles, W. C. 1956. *Land Survey Descriptions.* Los Angeles: Title Insurance and Trust Company, Revised by G. H. Wattles, 1974, Orange, California 92667: G. H. Wattles, P.O. Box 5702.

Chapter 19
Municipal and Subdivision Surveys

19-1. CONTROL MONUMENTS AND MAPS The value of a comprehensive city survey cannot be overestimated. All public agencies, utilities, land developers, engineers, architects, and others make constant use of the information furnished by city surveys. The money saved by the users of the information in planning, developing, engineering, and construction far exceeds the cost of the surveys.

Information furnished by well-planned and well-executed city surveys consists of horizontal-control and vertical-control monuments, together with their coordinates or elevations or both, large-scale base maps, topographic maps, property maps, and overall city maps to smaller scale. Horizontal-control monuments are established by triangulation and traverse of first- and second-order accuracy. Vertical-control monuments, or benchmarks, are established by first- and second-order lines of differential levels.

Base maps are compiled for a multiplicity of uses. In general, they are used to form a common basis for topographic maps, property maps, subdivision maps, and maps showing the positions of utility lines. For the compilation of base maps, scales of 1 in. = 50 ft, 1 in. = 100 ft, and 1 in. = 200 ft are most satisfactory, the choice depending on the intensity of land use. Topographic maps should be published at a scale of 1 in. = 200 ft; property maps, at a scale of 1 in. = 50 ft; and the overall city map, at a scale no smaller than 1 in. = 2000 ft.

19-2. STEPS IN A CITY SURVEY

Where it is possible to make a complete survey, the first step is to establish a network of triangulation control. The triangulation should be of first-order accuracy, should be tied to existing National Geodetic Survey triangulation stations, and should provide one control point for every 1 to 3 square miles throughout the metropolitan area. The next step is to run traverses connecting the triangulation stations. These traverses should furnish the equivalent of second-order control for the base maps and property surveys, and of third-order control for the topographic work. Leveling of the same three orders should follow the lines of horizontal control.

From this basic control, base maps are prepared as described in Section 19-6, and topographic maps are compiled either by photogrammetric methods or by plane-table methods, additional control being provided for the latter by plane-table traversing and plane-table leveling. Much of the information obtained in the topographic survey forms the basis of the property surveys that are made to locate the monuments and boundaries of all public property, supplementary information being obtained by a search of all public records and by additional field work.

19-3. TRIANGULATION FOR CITY SURVEYS

Although the principles governing the triangulation survey for a city are the same as those given in Chapter 10, the work is modified somewhat because of the shorter lengths of the sides and because of the special conditions encountered. In planning the network the tops of hills and of prominent buildings are utilized for stations, as far as possible, in order to keep the cost at a minimum. Towers are erected only when unavoidable. A line of sight that passes close to the ground along a hillside or bluff, or close to a building or chimney giving off heated air, is likely to produce uncertain results because of lateral refraction of the line of sight. Where such a sight is necessary, the observations should be made when the wind is blowing toward the object.

The triangulation net should be made up of either quadrilaterals or central-point figures, no single triangles being used. The diagram in Fig. 19-1 shows the main scheme first-order and second-order triangulation covering the city of Oakland, California, and including neighboring smaller cities. This is an excellent example of a strong triangulation network, which serves to control all subsequent municipal surveys in the area. The network is integrated directly into the triangulation network of the National Geodetic Survey, and the positions of all the stations are given in state plane coordinates.

To maintain high precision, strong figures are necessary. The value of R_1 for any of the basic figures should seldom exceed 5. The value of R_1 between

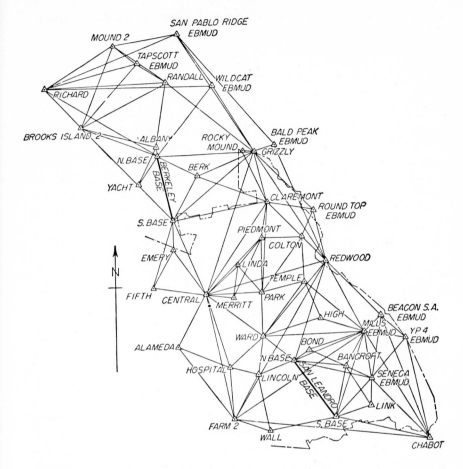

Figure 19-1. Triangulation diagram of the city of Oakland, California. Courtesy of city of Oakland.

adjacent bases should be kept less than 20, if possible, and should seldom be allowed to exceed 25. The lengths of these bases should approximate the average length of the sides of the triangles.

In order to keep temperature effects to a minimum and to control calibration errors, two or more standardized invar tapes are used in measuring the bases. If steel tapes must be used, the measurements should be made at night when the air temperature most truly indicates the actual temperature of the tape. The use of steel tripods, called *taping bucks*, for supporting the tape during measurement will be advantageous. These should weigh at least 15 pounds, and be from 15 to 30 in. high. The tape graduations are transferred to metal plates on top of the tripods by means of a pencil or steel awl.

A leveling party is necessary when measuring the baseline in order to run levels over the tripod heads. The levels furnish the differences of elevation between successive tripods needed to compute the individual horizontal distances. The standard error of the bases should be no larger than 1 : 700,000. After a least-squares adjustment of the network, the discrepancy between the measured and computed lengths of the bases should average no greater than 1 : 100,000.

If EDM is to be used on a relatively short base typical of city surveys, the instrument must be capable of high resolution and small inherent error, in order to maintain a high relative accuracy. Because of the ease of measuring distances using EDM, a mixture of triangulation and trilateration can help to strengthen an otherwise weak portion of the net.

The angles are measured with a 1″ direction instrument. The average triangle closure should not exceed 1″, and the maximum allowable error should be about 3″. To attain this precision, 8, 12, or 16 positions with the direction instrument (the number depending on the accuracy of the instrument and the conditions surrounding the measurements), will generally be required. Because of the relatively short lengths of the sides, the signals and the instrument must be centered within 0.01 ft. All angles should be measured at night, any one of a variety of triangulation signal lights being used. The instrument should be shielded from the wind during the observation period by means of a canvas enclosure or other device. Except when the net is connected with the first-order control of the National Geodetic Survey, astronomical observations for azimuths should be made at a few selected stations in the network.

The monuments should be of permanent character and carefully referenced. An iron pipe that is provided with a bronze cap and is set in a concrete post at least 12 in. in diameter makes a suitable monument if it extends below frost line and is so located as to minimize the chances of destruction by public or private construction. By using a bronze cap with a spherical top, the monument will serve also as a benchmark. The exact station mark can be designated by a small drill hole or an etched cross in the top of the cap. When the monument is set several inches below the surface of the ground, a small cast-iron box of the manhole type makes the monument more accessible.

In order to make the triangulation stations usable to a maximum number of people, the triangulation computations should be carried out on the state plane coordinate system. The least-squares adjustment is most easily made by the method of variation of plane coordinates described in Appendix A and in Special Publication No. 193 of the U.S. Coast and Geodetic Survey. If trilateration has been combined with the triangulation, the method of variation of coordinates in trilateration adjustment shown in Appendix A can be combined with the methods given in Special Publication No. 193 to perform a combined adjustment.

For the use of local surveyors and engineers, an index map showing the locations of all triangulation stations should be published. This map should be accompanied by a list of descriptions of the stations and their references, a tabulated list of coordinates, and the azimuths and lengths of the sides.

19-4. TRAVERSE FOR CONTROL OF CITY SURVEYS The use
of first-order traverse as the principal horizontal control in a city survey,
independent of triangulation, is advisable only for small areas where triangulation is unduly expensive because of topographic conditions. The
principal use of first-order traverse is in establishing additional control points
for originating second-order traverses, which control property surveys, and
third-order traverses, which control the topographic surveys.

A first-order traverse begins at one triangulation station and closes on
another triangulation station. This practice tends to eliminate the effects of
unknown systematic errors in measuring the traverse sides. The traverse
lines are marked at intervals not greater than a mile by pairs of permanent
intervisible monuments, 500 to 1000 ft apart. These marked lines provide
positions for subsequent traversing of lower order. Set in pairs, the monuments provide beginning azimuths for subsequent traverses. Intermediate
stations are semipermanent, are marked by iron pins or by lead or copper hubs
in pavements or sidewalks, and are referenced for future recovery. The stations
should be so located as to provide long clear sights free from horizontal
refraction. Where a series of short sights is necessary, azimuths should be
carried by means of cutoff lines discussed in Section 8-8 in order to maintain
accuracy in the directions of the lines. Long narrow loop circuits should be
avoided; for connections subsequently made across the narrow dimension
may show unallowable discrepancies, despite satisfactory closures obtained
in the original circuit.

The average angular error, in seconds, should not exceed $2\sqrt{N}$, in which
N is the number of sides or angles in the traverse, and the maximum error
should not exceed $4\sqrt{N}$. The angles may be measured by four, six, or eight
positions of a $1''$ direction instrument, the number depending on the conditions
surrounding the measurements. Triangulation lights or specially designed
traverse lights may be used for signals in both day and night operations.

The traverse sides are measured by EDM, and the measured lengths are
corrected for meteorological conditions and instrumental errors. Since the
lines in the traverse are relatively short, slope reductions can be made by the
methods outlined in Sections 2-11 and 2-12. The lines are then reduced to grid
lengths for computing on the state coordinate system. After the azimuth errors
are distributed, the average position closure should not exceed 1 : 50,000, and
the maximum closure should not exceed 1 : 25,000.

The second-order traverse is used to furnish coordinates, bearings, and
distances for street property monuments. The stations should coincide with
the street monuments wherever possible. When a $1''$ direction instrument is
used to measure angles, two or four positions will suffice. The average angular
closure, in seconds, should not exceed $6\sqrt{N}$, and the maximum closure should
not exceed $8\sqrt{N}$. After the angular error is distributed, the average position
closure should be 1 : 20,000 or better.

The diagram in Fig. 19-2 shows the traverse network covering the City of
Alameda, California. The entire network is tied to five triangulation stations

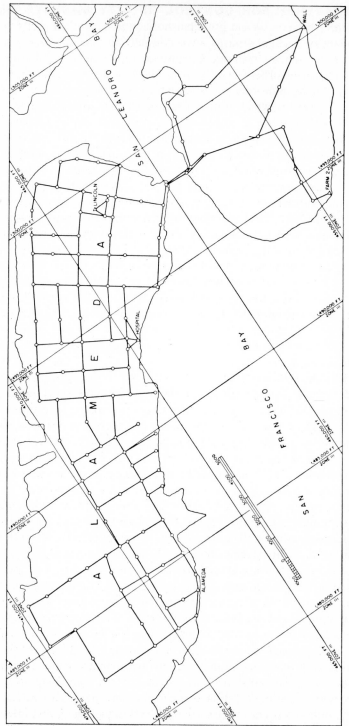

Figure 19-2. Traverse diagram of the city of Alameda, California. Courtesy of city of Alameda.

which appear in the lower left-hand part of Fig. 19-1. Nearly all the traverse lines are along the city streets (eliminated from the drawing for clarity), and the monuments are located at street corners. This is a flat-lying area of limited extent, and the traverse network is entirely adequate to control all subsequent surveys. The positions of the traverse stations are given by state plane coordinates.

The third-order traverse furnishes control for the topographic survey. Such a traverse should begin at a station of higher order and close on another station of higher order. Semipermanent monuments are used, since they can be replaced easily from the references or by retracing the survey. The average angular closure should not exceed 10″ per station. The angles can be measured with a 30″ engineer's transit or theodolite, each angle being measured once direct and once reversed. After the angular error is distributed, the position closure should not exceed 1 : 7500. The distances are most conveniently measured using EDM. This will assure a good linear closure at very little effort. If a tape is used, a spring balance and thermometer should be used in order to maintain the necessary closure accuracy.

The published data for the traverse stations should be the same as for the triangulation system, the order of the various stations being indicated. All rectangular coordinates should be referenced to the state plane coordinate system.

19-5. LEVELING FOR CITY SURVEYS   One of the difficulties facing engineers, surveyors, and others who use city benchmarks to control their operations is the multiplicity of vertical datums in existence in one given city. This is troublesome when trying to tie a line of levels to two networks on different datums. A city should endeavor to establish the elevations of all benchmarks with respect to a common level datum at the earliest possible time. This datum should preferably be the National Geodetic Vertical Datum of 1929 based on leveling operations of the National Geodetic Survey.

A first-order level net should be laid out along major streets and railroad lines if such a net has not already been established by the National Geodetic Survey. Traverse monuments should be incorporated into the net where possible. Additional benchmarks are set on bridge abutments or piers, masonry retaining walls, and masonry walls of buildings. These marks should be bronze tablets and should be carefully described and referenced for future use. The levels are run using three wires as described in Section 3-37 or else with a first-order level equipped with an optical micrometer as discussed in Section 3-21. When the latter is used, only a single wire is read. Thus an auxiliary method must be used to determine the lengths of the sights in order to provide the proper weights in the least-squares adjustment of the level net (see Appendix A). These distances can be paced, scaled from a map or obtained directly from values for the traverse that the lines happen to follow. The levels are run in both directions over a section. The resulting differences of elevation, in feet, should agree within $0.017\sqrt{M}$, in which M is the length of the section in miles.

The second-order levels are run in closed circuits that begin and end on the same first-order benchmark, or are run from one such benchmark to another. These lines of levels follow the lines of second-order traverse, the traverse stations serving as benchmarks. Additional benchmarks are established in such places as on fire hydrants, curb catch basins, and crosses chiseled in concrete sidewalks. They are referenced by steel-tape measurements. The maximum closure, in feet, of any line or circuit should not exceed $0.035\sqrt{M}$, in which M is the length of the line or circuit in miles. Three-wire leveling or a tilting level equipped with an optical micrometer should be used.

The final elevations, adopted after a least-squares adjustment of the level net, should be published for the use of local engineers and surveyors. The order of each benchmark should be indicated, together with its complete description and references and the vertical datum with respect to which its elevation is established.

19-6. BASE MAPS

As previously stated, a base map is prepared to provide a common basis for other types of maps. It should show a maximum amount of basic information regarding horizontal and vertical control, city property, street center lines, boundary lines, and related features. Base maps are most useful at scales of 50, 100, and 200 ft to the inch. They should be indexed in series by means of the state plane coordinates of the lower left-hand corner of each sheet. A series at a scale of 1 in. = 50 ft with a working area of 20 in. in the north-south direction, and 30 in. in the east-west direction will have index x coordinates in multiples of 1500 ft and index y coordinates in multiples of 1000 ft. A series at a scale of 1 in. = 200 ft on the same size sheet will have index coordinates in multiples of 6000 and 4000 ft in the x direction and y direction, respectively.

The base-map manuscript should be prepared in pencil on dimensionally stable translucent material. The manuscript should contain grid lines for state plane coordinates at 10-in. intervals. Triangulation and traverse stations are plotted by coordinates, and their names and coordinate values are shown directly on the map. The positions of benchmarks, together with their elevations, should be indicated. The map should contain all construction-line notes, dimensions, and references; right-of-way lines; street center lines with bearings; all points of curvature, tangency, and intersections along the center lines; and curve radii and central angles. The corners of streets and angle points should be clearly defined. Any information that cannot otherwise be made clear at the map scale should be shown in enlarged detail on the margin of the sheet. Streets should be plotted and their widths noted on the map.

The information shown on the base map should be reliable and complete, so that it will not be necessary to consult original records. The manuscripts should be kept up to date as more survey information is obtained. Results of survey activities conducted by state, federal, and private agencies, as well as by public utilities, should be shown on the base maps when the reliability of such information is established.

Reproductions of the base maps should be made available to surveyors and engineers operating in the city, so that all activities are coordinated on one common data base.

19-7. TOPOGRAPHIC MAP OF CITY   The topographic map should cover the built-up areas of the city and the development areas surrounding it. The map, to a scale of 1 in. = 200 ft and with a contour interval of 2 or 5 ft, should conform to the size of the base-map sheets. If the maps are to be published in color, the published sheets should show the topography by means of contours in brown. The drainage, comprising such features as streams, ponds, and lakes, is portrayed in blue. Black ink is used for the following features: structures, such as railroads, bridges, culverts, and curbs; benchmarks and other monuments; property lines, including street lines; boundaries of public property and political subdivisions; boundaries of recorded subdivisions; street names; names of parks, subdivisions, streets, and lakes; names of buildings; property dimensions; the height and character of buildings; grid lines or grid ticks; and lot numbers. Wooded areas and public property are shown in green.

The compilation of city topographic maps by means of aerial photogrammetry is recommended as superior to any other method of topographic mapping in accuracy, time, and cost. The control is obtained by second-order traverse and levels. Photogrammetric compilation is supplemented by plane-table mapping in areas where more information is necessary than can be obtained photogrammetrically.

In Fig. 19-3 is shown a portion of Sheet G-18 of the Metropolitan Topographic Survey of the City of San Diego, California. The published scale is 1 in. = 200 ft, and the contour interval is 5 ft. Elevations are based on the City of San Diego Vertical Datum. Grid lines of the California Coordinate System, Zone VI, are shown at 5-in. intervals. This series of maps was produced by aerial photogrammetry supplemented by field-completion and field-editing surveys.

A final topographic map is judged to be sufficiently accurate if elevations determined from the contours on the map are correct within one-half the contour interval. The horizontal position of any well-defined point should be correct within 0.02 in. at the published scale. Errors in excess of these limits should not occur in more than 10 % of the number of points tested. The mapping should be tested by running profile levels along traverse lines selected at random, and comparing these profiles with those obtained from the contours on the map.

Property lines located on the map are checked by assembling all available recorded plats of subdivisions. Distances between identified points and lines,

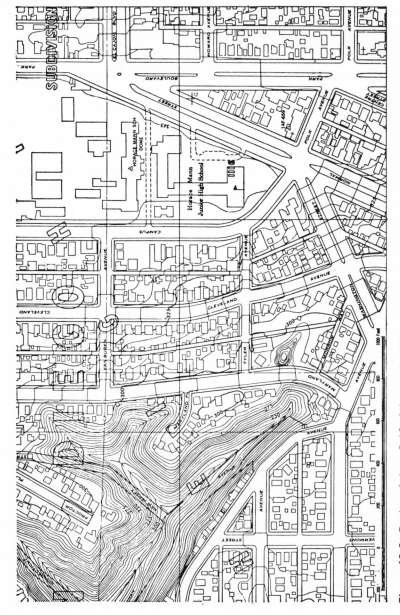

Figure 19-3. Portion of sheet G-18 of the topographic map series covering the city of San Diego, California. Original scale is 1 in. = 200 ft. Courtesy of city of San Diego.

such as monuments, fences, street lines, and block corners, are scaled from the manuscript sheets and compared with the corresponding distances shown on the recorded plats. Where there are many discrepancies, the manuscript sheet is checked in the field. When all identified lines have been checked, these lines are used as a basis for plotting the recorded information pertaining to the remaining lines and corners. The completed map should show street and alley lines, boundaries of recorded subdivisions, boundaries of public property, parks, playgrounds, schools, and other important areas.

19-8. CITY PROPERTY SURVEY The city property survey marks by suitable monuments all corner, angle, and curve points of the street lines, and establishes coordinate values for these and other critical points. When this is done, property monuments that become lost or obliterated may be conveniently and precisely reestablished from their known mathematical relations with other marked points. Thus each monument serves as a witness for every other monument.

The first aim of the property survey is the adequate location and recording of all street lines, which constitute the boundaries of the public's property. The property survey tends to stabilize the boundaries of private property by furnishing an adequate control for the use of those engaged in property surveying. It is usually desirable to complete the work of establishing authoritative dimensions for public property as rapidly as time and opportunity permit, especially in the older sections of a city where original surveys are often very faulty. The work in the newer sections is usually of a better grade, having been controlled through plat laws.

The work is begun by assembling all the available recorded information and the records of earlier surveys. When little or no information is available, this fact should furnish one of the strongest arguments for immediate action. The next step is the field location of all street intersection, angle, and curve points. Locating such points involves skill and judgment that is acquired only by years of experience with this class of work. The experienced man is familiar with the more important court decisions and realizes the weight that must be given to occupation lines. As soon as the location of a point has been definitely determined, a monument should be set either directly at the point or at a standard offset distance from it. Because of present-day traffic conditions, the use of standard offset monuments is usually preferable. These monuments should be located from the second-order traverse with a precision of 1 : 20,000, or with such precision that any point whose location is dependent on that of the monument can be located within $\frac{1}{4}$ in. of its true position.

The property maps should be drawn to a scale of 1 in. = 50 ft, and the layout of the sheets should conform with that used for the topographic map. The original sheets should be of the best quality paper, and from them tracings should be made on the best quality translucent material. The maps should show the bearings and lengths of street lines, alley lines, and boundaries of public property; the street widths and intersections; the coordinates of all intersection, angle, and curve points; and the locations of traverse stations. In addition there should appear all existing monuments or other witness markers; public, industrial, and commercially important buildings; private buildings in sparsely settled areas; and such features as railroads and bridges. The lettering should include street names; names of parks, subdivisions, streams, and lakes; names of public or semipublic buildings; and the numbers of benchmarks and other monuments.

19-9. UNDERGROUND MAP The purpose of the underground map is to show the locations of all sewers, water pipes, gas mains, electrical conduits, heating lines, subways, and tunnels, including their elevations at critical points. Much of the information to appear on such a map can be taken from the original construction plans and the records of public-service companies, this information being supplemented by such field surveys as may be required to complete the map.

The property maps can be used as a base for the underground map, many of the property dimensions being omitted to prevent overcrowding. The final maps, made on the best quality translucent material, should show street, alley, and easement lines, plotted to scale, with the various widths in figures; existing monuments and benchmarks; and surface structures, such as sidewalks, curbs, pavements, street railroads, transmission-line poles, and trees in parking areas. The underground structures should be differentiated by suitable symbols, and the sizes of the pipes, the percents of grade on gravity lines, and the elevations at critical points should be indicated. The leveling should be of second-order precision, or at least sufficiently accurate to permit the preparation of plans for new construction without fear of interference with existing structures. The horizontal locations should be shown with respect to the street and alley lines. As such a map soon becomes obsolete, some provision should be made for keeping it up to date.

19-10. TOWN, CITY AND VILLAGE PLATS A building lot that is to be conveyed by deed can be described either with respect to existing land lines or by reference to a plat. In the first method, the bearing and length of each side are given, and the description as a whole is tied to some section line or other well-defined property line. As the distance from the section line increases, the description becomes more and more complicated, particularly when curved boundaries are encountered and when the tie to the section line is a traverse of many sides. In the second method, the description is greatly simplified, consisting merely of the lot number and the name of the recorded plat of which it is a part.

Most states have laws that govern the platting or subdividing of land into city or town lots. While some of the details may differ, the main requirements in the various states are essentially the same. The first requirement is usually a well-defined boundary of the tract to be platted, with the corners marked by permanent monuments, to which reference is made on the plat. The tract is divided into streets and blocks by placing permanent monuments at all intersections of two streets, of two alleys, of a street and an alley, and of a street or alley with a boundary of the plat. The lot corners are marked by semi-permanent monuments, preferably iron pipe.

The survey is then platted on paper, the plat showing all blocks, lots, streets, alleys, parks, and monuments; the sizes of all lots, the lot numbers, and the names or numbers of all streets with their widths and courses; the directions and lengths of the boundaries of the tract, and all other information that will enable surveyors to retrace the lines accurately. The plat contains a written or printed description, which should be so complete that from it, without reference to the plat, the starting point of the tract can be determined and the outlines can be run. The starting point should be located with reference to some well-defined point, such as some certain point in the section when the land platted is in a state where the rectangular system of subdivision has been in effect.

The owners of the land sign a certificate on the plat, acknowledging that they have caused the plat to be made, and dedicating all streets, alleys, and parks to the use of the public. The surveyor certifies that the plat is a correct one and that the monuments described in it have been placed as shown on the plat.

Before a plat can be recorded, it must be approved by certain public officials or boards. In this manner, the locations and widths of the streets, the sizes of the lots, and the proper grading of the streets can be controlled by the municipality or other interested public boards. When the plat has been finally approved, one copy is placed on record in the office of the register of deeds or similar official. Additional copies may be required for city and state records.

19-11. SURVEY FOR PLAT   The first step in the subdivision of a tract into blocks and lots is an accurate survey of the boundaries of the parcel. This survey should be placed on a plane coordinate system. The next step is to prepare a topographic survey of the area. The blocks and lots together with coordinate grid ticks are then laid out on the topographic map.

Where zoning or planning boards are in existence, it is usually necessary to locate the principal streets in conformity with the general layout of the district adopted by such boards. In such cases, the widths of the streets are also specified by the board.

Rectangular blocks with streets at right angles to one another should not be adopted when the topography is such that curved streets and irregular-shaped lots would be more advantageous. By fitting the locations of the streets to the topography of the ground, the rates of grade will be comparatively low. Less grading will then be required for the streets, and the location of storm and sanitary sewers will generally be simplified. Some of the older portions of San Francisco can be pointed out as a horrible example of the adoption of a rectangular layout to topography that is utterly unsuited to such a plan. The result has been that the grades on some streets are so excessive as to render them practically worthless as thoroughfares.

Minimum lot frontage and lot sizes are usually controlled by local ordinances. These and other constraints must be taken into account when laying out the lots in the tract on the map. The coordinates of all points on the street center lines, including PI's, PC's and PT's of curves, as well as the coordinates of the centers of the curves, can be computed by scaling the coordinates of the beginning of the street center lines where they enter the tract (if these coordinates are not already known). Tangents and curves are laid out, and closed paper traverses are then computed along these lines between the fixed coordinates of the ends of the street center lines. Adjusting the closed traverse fixes the coordinates of the points on all of the streets. Alternatively, the coordinates of these points can be scaled directly from the paper location. Next the coordinates of the block and lot corners are computed on the coordinate system using the principles of analytical geometry discussed in Section 8-20.

After the paper layout has been completed, the field location is made. Street lines or block corners are first located from points on the boundaries of the tract by inversing the lines to these points. The individual lot corners are then established. As these lines are to serve as property boundaries, extreme care should be taken to ensure their accuracy. The work should be so conducted as to ensure a precision of about 1 : 10,000. The monuments marking important points should be as permanent as possible and referenced whenever possible. Some state plat laws now require for all important monuments the use of iron pipe that is at least 1 in. in diameter and at least 15 in. long, and is set in a concrete base at least 4 in. in diameter and 48 in. in depth.

19-12. LOT SURVEY The survey of a lot that is described with references to a certain plat may be simple or complicated, the requirements depending on the shape of the block and on the monuments that may be available. In Fig. 19-4 is illustrated an example of the simpler sort, where the monuments that were placed at the block corners can be found and identified. In locating the corners of Lot No. 40, the method of proportionate measurements must be used. Although the widths of the lots fronting on Ashland Avenue are shown on the plat as 50.00 ft, it is also evident that the total frontage has been equally divided among the seven lots. Consequently, if there is any excess or deficiency in the frontage, it should be equally distributed. For this reason, the corners a and b are set at distances from the northwest corner of the block equal, respectively, to $\frac{2}{7}$ and $\frac{3}{7}$ times the total block frontage. Temporary marks are set at e and f by measuring the south side of the block and locating these points by proportionate measurements. The

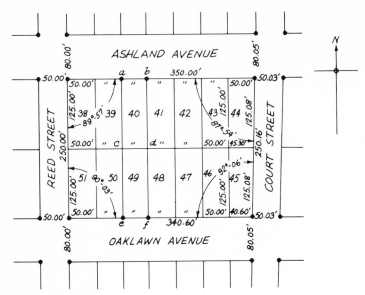

Figure 19-4. Lot survey.

other two lot corners, *c* and *d*, are located at the midpoints of the lines *ae* and *bf*.

If corners set on the original survey can be found and identified, such points are usually held to be the true corners, even though they have been erroneously located. When land has been held by an owner for a certain period of time, ordinarily 20 years, and adjacent owners are unable to reach an agreement on the boundary lines and bring the case into court, the occupation lines are usually held by the court to be the property lines. As pointed out in Section 18-24, the surveyor has no power to "establish" property lines. In the majority of cases, particularly where there is no serious disagreement, his findings will be accepted as final by the adjacent property owners. He should at all times so conduct his work as to be reasonably certain that his opinions will be supported if the question of location is brought before a court for final decision.

When many of the monuments are missing and when errors have been made in the original survey, the problem will be much more difficult than in the example shown in Fig. 19-4. Considerable skill and judgment will be required to weigh properly the conflicting evidence that may be disclosed as the survey progresses.

Some of the more important legal decisions relative to property surveys are cited in the books listed in the Bibliography at the end of Chapter 18.

19-13. METES AND BOUNDS The survey of a piece of ground by metes and bounds is the oldest type of land surveying. It consists of determining the length and direction of each side of the parcel and planting permanent monuments at the corners of the property. The lands in the thirteen original states were surveyed in this manner, but in many instances the descriptions are so vague and incomplete as to make the rerunning of the lines an impossibility. The following description, taken from the Hartford, Connecticut, Probate Court records for 1812 will illustrate what is often encountered.

147 acres, 3 rods, and 19 rods after deducting whatever swamp, water, rock and road areas there may be included therein and all other lands of little or no value, the same being part of said deceased's 1280 acre colony grant, and the portion hereby set off being known as near to and on the other side of Black Oak Ridge, bounded and described more in particular as follows, to wit:—Commencing at a heap of stone, about a stone's throw from a certain small clump of alders, near a brook running down off from a rather high part of said ridge; thence, by a straight line to a certain marked birch tree, about two or three times as far from a jog in a fence going around a ledge nearby; thence, by another straight line in a different direction, around said ledge and the Great Swamp, so called; thence, in line of said lot in part and in part by another piece of fence which joins on to said line, and by an extension of the general run of said fence to a heap of stone near a surface rock; thence, as aforesaid, to the "Horn," so called, and passing around the same as aforesaid, as far as the "Great Bend," so called, and from thence to a squarish sort of a jog in another fence, and so on to a marked black oak tree with stones piled around it; thence, by another straight line in about a contrary direction and somewhere about parallel with the line around by the ledge and the Great Swamp, to a stake and stone bounds not far off from the old Indian trail; thence, by another straight line on a course diagonally parallel, or nearby so, with "Fox Hollow Run," so called, to a certain marked red cedar tree out on a sandy sort of a plain; thence, by another straight line, in a different direction, to a certain marked yellow oak tree on the off side of a knoll with a flat stone laid against it; thence, after turning around in another direction, and by a sloping straight line to a certain heap of stone which is, by pacing, just 18 rods and about one half a rod more from the stump of the big hemlock tree where Philo Blake killed the bear; thence, to the corner begun at by two straight lines of about equal length, which are to be run by some skilled and competent surveyor, so as to include the area and acreage as herein before set forth.

For purposes of rerunning, a more favorable description would contain the length and bearing of each side and would include references to permanent objects which are still in existence. If the surveyor can find and identify the two ends of any one line, his work is greatly simplified, since that line affords him a comparison between his tape and that used in the original survey and makes unnecessary any consideration of magnetic declinations. The interior angles of the tract can be calculated from the given bearings and these computed angles turned off from the identified corners when searching for the remaining ones.

If only one of the original corners can be found, it will be necessary to know whether the bearings given in the description are magnetic or true ones. If they are magnetic, the surveyor must ascertain the value of the declination at the time the survey was made and what it is at present, if he is to use a compass in determining the approximate directions of the lines from that corner. What has been said in Section 18-24 relative to the relocation of lost corners applies also to the resurvey of a tract described by metes and bounds.

BIBLIOGRAPHY

Dracup, J. F. 1969. Control surveys for Akron, Ohio. *Surveying and Mapping* 29:243.

Gale, L. A. 1969. Control surveys for urban areas in Canada. *Surveying and Mapping* 29:669.

Gale, P. M. 1970. Control surveys for the City of Houston. *Surveying and Mapping* 30:95.

Gale, P. M. 1976. Cadastral surveys for small municipalities. *Proc. Fall Conv. Amer. Cong. Surv. & Mapp.* p. 387.

Grant, B. 1969. Survey record problems in the East Bay Area. *Surveying and Mapping* 29:469.

Pittman, R. Jr., and Tannen, G. 1971. Early, present, and future platting procedures. *Proc. Fall Conv. Amer. Cong. Surv. & Mapp.* p. 380.

Scher, M. B. 1969. Orthophoto maps for urban areas. *Surveying and Mapping* 29:413.

Urban, L. J. 1973. The use of geodetic control in a rapidly developing metropolitan area. *Proc. Fall Conv. Amer. Cong. Surv. & Mapp.* p. 48.

Appendix A
Adjustment of Elementary Surveying Measurements by the Method of Least Squares

A-1. GENERAL Whenever the engineer or surveyor conducts a field survey, no matter how simple or complex, he invariably makes more measurements than are absolutely necessary to locate the points in the survey. A line taped in two directions introduces one measurement more than is necessary to establish the length of the line. Measuring all three angles of a triangle also introduces one superfluous measurement. Closing a line of levels on a fixed benchmark introduces more measurements than are necessary to determine the elevations of the unknown set benchmarks. A closed traverse has more than enough measurements necessary to fix the positions of the intermediate traverse stations. These extra measurements are termed *redundant measurements*. They immediately impose conditions that must be satisfied in order to resolve disagreements or inconsistencies in the measurements.

The redundant measurements are made for the purpose of checking other measurements, uncovering mistakes, adjusting the measurements, and estimating the magnitudes of the random errors. The mean of several measurements, for example, is the adjusted value of the measured quantity. The residuals resulting from subtracting the mean from the measured values furnish the data necessary to evaluate the random errors by determining the standard error of the set and of the mean value. The adjusted values of intermediate benchmarks in a level line are obtained by first determining the closure on a known benchmark, and then distributing this closing error back through the line. This is the same procedure used for the adjustment of a traverse run between two fixed points.

In Chapter 5 procedures for adjusting some of the more elementary surveying measurements were discussed. These simple adjustments were based on one of the following principles of least squares: (1) In a set of measurements all of which are made with the same reliability, that is, all of which have equal weight, the most probable values that can be derived from the set are those which make the sum of the squares of the residuals or corrections a minimum. (2) In a set of measurements having unequal weights, the most probable values are those which make the sum of the products of the weights and the squares of the corrections a minimum. These principles may be stated briefly as follows: For equal-weight measurements,

$$\sum v^2 = \text{minimum} \tag{A-1}$$

or

$$[vv] = \text{minimum}$$

For unequal-weight measurements,

$$\sum (pv^2) = \text{minimum} \tag{A-2}$$

or

$$[pvv] = \text{minimum}$$

The adjustment of measurements by the method of least squares accomplishes the double purpose of satisfying the above conditions and resolving discrepancies or closure errors of the measurements. The least-squares adjustment can be accomplished by either of two general methods. The first method makes use of the formation of *observation equations* in which the corrections are stated as functions of indirectly determined values of parameters of the measurements. By way of example, the mean of the series of measurements in Section 5-3 was shown to be the most probable value by the formation of the following series of equations:

$$v_1 = M_1 - M$$

$$v_2 = M_2 - M$$

$$\vdots$$

$$v_n = M_n - M$$

These are observation equations. The corrections, or v's, appear on the left-hand side. These are expressed as functions of the measurements themselves and the mean, which is an indirectly determined value. Then the condition of least squares was applied to the corrections by getting the sum of the squares of the corrections, differentiating with respect to the mean (the indirectly determined parameter), and setting the result equal to zero. The result of these operations is an equation of the form

$$\frac{d(\sum v^2)}{dM} = M_1 - M + M_2 - M + \cdots + M_n - M$$

$$= M_1 + M_2 + \cdots + M_n - nM = 0$$

Such an equation is known as a *normal equation*. The solution of this normal equation gives the value of the indirectly determined parameter M. The residuals are then found by solving the observation equations for the v's. This least-squares adjustment is so simple that we do not bother to form observation equations, differentiate to get the normal equation, and then solve the normal equation to arrive at the mean value. We simply add up all the measured values and divide by the number of measurements.

For a more complex series of measurements, the observation equations must be formed, and then normal equations are developed by differentiation of $\sum v^2$ with respect to each parameter on the right-hand side of the observation equations and setting equal to zero. These are solved simultaneously in order to arrive at the best set of parameters that will make $\sum v^2$ a minimum. However, certain systematic procedures tend to simplify these operations, as will be shown later.

The second method for adjustment by least squares takes advantage of a priori conditions which must be imposed on the residuals. Such an a priori condition is the prior knowledge that the sum of three angles in a plane triangle must be $180°$. Also, in a traverse between two fixed points the sum of the latitudes must equal the difference between the Y coordinates of the two fixed points. This second method is the method of *condition equations*. For example, if three angles A_0, B_0, and C_0 of a triangle are observed and

$$A_0 + B_0 + C_0 = 179° \ 59' \ 56.7''$$

then we may state the condition by the equation

$$v_A + v_B + v_C = 3.3'' \quad \text{or} \quad v_A + v_B + v_C - 3.3'' = 0$$

in which v_A, v_B, and v_C are the corrections to be applied to the measured angles. The solution of this condition equation, after the condition of least squares is imposed, is

$$v_A = +1.1''; \qquad v_B = +1.1''; \qquad v_C = +1.1''$$

This simple adjustment was discussed in Section 5-9.

There are different approaches to the solution of the above condition equation. One approach is to express all the corrections as functions of only as many corrections as there are independent measurements. Thus in the preceding example there are only two independent angles. If A_0 and B_0 are considered independent, then C_0 must be considered dependent for the following reason: If there were no redundant measurements, then C would equal $180° - (A_0 + B_0)$. Therefore if the corrections v_A and v_B to A_0 and B_0 are independent, then

$$v_C = +3.3'' - (v_A + v_B)$$

In other words

$$v_A = v_A$$
$$v_B = v_B$$
$$v_C = +3.3'' - (v_A + v_B)$$

These are connecting equations, relating all the v's in terms of only the number of v's that are independent. On squaring we get

$$v_A^2 = v_A^2$$
$$v_B^2 = v_B^2$$
$$v_C^2 = [3.3'' - (v_A + v_B)]^2$$

Then

$$\sum v^2 = 2v_A^2 + 2v_B^2 + 3.3^2 - 6.6v_A - 6.6v_B + 2v_A v_B$$

In order to satisfy the condition of least squares, the partial derivative of $\sum v^2$ with respect to each of the independent quantities v_A and v_B must be made zero. Thus we get the following normal equations:

$$\frac{\partial(\sum v^2)}{\partial v_A} = 0$$

$$\frac{\partial(\sum v^2)}{\partial v_B} = 0$$

When the differentiations are performed, the results are

$$4v_A - 6.6 + 2v_B = 0$$
$$4v_B - 6.6 + 2v_A = 0$$

By solving these normal equations, we get

$$v_A = +1.1'' \quad \text{and} \quad v_B = +1.1''$$

Substitution of these values of v_A and v_B in the connecting equation

$$v_C = +3.3'' - (v_A + v_B)$$

gives

$$v_C = +1.1''$$

Another approach for solving condition equations is by the method of *correlates* or *Lagrange multipliers*. This approach, which is the more common, is as follows. Take the geometric condition equation and the condition of least squares

$$v_A + v_B + v_C - 3.3'' = 0 \tag{A-3}$$

$$v_A^2 + v_B^2 + v_C^2 = \text{minimum} \tag{A-4}$$

There are an infinite number of sets of values of the unknown v's that will satisfy Eq. (A-3), but only one set that will satisfy both Eq. (A-3) and (A-4). In order to determine that set, differentiate each equation, giving

$$dv_A + dv_B + dv_C = 0 \tag{A-5}$$

$$v_A \, dv_A + v_B \, dv_B + v_C \, dv_C = 0 \tag{A-6}$$

Obviously, an infinite number of sets of differentials given in Eq. (A-5) can satisfy that equation. If Eq. (A-5) is multiplied by an arbitrary multiplier C, the results are

$$C \, dv_A + C \, dv_B + C \, dv_C = 0 \tag{A-7}$$

If Eq. (A-7) is now compared with Eq. (A-6), term for term, the following equalities result:

$$v_A = C$$
$$v_B = C \tag{A-8}$$
$$v_C = C$$

Equations (A-8) are called *correlate* equations, and C is the Lagrange multiplier. If the values of v_A, v_B, and v_C given by Eq. (A-8) are substituted into the condition equation (A-3), the result is

$$C + C + C - 3.3'' = 0$$

or

$$3C - 3.3'' = 0 \tag{A-9}$$

Equation (A-9) is a normal equation whose solution gives $C = +1.1''$. If this value of the Lagrange multiplier is used in Eq. (A-8), then

$$v_A = +1.1'', \qquad v_B = +1.1'', \qquad v_C = +1.1''$$

as before.

The method of correlates will be more fully understood by reference to the adjustment of a quadrilateral given in Section A-3.

When adjusting the three angles of a triangle, we do not bother to go through either of the elaborate procedures given above. We simply subtract 180° from the sum of the three measured angles, divide the discrepancy by three, and apply a correction of opposite sign to each angle. However, if a quadrilateral or some other net is to be adjusted, many conditions must be satisfied simultaneously. Therefore the equations must be formed in an orderly fashion in order that they may be solved simultaneously.

If the observations to be adjusted have different weights, then the weights must be incorporated into the adjustment procedure. Let us consider three measurements of the length of a line designated as M_1, M_2, and M_3, with corresponding weights p_1, p_2, and p_3. Also, let the most probable value be designated as $\overline{M}$. Then, by forming observation equations, squaring, and multiplying by the weights, we get

$$v_1 = M_1 - \overline{M}; \qquad v_1^2 = (M_1 - \overline{M})^2; \qquad p_1 v_1^2 = p_1(M_1 - \overline{M})^2$$
$$v_2 = M_2 - \overline{M}; \qquad v_2^2 = (M_2 - \overline{M})^2; \qquad p_2 v_2^2 = p_2(M_2 - \overline{M})^2$$
$$v_3 = M_3 - \overline{M}; \qquad v_3^2 = (M_3 - \overline{M})^2; \qquad p_3 v_3^2 = p_3(M_3 - \overline{M})^2$$

In order to satisfy the condition that $\sum (pv^2)$ is a minimum, it is necessary to differentiate $\sum (pv^2)$ with respect to $\overline{M}$ and to set the derivative equal to zero. Thus

$$\frac{d \sum (pv^2)}{d\overline{M}} = -2p_1(M_1 - \overline{M}) - 2p_2(M_2 - \overline{M}) - 2p_3(M_3 - \overline{M}) = 0$$

and

$$\overline{M} = \frac{p_1 M_1 + p_2 M_2 + p_3 M_3}{p_1 + p_2 + p_3} = \frac{\sum (pM)}{\sum p}$$

This result is the weighted mean given by Eq. (5-27).

In case of three angles of a triangle having different weights, the adjustment by the method of correlates is as follows: Let us assume that

$$A_0 + B_0 + C_0 = 179° \ 59' \ 30''$$

and that

$$p_A = 6, \qquad p_B = 3, \qquad p_C = 1$$

The condition is then

$$v_A + v_B + v_C - 30'' = 0 \tag{A-10}$$

and the least-squares condition is

$$6v_A^2 + 3v_B^2 + 1v_C^2 = \text{minimum} \tag{A-11}$$

Differentiation of these two equations gives

$$dv_A + dv_B + dv_C = 0 \tag{A-12}$$

$$6v_A \, dv_A + 3v_B \, dv_B + 1v_C \, dv_C = 0 \tag{A-13}$$

Multiplying Eq. (A-12) by the Lagrange multiplier C gives

$$C \, dv_A + C \, dv_B + C \, dv_C = 0 \tag{A-14}$$

Equating coefficients between Eqs. (A-13) and (A-14) gives the correlate equations

$$6v_A = C \qquad \text{or} \qquad v_A = \frac{C}{6}$$

$$3v_B = C \qquad \text{or} \qquad v_B = \frac{C}{3} \tag{A-15}$$

$$1v_C = C \qquad \text{or} \qquad v_C = \frac{C}{1}$$

Substituting the values of the v's from Eq. (A-15) into the condition equation (A-10) gives the following normal equation:

$$\frac{C}{6} + \frac{C}{3} + \frac{C}{1} - 30'' = 0 \qquad \text{(A-16)}$$

whose solution gives

$$C = 20''$$

This value of C is substituted into the correlate equations (A-15) to give the solution for the v's. Thus

$$v_A = +3.3'', \qquad v_B = +6.7'', \qquad v_C = +20.0''$$

Compare these results with those of the example given in Section 5-10.

The concept of least-squares adjustment by the methods of observation equations and condition equations given here is very rudimentary. In fact observation and condition equations can be combined to form one system which is solved in such fashion as to make the weighted sums of squares of the corrections to the observations a minimum. The readings should be consulted for further investigation of these methods.

A-2. LEAST-SQUARES ADJUSTMENT BY MATRIX METHODS

The examples to follow in this appendix will be solved using either the method of observation equations or the method of condition equations as developed up to this point. These solutions can be simplified considerably by employing matrix algebra. Consequently, the examples to follow will also be solved by matrix methods.

The matrix expression for the condition of least squares given in Eq. (A-1) in which the observation equations are all of equal weight is

$$\mathbf{V}^T\mathbf{V} = \text{minimum} \qquad \text{(A-17)}$$

For observations of different weights, the least-squares condition given by Eq. (A-2) is expressed in matrix form as follows:

$$\mathbf{V}^T\mathbf{P}\mathbf{V} = \text{minimum} \qquad \text{(A-18)}$$

in which $\mathbf{V}$ is a column matrix of n residuals with dimensions $n \times 1$, $\mathbf{V}^T$ is the transpose of $\mathbf{V}$ with dimensions $1 \times n$, and $\mathbf{P}$ is a square diagonal matrix of weights with dimensions $n \times n$ or

$$\mathbf{P} = \begin{bmatrix} p_1 & 0 & \cdots & 0 \\ 0 & p_2 & & \vdots \\ \vdots & & & 0 \\ 0 & 0 & \cdots & p_n \end{bmatrix}$$

n being the number of observations or measurements. This weight matrix assumes that the measurements to be adjusted are not correlated.

The matrix method of satisfying the condition of least squares using *observation equations* will be developed in Section A-3(b). The matrix method used to handle *condition equations* will be developed in Section A-4(b).

A-3(a). ADJUSTMENT OF A LEVEL NET BY OBSERVATION EQUATIONS

The problem of adjusting a level net involves the tying of the net to previously established benchmarks, and the adjustment of the differences in elevations (DE's) between various junction points in the net so that no matter what route is taken to a point, the same elevation will be obtained. Furthermore, in order to satisfy the condition of least squares the corrections to the observed DE's must be such that the sum of the weighted squares of the corrections will be a minimum. The weights of the DE's between points are inversely proportional to the lengths of the routes along which the DE's are obtained. For instance, if the DE between A and B is determined along two routes, the first one being 3 miles long and the second being 4 miles long, the weight of the first DE is $\frac{1}{3}$ and that of the second is $\frac{1}{4}$; or the first has weight 1, whereas the second has weight 0.75. Weights may also be assumed to be inversely proportional to the number of setups of the instrument between two points.

In the level net shown in Fig. A-1, there are 11 independent lines, along which the differences in elevation have been obtained as a result of three-wire leveling. The values shown in Table A-1 have been corrected for systematic errors discussed in Section 3-38. The subscripts for the v's correspond to the numbers assigned to the lines in Fig. A-1. The elevations of the three benchmarks A, D, and K are fixed, as follows:

$$A = 856.425 \text{ ft}, \qquad D = 878.337 \text{ ft}, \qquad K = 867.271 \text{ ft}$$

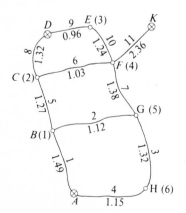

Figure A-1. Level net.

TABLE A-1

LINE	LENGTH (miles)	FROM TO	OBSERVED DE	CORRECTION TO DE
1	1.49	$A–B$	− 3.278	v_1
2	1.12	$B–G$	− 1.467	v_2
3	1.32	$G–H$	+ 9.756	v_3
4	1.15	$H–A$	− 4.975	v_4
5	1.27	$B–C$	+ 8.429	v_5
6	1.03	$C–F$	− 6.528	v_6
7	1.38	$F–G$	− 3.352	v_7
8	1.32	$C–D$	+16.724	v_8
9	0.96	$D–E$	−12.563	v_9
10	1.29	$E–F$	−10.721	v_{10}
11	2.36	$F–K$	+12.249	v_{11}

The most common procedure for adjusting a set of measurements by the method of observation equations is first to compute an approximate set of values of the quantities sought, and then to compute small corrections to these initial values. In the case of a level net, an approximate elevation is computed for each junction point by adding the DE of a line from a known elevation to the junction point. Thus

$$\text{approx. elevation } B = \text{elevation } A + \text{DE}_{AB}$$

$$\text{approx. elevation } B = 856.425 − 3.278 = 853.147 \text{ ft}$$

$$\text{approx. elevation } C = \text{approx. elevation } B + \text{DE}_{BC}$$

$$\text{approx. elevation } C = 853.147 + 8.429 = 861.576 \text{ ft}$$

Similarly, the approximate elevations of the other junction points are obtained as follows:

POINT	FROM LINE	APPROXIMATE ELEVATION
B	AB	853.147
C	BC	861.576
E	CF and FE	865.769
F	CF	855.048
G	BG	851.680
H	GH	861.436

These approximate elevations must be adjusted by adding small corrections, which will be designated by x's with subscripts that correspond to the numbers of the junction points shown in parentheses in Fig. A-1. Thus the adjusted elevations may be represented as follows:

POINT	ADJUSTED ELEVATION
B	$853.147 + x_1$
C	$861.576 + x_2$
E	$865.769 + x_3$
F	$855.048 + x_4$
G	$851.680 + x_5$
H	$861.436 + x_6$

We can now derive observation equations by applying the following general relationship:

$$\text{(observed DE}_{1-2}) + \text{(corrections to DE}_{1-2})$$
$$= \text{(adjusted elevation of 2)} - \text{(adjusted elevation of 1)}$$

For example, between A and B,

$$-3.278 + v_1 = (853.147 + x_1) - 856.425$$

Between B and G,

$$-1.467 + v_2 = (851.680 + x_5) - (853.147 + x_1)$$

Between G and H,

$$+9.756 + v_3 = (861.436 + x_6) - (851.680 + x_5)$$

Between H and A,

$$-4.975 + v_4 = 856.425 - (861.436 + x_6)$$

Between B and C,

$$+8.429 + v_5 = (861.576 + x_2) - (853.147 + x_1)$$

By writing an observation equation for each DE and combining numbers, we obtain 11 equations that express the v's, or the corrections to the DE's, in terms of the x's, or the corrections to the approximate elevations of the points.

The results are

$$v_1 = x_1 + 0$$

$$v_2 = -x_1 + x_5 + 0$$

$$v_3 = -x_5 + x_6 + 0$$

$$v_4 = -x_6 - 0.036$$

$$v_5 = -x_1 + x_2 + 0$$

$$v_6 = -x_2 + x_4 + 0$$

$$v_7 = -x_4 + x_5 - 0.016$$

$$v_8 = -x_2 + 0.037$$

$$v_9 = x_3 - 0.005$$

$$v_{10} = -x_3 + x_4 + 0$$

$$v_{11} = -x_4 - 0.026$$

If each DE had equal or unit weight, the condition to be satisfied would be that $\sum v^2$ = minimum, or

$$v_1^2 + v_2^2 + v_3^2 + \cdots + v_{11}^2 = \text{minimum}$$

However, since each DE is obtained either through a different distance or by a different number of setups (the distance is used in this example), each DE has a weight that is inversely proportional to the distance involved. Consequently, the condition to be satisfied is $\sum (pv^2)$ = minimum, or

$$p_1 v_1^2 + p_2 v_2^2 + \cdots + p_{11} v_{11}^2 = \text{minimum}$$

The weights determined by taking the reciprocals of the lengths of the lines are tabulated as shown in Table A-2. It is usually desirable to adjust each

TABLE A-2

LINE	FROM TO	LENGTH (miles)	WEIGHT
1	A–B	1.49	0.672
2	B–G	1.12	0.893
3	G–H	1.32	0.758
4	H–A	1.15	0.870
5	B–C	1.27	0.787
6	C–F	1.03	0.971
7	F–G	1.38	0.725
8	C–D	1.32	0.758
9	D–E	0.96	1.043
10	E–F	1.29	0.775
11	F–K	2.36	0.424
		$\sum =$	$\overline{8.676}$

weight by a factor in order to make the average weight equal to unity. In this example it would be necessary to have the weights add up to 11. Since the sum of the tabulated weights is 8.676, each such weight would have to be multiplied by 11/8.676. However, it would be an unnecessary refinement to make the average weight equal to unity in the present example. When the square of each value of v is multiplied by the tabulated weight, the results are as follows:

$$0.672v_1^2 = 0.672(+x_1 + 0)^2$$

$$0.893v_2^2 = 0.893(-x_1 + x_5 + 0)^2$$

$$0.758v_3^2 = 0.758(-x_5 + x_6 + 0)^2$$

$$0.870v_4^2 = 0.870(-x_6 - 0.036)^2$$

$$0.787v_5^2 = 0.787(-x_1 + x_2 + 0)^2$$

$$0.971v_6^2 = 0.971(-x_2 + x_4 + 0)^2$$

$$0.725v_7^2 = 0.725(-x_4 + x_5 - 0.016)^2$$

$$0.758v_8^2 = 0.758(-x_2 + 0.037)^2$$

$$1.043v_9^2 = 1.043(+x_3 - 0.005)^2$$

$$0.775v_{10}^2 = 0.775(-x_3 + x_4 + 0)^2$$

$$0.424v_{11}^2 = 0.424(-x_4 - 0.026)^2$$

Six normal equations can be formed by adding the right-hand sides of the above equations to get $\sum (pv^2)$ and then taking the partial derivative of $\sum (pv^2)$ with respect to each of the x's and setting each partial derivative equal to zero. Thus

$$\frac{\partial \sum (pv^2)}{\partial x_1} = 0; \qquad \frac{\partial \sum (pv^2)}{\partial x_2} = 0; \qquad \dots; \qquad \frac{\partial \sum (pv^2)}{\partial x_6} = 0$$

The solution of these equations would give the values of the six parameters (the x's) that would satisfy the condition of least squares.

The formation of the normal equations can be made quite mechanical by realizing that they take the following form:

$$[paa]x_1 + [pab]x_2 + [pac]x_3 + [pad]x_4 + [pae]x_5 + [paf]x_6 + [pak] = 0$$

$$[pba]x_1 + [pbb]x_2 + [pbc]x_3 + [pbd]x_4 + [pbe]x_5 + [pbf]x_6 + [pbk] = 0$$

$$[pca]x_1 + [pcb]x_2 + [pcc]x_3 + [pcd]x_4 + [pce]x_5 + [pcf]x_6 + [pck] = 0$$

$$[pda]x_1 + [pdb]x_2 + [pdc]x_3 + [pdd]x_4 + [pde]x_5 + [pdf]x_6 + [pdk] = 0$$

$$[pea]x_1 + [peb]x_2 + [pec]x_3 + [ped]x_4 + [pee]x_5 + [pef]x_6 + [pek] = 0$$

$$[pfa]x_1 + [pfb]x_2 + [pfc]x_3 + [pfd]x_4 + [pfe]x_5 + [pff]x_6 + [pfk] = 0$$

TABLE A-3. Tabulated Observation Equations

v	p	x_1	x_2	x_3	x_4	x_5	x_6	k
1	0.672	+1						0
2	0.893	−1				+1		0
3	0.758					−1	+1	0
4	0.870						−1	−0.036
5	0.787	−1	+1					0
6	0.971		−1		+1			0
7	0.725				−1	+1		−0.016
8	0.758		−1					+0.037
9	1.043			+1				−0.005
10	0.775			−1	+1			0
11	0.424				−1			−0.026

In these equations the p's are the weights, the a's are the coefficients of the x_1's in the observation equations, the b's are the coefficients of the x_2's, and so on, and the k's are the constant terms of the observation equations. The quantity $[pbd]$, for example, is the sum of six products, or $p_1 b_1 d_1 + p_2 b_2 d_2 + p_3 b_3 d_3 + p_4 b_4 d_4 + p_5 b_5 d_5 + p_6 b_6 d_6$. The student should verify this formulation.

Arranging the observation equations in tabular form as shown in Table A-3 simplifies the formation of the normal equations. The values under x_1 are the a's, those under x_2 are the b's, and so on. The coefficients all happen to be plus or minus unity in this adjustment. The normal equations, which are formed as indicated, can also be tabulated as shown in Table A-4. Because $[pab] = [pba]$, $[pac] = [pca]$, and so on, the coefficients of the x's in the table are symmetrical about the diagonal running through the coefficients $[paa]$, $[pbb]$, ..., $[pff]$. Thus the coefficients in the row i are equal to the coefficients in the column i. Being familiar with this symmetry obviously permits a great saving of time in forming the normal equations.

The solution of the preceding set of six normal equations gives the values of the corrections to be added to the approximate elevations of the junction points to get their adjusted values. Solving the original observation equations

TABLE A-4. Tabulated Normal Equations

x_1	x_2	x_3	x_4	x_5	x_6	k	
+2.352	−0.787			−0.893		0	= 0
−0.787	+2.516		−0.971			−0.02805	= 0
		+1.818	−0.775			−0.00521	= 0
	−0.971	−0.775	+2.895	−0.725		+0.02262	= 0
−0.893			−0.725	+2.376	−0.758	−0.01160	= 0
				−0.758	+1.628	+0.03132	= 0

for the v's gives the quantities needed to determine the standard errors of the measured values. The standard error of unit weight is

$$\sigma_0 = \sqrt{\frac{\sum pv^2}{n - u}} \qquad \text{(A-19)}$$

in which n is the number of lines and u is the number of junction points. The quantity $(n - u)$ is the number of redundant measurements or the number of degrees of freedom (see Section 5-4). In the case at hand, $n = 11$ and $u = 6$. Thus

$$\sigma_0 = \sqrt{\frac{\sum pv^2}{5}}$$

The standard error of each line is then obtained by Eq. (5-30).

After the approximate elevations of the junction points have been adjusted, the corrected values are considered as fixed elevations. The elevations of any intermediate benchmarks established along the various lines are then adjusted between two fixed junction points by the method discussed in Section 5-11.

A-3(b). ADJUSTMENT OF LEVEL NET BY MATRIX METHODS

The solution of the adjustment of a level net by the method of observation equations will now be performed by matrix methods. First, the matrix algebra involved in the solution must be developed. The matrix expression for a set of observation equations is given as follows: If there are n observations and u unknown parameters, then

$$\mathbf{V} = \mathbf{BX} + \mathbf{L} \qquad \text{(A-20)}$$

in which $\mathbf{V}$ is a column vector of n residuals with dimensions $n \times 1$, $\mathbf{X}$ is a column vector of unknown parameters with dimensions $u \times 1$, $\mathbf{B}$ is a coefficient matrix of dimensions $n \times u$, and $\mathbf{L}$ is a column vector of constant terms of dimensions $n \times 1$.

In Eq. (A-20) the coefficients of the residuals are all assumed to be unity. This is the usual case in the adjustment of elementary surveying measurements by observation equations.

Taking the general case of observations of different weights, the value of $\mathbf{V}$ from Eq. (A-20) is substituted into Eq. (A-18), giving

$$\mathbf{V}^T\mathbf{PV} = (\mathbf{BX} + \mathbf{L})^T\mathbf{P}(\mathbf{BX} + \mathbf{L}) = \text{minimum}$$

or

$$\mathbf{V}^T\mathbf{PV} = (\mathbf{X}^T\mathbf{B}^T + \mathbf{L}^T)\mathbf{P}(\mathbf{BX} + \mathbf{L})$$

Expanding gives

$$\mathbf{V}^T\mathbf{PV} = \mathbf{X}^T\mathbf{B}^T\mathbf{PBX} + \mathbf{X}^T\mathbf{B}^T\mathbf{PL} + \mathbf{L}^T\mathbf{PBX} + \mathbf{L}^T\mathbf{PL}$$

or since

$$\mathbf{L}^T\mathbf{PBX} = \mathbf{X}^T\mathbf{B}^T\mathbf{P}^T\mathbf{L} \quad \text{and} \quad \mathbf{P}^T = \mathbf{P}$$

then

$$\mathbf{L}^T\mathbf{PBX} = \mathbf{X}^T\mathbf{B}^T\mathbf{PL}$$

Finally,

$$\mathbf{V}^T\mathbf{PV} = \mathbf{X}^T\mathbf{B}^T\mathbf{PBX} + 2\mathbf{X}^T\mathbf{B}^T\mathbf{PL} + \mathbf{L}^T\mathbf{PL} \tag{A-21}$$

In order to make the weighted sums of squares of the residuals a minimum, we take the partial derivative of $\mathbf{V}^T\mathbf{PV}$ with respect to the vector of unknown parameters $\mathbf{X}$ and set equal to zero. Thus from Eq. (A-21)

$$\frac{\partial \mathbf{V}^T\mathbf{PV}}{\partial \mathbf{X}} = 2\mathbf{B}^T\mathbf{PBX} + 2\mathbf{B}^T\mathbf{PL} = 0$$

Giving the set of normal equations in matrix form as

$$\mathbf{B}^T\mathbf{PBX} + \mathbf{B}^T\mathbf{PL} = 0 \tag{A-22}$$

in which $\mathbf{X}$ is a column vector of unknown parameters with dimensions $u \times 1$, $\mathbf{B}^T\mathbf{PB}$ is a coefficient matrix of dimensions $u \times u$, and $\mathbf{B}^T\mathbf{PL}$ is a column matrix of constant terms of dimensions $u \times 1$.

The normal equations of Eq. (A-22) are solved for the $\mathbf{X}$ vector by matrix algebra giving

$$\mathbf{X} = -(\mathbf{B}^T\mathbf{PB})^{-1}\mathbf{B}^T\mathbf{PL} \tag{A-23}$$

The vector $\mathbf{V}$ of residuals can then be obtained by substitution of the value of $\mathbf{X}$ from Eq. (A-23) into the original observation equations of (A-20).

In the level net adjustment there are 11 observations and therefore $n = 11$. Since there are six junction points, there are six parameters to be computed, $x_1, x_2, \ldots, x_6$. Thus $u = 6$. The matrices making up the observation equation given by Eq. (A-20) are then as follows:

$$
V = \begin{bmatrix} v_1 \\ v_2 \\ v_3 \\ v_4 \\ v_5 \\ v_6 \\ v_7 \\ v_8 \\ v_9 \\ v_{10} \\ v_{11} \end{bmatrix} \quad
B = \begin{bmatrix}
+1 & 0 & 0 & 0 & 0 & 0 \\
-1 & 0 & 0 & 0 & +1 & 0 \\
0 & 0 & 0 & 0 & -1 & +1 \\
0 & 0 & 0 & 0 & 0 & -1 \\
-1 & +1 & 0 & 0 & 0 & 0 \\
0 & -1 & 0 & +1 & 0 & 0 \\
0 & 0 & 0 & -1 & +1 & 0 \\
0 & -1 & 0 & 0 & 0 & 0 \\
0 & 0 & +1 & 0 & 0 & 0 \\
0 & 0 & -1 & +1 & 0 & 0 \\
0 & 0 & 0 & -1 & 0 & 0
\end{bmatrix} \quad
X = \begin{bmatrix} x_1 \\ x_2 \\ x_3 \\ x_4 \\ x_5 \\ x_6 \end{bmatrix} \quad
L = \begin{bmatrix} 0 \\ 0 \\ 0 \\ -0.036 \\ 0 \\ 0 \\ -0.016 \\ +0.037 \\ -0.005 \\ 0 \\ -0.026 \end{bmatrix}
$$

$11 \times 1 \qquad\qquad\qquad 11 \times 6 \qquad\qquad\qquad 6 \times 1 \qquad\qquad 11 \times 1$

These matrices can be identified in the tabulated observation equations in Table A-3.

The weight matrix for the level-net adjustment is

$$
P = \begin{bmatrix}
0.672 & 0 & \cdot & \cdot & \cdot & \cdot & \cdot & \cdot & \cdot & \cdot & 0 \\
0 & 0.893 & & & & & & & & & \cdot \\
\cdot & & 0.758 & & & & & & & & \cdot \\
\cdot & & & 0.870 & & & & & & & \cdot \\
\cdot & & & & 0.787 & & & & & & \cdot \\
\cdot & & & & & 0.971 & & & & & \cdot \\
\cdot & & & & & & 0.725 & & & & \cdot \\
\cdot & & & & & & & 0.758 & & & \cdot \\
\cdot & & & & & & & & 1.043 & & \cdot \\
\cdot & & & & & & & & & 0.775 & 0 \\
0 & \cdot & \cdot & \cdot & \cdot & \cdot & \cdot & \cdot & \cdot & 0 & 0.424
\end{bmatrix}
$$

in which all the off-diagonal terms are zeros since no correlation is assumed to exist between the measured DE's. If the matrix multiplications indicated in Eq. (A-22) are performed, the resulting normal equations are then given in matrix form as follows:

$$
\begin{bmatrix}
+2.352 & -0.787 & 0 & 0 & -0.893 & 0 \\
-0.787 & +2.516 & 0 & -0.971 & 0 & 0 \\
0 & 0 & 0 & +1.818 & -0.775 & 0 \\
0 & -0.971 & -0.775 & +2.895 & -0.725 & 0 \\
-0.893 & 0 & 0 & -0.725 & +2.376 & -0.758 \\
0 & 0 & 0 & 0 & -0.758 & +1.628
\end{bmatrix}
\begin{bmatrix} x_1 \\ x_2 \\ x_3 \\ x_4 \\ x_5 \\ x_6 \end{bmatrix}
+ \begin{bmatrix} 0 \\ -0.02805 \\ -0.00521 \\ +0.02262 \\ -0.01160 \\ +0.03132 \end{bmatrix} = 0
$$

$$\qquad\qquad B^T P B \qquad\qquad\qquad\qquad X \qquad\qquad B^T P L$$

The solution for the vector of parameters given by Eq. (A-23), and the adjusted elevations are as follows:

SOLUTION (ft)		ADJUSTED ELEVATION (ft)
		$B = 853.147 + 0.003 = 853.150$
		$C = 861.576 + 0.010 = 861.586$
		$E = 865.769 + 0.001 = 865.770$
		$F = 855.048 - 0.005 = 855.043$
		$G = 851.680 - 0.002 = 851.678$
		$H = 861.436 - 0.020 = 861.416$

$$\mathbf{X} = \begin{bmatrix} x_1 \\ x_2 \\ x_3 \\ x_4 \\ x_5 \\ x_6 \end{bmatrix} = \begin{bmatrix} +0.003 \\ +0.010 \\ +0.001 \\ -0.005 \\ -0.002 \\ -0.020 \end{bmatrix}$$

The corrections to the measured DE's can now be obtained by substituting the value of **X** into Eq. (A-20) giving

$$v_1 = +0.003 \text{ ft} \qquad v_5 = +0.008 \text{ ft} \qquad v_9 = -0.004 \text{ ft}$$

$$v_2 = -0.005 \text{ ft} \qquad v_6 = -0.015 \text{ ft} \qquad v_{10} = -0.006 \text{ ft}$$

$$v_3 = -0.018 \text{ ft} \qquad v_7 = -0.013 \text{ ft} \qquad v_{11} = -0.021 \text{ ft}$$

$$v_4 = -0.016 \text{ ft} \qquad v_8 = +0.027 \text{ ft}$$

The standard error of unit weight for the network is given as

$$\sigma_0 = \sqrt{\frac{\mathbf{V}^T\mathbf{P}\mathbf{V}}{n - u}} \tag{A-24}$$

For this level network, the standard error of unit weight is ± 0.018 ft. The standard error of the measured DE's can then be found by Eq. (5-30), resulting in the following:

$$\sigma_1 = \pm 0.022 \text{ ft} \qquad \sigma_5 = \pm 0.020 \text{ ft} \qquad \sigma_9 = \pm 0.018 \text{ ft}$$

$$\sigma_2 = \pm 0.019 \text{ ft} \qquad \sigma_6 = \pm 0.018 \text{ ft} \qquad \sigma_{10} = \pm 0.020 \text{ ft}$$

$$\sigma_3 = \pm 0.021 \text{ ft} \qquad \sigma_7 = \pm 0.021 \text{ ft} \qquad \sigma_{11} = \pm 0.028 \text{ ft}$$

$$\sigma_4 = \pm 0.019 \text{ ft} \qquad \sigma_8 = \pm 0.021 \text{ ft}$$

A-4(a). ADJUSTMENT OF A QUADRILATERAL BY CONDITION EQUATIONS

The conditions to be satisfied in a completed quadrilateral are as follows: (1) The angles in three of the four triangles must add up to 180° plus spherical excess, and (2) the length of the closing side of the quadrilateral must be the same, no matter which way it is computed. Thus there are three angle conditions and one side condition in a completed quadrilateral.

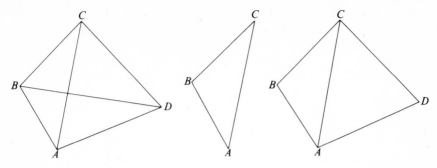

Figure A-2. Conditions in a quadrilateral.

To amplify the above conditions, consider the quadrilateral shown in Fig. A-2. On the left is shown the complete quadrilateral. The middle diagram shows C located from A and B by the lines AC and BC. These lines introduce an angle condition but do not create a condition relating to the lengths of the sides. The right-hand diagram shows D located from C and A by the lines CD and AD. These two lines introduce a second angle condition because they complete the second triangle, but there is still no condition relating to the lengths of the sides. Now, consider the effect of adding the line BD. This line forms two additional triangles, and the angles of each must add up to 180°. However, if the angles of one of the two triangles sum up correctly, and the first two angle conditions are satisfied, the angle condition in the other triangle will be satisfied. Therefore the addition of the line BD adds one more angle condition, making three in all. The addition of BD permits the development of a single side condition which is based on the following requirement: Beginning with side AB, four different routes may be taken through the triangles to calculate side CD, each of which must give the same value to the computed length of CD.

Following is an abstract of directions measured at the four stations shown in Fig. A-3. For simplicity, all angles are assumed to be plane angles with no spherical excess to contend with. Also, each angle has unit weight.

INSTRUMENT AT 20		INSTRUMENT AT CORNING	
18	0° 00′ 00.0″	Peters Rock	0° 00′ 00.0″
Corning	40° 08′ 17.9″	20	61° 29′ 34.3″
Peters Rock	112° 04′ 06.9″	18	103° 20′ 44.2″

INSTRUMENT AT 18		INSTRUMENT AT PETERS ROCK	
Corning	0° 00′ 00.0″	20	0° 00′ 00.0″
Peters Rock	53° 11′ 23.7″	18	23° 06′ 37.3″
20	98° 00′ 38.4″	Corning	46° 34′ 28.5″

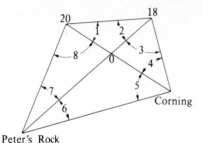

Figure A-3. Observed quadrilateral.

The resulting angles including the corrections are

1. $40° 08' 17.9'' + v_1$ 2. $44° 49' 14.7'' + v_2$
3. $53° 11' 23.7'' + v_3$ 4. $41° 51' 09.9'' + v_4$
5. $61° 29' 34.3'' + v_5$ 6. $23° 27' 51.2'' + v_6$
7. $23° 06' 37.3'' + v_7$ 8. $71° 55' 49.0'' + v_8$

When the angles in three of the four triangles are considered, the results are as follows:

Corning	$41° 51' 09.9'' + v_4$
20	$40° 08' 17.9'' + v_1$
18	$98° 00' 38.4'' + v_2 + v_3$

$180° 00' 06.2'' + v_1 + v_2 + v_3 + v_4 = 180° 00' 00.0''$

Peters Rock	$23° 06' 37.3'' + v_7$
20	$112° 04' 06.9'' + v_1 + v_8$
18	$44° 49' 14.7'' + v_2$

$179° 59' 58.9'' + v_1 + v_2 + v_7 + v_8 = 180° 00' 00.0''$

Corning	$61° 29' 34.3'' + v_5$
Peters Rock	$46° 34' 28.5'' + v_6 + v_7$
20	$71° 55' 49.0'' + v_8$

$179° 59' 51.8'' + v_5 + v_6 + v_7 + v_8 = 180° 00' 00.0''$

These relationships give rise to the following three angle-condition equations (see Section A-1):

$$v_1 + v_2 + v_3 + v_4 + 06.2'' = 0 \qquad (1)$$

$$v_1 + v_2 + v_7 + v_8 - 1.1'' = 0 \qquad (2)$$

$$v_5 + v_6 + v_7 + v_8 - 8.2'' = 0 \qquad (3)$$

The one side-condition equation can be arrived at in a number of different ways by starting with any one of the four sides of the quadrilateral. The necessary relationships for each are as follows:

1.
$$\frac{\text{Peters Rock--18}}{\text{Peters Rock--20}} = \frac{\sin(1 + 8)}{\sin 2}$$

$$\frac{\text{Peters Rock--Corning}}{\text{Peters Rock--18}} = \frac{\sin 3}{\sin(4 + 5)}$$

$$\frac{\text{Peters Rock--20}}{\text{Peters Rock--Corning}} = \frac{\sin 5}{\sin 8}$$

Here, angles 6 and 7 are omitted.

2.
$$\frac{\text{20--Corning}}{\text{20--18}} = \frac{\sin(2 + 3)}{\sin 4}$$

$$\frac{\text{20--Peters Rock}}{\text{20--Corning}} = \frac{\sin 5}{\sin(6 + 7)}$$

$$\frac{\text{20--18}}{\text{20--Peters Rock}} = \frac{\sin 7}{\sin 2}$$

Here, angles 1 and 8 are omitted.

3.
$$\frac{\text{18--Peters Rock}}{\text{18--Corning}} = \frac{\sin(4 + 5)}{\sin 6}$$

$$\frac{\text{18--20}}{\text{18--Peters Rock}} = \frac{\sin 7}{\sin(1 + 8)}$$

$$\frac{\text{18--Corning}}{\text{18--20}} = \frac{\sin 1}{\sin 4}$$

Here, angles 2 and 3 are omitted.

4.
$$\frac{\text{Corning--20}}{\text{Corning--Peters Rock}} = \frac{\sin(6 + 7)}{\sin 8}$$

$$\frac{\text{Corning--18}}{\text{Corning--20}} = \frac{\sin 1}{\sin(2 + 3)}$$

$$\frac{\text{Corning--Peters Rock}}{\text{Corning--18}} = \frac{\sin 3}{\sin 6}$$

Here, angles 4 and 5 are omitted.

It is also possible to develop the side-condition equation by considering the point O in Fig. A-3 in the following manner:

5. $\dfrac{O-20}{O-\text{Peters Rock}} = \dfrac{\sin 7}{\sin 8}$

$\dfrac{O-18}{O-20} = \dfrac{\sin 1}{\sin 2}$

$\dfrac{O-\text{Corning}}{O-18} = \dfrac{\sin 3}{\sin 4}$

$\dfrac{O-\text{Peters Rock}}{O-\text{Corning}} = \dfrac{\sin 5}{\sin 6}$

In this case all the measured angles are included.

The selection of some one of the five sets of triangles depends on the size of the angles involved. In order to make the constant term in the side-condition equation as large as possible, the smallest angles are selected. If all the angles involved are of the same general magnitude, then set (5) is selected because it involves all the angles. We will select set (2), since the sum of angles 1 and 8 is larger than the sum of the two angles at any other station.

Using the three equations in (2), we multiply the left-hand members, multiply the right-hand members, and equate the products. The result is

$$\frac{(20-\text{Corning})}{(20-18)} \cdot \frac{(20-\text{Peters Rock})}{(20-\text{Corning})} \cdot \frac{(20-18)}{(20-\text{Peters Rock})}$$

$$= \frac{\sin(2+3)}{\sin 4} \cdot \frac{\sin 5}{\sin(6+7)} \cdot \frac{\sin 7}{\sin 2}$$

or

$$1 = \frac{\sin(2+3)}{\sin 4} \cdot \frac{\sin 5}{\sin(6+7)} \cdot \frac{\sin 7}{\sin 2}$$

Clearing fractions gives

$$\sin 4 \cdot \sin(6+7) \cdot \sin 2 = \sin(2+3) \cdot \sin 5 \cdot \sin 7$$

It is difficult to handle this trigonometric equation. We therefore make a linear equation by taking the logarithms of both sides, as follows:

$$\log \sin 4 + \log \sin(6+7) + \log \sin 2 = \log \sin(2+3) + \log \sin 5 + \log \sin 7$$

In order to make a condition equation, we include the corrections to the angles. Thus

$$\log \sin(4 + v_4) + \log \sin(6 + 7 + v_6 + v_7) + \log \sin(2 + v_2)$$
$$= \log \sin(2 + 3 + v_2 + v_3) + \log \sin(5 + v_5) + \log \sin(7 + v_7)$$

If the v's are corrections in seconds, we can express each logarithm in a more convenient form. For example, the function for angle 4 would be

$$\log \sin(4 + v_4) = \log \sin 4 + (\text{change in value of log sin 4 per second})v_4$$

in which v_4 is in seconds. The change in value of the log sine of the angle obtained in the "difference" column in the logarithm tables is usually used with the decimal point after the sixth place. The hand calculator can be used to obtain both the logarithms and the differences.† The side condition equation can then be laid out as follows:

$$
\begin{array}{ll}
1 \sin 4 & = 9.824\,2681 + 2.35v_4 \\
1 \sin(6 + 7) & = 9.861\,0980 + 2.00v_6 + 2.00v_7 \\
1 \sin 2 & = 9.848\,1221 + 2.12v_2 \\
\hline
& \quad 9.533\,4882 + 2.35v_4 + 2.00v_6 + 2.00v_7 + 2.12v_2
\end{array}
$$

$$
\begin{array}{ll}
1 \sin(2 + 3) & = 9.995\,7414 - 0.30v_2 - 0.30v_3 \\
1 \sin 5 & = 9.943\,8691 + 1.14v_5 \\
1 \sin 7 & = 9.593\,8435 + 4.93v_7 \\
\hline
& = 9.533\,4540 - 0.30v_2 - 0.30v_3 + 1.14v_5 + 4.93v_7
\end{array}
$$

Transposing all terms to the left, we have

$$9.533\,4882 - 9.533\,4540 + 2.35v_4 + 2.00v_6 + 2.00v_7$$
$$+ 2.12v_2 + 0.30v_2 + 0.30v_3 - 1.14v_5 - 4.93v_7 = 0$$

Collecting terms, simplifying, and arranging the v's in order, we get the fourth condition equation

$$+2.42v_2 + 0.30v_3 + 2.35v_4 - 1.14v_5 + 2.00v_6 - 2.93v_7 + 34.2 = 0 \quad (4)$$

The constant term 34.2 is the difference between the sums of the logarithms on both sides, with the decimal point after the sixth place. The form for adding the logarithms on both sides shown above is usually arranged in tabular form as shown in Table A-5.

† See footnote to Table 10-6 on page 393.

TABLE A-5

ANGLE	LOG SIN $(+)$	DIFFERENCE $1''$	ANGLE	LOG SIN $(-)$	DIFFERENCE $1''$
4	9.824 2681	$+2.35$	2 + 3	9.995 7414	-0.30
6 + 7	9.861 0980	$+2.00$	5	9.943 8691	$+1.14$
2	9.848 1221	$+2.12$	7	9.593 8435	$+4.93$
	9.533 4882			9.533 4540	
	$-$ 540				
	342				

The four condition equations are:

$$v_1 + v_2 + v_3 + v_4 + 6.2 = 0$$

$$v_1 + v_2 + v_7 + v_8 - 1.1 = 0$$

$$v_5 + v_6 + v_7 + v_8 - 8.2 = 0$$

$$2.42v_2 + 0.30v_3 + 2.35v_4 - 1.14v_5 + 2.00v_6 - 2.93v_7 + 34.2 = 0$$

If it is assumed that all the angles have been measured with the same degree of precision, then the weight of each measurement is unity. Consequently, $\sum v^2$ must be a minimum in order to arrive at the most probable values for the angles in the solution. That is,

$$v_1^2 + v_2^2 + v_3^2 + v_4^2 + v_5^2 + v_6^2 + v_7^2 + v_8^2 = \text{minimum} \qquad \text{(A-25)}$$

Since the four condition equations contain eight unknowns, there is no unique solution as they stand. Also, there are only four independent v's in these equations. However, the condition of least squares allows us to find a set of v's that satisfy the four condition equations, while at the same time forcing $\sum v^2$ to be a minimum.

If each of the four condition equations are multiplied by a Lagrange multiplier, the results are

$$C_1(v_1 + v_2 + v_3 + v_4 + 6.2) = 0$$

$$C_2(v_1 + v_2 + v_7 + v_8 - 1.1) = 0$$

$$C_3(v_5 + v_6 + v_7 + v_8 - 8.2) = 0$$

$$C_4(2.42v_2 + 0.30v_3 + 2.35v_4 - 1.14v_5 + 2.00v_6 - 2.93v_7 + 34.2) = 0$$

and since the left-hand side of each of the condition equations is equal to zero, then the sum of all the left-hand sides is also equal to zero. Thus

$$C_1(v_1 + v_2 + v_3 + v_4 + 6.2)$$

$$+ C_2(v_1 + v_2 + v_7 + v_8 - 1.1)$$

$$+ C_3(v_5 + v_6 + v_7 + v_8 - 8.2)$$

$$+ C_4(2.42v_2 + 0.30v_3 + 2.35v_4 - 1.14v_5 + 2.00v_6 - 2.93v_7 + 34.2) = 0$$

$$\text{(A-26)}$$

Differentiation of Eq. (A-26) gives

$$C_1(dv_1 + dv_2 + dv_3 + dv_4)$$
$$+ C_2(dv_1 + dv_2 + dv_7 + dv_8)$$
$$+ C_3(dv_5 + dv_6 + dv_7 + dv_8)$$
$$+ C_4(2.42\, dv_2 + 0.30\, dv_3 + 2.35\, dv_4 - 1.14\, dv_5 + 2.00\, dv_6 - 2.93\, dv_7) = 0$$

$$(A\text{-}27)$$

Now since $\sum v^2$ has to be minimum, differentiation of Eq. (A-25) and setting equal to zero gives

$$v_1\, dv_1 + v_2\, dv_2 + v_3\, dv_3 + v_4\, dv_4 + v_5\, dv_5 + v_6\, dv_6 + v_7\, dv_7 + v_8\, dv_8 = 0$$

$$(A\text{-}28)$$

Multiplying the terms in parentheses in Eq. (A-27) by the appropriate Lagrange multipliers gives

$$C_1\, dv_1 + C_1\, dv_2 + C_1\, dv_3 + C_1\, dv_4 + C_2\, dv_1 + C_2\, dv_2 + C_2\, dv_7$$
$$+ C_2\, dv_8 + C_3\, dv_5 + C_3\, dv_6 + C_3\, dv_7 + C_3\, dv_8 + 2.42 C_4\, dv_2$$
$$+ 0.30 C_4\, dv_3 + 2.35 C_4\, dv_4 - 1.14 C_4\, dv_5 + 2.00 C_4\, dv_6 - 2.93 C_4\, dv_7 = 0$$

Collecting coefficients, we get

$$(C_1 + C_2)\, dv_1 + (C_1 + C_2 + 2.42 C_4)\, dv_2 + (C_1 + 0.30 C_4)\, dv_3$$
$$+ (C_1 + 2.35 C_4)\, dv_4 + (C_3 - 1.14 C_4)\, dv_5 + (C_3 + 2.00 C_4)\, dv_6$$
$$+ (C_2 + C_3 - 2.93 C_4)\, dv_7 + (C_2 + C_3)\, dv_8 = 0$$

$$(A\text{-}29)$$

Comparison of Eq. (A-28) with Eq. (A-29) gives the following relationships:

$$v_1 = C_1 + C_2$$
$$v_2 = C_1 + C_2 \qquad + 2.42 C_4$$
$$v_3 = C_1 \qquad\qquad + 0.30 C_4$$
$$v_4 = C_1 \qquad\qquad + 2.35 C_4$$
$$v_5 = \qquad\qquad C_3 - 1.14 C_4$$
$$v_6 = \qquad\qquad C_3 + 2.00 C_4$$
$$v_7 = \qquad C_2 + C_3 - 2.93 C_4$$
$$v_8 = \qquad C_2 + C_3$$

These are the *correlate equations* discussed in Section A-1. Note that there are only four variables on the right-hand sides of the equations expressing the values of the v's. Substitution of these values in the condition equations gives:

$$(C_1 + C_2) + (C_1 + C_2 + 2.42C_4)$$
$$+ (C_1 + 0.30C_4) + (C_1 + 2.35C_4) + 6.2 = 0$$
$$(C_1 + C_2) + (C_1 + C_2 + 2.42C_4)$$
$$+ (C_2 + C_3 - 2.93C_4) + (C_2 + C_3) - 1.1 = 0$$
$$(C_3 - 1.14C_4) + (C_3 + 2.00C_4)$$
$$+ (C_2 + C_3 - 2.93C_4) + (C_2 + C_3) - 8.2 = 0$$
$$2.42(C_1 + C_2 + 2.42C_4) + 0.30(C_1 + 0.30C_4)$$
$$+ 2.35(C_1 + 2.35C_4) - 1.14(C_3 - 1.14C_4)$$
$$+ 2.00(C_3 + 2.00C_4) - 2.93(C_2 + C_3 - 2.93C_4) + 34.2 = 0$$

When the last four equations are reduced to their simplest forms, they become

$$4C_1 + \quad 2C_2 \qquad\qquad + \quad 5.07C_4 + \quad 6.2 = 0 \qquad \text{(A-30)}$$
$$2C_1 + \quad 4C_2 + \quad 2C_3 - \quad 0.51C_4 - \quad 1.1 = 0 \qquad \text{(A-31)}$$
$$2C_2 + \quad 4C_3 - \quad 2.07C_4 - \quad 8.2 = 0 \qquad \text{(A-32)}$$
$$5.07C_1 - 0.51C_2 - 2.07C_3 + 25.30C_4 + 34.2 = 0 \qquad \text{(A-33)}$$

Equations (A-30), (A-31), (A-32), and (A-33) are the normal equations derived from the original condition equations through the correlate equations.

When the C's are determined by solving the normal equations, they are substituted in the correlate equations to obtain the v's. The v's are then applied to the observed angles to obtain the adjusted angles.

To make the preceding operations more systematic, the condition equations are first tabulated as shown in Table A-6. Then the rows of coefficients of the condition equations are arranged as columns. The tabulated

TABLE A-6. Tabulated Condition Equations

v_1	v_2	v_3	v_4	v_5	v_6	v_7	v_8	K
1	1	1	1					$+ \ 6.2 = 0$
1	1					1	1	$- \ 1.1 = 0$
				1	1	1	1	$- \ 8.2 = 0$
	2.42	0.30	2.35	-1.14	2.00	-2.93		$+34.2 = 0$

TABLE A-7. Tabulated Correlate Equations

	C_1	C_2	C_3	C_4
$v_1 =$	1	1		—
$v_2 =$	1	1		2.42
$v_3 =$	1			0.30
$v_4 =$	1			2.35
$v_5 =$			1	−1.14
$v_6 =$			1	2.00
$v_7 =$		1	1	−2.93
$v_8 =$		1	1	—

values represent the coefficients of the C's in the correlate equations, as indicated in Table A-7. This should be verified by the student. From the tabular arrangement for the correlate equations, the normal equations are formed by a cyclic and repetitive process. Let the coefficients of the C's be $a, b, c,$ and d, respectively. Then the normal equations would be represented as follows:

$$[aa]C_1 + [ab]C_2 + [ac]C_3 + [ad]C_4 + K_1 = 0$$
$$[ba]C_1 + [bb]C_2 + [bc]C_3 + [bd]C_4 + K_2 = 0$$
$$[ca]C_1 + [cb]C_2 + [cc]C_3 + [cd]C_4 + K_3 = 0$$
$$[da]C_1 + [db]C_2 + [dc]C_3 + [dd]C_4 + K_4 = 0$$

in which

$$[aa] = a_1a_1 + a_2a_2 + a_3a_3 + a_4a_4$$
$$[ab] = a_1b_1 + a_2b_2 + a_3b_3 + a_4b_4$$

and so on. Also $K_1, K_2, K_3,$ and K_4 are the same as the constant terms of the condition equations. This operation performs the same function as does the substitution of the correlate equations into the condition equations as shown on page 770.

Note that the coefficients of the C's about the diagonal terms are symmetrical. Thus the coefficients in row i are the same as the coefficients in column i. The normal equations could be shown in tabular form as in Table A-8.

TABLE A-8. Tabulated Normal Equations

C_1	C_2	C_3	C_4	K
$[aa]$	$[ab]$	$[ac]$	$[ad]$	$K_1 \;\; = 0$
$[ba]$	$[bb]$	$[bc]$	$[bd]$	$K_2 \;\; = 0$
$[ca]$	$[cb]$	$[cc]$	$[cd]$	$K_3 \;\; = 0$
$[da]$	$[db]$	$[dc]$	$[dd]$	$K_4 \;\; = 0$

TABLE A-9. Tabulated Normal Equations

C_1	C_2	C_3	C_4	K	
4	2	0	5.07	+ 6.2	= 0
2	4	2	− 0.51	− 1.1	= 0
0	2	4	− 2.07	− 8.2	= 0
5.07	−0.51	−2.07	+25.30	+34.2	= 0

In the present example, the tabulated normal equations would be as in Table A-9. The four normal equations are now solved simultaneously. Then the values of the C's are substituted in the correlate equations, from which the v's are obtained. A check on the calculations can be made by substituting the v's in the condition equations to see whether all four condition equations are satisfied. When the normal equations in the present example are solved, the results are

$$C_1 = +0.89; \qquad C_2 = -1.35; \qquad C_3 = +2.01; \qquad C_4 = -1.39$$

Substitution in the correlate equations gives

$$v_1 = -0.46'' \qquad v_5 = +3.59''$$

$$v_2 = -3.82'' \qquad v_6 = -0.77''$$

$$v_3 = +0.47'' \qquad v_7 = +4.73''$$

$$v_4 = -2.38'' \qquad v_8 = +0.66''$$

When this set of v's is substituted in the condition equations, all four of them are satisfied. Also, we are assured that $\sum v^2$ is a minimum.

Since each angle is assumed to have unit weight, the standard error of the measured angles can be determined from the v's. Thus

$$\sigma_0 = \sqrt{\frac{\sum v^2}{n - u}} \tag{A-34}$$

In this case, n is the number of angles and u is the number of conditions. The quantity $n - u$ again is the number of redundant observations, or the number of degrees of freedom. For a fully observed quadrilateral such as that in Fig. A-3, $n = 8$ and $u = 4$. Hence,

$$\sigma_0 = \sqrt{\frac{\sum v^2}{4}}$$

A-4(b). QUADRILATERAL ADJUSTMENT BY CONDITION EQUATIONS USING MATRICES

The solution of the adjustment of a quadrilateral by the method of condition equations will now be performed using matrices. First, the matrix algebra involved in the solution must be developed. The reader should correlate this development with the method just presented in order to satisfy himself that the two approaches are identical.

The matrix expression for a set of u condition equations containing corrections (v's) to n observations is given by

$$\mathbf{AV} - \mathbf{L} = 0 \qquad \text{(A-35)}$$

in which $\mathbf{V}$ is a column vector of residuals of dimensions $n \times 1$, $\mathbf{A}$ is a matrix of coefficients of the residuals of dimensions $u \times n$, and $\mathbf{L}$ is the column vector of constant terms of dimension $u \times 1$. Assuming weighted observations, whether or not all the weights are equal, the function F to be minimized is given by Eq. (A-18). Thus we wish to minimize

$$F = \mathbf{V}^T \mathbf{P} \mathbf{V}$$

Let $\mathbf{C}$ represent a vector of Lagrange multipliers of dimension $u \times 1$, (that is, one multiplier for each condition equation). For example, in the case of the fully observed quadrilateral

$$\mathbf{C}^T = [C_1 C_2 C_3 C_4]$$

Now since

$$\mathbf{AV} - \mathbf{L} = 0$$

by Eq. (A-35), then $2\mathbf{C}^T(\mathbf{AV} - \mathbf{L})$ also is zero. Thus the original function F can be expressed as

$$F = \mathbf{V}^T \mathbf{P} \mathbf{V} - 2\mathbf{C}^T(\mathbf{AV} - \mathbf{L}) \qquad \text{(A-36)}$$

This function is minimized by taking the partial derivative with respect to $\mathbf{V}$ and setting equal to zero. Thus

$$\frac{\partial F}{\partial \mathbf{V}} = 2\mathbf{V}^T \mathbf{P} - 2\mathbf{C}^T \mathbf{A} = 0 \qquad \text{(A-37)}$$

If Eq. (A-37) is divided by 2 and the transpositions are performed, then

$$\mathbf{P}^T \mathbf{V} - \mathbf{A}^T \mathbf{C} = 0 \qquad \text{(A-38)}$$

Examination of the weight matrix given in Eq. (A-18) shows that $\mathbf{P}^T = \mathbf{P}$. Thus Eq. (A-38) becomes

$$\mathbf{P} \mathbf{V} - \mathbf{A}^T \mathbf{C} = 0$$

Solving for **V** in terms of **C** gives

$$\mathbf{V} = \mathbf{P}^{-1}\mathbf{A}^T\mathbf{C} \tag{A-39}$$

Equation (A-39) is equivalent to the correlate equations shown on page 769 in which all the weights are assumed to be the same; that is, **P** is a unit matrix. Substituting this value of **V** into Eq. (A-35) gives

$$\mathbf{A}\mathbf{P}^{-1}\mathbf{A}^T\mathbf{C} - \mathbf{L} = 0$$

This is equivalent to the normal Eqs. (A-30), (A-31), (A-32), and (A-33), the solutions of which give

$$\mathbf{C} = (\mathbf{A}\mathbf{P}^{-1}\mathbf{A}^T)^{-1}\mathbf{L} \tag{A-40}$$

The vectors of residuals **V** is then obtained by substituting the value of **C** in Eq. (A-40) into Eq. (A-39), giving

$$\mathbf{V} = \mathbf{P}^{-1}\mathbf{A}^T(\mathbf{A}\mathbf{P}^{-1}\mathbf{A}^T)^{-1}\mathbf{L} \tag{A-41}$$

The reader should review this matrix algebra in order to satisfy himself of the equivalence of this derivation with that given in the quadrilateral adjustment.

In this adjustment there are four condition equations and $u = 4$. Also there are a total of eight observations and $n = 8$.

The values of the matrices given in Eq. (A-41) for this adjustment are as follows:

$$\mathbf{P} = \begin{bmatrix} 1 & 0 & . & . & . & . & . & 0 \\ 0 & 1 & & & & & & . \\ . & & 1 & & & & & . \\ . & & & 1 & & & & . \\ . & & & & 1 & & & . \\ . & & & & & 1 & & . \\ . & & & & & & 1 & 0 \\ 0 & . & . & . & . & . & 0 & 1 \end{bmatrix}$$

$$8 \times 8$$

$$\mathbf{A} = \begin{bmatrix} 1 & 1 & 1 & 1 & 0 & 0 & 0 & 0 \\ 1 & 1 & 0 & 0 & 0 & 0 & 1 & 1 \\ 0 & 0 & 0 & 0 & 1 & 1 & 1 & 1 \\ 0 & 2.42 & 0.30 & 2.35 & -1.14 & 2.00 & -2.93 & 0 \end{bmatrix} \quad \mathbf{L} = \begin{bmatrix} +6.2 \\ -1.1 \\ -8.2 \\ +34.2 \end{bmatrix}$$

$$4 \times 8 \qquad\qquad\qquad 4 \times 1$$

Reduction of Eq. (A-41) using these matrices will give the vector of residuals or corrections shown on page 772.

A-5. ADJUSTMENT OF TRIANGULATION BY VARIATION OF COORDINATES

A triangulation net of any degree of complexity can be adjusted by the method of observation equations commonly referred to as the *variation of coordinates*. If the work is being done on a plane coordinate system, the variations of the coordinates are designated ΔX and ΔY; and if on a geodetic system, the variation of coordinates are designated $\Delta\phi$ and $\Delta\lambda$. This section will deal with a plane coordinate system.

In the adjustment of the level net of Section A-3 by observation equations a set of approximate elevations was determined using selected measured DE's as shown on page 754. Then a set of observation equations was developed that expressed the corrections to the DE's (the v's) as functions of corrections to the approximate elevations. In solving the problem, the elevation of each point i was allowed to vary by an amount x_i in such a way that the values of the x's made the weighted sum of squares of the corrections to the measured DE's a minimum. This condition was assured by the method of formation of the normal equations; that is, the partial derivative of the weighted sum of squares of the residuals with respect to each of the x's was set equal to zero.

In the adjustment of a triangulation net by observation equations, a set of approximate X and Y coordinates is first computed for each new station. These computations are made using selected triangles with unadjusted angles, together with preliminary azimuths. These approximate coordinates are then allowed to vary by small amounts ΔX and ΔY in such a way that the computed values of the ΔX's and ΔY's make the weighted sum of squares of the corrections to the measured angles a minimum.

Each measured angle in the triangulation net involves three points in the net. These are the *at* station designated A, the *from* station designated F, and the *to* station designated T (see Section 4-15). If any of these are fixed stations, then the corrections to the coordinates of these stations will, of course, be zero.

The form of the observation equation that expresses the correction v to a measured angle as a function of the variations ΔX and ΔY of the triangulation stations can be developed in a number of different ways. The development presented here is geometric in nature. The angle FAT in Fig. A-4 is equal to the azimuth of the line AT minus the azimuth of the line AF. Stated as an observation equation

$$\angle FAT + v_A = \alpha_{AT} - \alpha_{AF} \tag{A-42}$$

in which $\angle FAT$ is the measured angle at A from F to T, v_A is the correction to the measured angle at A, α_{AF} is the adjusted azimuth of AF, and α_{AT} is the adjusted azimuth of AT.

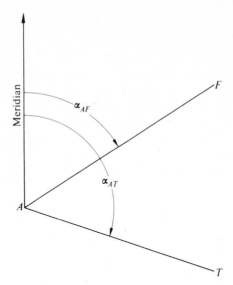

Figure A-4. Angle as difference in azimuths.

The adjusted azimuths of the two lines are obtained by applying corrections $\Delta\alpha_{AF}$ and $\Delta\alpha_{AT}$ to preliminary azimuths obtained from approximate coordinates of the stations by inversing according to Eq. (8-13). Thus

$$\alpha_{AF} = \alpha'_{AF} + \Delta\alpha_{AF}$$
$$\alpha_{AT} = \alpha'_{AT} + \Delta\alpha_{AT}$$

(A-43)

in which the primes (') denote preliminary azimuths. The observation equation given by Eq. (A-42) can then be written as follows:

$$v_A = \Delta\alpha_{AT} - \Delta\alpha_{AF} + [\alpha'_{AT} - \alpha'_{AF} - \angle FAT]$$

(A-44)

The terms inside the brackets make up the constant term of the observation equation. Equation (A-44) can then be expressed as

$$v_A = \Delta\alpha_{AT} - \Delta\alpha_{AF} + K_A$$

(A-45a)

in which

$$K_A = \alpha'_{AT} - \alpha'_{AF} - \angle FAT$$

(A-45b)

Expressions for $\Delta\alpha_{AT}$ and $\Delta\alpha_{AF}$ will now be developed geometrically. In Fig. A-5 points A' and F' are approximate positions of two triangulation stations obtained by preliminary position computations (see Section 10-25).

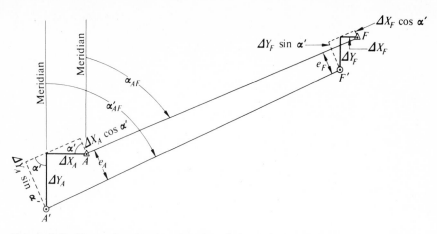

Figure A-5. Change in azimuth as function of change in coordinates.

The azimuth of this line α'_{AF} is obtained by Eq. (8-13) unless it has already been obtained for the position computation. The positions of the two points are allowed to vary by small amounts ΔX_A, ΔY_A and ΔX_F, ΔY_F in order to move them into the adjusted positions at A and F. The variation of the coordinates at each end of the line will change the preliminary azimuth of the line by a small amount $\Delta\alpha_{AF}$. The small distance e_A between the preliminary line $A'F'$ and the final line AF is expressed as

$$e_A = \Delta Y_A \sin\alpha'_{AF} - \Delta X_A \cos\alpha'_{AF} \qquad \text{(A-46a)}$$

and the small distance e_F at the other end of the line is

$$e_F = \Delta Y_F \sin\alpha'_{AF} - \Delta X_F \cos\alpha'_{AF} \qquad \text{(A-46b)}$$

If the line AF is imagined to be moved parallel with itself to make A coincide with A', the results are shown in Fig. A-6. The small angle between the two lines is $\Delta\alpha_{AF}$, and can be expressed as

$$\Delta\alpha_{AF} = \frac{e_A - e_F}{AF} \qquad \text{(A-47)}$$

in which the distance AF is obtained with sufficient accuracy from the approximate X and Y coordinates of A and F using Eq. (8-12). Since $\Delta\alpha_{AF}$ is a small angle, Eq. (A-47) can be expressed as

$$\Delta\alpha''_{AF} = \frac{e_A - e_F}{AF \text{ arc } 1''} \qquad \text{(A-48)}$$

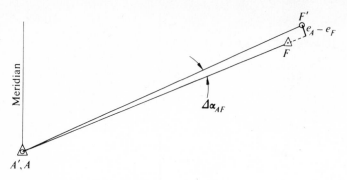

Figure A-6. Change in preliminary azimuth.

Substituting the value of e_A from Eq. (A-46a) and e_F from Eq. (A-46b) into Eq. (A-48) gives the expression for $\Delta\alpha_{AF}$ in seconds as a function of the variation in the coordinates of A and F, thus

$$\Delta\alpha''_{AF} = -\frac{\cos\alpha'_{AF}}{AF \text{ arc } 1''}\Delta X_A + \frac{\sin\alpha'_{AF}}{AF \text{ arc } 1''}\Delta Y_A$$

$$+ \frac{\cos\alpha'_{AF}}{AF \text{ arc } 1''}\Delta X_F - \frac{\sin\alpha'_{AF}}{AF \text{ arc } 1''}\Delta Y_F \qquad \text{(A-49)}$$

By the same geometrical analysis an expression for $\Delta\alpha_{AT}$ can be developed in terms of the variation of the coordinates of A and T, giving

$$\Delta\alpha''_{AT} = -\frac{\cos\alpha'_{AT}}{AT \text{ arc } 1''}\Delta X_A + \frac{\sin\alpha'_{AT}}{AT \text{ arc } 1''}\Delta Y_A$$

$$+ \frac{\cos\alpha'_{AT}}{AT \text{ arc } 1''}\Delta X_T - \frac{\sin\alpha'_{AT}}{AT \text{ arc } 1''}\Delta Y_T \qquad \text{(A-50)}$$

The observation equation for angle FAT is obtained in its final form by substituting the value of $\Delta\alpha_{AF}$ from Eq. (A-49) and $\Delta\alpha_{AT}$ from Eq. (A-50) into Eq. (A-45a), giving

$$v_A = \frac{\cos\alpha'_{AF}}{AF \text{ arc } 1''}\Delta X_A - \frac{\sin\alpha'_{AF}}{AF \text{ arc } 1''}\Delta Y_A - \frac{\cos\alpha'_{AF}}{AF \text{ arc } 1''}\Delta X_F + \frac{\sin\alpha'_{AF}}{AF \text{ arc } 1''}\Delta Y_F$$

$$- \frac{\cos\alpha'_{AT}}{AT \text{ arc } 1''}\Delta X_A + \frac{\sin\alpha'_{AT}}{AT \text{ arc } 1''}\Delta Y_A$$

$$+ \frac{\cos\alpha'_{AT}}{AT \text{ arc } 1''}\Delta X_T - \frac{\sin\alpha'_{AT}}{AT \text{ arc } 1''}\Delta Y_T + K_A \qquad \text{(A-51)}$$

The coefficients of the ΔX's and ΔY's in Eq. (A-51) are needed to only three or four significant figures at the most. However, the values of the prelim-

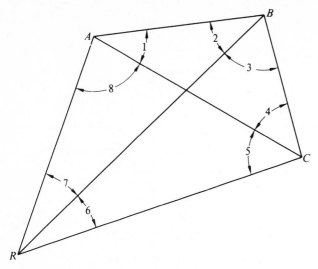

Figure A-7. Quadrilateral to be adjusted by variation of coordinates.

inary azimuths α'_{AF} and α'_{AT} contained in the constant term must be determined with a precision consistent with the precision of the measured angles. The value of the residual v_A and the constant term K_A are both expressed in seconds of angle.

Each independent measured angle in the triangulation net will introduce an observation equation of the form given by Eq. (A-51). These are then reduced by the methods given in Section A-3 in order to make the weighted sum of squares of the residuals a minimum.

The quadrilateral adjusted in Section A-4 by the method of condition equations will now be adjusted by the variation of coordinates technique. Figure A-7 is a redrawing of Fig. A-3, except that the triangulation stations are designated $A, B, C,$ and R for simplicity. Assume that the fixed coordinates of A and B are

$$X_A = 500,000.00, \qquad Y_A = 500,000.00$$
$$X_B = 525,000.00, \qquad Y_B = 502,000.00$$

Then, by Eq. (8-12)

$$AB = (25,000.00^2 + 2,000.00^2)^{1/2} = 25,079.87$$

and by Eq. (8-13)

$$\tan \alpha_{AB} = \frac{25,000.00}{2,000.00} = +12.50000000$$

TABLE A-10

STATION	ANGLE NUMBER	ANGLE VALUE	DISTANCE	SIDE
			25,079.87	AB
C	4	41° 51′ 07.8″	0.667 21094	
A	1	40° 08′ 15.8″	0.644 62710	
B	2 + 3	98° 00′ 36.4″	0.990 24369	
		180° 00′ 00.0″	24,230.96	BC
			37,222.39	CA

TABLE A-11

STATION	ANGLE NUMBER	ANGLE VALUE	DISTANCE	SIDE
			25,079.87	AB
R	7	23° 06′ 37.7″	0.392 50523	
A	1 + 8	112° 04′ 07.3″	0.926 73405	
B	2	44° 49′ 15.0″	0.704 89217	
		180° 00′ 00.0″	59,215.44	BR
			45,040.43	RA

and

$$\alpha_{AB} = 85° 25′ 33.9″$$

A set of approximate coordinates are now computed for stations C and R. The triangle CAB is computed and is then used for the position computation of station C. In this preliminary calculation the three angles are forced to add up to 180° by distributing the error of closure equally between the three angles. Thus from the layout of this triangle (Table A-10) each angle is seen to take a $-2.1″$ correction. The calculation of the triangle then follows (see Section 10-24).

Triangle RAB is used to compute an approximate set of coordinates for Station R. The triangle solution is as in Table A-11.

Calculations are laid out below for the position computation of C and R as given in Section 10-25. The azimuths are computed as follows:

azimuth AB	=	85° 25′ 33.9″	azimuth BA	=	265° 25′ 33.9″
+ angle 1	= +	40° 08′ 15.8″	− angle 2 + 3	−	98° 00′ 36.4″
azimuth AC	=	125° 33′ 49.7″	azimuth BC	=	167° 24′ 57.5″
				−	125° 33′ 49.7″
			angle 4	=	41° 51′ 07.8″ (check)

$$\text{azimuth } AB = 85° 25' 33.9'' \quad \text{azimuth } BA = 265° 25' 33.9''$$
$$+\text{angle } 1 + 8 = +112° 04' 07.3'' \quad -\text{angle } 2 = -44° 49' 15.0''$$
$$\text{azimuth } AR = 197° 29' 41.2'' \quad \text{azimuth } BR = 220° 36' 18.9''$$
$$-197° 29' 41.2''$$
$$\text{angle } 7 = 23° 06' 37.7'' \text{ (check)}$$

The position computations are shown in Table A-12.

The approximate coordinates of the new stations C and R, together with those of the fixed stations A and B, are now used to compute lengths and preliminary azimuths of all the lines in the figure. These are used to determine the coefficients of the ΔX's and ΔY's in the observation equations. The preliminary azimuths are also used in computing the constant terms of the observation equations. In this very simple net the only line that must be inversed is the line CR because all the other lines in the net were used to determine the approximate coordinates in the first place. In a more complex net, however, it will generally be necessary to inverse a substantial number of lines before an observation equation can be formed for each measured independent angle.

The calculations necessary to inverse the line CR are shown here.

$$X_R = 486,460.00 \qquad Y_R = 457,042.94$$
$$X_C = 530,279.24 \qquad Y_C = 478,351.12$$
$$\Delta X_{CR} = -43,819.24 \qquad \Delta Y_{CR} = -21,308.18$$

$$CR = \sqrt{43,819.24^2 + 21,308.18^2} = 48,725.40$$

$$\tan \alpha_{CR} = \frac{-43,819.24}{-21,308.18} = 2.056\,45156$$

$$\text{azimuth } CR = 244° 04' 03.1''$$

TABLE A-12

STATION	LENGTH	COS AZIMUTH	SIN AZIMUTH	Y	X
A				500,000.00	500,000.00
	37,222.39	−0.581 60921	+0.813 46834	−21,648.88	+30,279.24
C				478,351.12	530,279.24
	24,230.96	+0.975 97754	−0.217 87117	+23,648.87	−5,279.23
B				501,999.99	525,000.01
			fixed	502,000.00	525,000.00
A				500,000.00	500,000.00
	45,040.43	−0.953 74436	−0.300 61887	−42,957.06	−13,540.00
R				457,042.94	486,460.00
	59,215.44	+0.759 21167	+0.650 84378	+44,957.05	+38,540.00
B				501,999.99	525,000.00
			fixed	502,000.00	525,000.00

TABLE A-13

LINE	PRELIMINARY LENGTH	PRELIMINARY AZIMUTH	COS α'	SIN α'
AB	—	85° 25′ 33.9″	—	—
AC	37,222	125° 33′ 49.7″	−0.58161	+0.81347
AR	45,040	197° 29′ 41.2″	−0.95374	−0.30062
BA	—	265° 25′ 33.9″	—	—
BC	24,231	167° 24′ 57.5″	−0.97598	+0.21787
BR	59,215	220° 36′ 18.9″	−0.75921	−0.65084
CR	48,725	244° 04′ 03.1″	−0.43731	−0.89931
CA	37,222	305° 33′ 49.7″	+0.58161	−0.81347
CB	24,231	347° 24′ 57.5″	+0.97598	−0.21787
RA	45,040	17° 29′ 41.2″	+0.95374	+0.30062
RB	59,215	40° 36′ 18.9″	+0.75921	+0.65084
RC	48,725	64° 04′ 03.1″	+0.43731	+0.89931

A listing is now made of all the lines needed for the formation of the observation equations. Neither the length of AB nor the functions of the azimuth are needed and are not shown. (See Table A-13.)

The observation equation for angle 1 involves stations A, B, and C. An examination of Eq. (A-51) shows that since A and B are fixed points, there will be only two unknowns on the right side of the equation, that is, ΔX_C and ΔY_C. The constant term K_1 is computed by Eq. (A-45b) as follows:

$$
\begin{aligned}
\alpha'_{AC} &= 125° 33′ 49.7″ \\
\alpha'_{AB} &= -85° 25′ 33.9″ \\
\hline
&40° 08′ 15.8″ \\
-\text{measured} \angle 1 &= -40° 08′ 17.9″ \quad \text{(see page 763)} \\
\hline
K_1 &= -2.1″
\end{aligned}
$$

Then, by Eq. (A-51) the observation equation for angle 1 is

$$
v_1 = \frac{\cos \alpha'_{AC}}{AC \text{ arc } 1''} \Delta X_C - \frac{\sin \alpha'_{AC}}{AC \text{ arc } 1''} \Delta Y_C - 2.1''
$$

or

$$
v_1 = \frac{-0.58161}{37{,}222 \times 4.85 \times 10^{-6}} \Delta X_C - \frac{0.81347}{37{,}222 \times 4.85 \times 10^{-6}} \Delta Y_C - 2.1''
$$

or

$$
v_1 = -3.22 \Delta X_C - 4.50 \Delta Y_C - 2.1''
$$

The constant term for the equation for angle 2 is as follows:

$$\alpha'_{BA} = 265° \ 25' \ 33.9''$$
$$\alpha'_{BR} = -220° \ 36' \ 18.9''$$
$$\overline{\phantom{\alpha'_{BR} = -2} 44° \ 49' \ 15.0''}$$
$$-\text{measured} \ \angle 2 = -44° \ 49' \ 14.7''$$
$$K_2 = +0.3''$$

The observation equation for angle 2 is then

$$v_2 = -\frac{\cos \alpha'_{BR}}{BR \ \text{arc} \ 1''} \Delta X_R + \frac{\sin \alpha'_{BR}}{BR \ \text{arc} \ 1''} \Delta Y_R + 0.3''$$

$$v_2 = -\frac{-0.75921}{59,215 \times 4.85 \times 10^{-6}} \Delta X_R + \frac{-0.65084}{59,215 \times 4.85 \times 10^{-6}} \Delta Y_R + 0.3''$$

$$v_2 = +2.64 \ \Delta X_R - 2.23 \ \Delta Y_R + 0.3''$$

The constant term for the equation for angle 3 is

$$\alpha'_{BR} = 220° \ 36' \ 18.9''$$
$$-\alpha'_{BC} = -167° \ 24' \ 57.5''$$
$$\overline{\phantom{-\alpha'_{BC} = -1} 53° \ 11' \ 21.4''}$$
$$-\text{measured} \ \angle 3 = -53° \ 11' \ 23.7''$$
$$K_3 = -2.3''$$

The observation equation for angle 3 is then

$$v_3 = -\frac{\cos \alpha'_{BC}}{BC \ \text{arc} \ 1''} \Delta X_C + \frac{\sin \alpha'_{BC}}{BC \ \text{arc} \ 1''} \Delta Y_C$$

$$+ \frac{\cos \alpha'_{BR}}{BR \ \text{arc} \ 1''} \Delta X_R - \frac{\sin \alpha'_{BR}}{BR \ \text{arc} \ 1''} \Delta Y_R - 2.3''$$

$$v_3 = -\frac{-0.97598}{24,231 \times 4.85 \times 10^{-6}} \Delta X_C + \frac{0.21787}{24,231 \times 4.85 \times 10^{-6}} \Delta Y_C$$

$$+ \frac{-0.75921}{59,215 \times 4.85 \times 10^{-6}} \Delta X_R - \frac{-0.65084}{59,215 \times 4.85 \times 10^{-6}} \Delta Y_R - 2.3''$$

$$v_3 = +8.30 \ \Delta X_C + 1.85 \ \Delta Y_C - 2.64 \ \Delta X_R + 2.23 \ \Delta Y_R - 2.3''$$

Note that the numerical values of the coefficients are repeated each time a line is used. The algebraic sign of the coefficients must be determined very carefully in forming the observation equations.

The observation equations for the eight measured angles are grouped here for convenience.

$$v_1 = -3.22\,\Delta X_C - 4.50\,\Delta Y_C \qquad\qquad\qquad\qquad\quad -2.1''$$

$$v_2 = \qquad\qquad\qquad\qquad +2.64\,\Delta X_R - 2.23\,\Delta Y_R + 0.3''$$

$$v_3 = +8.30\,\Delta X_C + 1.85\,\Delta Y_C - 2.64\,\Delta X_R + 2.23\,\Delta Y_R - 2.3''$$

$$v_4 = -5.08\,\Delta X_C + 2.65\,\Delta Y_C \qquad\qquad\qquad\qquad\quad -2.1''$$

$$v_5 = -5.07\,\Delta X_C - 0.69\,\Delta Y_C + 1.85\,\Delta X_R - 3.81\,\Delta Y_R + 12.3''$$

$$v_6 = +1.85\,\Delta X_C - 3.80\,\Delta Y_C + 0.79\,\Delta X_R + 1.57\,\Delta Y_R - 7.0''$$

$$v_7 = \qquad\qquad\qquad\qquad +1.72\,\Delta X_R + 0.85\,\Delta Y_R + 0.4''$$

$$v_8 = +3.22\,\Delta X_C + 4.50\,\Delta Y_C - 4.36\,\Delta X_R + 1.38\,\Delta Y_R + 2.5''$$

The matrices making up the above observation equations given by Eq. (A-20) are as follows:

$$
\mathbf{V} =
\begin{bmatrix}
v_1 \\ v_2 \\ v_3 \\ v_4 \\ v_5 \\ v_6 \\ v_7 \\ v_8
\end{bmatrix}
\qquad
\mathbf{B} =
\begin{bmatrix}
-3.22 & -4.50 & 0 & 0 \\
0 & 0 & +2.64 & -2.23 \\
+8.30 & +1.85 & -2.64 & +2.23 \\
-5.08 & +2.65 & 0 & 0 \\
-5.07 & -0.69 & +1.85 & -3.81 \\
+1.85 & -3.80 & +0.79 & +1.57 \\
0 & 0 & +1.72 & +0.85 \\
+3.22 & +4.50 & -4.36 & +1.38
\end{bmatrix}
$$

$$8 \times 1 \qquad\qquad\qquad\qquad 8 \times 4$$

$$
\mathbf{X} =
\begin{bmatrix}
\Delta X_C \\ \Delta Y_C \\ \Delta X_R \\ \Delta Y_R
\end{bmatrix}
\qquad
\mathbf{L} =
\begin{bmatrix}
-2.1'' \\ +0.3'' \\ -2.3'' \\ -2.1'' \\ +12.3'' \\ -7.0'' \\ +0.4'' \\ +2.5''
\end{bmatrix}
$$

$$4 \times 1 \qquad\qquad\qquad\qquad 8 \times 1$$

Since all the angles in this triangulation are assumed to have equal weights, the weight matrix P appearing in the solution for X given by Eq. (A-23) is the unit matrix. The solution for X gives

$$\Delta X_C = -0.10 \text{ ft}; \qquad \Delta Y_C = -0.30 \text{ ft}$$
$$\Delta X_R = +1.00 \text{ ft}; \qquad \Delta Y_R = +3.00 \text{ ft}$$

These values are added to the approximate coordinates of C and R to give the final adjusted values.

approx. X_C = 530,279.24   approx. Y_C = 478,351.12

        −0.10   −0.30

adjusted X_C = 530,279.14   adjusted Y_C = 478,350.82

approx. X_R = 486,460.00   approx. Y_R = 457,042.94

        +1.00   +3.00

adjusted X_R = 486,461.00   adjusted Y_R = 457,045.94

The values of X are substituted into the original observation equation (A-20) in order to obtain the vector of residuals. The results are

$$v_1 = -0.51'' \qquad v_5 = +3.44''$$
$$v_2 = -3.75'' \qquad v_6 = -0.64''$$
$$v_3 = +0.37'' \qquad v_7 = +4.67''$$
$$v_4 = -2.31'' \qquad v_8 = +0.69''$$

These results are essentially the same as the set obtained by the adjustment of the quadrilateral using condition equations shown on page 772.

In the formation of the observation equations there will be one equation for each measured independent angle. In the quadrilateral used in this example, all eight angles are independent. There are two general cases in which a measured angle is not an independent angle. The first case involves angles measured between fixed lines, and the second case arises when all angles are measured to close the horizon. In Fig. A-8, A, B, and C are fixed points. If angles CBF and FBA are both measured, then only one is an independent angle since the angle CBA is a fixed angle. Thus only one of the two is used to form an observation equation.

If all five of the angles are measured at the new station F, then one of these is dependent on the other four since all five must add up to the a priori value of $360°$. Thus one of the angles must be omitted from the formation of the observation equations.

After the observation equations are formed, the coefficients and constant terms can be checked for consistency. The coefficients of the observation

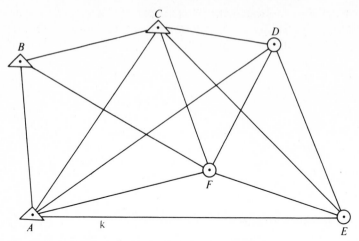

Figure A-8. Triangulation net with dependent angles.

equations involved in any one triangle must add up to zero. For example, in triangle ACR of Fig. A-7 the coefficients of observation equations 5, 6, 7, and 8 must sum up to zero. Also, the sum of the constant terms in the equations involved in any one triangle must equal the error of closure of the triangle with opposite algebraic sign. As an example, the constant terms in observation equations 5, 6, 7, and 8 shown on page 784 must add up to $+8.2''$ which in fact they do.

A-6. ADJUSTMENT OF TRILATERATION BY VARIATION OF COORDINATES
A trilateration net of measured lengths can be adjusted by first determining a set of preliminary coordinates of the stations in the net and then letting these coordinates vary in such a way as to make the weighted sum of squares of the corrections to the measured lengths a minimum. The overall method is the same as that used for the adjustment of angles by variation of coordinates.

In Fig. A-9, A' and T', signifying the *at* and the *to* stations, are the approximate positions of two stations having been determined by a preliminary trilateration computation. Points A and T are the adjusted positions of the two stations. The small distance ΔX_A, ΔY_A, ΔX_T, and ΔY_T are the amounts by which the approximate coordinates change to bring the points to their final positions. An observation equation is expressed giving the measured length, plus its correction, in terms of a preliminary computed length, plus a small change in the computed length. Thus

$$l_m + v = l' + \Delta l' \tag{A-52}$$

in which l_m is the measured length, reduced to the plane coordinate projection; v is the correction to measured length; l' is the length computed from approximate coordinates; and $\Delta l'$ is the correction to computed length.

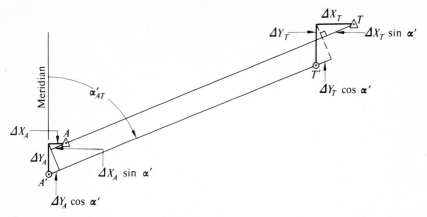

Figure A-9. Change in length as function of change in coordinates.

The value of $\Delta l'$ can be obtained in terms of the ΔX's and ΔY's of Fig. A-9. Thus

$$AT = A'T' + \Delta l'$$
$$= A'T' + (\Delta Y_T \cos \alpha'_{AT} + \Delta X_T \sin \alpha'_{AT}) - (\Delta Y_A \cos \alpha'_{AT} + \Delta X_A \sin \alpha'_{AT})$$

and

$$\Delta l' = -\sin \alpha'_{AT} \Delta X_A - \cos \alpha'_{AT} \Delta Y_A + \sin \alpha'_{AT} \Delta X_T + \cos \alpha'_{AT} \Delta Y_T$$

$$\text{(A-53)}$$

Substituting the value of $\Delta l'$ from Eq. (A-53) into (A-52) and rearranging terms gives the final form of the observation equation.

$$v = -\sin \alpha'_{AT} \Delta X_A - \cos \alpha'_{AT} \Delta Y_A + \sin \alpha'_{AT} \Delta X_T + \cos \alpha'_{AT} \Delta Y_T + K$$

$$\text{(A-54a)}$$

in which

$$K = l' - l_m \qquad \text{(A-54b)}$$

Just as in the case of observation equations in triangulation, the coefficients of the ΔX's and ΔY's are needed to only a few places after the decimal in most instances. However, the preliminary value of the length l' contained in the constant term must be computed by inversing with a precision consistent with the precision of the measured length l_m.

After all the observations have been formed, they are then reduced by the methods given in Section A-3 in order to make the weighted sum of squares of the residuals a minimum.

One difficulty that arises in the adjustment of a trilateration net is the assignment of proper weights in the **P** matrix of Eq. (A-23). Different types of EDM's have different accuracies. Long lines are liable to be affected by environmental conditions more than shorter lines. On the other hand the relative accuracy of short lines is usually lower than that of longer lines because of the constant error or the limiting resolution of the phase meter. Thus the a priori assignment of elements in the **P** matrix is always subject to error.

A fairly acceptable rationale for the assignment of weights to length observation is to consider a constant error and a proportional error to affect each measurement. These errors are somewhat subjective, being based partly on experience with a given EDM and partly on the manufacturer's statement of the magnitude of the constant error. Assume that experience indicates consistent agreement between measured and "known" distances to be approximately 1/100,000 or 10 ppm. If the so-called constant error is, say, 0.02 ft, then the total assumed error in a distance of 10,000 ft would be $0.02 + 10,000/100,000 = 0.12$ ft. Letting this combination represent the uncertainty expressed as a standard error, the weight to be assigned to a measurement is then the inverse of the square of the standard error. In the case of the 10,000-ft length, the weight would be $1/0.12^2 = 69.5$. The assumed standard error of a 30,000-ft length would be 0.32 ft with a corresponding weight of 9.8. The weights for these two lengths could be expressed also in relative terms as 6.95 and 0.98, respectively.

If the reader wishes to pursue an investigation of weighted observations in EDM, he should study current literature on this subject. Since the use of EDM in the field of surveying is relatively new, there is not at present enough supporting evidence to substantiate any one particular method for assigning proper weights to the direct measurement of distance by EDM.

The quadrilateral adjusted in Section A-5 as a simple triangulation net and shown in Fig. A-7 will now be used as a trilateration net to illustrate the adjustment by variation of coordinates. In Table A-14 the lengths have been measured by EDM, corrected for environmental conditions, reduced to their sea-level lengths, and then to their corresponding plane coordinate projection or grid length. The relative weights were computed as k/σ^2, in which the arbitrary constant of proportionality $k = 0.10$ and σ is the assumed standard

TABLE A-14

LINE NUMBER	LINE	MEASURED GRID LENGTH (ft)	ASSUMED WEIGHT
9	AC	37,221.98	0.66
10	BC	24,231.15	1.48
11	AR	45,040.08	0.45
12	BR	59,215.10	0.27
13	CR	48,726.18	0.40

error based on the length of the line, a constant error of 0.02 ft, and 10 ppm proportionate error. The significance of the numbers assigned to the lines will become apparent in Section A-7. The coordinates of A and B are as given in the example in Section A-5 from which AB is computed to be 25,079.87 ft. Only two triangles need be computed to determine approximate coordinates of C and R. First, the angles in each triangle are computed using the law of cosines. These are checked to verify that they sum up to $180°$ in each triangle.

In triangle CAB

$$\cos C = \frac{\overline{CA}^2 + \overline{CB}^2 - \overline{AB}^2}{2 \times CA \times CB} = \frac{37,221.98^2 + 24,231.15^2 - 25,079.87^2}{2 \times 37,221.98 \times 24,231.15}$$
$$= 0.744\,85963$$

$$\cos A = \frac{\overline{AB}^2 + \overline{AC}^2 - \overline{BC}^2}{2 \times AB \times AC} = \frac{25,079.87^2 + 37,221.98^2 - 24,231.15^2}{2 \times 25,079.87 \times 37,221.98}$$
$$= 0.764\,48461$$

$$\cos B = \frac{\overline{BA}^2 + \overline{BC}^2 - \overline{AC}^2}{2 \times BA \times BC} = \frac{25,079.87^2 + 24,231.15^2 - 37,221.98^2}{2 \times 25,079.87 \times 24,231.15}$$
$$= -0.139\,31492$$

$$
\begin{aligned}
C &= 41° 51' 10.6'' \\
A &= 40° 08' 19.8'' \\
B &= 98° 00' 29.5'' \\
\hline
&179° 59' 59.9'' \quad \text{(check)}
\end{aligned}
$$

In triangle RAB

$$\cos R = \frac{\overline{RA}^2 + \overline{RB}^2 - \overline{AB}^2}{2 \times RA \times RB} = \frac{45,040.08^2 + 59,215.10^2 - 25,079.87^2}{2 \times 45,040.08 \times 59,215.10}$$
$$= 0.919\,74874$$

$$\cos A = \frac{\overline{AB}^2 + \overline{AR}^2 - \overline{BR}^2}{2 \times AB \times AR} = \frac{25,079.87^2 + 45,040.08^2 - 59,215.10^2}{2 \times 25,079.87 \times 45,040.08}$$
$$= -0.375\,71701$$

$$\cos B = \frac{\overline{BA}^2 + \overline{BR}^2 - \overline{AR}^2}{2 \times BA \times BR} = \frac{25,079.87^2 + 59,215.10^2 - 45,040.08^2}{2 \times 25,079.87 \times 59,215.10}$$
$$= 0.709\,31560$$

$$
\begin{aligned}
R &= 23° 06' 38.2'' \\
A &= 112° 04' 07.1'' \\
B &= 44° 49' 14.7'' \\
\hline
&180° 00' 00.0'' \quad \text{(check)}
\end{aligned}
$$

The angles in the triangles are now used to compute the azimuths of the lines, from which preliminary position computations can be made. The azimuths are computed as follows:

azimuth AB =	85° 25′ 33.9″	azimuth BA =	265° 25′ 33.9″	
+ angle at A =	+40° 08′ 19.8″	− angle at B =	−98° 00′ 29.5″	
azimuth AC =	125° 33′ 53.7″	azimuth BC =	167° 25′ 04.4″	
			−125° 33′ 53.7″	
		angle at C =	41° 51′ 10.7″	(check)

azimuth AB =	85° 25′ 33.9″	azimuth BA =	265° 25′ 33.9″	
+ angle at A =	+112° 04′ 07.1″	− angle at B =	−44° 49′ 14.7″	
azimuth AR =	197° 29′ 41.0″	azimuth BR =	220° 36′ 19.2″	
			−179° 29′ 41.0″	
		angle at R =	23° 06′ 38.2″	(check)

The position computations are shown in Table A-15.

The approximate coordinates of the new stations C and R, together with those of the fixed stations A and B, are now used to compute the lengths and azimuths of all the lines in the figure. The azimuths are used to determine the coefficients of the ΔX's and ΔY's in the observation equations. The preliminary lengths are used in computing the constant term of the observation equation. In this very simple net, line CR is the only line that must be inversed because the other lines were used to determine the approximate coordinates in the first place. In a more complex net, however, it will generally be necessary to inverse a substantial number of lines before an observation equation can be formed for each measured distance.

TABLE A-15

STATION	LENGTH	COS AZIMUTH	SIN AZIMUTH	Y	X
A				500,000.00	500,000.00
	37,221.98	−0.58162491	+0.81345706	−21,649.23	+30,278.48
C				478,350.77	530,278.48
	24,231.15	+0.97598482	−0.21783853	+23,649.23	− 5,278.48
B				502,000.00	525,000.00
			fixed	502,000.00	525,000.00
A				500,000.00	500,000.00
	45,040.08	−0.95374465	−0.30061795	−42,956.74	−13,539.86
R				457,043.26	486,460.14
	59,215.10	+0.75921073	+0.65084489	+44,956.74	+38,539.85
B				502,000.00	524,999.99
			fixed	502,000.00	525,000.00

TABLE A-16

LINE	PRELIMINARY LENGTH	PRELIMINARY AZIMUTH	COS α'	SIN α'
AC	37,221.98	125° 33′ 54″	−0.58162	+0.81346
BC	24,231.15,	167° 25′ 04″	−0.97598	+0.21784
AR	45,040.08	197° 29′ 41″	−0.95374	−0.30062
BR	59,215.10	220° 36′ 19″	−0.75921	−0.65084
CR	48,724.29	244° 04′ 04″	−0.43731	−0.89931

The calculations necessary to inverse the line CR are shown below.

$$X_R = \ 486{,}460.14 \qquad Y_R = \ 457{,}043.26$$
$$X_C = \ 530{,}278.48 \qquad Y_C = \ 478{,}350.77$$
$$\Delta X_{CR} = -43{,}818.34 \qquad \Delta Y_{CR} = -21{,}307.51$$
$$CR = \sqrt{43{,}818.34^2 + 21{,}307.51^2} = 48{,}724.29$$

$$\tan \alpha_{CR} = \frac{-43{,}818.34}{-21{,}307.51} = +2.056\ 47398$$

azimuth $CR = 244° 04′ 04″$ (needed only to nearest minute)

A listing is now made (Table A-16) of all the lines needed for the formation of the observation equations.

The observation equation for line 9 (AC) contains the two unknown parameters ΔX_C and ΔY_C on the right side. Thus by Eq. (A-54a)

$$v_9 = \sin \alpha'_{AC} \, \Delta X_C + \cos \alpha'_{AC} \, \Delta Y_C + K_9$$

in which

$$K_9 = 37{,}221.98 - 37{,}221.98 = 0$$

thus

$$v_9 = +0.813 \, \Delta X_C - 0.582 \, \Delta Y_C + 0$$

Similarly, the other four observation equations are formed. For line 13 (CR),

$$v_{13} = -\sin \alpha'_{CR} \, \Delta X_C - \cos \alpha'_{CR} \, \Delta Y_C + \sin \alpha'_{CR} \, \Delta X_R + \cos \alpha'_{CR} \, \Delta Y_R + K_{13}$$

in which

$$K_{13} = 48{,}724.29 - 48{,}726.18 = -1.89$$

The observation equations for the five measured lines are grouped here for convenience.

$$v_9 = +0.813 \, \Delta X_C - 0.582 \, \Delta Y_C \qquad\qquad\qquad\qquad + 0$$

$$v_{10} = +0.218 \, \Delta X_C - 0.976 \, \Delta Y_C \qquad\qquad\qquad\qquad + 0$$

$$v_{11} = \qquad\qquad\qquad\quad - 0.301 \, \Delta X_R - 0.954 \, \Delta Y_R + 0$$

$$v_{12} = \qquad\qquad\qquad\quad - 0.651 \, \Delta X_R - 0.759 \, \Delta Y_R + 0$$

$$v_{13} = +0.899 \, \Delta X_C + 0.437 \, \Delta Y_C - 0.899 \, \Delta X_R - 0.437 \, \Delta Y_R - 1.89$$

The matrices making up the observation equations given by Eq. (A-20) are as follows:

$$
\mathbf{V} = \begin{bmatrix} v_9 \\ v_{10} \\ v_{11} \\ v_{12} \\ v_{13} \end{bmatrix} \qquad
\mathbf{B} = \begin{bmatrix}
+0.813 & -0.582 & 0 & 0 \\
+0.218 & -0.976 & 0 & 0 \\
0 & 0 & -0.301 & -0.954 \\
0 & 0 & -0.651 & -0.759 \\
+0.899 & +0.437 & -0.899 & -0.437
\end{bmatrix}
$$

$$5 \times 1 \qquad\qquad\qquad\qquad\qquad 5 \times 4$$

$$
\mathbf{X} = \begin{bmatrix} \Delta X_C \\ \Delta Y_C \\ \Delta X_R \\ \Delta Y_R \end{bmatrix} \qquad
\mathbf{L} = \begin{bmatrix} 0 \\ 0 \\ 0 \\ 0 \\ -1.89 \end{bmatrix}
$$

$$4 \times 1 \qquad\qquad 5 \times 1$$

The weight matrix $\mathbf{P}$ used in the solution for $\mathbf{X}$ given by Eq. (A-23) is

$$
\mathbf{P} = \begin{bmatrix}
0.66 & 0 & 0 & 0 & 0 \\
0 & 1.48 & 0 & 0 & 0 \\
0 & 0 & 0.45 & 0 & 0 \\
0 & 0 & 0 & 0.27 & 0 \\
0 & 0 & 0 & 0 & 0.40
\end{bmatrix}
$$

$$5 \times 5$$

Solving Eq. (A-23) for **X** gives

$$\Delta X_C = +0.35 \text{ ft} \qquad \Delta X_R = -1.81 \text{ ft}$$
$$\Delta Y_C = +0.16 \text{ ft} \qquad \Delta Y_R = +0.78 \text{ ft}$$

These values are added to the approximate coordinates of C and R to get the adjusted coordinates of the two stations. Thus

approx. $X_C = 530,278.48$	approx. $Y_C = 478,350.77$
$+0.35$	$+0.16$
adjusted $X_C = \overline{530,278.83}$	adjusted $Y_C = \overline{478,350.93}$
approx. $X_R = 486,460.14$	approx. $Y_R = 457,043.26$
-1.81	$+0.78$
adjusted $X_R = \overline{486,458.33}$	adjusted $Y_R = \overline{457,044.04}$

Substituting the values of **X** into Eq. (A-20) gives the vector of residuals as

$$v_9 = +0.19 \text{ ft} \qquad v_{12} = +0.59 \text{ ft}$$
$$v_{10} = -0.08 \text{ ft} \qquad v_{13} = -0.22 \text{ ft}$$
$$v_{11} = -0.19 \text{ ft}$$

The standard error of unit weight for the net, given by Eq. (A-24), in which n is 5 and u is 4, is ± 0.403 ft. The standard error of each of the lines is then computed by Eq. (5-30), giving

$$\sigma_9 = +0.49 \text{ ft} \qquad \sigma_{12} = \pm 0.78 \text{ ft}$$
$$\sigma_{10} = \pm 0.33 \text{ ft} \qquad \sigma_{13} = \pm 0.64 \text{ ft}$$
$$\sigma_{11} = \pm 0.60 \text{ ft}$$

A-7. COMBINED TRIANGULATION-TRILATERATION ADJUSTMENT
When a horizontal control network has been surveyed by measuring both angles and distances, the method of variation of coordinates can be used very effectively to adjust the measurements. The most frequent situation is that in which a triangulation net has been strengthened by measuring the lengths of selected lines directly, using EDM. The angle observations given in Section A-5 and the distance measurements given in Section A-6 will be used to perform a single adjustment. It is to be noted that a complete observation of both angles and distances in a net is usually not justified, and is performed only for special studies.

To begin the adjustment an approximate set of coordinates are computed for each new station in the net. This can be done as shown in Section A-5, in which the angles and a starting side are used to compute the triangles, or else as shown in Section A-6 using the measured lengths and application of the law of cosines. Using the results of the computations performed in Section A-5, the only additional work needed to form the observation equations of this example is a recomputation of the constant terms in the length equations. The values of the sines and cosines of the preliminary azimuths shown in the tabulation in Table A-13 are more than sufficient to use as coefficients in the length equations. Referring to the lengths obtained in the triangle computations in Tables A-10 and A-11, and the inverse of CR shown on page 781, together with the measured lengths given in Table A-14, the values of the constant terms for the length equations are

$$(AC) K_9 = 37{,}222.39 - 37{,}221.98 = +0.41$$

$$(BC) K_{10} = 24{,}230.96 - 24{,}231.15 = -0.19$$

$$(AR) K_{11} = 45{,}040.43 - 45{,}040.08 = +0.35$$

$$(BR) K_{12} = 59{,}215.44 - 59{,}215.10 = +0.34$$

$$(CR) K_{13} = 48{,}725.40 - 48{,}726.18 = -0.78$$

A summary of the observation equations is given here.

$$v_1 = -3.22\,\Delta X_C - 4.50\,\Delta Y_C \qquad\qquad\qquad - 2.1''$$

$$v_2 = \qquad\qquad\qquad +2.64\,\Delta X_R - 2.23\,\Delta Y_R + 0.3''$$

$$v_3 = +8.30\,\Delta X_C + 1.85\,\Delta Y_C - 2.64\,\Delta X_R + 2.23\,\Delta Y_R - 2.3''$$

$$v_4 = -5.08\,\Delta X_C + 2.65\,\Delta Y_C \qquad\qquad\qquad - 2.1''$$

$$v_5 = -5.07\,\Delta X_C - 0.69\,\Delta Y_C + 1.85\,\Delta X_R - 3.81\,\Delta Y_R + 12.3''$$

$$v_6 = +1.85\,\Delta X_C - 3.80\,\Delta Y_C + 0.79\,\Delta X_R + 1.57\,\Delta Y_R - 7.0''$$

$$v_7 = \qquad\qquad\qquad +1.72\,\Delta X_R + 0.85\,\Delta Y_R + 0.4''$$

$$v_8 = +3.22\,\Delta X_C + 4.50\,\Delta Y_C - 4.36\,\Delta X_R + 1.38\,\Delta Y_R + 2.5''$$

$$v_9 = +0.813\,\Delta X_C - 0.582\,\Delta Y_C \qquad\qquad\qquad + 0.41\ \text{ft}$$

$$v_{10} = +0.218\,\Delta X_C - 0.976\,\Delta Y_C \qquad\qquad\qquad - 0.19\ \text{ft}$$

$$v_{11} = \qquad\qquad\qquad -0.301\,\Delta X_R - 0.954\,\Delta Y_R + 0.35\ \text{ft}$$

$$v_{12} = \qquad\qquad\qquad -0.651\,\Delta X_R - 0.759\,\Delta Y_R + 0.34\ \text{ft}$$

$$v_{13} = +0.899\,\Delta X_C + 0.437\,\Delta Y_C - 0.899\,\Delta X_R - 0.437\,\Delta Y_R - 0.78\ \text{ft}$$

The matrices making up the observation equations given by Eq. (A-20) are as follows:

$$
V = \begin{bmatrix} v_1 \\ v_2 \\ v_3 \\ v_4 \\ v_5 \\ v_6 \\ v_7 \\ v_8 \\ v_9 \\ v_{10} \\ v_{11} \\ v_{12} \\ v_{13} \end{bmatrix}_{13 \times 1}
\quad
B = \begin{bmatrix}
-3.22 & -4.50 & 0 & 0 \\
0 & 0 & +2.64 & -2.23 \\
+8.30 & +1.85 & -2.64 & +2.23 \\
-5.08 & +2.65 & 0 & 0 \\
-5.07 & -0.69 & +1.85 & -3.81 \\
+1.85 & -3.80 & +0.79 & +1.57 \\
0 & 0 & +1.72 & +0.85 \\
+3.22 & +4.50 & -4.36 & +1.38 \\
+0.813 & -0.582 & 0 & 0 \\
+0.218 & -0.976 & 0 & 0 \\
0 & 0 & -0.301 & -0.954 \\
0 & 0 & -0.651 & -0.759 \\
+0.899 & +0.437 & -0.899 & -0.437
\end{bmatrix}_{13 \times 4}
\quad
X = \begin{bmatrix} \Delta X_C \\ \Delta Y_C \\ \Delta X_R \\ \Delta Y_R \end{bmatrix}_{4 \times 1}
\quad
L = \begin{bmatrix} -2.1 \\ +0.3 \\ -2.3 \\ -2.1 \\ +12.3 \\ -7.0 \\ +0.4 \\ +2.5 \\ +0.41 \\ -0.19 \\ +0.35 \\ +0.34 \\ -0.78 \end{bmatrix}_{13 \times 1}
$$

These equations contain two different kinds of units, that is, seconds of arc and feet of distance. It thus becomes very important to assign the correct weights to the weight matrix. The weight of an observation is inversely proportionate to the square of the standard error. If a priori standard errors can be assigned to the measured angles and the measured lengths, the weight matrix will then contain diagonal elements equal to $1/\sigma^2$ in the proper units. In this example a standard error of $1''$ will be assumed for all eight measured angles, whereas a standard error of 0.02 ft $+ D/100{,}000$ will be assumed for the measured distances in which D is the measured distance. For example line AC is 37,222 ft long, and

$$\sigma_{AC} = 0.02 + 0.37 = 0.39 \text{ ft}$$

The weight is then $1/0.39^2 = 6.6$, and the weight matrix is

$$
P = \begin{bmatrix}
1 & 0 & & & & & & & & & & & 0 \\
0 & 1 & & & & & & & & & & & \\
& & 1 & & & & & & & & & & \\
& & & 1 & & & & & & & & & \\
& & & & 1 & & & & & & & & \\
& & & & & 1 & & & & & & & \\
& & & & & & 1 & & & & & & \\
& & & & & & & 1 & & & & & \\
& & & & & & & & 6.6 & & & & \\
& & & & & & & & & 14.8 & & & \\
& & & & & & & & & & 4.5 & & \\
& & & & & & & & & & & 2.7 & 0 \\
0 & & & & & & & & & & & 0 & 4.0
\end{bmatrix}_{13 \times 13}
$$

Solving Eq. (A-23) for **X** gives

$$\Delta X_C = +0.06 \text{ ft} \qquad \Delta X_R = +0.11 \text{ ft}$$

$$\Delta Y_C = -0.46 \text{ ft} \qquad \Delta Y_R = +1.75 \text{ ft}$$

The adjusted coordinates are then obtained by adding the above values to the approximate values (see page 781), thus

$$
\begin{array}{ll}
\text{approx. } X_C = 530{,}279.24 & \text{approx. } Y_C = 478{,}351.12 \\
\qquad\qquad\quad +0.06 & \qquad\qquad\quad -0.46 \\
\text{adjusted } X_C = \overline{530{,}279.30} & \text{adjusted } Y_C = \overline{478{,}350.66}
\end{array}
$$

$$
\begin{array}{ll}
\text{approx. } X_R = 486{,}460.00 & \text{approx. } Y_R = 457{,}042.94 \\
\qquad\qquad\quad +0.11 & \qquad\qquad\quad +1.75 \\
\text{adjusted } X_R = \overline{486{,}460.11} & \text{adjusted } Y_R = \overline{457{,}044.69}
\end{array}
$$

The values of the residuals are obtained by substituting the value of **X** into Eq. (A-20), giving

$$
\begin{array}{lll}
v_1 = -0.24'' & v_5 = +5.85'' & v_9 = +0.72 \text{ ft} \\
v_2 = -3.31'' & v_6 = -2.33'' & v_{10} = +0.27 \text{ ft} \\
v_3 = +0.98'' & v_7 = +2.07'' & v_{11} = -1.35 \text{ ft} \\
v_4 = -3.62'' & v_8 = +2.59'' & v_{12} = -1.05 \text{ ft} \\
& & v_{13} = -1.78 \text{ ft}
\end{array}
$$

The standard error of unit weight given by Eq. (A-24), in which n is 13 and u is 4, is ± 3.40 units. Then by Eq. (4-23)

$$
\begin{array}{lll}
\sigma_1 = \pm 3.40'' & \sigma_5 = \pm 3.40'' & \sigma_9 = \pm 1.32 \text{ ft} \\
\sigma_2 = \pm 3.40'' & \sigma_6 = \pm 3.40'' & \sigma_{10} = \pm 0.88 \text{ ft} \\
\sigma_3 = \pm 3.40'' & \sigma_7 = \pm 3.40'' & \sigma_{11} = \pm 1.60 \text{ ft} \\
\sigma_4 = \pm 3.40'' & \sigma_8 = \pm 3.40'' & \sigma_{12} = \pm 2.10 \text{ ft} \\
& & \sigma_{13} = \pm 1.70 \text{ ft}
\end{array}
$$

A-8. ADJUSTMENT OF A TRAVERSE NET BY OBSERVATION EQUATIONS

In Section A-3 a level net was adjusted by assuming an initial set of junction-point elevations and then letting these values be changed by small amounts in order to cause the sum of the products of the weights of the lines of levels and the squares of the corrections to the

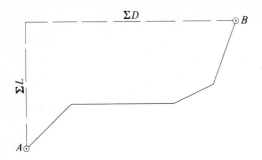

Figure A-10. Traverse section.

measured DE's of these lines to be a minimum. A traverse net consisting of several lines joined at junction points can be adjusted in a similar manner. In theory, however, there is one important difference. The corrections which we refer to as v's are supposed to be applied to quantities that are measured directly. In the level-net adjustment, we consider the DE's to be measured directly, since they represent a series of additions and subtractions of directly observed readings of the leveling rod. The corrections that will be referred to in this section as v_x's and v_y's are applied to derived quantities and not to direct observations.

The basis for the adjustment of a traverse net by the method of observation equations as set forth in this section is the compass rule discussed in Section 8-15. This rule states that the correction to the latitude or departure of a line is directly proportional to the length of that line. Let us consider a series of traverse lines joining two points A and B, as shown in Fig. A-10. Such a series of lines is referred to as a section. A traverse net is composed of several sections, each of which joins either two junction points or a junction point and a fixed point, as indicated in Fig. A-11. For each section we compute the sum of the latitudes, called $\sum L$, and the sum of the departures, called $\sum D$. Each sum is analogous to the DE of a level line. The corrections to

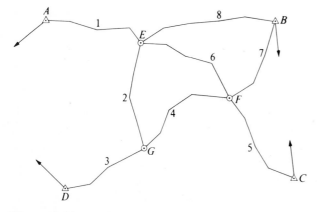

Figure A-11. Traverse net.

these two derived quantities are the v_y's and the v_x's, respectively. The quantities $\sum L$ and $\sum D$ are derived, since they are products of measured distances and functions of angles that have been obtained from measured angles, as explained in Chapter 8 (see also Section 5-7).

In Fig. A-11 traverse sections have been run from each of the four control points A, B, C, and D to each of the other three through the junction points E, F, and G. The fixed coordinates of the four control points in feet are as follows:

POINT	Y	X
A	28,696.58	51,582.20
B	28,878.62	56,454.81
C	25,674.32	56,890.80
D	25,256.32	52,064.40

The lines with the arrowheads at the four control points indicate azimuth backsights. A traverse can be carried, for example, by backsighting at A, running in succession to E, to F, and to C, and then closing in azimuth at C. When all the traversing shown in Fig. A-11 has been executed, either the angular closures may be adjusted to give adjusted azimuths or else the computed azimuths may be used as unadjusted azimuths. The latter procedure is perhaps more valid. The latitudes and departures of all the lines are next computed by applying Eqs. (8-1) and (8-2) or Eqs. (8-3) and (8-4). The values of $\sum L$ and $\sum D$ for the sections are then determined and listed in tabular form. The measured and computed values for the traverse in Fig. A-11 are shown in Table A-17. The weights of the sections are assumed to be inversely proportional to their lengths, which is the basis of the compass rule. These weights have been modified so that their average is unity.

An approximate set of values for the coordinates of the junction points are now computed by adding the proper values of $\sum L$ to known Y co-

TABLE A-17

SECTION	LENGTH 1000 ft	FROM TO	$\sum L$	$\sum D$	CORRECTION TO $\sum L$	CORRECTION TO $\sum D$	p
1	2.24	A–E	$-$ 357.30	$+2076.02$	v_{y_1}	v_{x_1}	1.04
2	2.20	E–G	-2196.62	$+$ 96.08	v_{y_2}	v_{x_2}	1.06
3	1.88	G–D	$-$ 886.02	-1689.68	v_{y_3}	v_{x_3}	1.24
4	2.74	G–F	$+1113.50$	$+1725.88$	v_{y_4}	v_{x_4}	0.85
5	2.44	F–C	-1609.23	$+1410.42$	v_{y_5}	v_{x_5}	0.96
6	2.51	E–F	-1082.82	$+1822.00$	v_{y_6}	v_{x_6}	0.93
7	2.00	F–B	$+1622.48$	$+$ 974.60	v_{y_7}	v_{x_7}	1.17
8	3.02	B–E	$-$ 539.59	-2796.74	v_{y_8}	v_{x_8}	0.77

ordinates and adding the proper values of $\sum D$ to known X coordinates. Thus for point E

$$Y_E = Y_A + \sum L_{AE} = 28,696.58 - 357.30 = 28,339.28$$

$$X_E = X_A + \sum D_{AE} = 51,582.20 + 2076.02 = 53,658.22$$

The values of the approximate coordinates of E, F, and G are as follows:

POINT	FROM SECTION	APPROX. Y	APPROX. X
E	AE	28,339.28	53,658.22
F	BF	27,256.14	55,480.21
G	DG	26,142.34	53,754.08

We obtain the adjusted coordinates of each junction point by adding an as-yet unknown correction ΔY to the approximate Y coordinate and a correction ΔX to the approximate X coordinate. The adjusted coordinates may therefore be represented as follows:

POINT	ADJUSTED Y	ADJUSTED X
E	$28,339.28 + \Delta Y_E$	$53,658.22 + \Delta X_E$
F	$27,256.14 + \Delta Y_F$	$55,480.21 + \Delta X_F$
G	$26,142.34 + \Delta Y_G$	$53,754.08 + \Delta X_G$

We can now derive an observation equation in Y and in X for each section. For example, for the section between A and E,

$$\sum L_{AE} + v_{y_1} = \text{adjusted } Y_E - Y_A$$

or

$$-357.30 + v_{y_1} = 28,339.28 + \Delta Y_E - 28,696.58$$

Hence,

$$v_{y_1} = +\Delta Y_E + 0$$

Also,

$$\sum D_{AE} + v_{x_1} = \text{adjusted } X_E - X_A$$

or

$$+2076.02 + v_{x_1} = 53,658.22 + \Delta X_E - 51,582.20$$

Hence,

$$v_{x_1} = +\Delta X_E + 0$$

The complete set of observation equations is as follows:

$$v_{y_1} = +\Delta Y_E + 0 \qquad\qquad x_{x_1} = +\Delta X_E + 0$$
$$v_{y_2} = -\Delta Y_E + \Delta Y_G - 0.32 \qquad v_{x_2} = -\Delta X_E + \Delta X_G - 0.22$$
$$v_{y_3} = -\Delta Y_G + 0 \qquad\qquad v_{x_3} = -\Delta X_G + 0$$
$$v_{y_4} = +\Delta Y_F - \Delta Y_G + 0.30 \qquad v_{x_4} = +\Delta X_F - \Delta X_G + 0.25$$
$$v_{y_5} = -\Delta Y_F + 0.41 \qquad\qquad v_{x_5} = -\Delta X_F + 0.17$$
$$v_{y_6} = -\Delta Y_E + \Delta Y_F - 0.32 \qquad v_{x_6} = -\Delta X_E + \Delta X_F - 0.01$$
$$v_{y_7} = -\Delta Y_F + 0 \qquad\qquad v_{x_7} = -\Delta X_F + 0$$
$$v_{y_8} = +\Delta Y_E + 0.25 \qquad\qquad v_{x_8} = +\Delta X_E + 0.15$$

The matrix form of these equations is

$$
\begin{bmatrix} v_{y_1} \\ v_{y_2} \\ v_{y_3} \\ v_{y_4} \\ v_{y_5} \\ v_{y_6} \\ v_{y_7} \\ v_{y_8} \end{bmatrix} =
\begin{bmatrix} +1 & 0 & 0 \\ -1 & 0 & +1 \\ 0 & 0 & -1 \\ 0 & +1 & -1 \\ 0 & -1 & 0 \\ -1 & +1 & 0 \\ 0 & -1 & 0 \\ +1 & 0 & 0 \end{bmatrix}
\begin{bmatrix} \Delta Y_E \\ \Delta Y_F \\ \Delta Y_G \end{bmatrix} +
\begin{bmatrix} 0 \\ -0.32 \\ 0 \\ +0.30 \\ +0.41 \\ -0.32 \\ 0 \\ +0.25 \end{bmatrix}
$$
$$\mathbf{V}_y \qquad\qquad \mathbf{B} \qquad\qquad \mathbf{X}_y \qquad\qquad \mathbf{L}_y$$

$$
\mathbf{P}_y =
\begin{bmatrix}
1.04 & . & . & . & . & . & . & 0 \\
0 & 1.06 & & & & & & . \\
. & & 1.24 & & & & & . \\
. & & & 0.85 & & & & . \\
. & & & & 0.96 & & & . \\
. & & & & & 0.93 & & . \\
. & & & & & & 1.17 & . \\
0 & . & . & . & . & . & . & 0.77
\end{bmatrix}
$$
$$8 \times 8$$

$$
\begin{bmatrix} v_{x_1} \\ v_{x_2} \\ v_{x_3} \\ v_{x_4} \\ v_{x_5} \\ v_{x_6} \\ v_{x_7} \\ v_{x_8} \end{bmatrix} =
\begin{bmatrix} +1 & 0 & 0 \\ -1 & 0 & +1 \\ 0 & 0 & -1 \\ 0 & +1 & -1 \\ 0 & -1 & 0 \\ -1 & +1 & 0 \\ 0 & -1 & 0 \\ +1 & 0 & 0 \end{bmatrix}
\begin{bmatrix} \Delta X_E \\ \Delta X_F \\ \Delta X_G \end{bmatrix} +
\begin{bmatrix} 0 \\ -0.22 \\ 0 \\ +0.25 \\ +0.17 \\ -0.01 \\ 0 \\ +0.15 \end{bmatrix} \qquad \mathbf{P}_x = \mathbf{P}_y
$$
$$\mathbf{V}_x \qquad\qquad \mathbf{B} \qquad\qquad \mathbf{X}_x \qquad\qquad \mathbf{L}_x$$

Two sets of normal equations are formed, one in Y and one in X. The procedure for the formation of the normal equations is that described in Section A-3. The coefficients $[\mathbf{B}^T\mathbf{P}\mathbf{B}]$ in both sets of equations will be identical since $\mathbf{B}$ and $\mathbf{P}$ are identical for both sets. If the adjustment is being performed on a desk calculating machine, the values used in the solution of both sets of normal equations can be placed side by side, because of the equality of the coefficients.

The proper values of ΔY are added to the approximate Y coordinates of the junction points, and the proper values of ΔX are added to the approximate X coordinates. These results are the adjusted coordinates. The standard errors of the $\sum L$'s and $\sum D$'s are determined as discussed in Section A-3 for determining the standard errors of the DE's.

After the adjusted coordinates of the junction points have been determined, the coordinates of the intermediate traverse stations in each section are made to be consistent with those of the junction points. This adjustment is an application of the compass rule, as outlined in Section 8-15 and as illustrated by the example in Section 8-19.

A-9. ADJUSTMENT OF INTERSECTION OBSERVATIONS

In the example of an intersection solution given in Section 10-34, the minimum of two measured angles α and β were used to compute the X- and Y-coordinates of P. If the third angle γ as shown in Fig. 10-18 is also measured, the position of P can be determined by the method of variation of coordinates. The coordinates of known point C in Fig. 10-18 are $X_C = 1,465,872.89$ and $Y_C = 639,864.01$. Angle $\gamma = 87° 32' 02''$. Since an approximate position for P has already been computed in Example 10-5, the three lines AP, BP, and CP can be inversed to determine their preliminary lengths and azimuths as shown in Table A-18. The azimuths of AB and BA were computed in Example 10-5. The azimuth of BC is determined by inversing.

Using the scheme shown in Fig. A-4 to designate the *at, from,* and *to* stations, in forming the observation equation given by Eq. (A-51) for angle α, A is the *at* station, P is the *from* station, and B is the *to* station. The constant K_α is computed from Eq. (A-45b) and Table A-18.

$$
\begin{aligned}
\alpha'_{AB} &= 108° 49' 36'' \\
-\alpha'_{AP} &= -34° 20' 20'' \\
\hline
&74° 29' 16''
\end{aligned}
$$

$$
\begin{aligned}
-\text{measured} \angle \alpha \quad & -74° 29' 16'' \\
\hline
K_\alpha &= 0
\end{aligned}
$$

Then, since ΔX_A, ΔY_A, ΔX_B, and ΔY_B are all zero,

$$
v_\alpha = -\frac{\cos \alpha'_{AP}}{AP \text{ arc } 1''} \Delta X_P + \frac{\sin \alpha'_{AP}}{AP \text{ arc } 1''} \Delta Y_P + 0
$$

TABLE A-18

LINE	PRELIMINARY LENGTH	PRELIMINARY AZIMUTH	cos α′	sin α′
AB		108° 49′ 36″		
AP	5886	34° 20′ 20″	+0.82572	+0.56409
BP	9258	326° 36′ 04″	+0.83486	−0.55046
CP	8837	306° 37′ 23″	+0.59655	−0.80258
BC		39° 05′ 15″		

For angle β, in Eq. (A-51) B is the *at* station, A is the *from* station, and P is the *to* station. The constant K_β is computed

$$\alpha'_{BP} = \quad 326°\ 36'\ 04''$$
$$-\alpha'_{BA} = -288°\ 49'\ 36''$$
$$\overline{\qquad\qquad 37°\ 46'\ 28''}$$

$$-\text{measured} \angle \beta \quad -37°\ 46'\ 28''$$
$$\overline{\qquad\qquad K_\beta = 0}$$

Then, since ΔX_A, ΔY_A, ΔX_B, ΔY_B are all zero,

$$v_\beta = +\frac{\cos \alpha'_{BP}}{BP \text{ arc } 1''}\Delta X_P - \frac{\sin \alpha'_{BP}}{BP \text{ arc } 1''}\Delta Y_P + 0$$

For angle γ, in Eq. (A-51), C is the *at* station, B is the *from* station, and P is the *to* station. The constant K_γ is

$$\alpha'_{CP} = \quad 306°\ 37'\ 23''$$
$$-\alpha'_{CB} = -219°\ 05'\ 15''$$
$$\overline{\qquad\qquad 87°\ 32'\ 08''}$$

$$-\text{measured} \angle \gamma \quad -87°\ 32'\ 02''$$
$$\overline{\qquad\qquad K_\gamma = +06''}$$

Then since ΔX_B, ΔY_B, ΔX_C, and ΔY_C in Eq. (A-51) are all zero,

$$v_\gamma = \frac{\cos \alpha'_{CP}}{CP \text{ arc } 1''}\Delta X_P - \frac{\sin \alpha'_{CP}}{CP \text{ arc } 1''}\Delta Y_P + 6''$$

The three resulting observation equations are then

$$v_\alpha = -28.94\,\Delta X_P + 19.77\,\Delta Y_P + 0$$
$$v_\beta = +18.60\,\Delta X_P + 12.26\,\Delta Y_P + 0$$
$$v_\gamma = +13.92\,\Delta X_P + 18.73\,\Delta Y_P + 6''$$

$$\mathbf{V} = \begin{bmatrix} v_\alpha \\ v_\beta \\ v_\gamma \end{bmatrix} \quad \mathbf{B} = \begin{bmatrix} -28.94 & +19.77 \\ +18.60 & +12.26 \\ +13.92 & +18.73 \end{bmatrix} \quad \mathbf{X} = \begin{bmatrix} \Delta X_P \\ \Delta Y_P \end{bmatrix} \quad \mathbf{L} = \begin{bmatrix} 0 \\ 0 \\ +6'' \end{bmatrix}$$

Assuming that all three angles have the same weight, P is the unit matrix. Then by Eq. (A-23)

$$\mathbf{X} = -(\mathbf{B}^T\mathbf{B})^{-1}\mathbf{B}^T\mathbf{L}$$

The solution of this equation gives $\Delta X_P = -0.07$ and $\Delta Y_P = -0.13$. The adjusted coordinates of P are then

$$
\begin{array}{lll}
X'_P = 1,458,780.23 & Y'_P = 645,135.92 & \text{(from Example 10-5)} \\
-0.07 & -0.13 & \\
\hline
X_P = 1,458,780.16 & Y_P = 645,135.79 &
\end{array}
$$

A-10. ADJUSTMENT OF CONDITIONED RESECTION
In the example of a resection solution given in Section 10-35, the minimum of two measured angles p' and p'' were used to compute the X- and Y-coordinates of P. If a third angle p''' is also measured as shown in Fig. A-12, the position of P can be obtained by the method of variation of coordinates. Table A-19 lists the coordinates of the four control points in Fig. A-12.

The measured angles are shown in Fig. A-12. The length and azimuth of AB needed to compute the preliminary coordinates of P are 1642.83 ft and 261° 40′ 52″, respectively, obtained by inversing AB. The position computation for the preliminary position of P (see Section 10-25) using the results of triangle PBA of the example shown in Table 10-11 is laid out in Table A-20.

The lines needed for the adjustment are inversed to obtain the distances and preliminary azimuths of these lines. This is shown in Table A-21. Some of the lines are obtained from the triangle solution and the position solution

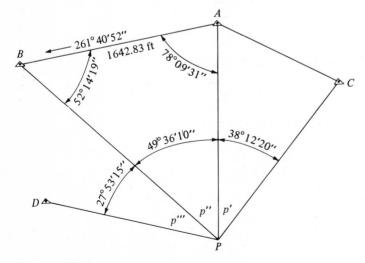

Figure A-12. Resection on four control points.

TABLE A-19

STATION	X	Y
A	118,476.24	267,935.80
B	116,850.70	267,698.11
C	119,406.52	267,394.23
D	117,012.87	266,623.98

given in Section 10-35 and Table A-20 but are shown in Table A-21 for convenience.

In forming the observation equations given by Eq. (A-51), the unknown point P is the *at* station for each angle. With reference to Fig. A-12, for angle p''', D is the *from* station, and B is the *to* station. The constant $K_{p'''}$ is obtained by Eq. (A-45b).

$$
\begin{aligned}
\alpha'_{PB} &= 313°\,55'\,11'' \\
-\alpha'_{PD} &= -286°\,01'\,48'' \\
\hline
&27°\,53'\,23'' \\
-\text{measured} \angle p''' &= -27°\,53'\,15'' \\
\hline
K_{p'''} &= +8''
\end{aligned}
$$

Then, since ΔX_B, ΔY_B, ΔX_D, and ΔY_D are all zero,

$$
v_{p'''} = \frac{\cos \alpha'_{PD}}{PD \text{ arc } 1''} \Delta X_P - \frac{\sin \alpha'_{PD}}{PD \text{ arc } 1''} \Delta Y_P - \frac{\cos \alpha'_{PB}}{PB \text{ arc } 1''} \Delta X_P
$$

$$
+ \frac{\sin \alpha'_{PB}}{PB \text{ arc } 1''} \Delta Y_P + 8''
$$

TABLE A-20

azimuth $AB =$	261° 40′ 52″		azimuth $BA =$	$+81°40′52″$	
$- \angle A$	$-78°09′31″$		$+ \angle B$	$+52°14′19″$	
azimuth $AP =$	183° 31′ 21″		azimuth $BP =$	133° 55′ 11″	
	$-133°55′11″$				
$\angle P$	49° 36′ 10″ (checks)				

STATION	LENGTH	COS AZIMUTH	SIN AZIMUTH	Y	X
A				267,935.80	118,476.24
	1705.38	−0.998 11075	−0.061 44050	−1,702.16	−104.78
P				266,233.64	118,371.46
	2111.26	+0.693 64981	−0.720 31239	+1,464.48	−1,520.77
B				267,698.12	116,850.69
			fixed	267,698.11	116,850.70

TABLE A-21

LINE	PRELIMINARY LENGTH	PRELIMINARY AZIMUTH	cos α'	sin α'
PA	1705	3° 31′ 21″	+0.99811	+0.06144
PB	2111	313° 55′ 11″	+0.69365	−0.72031
PC	1555	41° 43′ 41″	+0.74631	+0.66559
PD	1414	286° 01′ 48″	+0.27614	−0.96112

For angle p'', B is the *from* station, and A is the *to* station. The constant $K_{p''}$ is computed

$$
\begin{aligned}
\alpha'_{PA} &= 363° \, 31' \, 21'' \\
-\alpha'_{PB} &= -313° \, 55' \, 11'' \\
\hline
&49° \, 36' \, 10''
\end{aligned}
$$

$$
\begin{aligned}
-\text{measured} \angle p'' &\quad -49° \, 36' \, 10'' \\
\hline
K_{p''} &= 0
\end{aligned}
$$

Then, since ΔX_A, ΔY_A, ΔX_B, and ΔY_B are all zero,

$$
v_{p''} = \frac{\cos \alpha'_{PB}}{PB \text{ arc } 1''} \Delta X_P - \frac{\sin \alpha'_{PB}}{PB \text{ arc } 1''} \Delta Y_P - \frac{\cos \alpha'_{PA}}{PA \text{ arc } 1''} \Delta X_P
$$

$$
+ \frac{\sin \alpha'_{PA}}{PA \text{ arc } 1''} \Delta Y_P + 0
$$

For angle p', A is the *from* station, and C is the *to* station.

$$
\begin{aligned}
\alpha'_{PC} &= 41° \, 43' \, 41'' \\
-\alpha'_{PA} &= -3° \, 31' \, 21'' \\
\hline
&38° \, 12' \, 20''
\end{aligned}
$$

$$
\begin{aligned}
-\text{measured} \angle p' &\quad -38° \, 12' \, 20'' \\
\hline
K_{p'} &= 0
\end{aligned}
$$

Then since ΔX_A, ΔY_A, ΔX_C, and ΔY_C are all zero,

$$
v_{p'} = \frac{\cos \alpha'_{PA}}{PA \text{ arc } 1''} \Delta X_P - \frac{\sin \alpha'_{PA}}{PA \text{ arc } 1''} \Delta Y_P - \frac{\cos \alpha'_{PC}}{PC \text{ arc } 1''} \Delta X_P
$$

$$
+ \frac{\sin \alpha'_{PC}}{PC \text{ arc } 1''} \Delta Y_P + 0
$$

The reduced observation equations are then

$$v_{p'''} = -27.50 \, \Delta X_P + 69.83 \, \Delta Y_P + 8''$$

$$v_{p''} = -52.97 \, \Delta X_P + 77.81 \, \Delta Y_P + 0$$

$$v_{p'} = +21.75 \, \Delta X_P + 80.86 \, \Delta Y_P + 0$$

$$\mathbf{V} = \begin{bmatrix} v_{p'''} \\ v_{p''} \\ v_{p'} \end{bmatrix} \quad \mathbf{B} = \begin{bmatrix} -27.50 & +69.83 \\ -52.97 & +77.81 \\ +21.75 & +80.86 \end{bmatrix} \quad \mathbf{X} = \begin{bmatrix} \Delta X_P \\ \Delta Y_P \end{bmatrix} \quad \mathbf{L} = \begin{bmatrix} 8'' \\ 0 \\ 0 \end{bmatrix}$$

Assuming all three angles have equal weights, **P** is the unit matrix. Then by Eq. (A-23)

$$\mathbf{X} = -(\mathbf{B}^T\mathbf{B})^{-1}\mathbf{B}^T\mathbf{L}$$

The solution of this equation gives $\Delta X_P = +0.03$ and $\Delta Y_P = -0.03$. The adjusted coordinates of P are then

$$X'_P = 118{,}371.46 \qquad Y'_P = 266{,}233.64 \qquad \text{(from Table A-20)}$$

$$\frac{+0.03}{X_P = 118{,}371.49} \qquad \frac{-0.03}{Y_P = 266{,}233.61}$$

A-11. REMARKS ON ADJUSTMENT OF OBSERVATIONS

Since this appendix is intended only to introduce the reader to the concepts of least-squares adjustments, no attempt has been made to develop the subject in depth. This is the proper function of a separate textbook. A study of this appendix should raise several questions. For example, in the adjusted level net, how reliable are the adjusted elevations of the points? Or in the triangulation or trilateration nets, how reliable are the adjusted plane coordinates? And do different parts of the net bear any relation to other parts of the net as far as dependability is concerned? Answers to these questions can be answered only by a rather thorough study of random-error theory and error propagation in a framework of modern statistics.

For a more complete study of the adjustment of surveying measurements, the readings given at the end of this appendix should be consulted.

The matrix formulation of the solution to the problem of the adjustment of weighted observations is the most efficient method because modern high-speed computers and programs for manipulating matrices are generally available. Before applying the matrix methods, however, a sound understanding is needed of what the least-squares adjustment does for a set of measurements.

BIBLIOGRAPHY

Adams, O. S., and Claire, C. N. *Manual of Plane Coordinate Computation.* Special Publication No. 193. U. S. Coast and Geodetic Survey. U. S. Government Printing Office.

Brown, D. C. 1957. *A Treatment of the General Problem of Least Squares and the Associated Error Propagation.* App. A, R.C.A. Date Reduction Technical Report No. 39.

Deming, W. E. 1964. *Statistical Adjustment of Data.* New York: Dover Publications.

Derenyi, E., and Chrzanowski, A. 1968. Preanalysis of trilateration nets for engineering surveys. *Surveying and Mapping* 28:615.

Dracup, J. F. 1969. Trilateration—a preliminary evaluation. *Proc. Fall Conv. Amer. Cong. Surv. & Mapp.* p. 95.

Lafferty, M. E. 1980. Simple L. S. adjustment with the TI 59 printing calculator. *Proc. Ann. Meet. Amer. Cong. Surv. & Mapp.* p. 444.

Mikhail, E. M. 1976. *Observations and Least Squares.* New York: Harper & Row.

Mikhail, E. M., and Gracie, G. 1981. *Analysis and Adjustment of Survey Measurements.* New York: Van Nostrand Reinhold.

Rainsford, H. F. 1957. *Survey Adjustments and Least Squares.* London: Constable.

Richardus, P. 1966. *Project Surveying.* New York: Wiley.

Stipp, D. W. 1962. Trilateration adjustment. *Surveying and Mapping* 22:575.

Stoughton, H. W. 1977. An algorithm to adjust a quadrilateral. *Surveying and Mapping* 37:201.

Stoughton, H. W. 1977. Introduction to least squares adjustment for the land surveyor. *Proc. Fall Conv. Amer. Cong. Surv. & Mapp.* p. 375.

Veress, S. A. 1974. Selection of least squares adjustment for control survey. *Proc. Ann. Meet. Amer. Cong. Surv. & Mapp.* p. 361.

Vincenty, T. 1975. Length ratios and scale unknowns in trilateration. *Surveying and Mapping* 35:245.

Wolf, P. R. 1969. Horizontal position adjustment. *Surveying and Mapping* 29:635.

Appendix B
The Adjustment of Instruments

B-1. REMARKS No matter how perfectly an instrument may be adjusted when it leaves the instrument maker, it seldom remains in that condition for any considerable length of time, particularly when in daily use, subjected to rough handling, and transported by car or truck over all kinds of roads. Although the field operations can be conducted so as to eliminate errors due to imperfect adjustment, more time is required when the work is done in this manner. For this reason every engineer and surveyor should be able to determine whether or not his instruments are in proper adjustment and if they are not, how to adjust them.

Nearly all the adjustments of the transit and the level depend on the principle of reversal. By reversing the position of the instrument, the effect of the error is doubled. The adjustments are made by turning the capstan-headed screws or nuts that control the positions of the cross hairs and the bubble tubes. The beginner should be very careful not to put too much tension on these screws, as otherwise the threads may be stripped or the screw broken. The adjusting pins should be of the proper size to prevent damaging the holes in the screws and nuts.

When an instrument is badly out of adjustment, it may be necessary to repeat the adjustments several times. Time will usually be saved in such cases by not attempting to perfect each adjustment the first time, since the adjustment of the other parts of the instrument may affect the earlier adjustments.

Modern-day transportation provides the means to ship an instrument to an instrument maker for routine cleaning, adjusting, and lubrication, and to put it back into operation in a short period of time. Although the various adjustments to be discussed in this chapter can be performed by the engineer or surveyor, they should be performed by specialists. No attempt should be made by the surveyor to adjust an optical reading theodolite or a self-leveling level unless he has been trained to do so. These adjustments must be performed in a properly equipped shop or laboratory.

B-2. CROSS HAIRS OF DUMPY LEVEL The first adjustment of the dumpy level is to make the horizontal cross hair truly horizontal when the instrument is leveled. With the level firmly set up, although not necessarily leveled, the clamp screw is clamped. A well-defined point is selected which lies on, say, the left edge of the horizontal cross hair. The telescope is now slowly rotated in azimuth about the vertical axis of the level by means of the tangent screw. If the point stays on the horizontal cross hair all the way over to the right edge, no adjustment is required. If the point departs from the horizontal cross hair, two adjacent capstan screws (shown in Fig. 3-9) are loosened, and the cross-hair ring is rotated to bring the point halfway back to the horizontal cross hair. The capstan screws are then tightened, and the test is repeated until the point stays on the cross hair.

B-3. BUBBLE TUBE OF DUMPY LEVEL The second adjustment of the dumpy level is to make the axis of the bubble tube perpendicular to the vertical axis. With the level firmly set up, it is leveled as best as can be using the leveling screws. The telescope is turned to bring the level bubble over one pair of leveling screws, and the bubble is carefully centered. The telescope is now rotated 180° in azimuth about the vertical axis. If the bubble returns to the center, the bubble tube is in adjustment. If the bubble does not return to the center, it is brought halfway back to center by raising or lowering the capstan end of the bubble tube.

Some level tubes contain two capstan-headed screws, one on top and one on the bottom. These are worked in opposition to one another; that is, one screw is loosened and the other one is tightened to raise or lower the end of the tube. Still other dumpy levels contain adjusting screws at both ends of the bubble tube. Either end can then be manipulated.

After the bubble has been brought halfway back to the center using the adjusting screws, it is then centered the remainder of the way by means of the leveling screws. Now again on rotation through 180°, the bubble should remain centered. If it does not, the adjustment is repeated until the bubble does remain centered.

B-4. LINE OF SIGHT OF DUMPY LEVEL

The final adjustment of the dumpy level is to make the line of sight parallel to the axis of the bubble tube, or to make the line of sight horizontal when the bubble is in the center of the tube. This is accomplished by the method referred to as the "two-peg adjustment" and is a form of reciprocal leveling described in Section 3-36. Two pegs perhaps 100 ft apart are set into the ground as shown in Fig. B-1. The level is first set up so close to point A that the eyepiece barely clears a leveling rod held on the point. By sighting through the wrong end of the telescope, the reading of a rod held on A can be obtained. The reading of a rod held on B is obtained in the usual way. If the level is in adjustment, the difference between these two readings will be the true difference in elevation. The level is next set up very close to B and readings are taken of rods held on B and A. If the difference in elevation obtained from this position of the level does not agree with the value obtained with the level near A, the instrument is out of adjustment. The true difference in elevation will be the mean of the two differences. From this value and from the rod reading on B, the correct rod reading on A can be calculated. The cross hair is brought to this calculated reading by moving the cross-hair ring vertically up or down by backing off on the top capstan screw and tightening the bottom screw, or vice versa. This must be done in small increments, as it is a very sensitive operation. After the adjustment has been made, the level bubble should be checked to make sure it is still centered. The reading at A should be then verified.

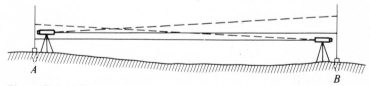

Figure B-1. Adjustment by two-peg method.

B-5. CROSS HAIR OF TILTING DUMPY LEVEL

The first adjustment of the tilting dumpy level is to make the horizontal cross hair truly horizontal when the bull's-eye bubble is centered. The test is the same as for the dumpy level. A point is chosen that coincides with one end of the cross hair. If the point does not remain on the hair as the telescope is slowly turned about its vertical axis, the adjustment is made by loosening two adjacent capstan screws and rotating the cross-hair ring.

B-6. BULL'S-EYE BUBBLE OF TILTING DUMPY LEVEL

The second adjustment of the tilting level is that of making the circular bubble stay in the center as the telescope is rotated about the vertical axis. If the bubble moves off center when the telescope is turned 180° in azimuth, it is brought halfway back to center in both directions by raising or lowering the bubble mount by means of either capstan screws or spring screws, the method depending on the make of level, and the remaining distance with the leveling screws. This adjustment is not critical because of the tilting feature of the level.

B-7. BUBBLE TUBE OF SENSITIVE LEVEL OF TILTING LEVEL

This adjustment gives a horizontal line of sight when the main sensitive bubble is centered. If the level bubble is of the coincidence type, the coincidence of the two ends of the bubble indicates centering. The two-peg method described in Section B-4 is used to make this adjustment. When the rod reading that will give a horizontal line of sight has been calculated, the telescope is tilted up or down by means of the tilting knob until the horizontal cross hair is on the correct reading. The main bubble is brought to center or to coincidence, by raising or lowering one end of the bubble vial.

B-8. PLATE BUBBLES OF TRANSIT

The first adjustment of the transit is to make the axes of the plate levels perpendicular to the vertical axis of the instrument. The transit is set up and each bubble is brought to the center of its tube by means of the leveling screws. The instrument is then revolved on its vertical axis through 180°. If either bubble moves toward one end of the tube, it is brought halfway back by means of the capstan-headed screw at the end of the tube, and the remaining distance by means of the leveling screws. This operation is repeated until both bubbles remain in the centers of the tubes as the telescope is turned in azimuth.

B-9. CROSS HAIRS OF TRANSIT

The first adjustment of the cross hairs of a transit is to make the vertical cross hair perpendicular to the horizontal axis of the telescope, so that when the other adjustments have been made it will be truly vertical. This adjustment is made in a manner similar to the one used in making the horizontal cross hair in a level truly horizontal. One end of the vertical hair is brought to some well-defined point and the telescope is revolved on its transverse axis, so that the point appears to move along the hair. If it does not remain on the hair throughout the motion, two adjacent screws, which control the cross-hair ring are loosened, and the ring is rotated.

The purpose of the second adjustment of the vertical cross hair is to make the line of sight perpendicular to the horizontal axis. Whether or not an adjustment is necessary can be determined by prolonging a straight line by the method of double centering, as described in Section 6-6. If, as shown in Fig. B-2, the two points B and C do not coincide, the cross-hair ring should be shifted laterally until the vertical cross hair appears to strike D, which is at one-fourth of the distance from the second point C to the first point B. Since a double reversing is involved, the apparent error CB is four times the real error.

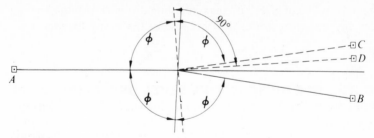

Figure B-2. Second adjustment of vertical cross hair.

The third adjustment of the cross hairs is necessary when the transit is used as a level, or when it is used in measuring vertical angles. The purpose of the adjustment is to bring the horizontal hair into the plane of motion of the optical center of the object glass, so that the line of sight will be horizontal when the telescope bubble is in adjustment and the bubble is centered. In making the adjustment the transit is set up at A, Fig. B-3, and two stakes are driven at B and C on the same straight line and at distances of about 30 and 300 ft from A. With the telescope clamped in the normal position and the line of sight represented by cd, rod readings Be and Cd are obtained by holding leveling rods on the two stakes. The telescope is then plunged and the line of sight fg is brought to the former rod reading Be. If the rod reading on C agrees with the reading previously obtained, no adjustment is necessary. If the two readings do not agree, the target is set at Cb, which is the mean of the two readings Cd and Cg, and the horizontal cross hair is brought to the position a where it appears to bisect the target, by shifting the cross-hair ring vertically by means of the capstan-headed screws on the top and bottom of the telescope. If the instrument is badly out of adjustment, the reading Be will be changed perceptibly as the cross-hair ring is shifted, and the adjustment may have to be repeated several times.

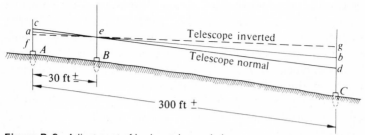

Figure B-3. Adjustment of horizontal cross hair.

B-10. STANDARDS If the standards of the transit are in adjustment, the horizontal axis of the telescope is perpendicular to the vertical axis of the instrument. When the plates are leveled, the horizontal axis is truly horizontal, and the line of sight moves in a vertical plane as the telescope is raised and

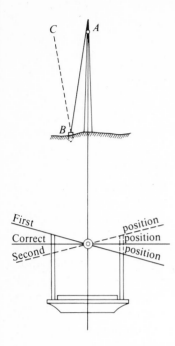

Figure B-4. Adjustment of standards.

lowered. The test of the standards is made by sighting, with the telescope normal, to some well-defined point on a high object, such as a flagpole or church spire. In Fig. B-4, *A* is such a high point. With the horizontal motion clamped, the telescope is depressed and a point *B* is set on the ground in the line of sight. The telescope is then plunged, the lower clamp is released, and a sight is taken on *B*. With the lower clamp tight, the telescope is again elevated. If the line of sight strikes the high point *A*, no adjustment is necessary. If, as in Fig. B-4, the line of sight now strikes at *C*, one end of the horizontal axis must be raised or lowered to bring the line of sight midway between *C* and *A*. This is done by loosening the upper screws that hold the horizontal axis in place and then turning the small capstan-headed screw that controls the position of the block on which the horizontal axis rests. In the example shown in the figure, the left-hand side should be raised or the right-hand side lowered when the telescope is in the reversed position. When the axis has been properly adjusted, the upper screws are tightened just sufficiently to prevent looseness of the bearing. If the line of sight is now lowered, it should strike point *B*.

B-11. TELESCOPE BUBBLE   The purpose of the adjustment of the telescope bubble is to make the axis of its tube parallel to the line of sight. This adjustment is necessary whenever the transit is to be used as a level. When the direction of a horizontal line of sight has been determined by the peg method, described in Section B-5, the line of sight is brought to the correct reading by raising or lowering the telescope with the vertical tangent screw. The bubble tube is then adjusted to bring the bubble to the center of the tube.

B-12. VERTICAL-CIRCLE VERNIER

The vertical circle should preferably read 0° when the telescope is horizontal. With the instrument carefully leveled and the telescope bubble in the center of the tube, the screws holding the vernier in place are loosened slightly and the zeros are made to coincide by tapping the vernier lightly.

B-13. CENTERING THE EYEPIECE

If the intersection of the cross hairs appears to be out of the center of the field of view when the instrument is otherwise in adjustment, the end of the eyepiece must be out of position. Erecting eyepieces are usually held in place by a ring, similar to the cross-hair ring, which is controlled by a set or adjusting screws between the eyepiece and the cross-hair ring. By moving this ring, the intersection of the cross hairs can be brought to the center of the field. As the inverting eyepiece is very short, its inner end needs no support.

B-14. PLANE-TABLE ALIDADE

The method of adjustment of the plane table alidade varies somewhat with the type of alidade. If the telescope is equipped with a removable striding level, the telescope is of the tube-in-sleeve type, and can be rotated around its longitudinal axis. The first adjustment to make is the proper positioning of the cross hairs. With the alidade sitting on a solid flat surface, a point is selected at the intersection of the horizontal and vertical cross hairs. The telescope is now rotated 180° about its longitudinal axis. If the point remains on the cross-hair intersection, the line of sight is parallel with the telescope tube. If the point departs from the intersection of the cross hairs, it is brought halfway back, both horizontally and vertically, by means of the opposite horizontal and vertical adjusting screws on the cross-hair ring.

The striding level is adjusted next. It is placed in position on the collars of the telescope. The vertical tangent screw is used to center the striding-level bubble. The striding level is now picked up, reversed end for end, and replaced on the telescope. If the bubble comes back to the center, then the axis of the bubble tube is parallel with the axis of the collars, which are in turn assumed to be coincident with the longitudinal axis of the telescope. Thus the bubble tube is parallel with the line of sight because of the previous adjustment. If the bubble does not return to the center, it is brought halfway back to center by the adjusting screws of the level tube. It is then centered using the vertical tangent screw. The bubble is checked by reversing it end for end.

Finally, the control or index bubble is adjusted by first centering the striding-level bubble with the telescope vertical tangent screw. Then the V-scale index line is set or read 50 (or the vertical circle is set to read 30° 00′) by means of the control-bubble tangent screw. The control bubble is now brought to the center by means of its adjusting screws.

The microptic alidade shown in Fig. 14-16 is not equipped with a striding level. The adjustment procedure is therefore different than for the tube-in-sleeve type. A two-peg test is performed to make the line of sight horizontal when the control bubble is centered and the V scale reads zero. At each setup, the control bubble is carefully centered using the control-bubble tangent screw, and then the telescope is moved using the telescope tangent screw until the V scale reads zero. Backward and forward readings are taken from setups at *A* and *B* as described in Section B-4. After the correct reading is calculated at the second setup at *B*, the telescope is raised or lowered using the telescope tangent screw until the line of sight is on the correct reading at *A*. The V-scale reading is then set to zero using the control-bubble tangent screw. Finally, the control bubble is brought to center using the adjusting nuts on the end of the level tube.

B-15. LEVELS ON STRAIGHTEDGE The levels attached to the straightedge are adjusted by leveling the board and then lifting the alidade from the board and turning it end for end. The adjustment is made, when necessary, by bringing the bubble halfway back to the center by means of the adjusting screws. Since the board may be warped, the first position of the alidade must be carefully marked so that the alidade will occupy exactly the same position when it is turned end for end on the board.

B-16. VERTICAL CROSS HAIR OF ALIDADE The vertical cross hair of an alidade is adjusted according to the method described for the transit in Section B-9, that is, by noting whether or not the cross hair remains on a fixed point as the telescope is raised or lowered. If an adjustment is necessary, two adjacent screws which hold the cross-hair ring are loosened, and the ring is then turned until the hair will remain on a point as the telescope is rotated.

B-17. PROTECTION AND CLEANING OF INSTRUMENTS The instrumentman can do much to render unnecessary the frequent adjustment of any surveying instrument. The workmanship on a high-grade transit or level is comparable to that of the finest watch, and the instrument should be handled accordingly. When it is carried from point to point on the shoulder, the clamps should be tight enough to prevent needless wear, yet loose enough to yield if the parts are accidentally bumped. When the transit is not in use or is being carried, the needle should be raised. As the instrument is taken from the shoulder, it should be carefully set down. Otherwise a cross hair may be broken or the instrument thrown out of adjustment. When being transported by car or truck, it should be held in the lap, if possible. When being transported long distances, it should be packed in its box, crumpled newspapers being used to hold it securely in place. The instrument box should then be packed in a box large enough to permit the use of several inches of excelsior padding on all sides.

It should be remembered that repairs to a damaged instrument are costly. In addition, it is almost impossible to restore a badly damaged instrument to its original condition. For these reasons, a transit, level, or plane table should not be left unguarded when a setup on a sidewalk or pavement is unavoidable. The tripod legs should be spread and pushed into the ground sufficiently to prevent its being blown or knocked over.

The instrument should be protected from the rain by a waterproof cover. If such a cover is not at hand, the dust cap should be placed on the objective and the object glass turned upward. When brought in wet from the field, the instrument should be thoroughly dried before it is replaced in the instrument box. The lenses should be wiped with a piece of the softest chamois skin, silk, or linen. If objects appear blurred or indistinct, the cause probably is a dirty eyepiece. Any dust should be removed with a fine camel's-hair brush. If very dirty, the eyepiece may be cleaned with alcohol and wiped with a clean soft cloth. The lenses should not be cleaned too frequently, as there is danger of their being scratched.

Extreme care should be used in wiping the vertical circle and vernier. The edges should not be rubbed, and the hands should be kept off the vertical circle to prevent tarnishing. If any of the graduated circle become tarnished, the dust should first be removed with a camel's-hair brush. A thin film of watch oil is then spread on the tarnished spots. The oil is allowed to remain for a few hours and is then wiped off with a soft piece of linen.

Although an instrument should not be taken apart unnecessarily, any part that does not work freely should not be forced, but cleaned. Old oil and accumulated dirt should be thoroughly removed with alcohol, or, in the case of screws and screw holes, by the use of soap and water. No oil should be used on the exposed parts, as it soon accumulates enough dust to prevent the part from working freely. On interior bearings, only the finest grade of watch oil should be used, and this very sparingly since it tends to collect dust and in cold weather may become stiff and prevent the free movement of the part. Usually, enough oil will remain if the oiled part is wiped dry with a soft cloth before it is replaced in the instrument.

Appendix C
Tables

TABLE C-1. Lengths of Circular Arcs for Radius = 1

DEG.	LENGTH	DEG.	LENGTH	DEG.	LENGTH	DEG.	LENGTH
1	0.017 45 329	31	0.541 05 207	61	1.064 65 084	91	1.588 24 962
2	0.034 90 659	32	0.558 50 536	62	1.082 10 414	92	1.605 70 291
3	0.052 35 988	33	0.575 95 864	63	1.099 55 743	93	1.623 15 620
4	0.069 81 317	34	0.593 41 195	64	1.117 01 072	94	1.640 60 950
5	0.087 26 646	35	0.610 86 524	65	1.134 46 401	95	1.658 06 279
6	0.104 71 976	36	0.628 31 853	66	1.151 91 731	96	1.675 51 608
7	0.122 17 305	37	0.645 77 182	67	1.169 37 060	97	1.692 96 937
8	0.139 62 634	38	0.663 22 512	68	1.186 82 389	98	1.710 42 267
9	0.157 07 963	39	0.680 67 841	69	1.204 27 718	99	1.727 87 596
10	0.174 53 293	40	0.698 13 170	70	1.221 73 048	100	1.745 32 925
11	0.191 98 622	41	0.715 58 499	71	1.239 18 377	101	1.762 78 254
12	0.209 43 951	42	0.733 03 829	72	1.256 63 706	102	1.780 23 584
13	0.226 89 280	43	0.750 49 158	73	1.274 09 035	103	1.797 68 913
14	0.244 34 610	44	0.767 94 487	74	1.291 54 365	104	1.815 14 242
15	0.261 79 939	45	0.785 39 816	75	1.308 99 694	105	1.832 59 571
16	0.279 25 268	46	0.802 85 146	76	1.326 45 023	106	1.850 04 901
17	0.296 70 597	47	0.820 30 475	77	1.343 90 352	107	1.867 50 230
18	0.314 15 927	48	0.837 75 804	78	1.361 35 682	108	1.884 95 559
19	0.331 61 256	49	0.855 21 133	79	1.378 81 011	109	1.902 40 888
20	0.349 06 585	50	0.872 66 463	80	1.396 26 340	110	1.919 86 218
21	0.366 51 914	51	0.890 11 792	81	1.413 71 669	111	1.937 31 547
22	0.383 97 244	52	0.907 57 121	82	1.431 16 999	112	1.954 76 876
23	0.401 42 573	53	0.925 02 450	83	1.448 62 328	113	1.972 22 205
24	0.418 87 902	54	0.942 47 780	84	1.466 07 657	114	1.989 67 535
25	0.436 33 231	55	0.959 93 109	85	1.483 52 986	115	2.007 12 864
26	0.453 78 561	56	0.977 38 438	86	1.500 98 316	116	2.024 58 193
27	0.471 23 890	57	0.994 83 767	87	1.518 43 645	117	2.042 03 522
28	0.488 69 219	58	1.012 29 097	88	1.535 88 974	118	2.059 48 852
29	0.506 14 548	59	1.029 74 426	89	1.553 34 303	119	2.076 94 181
30	0.523 59 878	60	1.047 19 755	90	1.570 79 633	120	2.094 39 510

TABLE C-1. (continued)

MIN.	LENGTH	MIN.	LENGTH	SEC.	LENGTH	SEC.	LENGTH
1	0.000 29 089	31	0.009 01 753	1	0.000 00 485	31	0.000 15 029
2	0 58 178	32	9 30 842	2	00 970	32	15 514
3	0 87 266	33	9 59 931	3	01 454	33	15 999
4	1 16 355	34	9 89 020	4	01 939	34	16 484
5	1 45 444	35	10 18 109	5	02 424	35	16 969
6	0.001 74 533	36	0.010 47 198	6	0.000 02 909	36	0.000 17 453
7	2 03 622	37	10 76 286	7	03 394	37	17 938
8	2 32 711	38	11 05 375	8	03 879	38	18 423
9	2 61 799	39	11 34 464	9	04 363	39	18 908
10	2 90 888	40	11 63 553	10	04 848	40	19 393
11	0.003 19 977	41	0.011 92 642	11	0.000 05 333	41	0.000 19 877
12	3 49 066	42	12 21 730	12	05 818	42	20 362
13	3 78 155	43	12 50 819	13	06 303	43	20 847
14	4 07 243	44	12 79 908	14	06 787	44	21 332
15	4 36 332	45	13 08 997	15	07 272	45	21 817
16	0.004 65 421	46	0.013 38 086	16	0.000 07 757	46	0.000 22 301
17	4 94 510	47	13 67 175	17	08 242	47	22 786
18	5 23 599	48	13 96 263	18	08 727	48	23 271
19	5 52 688	49	14 25 352	19	09 211	49	23 756
20	5 81 776	50	14 54 441	20	09 696	50	24 241
21	0.006 10 865	51	0.014 83 530	21	0.000 10 181	51	0.000 24 726
22	6 39 954	52	15 12 619	22	10 666	52	25 210
23	6 69 043	53	15 41 708	23	11 151	53	25 695
24	6 98 132	54	15 70 796	24	11 636	54	26 180
25	7 27 221	55	15 99 885	25	12 120	55	26 665
26	0.007 56 309	56	0.016 28 974	26	0.000 12 605	56	0.000 27 150
27	7 85 398	57	16 58 063	27	13 090	57	27 634
28	8 14 487	58	16 87 152	28	13 575	58	28 119
29	8 43 576	59	17 16 240	29	14 060	59	28 604
30	8 72 665	60	17 45 329	30	14 544	60	29 089

TABLE C-2. Trigonometric Formulas for the Solution of Right Triangles

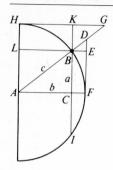

Let A = angle BAC = arc BF, and let radius AF = $AB = AH = 1$. Then,

$\sin A = BC$	$\csc A = AG$
$\cos A = AC$	$\sec A = AD$
$\tan A = DF$	$\cot A = HG$
$\text{vers } A = CF = BE$	$\text{covers } A = BK = LH$
$\text{exsec } A = BD$	$\text{coexsec } A = BG$
$\text{chord } A = BF$	$\text{chord } 2A = BI = 2BC$

In the right-angled triangle ABC, let $AB = c$, $BC = a$, $CA = b$. Then,

1. $\sin A = \dfrac{a}{c}$

2. $\cos A = \dfrac{b}{c}$

3. $\tan A = \dfrac{a}{b}$

4. $\cot A = \dfrac{b}{a}$

5. $\sec A = \dfrac{c}{b}$

6. $\csc A = \dfrac{c}{a}$

7. $\text{vers } A = 1 - \cos A = \dfrac{c - b}{c} = \text{covers } B$

8. $\text{exsec } A = \sec A - 1 = \dfrac{c - b}{b} = \text{coexsec } B$

9. $\text{covers } A = \dfrac{c - a}{c} = \text{vers } B$

10. $\text{coexsec } A = \dfrac{c - a}{a} = \text{exsec } B$

11. $a = c \sin A = b \tan A$

12. $b = c \cos A = a \cot A$

13. $c = \dfrac{a}{\sin A} = \dfrac{b}{\cos A}$

14. $a = c \cos B = b \cot B$

15. $b = c \sin B = a \tan B$

16. $c = \dfrac{a}{\cos B} = \dfrac{b}{\sin B}$

17. $a = \sqrt{c^2 - b^2} = \sqrt{(c - b)(c + b)}$

18. $b = \sqrt{c^2 - a^2} = \sqrt{(c - a)(c + a)}$

19. $c = \sqrt{a^2 + b^2}$

20. $C = 90° = A + B$

21. $\text{area} = \frac{1}{2}ab$

TABLE C-3. Trigonometric Formulas for the Solution of Oblique Triangles

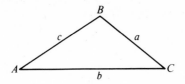

NO.	GIVEN	SOUGHT	FORMULA
22	A, B, a	C, b, c	$C = 180° - (A + B)$
			$b = \dfrac{a}{\sin A} \times \sin B$
			$c = \dfrac{a}{\sin A} \times \sin(A + B) = \dfrac{a}{\sin A} \times \sin C$
		area	$\text{area} = \tfrac{1}{2}ab \sin C = \dfrac{a^2 \sin B \sin C}{2 \sin A}$
23	A, a, b	B, C, c	$\sin B = \dfrac{\sin A}{a} \times b$
			$C = 180° - (A + B)$
			$c = \dfrac{a}{\sin A} \times \sin C$
		area	$\text{area} = \tfrac{1}{2}ab \sin C$
24	C, a, b	c	$c = \sqrt{a^2 + b^2 - 2ab \cos C}$
25		$\tfrac{1}{2}(A + B)$	$\tfrac{1}{2}(A + B) = 90° - \tfrac{1}{2}C$
26		$\tfrac{1}{2}(A - B)$	$\tan \tfrac{1}{2}(A - B) = \dfrac{a - b}{a + b} \times \tan \tfrac{1}{2}(A + B)$
27		A, B	$A = \tfrac{1}{2}(A + B) + \tfrac{1}{2}(A - B)$
			$B = \tfrac{1}{2}(A + B) - \tfrac{1}{2}(A - B)$
28		c	$c = (a + b) \times \dfrac{\cos \tfrac{1}{2}(A + B)}{\cos \tfrac{1}{2}(A - B)} = (a - b) \times \dfrac{\sin \tfrac{1}{2}(A + B)}{\sin \tfrac{1}{2}(A - B)}$
29		area	$\text{area} = \tfrac{1}{2}ab \sin C$

(continued)

TABLE C-3. (continued)

30	a, b, c	A	Let $s = \dfrac{a + b + c}{2}$
31			$\sin \tfrac{1}{2}A = \sqrt{\dfrac{(s - b)(s - c)}{bc}}$
			$\cos \tfrac{1}{2}A = \sqrt{\dfrac{s(s - a)}{bc}}$
			$\tan \tfrac{1}{2}A = \sqrt{\dfrac{(s - b)(s - c)}{s(s - a)}}$
			$\sin A = \dfrac{2\sqrt{s(s - a)(s - b)(s - c)}}{bc}$
32			$\cos A = \dfrac{b^2 + c^2 - a^2}{2bc}$
33		area	$\text{area} = \sqrt{s(s - a)(s - b)(s - c)}$

Index